The Commercial Greenhouse

THIRD EDITION

Join us on the web at

agriculture.delmar.com

The Commercial Greenhouse

THIRD EDITION

James W. Boodley
Steven E. Newman

DELMAR
CENGAGE Learning

Australia • Brazil • Japan • Korea • Mexico • Singapore • Spain • United Kingdom • United States

DELMAR
CENGAGE Learning

The Commercial Greenhouse,
Third Edition
James W. Boodley, Steven E. Newman

**Vice President, Career and
Professional Editorial:**
Dave Garza

Director of Learning Solutions:
Matthew Kane

Senior Acquisitions Editor:
David Rosenbaum

Managing Editor:
Marah Bellegarde

Product Manager:
Christina Gifford

Editorial Assistant:
Scott Royael

**Vice President, Career, and
Professional Marketing:**
Jennifer McAvey

Marketing Director:
Debbie Yarnell

Marketing Manager:
Erin Brennan

Marketing Coordinator:
Jonathan Sheehan

Production Director:
Carolyn Miller

Production Manager:
Andrew Crouth

Content Project Manager:
Katie Wachtl

Senior Art Director:
Dave Arsenault

Technology Project Manager:
Ben Knapp

Production Technology Analyst:
Thomas Stover

For product information and technology assistance, contact us at
Professional & Career Group Customer Support, 1-800-648-7450
For permission to use material from this text or product,
submit all requests online at **cengage.com/permissions**
Further permissions questions can be emailed to
permissionrequest@cengage.com

Library of Congress Control Number: 2008925689

ISBN-13: 978-1-4180-3079-7

ISBN-10: 1-4180-3079-1

Delmar
5 Maxwell Drive
Clifton Park, NY 12065-2919
USA

Cengage Learning is a leading provider of customized learning solutions with office locations around the globe, including Singapore, the United Kingdom, Australia, Mexico, Brazil, and Japan. Locate your local office at: **international.cengage.com/region**

Cengage Learning products are represented in Canada by Nelson Education, Ltd.

For your course and learning solutions, visit **academic.cengage.com**

Notice to the Reader

Publisher does not warrant or guarantee any of the products described herein or perform any independent analysis in connection with any of the product information contained herein. Publisher does not assume, and expressly disclaims, any obligation to obtain and include information other than that provided to it by the manufacturer. The reader is expressly warned to consider and adopt all safety precautions that might be indicated by the activities described herein and to avoid all potential hazards. By following the instructions contained herein, the reader willingly assumes all risks in connection with such instructions. The publisher makes no representations or warranties of any kind, including but not limited to, the warranties of fitness for particular purpose or merchantability, nor are any such representations implied with respect to the material set forth herein, and the publisher takes no responsibility with respect to such material. The publisher shall not be liable for any special, consequential, or exemplary damages resulting, in whole or part, from the readers' use of, or reliance upon, this material.

Printed in the USA
3 4 5 6 7 11

CONTENTS

Preface xv
Acknowledgments xvi

SECTION 1 — THE INDUSTRY—SCOPE AND DEVELOPMENT 1

Chapter 1: Geographical Location in the United States 2
  The Floriculture Business 2
  Plant Production Areas 6

Chapter 2: Foreign Competition with U.S. Flower Crops 17
  Impact of Offshore Flower Production on U.S. Production 17
  Major Exporters of Flower Crops 20
  Hazards of Importing Plant Materials 26

Chapter 3: Opportunities and Occupations in the Floriculture
  Production Industry 28
  Expert Knowledge Required by Plant Growers 28
  Types of Growers 30
  Job Opportunities 31

SECTION 2 — TYPES OF PLANT GROWING STRUCTURES 37

Chapter 4: Greenhouses 38
  Types of Greenhouses 38
  Construction of the Greenhouse 46
  Framing Materials 49
  Methods of Framing to Achieve Maximum Strength 51
  Foundations and Curtain Wall Construction 51
  Wooden Framing for Greenhouses 52
  Framing for Plastic-Covered Greenhouses 53
  Glazing Materials 55
  Other Growing Structures 66

Chapter 5: Financing, Locating, and Sizing a Greenhouse Facility ... 73
  Getting Started—Financing the Business 74
  Site Selection 76
  Greenhouse Structures 79
  Orientation and Arrangement of Greenhouses 80
  Greenhouse Benches 81
  Service Buildings 85

Chapter 6: Control of the Greenhouse Environment 89
  Heating 89
  Types of Fuel 90
  Types of Heating Systems for Glass Houses 91
  Distribution of Heat 94
  Calculation of Heat Requirements 98
  Piping for Hot Water and Steam Heated Greenhouses 101
  Heating Plant Boilers 103
  Emergency Heating Equipment 105
  Heating Systems for Plastic-Covered Greenhouses 106
  Heat Conservation Methods 108
  Solar Heating 111
  Ventilation and Cooling 111
  Types of Ventilation Systems 112
  Cooling Systems 119

SECTION 3 · EFFECTS OF ENVIRONMENTAL FACTORS ON PLANT GROWTH 131

Chapter 7: The Plant 132
  Cells 132
  Tissues 133
Chapter 8: Light 141
  Photosynthesis 141
  Characteristics of Light 142
  Photoperiod 147
  Light Quality 152
Chapter 9: Temperature 156
  Responses of Plants to Temperature at Various Stages
    of Growth 157
  Night Temperature 157
  DIF Day–Night Temperature Relationships 157
  Seed Germination 159
  Vegetative Growth 159
  Flowering Growth 163
Chapter 10: Gases 166
  Oxygen 167
  Carbon Dioxide 168
  Harmful Gases 171
  Wood Preservatives 173
  Other Air Pollutants 174
  Soil Fumigants 175
  Herbicides 176

SECTION 4 · GROWING MEDIA 179

Chapter 11: Soils: Properties and Makeup 180
  Growing Media 180
  Functions of a Growing Medium 181
  Composition of the Growing Medium 182
  Soil Basics 182
Chapter 12: Artificial Soils 201
  Need for Artificial Soils 201
  Standard Growing Media 201
  Industry Regulation 208
Chapter 13: Sterilization (Pasteurization) of Growing Media 211
  Steam Pasteurization 211
  Chemical Soil Treatment 216

SECTION 5 · NUTRITION AND WATERING 221

Chapter 14: Essential Elements for Plant Growth 222
  Types of Mineral Elements 222
  Distribution of Mineral Elements in Plants 223
  Mineral Element Absorption 223
  Essential Macro Elements 228
  Essential Trace or Micro Elements 235

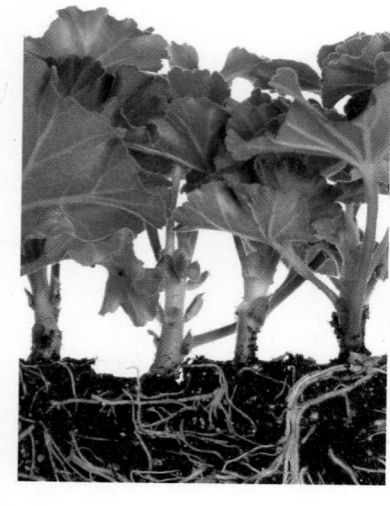

Chapter 15: Fertilizers and Methods of Application ----- 242

Organic Fertilizers 242
Inorganic Fertilizers 243
Complete Fertilizers 248
Trace Elements 250
Methods of Fertilizer Application 252
Adjusting Soil Acidity and Alkalinity 261
Determining Plant Nutrient Needs 264
Total Soluble Salts 265
Measuring Soluble Salt Content 267

Chapter 16: Water: Plant Requirements and Methods
of Application ----- 272

Role of Water 272
Application of Water 277

SECTION 6   PLANT PROPAGATION   299

Chapter 17: Sexual (Seed) Propagation ----- 300

Life Cycle of Plants 301
Annual Plants 301
Biennial Plants 302
Perennial Plants 302
Seeds 302
Seed Germination 304
Germination Media 307
Soil Pasteurization 308
Germination Environment 310
Seed Sowing 312
Fertilizer Requirements 317
Transplanting 317

Chapter 18: Asexual (Vegetative) Propagation ----- 323

Basis for Vegetative Propagation 324
Methods of Asexual Propagation 325
Propagation Environment 328
Other Vegetative Propagation Methods 335

SECTION 7   PEST MANAGEMENT AND CHEMICAL PLANT GROWTH CONTROL   347

Chapter 19: Insects, Mites, and Other Invertebrate
Greenhouse Pests ----- 348

Pest Prevention 350
Sanitation 350
Exclusion 351
Greenhouse Environment 353
Monitoring 354
Management 356
Pesticide Selection 359
Insects, Mites, and Other Invertebrate Pests 370

Chapter 20: Greenhouse Diseases ----- 382

Plant Disease 383
Fungi 383
Bacteria 392

Viruses   392
Phytoplasmas   395

Chapter 21:   Regulating Plant Growth - - - - - - - - - - - - - - - - - - - **398**
Chemical Plant Growth Regulators   399
Chemical Plant Growth Retardants   399
Other Chemical Plant Growth Regulators   404
Mechanical Conditioning   405

SECTION 8   CONTAINER GROWN CROPS   409

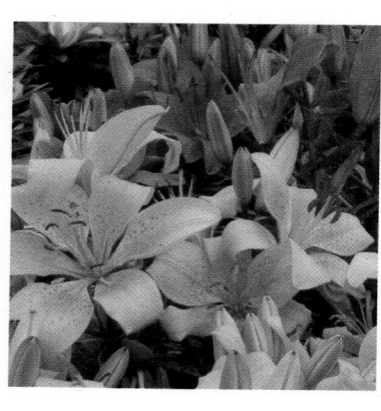

Chapter 22:   Containers and the General Handling of Container Grown Crops - - - - - - - - - - - **410**
Clay Pots   411
Plastic Pots   411
Peat Pots and Multicell Paks   412
Pot Sizes and Shapes   413
Growing Media   413
Transplanting, Potting, and Repotting   414
Preparation for Marketing   416

Chapter 23:   Flowering Plants I - **419**
Abutilon (Flowering Maple)   420
Achimenes (Magic Flower)   420
Azalea   421
Begonia   432
Bromeliads   437
Bougainvillea   438
Caladium (Fancy-Leaved Caladium)   439
Calceolaria (Pocketbook Plant)   440

Chapter 24:   Flowering Plants II - **444**
Christmas Cactus   444
Christmas Cherry   446
Cineraria   446
Clerodendrum   448
Cyclamen   449
Chrysanthemum   454
Exacum (Arabian Violet)   463
Fuchsia   463
Gardenia   464
Genista   466

Chapter 25:   Flowering Plants III - **470**
African Violet   470
Geraniums   474
Gloxinia   483
Hydrangea (Snow Ball Plant)   484
Kalanchoe   488
Poinsettia   491
Production, Regular Program   496
Primrose   507
Roses   508

Chapter 26:   Flowers from Bulbous Species - - - - - - - - - - - - - - - **516**
Bulbs for Forcing   516
Anemones   521
Calla (Calla Lily)   522
Crocus   524

Freesia   525
Gladiolus   526
Hyacinth   530
Iris   532
Lily   533
Narcissus (Daffodils)   546
Tulips   552

Chapter 27:   **Tropical Foliage Plants** - **574**

Production Areas   574
World Distribution   575
General Culture   576
Light and Temperature   579
Moisture and Humidity   580
Acanthaceae   580
Agavaceae   580
Amarathaceae   582
Araceae   582
Araliaceae   586
Araucariaceae   588
Asteraceae (Compositae)   588
Bromeliaceae   589
Commelinaceae   589
Euphorbiaceae   589
Gesneriaceae   589
Lamiaceae (Labiatae)   590
Liliaceae   590
Marantaceae   591
Moraceae   591
Myrsinaceae   593
Palmae (Arecaceae)   593
Pandanaceae   594
Piperaceae   594
Pittosporaceae   595
Polypodiaceae   595
Rubiaceae   596
Rutaceae   596
Saxifragaceae   597
Urtricaceae   597
Vitaceae   597
Foliage Plant Acclimatization   597
Ferns   600
Palms   600

Chapter 28:   **Bedding Plants** - **605**

The Production of Bedding Plants   605
Popular Crops   614
Ageratum (Floss Flower)   614
Alyssum (Sweet Alyssum)   614
Asters   615
Begonias   615
Browallia   615
Caladium   615
Celosia (Cockscomb)   615
Coleus   616
Geraniums   617
Pot Plants from Seed   617
Impatiens   620

Lobelia (Fairy Wings)   621
Marigolds   622
Pansy   622
Petunia   624
Phlox (Annual Phlox)   626
Salvia (Scarlet Sage)   627
Snapdragons   627
Verbena   628
Zinnia   629
Vegetable Crops   629

SECTION 9   CUT FLOWER CROPS   635

Chapter 29:   Major Cut Flowers: Antirrhinum, Dendranthema, Dianthus --------------------- 636
Antirrhinum (Snapdragons)   638
Chrysanthemums   643
Dianthus (Carnation)   658

Chapter 30:   Orchids and Roses ------------------------------- 679
Orchids   679
Roses   690

Chapter 31:   Minor Cut Flower Crops ------------------------- 712
Alstroemeria   712
Anthurium   714
Calendula (Pot Marigold)   714
Callistephus (China Aster)   715
Centaurea (Bachelor's Button)   718
Dahlias   720
Freesias   722
Gerbera (Transvaal Daisy)   722
Gypsophila (Baby's Breath)   724
Iris   724
Matthiola (Florists' Stock)   724
Narcissus   726
Peonies   726
Ranunculus   728
Schizanthus (Butterfly Flower)   728
Stephanotis   729
Tulips   730
Zantedeschia (Calla Lily)   730

SECTION 10   POSTHARVEST HANDLING AND MARKETING OF POTTED PLANTS AND CUT FLOWERS   733

Chapter 32:   Harvesting and Storage Methods -------------------- 734
Factors That Influence Cut Flower Life   734
Production and Harvesting Methods   735
Preharvest Handling   736
Harvesting   736
Cold Storage Treatments   737

Chapter 33:   The Role of the Wholesaler in the Marketing and Distribution of Floriculture Crops -------------- 746
Wholesale Distributor   746

Chapter 34:   The Retail Florist, Mass Market, and Other Outlets   757
          Types of Retail Outlets   758
          Other Types of Retail Outlets   759

Appendix I: A Partial List of Boom Irrigation Manufacturers and Suppliers   766

Appendix II: Association of Natural Biocontrol Products List of US ANBP Supplier Members   767

Appendix III: University Plant Disease Laboratories in the United States and Canada   769

Appendix IV: Formula for Preparing a Citric Acid Hydrating Solution   773

GLOSSARY   775

INDEX   787

A $8.65 billion dollar industry, commercial greenhouse and nursery plant production and sales play an important role in the dynamics of modern agribusiness. This growing industry has an increasing need for personnel trained in sound business practice and basic horticultural principles.

In this text, authors James W. Boodley and Steven E. Newman develop the basic outlines of the economics of the industry and present the individual's role within this environment. Principles and practical applications of commercial plant and flower production are taught in the sections on Plant Growing Structures; Effects of Environmental Factors; Growing Media; Nutrition and Watering; Plant Propagation; Container Grown Crops; and Cut Flower Crops. The concluding section on Postharvest Handling and Marketing of Pot Plants and Cut Flowers discusses the harvesting and storing of crops preparatory to marketing through wholesale and retail distribution channels.

In keeping with modern educational practice, the text has been written to develop the skills and knowledge required by present-day students in vocational, technical, and adult education. In addition to chapter-by-chapter learning objectives, the text includes unit-end material that not only reinforces the information presented in the chapter but also serves as a departure point for further student inquiry.

ABOUT THE AUTHORS

Dr. James W. Boodley is a Professor Emeritus of Cornell University, Ithaca, New York, where he served for almost 25 years in teaching, research, and extension activities of commercial floriculture interests. He co-developed the Cornell peat-lite soilless growing media, which revolutionized the methods used in the production of greenhouse flower crops. His research in postharvest physiology led to the development of a floral preservative that is used to prolong shelf life of cut flowers as well as a bud opening solution.

He is the author of over 75 scientific articles and 200 popular articles relating to flower crop production and handling. His textbook *The Commercial Greenhouse* is widely used in industry and college teaching programs. He received his B.S., M.S., and Ph.D. degrees at Pennsylvania State University, where he served as Assistant Professor of Horticulture for 2 years prior to his Cornell University career.

In 1983, Dr. Boodley took early retirement from Cornell University to become Manager of Horticultural and Post-Harvest Research at the Smithers Oasis Company in Kent, Ohio. In this position he has traveled extensively in the United States and Canada giving presentations to commercial flower growers, wholesale florists, and retail florists on the use of the Smithers Oasis Company line of floral products. He has also traveled to Australia, New Zealand, Japan, Hawaii, Mexico, Colombia, Peru, and Puerto Rico, and has made several trips to various countries in Europe evaluating the florist industry and giving guest lectures. Dr. Boodley is now retired from Smithers Oasis Company.

Dr. Steven E. Newman is a professor at Colorado State University, Fort Collins, Colorado, where he has served since 1994. There he conducts extension, teaching, and research programs for the commercial greenhouse industries of Colorado and the intermountain west. His work centers around water quality and water sanitation in the greenhouse.

He is the author of over 65 scientific articles and 100 popular articles relating to flower and nursery crop production. He received his B.S. at Montana State University, his M.S. at the University of Nebraska–Lincoln, and his Ph.D. at Texas A&M University. He served at Mississippi State University as Associate Professor for 9 years prior to his Colorado State University career.

In 2004, Dr. Newman was inducted into the Society of American Florists American Academy of Florists. He has traveled extensively in the United States giving presentations to commercial flower and nursery growers. He has also traveled to Colombia and Costa Rica where he assisted local grower groups with educational programs.

ACKNOWLEDGMENTS

My thanks ultimately must start with a recognition of Dr. Boodley who first prepared this text. It is an honor and a privilege to have the opportunity to provide an update to his work.

Many thanks go out to my teachers, associates, and mentors who throughout my career provided me with vast experiences with the growing of plants. I especially want to include a special acknowledgment to Dr. Fred T. Davies of Texas A&M University who helped me have faith in my own abilities.

Special thanks go to John A. Ray and many student workers and graduate students whose dedicated efforts contributed to many of the images and updated information included in this new edition.

Thanks go to the editorial and graphics staff of Delmar, Cengage Learning and S4Carlisle Publishing Services. If it were not for the gentle and continuous coercion of Christina Gifford, this work would not have been completed.

Appreciation is also extended to the many individuals, companies, and organizations who allowed the use of photographs.

My deepest appreciation goes to my family whose patience and devotion was tested during the preparation of this textbook: My wife and best friend Debra and my two sons Russell and Thomas.

Finally, I must acknowledge my mother, father, and stepmother: Kay Newman and Drs. Walter and Rosemary Newman. Their unceasing devotion to learning and to the academy has been my guiding light.

Illustrations were provided by the following:
Dr. Scott Bauer, USDA-ARS
Blackmore Company, Belleville, Michigan
Bouldin & Lawson, McMinnville, Tennessee
Colombian Centre for Innovation in Floriculture
Chapin Watermatics, Watertown, New York
Dr. Whitney Cranshaw, Colorado State University, Fort Collins, Colorado

Cravo Equipment, Limited, Brantford, Canada
Dan Gerace, Welby Gardens, Inc., Denver, Colorado
Grower Talks, Batavia, Illinois
Dr. Allen Hammer, Purdue University, West Lafayette, Indiana
Dr. Joe J. Hannan, Colorado State University, Fort Collins, Colorado
Mr. John Hughes, Guelph University, Guelph, Canada
Jiffy Products of America, W. Chicago, Illinois
Dr. Bill Miller, Cornell University, Ithaca, New York
The Paul Ecke Ranch, Encinitas, California
Laura Pottorff, Colorado Department of Agriculture, Lakewood, Colorado
Premier Horticulture, Inc., Dorval, Quebec, Canada
Professional Landcare Network (PLANET)
Dr. David W. Reed, Texas A&M University, College Station, Texas
Dr. Jack T. Reed, Mississippi State University, Starkville, Mississippi
Smithers Oasis Corporation, Cuyahoga Falls, Ohio
H. Edward Reiley
Dr. Brian Whipker, North Carolina State University, Raleigh, North Carolina

For permission to take photographs in their facilities, our thanks to:
Agri-Starts, Inc., Apopka, Florida
Aldershot of New Mexico, Las Cruces, New Mexico
Bath Garden Center, Fort Collins, Colorado
Clackamas Greenhouse, Aurora, Oregon
Colorado State University, Fort Collins, Colorado
Dramm & Echter Greenhouses, Encinitas, California
Denver Wholesale Florists, Denver, Colorado
Green Circle Growers, Oberlin, Ohio
Gulley Greenhouses, Inc., Fort Collins, Colorado
Johnson's House of Flowers, Grand Junction, Colorado
Jordan's Floral Gardens, Fort Collins, Colorado
Kitayama Brothers, Inc., Brighton, Colorado
Leider Greenhouse, Buffalo Grove, Illinois
Majorie McNeely Conservatory, Como Park and Conservatory, Saint Paul,
   Minnesota
Mellano & Company, San Luis Rey, California
Orchids by Hausermann, Inc., Villa Park, Illinois
Pacific Plug and Liner, Watsonville, California
Paul Wood Florist, Fort Collins, Colorado
The Paul Ecke Ranch, Encinitas, California
Priola Greenhouse, Denver, Colorado
Tagawa Greenhouses, Brighton, Colorado
Welby Gardens, Inc., Denver, Colorado
Mr. Donald Bradley, Elmira, New York
Mr. George Kobylarz, Newark Valley, New York
Mr. Marshall Lowman, Jr., Elmira, New York
Cleveland Plant Wholesale, Cleveland, Ohio
FINAST Stores, Cuyahoga Falls, Ohio
Ingliss Greenhouses, Youngstown, Ohio

Special request to use information from video tape:
   Yoder Brothers Incorporated, Barberton, Ohio

The authors and Delmar, Cengage Learning wish to express their appreciation to the following reviewers:

Robert Dockery
Western Piedmont Community College
Morganton, NC

Lauri Carey
Illinois Valley Community College
Oglesby, IL

Paul A. Thomas, Ph.D.
University of Georgia
Athens, GA

Valerie Simmens
W.B. Saul High School for Agricultural Sciences
Philadelphia, PA

Kathryn J. Johnson
Gateway Technical College
Kenosha, WI

Luann Jones
Troup Co. Comprehensive High School
LaGrange, GA

Peter Shaw
Spokane Community College
Spokane, WA

Special thanks go to David Rosenbaum, Acquisitions Editor, and Katie Wachtl, Content Project Manager.

DEDICATION

This third edition of *The Commercial Greenhouse* is dedicated to Homer Metcalf, Professor Emeritus, Montana State University, Bozeman, Montana. Educator and mentor. Without Homer, I would not have continued my education in graduate school.

The Industry-Scope and Development

Geographical Location in the United States

OBJECTIVES:

After studying this chapter, the student should have learned:

■ the areas of major flower crop production in the United States.

■ the 15 highest value florist crops.

■ the top seven states in floriculture production.

The Floriculture Business

Growing flowers on a commercial basis is big business in the United States. From a small start around the major population centers in the northeastern United States in the early 1800s, flower crop production (floriculture) now takes place in every state. According to the 2005 USDA Summary of Floriculture Crops for 2004, the total wholesale value of flower crops sold in the United States was $5.18 billion. The sales value for the top 15 crops is given in Table 1-1 and the sales value for the top 10 cut flower crops is given in Table 1-2.

The sales figures in Tables 1-1 and 1-2 are known as the farm gate value or the value to the wholesale grower. The retail value of floriculture is said to be three times as large. Thus, the retail floriculture industry is estimated to be worth more than $15.5 billion. This would bring the total floriculture industry value to over $20 billion, an increase of over six times the value reported in the 1976 Census of Horticultural Specialties by the USDA (Figure 1-1).

Over the past 30 years, major changes in the types of crops grown have taken place. Bedding and garden plants are now the major crops being produced, representing more than 50 percent of wholesale production (Figure 1-2). This change reflects the general population's increasing interest in gardening and growing plants outdoors. The influence of offshore producers of roses, chrysanthemums, and carnations has taken a heavy toll on the production of these crops in the United States. At one time chrysanthemums and carnations ranked number one and two, respectively, in dollar value; now chrysanthemums rank seventh among domestically grown cut flowers and carnations do not even make the top 10 (Table 1-2).

Although cut flowers are a major segment of the industry, offshore producers have forced many carnation, chrysanthemum, and rose growers to

Table 1-1

The Top 15 Florist Crops Grown for 36 States		
Crop	Ranking	Sales ($1,000)
Bedding and Garden Plants	1	2,532,545
Potted Foliage	2	638,979
Poinsettias	3	247,845
Potted Orchids	4	127,608
Florist Greens	5	92,445
Cut Lilies	6	78,169
Potted Chrysanthemums	7	75,018
Potted Spring Bulbs	8	53,268
Cut Roses	9	43,111
Azaleas	10	41,130
Easter Lilies	11	38,488
Cut Tulips	12	36,217
Cut Gerber Daisies	13	31,280
Cut Gladioli	14	25,957
Potted Miniature Roses	15	17,977

(*Source:* United States Department of Agriculture, Floriculture Crops Summary, 2004)

Table 1-2

The Top 10 Florist Cut Flower Crops Grown in 36 Selected States		
Cut Flower Crop	Ranking	Total Wholesale Value ($1,000)
Lilies	1	78,169
Roses	2	43,111
Tulips	3	36,217
Gerber Daisies	4	31,280
Gladoli	5	25,957
Iris	6	21,708
Chrysanthemums	7	20,030
Snapdragons	8	15,001
Delphiniums	9	9,891
Orchids	10	8,195

(*Source:* United States Department of Agriculture, Floriculture Crops Summary, 2004)

switch to other crops. This has resulted in increased demand such as cuttings and plug started seedlings. As a result, numerous specialty propagator/shippers have started in business. This sector of floriculture represents 8 percent of the total wholesale production for the United States.

Potted flowering and foliage plants are still highly profitable as these crops are not being challenged by their offshore producers. Plant Quarantine Act 37 has been relaxed for a few potted crops from Europe and that impact continues to be debated. Any relaxing of the rulings would allow importation of these crops and could have a devastating effect on U.S. growers. Lin Schmale with the Society of American Florists describes the Quarantine 37 (Q-37) regulations (properly referred to as 7 CFR 319.37) as the federal regulations governing the shipping of "plants for planting" (including seeds, bareroot material, cuttings) into the United States. Q-37 also covers "plants in approved growing media," which will always be regarded as high-risk material. Current regulations permit ferns, African violets, gloxinia, Begonia, Peperomia, hyacinth Alstroemeria, Ananas, Anthurium, Nidularium, Rhododendron, Phalaenopsis orchids, Penjing from China, and Christmas and Easter Cactus to enter the United States when grown in special conditions and shipped in approved media other than soil. In 2004, the USDA's Animal and Plant Health Inspection Service (APHIS) began the multiyear process of revising the Q-37 regulations. The outdated regulations do not provide U.S. agriculture and the environment with adequate protection against introducing plant pests, including noxious weeds. When the regulations were originally put into place in the 1930s, plants came to the United States from only a few foreign locations, were imported by nurseries in low numbers as germplasm "starter" material, and required treatment for pests. National plant distribution was slow. Today, plants from worldwide locations come to the United States in unlimited "consumer ready" quantities, and are treated only when quarantine pests are found. Border inspections, upon which the United States has historically relied, are increasingly ineffective.

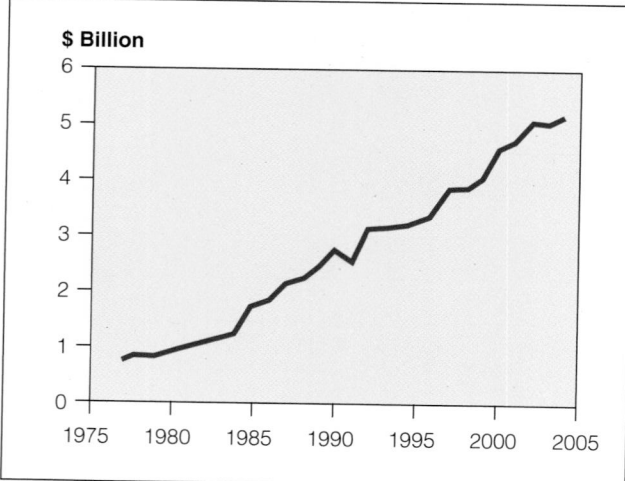

FIGURE 1-1 Value of floriculture sales at wholesale, 1975 through 2004. *(Source: United States Department of Agriculture, Floriculture Crops Summary, 1976–2005: 1979–1989, Data from 28 states. 1994–2004, Data from 36 states.)*

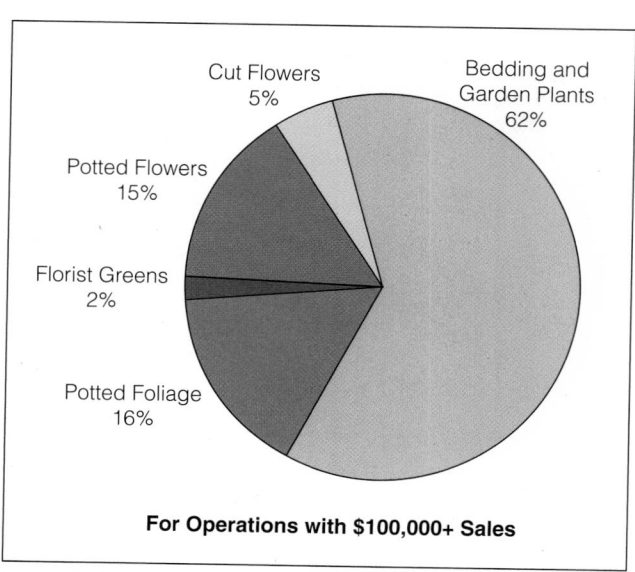

FIGURE 1-2 Wholesale value of floriculture, percentage by type of crop, United States 2004. *(Source: United States Department of Agriculture, Floriculture Crops Summary, 2004.)*

Table 1-3 lists the top eight potted florist crops and their dollar value. Hanging baskets are not considered in this category because they are made up using a mixture of bedding plants, small foliage, and other types of plants.

As stated earlier, bedding and garden plants are the major crops being produced in the United States (Table 1-1). California dominates this production, followed by Michigan, as shown in Table 1-4. However, this sector of floriculture is not without change. Since 2000, floriculture census data have divided bedding and garden plants into annuals and herbaceous perennial plants. From 2000 to 2003, the percentage of the total production of bedding and garden plants has increased from about 21 percent of the total to more than 25 percent (Figure 1-3). This trend is expected to continue as the interest in herbaceous perennial plants expands.

Table 1-3

The Top Eight Florist Potted Crops Grown in 36 Selected States		
Florist Potted Crop	Ranking	Total Wholesale Value ($1,000)
Poinsettias	1	247,845
Orchids	2	127,608
Chrysanthemums	3	75,018
Spring Bulbs	4	53,268
Azaleas	5	41,130
Easter Lilies	6	38,488
Miniature Roses	7	17,977
African Violets	8	16,308

(*Source:* United States Department of Agriculture, Floriculture Crops Summary, 2004)

Table 1-4

Top 10 States Reporting Sales by Growers of Annual Bedding and Garden Plants and Herbaceous Perennial Plants, $100,000 or More				
State	Ranking Based on Total Values	Annual Bedding and Garden Plants ($1,000)	Herbaceous Perennial Plants ($1,000)	Total Bedding and Garden Plants ($1,000)
California	1	254,145	71,409	325,554
Michigan	2	188,030	61,723	249,753
Texas	3	167,172	32,498	199,670
Ohio	4	94,399	39,669	134,068
Florida	5	94,121	22,883	117,004
New York	6	79,853	28,021	107,874
North Carolina	7	82,559	23,735	106,294
New Jersey	8	53,422	35,936	89,358
South Carolina	9	15,071	73,448	88,519
Pennsylvania	10	68,700	16,244	84,944

(*Source:* United States Department of Agriculture, Floriculture Crops Summary, 2004)

FIGURE 1-3 Increase in percentage of herbaceous perennials of the total bedding and garden plants. *(Source: United States Department of Agriculture, Floriculture Crops Summary, 2004.)*

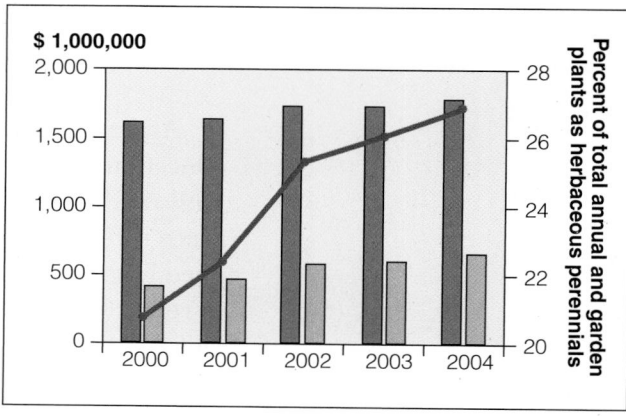

Table 1-5

Top 10 States Reporting Sales by Growers of Floriculture Crops, $100,000 or More		
State	Ranking	Total Wholesale Value of Reported Crops ($1,000)
California	1	1,018,247
Florida	2	825,672
Michigan	3	371,209
Texas	4	289,873
New York	5	209,015
Ohio	6	203,187
Pennsylvania	7	186,123
North Carolina	8	161,871
New Jersey	9	152,647
Washington	10	123,931

(*Source:* United States Department of Agriculture, Floriculture Crops Summary, 2004)

Plant Production Areas

Table 1-5 lists the top 10 states reporting sales by growers in excess of $100,000. California continues to be the leader; however, Florida has become a close second. Both states produce large numbers of foliage plants as does Texas (Table 1-6).

Tables 1-7 through 1-14 list the individual crops as reported by the 10 leading producing states for these crops.

The early development of the greenhouse industry centered around the larger cities. The major reason was the close proximity to the market. Yet, the availability of manure, which growers obtained from farmers on the outskirts of the cities, also had its impact. As the cities expanded and taxes increased, growers were forced to move to cheaper land.

The geographic location of flower production near urban centers is not as important as it once was. Jet aircraft now transport products quickly from one

Table 1-6

Top 10 States Reporting Sales by Growers of Foliage Plants for Indoor or Patio Use, $100,000 or More		
State	Ranking	Total Wholesale Value of Foliage for Indoor or Patio Use ($1,000)
Florida	1	415,967
California	2	96,607
Texas	3	28,622
Hawaii	4	15,443
Ohio	5	10,542
New York	6	6,740
North Carolina	7	5,201
Illinois	8	4,759
Louisiana	9	4,500
Michigan	10	4,152

(*Source:* United States Department of Agriculture, Floriculture Crops Summary, 2004)

Table 1-7

Top 10 States Reporting Sales by Growers of Poinsettias, $100,000 or More		
State	Ranking	Total Wholesale Value of Poinsettias ($1,000)
California	1	31,645
New York	2	19,072
North Carolina	3	17,496
Florida	4	15,468
Texas	5	15,141
Pennsylvania	6	13,528
Ohio	7	13,005
Michigan	8	12,743
Illinois	9	9,605
New Jersey	10	9,437

(*Source:* United States Department of Agriculture, Floriculture Crops Summary, 2004)

coast to another. The interstate highway system has also reduced the time needed to move heavy, bulky items such as lily bulbs.

Weather has always been a factor in the location of a flower growing business. Certain types of climate make one area of the country more suited to growing one type of crop than another. For example, the need for high light intensity during the winter months resulted in the growth of the carnation industry in New England, along the middle Atlantic coastal region in the east, and in Colorado in the west.

Table 1-8

United States Sales by State of Potted Orchids, $100,000 or More		
State	Ranking	Total Wholesale Value of Potted Orchids ($1,000)
California	1	49,597
Florida	2	46,965
Hawaii	3	17,133
Illinois	4	2,453
New Jersey	5	1,896
New York	6	1,746
Virginia	7	952
Pennsylvania	8	459
Texas	9	81
All other states		127,608

(*Source:* United States Department of Agriculture, Floriculture Crops Summary, 2004)

Table 1-9

Top 10 States Reporting Sales by Growers of Potted Chrysanthemums, $100,000 or More		
State	Ranking	Total Wholesale Value of Potted Chrysanthemums ($1,000)
California	1	12,503
Texas	2	9,391
Florida	3	6,850
New York	4	5,628
Pennsylvania	5	4,321
North Carolina	6	4,283
New Jersey	7	2,978
Illinois	8	2,199
Iowa	9	1,222
Missouri	10	1,127

(*Source:* United States Department of Agriculture, Floriculture Crops Summary, 2004)

The ability to control chrysanthemum flowering year round led to the start of that industry in Florida. An ideal year-round climate and the availability of a large labor force led to the start of the industry in California. Offshore production of carnations, roses, and chrysanthemums has forced growers to switch their production to other crops.

Increased energy and fuel costs have caused growers to adapt energy-saving measures and/or move to the Sun Belt region of the United States. Energy will continue to be a major issue for growers.

Table 1-10

Top 10 States Reporting Sales by Growers of Potted Spring Bulbs, $100,000 or More		
State	Ranking	Total Wholesale Value of Potted Spring Bulbs ($1,000)
New Jersey	1	8,671
California	2	6,875
Michigan	3	6,011
Pennsylvania	4	4,122
New York	5	3,393
Texas	6	2,110
Illinois	7	1,376
Florida	8	1,348
North Carolina	9	1,312
Massachusetts	10	1,303

(*Source:* United States Department of Agriculture, Floriculture Crops Summary, 2004)

Table 1-11

Top 10 States Reporting Sales by Growers of Azaleas, $100,000 or More		
State	Ranking	Total Wholesale Value of Azaleas ($1,000)
Oregon	1	14,085
California	2	5,387
Texas	3	1,489
Illinois	4	1,384
Wisconsin	5	797
Minnesota	6	860
Iowa	7	769
Michigan	8	727
North Carolina	9	605
Ohio	10	413

(*Source:* United States Department of Agriculture, Floriculture Crops Summary, 2004)

Pacific Northwest

Bulbs are produced in large quantities in northern California, Washington, and Oregon. The mild winters and the cool summers in this area produce the correct environment for lily, iris, and narcissus bulbs.

Lily bulbs are grown in northern California and southwestern Oregon for Easter forcing. Wedgewood iris bulbs are also used for forcing. Growers on the Isle of Jersey off the English coast prefer American iris to those produced in Holland (despite the fact that transportation costs from Holland are much less expensive).

Table 1-12

Top 10 States Reporting Sales by Growers of Easter Lilies, $100,000 or More		
State	Ranking	Total Wholesale Value of Easter Lilies ($1,000)
California	2	6,164
Michigan	1	4,878
New York	3	3,580
Pennsylvania	4	2,675
Ohio	5	2,190
New Jersey	6	2,142
Texas	7	1,511
Illinois	8	1,455
North Carolina	9	1,164
Florida	10	1,067

(*Source:* United States Department of Agriculture, Floriculture Crops Summary, 2004)

Table 1-13

Top 10 States Reporting Sales by Growers of Potted Miniature Roses, $100,000 or More		
State	Ranking	Total Wholesale Value of Potted Miniature Roses ($1,000)
California	1	9,005
Florida	2	733
Texas	3	567
New York	4	276
Michigan	5	253
Hawaii	6	248
Minnesota	7	126
Ohio	8	83
Illinois	9	77
Tennessee	10	74

(*Source:* United States Department of Agriculture, Floriculture Crops Summary, 2004)

Many narcissus bulbs are sent to eastern growers to be forced in greenhouses. Other bulbs are brought to the bud stage and then shipped for cut flowers. Millions of narcissus are produced in the Northwest for this purpose.

The climatic conditions that favor bulb growing and forcing in this area also benefit other crops. For example, azaleas from Oregon are grown until they can be shipped to other areas of the country for forcing directly into bloom or for growing on (Table 1-11). The center of production for budded and grafted roses is in California, Washington, and Oregon. These plants are used by rose growers in the United States to produce cut flowers year round. This region

Table 1-14

U.S. Sales by State of African Violets, $100,000 or More		
State	Ranking	Total Wholesale Value of African Violets ($1,000)
California	1	2,164
New York	2	1,109
Florida	3	824
Pennsylvania	4	423
Illinois	5	292
Minnesota	6	64
Colorado	7	51
Indiana	8	15
Louisiana	9	9
All other states		11,357

(*Source:* United States Department of Agriculture, Floriculture Crops Summary, 2004)

also supplies Christmas holly and other cut greens used by retailers for flower arrangements.

California

California is the leading agricultural state in the United States and this includes floriculture. With a coastline extending a thousand miles from north to south, California provides nearly any climate needed to grow flower crops. Chrysanthemums and roses are produced in record numbers. These and other specialty cut flowers are shipped to all parts of the country.

California has unique mountain ranges with valleys that open to the sea. The conditions in these valleys are ideal for the production of flower and vegetable seeds. Plant growth is stimulated by the cool nights caused by fog. The fog comes off the ocean in the late afternoon and disappears by midmorning. Little or no rain at the time of seed harvest ensures a good crop. Widely known California seed producers include Bodger Seeds, Ltd., W. Atlee Burpee Company, Goldsmith Seeds, and PanAmerican Seed, as well as many more.

Specialty propagators are attracted to the unique climate of California. The Paul Ecke Ranch of Encinitas is the world's leading propagator of poinsettias. A number of companies have main offices in other sections of the country but maintain propagation facilities in California. For example, Yoder Brothers is the primary supplier of chrysanthemum cuttings in the United States. The main office and headquarters of this company are in Barberton, Ohio. Ball Horticulture, Pacific, headquartered in West Chicago, Illinois, is another large horticultural supplier. This company is known in horticultural circles throughout the world.

Producers prominent in the production of new cultivars of bedding and other plants from seed are Bodger Seeds, Ltd., Goldsmith Seeds, PanAmerican Seed, American Takii, and Sakata (Figures 1-4 through 1-8). American Takii and Sakata are well-known Japanese seed companies that have production and processing facilities in California.

FIGURE 1-4 Gerber daisies grown for cut flowers hydroponically in southern California.

FIGURE 1-5 Vegetative annuals under evaluation by The Paul Ecke Ranch of Encinitas, California.

FIGURE 1-6 PanAmerican Seed Company's new 50,000 square foot, positive pressure trial greenhouses. 1995 California Pack Trials. *(Courtesy Grower Talks, June 1995.)*

FIGURE 1-7 Overview of 1995 California 1995 California Pack Trials. *(Courtesy Grower Talks, June 1995.)*

FIGURE 1-8 Another overview of the 1995 California Pack Trials. Note signs identify each cultivar. *(Courtesy Grower Talks, June 1995.)*

Hawaii

The floriculture industry in Hawaii has increased more than tenfold from the 1970 Horticultural Specialty Census dollar value of $5,663,323 to the 2004 Floriculture Crops Summary dollar value of $68,916,000. Potted foliage, other cut flowers, orchids (both cut flowers and potted plants), and anthuriums are a

FIGURE 1-9 Globe amaranth grown in Hawaii for leis.

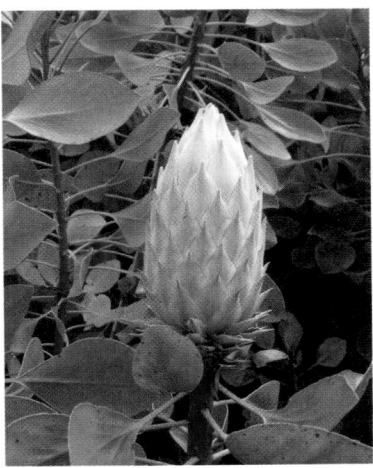

FIGURE 1-10 Protea flowers are widely grown on the island of Maui in Hawaii.

FIGURE 1-11 The pseudobulbs of these cattleya orchids are being grown in volcanic rock.

FIGURE 1-12 These *Oncidium* orchids are being grown for corsages and leis.

few of the top dollar crops grown. Flower production in Hawaii is shown in Figures 1-9 through 1-12.

Colorado

Denver, Colorado, for many years was called "The Carnation Capital of the World." This was due to the high light intensity that prevailed during the winter, which made ideal growing conditions for this crop. That title has long been lost to Bogota, Colombia. As a result of this offshore competition, carnation growers along the front range of the Colorado Rocky Mountains have shifted their production to bedding plants and some potted florist crops. For a period, many growers shifted to cut roses, but as of 2008, no large scale growers of cut roses exist in Colorado. There are a few successful specialty propagator/shippers that root geranium cuttings and produce bedding plant plug seedlings.

Texas

Tyler, Texas, is called the "Rose Capital of the State." Acres of field-grown roses continue to be produced in Tyler for sale to home owners as garden plants.

Texas also produces some commercial stock and grows more potted miniature roses than any other state. Growers in Texas have made a major shift to the production of foliage plants. There is a large concentration of foliage plant producers in the lower Rio Grande Valley near McAllen, Texas. Texas now ranks third behind Florida and California in the dollar value of sales for foliage crops (Table 1-6).

Florida

Florida leads other states in the production of tropical foliage plants (Table 1-6). Chrysanthemums are grown in quantity outdoors under shade cloth (Figure 1-13), but not to the extent they once were. Offshore production of these crops has all but eliminated the large chrysanthemum grower in Florida. Miami is a port of entry for flowers shipped from Puerto Rico and Central and South America.

Other Areas of the United States

Throughout the rest of the country, a large number of different crops are grown in glass- and plastic-covered greenhouses. During frost-free periods, a few special crops are grown outdoors.

Peonies are a major outdoor crop in parts of Illinois. However, they are less important now as a wholesale crop. Gladioli are grown in large numbers outdoors. Plantings from Florida to New England are made as soon as the frost-free date arrives for the area.

Specialty cut flowers are being produced in many areas. Research carried on in the horticultural departments at Oklahoma State University and now the University of North Carolina has provided growers with detailed cultural information for many of these crops.

Changes will continue to take place in the floriculture industry as the costs of production increase without an equal increase in the return to growers. Fuel and labor continue to dominate the expenses for any grower. Cheaper production in countries outside the United States has already forced major production changes in areas of this country. Foreign competition is discussed in Chapter 2.

FIGURE 1-13
Chrysanthemum stock plants grown under saran cloth for cuttings. The lights are included to prevent the plants from flowering.

ACHIEVEMENT REVIEW

Select the best answer or answers to complete each statement. List the appropriate letter(s).

1 The value of the flower crops sold in 2004 according to the USDA Summary of Floriculture Crops was:
a. $2,750,000,000.
b. $3,230,000,000.
c. $5,179,000,000.
d. $9,690,000,000.

2 Place the following potted crops in the order of sales, first to fourth, as reported in the 2004 Summary.
a. Poinsettias
b. Potted foliage
c. Chrysanthemums
d. Azaleas

3 Over the past 30 years, bedding and garden plant sales:
a. stayed the same.
b. increased by doubling.
c. became dominant.
d. decreased.

4 Place the following four states in order of their importance, first to last, in the production of flower crops in the United States.
a. California
b. Florida
c. Michigan
d. Texas

5 The chrysanthemum industry started in Florida as a result of:
a. the invention of plastic shade cloth.
b. the ability to control flowering year round.
c. a migration of northern growers to Florida.
d. the development of grades and standards.

6 The geographical area where bulb crops are grown is the:
a. middle Atlantic states.
b. southeastern United States.
c. southwestern United States.
d. Pacific Northwest.

7 Fog coming from the ocean to the coastal valleys of California at night provides an ideal climate for growing:
a. crops for seed.
b. cut stock.
c. outdoor roses.
d. gladioli.

8 The three main flower crops exported from Hawaii are:
a. other cut flowers.
b. potted foliage.
c. protea.
d. orchids.

9 For years Denver, Colorado, was called "The _____ Capital of America."
a. Carnation
b. Chrysanthemum
c. Foliage plant
d. Rose

10 Florida is first in the production of:
a. chrysanthemums.
b. foliage plants.
c. poinsettias.
d. roses.

Foreign Competition with U.S. Flower Crops

OBJECTIVES:

After studying this chapter, the student should have learned:

■ the major cut flower crops imported into the United States.

■ the countries with the largest number of flower exports to the United States.

■ the importance of imports and exports to the floriculture industry.

Flower and ornamental crops imported into the United States are valued at millions of dollars. Both the quantity and the dollar value of these imports increase each year. Although the United States has been importing these crops for many years, it was not until 1970 that complete records were kept on imports.

Impact of Offshore Flower Production on U.S. Production

In the last 35 years, offshore producers of cut flower crops have made significant increases in the number of cut flowers shipped into the United States. Colombia is the country primarily responsible for these changes, in that more than 74 percent of the cut flowers imported into the United States are grown in that country (Table 2-1).

Table 2-2 lists a selection of the most actively imported crops as they were reported from 1989 through 2003 by country. As an example of these changes, in 1989, Colombia shipped $169 million worth of cut flowers into the United States. Fourteen years later, shipments of cut flowers increased over 1.5 times to $267 million. Other examples of similar changes can be calculated by the reader.

These imports of cut flowers have had a significant impact on the numbers of flower growers of carnations, chrysanthemums, and roses who have

Table 2-1

Imports of Flowers, Value of Cut Stems by Countries—1989 through 2003								
Country	1989	1991	1993	1995	1997	1999	2001	2003
	Value in $1,000							
Canada	1,558	1,330	1,737	2,522	3,496	3,033	2,939	2,545
Colombia	169,199	184,261	222,782	276,995	311,569	297,464	246,429	266,574
Costa Rica	5,692	7,028	7,702	6,698	6,987	6,154	4,273	3,257
Ecuador	5,793	7,894	12,251	30,782	57,027	65,274	67,536	60,644
Guatemala	2,568	3,689	5,044	7,110	7,719	5,266	3,249	3,776
Mexico	8,492	13,301	10,998	17,942	16,940	17,405	18,303	8,502
Netherlands	7,928	6,156	6,174	7,230	7,609	6,593	4,403	3,093
Thailand	3,928	4,465	4,178	4,577	3,765	2,562	2,698	2,982
All other countries	11,589	7,302	6,516	9,402	7,167	6,943	6,749	6,660
Total cut flowers imported into U.S.	216,747	235,426	277,382	363,258	422,279	410,694	356,579	358,033

(*Source*: Agricultural Import Commodity Aggregations, Foreign Agriculture Service—United States Department of Agriculture)

Table 2-2

Imports of Selected Flowers, Value of Cut Stems by Countries—1989 through 2003									
Country	Flower Commodity	1989	1991	1993	1995	1997	1999	2001	2003
		Value in $1,000							
Canada	Total Cut Flowers	1,558	1,330	1,738	2,522	3,496	3,033	2,939	2,545
	Carnations (all)	124	9	0	6	19	50	47	35
	Pompon Chrysanthemums	104	19	1	7	27	45	189	419
	Standard Chrysanthemums	150	0	13	16	9	41	493	474
	Dried or Dyed Cut Flowers	530	469	942	672	716	499	404	253
	Roses	429	436	524	1,398	1,999	1,096	1,062	364
	Sweetheart Roses	218	390	258	411	721	1,295	720	729
Colombia	Total Cut Flowers	169,199	184,261	222,782	276,995	311,570	293,465	246,429	266,574
	Miniature Carnations	12,934	20,213	20,235	31,304	35,859	39,200	25,560	23,109
	Standard Carnations	55,741	47,708	61,466	78,167	73,755	64,947	43,858	36,124
	Pompon Chrysanthemums	37,388	40,439	47,641	59,954	60,262	61,409	46,067	51,822
	Standard Chrysanthemums	4,334	8,277	13,033	7,939	9,402	7,352	8,239	10,341
	Dried or Dyed Cut Flowers	2,339	68	51	36	37	372	91	48
	Roses	51,404	65,808	79,859	99,434	132,077	120,620	118,058	138,387
	Sweetheart Roses	5,011	1,735	447	151	170	3,447	4,302	6,428
	Orchids	46	12	44	4	8	118	255	304
Costa Rica	Total Cut Flowers	5,692	7,028	7,702	6,698	6,987	6,154	4,354	3,257
	Carnations (all)	397	165	0	0	2	0	0	0
	Pompon Chrysanthemums	4,468	5,465	6,416	4,694	5,528	5,252	3,781	2,591
	Standard Chrysanthemums	9	176	420	332	199	245	257	210
	Dried or Dyed Cut Flowers	21	412	21	24	11	94	10	25

(*continues*)

Table 2-2 *(continued)*

	Roses	783	789	756	1,435	1,025	186	57	57
	Sweetheart Roses	0	3	9	5	0	3	6	350
	Orchids	7	0	70	204	214	363	243	17
Ecuador	Total Cut Flowers	5,793	7,894	12,251	30,782	57,027	65,274	67,536	60,644
	Miniature Carnations	594	1,423	583	1,057	955	1,351	579	390
	Standard Carnations	77	262	311	942	3,747	4,264	474	280
	Pompon Chrysanthemums	486	195	368	663	173	89	93	137
	Standard Chrysanthemums	284	129	45	140	308	107	90	35
	Dried or Dyed Cut Flowers	2	0	27	22	118	274	191	18
	Roses	3,145	5,864	10,899	27,790	51,693	58,502	64,780	57,924
	Sweetheart Roses	1,204	19	18	162	33	687	1,328	1,858
Guatemala	Total Cut Flowers	2,568	3,689	5,044	7,111	7,719	5,266	3,249	3,776
	Miniature Carnations	0	0	0	103	33	88	246	0
	Standard Carnations	20	189	118	260	1,539	1,034	72	0
	Chrysanthemums (all)	0	3	0	0	6	0	4	0
	Dried or Dyed Cut Flowers	1	5	20	101	47	45	54	13
	Roses (all)	2,540	3,492	4,866	6,647	6,093	4,091	2,873	3,763
Mexico	Total Cut Flowers	8,492	13,301	10,998	17,942	16,940	17,405	18,303	8,503
	Miniature Carnations	140	3	276	249	8	43	29	3
	Standard Carnations	415	241	424	460	143	86	317	100
	Chrysanthemums (all)	96	3	136	0	0	3	0	0
	Dried or Dyed Cut Flowers	652	2,274	1,900	4,458	7,170	8,784	8,365	4,821
	Roses	7,170	10,729	8,140	12,250	9,378	8,224	9,238	3,367
	Sweetheart Roses	17	46	32	418	240	130	173	212
Netherlands	Total Cut Flowers	7,928	6,517	6,174	7,230	7,613	6,593	4,429	3,093
	Miniature Carnations	87	87	50	140	200	68	17	60
	Standard Carnations	22	16	1	30	40	9	4	16
	Pompon Chrysanthemums	85	178	227	621	730	634	487	44
	Standard Chrysanthemums	575	239	876	718	865	1,187	967	289
	Dried or Dyed Cut Flowers	4,558	4,422	3,898	3,969	3,016	2,134	463	229
	Roses	2,087	1,071	782	1,295	2,191	1,839	1,938	1,661
	Sweetheart Roses	9	25	7	60	192	339	144	200
	Orchids	502	119	333	387	379	382	409	592
Thailand	Total Cut Flowers	3,928	4,465	4,178	4,577	3,765	2,562	2,698	2,982
	Carnations (all)	1	0	0	0	0	0	0	2
	Chrysanthemums (all)	0	0	0	0	1	3	0	0
	Dried or Dyed Cut Flowers	949	755	207	295	205	67	116	520
	Roses (all)	4	0	0	0	0	0	0	0
	Orchids	1,326	1,000	1,031	1,203	598	579	695	334
	Dendrobium Orchids	1,648	2,709	2,940	3,074	2,960	1,913	1,887	2,126

(*Source:* Agricultural Import Commodity Aggregations, Foreign Agriculture Service—United States Department of Agriculture)

Table 2-3

Summary of United States Cut Flower Growers Decline in Numbers, 1979 through 2004			
Flower Commodity	Year	Total U.S. Growers (no.)	Percent Loss from 1979
Carnations	1979	618	—
	1984	410	33.7
	1989	471	23.8
	1994	185	70.1
	1999	109	82.4
	2004	28	95.5
Chrysanthemums	1979	1,828	—
	1984	1,221	33.2
	1989	874	52.2
	1994	261	85.7
	1999	162	91.1
	2004	61	96.7
Roses	1979	415	—
	1984	378	8.9
	1989	480	+15.7
	1994	307	26.0
	1999	173	58.3
	2004	66	84.1

(*Source*: United States Department of Agriculture, Floriculture Crops Summary, 1979 through 2004. Data from 28 states, 1979–1989. Data from 36 states, 1994–2004)

been forced to go out of business in the United States. The data presented in Table 2-3 give a comparison for the years 1979 through 2003 on the decrease in the number of growers and the percentage of grower loss. Producers of carnations, pompons, and standard mums have had the greatest decline. The number of rose growers did not decline so greatly, but the major impact of offshore producers of roses has only begun to have a negative influence in the last five years.

It should be obvious to the reader that producers of these crops have either switched to growing other crops or have gone out of business. Rose growers in particular, many of them formerly located close to large cities, have been encroached upon by the growth of suburban areas. Increasing tax burdens, owners getting older, sales of properties to real estate speculators and developers, in addition to being unable to compete financially with imports have all been reasons for rose growers going out of business.

Major Exporters of Flower Crops

The development of the floriculture industry in Central and South America was the result of farsighted growers in the United States and other countries joining forces with the local growers. Natural advantages such as favorable climatic conditions and national advantages such as government low interest loans and other subsidies favored the development of the industry. Rapid means of getting one's product to market by jet aircraft has also been a stimulus. These same kinds of influences are now at work in South Africa as well as in other countries. Yet, Central and South America dominate the volume of cut flowers imported into the United States.

Tropical regions experience a wide variation in climatic conditions. This is due to proximity to the equator and also to differences in elevation where crops are grown in many countries. It may be too warm at sea level for a particular crop to be successfully grown, but 3,000 feet (1,000 m) up the mountain the climate may be more favorable.

Colombia

Bogota, the capital city of Colombia, South America, is a major source of flower crops. The elevation of Bogota is nearly 8,600 feet (2,620 meters). At this altitude, the nights are cool and the days are comfortably warm. Generally the weather conditions are ideal for growing a number of different flower crops. Most of the larger growers provide extra benefits for their employees, such as medical care, child care, breakfast at a greatly reduced price, transportation to and from work, and paid vacations. Despite these benefits, the wages are so low that American growers cannot compete in labor costs. High fuel prices in the United States drive this issue as well.

Because the work is not complicated and the wages are low, Colombian growers can hire more workers than American growers normally employ. Most of these workers are women. In general, the greater number of workers means that the planting, maintenance, and harvesting of the crops will require less time.

The climate around Bogota is ideal for growing carnations and roses (Figure 2-1). Although there are frosts in the winter, no heat is required in the greenhouses where the carnations are grown (Figures 2-2 and 2-3). There are over 16,171 acres (6.544 hectares [ha]) of greenhouses in Colombia producing carnations, roses, and other crops. Some of these greenhouse farms were started by Americans

FIGURE 2-1 Bogotán worker inspects roses ready for harvest in Colombian greenhouse for shipment to the United States and other countries.

FIGURE 2-2 Minimal cost plastic-covered structures are all that is needed to protect crops from rain and cool nights.

FIGURE 2-3 Newly planted carnation cuttings in Colombian greenhouse. Sanitation is crucial to maintain the plantings disease free.

FIGURE 2-4 Daisies growing under protected culture in Colombia. *(Courtesy Colombian Centre for Innovation in Floriculture.)*

who recognized the advantages of this area for flower production. Many of the cut flowers are used in Colombia, yet most are grown for export. Flowers represent 5.67 percent of the total volume of all products exported from Colombia.

Marguerite daisies are grown in open fields near Bogota (Figure 2-4). The daisies are harvested in the field and are dyed pink, yellow, blue, green, and other colors. Then they are shipped to markets in the United States and Europe.

Chrysanthemums and roses are also grown on farms around Bogota (Figure 2-5). These crops need a minimum night temperature of 60°F (15.5°C). As a result, the greenhouses where these crops are grown must be heated. The greenhouses consist of a wooden framework covered with a single layer of polyethylene sheeting. These structures must be recovered every year because the high light intensity breaks down the plastic.

FIGURE 2-5
Chrysanthemums in Bogotá greenhouse.

Medellin (pronounced *med-a-yeen*) is a large city about one hour west of Bogota by airplane. Just over the mountain range from the city is a valley where the night temperatures seldom go below 60°F (15.5°C). Chrysanthemums are grown here under plastic-covered structures. Artificial heat is not required. The plastic protects the plants from the heavy rainfall during the rainy season. In addition the polyethylene protects the crop from a number of potentially serious diseases.

Despite the low labor costs and the favorable climate, crop growers in Colombia have a number of problems. Political instability and pressures from the illegal drug trade dominate these issues. Most supplies are brought into the United States by air. The cost of airfreight is very high. A single import duty on a large item increases the cost to five times that of the original price. The Andean Free Trade Act of 2002, which was updated in 2005, eliminates many of these duties.

The cut flower crops are exported from Colombia by air. Cut flowers are perishable and can be damaged easily. Part or all of a shipment can be lost by overheating while waiting to be loaded on the plane, or by freezing in the storage compartment of the plane during the flight. Yet advances in postharvest treatment of cut flowers have reduced the impact of many of these issues. Delays in shipment remain common due to flight cancellations.

Other American Countries as Exporters

Other areas in tropical America have become increasingly important in the production of cut flowers and unrooted cuttings of flowering and foliage plants. Brazil, Costa Rica, Dominican Republic, Ecuador, Guatemala, and Mexico have all increased production of various cut flower crops (Figure 2-6). Some of these crops are used internally, but the vast majority of the products grown are exported to the United States and other countries of the world.

One of the oldest floriculture firms in Central America is located in Cartago, Costa Rica. One of the outstanding plant breeders in the world was Mr. Claude Hope of Costa Rica. His farm has produced some of the best hybrid petunia seeds ever developed. Even though Mr. Hope has died, Guatemala continues to produce flower seeds that are sold worldwide.

Several U.S. propagators of herbaceous annual bedding plants have recently established farms in Guatemala to grow stock plants for unrooted liners for

FIGURE 2-6 Gerber daisies in Costa Rican greenhouse. Note that this is a sawtooth structure covered with fiberglass-reinforced plastic to protect the crop from rain.

FIGURE 2-7 A 100-acre (42 ha) field of mini carnations as they are grown in Kenya, Africa. *(Courtesy of Mr. John Hughes, Ontario Advisory Service, Canada.)*

shipment worldwide. These crops primarily include geraniums, poinsettias, and New Guinea impatiens, but the list of species is growing.

Mexico is one of the countries where greenhouses, mostly polyethylene covered, have increased in acreage for the production of roses. Yet, the greenhouse vegetable industry in Mexico has increased from 123 acres (50 ha) in 1991 to an estimated 6,670 acres (2,700 ha) in 2004. This is primarily for the production of hydroponic tomatoes.

The offshore production of roses in Mexico and Central and South America has taken its toll on the number of rose growers remaining in the United States. The number of U.S. rose growers continue to decline (Table 2-2).

Kenya

A relative newcomer to the industry, Kenya is a flower producer and exporter only since the early 1970s, yet very little is imported into the United States. Kenya is located in east Africa, just south of Ethiopia, on the equator. Direct flights to Europe serve this market, as it is only an eight-hour direct flight to Holland, Germany, or England.

The country has a benign climate with a variety of temperature ranges for different crops. Greenhouse structures are of simple construction (Figure 2-7). They serve merely to keep rain off sensitive crops such as roses. Most of the flower farms are around Lake Naivasha where no auxiliary heat is needed.

There is abundant, cheap local labor. Workers are paid approximately $2.65 U.S. per day for an eight-hour workday. On large farms where several hundred to several thousand people are employed, housing is provided. In addition, the employees and their families receive medical care, schooling for their young children, and usually one meal a day.

One farm alone has over 740 acres (300 ha) in production of flower crops. It is said to be the largest cut flower producer in the world. Other farms are lesser in size, down to as little as 6 acres (2.5 ha). Many of these farms are undergoing expansion.

Roses are a predominant crop grown. These are grown mostly under cover, but there is some field production of spray type roses.

Spray carnations are grown in open fields without greenhouse covering. Standard carnations are also field grown. Other crops grown seasonally are *Amaranthus, Celosia, Delphinium Euphorbia, Gypsophila,* and statice. *Alstroemeria* is grown in lesser amounts due to competition from Dutch growers in Holland.

Practically all of the cut flowers are shipped by air to Europe with the majority going to Holland. There they are sold through the import auction market providing more than 25 percent of the flowers sold in the European Union. A small number of roses are also shipped to Canada. New markets for flowers from Kenya are opening in Russia and Japan.

Two American companies, Goldsmith Seeds of California and Yoder Kenya Ltd. have been in operation in the country for several years. Goldsmith's farm produces seed for geraniums, impatiens, and petunias. This seed is shipped to California for testing, processing, and ultimately sale to the industry.

Yoder Kenya Ltd. produces unrooted *Dendranthema* (chrysanthemum) cuttings, which are shipped daily to the United Kingdom and Europe. Yet, disease pressure has made the future of these production facilities tenuous.

There is little doubt that the floriculture world will hear more of Kenya as flower production in that country increases.

Other Exporters of Floriculture Crops

Countries around the world ship flowers and plant materials to the United States. Almost 2.4 million orchids arrive each year from Thailand. Both Israel and the Netherlands export cut flowers to the United States. The Netherlands also ship several million cut tulip flowers to the United States (Figure 2-8). In addition, flower crops and foliage propagation materials are imported from various countries in Europe and from South Africa.

FIGURE 2-8 Cut tulips ready for auction in the clock auction market in Aalsmeer, Holland. *(Courtesy Bill Miller, Cornell University.)*

Not all flower and plant production in Europe and in other countries should be looked upon as competition for American products. Without the millions of tulip bulbs imported from the Netherlands, the bulb forcing industry in the United States would be very small. Bulbs, corms, roots, and similar products account for much of the multimillion dollar value of greenhouse and nursery products imported into the United States each year.

Shipping to Foreign Markets

Large numbers of plants and flowers are shipped from the United States to other countries. Gladioli are regularly shipped to markets in Europe, the Caribbean, and Canada. Canada not only exports plant materials to the United States, but also it is one of the largest foreign markets for floriculture products sent out of the United States.

Hazards of Importing Plant Materials

The plant materials imported into the United States arrive at a number of cities around the country. For example, most of the carnations, roses, and chrysanthemums from South America enter the United States through Miami, Florida. At each point of entry, the materials must be inspected by customs agents to ensure that diseases and insects are not brought into the country. The Animal and Plant Health Inspection Service of the U.S. Department of Agriculture is responsible for this effort. Since September 11, 2001, the mission that APHIS has been charged to deliver—the prevention of accidental or deliberate introduction of the diseases and other pests to U.S. agriculture—has increased in importance. On occasion, however, plant diseases do pass undetected through such inspections. White rust of chrysanthemums has been a serious problem for years in Europe. In the summer of 1977, an outbreak of this disease was discovered in Pennsylvania and it spread to California. This disease threatened the entire chrysanthemum industry in the United States and was eradicated. Since that time, only occasional outbreaks have occurred in southern California and more recently in Canada.

Other more serious outbreaks of diseases have occurred recently. *Ralstonia solanacearum* race 3 biovar 2, a bacterial pathogen causing a wilt disease in many agricultural crops, was previously unknown in the United States. During 2003, several U.S. greenhouses received imported geranium cuttings from Kenya and were subsequently eradicated. This disease has the potential to cause major disruption to U.S. agriculture and was placed on the list of select agents on the Agricultural Bioterrorism Act of 2002. The introduction into the United States from infected geraniums was not considered to be intentional, but the result of unsanitary greenhouse practices from a foreign source facility.

As faster means of travel and communication are developed, a continued increase is expected in the two-way movement of flowers and plant materials between the United States and other countries.

ACHIEVEMENT REVIEW

Select the best answer or answers to complete each statement. List the appropriate letter(s).

1 List in order of importance the four major cut flowers imported into the United States.
  a. Roses
  b. Chrysanthemums
  c. Pompons
  d. Carnations

2 Of the following countries, which one ships the greatest number of cut flowers to the United States?
  a. Costa Rica
  b. Colombia
  c. Guatemala
  d. Mexico

3 The development of the cut flower industry in South America is due to:
  a. a favorable climate.
  b. a large labor force.
  c. a low pay scale.
  d. all of the above conditions.

4 Most of the carnations shipped from South America are grown on farms located around:
  a. Bogota, Colombia.
  b. Medellin, Colombia.
  c. Quito, Ecuador.
  d. Mexico City, Mexico.

5 A deadly chrysanthemum disease was discovered in 1977 in the United States. It is called:
  a. damping off.
  b. *Pythium.*
  c. *Rhizoctonia.*
  d. white rust.

6 The tulip industry in the United States depends upon bulbs imported from:
  a. South America.
  b. Australia.
  c. the Netherlands.
  d. West Germany.

7 One of the largest customers for flowers grown in the United States is:
  a. South America.
  b. Africa.
  c. Canada.
  d. West Germany.

Opportunities and Occupations in the Floriculture Production Industry

OBJECTIVES:

After studying this chapter, the student should have learned:

- several types of growers.
- a number of skills a grower should have to ensure a successful business.
- several types of jobs in the floriculture industry and the tasks required in these jobs.

Growing greenhouse crops is probably the most advanced type of modern production agriculture. For example, the growth of holiday crops, such as poinsettias and lilies, must be controlled exactly if they are to be at the peak of their perfection and in bloom at the proper time. Such control requires all of the knowledge and skill of the grower. No other type of plant production is so demanding.

Expert Knowledge Required by Plant Growers

Growing Skills

A commercial producer of flower crops must be an expert in many areas to be successful. A grower must know a great deal about soils and container media because the proper growing medium for a particular crop is obtained by mixing various types of materials. It is important that growing plants be irrigated and fertilized on a regular schedule. Too much water or fertilizer may cause the plants to die and too little water or fertilizer results in low-quality plants, which cannot compete in the marketplace.

Chrysanthemums can be made to flower year round if the grower knows how temperature and day length control their growth. The use of artificial lights at night permits growers to maintain several other crops in a nonflowering or vegetative stage of growth. A change in temperature of a few degrees warmer or colder during the growth of a crop can either hasten or delay blooming by a few days, a week, or more.

The grower must be able to control plant diseases and insect pests without injuring the flower blooms or spotting or staining the foliage with spray residue. The entire area of disease control is a great challenge to growers. Many insects build up a resistance to various pesticides. Thus, it is often necessary to find alternative methods of control including integrated pest management.

Diseases such as *Botrytis* and mildews are kept at a low level in the greenhouse if the ventilation and heating systems are handled properly. If these factors are improperly managed, the flower crop can be ruined quickly.

Bringing the crop to the point of harvest is only half of the grower's job. The object of growing flowers and plants is to sell them at a profit. The grower must understand the factors that affect the postharvest physiology of plants. For example, what happens inside the cut flower once it is harvested? What can be done to keep the cut flowers and flowering plants alive and saleable? If the grower is knowledgeable, days can be added to the vase life of a cut flower as well as improve the longevity of potted flowering plants for the consumer.

Mechanical Skills

The successful grower must have other abilities besides growing skills. Knowledge of carpentry and welding will enable the grower to make repairs or build benches or other structures for greenhouse use. Irrigation systems may require basic plumbing skills. Basic electrical knowledge is required to maintain the lighting, heating, ventilation, and other systems necessary in a greenhouse. The skills of a heating engineer or boiler technician may be required as well.

Marketing Skills

Crops must be brought to market in good condition. To accomplish this, they are first graded according to certain standards of quality. Then they are packed carefully in containers that will protect them from overheating in the summer or freezing in the winter and from being damaged by careless handling. A properly packed box of cut flowers should be able to withstand an accidental 6-foot fall without damage to the flowers inside.

Administrative Skills

A grower must hire labor to produce crops for sale. Thus, the grower becomes an administrator, or "the boss." As the person in charge, it is the grower's responsibility to assign the various tasks that will result in a good plant crop. Crops must be planted, pinched, tied up, watered, fertilized, pruned, disbudded, given the black cloth treatment, treated with growth control chemicals, sprayed or fumigated for insects and diseases, harvested, refrigerated, graded, packed, and sold. The supervisor in charge must see that these tasks are performed on schedule, correctly, and in as little time as possible. Time is money, and labor is costly.

Computer Skills

The use of computers is involved in every aspect of growing flowering crops. Computers are used to maintain the greenhouse environment as well as to maintain good records. Skills with using spreadsheets, databases, electronic communications, and the Internet are required of all growers.

Laws, Regulations, and Taxes

To operate a business today, a grower must satisfy the rules and regulations of various local, state, and federal agencies. For example, the grower is required to keep accurate financial records of income and expenses. In many states, before a flower grower can buy pesticides and use them on the crops, the grower must be a certified or licensed applicator.

Types of Growers

A person who decides to become a grower should realize that this is a full-time job 365 days a year. Plants are living organisms and need daily attention. Although there are many automatic and semiautomatic methods and devices to do the required work in a greenhouse, people are needed to direct the operation of these devices. There are several options for a person who wishes to be a grower.

Wholesale Grower

The wholesale grower produces one or more crops for sale at the wholesale level only. Wholesale growers typically do not make retail sales. A grower may specialize in just one crop, such as roses; or a grower may produce two or three cut flower crops. For example, carnations and snapdragons are cool temperature crops that can be grown in the same greenhouse. Chrysanthemums and carnations may be grown by the same producer; however, they have different temperature needs and must be planted to separate greenhouses. A wholesale grower usually is responsible for a greenhouse range, or the area covered by one or more greenhouses, that is not less than 0.5 acre (0.2 ha) in size. The range may be as large as 10 to 20 acres (4.05 to 8.1 ha).

Many wholesale growers produce potted or bedding plants only. Flowering plants such as poinsettias or tropical green plants may be grown. Tropical green plants are grown and sold year round. Most growers of flowering potted plants raise more than one type of plant. Poinsettias are grown for Christmas. Lilies and azaleas are important Easter and Mother's Day crops. Pot mums (chrysanthemums) and many other types of pot plants can be grown year round.

The growing of potted plants uses more labor and is more costly than cut flower production. However, cut flower crops cannot be adapted to mechanical laborsaving methods as easily as potted plant production. Some growers in the United States produce both potted plants and cut flower crops.

In areas where flowers can be grown outdoors year round, single-crop specialty growers are common. For example, Florida produces gladioli and chrysanthemums. California produces large numbers of specialized crops outdoors.

Specialty Producers

In addition to growers of cut flowers and potted plants, there are a number of specialty producers of plant propagation materials. Millions of rooted and unrooted chrysanthemum and carnation cuttings are propagated by Yoder Brothers of Barberton, Ohio; Salinas, California; and Fort Myers, Florida. Van ZanTen California Plant Company also propagates large numbers of rooted chrysanthemum cuttings.

The Paul Ecke Ranch of Encinitas, California, propagates poinsettia cuttings for shipment around the world. Oglevee Ltd., Fischer USA, Dümmen, and Selecta are also large propagators of poinsettias. All of these companies have diversified their offerings to include many other vegetatively propagated herbaceous annuals. Many other specialty growers are located in various parts of the United States.

Seed Suppliers

Seed producers form another group of specialty growers. Many of the crops grown for sale are started from seed. The most rapidly expanding area of the floriculture industry is bedding plants. In general, nearly all bedding plants are started from seed.

Well-known growers and companies supplying seeds for the florist industry are as follows:

- American Takii, Inc., Salinas, California
- Bodger Seeds, Ltd., El Monte, California
- DeRuiter Seeds, Inc., Columbus, Ohio
- Geo Seed, Hodges, South Carolina
- Goldsmith Seeds, Inc., Gilroy, California
- G.S. Grimes Seeds, Concord, Ohio
- Harris Seeds, Rochester, New York
- Jelitto Perennial Seeds, Louisville, Kentucky
- PanAmerican Seed, West Chicago, Illinois
- Park Seed Wholesale, Greenwood, South Carolina
- Sakata Seed America, Inc., Morgan Hill, California
- S & G Seeds, Inc. (Syngenta Seeds), Downers Grove, Illinois
- Stokes Seeds, Buffalo, New York

Bulb Suppliers

Bulb producers usually are specialty growers. Most of the tulips used in the industry are imported from the Netherlands. A fairly large number of tulips are grown in Holland, Michigan. Lily, iris, daffodil, and narcissus bulbs and rhizomes are grown in the northwestern part of the United States.

Job Opportunities

The variety of organizations and businesses in the commercial greenhouse industry means that there are many different kinds of job opportunities. Descriptions of only a few of these jobs are given because it is not practical to list here all of the jobs available. A very large company may be so specialized that one person is in charge of a particular job. In a small company, the same job may be only one of several jobs assigned to one person. Figure 3-1 is an organizational chart of a large greenhouse range.

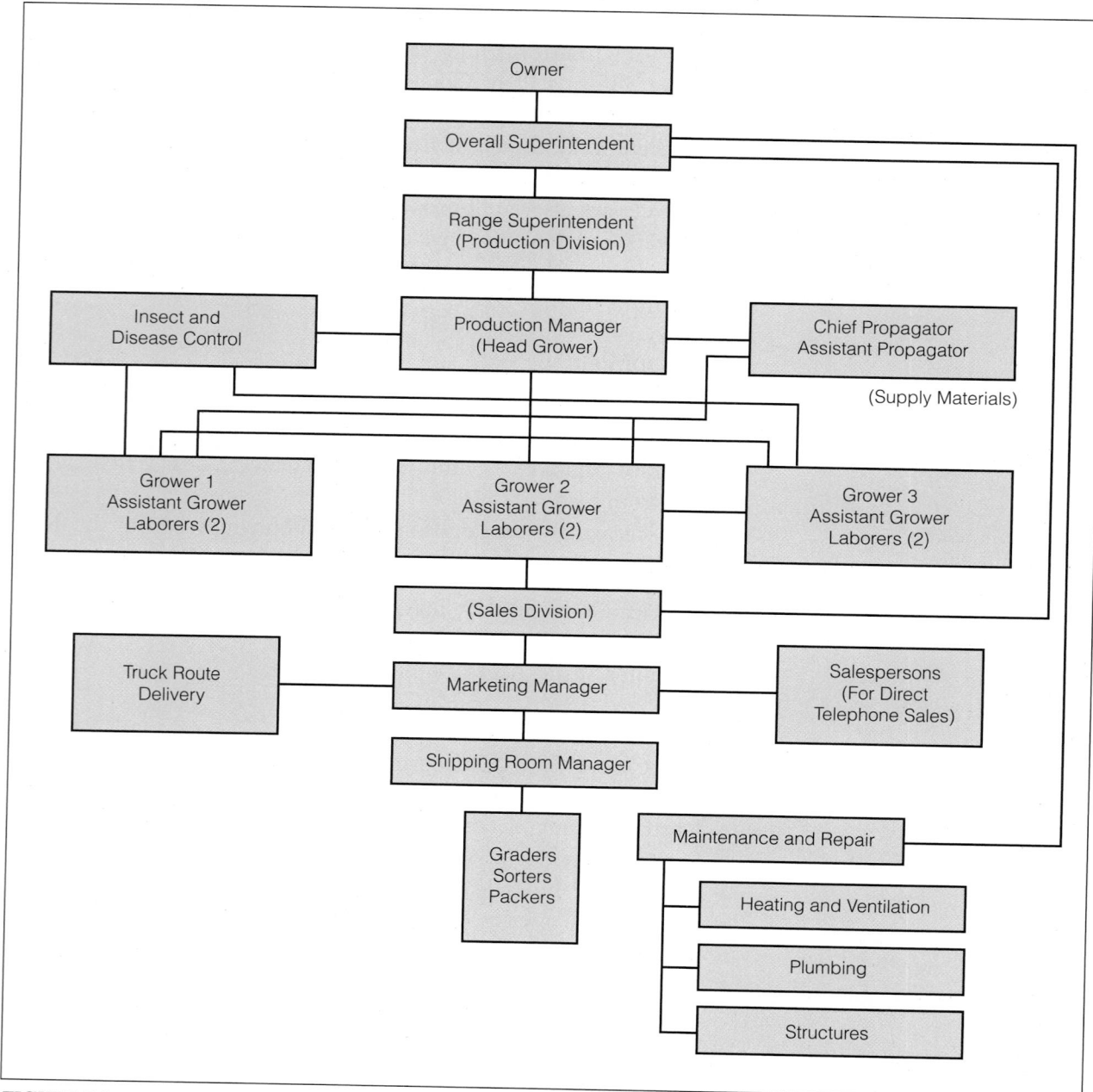

FIGURE 3-1 Organizational chart for a large greenhouse operation consisting of several ranges.

Greenhouse Production

Propagator

The propagator is in charge of all phases of propagation. It is the responsibility of this person to program the cuttings to be taken and when they are to be propagated. The propagator determines and maintains the light, temperature, and moisture conditions needed to root the cuttings in the shortest time. A propagator also uses seed, tissue culture, and other advanced propagation techniques to produce young plants that support the flower industry. The propagator determines all insect and disease control methods that must be used to ensure a good crop.

A propagator for a large specialty grower, such as Yoder Brothers, Inc., may supervise 30 to 50 or more employees. The propagator may be responsible for several hundred thousand root cuttings or several million seedling plugs.

Grower

The grower is usually placed in charge of at least one large greenhouse and sometimes two or more. One grower alone can care for nearly 10,000 rose plants. The grower is responsible for all cultural jobs in the assigned area, including planting, watering, fertilizing, tying up, and harvesting. A grower may have one, two, or many greenhouse assistants. The grower usually trains these assistants to prepare them to become growers who can take charge of their own areas.

In large greenhouse ranges, one or two people generally are assigned to insect and disease control. It is their job to maintain a regular schedule of spraying and fumigation to keep the plants pest free.

Head Grower

The head grower supervises several growers. In a small greenhouse operation, the head grower may be the only grower employed. In a very large operation, the head grower may have several growers to supervise. The head grower meets with these growers to assign tasks and review with them the jobs to be done for a particular crop or crops. A head grower may report to a range superintendent or to the production manager.

Range Superintendent

In a very large business, there may be several greenhouse ranges in different locations. Some ranges may be small enough to be managed by a grower or head grower alone. Other ranges may be so large that a range superintendent is required to carry out the orders of the owner. The range superintendent may be in charge of several head growers, plus growers, greenhouse assistants, and laborers.

Production Manager

A production manager is responsible for all stages of a single or group of greenhouse crops. The production manager must be competent in all phases of growing and producing greenhouse crops as well as possess the ability to communicate, train, lead, and manage production employees. The manager assigns jobs to others but has the final responsibility for producing the crop. The production manager orders all plant materials, supplies, and fertilizers that are needed to grow the crop.

Sales and Marketing

After the crop is grown, it must be sold. In a small greenhouse business, the production manager and sales manager may be the same person. In a large operation, the functions of sales and marketing are often separated and assigned to specialists. Individuals responsible for sales and marketing also produce marketing brochures and product catalogs, develop sales projections, manage product lines, as well as train and motivate sales personnel.

Marketing Manager

The marketing manager is in charge of grading, postharvest handling and storage, packaging, and shipment of plants and flowers. This person is responsible for ordering and receiving the cartons, containers, and packaging materials needed to service the crop. The shipping room manager reports to the marketing manager.

Shipping Room Manager

The shipping room manager supervises the handling of the crop from the time it comes from the greenhouse until it is packed for sale. This person directs the work of the sorters and graders who trim the foliage and grade the flowers by comparing them to certain standards. A large greenhouse range specializing in roses may employ 10 or more people to sort and grade. A packer takes the graded flowers and packs them into boxes for shipment. Skill and care are needed to package the cut flowers properly to prevent damage in transit. In greenhouses specializing in spring blooming plants, proper coordination in loading trucks can guarantee a successful relationship between the wholesale greenhouse and the retailer.

Additional jobs in the floriculture industry include the following:

- People are needed to prepare soils or container media for planting.
- Fork lift operators and tractor and truck drivers are needed to move supplies, plants, and materials. Trained drivers must be fully competent to manage perishable floral crops during a delivery.
- Workers are needed to pot plants and move them to their site in the greenhouse.
- People are needed to wrap and package potted plants at the time of sale.

It is important that all greenhouse equipment be repaired and maintained in good operating condition. Therefore, maintenance people are required to repair defective equipment and also service the equipment on a regular basis to prevent problems. Individuals who are in charge of maintenance must be skilled in all trades including carpentry, plumbing, and electrical. Many large greenhouse operations employ fully certified boiler technicians. Mention should also be made of the salespeople who sell seeds, bulbs, rooted cuttings, small plants, equipment, chemicals, and fertilizers. All of these people are needed to make the industry what it is today.

In addition to the growing side of the business, there are a number of job opportunities at both the wholesale and the retail level of the industry. Details about these opportunities can be found from other sources of information.

ACHIEVEMENT REVIEW

Select the best answer or answers to complete each statement. List the appropriate letter(s).

1 Growing florist crops today requires:
   a. very little special knowledge.
   b. a small dollar investment.
   c. knowledge of many special skills.
   d. mostly hand labor.

2 Areas of knowledge in which a grower must be an expert are:
   a. light and temperature control.
   b. soil preparation.
   c. insect and disease control.
   d. all of the above.

3 Cut flowers that are not handled properly after cutting:
   a. may be worthless for sale in one hour.
   b. are sold at lower prices than normal.
   c. are given away to peddlers.
   d. should be thrown out.

4 A wholesale grower may:
   a. grow cut flowers.
   b. sell flowers separately to walk-in customers.
   c. grow both cut flowers and potted plants.
   d. produce a specialty crop.

5 Chrysanthemums and carnations cannot be grown in the same greenhouse because:
   a. they produce different kinds of flowers.
   b. they have different day length requirements.
   c. they are attacked by the same insects.
   d. they have different temperature needs.

6 The production of potted plants:
   a. takes more labor than growing cut flowers.
   b. costs less than growing cut flowers.
   c. may be highly mechanized.
   d. costs the same and uses as much labor as growing cut flowers.

7 A propagator for a large specialty producer:
   a. only inserts the cuttings in the rooting medium.
   b. is responsible for all phases of root cuttings.
   c. may be in charge of a large group of employees.
   d. works only on weekends.

8 The range superintendent:
   a. carries out the orders of the owner(s) in getting the crop produced.
   b. is in charge of one or two greenhouses.
   c. directs the work of head growers, growers, assistants, and laborers.
   d. replaces broken glass in the greenhouse range.

9 The shipping room manager supervises:
   a. all of the workers in the shipping room.
   b. the graders and sorters who work in the shipping room.
   c. the flower packers.
   d. the ordering of boxes and other containers.

10 There are many job opportunities in the allied trades that support the grower. These companies sell:
   a. seeds, bulbs, and plants.
   b. pots and containers.
   c. fertilizers and insect killers.
   d. all of the above items and much more.

Types of Plant Growing Structures

CHAPTER

4

Greenhouses

OBJECTIVES:

After studying this chapter, the student should have learned:

■ different styles of greenhouse structures.

■ various components of a greenhouse.

■ the glazing materials used to cover greenhouses, including advantages and disadvantages of each material.

■ the economic factors to be considered in constructing a greenhouse.

The basic function of a greenhouse is to provide a protective environment for crop production while allowing natural light transmission. Greenhouse crops are either vegetable or ornamental. Ornamental crops are grown for use as cut flowers, as flowering potted plants, and as bedding plants. Green plants, such as tropical foliage plants, are grown for their aesthetic value.

Types of Greenhouses

Detached Houses

A detached house (Figure 4-1) is a freestanding greenhouse that may be constructed in one of several different styles: even span, uneven span, lean-to, Quonset, Gothic arch, curvilinear, curved eave, and dome shaped. It is detached in that the greenhouse is not attached to another greenhouse structure, but may be attached to a headhouse.

Even Span

The even span detached greenhouse is the style most commonly used for single houses or for several houses connected together to form one large

FIGURE 4-1 Detached freestanding greenhouse. The greenhouse is covered with single-pane glass.

structure. Even span greenhouses usually have clear spans with truss supports rather than supporting columns. The absence of columns means that heat-saving closures can be used to conserve energy. Conservation is very important now because of the increasing cost of fuel. Clear span greenhouses also make it easier to use semiautomatic methods to control the day length for chrysanthemums and other photoperiodic or day-length responsive crops.

The roof of an even span greenhouse is equal in width and pitch on both sides. **Pitch** is the angle the roof makes with the horizontal. The rise in the roof pitch compared to the horizontal length is typically 6 feet (1.8 m) by 12 feet (3.7 m). In other words, for every 12 feet of horizontal distance, the roof angle is 6 feet high. Even span greenhouses range in width from 10 feet (3 m) to 76 feet (23.1 m).

The length of the greenhouse is typically determined by the grower and usually depends upon the type of crop to be grown. For example, when many different greenhouse crops are grown, smaller houses are used. The smaller areas mean that different temperatures can be maintained to meet the requirements of specific crops. Cut flower crops and many greenhouse vegetables are typically grown in longer greenhouses for labor and space efficiency.

Greenhouses that extend more than 150 feet (45.8 m) from a central aisle were once not recommended. This was due to the amount of the workers' time spent walking from point to point when the greenhouse length is greater than this value. Yet, with modern plant moving systems, conveyors, monorails, and the like, this is not the issue today that it once was.

The **eave height** of a greenhouse should be not less than 7 feet (2.1 m), including the **curtain wall** of 2 feet (0.6 m) to 3 feet (0.9 m). (The curtain wall is the solid part of the sidewall and is often called a knee wall.) This is to allow a worker to stand upright when working close to the eave for better worker ergonomics. Many modern greenhouses now use an eave height of 10 feet (3 m) to 14 feet (4.3 m) to allow for a more uniform temperature environment in the greenhouse and to allow for efficient placement of greenhouse equipment overhead. Many specialized greenhouses, which are used for large tropical plants or woody ornamentals, also may have an eave height of 10 to 14 feet (3 to 4.3 m).

FIGURE 4-2 A large garage-type door makes it easy for vehicles and trailers to enter the greenhouse.

The gable end of a large greenhouse should have a large, garage type or sliding door so that plant carts, tilling equipment, production equipment, and small trucks can move in and out readily (Figure 4-2). Ideally, this door should be placed on the north end of the greenhouse. In a greenhouse of intermediate size, the door may be centered in the end wall. In a large house, the door may be placed to one side of the end wall. Small houses require only a standard, room type door for employee use. Yet, if this is the only access in and out of the greenhouse, size the door based on the equipment that is required to move into and out of the greenhouse.

Uneven Span

In an **uneven span greenhouse**, one side of the roof is longer than the other side. That is, it makes a smaller angle with the horizontal and has a lower pitch.

Uneven span greenhouses were popular years ago. At that time, greenhouses were placed on hillsides having southern exposures. When the long side of the greenhouse faced south (i.e., with the ridge line running east and west), the sun was able to provide it with maximum heating power for little cost. The present high fuel costs are causing some growers to consider returning to this type of greenhouse. A construction technique that has been proposed for uneven span greenhouses is the use of a solid north wall and the northern, smaller roof area covered with a solid opaque material as well. When the wall and roof are covered inside with an efficient reflecting and insulating material, a great deal of energy can be conserved. Uneven span greenhouses are practical if the slope of the land is not great enough to cause difficult working conditions.

Lean-To

Lean-to greenhouses are built against other greenhouses or buildings (Figure 4-3). The roof slopes in one direction only; generally, it faces south. Lean-to greenhouses are small and may be added to other structures. They are used frequently for bulb forcing or seed starting operations. Lean-to greenhouses are also common at high latitude locations. They are also useful for institutional installations including schools and universities (Figure 4-4).

FIGURE 4-3 A lean-to greenhouse in which seeds are started.

FIGURE 4-4 An institutional lean-to greenhouse facing south for research and teaching at Montana State University. High intensity discharge lamps are often used for supplemental light during winter months at high latitude sites.

FIGURE 4-5A Uncovered Quonset style greenhouse showing structural pieces.

FIGURE 4-5B Quonset style greenhouses. The front two greenhouses are covered with double wall polycarbonate and the greenhouse to the rear is covered with fiberglass-reinforced plastic.

Quonset

The concept of the Quonset hut was developed during World War II when inexpensive, easy to build structures were needed for warehouses and barracks. In 1941, Peter Dejongh and Otto Brandenberger of the George A. Fuller Construction Company in New York went to work and within a month they set up a production facility near Quonset, Rhode Island, hence the name Quonset.

The framework for the Quonset greenhouse consists of curved **roof bars** in the shape of semicircles (Figure 4-5A). Note that the structure is a clear span type. The framework can be covered with **sheet film plastic** or rigid plastic panels (Figure 4-5B).

Gothic Arch

Greenhouses with Gothic arch construction (Figure 4-6) are not widely used. The framework for this type of greenhouse was once formed from laminated wooden trusses. Each truss consists of thin layers of wood glued together to

FIGURE 4-6 Greenhouse with Gothic arch construction. Interior view of the Marjorie McNeely Conservatory at the Como Park Zoo and Conservatory, Saint Paul, Minnesota.

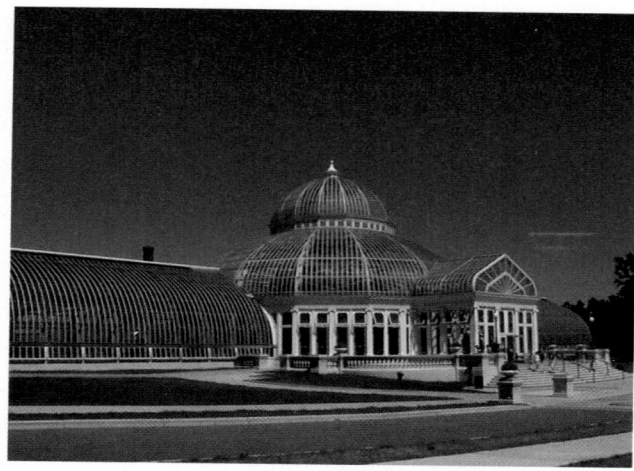

FIGURE 4-7 Marjorie McNeely Conservatory at the Como Park Zoo and Conservatory, Saint Paul, Minnesota.

form one large beam. The Gothic arch achieves a greater span than is possible when using standard wood framing members. Spring flowering annuals and other potted crops were generally grown in Gothic arch structures. Ridge ventilators cannot be used because of the curved roof bar construction. Thus, fans must be used for ventilation. Modern construction materials, including steel and aluminum, have replaced the laminated wood trusses, yet the Gothic arch frame is now used by botanical gardens and conservatories and more expensive hobby or home greenhouses for its architectural appeal (Figure 4-7).

Curvilinear and Curved Eave

The curvilinear and curved eave type of construction was popular when very large conservatories were built, but is rarely used today because of the cost of curved glass. Many city parks, botanical gardens, and even a few wealthy individuals maintain conservatory-type structures (Figure 4-7). Many of these conservatories are used for seasonal displays and exhibitions of floral crops. They are too small for commercial use, but are often found at technical institutes and vocational schools where they are used for instruction.

Advantages of the Detached Greenhouse

A detached greenhouse has certain advantages and disadvantages. One advantage of a separate house is that it is easier to program and maintain the temperature to meet specific crop requirements. In addition, detached greenhouses are easier to ventilate without exposing the plants to blasts of cold air. A greenhouse contains a large volume of air that buffers temperature changes. This large air volume also means that **carbon dioxide** does not quickly become a limiting factor of growth in the winter when the house is closed.

The light that enters a detached greenhouse is rather uniformly distributed over the entire growing area. In connected houses, the roof gutters cut down on the light admitted and shade the crops, affecting their growth.

Detached houses are more easily cared for and maintained than connected houses. The potential for cross contamination of insects and diseases is also reduced by using detached houses. This makes detached houses ideal for quarantining plants that may be contaminated. Properly spaced detached greenhouses are easily cleared of snow without damaging the structure. Finally, single detached greenhouses require fewer ventilators and less ventilating machinery.

Disadvantages of Single Units

The greater height of a single detached house, as well as all walls exposed means that more energy is required to heat the large volume of contained air. The height also means that more of the surface is exposed to winter winds. As a result, there is a greater loss of heat than in low houses.

Large detached greenhouses have a higher initial cost. They are less easily painted and reglazed. Because more ground area is needed with a number of single detached structures, a large greenhouse range requires more land area than connected houses need.

Connected Houses

Many of the same types of construction shown for detached houses can be used for connected houses. Ridge and furrow or gutter connected houses consist of even span structures placed one after the other with no sidewalls between individual units (Figure 4-8). The grower decides how many houses are needed based on the requirements of the crop or crops to be grown. A grower may have from two ridge and furrow houses to as many as are needed to cover several acres of ground. A range of ridge and furrow greenhouses is easily expanded in size by simply adding an additional connected house.

Sawtooth

Sawtooth greenhouse construction is used in areas such as California, Texas, Florida, and maritime regions where the climate is mild and a prevailing wind is common. Sawtooth greenhouses consist of a series of lean-to greenhouses connected together. Where sawtooth greenhouses were once considered an inexpensive construction design and covered with sheet plastic, they are now common to coastal regions and covered with more permanent glazing materials. The elevated portion of the sawtooth opening is directed leeward to the prevailing

FIGURE 4-8 A range of connected greenhouses.

winds to aid in ventilation. Open sidewalls facilitate this energy-efficient cooling strategy (Figure 4-9).

Multichamber Houses

One style of multichamber greenhouse is a modification of the Quonset type detached greenhouse. As shown in Figure 4-10, the barrel vault greenhouse consists of Quonset type structures set on sidewalls. Figure 4-11 shows connected barrel vault greenhouses. The height of the sidewall depends upon the crop to be grown. One unit offered to the industry has sidewalls ranging in height from 7 feet (2.1 m) to 10 feet (3 m) or more.

In the last 10 years, another type of multichamber greenhouse has become popular in the United States. The Dutch Venlo type of connected greenhouse has a ridge and furrow construction. (Venlo refers to a municipality in the southeastern Netherlands.) Venlo greenhouses use 2-foot (0.6-m) by 4-foot (1.2-m) glass lights with lightweight roof bars and supporting structures (Figure 4-12). The roof pitch is also less than the typical greenhouse design, providing a lower

FIGURE 4-9 Sawtooth greenhouses are commonly found in warm areas such as Florida and California, where prevailing winds and mild winter temperatures make the most of its built-in ventilation. *(Courtesy Chris Beytes, Ball Publishing.)*

FIGURE 4-10 An end view of barrel vault greenhouses covered with double wall polycarbonate in Fort Collins, Colorado.

FIGURE 4-11 A range of barrel vault greenhouses covered with double wall polycarbonate end walls and a double layer polyethylene roof. Note the tank for collecting runoff from the greenhouse for irrigation water.

FIGURE 4-12 Venlo or low profile glass greenhouse.

profile. This type of greenhouse has the highest percentage of light transmission of any greenhouse presently on the market as well as the lowest roof surface area exposed, thus reducing the heating demand. In the Netherlands, Dutch Venlo type greenhouses are typically used to produce tomatoes and cucumbers. These greenhouses are adapted in the United States to the production of all types of ornamental horticulture crops. Yet, due to the lighter weight roof bars and supporting structures, the Venlo greenhouse design may not comply with many municipal building codes for snow and wind loading.

Saran-Covered Houses

Plastic Saran-covered structures are used in the production of florist crops in milder climates (Figure 4-13). They are not considered to be true greenhouses. In the northern part of the United States, summer crops such as chrysanthemums and asters are grown under Saran-covered houses. In areas where frost is not a problem, Saran-covered houses are used to reduce the intensity of bright sunlight and to provide some protection from heavy rains and certain insects. For example, tropical foliage plants are grown under Saran-covered houses in southern Florida, Texas, California, and Hawaii.

Lath Houses

Lath houses generally are made of wood or wood with a metal framework (Figure 4-14). They are used in tropical and semitropical areas to provide shade for various ornamental plants. The degree of shade is determined by the closeness of the lath spacing. In northern areas, lath houses are used in the production of ground covers and other crops that require shade or broken sunlight. Lath houses are not as common as in the past due to the high degree of maintenance required compared to a Saran-covered house.

Advantages of Connected Greenhouses

Connected greenhouses require less land area than detached greenhouses for the same unit of covered space. The absence of sidewalls between the houses means that fewer construction materials are required. It is more difficult in

FIGURE 4-13 Hoop house with saran cloth covering providing shade for a shipping and receiving area for a southern California greenhouse.

FIGURE 4-14 Lath house.

connected greenhouses to zone heat for different crops and to apply fumigant type insecticides as well as to prevent the spread of insects and diseases unless internal walls are erected. However, less heat is required because there are fewer exposed wall surfaces where heat exchange can occur. Work is done more efficiently because there is easy passage between houses.

Disadvantages of Connected Greenhouses

Snow cannot be cleared readily from connected greenhouses. The buildup of snow in the furrows (gutters) between houses can become great enough to cause the house to collapse. To prevent this, additional heat lines must be located under the gutters to melt the snow. These same gutters produce shadows that reduce the light intensity and often delay the flowering of a crop.

Soil and other materials must be carried into the greenhouse at one end or the other. It is easier to bring materials through a side entrance in a long narrow greenhouse.

A smaller volume of air above the crops in connected greenhouses limits the amount of carbon dioxide available for plant growth in the winter. This can be overcome by injecting carbon dioxide into the greenhouse from burning fuel, compressed gas, or other methods. In addition, the lower air volume cannot provide as much buffering of temperature changes at other times of the year. Taller sidewalls, if incorporated in the early design and construction, can alleviate some of these issues.

The same amount of roof area is required for several small connected greenhouses as for one large house covering the same ground area if the roof pitches (roof angles) are equal. However, the costs of construction for a single large house are higher because the supporting trusses must have greater strength and are heavier in weight.

Construction of the Greenhouse

Framework

The greenhouse framework provides support for the covering. The earliest greenhouses used wood for all framing members. The combination of bulky structural members and narrow panes of glass resulted in structures with dark interiors. Such conditions were unsuited for growing crops requiring a high light intensity.

Wood is still used in modern construction. However, it is used primarily for small hobby type greenhouses and for greenhouses that are to be covered with sheet plastic, thus requiring a minimal amount of framework. All aluminum or aluminum and steel framing are used in large commercial greenhouses.

In the following sections, each part of the framing used in a typical greenhouse is described and its function explained. Refer to Figure 4-15 for the location of each component in the framework.

Framework Components

Sideposts

Buildings have **foundations** that bear the weight of the structure. In greenhouses, however, **sideposts** bear the weight and side strain of the roof. These

FIGURE 4-15 Structural members of the greenhouse framework.

sideposts are usually encased in concrete and set on concrete footings or caissons that are placed below the frost line. If wooden sideposts are used, the wood must be pressure treated to resist decay.

Glazing Sill or Sash Sill

The **glazing sill** provides the support for the base of the glass or other glazing material. It is mounted on top of the curtain wall. When side vents are used in the greenhouse, this member is called a sash sill. Wooden or metal sills may be used at the ends of the greenhouse as well as along the sides.

Eave Plate

The eave plate rests on the sideposts. It forms the support for the roof members.

Gutters

Gutters are used to collect the runoff water and channel it away from the structure. In ridge and furrow houses, a gutter is used instead of an eave plate. Gutters should have a fall of at least 4 inches (10 cm) for each 100 feet (30.5 m) of length for proper drainage. Leaves and debris must be removed from gutters before the freezing temperatures of winter begin. The width of a gutter should be wide enough to accommodate a worker's foot for easier maintenance.

Detached houses typically do not use gutters because they interfere with the clearing of snow from the roof. Rain and melting snow are allowed to fall freely from the roof edge.

Drip Gutter

The **drip gutter** is located at the level of the underside of a gutter. Any condensation that forms on the inside of the glazing material is channeled to the ground and away from the crop by the drip gutter.

Sash Bars

The **sash bar** is one of the most important structural members of a greenhouse. Its shape and size must be such that it casts the least shade possible. At the same time, it must be strong enough to carry the weight of the glazing material, the weight of snow, and wind load forces. Sash bars have built-in channels. When sash bars are installed with the channels on the inside of the greenhouse, the channels direct the condensation that forms to a drip gutter. Sash bars were once constructed from cypress or redwood, but modern sash bars are typically extruded aluminum. Metal sash bars are uniform in shape and are manufactured in standard sizes. Variations result from differing manufacturers or from different glazing attachment requirements.

Bar Caps

Bar caps are attached to the outside of the greenhouse sash bars to hold the glazing material in place (Figure 4-16). The bar caps also hold the glazing compound used to seal around the glass (or other glazing material) to prevent leaks. When glazing compound is not protected from the sunlight, the ultraviolet rays of the sun hasten the deterioration of the compound. The bar caps shield the glazing compound from the ultraviolet rays. Each bar cap is easily attached to the sash bar by two or more screws. A greenhouse fitted with bar caps is almost maintenance free.

Purlins

Purlins run the length of the greenhouse. They are used to support the sash bars on a wide house. Purlins are supported by purlin posts, purlin braces, rafters, or other truss members.

Truss

A **truss** is a supporting member that is used to provide structural strength. Clear span greenhouses (Figure 4-17) require complex truss shapes to support the weight of the glazing materials and an estimated snow and wind load. The largest clear span greenhouse offered to the trade in the United States is approximately 50 feet (15.3 m) wide. In England, clear span greenhouses up to 76 feet (23.1 m) in width are in use for commercial crops.

Ridge

The **ridge** is the top of the greenhouse. The upper end of each sash bar is fastened to the ridge. A ridge cap is installed in the top of the ridge. The ridge also supports the ridge ventilators. In earlier wood framed greenhouses, the ridge was often used as a catwalk providing access to the top of the greenhouse for inspection and minor repairs. Good balance was required to walk along the ridge cap. Modern greenhouses using steel trusses and aluminum sash bars typically use an extruded aluminum ridge. These designs are typically narrower and

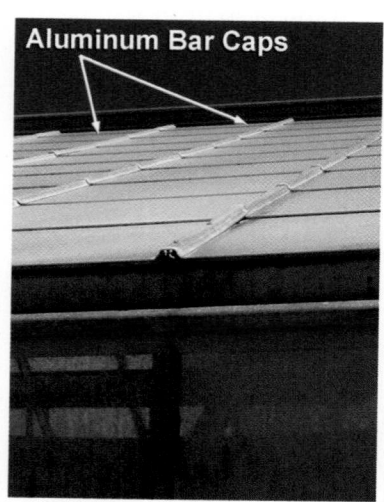

FIGURE 4-16 Aluminum bar caps over sash bars for fastening glass panes. The glass on this greenhouse has had whitewash applied for shade.

Two Methods of Framing a Semi-Iron
Column Pipe Frame Greenhouse

Barrel Vault Roper IBG

Clear Span

FIGURE 4-17 Methods of framing a greenhouse.

will not easily accommodate a worker. Modern safety regulations prohibit rooftop worker access without proper training and safety harnesses.

Ventilator

A **ventilator** is a movable unit that is fastened to the ridge. It can be opened to ventilate the greenhouse when the temperature inside becomes too high. Ridge ventilators may run the entire length of the house, or they may be installed in sections. They range in width from 2 to 4 feet (0.61 to 1.22 meters). Ridge ventilators once equaled 10 percent of the floor space, whereas modern ventilators are 20 percent or more of the total floor space for passive greenhouse cooling. Ideal construction places a roof ventilator on each side of the ridge, thus allowing the grower to open the ventilator to the leeward wind direction.

Sidewall ventilators are also used in greenhouses. These units increase the airflow through the greenhouse. They function like a damper in the bottom of the stove. When the sidewall ventilators are wide open, a maximum **chimney effect** is created in the greenhouse. As a result, hot air is pulled to the outside through the ridge ventilators.

Ventilator Header

The ventilator rests on the **ventilator header**. The header is mortised (grooved) on its lower side so that the glazing compound can be applied and the edge of the roof glass secured.

Framing Materials

There are two goals in designing greenhouse framing: (1) It must be strong and light, and (2) it must cast little shade. Other factors to be considered are the initial cost, durability, and maintenance costs. The attractiveness of low initial

FIGURE 4-18 Wood framed greenhouse with fiberglass-reinforced plastic roof.

cost may be offset by excessive maintenance costs. An initial cost that seems to be high may actually represent a savings if annual maintenance costs are low.

Greenhouses once were framed using wood alone (Figure 4-18). Modern construction techniques evolved greenhouses from wood alone to wood and iron (**semi-iron**), iron alone (steel), steel and aluminum, and finally aluminum. Metal framing for large greenhouses is described here. Wooden framing is described under "Wooden Framing for Greenhouses."

Semi-Iron Pipe Framework

In semi-iron pipe framing, iron or steel is used for the support structures and wood is used for the sash bars that hold the glass. Any steel used should be hot dip galvanized after fabrication to protect the metal from rust. If cuts or welds are made through the galvanized skin of the metal, the areas must be wire brushed thoroughly, coated with a primer, and then finished with an appropriate paint. Inadequate treatment of such areas makes them more liable to rust.

Aluminum Framework

At one time, greenhouse construction using steel supporting members and aluminum sash bars was considered to be ideal. Now, greenhouses can be constructed entirely of aluminum members. Aluminum is a soft metal, which means that pure aluminum is unsatisfactory for greenhouse construction. Other materials are added to aluminum to give the required strength, flexibility, and corrosion resistance (necessary in the humid environment of the greenhouse). These combinations of aluminum with other materials are called **alloys**. Aluminum alloys can be extruded in any required size and shape. The extrusions are equal in strength to similar steel forms and surpass steel in their resistance to corrosion. The most used aluminum extrusion is known as mill finish, with heat treatment of alloy 6063-T6 or 6063-T5.

Greenhouse manufacturers supply aluminum framing members in prefabricated form. Unskilled labor can assemble the framework easily. The framing materials are supplied in the form of modular sections known as bays. Each section is approximately 10 feet (3 m) long on center. This spacing permits aluminum glazing bars to be placed to accommodate 24-inch (60-cm) wide glass lights (panes). Each bay consists of five lights of glass. Any number of bays may be used to obtain the desired greenhouse length.

Methods of Framing to Achieve Maximum Strength

Several methods of framing semi-iron, steel, and aluminum greenhouses are shown in Figure 4-17. Each type of framework provides support for the structure for both normal and unusual loads.

Minimum loading and structural design guidelines are typically mandated by local building authorities through zoning codes. The National Greenhouse Manufacturers Association (NGMA) has worked closely with national groups to establish a standard set of codes for the United States. The following three types of loads must be considered in the design of the greenhouse structure.

1. The dead load is the weight of the structure and the glass covering the framework.
2. The live load is due to the weight of snow. Snow loads are estimated on the basis of 15 pounds per square foot (73.4 kg/m^2) of snow on a horizontal surface.
3. The wind load is the load caused by wind pressure. It is calculated on the basis of 20 pounds per square foot (98 kg/m^2) on a vertically projected area.

The three loads described can act together in several ways. Greenhouse manufacturers use the following load combinations when designing structures.

Dead load + live load
Dead load + wind load
Dead load + wind load + one-half of the live load
Dead load + wind load + one-half of the live load + an additional 100-pound (45.4-kg) load applied at the center of any span

Greenhouses are designed to withstand heavy stresses. However, extremely high winds may distort even the strongest greenhouse. The snow loads should not be a problem in a heated greenhouse. The heat radiating from the glass will melt and clear snow as fast as it falls. If the greenhouse is not heated, the structure will collapse under heavy snow loads. If a greenhouse is under construction and is covered but lacks a heating system, portable heaters must be used inside the structure to melt the snow and prevent a buildup.

A partially completed greenhouse is particularly subject to structural failure under stress. It is the total structure that is designed to withstand stresses. When a house is incomplete, the support each component supplies to another is missing. Thus, the structure lacks its total design strength. A carefully designed and engineered greenhouse is a perfectly balanced structure of metal and glass, or plastic.

Foundations and Curtain Wall Construction

Foundations

Greenhouses require a foundation to support the structure. Unlike other buildings, greenhouse foundations are typically not continuous around the perimeter of the building. Small hobby type greenhouses are an exception in that they may have a complete foundation.

FIGURE 4-19A Concrete curtain wall on Lord and Burnham greenhouse structure.

FIGURE 4-19B Concrete curtain wall on fiberglass-reinforced plastic greenhouse.

A typical foundation consists of concrete footings located below the frost line. The footings are spaced at intervals of 8 feet (2.4 m) to 10 feet (3 m) or more. The spacing of the footings is equal to the distance between the supporting column posts. These posts are embedded in concrete piers that rest on the footings. This system provides the main structural support for the greenhouse.

Curtain Wall

In some greenhouses, usually of older construction or institutional, a sidewall, known as a curtain wall or knee wall, is attached between the support posts and extends around the perimeter of the greenhouse (Figures 4-19A and 4-19B). This wall does not provide support for the greenhouse. In new construction, the greenhouse wall usually extends down to the ground to let as much light in to the crop as possible.

Curtain walls are made of poured concrete, cinder or cement block, corrugated fiberglass-reinforced plastic, wood, or any other solid building materials. The curtain wall in many plastic-covered greenhouses consists of a 1-inch (2.5-cm) by 6-inch (15-cm) board to which the plastic is attached.

The curtain wall should extend at least 6 inches (15 cm) below the surface of the ground. In this way, rodents and other small animals are prevented from burrowing into the greenhouse. The wall should be well insulated to prevent the loss of a large amount of heat. Heat also leaves the greenhouse through the ground underneath the curtain wall. To reduce this heat loss, a 2-inch (5-cm) thick ball of polyurethane insulating foam can be buried vertically, 2 to 3 feet (0.61 to 0.91 meters) below the curtain wall.

A curtain wall is generally 2 or 3 feet (0.61 to 0.91 meters) high. The remainder of the distance to the eave is taken up by a sidewall of glass or plastic.

Wooden Framing for Greenhouses

Greenhouses with all wooden construction have no metal structural members. Redwood, cedar, and bald cypress are the most commonly used woods because each has a high degree of decay resistance. Framing for plastic-covered greenhouses is often made of fir, spruce, or pine. Regardless of the type of wood used, only the heartwood has a natural decay resistance. Untreated sapwood of all species of wood has a low resistance to decay. Even when the sapwood is treated with **preservatives**, it usually has a short life under greenhouse conditions.

Wood decay is caused by fungi that feed on the wood. These fungi require air, warmth, food, and moisture for growth. They grow most rapidly at temperatures of 70°F to 85°F (21°C to 29.5°C) and in air with moisture content above 20 percent. These are the conditions normally maintained in a greenhouse for plant growth. Wood should not be used in a greenhouse unless it is treated with a preservative.

Preservatives

There are many wood preservatives on the market, but only a few should be used in a structure intended for plant crops. Many growers have lost plants valued at thousands of dollars by using the wrong wood preservative in the greenhouse.

The use of preservatives containing creosote and pentachlorophenol must be avoided when treating wooden structural members of the greenhouse, the benches, or wooden flats (containers) used for growing plants. Fumes from these chemicals are toxic to plants.

Wood preservatives that are safe for greenhouse use are copper **naphthenate** and zinc naphthenate. The wooden parts may be brushed, dipped, or soaked in the preservative. The greatest penetration of the chemical into the wood cells occurs when the parts are soaked. However, this method uses the most material.

A commercial process known as Wolmanizing infuses chromated copper arsenate salts into the wood under high pressure as a preservative. After the wood is treated with the preservative salts, it is placed in a high-pressure chamber where the salts are forced deep into the wood tissues and cells. Wolmanized lumber has a greater resistance to decay than hand-treated wood. The Environmental Protection Agency has removed many residential and commercial uses of chromated copper arsenate salts while farm use of chromated copper salts are still registered for barns and fencing, which loosely includes commercial greenhouses.

Framing for Plastic-Covered Greenhouses

The type of framing used for a greenhouse covered with plastic depends upon the size of the structure and the plastic used. Acrylic, polycarbonate, flexible film plastic, or rigid fiberglass-reinforced plastic (FRP) can be selected.

Methods of Framing

Several methods of framing can be used for greenhouses consisting of a wooden framework covered with plastic. Popular framing methods are shown in Figure 4-20. These methods include the A-frame, scissors truss, sawtooth, truss rafter, laminated beam, and external gusset.

The structural posts for wood framed greenhouses are set in the ground just as in metal construction. The posts are spaced either 4 feet (1.2 m) or 8 feet (2.4 m) on centers. The recommended maximum rafter spacing is 6 feet (1.8 m).

FIGURE 4-20 Framing methods used for greenhouses covered with plastic.

Large Greenhouses Covered with Rigid Plastic

The framing for large greenhouses covered with rigid plastic is similar to the framing for glass-covered greenhouses. The structural members and roof bars are made of aluminum alloys. The roof bars used on rigid plastic-covered greenhouses are simpler in design than those required on glass-covered greenhouses. (Recall that the sash bars must be shaped to receive the glass bedding compound, the glass light, and then the bar cap.) Fewer roof (sash) bars are used on a plastic-covered greenhouse, thus reducing the shadows inside the greenhouse.

Roof bars on glass-covered structures are spaced up to 28.75 inches (73 cm) apart. Roof bars for tempered glass structures are spaced 6.6 feet (2 m) apart. Greenhouses with a rigid plastic cover use roof bars spaced 12 feet (3.6 m) on centers. This wide spacing is possible for two reasons: (1) Corrugated rigid plastics, such as acrylic, polycarbonate, and FRP, have an inherent strength; and (2) corrugated rigid plastic pieces can be fastened together directly. Glass lights cannot be fastened in this manner.

Quonset and multiarch greenhouses also use preformed aluminum alloy or galvanized steel trusses. Some growers make their own metal framework for Quonset type greenhouses by bending metal tubing. However, these home-made structures may not withstand large snow or wind loads because the tubing used is not designed for greenhouse use.

Small to Medium Houses Covered with Plastic Film

Small to medium plastic-covered greenhouses commonly use wood framing. All wooden members must be weatherized before construction begins. Supporting posts installed below grade level must be treated with a preservative. Wooden members used above ground should receive at least one coat, or preferably two coats, of quality greenhouse paint.

Greenhouses covered with plastic film are often considered to be temporary structures. The designation of a structure as temporary by tax assessors is an advantage to the grower. Taxes are lower on temporary buildings than on permanent structures.

Glazing Materials

The covering of a greenhouse is known as the glazing or cladding, or the skin. The NGMA has placed glazing materials into three different categories. Type I glazing are thin films, which include polyethylene, ethylene vinyl acetate (EVA), polyvinyl chloride (PVC), polyvinyl fluoride (PVF), and polyester films. Type II glazing are flexible plastics, which include fiberglass-reinforced plastic (FRP), acrylic, polycarbonate, polypropylene, PVC, and PETG (glycol-modified polyethylene terephthalate). Type II glazing includes corrugated single wall sheeting as well as multiwall structured composites. Type III glazing includes glass such as float glass, annealed, chemically strengthened, tempered and laminated forms of float, annealed, chemically strengthened, and tempered and rolled pattern glass. An important property for both glass and plastic is the ability to transmit natural light.

Light Transmittance

The most important function of a greenhouse covering is to allow the maximum amount of available light to reach the crop. Therefore, the percentage of light transmitted by the covering material is an important factor in selecting the material to be used. Light transmittance is defined as the amount of light that strikes a surface and passes through it. No material can transmit 100 percent of the light that strikes it. What happens to the light that strikes a surface? Part of the light is reflected, part of the light is absorbed by the material it strikes, and the remainder of the light is transmitted through the material. The amount of light reflected, absorbed, or transmitted depends upon the material and the angle of incidence (Figures 4-21 and 4-22).

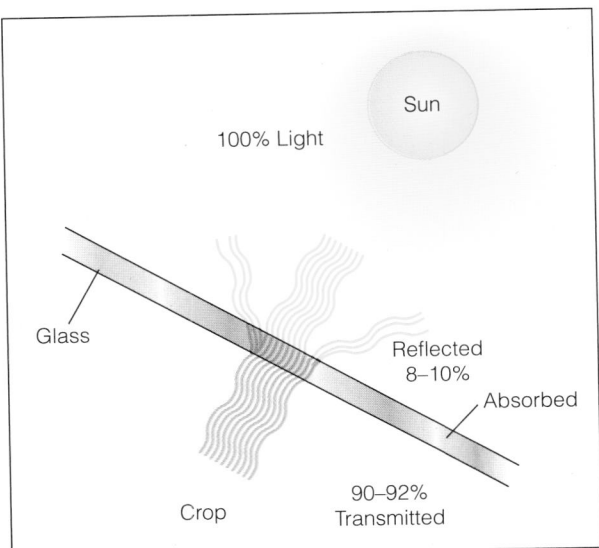

FIGURE 4-21 The percentages of light absorbed and transmitted by greenhouse coverings varies with the type of glazing.

FIGURE 4-22 Mid-winter view of a freestanding fiberglass-reinforced plastic greenhouse in Colorado reflecting late afternoon sun.

Table 4-1

	Spectral Transmittance (%) (in nanometers)*		
Glazing material	Ultraviolet (200–400 nm)	Visible (PAR)** (400–700 nm)	Infrared (700–2500 nm)
Clear float glass (0.125 in)	–	89.3	–
Double layer polyethylene (6 mil)	47.9	77.8	79.6
Twin wall acrylic (8 mm)	44.7	84.5	73.9
Twin wall polycarbonate (8 mm)	18.1	79.3	76.3
Co-Extruded Polycarbonate (UV coated)	10.0	86.9	85.0
Fiberglass reinforced plastic	19.6	90.5	86.6

Percentage of Sunlight Transmitted at Selected Wavelengths Through Various Types of Greenhouse Glazing Materials

*1 nanometer equals one-billionth of a meter (39.37 in.)
**PAR—photosynthetically active radiation

When light strikes a sheet of black plastic, for example, a small amount is reflected. The greatest amount of light is absorbed by the plastic. No light, or very little, is transmitted. The greatest amount of light transmitted is obtained with clear glass. In addition, a small amount of light is absorbed and a small amount is reflected. The percentage of light transmitted at various wavelengths depends upon the material. Photosynthetically active radiation (PAR) is primarily within the visible portion of the light spectrum and ranges from 400 to 700 nanometers (nm) and is critical for plant growth and development. Light transmission values are given in Table 4-1 for common glazing materials.

Type I Thin Films

Glass was used by the Romans and ancient Greeks to build conservatories. However, it was not until 1945 that researchers successfully produced plastic films. Since then plastic has been accepted worldwide as a greenhouse glazing material.

Polyethylene

Polyethylene is probably the best known of all plastic materials in use today. Sales of polyethylene worldwide account for 80 percent of the plastics market. Polyethylene is familiar to almost everyone in the form of plastic bags and food wraps. The majority of greenhouses covered with plastic use polyethylene (Figure 4-23).

Research at Clemson University, Clemson, South Carolina, Department of Horticulture determined that pigmentation in greenhouse film plastics can alter the effects on light reflectivity and transmission of specific wavelengths. This has the potential to eliminate or reduce the need for growth regulators, reduce solar load gain, reduce spore germination for disease control, and provide other plant growth responses that may result from spectrum alteration.

Polyethylene is waxy to the touch, is chemically inert, and stays flexible even at low temperatures. It is permeable for certain gases. That is, oxygen and carbon dioxide can diffuse or pass through the plastic at different rates. The

FIGURE 4-23 A Quonset-style greenhouse. Notice the plastic tubing side walls that, when deflated, allow air to enter the structure.

permeability of polyethylene is three times greater for carbon dioxide than it is for oxygen, but the diffusion rate is slow.

The material can be purchased in widths ranging from 1 foot (0.3 m) to 50 feet (15.4 m). The thickness of the sheet ranges from 1 mil to 10 mils. One mil is equal to 1/1,000 inch.

Polyethylene is only one of many plastics that have been developed through research in petroleum chemistry. Until the 1973–1974 petroleum shortage, the cost of polyethylene was decreasing. Prior to 1973, the price range of polyethylene was 0.0025 to 1 cent per square foot. Now the cost is more than 14 cents per square foot for 6-mil plastic in lots of 5,000 square feet (464 m^2). As petroleum prices and supplies vary, the availability of polyethylene and other plastics will also vary, resulting in fluctuating costs.

Polyethylene film for outdoor use contains antioxidants and ultraviolet absorbers (inhibitors) to combat the effects of weathering. The ultraviolet radiation of the sun causes the plastic to deteriorate. When polyethylene was first developed, it was applied to greenhouses in October after the high light intensity of the summer had passed. The plastic would go through the winter but would be in tatters by June or July of the following summer.

Not all polyethylene products manufactured at present contain ultraviolet inhibitors. For greenhouse use, the grower must be sure to purchase only polyethylene with an ultraviolet inhibitor. However, the presence of an inhibitor does not guarantee that the polyethylene will last longer than a year. According to manufacturers, the average life expectancy for polyethylene can be up to 4 years, depending on the light intensity, wind, and other environmental conditions. It is expected that manufacturing refinements will increase the lifetime of polyethylene.

Agricultural or construction grade polyethylene is not suitable for greenhouse coverings. This type of plastic is folded for shipment, causing a 50 percent reduction in the strength of the material at the folds. As a result, the plastic tends to split along the folds during cold weather.

An insulating effect is obtained by using two layers of polyethylene separated by a dead air space. This air space is maintained with a squirrel cage fan using cooler and drier outside air to prevent condensation between the layers. A 30 to 40 percent reduction in heat requirements can be obtained using a double layer of polyethylene. The outside layer should be 4 to 6 mils thick, and the inside layer should be 2 to 4 mils thick. The heavier materials last longer and have a greater insulating effect. Double walled houses are more expensive and have a reduced light level.

The conservation of heat in a greenhouse is of prime importance due to the rising cost of fuel. One way growers can reduce the heat loss is to cover the inside of the greenhouse walls with clear polyethylene film. However, the relative humidity of the house increases, resulting in more disease. The use of a polyethylene lining also causes a slight delay in the flowering of chrysanthemum and poinsettia crops.

Some growers have completely covered their glass greenhouses with polyethylene. This method allows them to achieve heat savings of 38 percent with only a 4 to 15 percent reduction of light intensity. This is especially useful if the roof has been damaged as a temporary repair.

Woven Polyethylene

A novel concept in greenhouse structures is that of the retractable roof (Figure 4-24A). The greenhouse covering is a woven, flexible polyethylene that is ultraviolet resistant. It also has antifog and anticondensate qualities. The light transmission of the covering is approximately 80 percent when it is closed. The expected life of the covering is 5 to 7 years.

The greenhouse roof is retractable in sections when weather permits. Benefits claimed for the greenhouse include:

- Production schedules and higher plant quality are easier to achieve because plant growth is more predictable.
- Less water and fertilizer are needed because cooler daytime temperatures reduce transpiration losses.
- Water consumption may further be reduced by opening the roof when it is raining.
- The roof can be opened or closed to maximize the effectiveness of chemical applications.
- Growers can promote to retailers and other customers that plants are properly acclimated for transplanting and that potted plants have a longer shelf life.
- Plants grown in the open roof structures are naturally compact (Figure 4-24B).

FIGURE 4-24A Cravo "RETRACT-A-ROOF" greenhouse *(Courtesy Cravo Equipment Limited, Ontario, Canada.)*

FIGURE 4-24B Left, plants grown in closed greenhouse; right, plants grown in the Cravo "RETRACT-A-ROOF" greenhouse. Notice the compact growth of the plants on the right with no growth retardant treatment. *(Courtesy Cravo Equipment Limited, Ontario, Canada.)*

Ethylene Vinyl Acetate (EVA) Copolymers

Films based on copolymers of ethylene with vinyl acetate are used as greenhouse covers. The advantages of EVA film are as follows:

- It is tough and clear.
- There is less radiant heat loss through EVA than there is through polyethylene films.
- It has greater sunlight resistance and thus a longer life.
- It is less brittle in cold weather and is easier to install. For example, EVA can be installed at −40°F (−40°C) as compared with polyethylene film, which cannot be handled below −13°F (−25°C).
- It has better light diffusion.

Manufacturers claim that EVA films have an average life expectancy of 3 years. EVA film is expensive. Because of its cost, it is not widely used in the United States. However, it is accepted and used extensively in Japan and Scandinavian countries.

Polyvinyl Chloride (PVC)

PVC film has many uses in horticulture. When it is considered for use as a greenhouse covering, it should be remembered that PVC reduces the light intensity transmitted because it has a static charge collecting dirt and dust. However, PVC has a number of properties that make it desirable as a covering. PVC has excellent resistance to wear. Oxidation has little effect on PVC, but heat and light break down PVC film in 2 to 3 years.

PVC film reduces the transmission of long wavelength infrared radiation, resulting in less heat loss at night using PVC as a covering than using polyethylene. Like glass, PVC film intercepts radiation from the soil and returns a part of this lost energy to the greenhouse. Temperatures under PVC coverings may be 3°F to 5°F (1.67°C to 2.78°C) higher than the outside temperature and up to 30°F (44.4°F) higher than the temperature in a polyethylene-covered house under the same conditions. PVC becomes brittle at very low temperatures and soft on very warm days.

In some localities, constant wind is a factor to be considered when selecting a greenhouse glazing material. Under such conditions, fatigue failure of PVC is a problem.

PVC costs more than other types of plastic, which limits its widespread use. In addition, PVC film is manufactured with a maximum width of only 63 inches (1.6 meters). The lack of larger sizes reduces its popularity in the United States. In Japan, PVC film is used more frequently than polyethylene. Japanese manufacturers have formulated and developed a number of special agricultural grades of PVC.

Polyvinyl Fluoride (PVF)

Fluorinated plastic (**PVF**) is higher in cost than polyethylene. However, PVF has several outstanding properties that recommend it for outdoor use. As a greenhouse covering, it has a high resistance to abrasion (wear) and weathering. It can withstand temperatures from below freezing to more than 356°F (180°C). In addition, it is transparent to ultraviolet radiation, resulting in a longer life. Tedlar7® is a well-known PVF film manufactured by the E.I. Dupont Corporation.

The estimated lifetime of PVF films is 8 years. Compare this value with the lifetimes of other types of plastic: 2 to 3 years for ultraviolet inhibited polyethylene, and 2 years for clear PVC film. Because of the superior aging characteristics of PVF film, it is applied to the surface of glass-reinforced polyester (fiberglass) to increase its useful life. PVF film costs almost as much as horticultural glass. It transmits 92 percent of the visible light, as compared to horticultural glass, which transmits 90 percent.

Type II Flexible Plastics

There are five types of flexible sheet plastics used for greenhouse glazing: glass fiber–reinforced polyester (fiberglass) (FRP), clear PVC, polymethyl methacrylate (acrylic), polycarbonate (polycarbonate of bisphenol A), and PETG (glycol-modified polyethylene terephthalate). To qualify as a glazing material, a plastic must (1) transmit PAR light, (2) be flexible, but rigid enough to provide structural stability, (3) have high impact strength, and (4) be resistant to weathering. Most single-layer flexible plastic sheets for greenhouses are corrugated for strength.

Glass Fiber–Reinforced Polyester (FRP)

The resins from which rigid sheet plastics are manufactured are known as thermosetting or heat setting materials. This means that once the plastic is formed, it will not soften at the normal temperature of a heated greenhouse. If the plastics are exposed to a very high temperature, such as that of a flame or a steam pipe, they will melt. FRP plastics exposed to a flame will ignite and burn rapidly. Extreme caution must be used when any device capable of producing a high temperature, such as welding equipment and portable heaters, is operated inside or near fiberglass greenhouses. It has been reported that a fiberglass-covered greenhouse measuring 30 feet (9 m) by 100 feet (30 m) burned completely in 5 minutes after ignition. A typical Quonset style of greenhouse covered with corrugated fiberglass is shown in Figure 4-25.

FIGURE 4-25 Quonset style greenhouse with corrugated fiberglass reinforced plastic panels. Note the severe yellowing and aging, which will reduce light transmission.

The clear PVF film generally used to coat FRP panels is patented as Tedlar7. This film is transparent to visible light. However, it has an ultraviolet screening component that protects the FRP panel when subjected to ultraviolet radiation. Unprotected FRP panels first turn yellow and then gradually darken to tan and brown. As the weathering of the panel increases, the amount of light it transmits is reduced. Badly weathered fiberglass can be refinished so that its light transmission value is almost equal to that of new Tedlar7-coated FRP.

Unlike other clear plastics used for greenhouse glazing, FRP can transmit and diffuse, or scatter, sunlight. FRP can scatter the light entering the greenhouse because of the large number of fibers in the panels. As a result, there is a uniform light intensity with few shadows throughout the greenhouse.

The visible light transmitted through FRP is about 80 to 90 percent of the light transmitted by glass. The roof bars on FRP glazed houses have a much wider spacing than do the sash bars that support the glass lights. Thus there are fewer shadows because fewer roof bars are used. This factor makes up about 10 percent of the light penetration loss through FRP.

FRP sheets are manufactured in corrugated, ribbed, or flat forms. Corrugated panels are the most widely used type of FRP for greenhouse glazing. Flat FRP panels are often used to glaze the end walls of polyethylene-covered greenhouses. Because this type of greenhouse must be recovered each year, the use of longer lasting FRP on the end walls reduces the labor required. It takes as much time to recover the end walls as it does to replace the roof of a polyethylene-covered greenhouse.

Because an FRP-covered greenhouse uses wider panels, there are fewer lap joints as compared to a greenhouse glazed with glass. This means that a greenhouse using FRP panels retains a greater amount of heat. On the other hand, the exterior surface area of such a greenhouse is greater because of the corrugations of the FRP panels. It has been estimated that the heat required is 30 to 40 percent higher than the amount needed for a structure covered with flat sheets. An unheated greenhouse covered with a corrugated material is much cooler than if it were covered with the same material in flat sheets.

FRP-glazed greenhouses can withstand severe impact without shattering. This factor is very important in areas where frequent hailstorms are a problem. On several occasions, FRP-covered greenhouses have withstood the impact of hailstones the size of baseballs that caused extensive damage to nearby glass-covered houses. FRP panels have the same impact resistance at $-100°F$ ($-73.3°C$). Other plastics become brittle at such a low temperature.

Although FRP panels are classified as a rigid plastic, they are flexible enough to be bent in a curve to fit the framework of a Quonset type or arch type greenhouse.

The initial cost of glazing a greenhouse with FRP panels is much higher than if polyethylene were used and slightly lower than if glass were used. An FRP-covered house can be maintained at a much lower cost than a glass-covered house. One of the greatest expenses in maintaining a glass house is the replacement of broken lights and slipped glass. FRP sheets are the least expensive of the rigid plastic materials used to cover greenhouses.

FRP was first used as a greenhouse covering in the United States in 1947. Poor light transmission and weathering were problems when fiberglass was introduced. Improved manufacturing techniques increased the life span of FRP. With improvements in other rigid plastics, FRP is not as widely used as it once was. Over time, the clear film or resin binding the glass fibers erodes and degrades, allowing the fibers to become exposed or bloom. These exposed fibers will then yellow and decay as well as collect atmospheric debris and other pollutants reducing light transmission.

Rigid Polyvinyl Chloride (PVC)

Rigid polyvinyl chloride (PVC) is similar in composition to flexible PVC film. Rigid PVC is used in the form of flat or corrugated sheets. Both rigid PVC and FRP sheets can be used for the same type of horticultural installations. However, they cannot be interchanged on a structure because the methods of installation for the two materials are different.

A stabilizer must be added to rigid PVC to reduce the deteriorating effects of ultraviolet light. Rigid PVC without a stabilizer will darken and lose much of its light-transmitting ability within a few years after installation. Unfortunately, the stabilizer affects the breaking or shattering strength of PVC. This can be a serious problem in areas where hail is common during spring and summer storms.

The light transmission curves for rigid PVC, FRP, and glass are nearly the same when comparing the amount and quality of light reaching plants growing in houses covered with these materials. However, FRP panels require fewer supports because they can be placed at 4-foot intervals. This means that more light is admitted to the greenhouse. The supports for rigid PVC must be spaced closer together, reducing the amount of light entering the greenhouse.

Rigid PVC scatters less light than FRP. Clear PVC transmits 70 to 80 percent of the visible light that strikes the material. Clear FRP transmits 80 to 90 percent of the incident light. The light transmission values are much lower for colored panels. For example, a yellow panel has a percentage transmission of 75 percent; green is 57 percent; blue is 54 percent; and red transmits only 52 percent of the total light striking it.

Japan is one of the largest manufacturers and users of PVC in both the flexible and rigid forms. PVC users in the United States have found that the product may become brittle when exposed to low temperatures. In a brittle state, PVC can be punctured easily by hail.

Rigid PVC panels expand and contract more than FRP panels during temperature changes. Special installation methods must be used to fasten PVC panels to the greenhouse structure to prevent problems in extreme temperatures. Prolonged exposure to high temperatures causes rigid PVC to change from clear to a smoky gray. As this change occurs, the light transmitted for plant growth is greatly reduced.

Polymethyl Methacrylate (PMMA) or Acrylic

The rigid transparent plastic material PMMA is commonly known as acrylic. (The term *acrylic* is frequently applied to a large number of polymers including PMMA.) Acrylic has been used for many years and is considered to be the most suitable rigid transparent plastic for greenhouse glazing. It is lighter and tougher than glass and has excellent outdoor weathering properties. Manufacturing improvements increase the life span of acrylic plastics by preventing yellowing. There will also be a reduction in the impact strength, yet manufacturing advancements have increased impact resistance as well.

Acrylic has a better light transmission rating than PVC. Clear acrylic transmits approximately 92 percent of the light that strikes it. Acrylic also transmits more infrared radiation than PVC.

When exposed to the heat of the sun, acrylic has a slightly higher coefficient of expansion than PVC. Thus, the anchor points where the acrylic sheets are attached should be kept to a minimum to allow for the stretching and shrinking due to the temperature changes.

Acrylic is the least expensive of all the clear rigid plastics used in sheet form for greenhouse glazing. Acrylic has a fire hazard similar to FRP.

Polycarbonate

Extruded polycarbonate sheets are a popular long-term option for greenhouse glazing. Many products have a UV-protected surface on the exterior side and a factory-applied condensate control to help prevent plant diseases caused by dripping condensation on the interior side. Polycarbonate has slightly less light transmission compared to acrylic, but is considerably stronger and impact resistant, more flexible, and only flammable when an active flame is maintained in contact with the material. It is attached to the greenhouse in a fashion similar to acrylic.

PETG (Glycol-Modified Polyethylene Terephthalate)

PETG is an extruded plastic with characteristics similar to acrylic and polycarbonate. It has a higher impact resistance than acrylic, but requires framing. It has a coefficient of expansion similar to acrylic and polycarbonate. The light transmission properties of PETG are equivalent to those of polycarbonate and acrylic. Most manufacturers co-extruded a UV protective layer for greenhouse applications.

Acrylic and Polycarbonate Compared

Acrylic and polycarbonate sheets are manufactured in the following colors: colorless, solar bronze, and two forms of translucent white. Only the colorless form should be used as greenhouse glazing. Acrylic transmits nearly 83 percent of the light that strikes it.

Acrylic burns readily. If sufficient air is present, the products of combustion are carbon dioxide and water. If sufficient air is not available during combustion, toxic carbon monoxide will form.

Polycarbonate was initially designed for use in commercial buildings other than greenhouses. This material transmits 73 percent of the light striking it. It is not as strong as the acrylic as far as the useful load is concerned, yet is more impact resistant. The material burns when ignited but normally extinguishes itself when the flame is removed. The combustion products in the presence or absence of sufficient air are the same as for acrylic.

Multilayer Flexible Plastics

Most plastic material is applied in a single layer for greenhouse glazing. One exception is the plastic film used in a double layer, air inflated greenhouse. This type of construction uses two layers of plastic held apart by air pressure. The trapped air acts as insulation and helps to reduce the heat loss.

FIGURE 4-26 DEGLAS® 16 mm double skinned high impact acrylic glazing material.

FIGURE 4-27 DEGLAS® 16 mm double skinned high impact acrylic glazing on greenhouse roof.

FIGURE 4-28 DEGLAS® 16 mm double skinned high impact acrylic glazing. Large panels reduce the number of sash bars required, which reduces shadows and increases light transmission.

FIGURE 4-29 DEGLAS® 8 mm double skinned high impact acrylic glazing. New acrylic products are flexible enough to shape to Quonset structures.

Rigid, double, and triple layer acrylic and polycarbonate materials are available for greenhouse glazing. Manufacturers claim that these materials, shown in Figure 4-26, have the following properties: outstanding heat insulating ability, high stiffness, excellent resistance to impact stress, and exceptional weatherability. Double layer products are often used for greenhouse roof applications where triple wall products are typically used for side and end walls. Figures 4-27, 4-28, and 4-29 show various applications of multiwalled glazing products.

Acrylic and polycarbonate can be purchased in sheet form, 4 feet (1.2 m) wide by 8 (2.4 m), 10 (3 m), or 12 feet (3.6 m) in length. The two sheets of plastic are separated by internal ribs spaced roughly 0.6 inch (15 mm) apart. The double layer sheet weighs nearly 1 pound per square foot (4.9 kg/m^2). The useful roof load for this material is related to the spacing of the roof sash bars. For example, sash bars spaced 24 inches (60 cm) apart will support a roof load of 60 pounds per square foot (293.7 kg/m^2). When the sash bars are spaced 5 feet (1.5 m) apart, the useful roof load drops to 10 pounds per square foot (49 kg/m^2).

Type III Glass

Composition

The ingredients in window glass are listed in Table 4-2. The iron content of glass affects the amount of light that passes through the glass. As the percentage of iron increases in glass, less light is transmitted. Glass intended for horticultural use does not transmit light energy in the ultraviolet region of the spectrum. Infrared radiation up to a wavelength of 2.7 microns is transmitted. There is no transmission of infrared radiation at 4.7 microns.

Grades and Weights

Glass is manufactured in many grades and weights. Common types of glass fused include single strength, double strength, heavy sheet, polished plate (also known as "crystal" sheet), and heavy plate.

Greenhouse glazing is double strength glass with a quality of AA, A, or B. The AA and A grades are a better quality than the B grade. However, they are too expensive for greenhouse glazing. B grade glass has satisfactory light transmittance, and is not too expensive.

In the Western Hemisphere, the largest size greenhouse glass available for covering a greenhouse is 6.6 feet (2 m) wide by 12 feet (3.65 m) long. This is a specially tempered glass that may be slightly curved. The use of large glass panes reduces the number of roof bars needed. This means that more light enters the greenhouse. This feature is important in the winter when cloudy weather lowers the light intensity.

British greenhouse manufacturers use glass lights measuring 28.75 inches (0.7 m) by 65 inches (1.6 m). Dutch Venlo style greenhouses are glazed with panes that are 28 inches (71 cm) wide by 60 inches (152 cm) long. There are many Dutch type greenhouses in the United States. Excellent crops are produced in these houses because there are fewer roof bars to shade the plants.

Some manufacturers object to large glass panes because they can be more easily broken by hail, snow loads, and high wind forces. Under normal conditions, Holland and Great Britain do not receive the same heavy snowfall that is

Table 4-2

Composition of Glass		
Compound	Chemical Symbol	Percentage
Silica dioxide	SiO_2	70-73
Sodium oxide	Na_2O	12-15
Calcium oxide	CaO	9-14
Magnesium oxide	MgO	0-3
Aluminum trioxide	Al_2O_3	0-1.5
Iron trioxide	Fe_2O_3	0-1.15

common in some areas of the United States. However, gale force winds frequently blow in from the North Sea for extended periods. The wide pane glass houses used in these countries withstand the winds successfully. It is possible that larger glass lights would be an advantage in greenhouses in the United States.

There are a number of advantages to using large glass lights. Larger lights mean that there are fewer joints to seal and less glazing compound is needed. Large panes reduce the number of places where heat loss can occur if the glass slips. The time and labor required to glaze a house are reduced when large panes of glass are used.

When the cost of construction is not a primary consideration, large glass panes are often used. To reduce the heat loss through the glass, thermopane glass is used. Thermopane consists of two sheets of glass separated by an air space of 0.25 inch (0.6 cm) to 0.5 inch (1.2 cm). The glass sandwich is hermetically sealed on all edges, and the entrapped air acts like insulation.

Saran Plastic Mesh

In areas of the country where the intensity of sunlight is high, some protection must be provided for growing plants. A mesh plastic film known as **Saran** or shade cloth is commonly used to shade crops. Saran has properties similar to PVC. Saran plastic is manufactured to provide various percentages of shade or light transmission. The closeness of the weave, the thickness, and the color of the individual threads of plastic determine the amount of shade provided. Unlike other plastics that must be stabilized to prevent ultraviolet breakdown, Saran performs well outdoors. Other products available include aluminized shade fabrics.

Saran-covered houses are used for the outdoor production of certain crops in the northeastern part of the United States during the summer months. Chrysanthemums and asters are two crops commonly grown under Saran shade houses. Tropical foliage plants are grown under Saran plastic covers in Florida, Texas, California, and Hawaii.

Saran shade houses are simple in construction. Sideposts are set to a depth of 3 feet (0.9 m). Interior supporting posts are spaced on 20-foot (6-m) centers. Crisscrossing the top of the house are wires used to support the Saran plastic.

Other Growing Structures

Cold Frames and Hotbeds

Cold frames and **hotbeds** are useful additions to a greenhouse operation. They may be used in the spring and summer only, or throughout the year. Bedding plants can be hardened off in cold frames before sale to the public. Bulb crops may be stored over winter in cold frames.

The framework of cold frames and hotbeds can be made of wood, poured concrete, or concrete or cinder blocks. Any building material can be used if it has good structural strength and some insulating ability. The framework is covered by a sash, which usually measures 3 feet (0.9 m) by 6 feet (1.8 m). The sash may be made of greenhouse quality (treated) wood or aluminum. It is glazed with glass, plastic film, or rigid fiberglass. If plastic film is used, the sash should be stored over summer out of direct sunlight. The ultraviolet radiation of the sun will cause the plastic to deteriorate if it is unprotected.

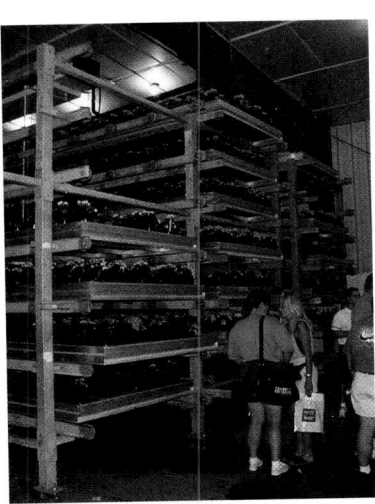

FIGURE 4-30 Hotbed construction.

Hotbeds are cold frames to which heat is added. Steam or hot water lines may be fastened around the perimeter of the bed (Figure 4-30). Electric heating cables can be buried beneath the soil to supply heat. Occasionally, hot air may be blown into a small hotbed from a large greenhouse. This method is not satisfactory for large areas because there is usually poor distribution of air.

Cold frames and hotbeds cost little when added to a greenhouse operation. However, the labor cost to operate them is high. During the warm days and cool nights of spring and fall, the sash must be raised each morning for ventilation. Then it must be closed at night to prevent the crops from freezing. Some growers have developed semiautomatic methods of ventilating cold frames. Instead of sashes, they use continuous covers that can be rolled on pipes or other devices. These methods can reduce the labor costs somewhat.

Bulb Cellars

Bulbs can be forced for spring sales. The old method of burying the bulbs in pots under layers of sawdust, soil, and salt hay has given way to **bulb cellars**. In the bulb cellar, the pots are checked periodically for moisture needs and growth rate. When the time for forcing arrives, it is an easy task to move the bulbs from the cellar into the greenhouse.

Bulb cellars are usually built underground to take advantage of the natural insulating effect of night. During the warm fall days, the ventilators and doors of the cellar are closed to prevent warm air from entering. In winter, the air temperature in the cellar is maintained at the range required to cool the bulbs, from 35°F to 41°F (2°C to 5°C).

Most growers now use aboveground storage areas for bulb forcing. Such structures are insulated (Figure 4-31). The cooling is provided by mechanical refrigeration units. In this case, there is no guesswork involved in determining if the bulbs are receiving the proper cool treatment required for uniform forcing.

FIGURE 4-31 Large walk-in cooler for forcing bulbs and other flowering crops.

Growing Rooms

Growing rooms have been used for many years in Europe. However, their use in the United States has grown with their increased use for germinating seed for plugs. Growing rooms are similar to the very expensive, controlled environment chambers used by research scientists, but on a more modest scale. Growing rooms are used to start seedlings (often called plugs) and accelerate (speed up) the growth of crops. For example, geraniums started from seed are placed in a growing room when they are planted in 2-inch (5-cm) diameter pots. Growing rooms are also used to accelerate the growth of tomatoes and lettuce before they are transplanted into greenhouses. By concentrating a maximum number of plants in each square foot of production area, the cost of operating a growing room can be held to reasonable levels.

A growing room is similar in construction to a refrigerator. The walls and ceilings are insulated. Artificial lights are installed to provide the required light intensity. Regardless of the claims made for special lamps developed for horticultural use, cool white fluorescent lamps provide the best illumination and are the least expensive to use. The illumination in the growing room should provide 1,000 **footcandles** (10.75K lux) of light energy at the surface of the plants. The light should be measured with an accurate light meter.

Foliage plants require 500 to 1,000 footcandles (5.37 to 10.75K lux) of light energy for best growth. There is an increasing interest in tropical foliage plants. As a result, facilities are needed in the northern part of the United States so that plants shipped from the South can be held before they are sent to local distributors. The process of holding plants is known as **acclimatization** and is discussed in Chapter 27.

Structures that were designed to hold meats and produce in cold storage can be converted for use in acclimatizing tropical foliage plants. To accommodate larger plants, ceiling heights of at least 15 feet (4.5 m) are desirable. Small plants can be grown on shelves. Cool white fluorescent lamps will provide sufficient light while the foliage plants are acclimatized.

ACHIEVEMENT REVIEW

Select the best answer or answers to complete each statement. List the appropriate letter(s).

1 The basic function of a greenhouse is to:
a. grow flowers.
b. raise vegetables and flowers.
c. provide a protected environment for crop production.
d. provide a protected environment for workers.

2 The maximum desirable length of a greenhouse from a central aisle is:
a. 75 feet (23 m).
b. 100 feet (30 m).
c. 150 feet (46 m).
d. 200 feet (60 m).

3 One length of the greenhouse roof is 24 feet (7.4 m), the other is 12 feet (3.7 m). This structure is called:
a. a clear span greenhouse.
b. an even span greenhouse.
c. an uneven span greenhouse.
d. a lean-to greenhouse.

4 Detached greenhouses have advantages and disadvantages as compared to attached houses. From the following choices, select the advantages.
a. Temperatures are more easily programmed for specific crops.
b. They are harder to ventilate.
c. Carbon dioxide quickly becomes a limiting factor for plant growth.
d. Carbon dioxide does not become a limiting factor for growth.

5 Crops that require heavy shade are often grown in:
a. glass-covered houses painted with whitewash.
b. Saran-covered houses.
c. wooden lath houses.
d. clear polyethylene-covered houses.

6 Draw and identify the various members of the greenhouse framework. Check this drawing against Figure 4-15.

7 Wooden greenhouse members should be treated with preservatives. Which of the following preservatives are recommended for use in a greenhouse?
a. Copper naphthenate
b. Creosote
c. Pentachlorophenol
d. Zinc naphthenate

8 Glass used to glaze a greenhouse is classified as:
a. crystal sheet.
b. heavy plate.
c. double strength.
d. single strength.

9 Large panes of glass used for greenhouse glazing have several advantages over smaller sizes. From the following choices, select the advantages.
   a. Fewer roof bars are needed, and more light enters the house.
   b. Fewer joints mean that less glazing compound is needed.
   c. Breakage is greater from high wind forces.
   d. Less time and labor are required to cover the greenhouse.

10 Polyethylene plastics break down from exposure to:
   a. infrared radiation.
   b. cosmic rays.
   c. ultraviolet radiation.
   d. gamma rays.

11 Two layers of polyethylene separated by air can save approximately how much heat as compared to a single layer of plastic?
   a. None
   b. 28 percent
   c. 40 percent
   d. 60 percent

12 Glass fiber–reinforced polyester (fiberglass):
   a. is completely resistant to fire.
   b. burns very slowly.
   c. is highly flammable.
   d. is unsafe to use.

13 The transmission of light through greenhouse glazing materials is:
   a. the same for all materials.
   b. different for all materials.
   c. greatest through glass.
   d. greatest through fiberglass.

14 Cold frames are used in greenhouse operations primarily to:
   a. harden off bedding plants in spring.
   b. produce cut flower crops.
   c. winter over bulb stock.
   d. start seedlings.

15 Hotbeds may be heated by several methods. The poorest way to supply heat is by means of:
   a. steam heating.
   b. hot water heat.
   c. hot air heat.
   d. electric heat.

16 Bulb cellars are built underground to:
   a. take advantage of the natural insulating effect of the soil.
   b. hide them from view.
   c. prevent rats and mice from entering.
   d. save on costs of heating.

Student Project 1—Constructing a Cold Frame

Objective

The student will lay out and assemble a standard section of cold frame. (Refer to Figure 4-30.)

Materials

4 Standard sash 3′ × 6′ (1.0 m × 1.8 m)

Lumber:

2 pieces, 2″ × 12″ × 12′ (5.0 cm × 30 cm × 3.6 m)
1 piece, 2″ × 6″ × 12′ (5.0 cm × 15 cm × 3.6 m)
2 pieces, 2″ ×12″ ×6′ (5.0 cm × 30 cm × 1.8 m)
1 piece, 2″ × 6″ × 6′ (5.0 cm × 15 cm × 1.8 m) (to be cut diagonally)
3 pieces, 1″ × 2″ × 6′ (2.5 cm × 5.0 cm × 1.8 m) (to be used as sash supports)
3 pieces, 1″ × 3″ × 6′ (2.5 cm × 7.5 cm × 1.8 m) (to be used as sash supports)
2 pieces, ¹/₂″ × 3″ ×6′ (1.2 cm × 7.5 cm × 1.8 m)
2 angle irons, 2″ × 2″ × 14″ × ¹/₄″ (5.0 cm×5.0 cm × 35 cm × 0.6 cm)
2 angle irons, 2″ × 2″ × 10″ × ¹/₄″ (5.0 cm × 5.0 cm × 25 cm × 0.6 cm)
2 flat metal straps, 2″ × 15″ × ¹/₄″ (5.0 cm × 37.5 cm × 0.6 cm)
2 flat metal straps, 2″ × 13″ × ¹/₄″ (5.0 cm × 32.5 cm × 0.6 cm)
2 flat metal straps, 2″ × 10″ × ¹/₄″ (5.0 cm × 25.0 cm × 0.6 cm)
42 machine bolts (round head), ³/₈″ × 2″ (1.1 cm × 5.0 cm) (with washers)

30 nails, 8-penny
$^1/_2$ gallon, exterior grade greenhouse paint, white
1 paintbrush, 3 inches wide
1 hammer
1 crosscut handsaw
1 6-foot (2 m) ruler
1 carpenter's square
1 drill with $^3/_8{''}$ (1.1 cm) wood bit

Procedure

1. Assemble all needed materials.
2. Notch the top of the frames on 37-inch (92.5-cm) centers for the sash supports. (See the notching detail in Figure 4-30.)
3. Paint all wooden members after cutting them to the required lengths and sizes.
4. Lay out the 12-foot (3.6-m) side boards and bolt them together. Place round-head bolts on the outside, using 15-inch (37.5-cm) flat straps to join the 6-inch (15.0-cm) and 12-inch (30.0-cm) wide boards together.
5. Attach the 14-inch (35-cm) angle irons at the ends to form corners.
6. Attach the 10-inch (25-cm) angle irons at the ends of the single 12-inch (30.0-cm) board to form corners for the foot of the frame.
7. Attach the 10-inch (25-cm) flat metal straps to the outside of the foot of the frame for support.
8. Lay out the 6-foot (1.8-m) long end boards (one cut diagonally). Bolt the end boards together using two 13-inch (32.5-cm) flat straps.
9. Bolt the two ends and the two sides together to form a rectangular box.
10. Nail the sash supports together (see the detail) using 1-inch (2.5-cm)× 2-inch (5.0-cm) boards for upright pieces. Use six nails per support.
11. Refer to the detail and notch the ends of the sash supports one-quarter inch deep to fit into the notches of the frames.
12. Fasten two one-half inch (1.2-cm) sash butt supports to each end of the frame.
13. Lay the sash on the frame.

CHAPTER

5

Financing, Locating, and Sizing a Greenhouse Facility

OBJECTIVES:

After studying this chapter, the student should have learned:

■ how to prepare a business plan to submit for financing to a lender.

■ how to determine when a location is suitable as a greenhouse site.

■ how to determine the type of structure best suited for the grower's needs.

■ how to design and lay out a greenhouse facility to obtain maximum production and efficient use.

Greenhouse operations are capital intense. The costs of acquiring land and then building the greenhouse structures can be intimidating to new and established growers alike. Yet, with good planning and accurate revenue projections, most lending institutions will work to help finance a greenhouse operation. The key is a sound business plan and good record keeping. Most greenhouse enterprises maintain financial records for their bank and the Internal Revenue Service. Yet, it is important to maintain accurate financial records of production costs in order to manage a business with continuous profitability. Whether you are financing a new business or financing an expansion, accurate record keeping will help you in developing a business plan.

Getting Started—Financing the Business

After you have finished your academic training, you probably feel that you are ready to start a business yourself by building a small greenhouse. Your classroom knowledge is one thing, but having real experience in the industry is something else. Before going into business yourself, unless you have had previous experience in a flower growing operation, you should seek a position with an established grower. A minimum of at least 1 year of employment is ideal, if for nothing more than to just learn the cycle of production for the various holidays and seasons. Growers are always planning ahead anywhere from 3 months for a short-term crop like pot mums to a year or more for crops such as azaleas.

After the decision has been made to become a plant grower, it is necessary to obtain the funding needed. Before any lending institution is going to loan you money, the loan officer will ask to see your business plan. This is a written statement that includes the factual information that you have gathered and your expectations of success in a new or existing business. The following is an outline of a business plan.

An Outline of a Business Plan

    I. The cover letter. The purpose of the cover letter is to describe for the reader the business goals you have and how you plan to obtain them.

   II. Business objectives and goals
- A. Explains why you are starting a business
- B. List of objectives and goals to be derived

  III. Business description
- A. Business name and location
- B. Is it to be wholesale only or a combination of wholesale and retail? What is your target market?
- C. What kinds of crops and/or services will you offer?
    1. Bedding plants only
    2. Potted plants only
    3. A combination of (1) and (2)
    4. Cut flowers only or a combination with other crops
    5. Landscape planning and installation
- D. What kinds of structures are needed?
    1. Greenhouses—describe type and size, cost
    2. Service buildings—estimated costs of construction
    3. Equipment—supplies of pots, packs, potting machines, etc.; list all equipment plus estimated
- E. Land area needed: How much improvement is needed to prepare for the structure's construction? Is there sufficient area for future expansion?
    1. Grading, leveling, and filling
    2. Installation of water lines, sewers, drainage pipes, electricity, liability, vehicles, gas
    3. Parking area for employees and customers
    4. Zoning issues

  IV. What is the target date for getting started?

V. How much money is needed to get started and carry on the business for 3 years? This assumes that you lose money the first year, break even the second year, and make a small profit the third year.
  A. Borrowed from a bank
  B. Borrowed from friends and/or relatives
  C. The amount you have yourself
  D. Borrowed from Small Business Association
  E. Borrowed from life insurance
  F. Other sources

VI. A breakdown of how funds are to be spent. Be as detailed and specific as possible.
  A. Title, searches, and appraisals
  B. Land purchase costs
  C. Attorney fees
  D. Improvements—grading, installation of drainage, electricity, gas
  E. Building costs, estimates from contractors
  F. Employee salaries, fringe benefits
  G. Taxes—property, school, and other for a year
  H. Fees for licenses and permits required
  I. Utility costs estimated for water, gas, electricity, other fuels for a year
  J. Insurance—structural, crop, personal liability, vehicles
  K. Equipment purchases for office, computers, supplies, etc.
  L. Growing supplies such as seeds, plants, started plugs, transplants, pots, packs, fertilizers, fertilizer injectors, fungicides, insecticides, growth control chemicals, potting media, miscellaneous tools, sprayers, etc.
  M. Plans for advertising, newspaper, radio, television, other

VII. Timetable for achieving
  A. Operating plan for 1 or 2 years
  B. Contingency fund

VIII. Plan of repayment of funds
  A. Length of time funds needed
  B. Source of repayment, monthly statements
  C. Break even analysis

IX. Collateral
  A. Assets available to secure the loan
  B. Appraised values of buildings, etc.
  C. The anticipated economic life of the asset

X. Historical financial analysis
  A. Financial statements from most recent 3 years
  B. Tax returns for last 3 years
  C. Most recent financial information

XI. Management and structure
  A. Organized as a sole proprietorship, partnership, or corporation
  B. Who are principals and what is their percentage of ownership?
  C. Personnel, list of officers, names, and their duties

After preparing the business plan it is then necessary to obtain an appointment with the lending officer to present your plan. There will be additional forms to fill out. Information that may require substantiation of what has been declared in the loan application is often needed. Obtaining financial support for starting a new business or expanding an older business takes time. It is not something that is done in a week or two, so start preparations months ahead of the target date. Also, consider approaching more than one funding institution for comparison.

Site Selection

Many factors must be considered when selecting the site and determining the size and arrangement of a greenhouse and auxiliary structures. Much more is involved in locating a greenhouse than simply buying a plot of land that is large enough for any future expansion of the business. The factors to be considered are examined separately in the sections that follow.

The Potential Market

Any grower planning to erect a greenhouse in a new location must ask several questions about the market to be served. First, where are the plants to be sold? What is the size of the market to be served? Second, what means of transportation are available to bring the products to this market? Today, motor freight transport is the principal means of carrying goods. Nearness to the interstate highway system can be an important factor in the success or failure of the business. A business located far off major highways will have serious problems in receiving supplies and serving customers in the winter in areas where the snowfall is heavy. For example, a critical situation may arise if a fuel supply cannot be delivered. If fuel trucks cannot reach the site, then trucks delivering the product to market will also be delayed; and do not forget the condition of bridges, overpasses, and underpasses. Although some cut flower crops such as roses, carnations, and chrysanthemums can be held in refrigerators for several days with no loss in quality, other plant crops must be sold immediately. Potted plant crops ready for market must be sold or there is a loss of income.

When a grower plans to sell in a local market, many of the factors to be considered are the same as those for distant markets. If the business is to be both wholesale and retail, then it is important to ensure easy access by the customer to the sales area. Adequate customer parking is another important consideration. All too often, the need for parking areas is an afterthought in site selection. Customer and employee parking must be segregated with appropriate signs.

Climatic Conditions

The increasing costs of fuel and electricity require that the existing climatic conditions and trends be thoroughly examined for the proposed construction area. The history of the weather in the area should be reviewed in detail.

Spring tornadoes, summer and fall hurricanes, and winter blizzards are all examples of conditions that can affect a decision on a region for building a greenhouse. A greenhouse located on a mountaintop may be well above any air pollution in the valley. However, the extra heat required to combat the wind chill factor at the higher elevation may offset the advantage of the increased light intensity due to the clearer air.

A greenhouse located in a valley may be protected from wind chill, but be subject to late spring frosts and early frosts in fall from trapped cold air.

The cost of the fuel required to heat a greenhouse in cold areas of the country may offset the cost of transporting flowers grown in warm climates to markets in all parts of the country.

The amount of heat that will be required for a proposed greenhouse can be estimated by reviewing weather bureau records. The grower should review the average annual temperatures for at least the last 50 years. Information can also be obtained from local records on the percentage of annual sunlight, wind velocity and direction, and amounts of precipitation. Areas of high precipitation usually have a low percentage of sunlight. The U.S. Coast and Geological Survey offices can provide information on the expected frequency of floods. The construction of a greenhouse facility on a river flood plain is risky and may not be allowed.

Topography of the Area

The differences in elevation of an area are shown on topographical maps. These maps are available from local soil conservation offices or from the U.S. Geological Survey. The grower can study these maps for details of the existing topography of the area. Greenhouses built on flat areas are more efficient. In addition, it is more economical to adapt mechanical systems of production if the greenhouse area is flat. However, as energy costs increase, it may become practical to construct greenhouses on slopes having a southern exposure.

If the underlying ground strata (layers) of the greenhouse site consist mainly of rock, then it may be very costly to grade the site so that it is uniformly level. EPA, state, or local requirements on preventing groundwater contamination from runoff as a result of irrigating and fertilizing greenhouse crops must be considered. Solid concrete floors that permit ebb and flood irrigation methods may have to be installed to meet these regulations. Other methods of water containment may be needed where ebb and flood benches are used for crop production.

Labor Availability

Any business requires a good source of labor. The skill requirements for greenhouse workers are not extreme. However, the success of the business depends upon the caliber of the workers, including their dependability and willingness to learn. The art of growing flower crops is not easily learned in a week or two. Most successful greenhouse enterprises rely on a reliable source of labor with farm-based skills. A low turnover of workers is desirable.

Zoning Laws

Many areas have zoning laws that prohibit certain types of businesses. Greenhouse flower crop production is an agricultural business. In general, greenhouses can be constructed wherever agriculture is permitted. The grower should check the local zoning laws before purchasing land as a greenhouse site.

Some localities restrict the type of building that can be constructed. Plastic-covered greenhouses generally are classified as temporary structures. However, some localities restrict the number of such greenhouses by taxing them as permanent structures. The grower should review any local tax laws that may affect the proposed greenhouse operation.

Types of Crops

The types of crops to be grown may be the deciding factor in locating the greenhouse. The distance the crop must be shipped may determine whether potted plant crops or cut flower crops are grown. Although the use of modern lightweight growing media reduces the weight of potted crops, they are not shipped as easily as cut flower crops. Trucks can effectively move pot crops and bedding plants 1,000 to 1,500 miles (1,609 to 2,414 km).

Cut flowers are more easily shipped over very long distances than pot crops. Orchids shipped from Australia and carnations from South America are good examples of the distances involved. Seedling plugs are easily shipped great distances as well using air freight.

Personal Desires

Personal preferences for certain areas of the country can be a very important factor in locating a greenhouse business and personal satisfaction. Consideration should be given to the desire to be close to the mountains, the seashore, flat plains, areas of high sunshine, a rural area, or a large urban center. Many consider the profession of operating a greenhouse a lifestyle as well as a career.

Availability of Services

The grower must look closely at the services available to the proposed greenhouse site. Can electricity be obtained at a reasonable rate? Is the electricity available nearby or must the grower pay the cost of installing poles to bring it to the site? What type of fuel can be supplied? If a choice of fuels can be made, the grower should calculate the fuel costs on the basis of units of heat delivered from each type of fuel. If natural gas is to be used, is the supply likely to be reduced or cut off completely if a shortage develops? Flower crops cannot be started and stopped in the same way an industrial operation can be turned on or off.

Is there an adequate supply of water? Is the quality of the water suitable for growing plants? Does the water have a high salt content? The process of desalinization (removal of salt) may be too expensive to make the water suitable for plant use. Water may contain excessive amounts of other mineral impurities, such as boron, fluoride, and sulphur. Some areas of the country may experience repeated or prolonged droughts during which water becomes scarce. In these areas, greenhouse operations and nurseries are among the first consumers to be restricted in water usage.

Acquiring an Established Business

If the grower is considering the purchase of an established business, a thorough appraisal of the total facility is needed. The following six points should be considered.

1. The age and condition of the buildings
2. The condition of the boiler and heating plant and its expected life
3. The economic condition of the neighborhood in which the business is located (i.e., is the neighborhood declining or developing?)
4. Rezoning issues (i.e., is the site rezoned to a nonagricultural use if the ownership changes hands?)

5. Expansion issues (i.e., is there room for expansion if, at some point, it is desirable to increase the size of the facility?)
6. Local laws regarding expansion

The prospective buyer should determine why the present owner is selling. A thorough investigation is called for to be sure that all of the reasons are known. For example, the owner may be selling because a superhighway is planned for the area or close by. The government can obtain the land under the right of eminent domain. However, the payment received through this process may be less than the business is worth.

Competition in the area may be a factor when considering the purchase of an existing property. Competition should be viewed as a stimulator, not as an adversary. The desire to be better than the competition has led many businesses to success.

After all of the necessary factors have been considered for either a new or an existing site, the decision can be made about the suitability of the site. The grower must consider the questions of the size and arrangement of the structures to be built.

Greenhouse Structures

Size

The types of crops to be grown are an important factor in determining the physical size of the structures. If many different crops are to be grown and the crops require different temperature ranges for best growth, small houses are preferred because the necessary temperatures can be controlled better in different areas. Growers of a single crop of cut flowers prefer large greenhouses where the plants will have enough headroom for growth.

Another advantage of a large house is that it buffers changes in the air temperature, such as those occurring in spring and fall. However, large houses require more heat than small houses to maintain desired temperatures. Large houses are taller than the small houses and are more exposed to the prevailing winds, resulting in greater heat loss.

Low houses are an advantage during storms with extremely high winds or heavy snowfall. Windbreaks provide more protection for low greenhouses than for high greenhouses.

Heat Conservation

An important consideration is the installation and use of heat conservation methods. For example, double skins, interior movable heat shades, and other methods of conserving heat should be examined before the size of the house is selected. It is easier to install semiautomatic black cloth insulating units in a clear span greenhouse of intermediate size than in a small house or a very large house.

Uneven Span Greenhouses

Uneven span greenhouses should be considered for new construction. During the low light conditions of winter, an uneven span greenhouse facing south will receive more light than a conventional even span greenhouse (roof angle of 35 degrees). A value of 11 percent more light is obtained if the south facing roof section is at an angle of 25 degrees with the horizontal and the north facing roof slope is at an angle of 55 degrees. If the angle at the peak of the roof is 90 degrees and the north facing roof slope is at an angle of 65 degrees, then even more light is admitted. The north facing wall can be covered with a reflective insulation material. As a result, there is a further increase in the amount of light reaching the crop. In addition, the heat loss from the structure is reduced.

Orientation and Arrangement of Greenhouses

Greenhouse orientation is an often debated subject. Some experts prefer a north–south orientation. Others believe an east–west orientation provides the maximum amount of sunlight during the winter months.

The orientation of the greenhouse was a more important factor when glass was the primary glazing material and the roof bars were spaced 10 inches (25 cm) or 12 inches (30 cm) apart. The decision to orient the house in a north–south direction or an east–west direction to obtain the maximum light is less critical now for two reasons: (1) Modern translucent plastic coverings scatter the light, and (2) roof bars are commonly spaced 8 feet apart.

The direction of the prevailing winds is an important factor to be considered in orienting a greenhouse. Winds affect the loss of heat in winter and the cooling of the greenhouse in the summer.

To conserve heat in winter, the smallest cross section of the greenhouse, usually an end wall, should face the prevailing winds. The end walls can be insulated easily. In the northeastern United States, the prevailing winter winds are generally from the northwest. In the summer, the prevailing winds are from the southwest, thus providing a natural advantage for summer cooling.

A greenhouse facility may consist of small single units, a single unit, large ridge and furrow units covering an acre or more, or small groups of units. A large greenhouse range with ridge and furrow houses and connected barrel vault houses is shown in Figure 5-1. The grouping of houses, also known as cluster construction, allows crops to be separated to provide different temperatures and eliminates the need for single units.

Common greenhouse arrangements are shown in Figure 5-2. The arrangement should provide for a smooth work flow. A covered central corridor connecting the greenhouses protects the plants as they are moved from one

FIGURE 5-1 Aerial view of a large greenhouse range. *(Courtesy Dr. Joe J. Hannan, Emeritus Professor of Floriculture, Colorado State University.)*

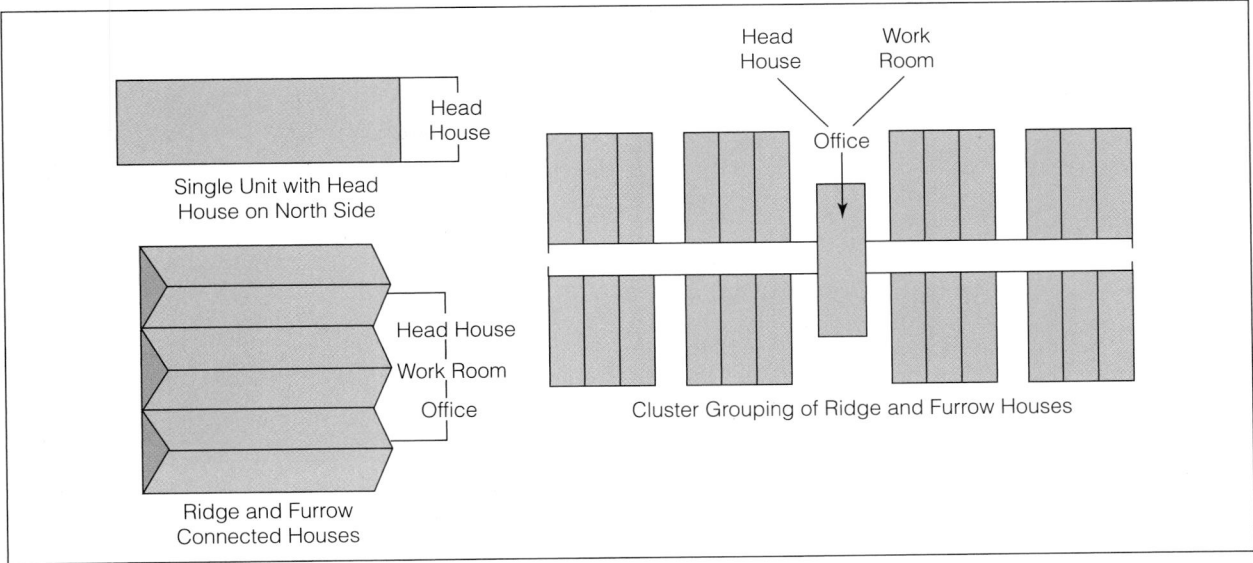

FIGURE 5-2 Common arrangements of greenhouse buildings.

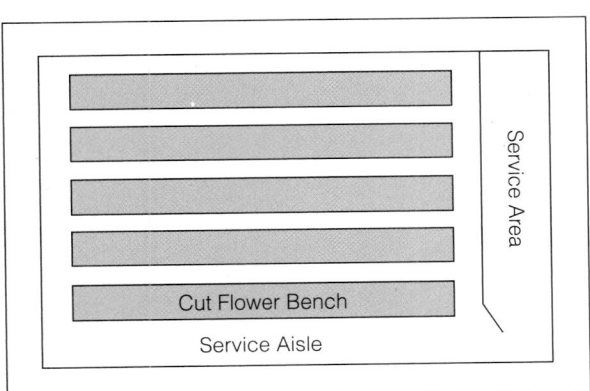

FIGURE 5-3A Cut flower benches.

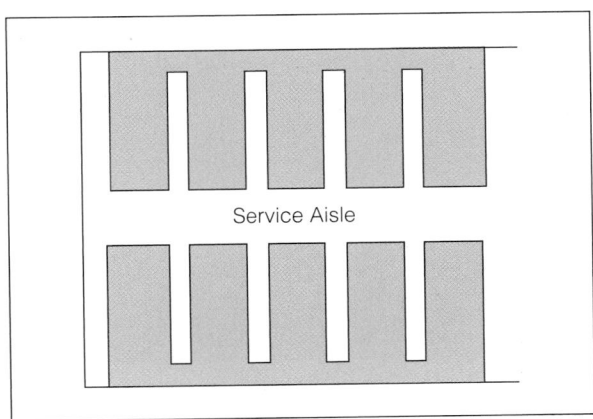

FIGURE 5-3B Peninsula bench layout for potted plants.

greenhouse to another, when they are received from suppliers, and as they are being readied for sales and shipment.

Greenhouse Benches

Bench Arrangements

The type of crop to be grown governs the bench arrangement and layout (Figures 5-3A and 5-3B). Benches for cut flower crops generally run along the length of the house. Potted plant crops and bedding plant crops may be grown on concrete floors (Figures 5-4A and 5-4B) or movable benches. Many potted plant and bedding plant crop growers use sectional benches. These can be moved around the greenhouse or outside the greenhouse at certain times of the year (Figures 5-5 and 5-6). The spring bedding plant crop is grown on the floor

FIGURE 5-4A Geraniums grown in large containers on the greenhouse floor where the concrete floor incorporates ebb and flood irrigation.

FIGURE 5-4B Bedding plants grown directly on the floor in the greenhouse, eliminating the need for benches.

FIGURE 5-5 Mobile benches in the greenhouse. These benches are supported on casters that roll on railing through the greenhouse.

FIGURE 5-6 Mobile benches in the greenhouse are adaptable to both bedding plant flats as well as pots.

FIGURE 5-7 Double cropping is done by growing one crop on the floor and the second crop on trays that can be moved outside during daylight hours.

FIGURE 5-8 Bedding plant production outside the greenhouse. Trays can be moved inside the greenhouse at night if frost threatens.

and also on the movable benches. In the morning, in early spring, the benches are moved outside into full light intensity and cool temperatures. This treatment conditions the plants to the outside environment. If no frost is predicted, the plants are left outdoors at night (Figures 5-7 and 5-8). In the event that frost threatens, the plants can be quickly moved into the greenhouses for protection. This grower has doubled the production area without having to build new greenhouses.

The grower can make the best use of the growing area in the greenhouse by using movable benches. Movable benches are used primarily for container crops because they are not easily adapted to cut flower crops. Movable benches are more expensive than fixed benches. A floating aisle provides access to the plants (Figures 5-9, 5-10, and 5-11).

A pyramid type bench is shown in Figure 5-12. This type of construction provides more production space in a given area than a flat bench.

FIGURE 5-9 Wire mesh rolling benches. The use of rolling benches maximizes use of greenhouse space in this propagation house.

FIGURE 5-10 Production greenhouse uses rolling top benches to maximize use of growing space. The pipes below the benches supply hot water heat for root zone heating.

FIGURE 5-11 Propagation house with rolling benches used for production needs after the propagation needs have ended. Notice the high intensity discharge (HID) lights overhead and the horizontal air flow (HAF) fans for moving the air. The HAF fans are usually on 24 hours a day.

FIGURE 5-12 A pyramid frame made from square tubing. This type of bench is useful for growing hanging baskets.

Bench Construction

Modern greenhouses use expanded metal as the material of choice for greenhouse bench construction. In older greenhouses, concrete may still be found, especially where cut flower crops such as roses are still grown (Figure 5-13). Ebb and flood benches are usually made of molded plastic (Figures 5-14A, 5-14B, and 5-14C). Some potted plant growers have begun to grow their crops in troughs. These are very effective in that they allow air to move around the plants, which helps to reduce foliage diseases. They also channel the irrigation solutions back to the central reservoir after the crop has been irrigated. This helps to prevent groundwater pollution.

FIGURE 5-13 Standard concrete bench designed for cut flowers, but now used for bedding plants.

FIGURE 5-14A Ebb and flood benches, drained.

FIGURE 5-14B Ebb and flood benches, fill pipe.

FIGURE 5-14C Ebb and flood benches, flooded.

The troughs are generally made of rigid molded plastic. Aluminum and other metals corrode too quickly from the fertilizer salts used. Concrete was once the best material for benches (Figure 5-13), but it too is used less because its cost is prohibitive.

Wire mesh benches are widely used for container grown crops (Figure 5-15). A wire mesh bench permits maximum air circulation around the plants. The greater the circulation, the less likelihood there is for the development and spread of foliage diseases. Galvanized wire mesh is subject to rust but still has a long, useful life as a bench material.

Some growers still use redwood. This wood resists decay and does not require treatment with wood preservatives. However, redwood is expensive. Many growers use other species of wood such as fir for benches. Pine is not used because it is too soft. Many growers build bench tops and support them on concrete cinder blocks to reduce expense and decrease deterioration (Figure 5-16).

All woods used should be treated with a preservative (with the exception of redwood). Preservatives that are not suitable for greenhouse use include creosote, pentachlorophenol, and other compounds that contain xylol (refer to Chapter 4). The fumes of these materials are toxic to plants.

Greenhouse benches are rarely painted. Some growers use whitewash on the sides of ground beds in rose houses. The whitewash is used to reflect the light and not to protect the wood.

Service Buildings

The service building, also known as the head house, is located on the northern side of the greenhouse. In a large greenhouse range, it may be located at the center of the range so that all of the houses are equidistant from it. The head house is connected to the greenhouse(s) to protect the plants as employees move them from one unit to another.

FIGURE 5-15 Wire mesh bench top placed on pipe framework.

FIGURE 5-16 Raised wooden bench tops supported by concrete cinder blocks. Open sides of cinder blocks are oriented in a way such as to not gather water and debris. *(Courtesy Dr. Joe J. Hannan, Emeritus Professor of Floriculture, Colorado State University.)*

The size and design of the head house is best determined by its intended use. Administrative offices should be located in one end of the building away from the dust and dirt of the soil storage area, the potting area, and the area where crops are prepared for shipment and sale.

The workroom area should be well lighted and heated. Rest rooms, a locker area, and lunch facilities must be provided for employees. Storage space is required for pots, boxes, soil mixtures, peat moss, and fertilizers. Chemicals used for insect, disease, and weed control are not stored in the head house. These materials must be stored in a separate, fireproof building that can be locked securely.

The head house will also have a large refrigerator or walk-in cooler to condition cut flowers. Grading tables and packing areas are designed so that enough space is provided even for the large shipments that occur at holiday times. Conveyors and hand trucks make it easier to move materials. The doors of the receiving area should be large and protected so that plants can be loaded and unloaded from trucks without damage from freezing.

ACHIEVEMENT REVIEW

Select the best answer or answers to complete each statement. List the appropriate letter(s).

1 The selection of a greenhouse site a great distance from a major highway is a poor choice because:
a. it is too far for workers to travel.
b. winter snowstorms may prevent fuel delivery.
c. there is less air pollution from highway vehicles.
d. it is too far from markets.

2 Weather bureau records for 50 years should be checked for long-term information on:
a. the percentage of annual sunshine.
b. the average annual temperatures.
c. the rise and fall of tides.
d. the percentage of annual precipitation.

3 Topographical maps provide information on:
a. the slope of the land.
b. the direction of underground streams.
c. the type of underground strata.
d. the type of vegetation in an area.

4 The type of crop to be grown may determine the location of a greenhouse. Which of the following items can be shipped more than 1,000 miles with the least difficulty?
a. Easter lilies
b. Petunias
c. Orchids
d. Roses

5 Which of the following conditions are most important when evaluating the water supply for a greenhouse?
a. Low soluble salt content
b. Availability in sufficient quantity year round
c. Low cost
d. High mineral content

6 Large greenhouses require more heat than small greenhouses because:
a. they contain a larger volume of air.
b. they are higher and more exposed to the prevailing winds.
c. they have a greater roof area than small houses.
d. they have more doorways.

7 During the low light conditions of winter, which type of greenhouse will allow the most light to enter?
a. Even span glass house
b. Quonset type plastic-covered house
c. Fiberglass multibarrel vault house
d. Uneven span glass house with the long slope facing south

8 For maximum light during winter months, the greenhouse should be facing:
a. north–south.
b. east–west.
c. 22° north of east and south of west.
d. none of these.

9 The preferred arrangement of benches for potted plant crops is:
a. not over 100 feet long by 4 feet wide.
b. a peninsula type arrangement.
c. peninsula benches that are 8 feet wide.
d. movable benches that are 4 feet by 20 feet.

10 The service building, or head house, should provide one or more of the following:
a. Space for administrative offices
b. Refrigerator space for holding cut flower crops
c. Storage space for chemicals used for insect, disease, and weed control
d. Rest rooms, locker area, and lunch facilities for workers

Control of the Greenhouse Environment

OBJECTIVES:

After studying this chapter, the student should have learned:

■ the basic differences between the types of heating units available to growers.

■ the most effective heating unit for a specific greenhouse installation.

■ how to calculate the heat requirements for a greenhouse.

■ the basic principles of greenhouse ventilation and cooling.

■ the various methods of greenhouse cooling.

Heating

After labor, the cost of heat in the greenhouse is typically the greatest expense a grower will have. Yet fluctuations in the energy market often alter this scenario from year to year. If it were not necessary to heat a greenhouse for a large part of the year, the grower could realize more profit from the operation than is normally the case.

Greenhouses are basically solar collectors. The sun provides the primary source of heat in the greenhouse. Unfortunately, there is a limited amount of sunshine. Even when the sun is shining, the amount of energy it provides as heat varies greatly. In the summer there is too much heat. On cloudy days in the winter, there is too little heat from the sun. Actually, sunny winter days fail to supply all of the heat required to maintain the greenhouse temperature at a desired level.

An important consideration in heating a greenhouse concerns the process known as the **greenhouse effect**. This effect is the way in which a greenhouse collects and stores heat from the sun. Greenhouse glazing passes **radiant energy** from the sun in the form of light and electromagnetic wavelengths in the range from 38 to 2,500 nanometers (nm). This energy warms up the plants, soil, and other objects in a greenhouse. The energy absorbed by these objects is then reradiated at comparatively low temperatures. The heat energy that is reradiated has longer wavelengths (from 500 to 3,500 nm or infrared energy), which cannot pass through the greenhouse glass. Heat supplied to the greenhouse by solar radiation during the day is trapped and largely retained at night.

The greenhouse effect varies for different plastic materials. For example, there is a high radiation heat loss at night through polyethylene. A greenhouse cannot retain all of the heat required to keep plants at the proper temperature. This is especially true during the winter months. As a result, additional heat must be provided. Some polyethylene films are available, however, that do not transmit infrared radiation, which will trap some of this heat energy.

Types of Fuel

The cost of energy varies with the type of fuel used, availability of that fuel, and the season of the year. This means that the type of heating system selected for a greenhouse must be compatible with the most economical fuel available in the region. Gas (natural or manufactured), fuel oil, electricity, and even coal or wood are fuels commonly used to heat greenhouses. Electricity is not discussed because it is not practical for commercial greenhouses. Wood is in limited use in some areas and practical where locally available. The use of **solar energy** alone as a means of heating has been thoroughly researched and found not practical in northern climates. Scientists at Rutgers University in New Jersey have shown that the size of conventional solar collectors needed must be as large in area as the greenhouse to be heated. Current solar collectors available at this time are an impractical situation for large scale production.

The choice of fuel used depends upon several factors. A prime consideration is the cost of fuel per **Btu (British thermal unit)**. One Btu is the quantity of heat required to raise the temperature of one pound of water one degree Fahrenheit.

Table 6-1 compares the cost of heat from three different fuels. Although coal appears to be the best buy for each dollar spent, one must take into consideration that coal has to be delivered and stockpiled. In winter, unless it is stored under cover, moisture from snow and rain may freeze the coal pile and make it difficult to use. Furnaces and boilers that use coal as fuel must be cleaned regularly of ashes. Disposal of ash becomes a problem when large amounts are handled.

In addition to these problems, the Clean Air Act has placed more strict controls on the amount of sulfur dioxide introduced into the atmosphere from burning coal. Besides the federal laws, many state and local community laws are written to protect neighbors from the nuisance of fly ash and soot that come from coal-fired heating systems.

Gas and fuel oil are much easier to control and provide no disposal problems. Natural gas is traditionally delivered through a network of gas lines.

Table 6-1

The Energy Output in Btus of One Dollar's Worth of Three Different Fuels		
Type of Fuel	Btus/Dollar of Fuel	Fuel Cost*
Bituminous (soft) coal	639,000	$14.50/ton
Fuel oil #2 grade	126,000	$2.80/gallon
Gas	229,200	$10.00/1,000 cu. ft.

*Fuel costs may be less than those quoted when large amounts are purchased.

FIGURE 6-1 An EPA approved outdoor fuel storage tank.

Propane is similar to natural gas, but is delivered to the site via a truck and stored on site. Fuel oil, common to the northeastern United States, is delivered via truck and must be stored on site as is propane.

■ The availability of the fuel is an important consideration, particularly in winter. Is the fuel readily available at all times? In recent years, it has become necessary for growers to have a second fuel source to take over the heating needs in the event that gas supplies are cut off due to temporary shortages. Some growers negotiate a reduced rate with natural gas suppliers and elect to be disconnected from the natural gas supply during times of heavy demand. This requires that the boilers be designed for dual fuels and that an alternate fuel source will be available on site.

■ Fuel transportation costs are a factor in selecting a fuel. Excessive costs for transportation may make a cheap fuel expensive. Will the fuel be supplied on a regular basis? If severe winter weather restricts fuel deliveries, the grower faces serious financial losses.

■ Can the fuel be stockpiled? Suppliers usually quote discounted prices if a large quantity of fuel can be purchased at one time. One method of storing liquid fuel is shown in Figure 6-1. However, local laws may prohibit the storage of large quantities of fuels. The grower must check any laws regulating fuel storage.

■ The grower must evaluate equipment costs and determine if part or all of the heating system can be automated. Simplified equipment and operation will reduce the labor (hence, the cost) required to maintain the desired greenhouse temperatures.

Types of Heating Systems for Glass Houses

Hot Water Heating System

Hot water heating systems were once only an option for small greenhouses having less than 15,000 square feet (1,500 m^2) of glazing. Larger greenhouses were once limited to steam. Hot water is also effective when a number of small compartments are to be heated to different temperatures. If a **gravity flow return system** is used, the boiler is placed below the ground level of the greenhouse (Figure 6-2). Gravity causes the cold water to flow back to the boiler.

FIGURE 6-2 Gravity flow heating system for which the boiler is located below the soil level of the greenhouse.

FIGURE 6-3 Accelerated heating system.

The initial cost of a gravity system is low, and the maintenance costs are also low. However, there are few gravity flow heating systems in use. They have been replaced by **motorized accelerated systems.**

An accelerated system has an electric pump located in the cold water return line near the boiler (Figure 6-3). The pump drives the return water to the boiler. Thus, the water moves through the heating system faster. As compared to a gravity system, the accelerated system requires less time to bring the greenhouse temperature to the desired level when the **thermostat** calls for heat.

Modern hot water systems are closed systems with forced water circulation. In a closed system, both the water temperature and pressure can be increased to improve the efficiency of the system. Fewer heating surfaces are required with this type of system to maintain the desired greenhouse temperatures.

Temperature Control

Greenhouse heating systems are effectively controlled by thermostats or greenhouse computerized environmental control systems (Figure 6-4). The thermostat or temperature sensor is to be installed in an aspirated enclosure. A small electric fan blows a continuous stream of air over the temperature sensor or the sensing unit of the thermostat. The air temperature is continuously sampled. As a result, a more uniform temperature can be maintained in the greenhouse.

It is recommended that the temperature sensor or thermostat be placed at the same level as the plant crop. If the crop is grown for cut flowers, the thermostat should be located one-third the way down from the top of the plants. For potted plants, the thermostat can be placed level with the top of the plants.

Zoning

A single temperature sensor or thermostat will not give accurate control of the temperature in a greenhouse. Factors affecting the temperature include the type of heating pipe used; the location of the heating pipes, such as on the sidewalls, below bench, or overhead, or a combination of sidewall and overhead locations; the type of ventilation system; and the type of greenhouse glazing material. **Infiltration** of cold air into the greenhouse due to wind effect can cause a 5°F to 10°F (2.78°C to 5.56°C) temperature difference in the house. The windward end may be 10°F colder than the end of the house away from the wind.

Zoned heating systems are used in large greenhouses to offset these temperature variations. With a zoned system, a separate thermostat is installed for each zone (Figure 6-5). One area can be maintained at a warmer or cooler temperature than other areas. Crops having different temperature requirements can be grown in the same house using zoned temperature control. It also allows the grower to make temperature adjustments to maintain a uniform temperature in the greenhouse.

If hot water is to be used to heat the greenhouse, one consideration in selecting the type of boiler is whether it allows **trimming for steam.** This is a procedure in which the level of water in the boiler is reduced to permit the production of steam. The steam is used to sterilize soils and pots. Trimming for steam is not possible on all hot water boilers. The grower should consult a heating engineer before attempting this procedure. The heating system must be shut off when a hot water boiler is steam trimmed. This means that steaming can be done only in the late spring, summer, and early fall when the greenhouse does not require heat to keep the crops at the proper temperature.

Steam Heating Systems

Steam heating systems were once used when the greenhouse has 15,000 square feet (1,500 m²) or more of glass area, the heating coils are continuous and over 200 feet (68 m) long, and the houses are more than 100 feet (34 m) distant from the boiler. Although steam heating systems continue to be popular, hot water heating systems are becoming more common for large scale greenhouse operations.

Modern steam heating plants are closed circuit systems. Water is heated to 212°F (100°C) or higher. The resulting steam is forced through the main pipes to the distribution pipes in the greenhouse (Figure 6-6). As the steam cools and gives up its heat, it condenses to water. The condensate (water) is returned, by pumps, to the boiler where it is reheated, and the process is repeated. The steam

FIGURE 6-4 Aspirated temperature and humidity sensor. This sensor is placed in the center of the greenhouse at or just above plant height.

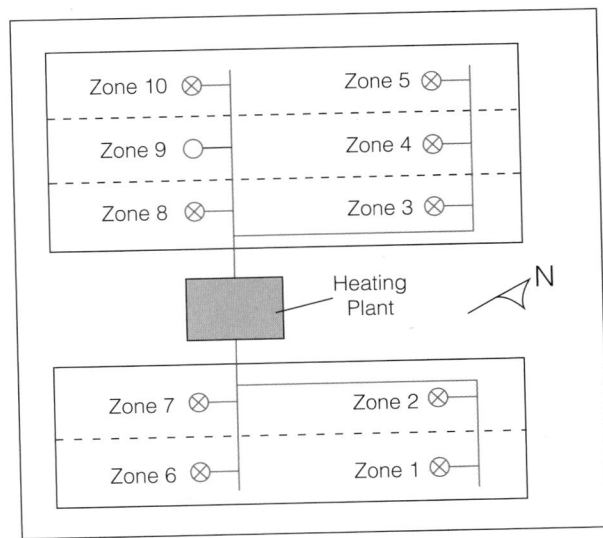

FIGURE 6-5 Diagram of a typical zoned heating system for a large greenhouse range.

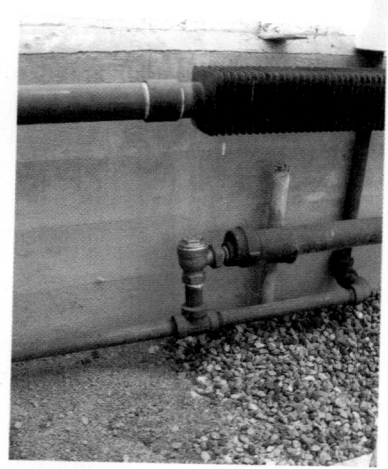

FIGURE 6-6 Typical steam pipes placed on curtain wall inside the greenhouse. The bronze steam trap is thermostatically controlled and does not allow steam to leave the section of pipe until it has condensed to water releasing 970 BTUs per pound of steam.

trap serves to prevent steam from returning to the boiler until it condenses to water. When steam condenses to water, 970 Btu per pound of steam are released.

In small greenhouse ranges, the steam is supplied at a low pressure of from 3 to 10 pounds per square inch (**psi**). In large ranges, the steam may be forced through several miles of main and distribution pipes. To meet the requirements of such a system, the steam pressure at the boiler is 60 psi or more. For very large greenhouse ranges, the pressure at the boiler may be as high as 120 psi.

A **unit steam heater** can be used to heat an entire greenhouse. Such a heater can also supplement conventional steam pipe installations (Figures 6-7A and 6-7B). A unit heater contains a **heat exchanger,** which is similar to an automobile radiator. A fan blows cold air through the steam-heated exchanger. The heated air is then distributed throughout the greenhouse.

Distribution of Heat

Natural Distribution

Greenhouse heating systems normally are designed to use the natural forces that affect the movement of air. Air moves by the process of natural **convection.** That is, as air is heated, it rises, and when it cools, it descends. In forced convection, fans aid in moving the hot air in the greenhouse.

The air currents due to natural convection cause the temperature of the greenhouse to vary at different locations. When heating pipes are placed along the sidewalls, the heated air in the vicinity rises and follows the slope of the greenhouse to the peak. As the air moves up the slope, it is cooled. At the peak, it meets the air rising from the other side of the house (Figure 6-8). The cooled air descends to the floor level. It then moves to the sides of the house where it is again heated by the pipes. Note that sidewall heating sets up a circular flow of air in the greenhouse.

A similar airflow pattern is formed if the heating pipes are placed on the end walls of the house. The force of the wind blowing on the greenhouse has a great

FIGURE 6-7A A steam fed unit heater (gas fired units are also common). The fan in front and to the right of the unit heater is a horizontal air flow (HAF) fan and is used to increase air circulation in the greenhouse.

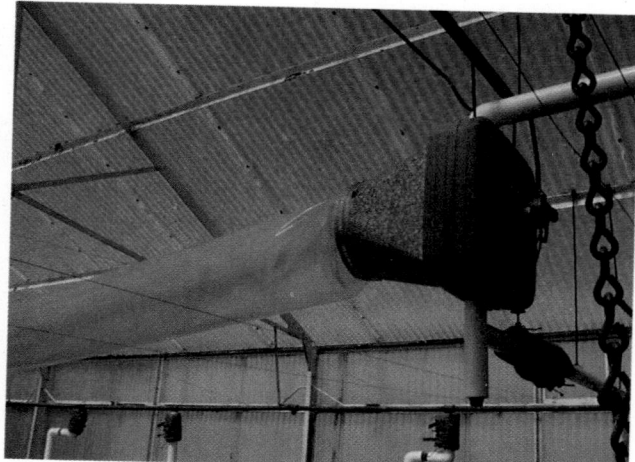

FIGURE 6-7B An overhead, steam fed heater with a perforated tube attached for heat distribution.

effect on the movement of air. This is especially true in glass-covered houses when air enters through loose lap joints between the glass lights.

Even more irregular convection currents are formed if heating pipes are placed overhead (in addition to those along the sidewalls) (Figure 6-9A). Because heat is supplied in several locations, there are no cold downdrafts on the plants. Yet, overhead heating pipes are rarely used in a modern greenhouse due to excessive shading, unless placed under a gutter to melt snow (Figure 6-9B).

Mechanical Methods of Distribution

In the past, growers may have installed large fans above the crops to improve the circulation of air in the greenhouse. These fans moved the warm air from the peak down to the plants. Studies demonstrated that such fans caused very distinct airflow patterns in greenhouses. This type of forced airflow causes greater temperature differences than those occurring from natural convection forces.

Perforated polyethylene tubing (Figure 6-10) is an advancement over large fans, because it moves air downward from the peak to the plants to provide a more effective means of improving the air (heat) distribution in the greenhouse. The tubing is installed above the crops. The tubing is from 18 to 24 inches (45 to 60 cm) in diameter and has small holes, 1.5 to 2 inches (4 to 6 cm) in diameter, spaced every 2 feet (50 cm) along the length. A fan-jet blows air into the tubing (Figure 6-11). The air may come from a heater (Figure 6-12) or from the

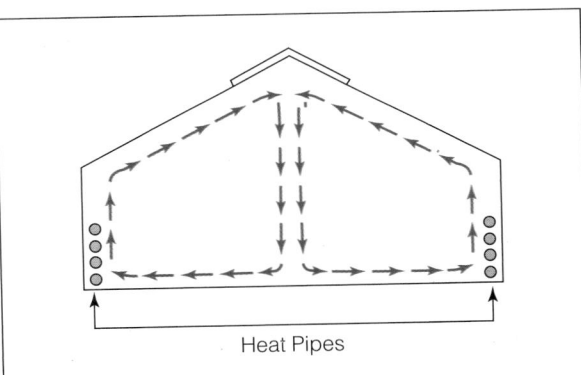

FIGURE 6-8 Airflow pattern in a greenhouse due to natural convection currents.

Heat Pipes

FIGURE 6-9A Perimeter arrangement of heating pipes.

FIGURE 6-9B Overhead steam pipe arrangement for melting snow in gutters.

greenhouse itself. The air blown through the tubing leaves it at the holes and mixes with the greenhouse air within a distance equal to 20 to 30 times the diameter of the hole. If the tubing has 2-inch (5-cm) diameter holes, the air is thoroughly mixed by the time it is 40 to 60 inches (100 to 150 cm) away from the tubing. This type of system sets up a continuous circulation in the greenhouse. This method of distribution is more important in winter ventilation when cold outdoor air is brought into the greenhouse for cooling purposes. A more detailed description of this method is given in the section Ventilation and Cooling in this chapter.

In recent years, the use of perforated polyethylene tubing for improving air circulation has decreased. Polyethylene tubing installed above the crops collects dirt and debris and casts shadows. An alternative solution to polyethylene tubing is the use of horizontal air flow (HAF) fans. These are small horizontal fans placed above the crops to circulate the air (Figure 6-13). They provide better temperature uniformity than tube systems. These fans are relatively small using 1/30 to 1/15 hp motors with a 16- to 18-inch blade diameter.

HAF fans typically have minimum to maximum velocities from 50 to 100 cfm (23.6 to 47.2 liters/sec). Place one fan, on average, every 50 feet (15.2 meters) of greenhouse. HAF fan placement should create a circular pattern of air

FIGURE 6-10 Perforated polyethylene tube is used to improve air distribution over long distances in a greenhouse.

FIGURE 6-11 ACME fan-jet ventilation and heating unit installed on polyethylene tubing.

FIGURE 6-12 ACME fan-jet system. The vent to the rear opens for first stage or winter cooling.

FIGURE 6-13 Overhead horizontal air flow (HAF) fan to improve air circulation. Note that the fan is mounted below the retractable shade system.

movement. They are placed typically 3 feet (0.9 meters) above the crop and below the eaves. Place the first fan 10 to 15 feet (3 to 4.6 meters) from the end wall and have no more than 50 feet (15.2 meters) between fans on the same side. The last fan (per side) should be 40 to 50 feet (12.2 to 15.2 meters) from the end wall toward which it is blowing. Mount fans one-fourth the house width from the sidewalls. When using horizontal unit heaters, they can serve as "corner" fans on the same side (first fan per side). When placing multiple fans in multispan greenhouses, direct the air movement in tandem rather than in opposition for maximum air movement.

Horizontal Unit Heaters

Horizontal unit heaters are commonly used in greenhouses. Perforated polyethylene tubing is attached to the heater to improve the distribution of heat. Compared to other types of heaters, horizontal unit heaters are low in cost and can be installed easily. In general, the circulation fans are connected to run continuously or in conjunction with HAF fans. Thus, the on–off cycling of the heater will not cause wide temperature fluctuations. The air is constantly moving. This condition reduces the amount of disease that can affect plants as a result of high humidity. The heat control thermostat is wired directly to provide heat on demand as the temperature decreases; however, most greenhouses wire unit heaters to a computerized climate control system. Unit heaters must be located properly if they are to operate efficiently and must be designed for use in a greenhouse. Greenhouse unit heaters typically have an aluminum manifold, which will not rust, reducing maintenance and replacement. In addition, most gas-fired unit heaters require a fresh air source as well as external venting. A typical hot air heater installation is shown in Figure 6-14.

Hot Air Heaters

Greenhouses can also be heated by hot air systems using residential type heating oil. This type of system provides dry heat in the greenhouse. In a small house, the built-in fan of the heater can distribute the heat. For large houses, 8- to 10-inch (20- to 25-cm) diameter polyethylene tubing is attached to the heater to distribute the heat throughout the house. Frequently, heaters are placed at several locations in large houses to maintain uniform temperatures. As with gas-fired unit heaters, hot air systems require a fresh air source and external venting.

FIGURE 6-14 Gas-fired unit heater.

Calculation of Heat Requirements

The amount of heat lost from a greenhouse depends on six factors: (1) the glazing material used, (2) the exposed surface area of the house, (3) the inside or crop temperature desired, (4) wind speed, (5) the condition of the house (that is, whether it is new or old with loose joints and slipped glass), and (6) the outside design temperature. The outside design temperature is the lowest average expected temperature during the heating season for the local area.

The heat requirements of a greenhouse can be determined using the formula:

$$H = Ka (t_1 - t_2)$$

where

H = the heat required, in Btu per hour
K = the **heat transfer coefficient** for the glazing material used
a = the exposed surface area, in square feet
t_1 = desired inside temperature, in °F
t_2 = outside temperature, in °F

This formula accounts for the heat lost through the covering of the house. It does not take into account the heat loss due to high winds, loose glass, or winter ventilation (when needed). To allow for these losses, an additional 10 percent can be added to the heat requirements.

Example

A typical problem will be solved to show how the heat formula is used to determine the heating requirements for a greenhouse (Figure 6-15).

Area of A =   32 ft × 6 ft (9.8 x 1.8 m)   = 192 ft² (17.6 m²) × 2 (Two End Walls)   =   384 ft² (35.28 m²)
Area of B = 100 ft × 6 ft (30.5 x 1.8 m) = 600 ft² (54.9 m²) × 2 (Two End Walls)   = 1,200 ft² (109.8 m²)
Area of C = $\dfrac{32\ ft \times 9\ ft}{2}$ $\dfrac{(9.8 \times 2.74\ m)}{2}$ = 144 ft² (13.4 m²) × 2 (Two Gable Ends)   =   288 ft² (26.8 m²)

Area of D = 18.3 ft × 100 ft (5.58 × 30.5 m) = 1,830 ft² (170.19 m²) × 2 (Two Roof Sections) = 3,660 ft² (340.4 m²)

Total   5,472 ft² (512.3 m²)

FIGURE 6-15 Calculating the exposed surface area of a greenhouse. The area measurement is to be used to calculate the heating requirements.

Glass as the Glazing Material

The greenhouse considered in this problem is 100 feet (30.5 m) long, 32 feet (9.8 m) wide with a 6-foot (1.8-m) high eave, and 15 feet (4.5 m) high at the ridge. The greenhouse is freestanding; that is, it does not have a workroom attached to one end. If a workroom is added, the heat loss for this area is not calculated.

The first step is to calculate the exposed surface area of the greenhouse. This area is term "a" in the formula.

Area of the walls: The area of a rectangle is equal to the length times the width.

End wall area (height to the eaves)

$$= 6 \text{ ft} \times 32 \text{ ft} (1.8 \text{ m} \times 9.8 \text{ m}) = 192 \text{ ft}^2 (17.6 \text{ m}^2)$$

$$192 \text{ ft}^2 (17.6 \text{ m}^2) \times 2 \text{ (two end walls)} = 384 \text{ ft}^2 (35.28 \text{ m}^2)$$

Side wall area (two walls to the eaves)

$$= 6 \text{ ft} \times 100 \text{ ft} (1.8 \text{ m} \times 30.5 \text{ m}) \times 2 = 1,200 \text{ ft}^2 (109.8 \text{ m}^2)$$

Area of the gable ends: The area of the gable ends can be calculated using the formula for a triangle:

Area of a triangle = $1/2 \times$ base $\times$ height

$$\text{Area} = 1/2 \times 32 \text{ ft} \times 9 \text{ ft} (1/2 \times 9.8 \text{ m} \times 2.74 \text{ m})$$

$$= 144 \text{ ft}^2 (13.4 \text{ m}^2)$$

$$144 \text{ ft}^2 (13.4 \text{ m}^2) \times 2 = 288 \text{ ft}^2 (26.8 \text{ m}^2), \text{ total area of two gable ends.}$$

Roof area: Another right triangle formula is used to calculate the length of the roof bar. This length is needed to determine the roof area.

$$a^2 + b^2 = c^2$$

In this formula, *a* is the base of the triangle, *b* is the height, and c is the hypotenuse of the triangle. (The hypotenuse is equal to the length of the roof bar.) The length of the base is 32 ft (9.8 m)/ 2 = 16 ft (4.9 m). The height is 9 ft (2.7 m). Using the formula $a^2 + b^2 = c^2$:

$$(16 \text{ ft})^2 + (9 \text{ ft})^2 = c^2$$

$$(4.9 \text{ m})^2 + (2.7 \text{ m})^2 = c^2$$

$$256 \text{ ft}^2 + 81 \text{ ft}^2 = 337 \text{ ft}^2$$

$$24 \text{ m}^2 + 7.3 \text{ m}^2 = 31.3 \text{ m}^2$$

Therefore the roof area is:

$$\sqrt{337} \text{ ft}^2 = 18.33 \text{ ft} (\sqrt{31.3} \text{ m}^2 = 5.58 \text{ m}), \text{ length of roof bar}$$

$$\text{length} \times \text{width: } 100 \text{ ft} \times 18.3 \text{ ft} (30.5 \text{ m} \times 5.58 \text{ m})$$

$$= 1,830 \text{ ft}^2 (170.19 \text{ m}^2)$$

The total area for two roof sections is:

$$1{,}830 \text{ ft}^2 \ (170.8 \text{ m}^2) \times 2 = 3{,}660 \text{ ft}^2 \ (341.6 \text{ m}^2).$$

The total exposed surface area is:

$$3{,}660 \text{ ft}^2 + 288 \text{ ft}^2 + 1{,}200 \text{ ft}^2 + 384 \text{ ft}^2 = 5{,}532 \text{ ft}^2$$

$$(340.4 \text{ m}^2 + 26.8 \text{ m}^2 + 109.8 \text{ m}^2 + 35.28 \text{ m}^2) = (512.3 \text{ m}^2)$$

For glass, the heat transfer coefficient (K) is 1.15. Assume that the crop requires a 60°F minimum temperature when the outdoor temperature is 0°F. Inserting these values in the heating formula gives the amount of heat required in Btu per hour.

$$H = 1.15 \times 5{,}532 \text{ ft}^2 \times (60 - 0)°F$$

$$H = 6{,}361.8 \times 60$$

$$H = 381{,}708 \text{ Btu/hr}$$

A heat input of 381,708 Btu per hour is required to heat this glass-covered greenhouse to 60°F (15.5°C) with an outdoor temperature of 0°F (−18°C). Heater output is rated in Btu per hour. Therefore, if one heater is to be used, it must be large enough to supply the required Btu value. Two smaller heaters can be used if their total Btu output equals the required value. For a house 100 feet (30.5 m) in length or longer, two smaller heaters are preferred to one large heater. There is better heat distribution from two small heaters.

Plastic-Covered Greenhouse

In the previous example, glass was used as the glazing material. Single layer polyethylene or rigid fiberglass glazing can also be used. These materials have nearly the same heat transfer coefficient as glass (1.15). Thus, the heat requirements will be the same for a greenhouse covered with either of these materials.

Double layer plastic has an insulating effect that reduces the amount of heat required. The heat transfer coefficient for double layer plastic is 0.80. Assuming that a greenhouse with the same exposed surface area, as determined in the previous example, is covered with double layer plastic, the heat requirements are:

$$H = Ka \ (t_1 - t_2)$$

$$H = 0.80 \times 5{,}532 \text{ ft}^2 \times (60 - 0)$$

$$H = 4{,}425.6 \times 60$$

$$H = 265{,}536 \text{ Btu/hr}$$

The heat requirement for a house covered with double layer plastic is 116,172 Btu per hour less than that for a house covered with glass or single layer plastic. This is a savings of about 30 percent in the amount of heat required. Single layer, plastic-covered greenhouses should not be constructed in areas with large winter heating requirements.

Piping for Hot Water and Steam Heated Greenhouses

The same formula is used to calculate the heat requirements for greenhouses using steam or hot water heating systems. First, the exposed area of the greenhouse is determined. Then the area, the design temperature, and the K value for glass are substituted in the formula to obtain the Btu per hour required. The steam and hot water move through pipes that act as heat exchangers or **radiators** in the greenhouse. Thus, the problem becomes one of calculating the total number of linear feet of pipe needed to radiate the required heat in Btu per hour.

In hot water heating systems, the water temperature is usually 180°F (82.1°C). A steam temperature of 215°F (101.6°C) is used for steam heated systems. This is the temperature of saturated steam at slightly more than atmospheric pressure. At atmospheric pressure, water turns to steam at 212°F (100°C).

Types of Pipe

Two types of pipe are used, plain or finned. **Finned pipe** is made by attaching closely spaced plates or tins (Figures 6-16A and 6-16B). The fins increase the surface area of the pipe where heat is radiated. The greater surface area means that more heat is given off. As a result, less pipe is required. One linear foot (30 cm) of 2-inch (5-cm) diameter finned pipe radiates the same amount of heat as 5 linear feet (1.8 m) of 2-inch (5-cm) diameter plain pipe. The heat transmission characteristics of finned pipe may vary. When selecting pipe, always consult the manufacturer's specifications to determine the exact rating of the pipe. One linear foot of 2-inch (5-cm) plain pipe carrying water heated to 180°F (82.1°C) will give off approximately 160 Btu per hour. Therefore, 1 linear foot (30 cm) of 2-inch (5-cm) firmed pipe carrying 180°F (82.1°C) water will give off 160 × 5 or 800 Btu per hour.

The pipe used in a steam heated greenhouse may be 1.25 inch (3.2 cm) or 1.5 inch (3.75 cm) in diameter. At a steam temperature of 215°F (101.6°C) and a

FIGURE 6-16A Finned steam pipe mounted on perimeter curtain wall.

FIGURE 6-16B Finned hot water pipe placed under the bench.

surrounding air temperature of 60°F (15.5°C), 1 linear foot (30 cm) of 1.5-inch (3.75-cm) pipe will radiate nearly the same amount of heat as a 2-inch (5-cm) pipe carrying hot water, or 180 Btu per hour. One and one-half inch (3.75-cm) iron pipe carrying steam at 215°F (101.6°C) will provide almost 210 Btu per hour when the surrounding air temperature is 60°F (15.5°C). One linear foot (30 cm) of 1.25-inch (3.2-cm) finned pipe is equivalent to 6 feet of 1.25-inch (3.2-cm) plain pipe and 4.2 feet (1.3 m) of 2-inch (5-cm) plain pipe.

Calculating the Amount of Piping Required

The glass-covered greenhouse described in the previous example requires 381,708 Btu of heat per hour. The amount of piping needed can be determined for both hot water and steam and for plain or finned pipe. For 2-inch (5-cm) plain pipe carrying hot water, the amount of pipe required in linear feet is equal to the total heat requirement (381,708 Btu) divided by the amount of heat radiated per foot of pipe (160 Btu), as follows:

$$381,708 \div 160 = 2,385.68 \text{ ft } (727.2 \text{ m})$$

For 2-inch (5-cm) finned pipe carrying hot water, the amount of pipe required in linear feet is equal to:

$$381,708 \div 800 = 477.13 \text{ ft } (145.4 \text{ m})$$

The same value can be obtained by dividing the required number of feet of plain pipe by 5. This number is the conversion factor for plain pipe to find the finned pipe equivalent.

$$2,385.68 \div 5 = 477.13 \text{ ft } (145.4 \text{ m})$$

Plain 1.25-inch (3.2-cm) diameter pipe carrying steam has the same Btu output as 2-inch (5-cm) diameter hot water pipe. This means that the number of feet of pipe required is the same, or 2,385.68 (727.2 m). If a 1.25-inch finned pipe is used, the amount needed is one-sixth that of plain pipe, or:

$$2,385.68 \div 6 = 397.61 \text{ linear feet } (727.2 \text{ m} \div 6 = 121.2 \text{ m})$$

If the greenhouse temperature is to be lower than 60°F (15.5°C), the length of pipe required will be different from the values calculated in the previous example. This change is due to the temperature differences between the lower temperature required and the hot water or steam temperatures. For a temperature less than 60°F (15.5°C), a given size of pipe will radiate more heat at the same water or steam temperature. A heating engineer should calculate the heating requirements for a greenhouse at various temperatures.

Piping Arrangement

The heating pipes in a greenhouse must be placed correctly to obtain the most uniform distribution of heat possible. The location of the pipes depends upon the type of crops to be grown, their heat requirements, the number of feet of pipe required, and the method of installing the pipes.

To avoid using valuable growing space, heating pipes are normally stacked or placed one on top of the other along the sidewalls and end walls of the greenhouse (Figure 6-17). The maximum heating efficiency is obtained only

FIGURE 6-17 Stacked finned steam heating pipe radiator.

when there is free airflow around the pipes. Therefore, the effect of stacking reduces the heating efficiency. Stacking reduces the airflow around the upper pipes. In addition, air is heated by the lower pipes, thus reducing the heating efficiency of the upper pipes. To make up for this loss in efficiency, additional heating pipes must be installed.

Many growers use bottom heat under their raised benches. The placement of heating pipes under benches also reduces the amount of airflow around the pipes. Again, the heating efficiency is reduced. Yet the gains in crop productivity through heating the root zone often offset the reduced heating efficiency.

When steam or hot water heating systems are the only means of heating a greenhouse, a **trombone,** or continuous, piping arrangement is used.

Such a piping system gives the most uniform heat distribution. To ensure that there is enough heat on the sides of the house, not less than 25 percent of the total radiating surface area of the pipe should be installed on the sidewalls or end walls.

Normally, finned pipe is installed on the sidewalls and end walls only. Plain pipe, once used overhead, is now more typically placed under bench.

Many greenhouses today are constructed with a heated floor. These are constructed by incorporating in-floor tubing as the concrete floor is poured. These heating systems are very efficient, especially for crops grown on the floor or when ebb and flood floor irrigation systems are used. In-floor heat will dry the floor quickly. Floor heat systems are typically not designed to provide the total heat requirement for the greenhouse and may need some sidewall and overhead heating. In-floor heating systems are very expensive to install, but are more efficient to operate as the large mass of the floor retains heat for longer periods.

Heating Plant Boilers

Modern heating plants can be purchased as complete packages that are easy to install in a greenhouse. There are two basic types of boilers used: the **fire tube boiler** and the **water tube boiler.** The fire tube boiler (Figures 6-18, 6-19A, and 6-19B) has a series of tubes through which heated air is forced. These tubes are surrounded by water. As the heated air moves through the cold water by way of the tubes, the water is heated.

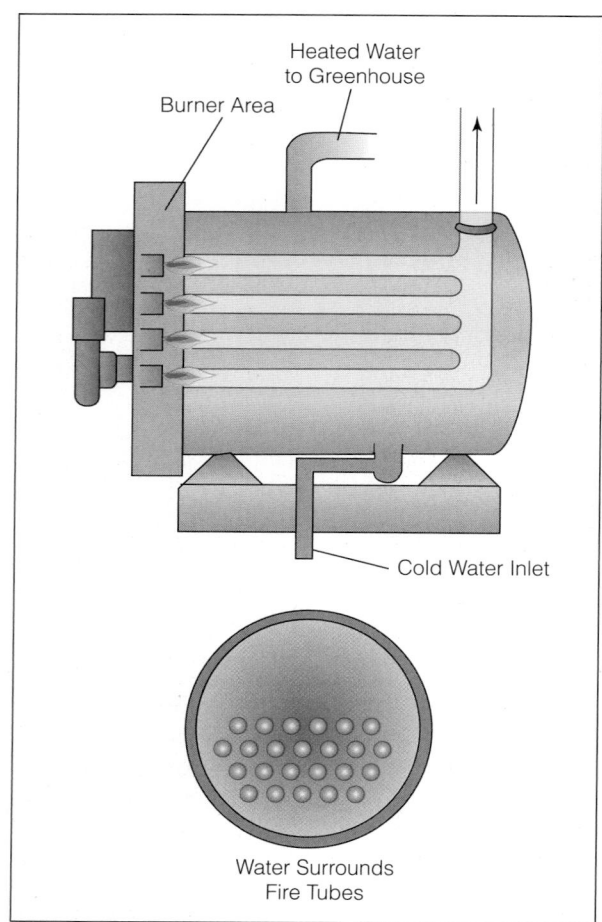

Heated Water
to Greenhouse

Burner Area

Cold Water Inlet

Water Surrounds
Fire Tubes

FIGURE 6-18 Fire tube boiler diagram.

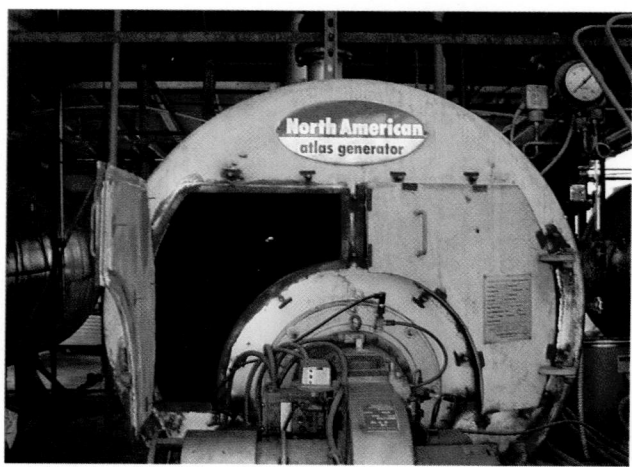

FIGURE 6-19A Fire tube boiler.

FIGURE 6-19B Fire tube boiler with fire tubes highlighted.

In a water tube boiler (Figure 6-20), water circulates in the tubes. Heat and flue gases pass around the tubes and heat the water. Each type of boiler has its advantages and fuel requirements, depending upon its specific use. The grower should consult with a heating expert to determine the type of boiler that will satisfy the requirements of the greenhouse.

Boiler Rating

Boilers are rated in horsepower. One **boiler horsepower** is equal to a heat output of 33,475 Btu per hour. The heating requirement of the greenhouse determines the boiler size. (Recall that the heating requirement depends upon the size of the area to be heated, the type of glazing material used, the desired inside temperature, and the outdoor design temperature.) In general,

FIGURE 6-20 Water tube boiler diagram.

a 200-hp boiler is sufficient to heat a 50,000 square foot (4,645.2 m²), glass-covered greenhouse to 60°F (15.5°C) at a design temperature of −10°F (−22.5°C). Most growers prefer to use two smaller boilers to carry the load. If one boiler fails for some reason, the second boiler will keep the greenhouse from freezing.

Emergency Heating Equipment

Most heating equipment cannot operate when power lines are down because of heavy snow or ice storms. An emergency source of power, such as a gasoline or diesel generator, should be available for use while the power company crews repair the wires. The emergency generating unit must have enough capacity to ensure that at least one-half of the boilers are operating. A qualified electrician can help the grower determine the size of the standby unit.

Heating Systems for Plastic-Covered Greenhouses

Many plastic-covered greenhouses are temporary structures or are used during part of the year only. This means that a centrally located hot water or steam heating system is often uneconomical for this type of structure. Instead, hot air furnaces or unit heaters are commonly used in plastic-covered greenhouses.

Hot Air Furnaces

In small greenhouses, a single hot air furnace with air deflectors is centrally located to maintain the desired temperature (Figure 6-21). Large greenhouses require more than one heater. In addition, the uniform distribution of heat is a problem in large houses when standard home type furnaces are used.

The heat distribution is improved if a perforated polyethylene distribution tube is used with fan-jet unit heaters. The fan can be wired to run continuously to circulate the air. The heat comes on when the thermostat calls for it.

Infrared Heating System

Greenhouses can be heated using an infrared heating system that supplies radiant energy (Figures 6-22A through 6-22C). Infrared heating transfers thermal energy as electromagnetic waves. One such infrared heating system that is readily adapted to greenhouse use is the CO-RAY-VAC® System. Gas-fired burners heat air that is distributed in an overhead pipe system, which serve as radiant tubes. The infrared radiant heat energy is reflected down to the crops. This energy is the same as the radiant energy from the sun. It heats in the same manner. The infrared heating system is satisfactory for uniformly low growing crops such as bedding plants and potted crops. However, it is not satisfactory for tall cut flower crops. In severe cold weather, crops located near the sides of the greenhouse do not receive enough heat and suffer cold damage. This system is not widely used to heat greenhouses other than those used to grow bedding plants or potted crops. Greenhouses heated with infrared heat often feel cold as the crop is heated and not the air.

FIGURE 6-21 A hot air heater is placed at plant height in a small plastic-covered greenhouse to provide uniform heat distribution.

FIGURE 6-22A CO-RAY-VAC® is an infrared radiant heating system.

FIGURE 6-22B Infrared radiant heating systems are gas fired and require a mechanical venting system to the exterior of the greenhouse. Make sure that the venting system is placed on the leeward end of the greenhouse to prevailing winds. Otherwise, exhaust gases may build up in the greenhouse.

FIGURE 6-22C Infrared heating system mounted near the roof of a greenhouse for growing bedding plants.

FIGURE 6-23 Properly vented gas-fired unit heaters with stack well above the roof line of the greenhouse.

Venting Heaters

Any heater that uses an open flame to burn the fuel must have a chimney pipe vented to the outside of the greenhouse (Figure 6-23). When compared to glass-covered greenhouses, which have many air leaks, plastic-covered greenhouses are almost airtight. All fuels need a good supply of oxygen to burn properly. If enough air is not available, the fuel will not burn completely. As a result, toxic (poisonous) gases may build up in the greenhouse. These gases can be hazardous to humans and severely damage plants.

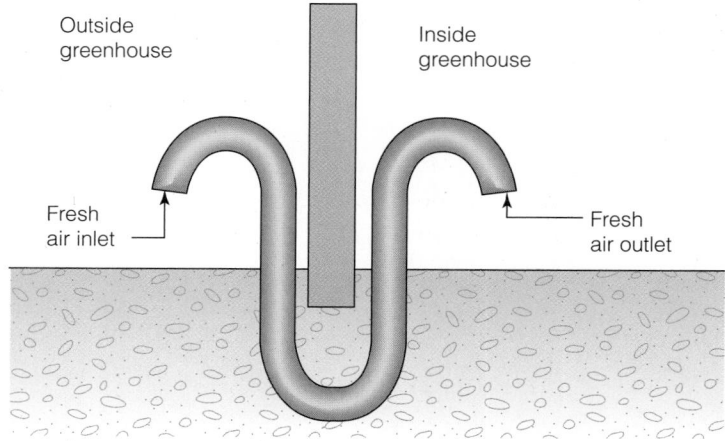

FIGURE 6-24 Fresh air inlet for gas-fired heater in a greenhouse. The configuration depicted eliminates the need for curtain wall mounting.

Outside greenhouse

Inside greenhouse

Fresh air inlet

Fresh air outlet

To ensure that there is enough air for the complete combustion of the fuel, **makeup air** is brought in from outside the greenhouse. A stovepipe is used to bring fresh air to the burner. The pipe should be screened to prevent mice and other small animals from entering (Figure 6-24). A rule of thumb can be used to decide the size of the pipe. An allowance of 50 square inches (322.6 cm²) of opening is made for each 100,000 Btu per hour furnace capacity. If the furnace is rated at 250,000 Btu per hour, then the opening for the makeup air pipe should be 50 × 2.5 = 125 in² (322.6 × 2.5 = 806.45 cm²). Thus, an opening of 10 by 12.5 inches (25 by 31.25 cm) will provide enough makeup air for the fuel to burn properly.

Carbon dioxide is used by plants during the daylight hours for **photosynthesis.** Some growers think that the carbon dioxide resulting from the fuel combustion will help the growth of the plants. Thus, they do not vent the heaters outside the building, and the carbon dioxide builds up in the house. However, other gases are also released when fuels are burned, such as carbon monoxide and ethylene. A buildup of these gases can be harmful to humans and plants. To safeguard personnel and the crop, heaters must be vented to the outside.

Fresh air venting is typically required by the local building code in many municipalities; however, there is an increasing selection of unvented heating systems designed for greenhouses. Unvented heating systems retain the heat energy otherwise lost to the outside within the greenhouse. Check local building codes before installation of any gas-fired heating system.

Heat Conservation Methods

As the cost of heating a greenhouse continues to increase, more attention should be given to ways of conserving or reducing the amount of heat needed to keep the greenhouse temperature at the desired level. There are a few commonly used methods of conserving heat, as well as several experimental methods being tested.

Double Layer Plastic

As mentioned in Chapter 4, a double layer plastic covering on the greenhouse can reduce the heat requirements by 30 to 40 percent. In this method, two

FIGURE 6-25 Squirrel cage fan used to inflate a double layer polyethylene roof.

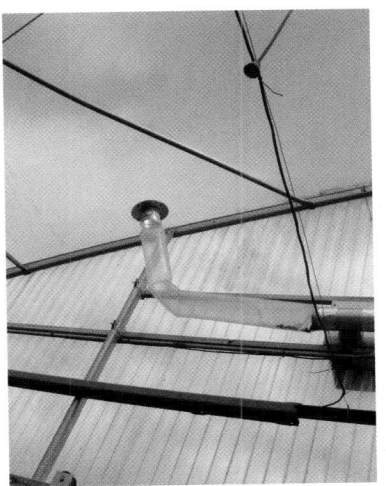

FIGURE 6-26 Fan bringing in outside air and ducted to the double polyethylene roof for inflation. Drier outside air is used to prevent condensation of water between the polyethylene sheets.

layers of plastic are applied to the greenhouse. The plastic layers are kept about 3 to 6 inches (7.62 to 15.24 cm) apart by means of a small fan that pumps air between them at low pressure (Figures 6-25 and 6-26).

Multiwall rigid plastic panels made of acrylic, polycarbonate, polyvinylchloride (PVC), and fiber-reinforced plastic (FRP) are widely used for greenhouse coverings in the United States. Twin-walled panels are common for roofing material and triple-walled panels are common for side and end walls. The cost of these materials is much higher than that of flexible plastic. The material must be used under actual growing conditions to determine if it will conserve as much heat as the manufacturer claims.

The gable end walls of the greenhouse can be covered with a second layer of plastic with an airspace between the layers to provide an insulating effect. Some growers use standard insulation materials with reflective backing to cover the end walls or north walls of the greenhouse. The insulation conserves heat, and the reflective backing reflects light into the crop so that growth is uniform.

Acrylic, polycarbonate, PVC, and FRP materials are also being used for the sides and end walls of large greenhouses clad with air-inflated polyethylene roofs.

North-facing greenhouse sidewalls can also be covered with the same type of insulation. The amount of light lost because of the insulation is more than made up by the savings in heat.

Interior Ceiling

An interior ceiling is widely used by growers. This method of insulating was first used by carnation growers when cooling the greenhouse by a **fan and pad cooling** system (Figure 6-27). The interior ceiling was installed just above the crop to keep the hot air in the peak of the house. In this way, the cooled air could be kept around the plants. Now, the interior ceiling is used to keep the heat around the crop, rather than being lost through the greenhouse roof (Figure 6-28). Both temporary and permanent interior ceilings can be installed.

Permanent Interior Ceiling

A permanent ceiling can be installed by hanging 4-mil polyethylene above the crop. Such an installation is easy in a clear span greenhouse. In

FIGURE 6-27 Mylar shade curtain material can be used as a false ceiling to keep cool air on plants in a fan and pad cooled house.

FIGURE 6-28 Retractable shade systems can be used as a ceiling for heat conservation.

FIGURE 6-29 Black cloth is pulled over plants to provide shade and act as a heat blanket. This photograph depicts a dual shading system. One curtain provides 50 percent shade and the second provides black cloth for extending the dark period in the greenhouse for photoperiodic plants as described in Chapter 8.

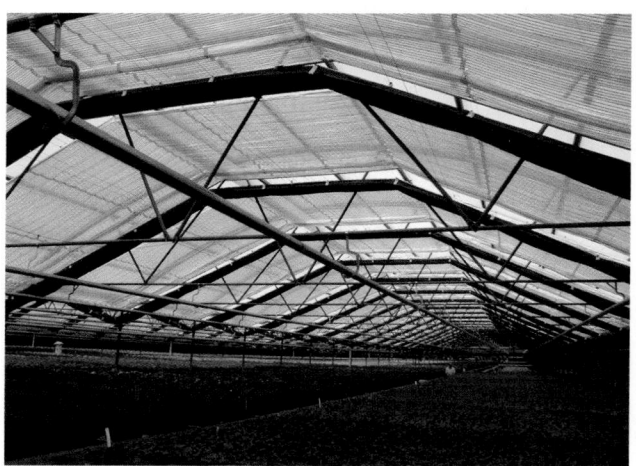

greenhouses with posts and other supports, it is more difficult to install an interior ceiling.

The double interior ceiling also makes it more difficult to vent the greenhouse when the interior temperature increases. Under these conditions, the best means of ventilation is automatic **exhaust fans**.

It has been calculated that an interior ceiling can reduce the heating requirements of a greenhouse by as much as 30 percent. If the ceiling is made of clear plastic, it may be left in place for the entire heating season.

Temporary Interior Ceiling

A temporary interior ceiling, often called a retractable curtain or retractable shade, can be installed to conserve heat (Figure 6-29). This type of ceiling is pulled into place each evening and removed each morning. It is similar to the methods used by growers to provide artificial short days so that chrysanthemums can be flowered out of season. Black sateen cloth is installed on wires, which is pulled over the plants to provide an insulating blanket above them. When the cloth is installed gutter to gutter and down the sidewalls, the heat requirement can be reduced as much as 50 percent or more. Semiautomatic black cloth shading machines simplify the task. The system can be made fully automatic by the use of a time clock or integration with the greenhouse computerized climate control system.

Experimental Methods

A number of experimental systems for conserving heat are being studied. For example, one system uses an inflatable interior ceiling to provide an insulating effect. In another system, the space between a double layer plastic roof is filled with foam to reduce the heat loss. During the daylight hours, the foam changes back to a liquid form so that it is easier to store and reuse. It has been suggested that the entire greenhouse be covered with 2-inch (5-cm) thick pieces of Styrofoam or some other type of rigid insulation. However, this method has not been used successfully on a commercial greenhouse.

As the energy shortage becomes more severe, many other methods of conserving energy will be studied and adapted to greenhouse use.

Solar Heating

Greenhouses were used as solar collectors long before scientists began the search for efficient methods of storing and using the sun's energy. As a solar collector, the greenhouse catches and stores solar energy. Unfortunately, the amount of heat retained in the greenhouse is not enough to maintain the desired temperature through the long nights of winter. However, it is very possible to store the solar energy collected during the day to meet part of the heat requirements at night. Two of the most commonly used storage materials are water and crushed gravel. If water is used, a tank must be constructed to serve as the heat reservoir. Complicated piping systems with heat exchangers are used. They transfer the heat energy from the sun into the storage area and then back to a usable form in the greenhouse.

If gravel is to be used to store the heat from the sun, a similar but less complicated system of heat exchangers is required. The mechanics of solar collectors and gravel storage have been worked out on a small scale, but such systems are not yet ready for application to large greenhouse installations. Part of the problem in using solar energy is the lack of control over the heating system. Much research is required before solar energy systems can be adapted to provide efficient greenhouse heating.

Ventilation and Cooling

There are three reasons for ventilating a greenhouse: (1) to reduce the air temperature when it becomes too high, (2) to exchange air to renew the supply of carbon dioxide for photosynthesis, and (3) to reduce the relative humidity in the greenhouse as a means of controlling diseases.

Ventilating to Reduce Air Temperatures

Plants cannot survive when temperatures exceed certain levels. For the majority of ornamental plants grown in greenhouses, the maximum temperature at which they will continue to function is around 95°F (35°C). Above this temperature, heat puts an unnatural stress on the internal structure of the plant and causes it to break down. As a result, greenhouses are ventilated to keep the temperature within the proper range for growing plants.

Ventilating to Supply Carbon Dioxide

The normal atmosphere contains approximately 0.033 percent carbon dioxide (CO_2), or 330 parts per million (ppm). Because plants use carbon dioxide for photosynthesis, a fresh supply of CO_2 must be available to plants at all times. If a greenhouse is closed for a long time, a deficiency of CO_2 can result, thus restricting plant growth.

Ventilating to Reduce Relative Humidity

Many fungus spores germinate only when the relative humidity is very high or in a film of water on the leaves of plants. High moisture conditions in the greenhouse can cause the water vapor in the air to condense on the plant leaves. Thus, ideal conditions for the start and spread of disease are provided. Ventilation controls the relative humidity and helps to reduce plant disease. This is especially effective by venting during the late afternoon or early evening in conjunction with operating the heating system, effectively drying out the greenhouse.

Types of Ventilation Systems

Natural Air Movement

As air heats up, it becomes less dense and rises. The grower uses this "chimney effect" to cool the greenhouse. When the greenhouse ridge ventilators are opened, hot air that rises to the top of the greenhouse exits through these vents. If sidewall ventilators are also used (Figure 6-30), airflow increases through the roof vents.

Whenever there is a wind, the roof ventilator that should be opened widest is the one opposite the direction of the wind. The flow of air over this vent creates a partial suction that helps to pull air from the greenhouse (Figure 6-31). It is risky to open the ventilator on the side facing the wind. If the wind is very strong, the ventilator may be torn from the greenhouse. Thus, if the weather

FIGURE 6-30 Double sidewall ventilators increase the chimney effect.

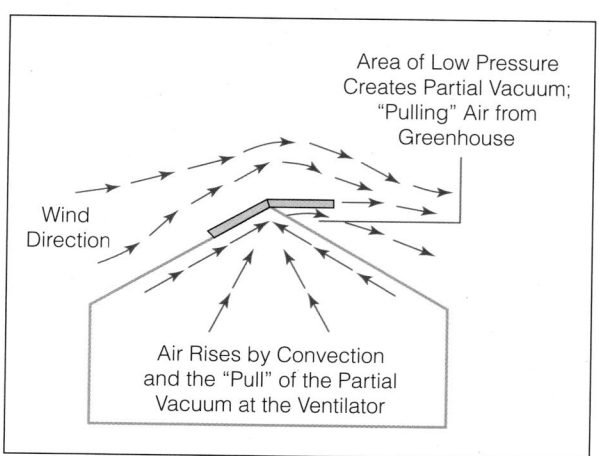

FIGURE 6-31 A partial vacuum is caused by wind currents over the ventilator. The low pressure pulls air from the greenhouse.

forecast predicts strong winds due to thunderstorm or gales, the ridge ventilators should be closed completely.

Natural ventilation methods are becoming more popular because of increased energy costs. Although fans and fan and pad cooling give more positive control over greenhouse ventilation, the cost of electricity is forcing growers to consider natural ventilation.

In small houses, less than 21 feet (7.2 m) wide, roof ventilators alone can keep the greenhouse temperature at the required level. Large freestanding greenhouses must be equipped with sidewall ventilators. Ridge and furrow greenhouses may require a double row of roof ventilators to ensure the proper amount of air exchange. Large houses 50 feet (17 m) or more in width require the maximum length of ventilator possible. To obtain the maximum ventilation, very large greenhouses should have a double row of 4-foot (1.4-m) wide roof ventilators and a double row of sidewall ventilators. Saw-tooth greenhouse designs also facilitate maximum ventilation of ridge and furrow greenhouses.

In Europe, 4-foot (1.4-m) wide ventilators are common. However, greenhouse manufacturers in the United States rarely use these wide ventilators. Instead, they provide 18-inch (46-cm) or 24-inch (61-cm) wide ventilators. Occasionally, 48-inch (122-cm) wide ventilators may be used.

Ventilating Sash Operation

Manual System

Hand cranks, chain pulls, or levers are used to open and close ventilators. Lines of ventilating sash on Dutch Venlo type greenhouses are operated by levers connected to push-pull rods on the vents.

Gear-driven ventilators are also commonly used. The gear unit is connected to the vent by an **elbow arm** (Figures 6-32 and 6-33), or a **rack and pinion** device (Figures 6-34A through 6-34C). The rack and pinion unit is more powerful than the elbow arm. In addition, it is easier to adapt to motorized operation, if desired. Many greenhouses also use roll-up walls (Figures 6-34D through 6-34F).

FIGURE 6-32 Crank and elbow arm apparatus used to open sidewall ventilators.

FIGURE 6-33 Close-up of elbow arm.

FIGURE 6-34A Rack and pinion vent machinery mounted outside the greenhouse to open side ventilators.

FIGURE 6-34B Close-up view view of rack and pinion vent mechanism.

FIGURE 6-34C
Exterior mounted rack and pinion vent mechanism. Note that these vents are contained within an insect screen enclosure with a gravel bed below the vent for weed control helping to keep the greenhouse insect free.

FIGURE 6-34D Drop wall side vent.

FIGURE 6-34E Roll-up wall side vent with hand crank.

FIGURE 6-34F Interior view of roll-up wall side vent with properly placed HAF fans.

Power System

In small greenhouses, manual operation of the ventilation sash is practical. In large greenhouses, however, power operation is more efficient and saves labor costs. In power operation, the greenhouse ventilation sash is automatically raised or lowered by electric motors. A thermostat can be connected to the sash system so that it is completely automatic; however, most systems integrate automatic sash operation with a computerized climate control system. Thus, the grower can concentrate on other tasks without having to worry about the vents. This is particularly important in the spring and fall months when the temperature can vary greatly as the sun plays hide-and-seek with the clouds.

Thermostat controls or climate control systems are set at a desired temperature. As the greenhouse warms up, the vent opens to allow the heated air to exit. Many controls have a staged or modulating feature whereby the vent opens only 2 inches (5.08 cm) when cooling is first required. If this setting does not reduce the temperature enough, the sash opens more. This procedure continues until the vent is wide open, or the maximum setting is reached. If the sash were to move from a fully closed position to a fully open position, there would be wide fluctuations of temperature in the greenhouse. The modulating control prevents such changes. Most greenhouse vent control systems have opening and closing limit controls. Such controls can limit the opening of the sash on windy or rainy days. They can also keep the sash slightly ajar so that the very humid air can exit the greenhouse at night.

If the ventilating sash is power operated, there should be some means of opening the vents by hand in case of a power failure (Figure 6-35). Most power failures occur in the winter when the ventilators normally are closed. However, power failures do occur at other times. For example, a severe windstorm may knock down power lines. If the ventilators were open in this case, it would be important to close them promptly to prevent them from blowing away. Ventilators should be closed before the storm arrives. Many climate control computer systems can be programmed to close ventilators when winds exceed a gale force limit.

Fan Ventilation

Fans alone are often used to ventilate greenhouses. Makeup air must come into the greenhouse to take the place of the hot air being exhausted by the fans. Thus, an opening or openings should be located in the side or end of the greenhouse wall opposite the fans (Figures 6-36 and 6-37). There is no problem

FIGURE 6-35 Chain fall apparatus for opening and closing roof ventilator.

FIGURE 6-36 Louvered openings for fan ventilation.

FIGURE 6-37 Electrically operated ventilation louver.

in bringing in cooler outside air in late spring, summer, or early autumn. Extremely cold air can be brought into the greenhouse in the late autumn, winter, and early spring causing severe injury to plants. For winter ventilation, the cold outside air should be brought in through a perforated polyethylene tube. The cold air leaving the tube thoroughly mixes with the warm greenhouse air within a distance equal to 20 to 30 times the diameter of the hole in the tube. This mixing of cold air with warm air in the greenhouse tempers the air and prevents sudden chilling of the plants.

A gravity operated louver is usually placed in the tube opening that is exposed to the weather. The louver prevents the snow and rain from blowing into the tube. A gravity operated or mechanically operated louver is placed on the outside, or delivery side, of the exhaust fan to prevent cold air from blowing into the greenhouse when the fan is turned off (Figure 6-38).

Fan Size

The exhaust fan is the heart of the ventilation system in a modern greenhouse. It is important to select the proper size of fan or fans. Greenhouse manufacturers recommend that there be at least one complete change of the greenhouse air every minute for summer ventilation.

Fans are rated on the basis of the amount of air they can move. This rating is given in cubic feet per minute (**cfm**). The cfm rating is usually determined for a static pressure of 0.125 inch (3.2 mm). Static pressure is a measure of the amount of force required to move the air. Many fan manufacturers provide fans with slant housing. Fans mounted on a slant provide increased protection to the motor from the weather and have a higher airflow capacity. The slant housing is at the same angle as the shutter, which directs air to fan. The cfm rating for fans mounted in slant housing uses a static pressure of 0.05 inch static pressure rather than 0.125. The fans should be able to deliver 15,000 to 18,000 cfm (7,079 to 8,495 liters/second) per horsepower rating. The size and number of fans required are determined by calculating the volume of air to be moved.

The volume of air is calculated by first determining the volume of the greenhouse. The volume is represented by the length, width, and height of the greenhouse to the eaves (Figure 6-39). If a greenhouse is 30 feet (9.1 m) wide, 100 feet (30.5 m) long, and 6 feet (1.8 m) high to the eaves, the volume is $30 \times 100 \times 6$ ($9.1 \times 30.5 \times 1.8$ m) = 18,000 ft^3 (499.59 m^3).

FIGURE 6-38 Exhaust fan properly screened for worker protection.

FIGURE 6-39 Calculating the volume of a greenhouse in cubic feet.

$$\text{Vol., Ft}^3 = \left[(W \times H_e) + \frac{W \times H_r}{2}\right] \times L$$

A. Even Span Greenhouse

$$\text{Vol., Ft}^3 = \left[(W \times H_e) + \frac{W_1 \times H_r}{2} + \frac{W_2 \times H_r}{2}\right] \times L$$

B. Uneven Span Greenhouse

$$\text{Vol., Ft}^3 = \left[(H_e \times W) + \frac{3.14 \times H_r \times H_r}{2}\right] \times L$$

C. Barrel Vault or Quonset-type Greenhouse

The next step is to calculate the volume of the greenhouse above the eave height. This calculation involves triangles and is a little more complicated. The area of a triangle is equal to one-half the length of the base times the height, or $1/2\ b \times h$. The volume of the area enclosed by the roof is equal to the area of the triangle times the length of the building.

For the greenhouse being considered, the base of the triangle formed by the roof is 30 feet (9.1 m) long. The height of the ridge peak above the eaves is 9 feet (2.7 m). The area of the roof above the eave height is $1/2$ 30 ft (9.1 m) $\times$ 9 ft (2.7 m) = 135 ft² (12.3 m²). Multiplying this area by the length of the roof gives a volume of 135 ft² (12.3 m²) $\times$ 100 ft (30.5 m) = 13,500 ft³ (375.15 m³). The total volume of the greenhouse is 18,000 ft³ (499.59 m³) + 13,500 ft³ (375.15 m³) = 31,500 ft³ (875 m³).

Therefore, a fan rated at 31,500 cfm (14,866 liters/second) is required to change the air once a minute. In a greenhouse of this size, one very large fan is impractical. For the same reasons that a single large heater does not give uniform heat distribution, one large fan will not give uniform cooling. More efficient cooling is obtained if several small fans are evenly spaced along one wall of the greenhouse (Figure 6-40). The cooling effect is more evenly distributed using this type of fan arrangement. The fans should not be spaced more than

FIGURE 6-40 Fans mounted opposing one another should be staggered. Note the barrier installed is to prevent fan exhaust from blowing into neighboring cooling pads.

25 feet (7.6 m) apart. For this greenhouse, four fans are required if each has a capacity of 8,000 cfm (3,776 liters/second).

The fans should be mounted on the side of the house away from the prevailing winds (Figure 6-41). If the fans are mounted so that they blow against the wind, their efficiency is reduced.

For plastic-covered greenhouses, a general rule of thumb is used to determine the size of fan needed. Most recommendations call for the movement of 7 to 10 cfm (3.3 to 4.7 liters/second) of air per square foot of floor area. Using a value of 7 cfm (3.3 liters/second), the required air movement for a house measuring 30 feet by 100 feet (9.1 m by 30.5 m) is 30 × 100 × 7 = 21,000 cfm (582.855 liters/second). This air volume is two-thirds the volume calculated previously. By using the 10 cfm (4.7 liters/second) value, the total air volume to be moved equals 30,000 cfm (14,130 liters/second). This value is closer to the previous calculation. Some experts say that the air in the peak of the greenhouse is not disturbed and should not enter the calculations. This may be true if there is an interior ceiling. However, with no ceiling, there is some turbulence due to the movement of air across the greenhouse. This turbulence reaches the hot air in the peak of the house. As a result, the hot air mixes with the cooler air entering the greenhouse. This mixture of air puts a greater load on the exhaust fans. For this reason, it is far better to have more fan capacity than is needed.

When the full cooling capacity of the fans is not needed, one or two of the fans should be capable of two-speed operation. This arrangement permits different levels of ventilation.

Thermostat Control of Fans

The operation of the fans may be controlled by a thermostat or preferably by a computerized climate control system. The control range of a thermostat should be from 45°F to 90°F (7°C to 32°C). A smaller control range is not recommended because the fans will cycle on and off more often than is necessary. This frequent cycling shortens the life of the motors. If the thermostat is sensitive enough, the fans will come on when the temperature is no more than 2°F (1.1°C) above the desired reading. They will go off when the temperature is no more than 2°F (1.1°C) below the thermostat setting.

The length of air travel from the inlet to the exhaust fan must be no longer than 100 feet (30.5 m). In very large ridge and furrow houses, this requirement may mean that the air inlets are located in the center of the house and the fans

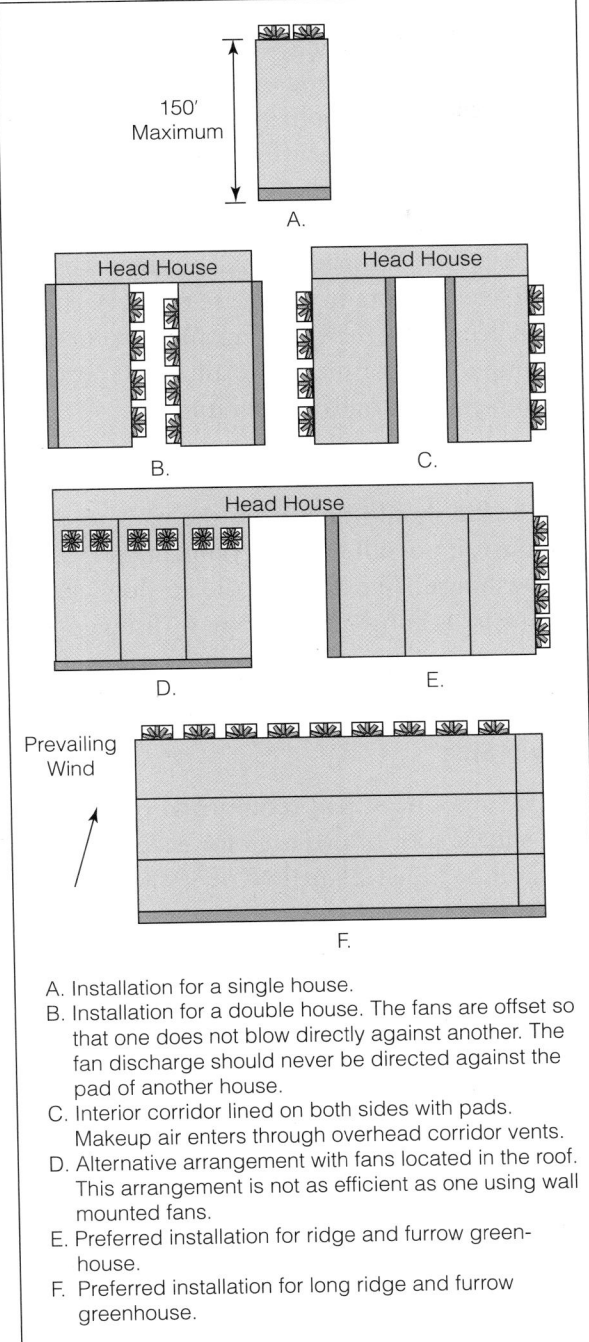

FIGURE 6-41 Various methods of installing fans and cooling pads for ventilation and cooling.

A. Installation for a single house.
B. Installation for a double house. The fans are offset so that one does not blow directly against another. The fan discharge should never be directed against the pad of another house.
C. Interior corridor lined on both sides with pads. Makeup air enters through overhead corridor vents.
D. Alternative arrangement with fans located in the roof. This arrangement is not as efficient as one using wall mounted fans.
E. Preferred installation for ridge and furrow greenhouse.
F. Preferred installation for long ridge and furrow greenhouse.

are located on all of the perimeter walls. Various methods of complying with this requirement are shown in Figure 6-41.

Cooling Systems

Fan and Pad Cooling

A wet pad can be added to the greenhouse ventilation system to increase the cooling effect of the fans. When the fans exhaust the air, a partial vacuum develops in the greenhouse. Makeup air is pulled into the greenhouse through a continually wetted pad that cools the air. The cooled air picks up heat as it moves through the crop area of the greenhouse. The exhaust fans then expel this heated air to the outside.

Fan and pad cooling systems work best in areas where the relative humidity is low. The cooling effect is due to the difference in the relative humidity of the air passing through the pad and the relative humidity of the pad, or 100 percent. The greater the difference in relative humidity, the greater the cooling effect. Fan and pad cooling systems work very well in the high, dry altitude of Denver, Colorado. They are not very effective in the humid regions of Florida. An accurately designed fan and pad cooling system should be able to lower the temperature of incoming air by 80 percent of the difference between the dry bulb and wet bulb temperatures.

The pads should be installed on the windward side of the house. This cooling system is more efficient if the natural winds do not blow against the fans. The air should flow from the pads to the fan or fans along the length of the house, if possible. This arrangement is more efficient than airflow across a narrow house. The distance the air flows should be less than 225 feet (77 m). The system is most efficient at a distance of 150 feet (51 m). As the distance increases above this value, the cooling efficiency decreases. Typical fan and pad installations for various situations are shown in Figure 6-41.

Pad Size

The amount of pad required to provide the desired cooling is approximately 1 square foot of pad area for each 150 cubic feet (4.2 cm^3) of air moved each minute. Thus, an airflow of 30,000 cfm (850 cm) will require 200 square feet (18.6 m^2) of pad area.

Correction factors are to be applied as the distance between the fan and the pad decreases from 100 feet (30.5 m), and for various elevations above sea level. The elevation correction is made because air gets lighter at higher altitudes. Thus, more air must be moved through the greenhouse to obtain the same cooling effect. These correction factors are given in Tables 6-2 and 6-3.

The greenhouse used in previous examples measures 30 feet by 100 feet (9.1 m by 30.5 m). The calculated airflow for this house is 30,000 cfm (850 cmm). Applying the correction factor of 1.56 from Table 6-2, the resulting fan requirement is 46,800 cfm (1,325 cmm). If the greenhouse is located at an elevation of 3,500 feet (1,064 m), the fan requirement remains the same because the correction factor of 1.56 for the distance is larger than the correction factor of 1.12 for the elevation. If the fan to pad distance is 100 feet (30.5 m) or more, the elevation correction factor of 1.12 is used. As a result, the fan requirement is 33,600 cfm (952 cmm).

Table 6-2

Correction Factor to Be Applied when the Fan-to-Pad Distance Is Less Than 100 ft (30.5 m)					
Meters	6–9	9–14	14–18	18–26	26–30.5	above 30.5
Feet	20–30	30–45	45–60	60–85	85–100	above 100
Factor	1.70	1.56	1.42	1.28	1.14	1.00

Table 6-3

Elevation in feet (meters)	Sea level to 1,000 (303 m)	1,000 (303 m) to 2,000 (606 m)	2,000 (606 m) to 3,000 (909 m)	3,000 (909 m) to 4,000 (1212 m)	4,000 (1212 m) to 5,000 (1515 m)	5,000 (1515 m) to 6,000 (1818 m)	6,000 (1818 m) to 7,000 (2121 m)
Factor	1.0	1.04	1.08	1.12	1.17	1.22	1.27

Correction Factor to Be Applied for Elevation Above Sea Level

FIGURE 6-42 Aspen fiber pads along a greenhouse wall.

If a greenhouse requires an air movement of 46,800 cfm (1,325 cmm), a pad area of 312 square feet (29.3 m²) is needed. For maximum efficiency, the greenhouse pad should be installed as a continuous unit. If it is installed in segments, with spaces between the pad segments, warm or hot spots may develop in these areas. When the required pad area is greater than the wall space available, the pad can be set up in a zigzag arrangement. The resulting pad area results in a greater surface area in the same linear distance.

Pad Materials

At one time **aspen fiber** pads were used in evaporative cooling systems (Figure 6-42). These usually were short lived due to algae growing in the pad as well as the breakdown of the fibers. These have been replaced by formed paper pads. One brand name is **Kool-cel**® (Figure 6-43). The most often used pads come in sections that are 4 inches (10 cm) thick. Other thicknesses are available if needed for a particular installation. The height and width of the pads vary according to the design of the pad area. They are easy to install and have a life of at least 10 years if properly maintained. Algae growth may still be a problem, but this can be controlled through chemicals placed in the water sump.

The pad system may be installed inside or outside the greenhouse. If the pads are installed outside the greenhouse in northern climates, they must be winterized in the off-cooling season.

Water Supply to Pad

The water used in the pads is recirculated so that it is not wasted. The amount of water flowing through the pad should be not less than 0.33 gallon (1.26 l)

FIGURE 6-43 Corrugated paper Kool-cel® evaporative cooling pad.

per minute for each linear foot (30.5 cm) of pad. This flow rate ensures the most efficient pad performance.

The water is supplied to the pads by an overhead pipe. A series of holes is evenly spaced along the pipe. To flush the pipe, an end plug is provided. If the pad is more than 75 feet (23 m) long, the water supply pipe should be connected near the midpoint of the distributor pipe. A filter is located in the line near the pump. The filter removes materials that otherwise would block the holes in the distributor pipe.

As stated previously, the system requires only 0.33 gallon (1.26 l) for each linear foot (30.5 cm) of pad length. However, the sump should have a storage capacity of 1.5 gallons (8.5 l) for each linear foot (30.5 cm) of pad. If the pad system is more than 75 feet (23 m) long, the sump is located near the middle of the pad.

This type of system loses water through evaporation. Makeup water is added from time to time by an automatic float valve in the sump.

Control of Insects

Large flying insects normally cannot penetrate the pads. Small insects, such as thrips, were once controlled by adding a nonvolatile insecticide to the water. This is no longer an acceptable practice. Children and pets should not be allowed near water treated with insecticides or other chemicals.

High-Pressure Fog Cooling

Fan and pad cooling is based on the principle that evaporating water takes heat from the air. This same principle is applied in high-pressure fog cooling (Figure 6-44). In this system, water is forced through small nozzles at a pressure of 500 to 600 psi. The nozzles are evenly spaced throughout the greenhouse about 10 feet (3 m) above the crops. The nozzles spray a very fine mist or *fog* that evaporates and cools the air in the greenhouse.

This system requires water that is free of silt, sand, or other materials; a pump or pumps that can deliver water at 500 to 600 psi; a control system, usually a **humidistat**, to activate the pump; and nozzles that can deliver high-pressure water at the rate of 0.6 gallon (2.25 l) per minute.

Water containing calcium and magnesium carbonates will soon block the fine holes in the nozzles. A water-softening device should be used to treat

FIGURE 6-44 High-pressure fog installation in a bedding plant plug greenhouse.

the water before it goes through the system. Unless hard water is treated, it may leave an undesirable film on the leaves of the plants. Soft water interferes with plant growth and should not be used to water the plants.

The control unit for this system is set to provide fog in a cycle of 30 seconds on and 30 seconds off. The air moisture will increase gradually until it reaches the desired level. A humidistat controls the misting cycle.

Stainless steel oil burner nozzles are installed on short "rattails" that can be adjusted in any direction. These rattails are fixed to copper pipes. Nozzles are installed at 5-foot intervals. One central line can serve a house up to 20 feet (6.1 m) wide. Two lines are needed in houses 25 to 45 feet (7.6 to 13.6 m) wide; three lines are used in houses 45 to 60 feet (13.6 to 18.2 m) wide; and four lines must be used in houses wider than 60 feet (18.2 m). The lines should be installed so that the nozzles can be cleaned periodically. If possible, both the ridge and side vents should be wide open during fogging. Air circulation fans, if installed, should be turned on as well.

If the evaporative conditions are such that the moisture being added to the air evaporates readily, cooling takes place with little or no moisture deposited on the leaves. If too much moisture is added, moisture may be deposited on the leaves, leading to disease.

An emergency cooling method should be available if the fogging system fails in hot weather. Crops grown under high-pressure fog cooling are very sensitive to low relative humidity and moisture stress. They are easily damaged if the water is turned off for even a few hours on hot days. A helicopter spraying whitewash over a greenhouse is one method of quickly shading the house to prevent serious crop losses.

The high-pressure fog system is used on roses and cymbidium orchids. It has also found use on foliage crops but is not widely used on other crops.

Shading as a Means of Cooling

Greenhouse temperatures can be reduced by applying shading materials to the glass (Figure 6-45). Whitewash, commercial shading compounds, and even mud can be used to shade crops. As the intensity of the light reaching the crops decreases, the temperature inside the greenhouse also decreases. If too much shade is applied, the temperature may be reduced to comfortable levels for

FIGURE 6-45 Shading or whitewash is applied to the exterior of the roof to lower the temperature. The apparatus on the roof is mounted on wheels and allows worker access along the length of the greenhouse for applying or removing shading compound.

workers, but the light intensity is so low that the quality of the crops is affected. That is, the plants grow too tall, have weak stems, and have a poor lasting quality.

Flowing Water Shade

There is another method of shading that is not in common use in the United States except in some greenhouse vegetable production. Some European growers use water flowing over the greenhouse roof as a shade. Water is pumped to the top of the greenhouse where it is distributed through a perforated pipe. It flows in a thin layer down the outside of the greenhouse glass. The water is dyed green because this color absorbs the heat of the infrared radiation and prevents it from entering the greenhouse. The water flows to a gutter where it is channeled to a sump for recycling. If the water becomes too warm as it flows over the greenhouse roof, it can be cooled mechanically in the sump. Some cooling of the water takes place even as it absorbs heat from the roof. This system of shading can be used only if the glass of the house is tightly sealed. Even a small leak through the glass means that water is lost from the system. This system has not been tried on plastic-covered greenhouses.

Mechanical Air Conditioning

It is impractical to use mechanical air conditioners to cool a greenhouse. The heat load built up in a greenhouse is too large for a mechanical unit. A 50- to 100-ton air-conditioning unit would be required to cool a medium size greenhouse. The size of such a unit can be imagined by noting that a home air conditioner may be rated at 0.25 or 0.5 ton.

Small hobby type greenhouses use package coolers called desert coolers (Figure 6-46). These units operate on the same principle as fan and pad cooling systems. In fact, they are a package type fan and pad cooling unit. They are not practical for large commercial greenhouses. However, high-efficiency positive-pressure cooling systems are available for greenhouses where crops sensitive to virus infection or other potential contamination warrant the increased costs.

Controlling the Greenhouse Environment

Control of the greenhouse environment, principally the air temperature, has moved a long way from the days when the grower controlled the heat by

FIGURE 6-46 Desert style swamp cooler unit mounted on a greenhouse. Cooling pads mounted internally are typically aspen or excelsior.

opening or closing a series of valves by hand. The temperature guide was a thermometer hanging at eye level in the middle of the greenhouse. In spring, summer, and fall, hand opening or closing overhead and/or side ventilators was the method of lowering greenhouse temperatures. When it got too hot in summer, whitewash shade, as has been mentioned, was applied to help reduce the temperature inside the greenhouse. When thermostats became popular as a means of sensing changes in temperature, this was a major step in more precise control of the greenhouse environment. The first units operated on an on–off basis. When the unit was in the on position, the heat control valve or ventilator was full open. When in the off position, the unit controlled was in the fully closed position. Later progress resulted in the development of a stage controller. Stage control made it possible for a temperature controlling device to be operated in stages (Figure 6-47). If only a little heat was called for, the valve controlling the flow of steam or water opened only partially. As the demand for heat increased, the valve opened farther until it was fully open. The same situation took place with motor-controlled ventilators. As more air movement was needed, the stage control opened the vent farther until it was fully open.

This kind of control was fine in late spring, summer, and early fall when outside temperatures were not too cold. However, in early spring and late fall, and even into early winter, to have the ventilator fully open would result in a rapid temperature change with possibly cold drafts of air hitting the plants. To overcome this, greenhouse manufacturers began to put limit controls on their ventilator control systems. Limit controls allowed the grower to override the full open response to the thermostat's call for more ventilation to control temperatures. Ventilator positions could be set at one-quarter, one-half, three-quarters, or full open as needed. In areas where frequent rain showers occurred, growers asked for another control device that would close the top ventilators on the greenhouse when a heavy shower developed unexpectedly. Top greenhouse vents when fully opened formed a V shape that acted as a funnel, which resulted in large amounts of water entering the greenhouse.

The next advancement in controlling the greenhouse environment was the introduction of the aspirated thermostat (as shown previously in Figure 6-4). It was learned that blowing an air stream across the thermostat control gave better sampling of the greenhouse air temperature than a nonaspirated unit. As a result, the variation between the high and low temperatures, around a set

FIGURE 6-47 Stage controller. This is a Wadsworth Control Systems, Inc. Step50®, which was installed in 1979 and has remained fully functional for more than 25 years. It controls one zone with two stages of heating and three stages of cooling.

point, is reduced. As an example, if the thermostat setting was for 60°F (15.5°C) in a greenhouse with a nonaspirated unit, the temperature could range plus or minus 4° or 5°. With the use of an aspirated controller, the temperature range may be only 2° or 3° plus or minus the desired setting. Closer control of the greenhouse air temperature gives better control and timing of the crop being grown.

More advanced than thermostats, the thermistor became the sensing unit most often used in greenhouse temperature controllers. These are units that are solid-state electronics. In response to temperature changes, an electrical signal is sent to the equipment controller that causes it to function.

The development and use of thermistors in the greenhouse allowed the introduction of more sophisticated equipment. Stage controllers and computerized climate control systems for monitoring the environment are now in common use. Computerized climate control systems provide the grower with a distinct advantage in precisely controlling not only the greenhouse temperature but also light intensity and carbon dioxide (CO_2) concentrations (Figure 6-48). Relative humidity (RH), which is closely related to vapor pressure deficit (VPD), can also be monitored and controlled. Knowledge of the relationship between RH, VPD, and light intensity in the greenhouse gives the grower the guidance of when to water and fertilize the crop. Exterior weather monitoring systems are also incorporated into the climate control system (Figure 6-49).

The photocell that measures light intensity relays the signal to the computer that controls turning on supplemental high intensity discharge (HID) lighting when needed. If the light intensity becomes too high for the crop, the computer actuates the shade cloth mechanism to reduce the light levels. This aspect of greenhouse environmental control is very important, especially in spring. It is at this time that dark, cloudy periods are often broken by short periods of high light intensity as the sun plays dodge with the clouds. Such high light intensity levels can quickly injure tender foliage that has developed during the cloudy period.

FIGURE 6-48 Wadsworth Control Systems, Inc. EnviroSTEP® climate control computer with attached vent control panel. This climate control computer manages one zone. It has the capacity to control 12 devices.

FIGURE 6-49 Weather station mounted on the exterior of a greenhouse. This unit measures wind speed, wind direction, temperature, relative humidity, light intensity, and it has a precipitation detector.

Through the use of multiple sensors placed at plant level throughout the greenhouse range, more precise control of temperature and other factors is possible. Factors affecting heat distribution in the greenhouse include (1) the type of heating system used, whether it is radiation from steam or hot water pipes located on the sides and/or overhead, from infrared heaters or from hot air units with forced air circulation; (2) the type of ventilation system; (3) the type of greenhouse glazing material; and (4) the size of the area being monitored. In glass-covered greenhouses, cold air infiltrating through loose laps in the glass due to wind effect can cause a 5° to 10°F (2.8° to 5.6°C) difference in temperature in the house. The windward end may be 10°F (5.6°C) colder than the end away from the wind in a large greenhouse. This is not so much a problem in smaller glass covered greenhouses or those covered with flexible or rigid plastic.

Zoned heating systems are used in large greenhouses to overcome temperature differences. With a zoned system, a separate temperature sensor is installed in each zone (as seen earlier in Figure 6-5), so that one area can be kept warmer or cooler than other areas. This permits growing crops that have different temperature requirements in the same house. Having zone heating also allows the grower to fine-tune temperature adjustments through the computer.

In addition to maintaining more accurate control of the greenhouse and plant growing environment, computers can assist in making crop production decisions. Research scientists at Michigan State University have developed what they call the CARE plant modeling software system. The CARE system is a computer decision support system, which has been further enhanced through a collaboration of scientists at the University of California at Davis and the University of New Hampshire. This system helps a grower to program the growth of a crop to meet market specifications for final plant height and flowering date.

The system is designed for personal computer use to provide graphical tracking of the crop's growth. The CARE programming suggests specific temperature set points and growth retardant applications for the crop. CARE modules exist for chrysanthemums, Easter lilies, and poinsettias. As more experience is gained with this new tool, other crop modules will become available.

The use of computer-controlled greenhouse environment and production systems is no longer out of the ordinary and is the greenhouse climate control system of choice. The use of robotics for sowing seed, transplanting, potting up, and moving the plants or the containers used for growing the crop to the growing location in the greenhouse and then back to the staging area for packing and sales is becoming more and more prevalent. Computers will continue to play an important role in the industry.

ACHIEVEMENT REVIEW

Select the best answer or answers to complete each statement. List the appropriate letter(s).

1 The greenhouse effect is a result of:
   a. the type of construction used for the greenhouse.
   b. short wavelengths of solar energy that pass through the glass and are converted to long wavelengths of heat that cannot be reradiated through the glass.
   c. long wavelengths of solar energy that are converted to short wavelengths that cannot pass through the glass.
   d. all of the above factors.

2 The choice of fuel used to heat the greenhouse depends on:
   a. the cost per Btu.
   b. the availability of the fuel.
   c. the equipment costs to burn it.
   d. all of the above factors.

3 Small greenhouses, with less than 15,000 square feet (1,393 m^2) of area, are best heated by means of:
   a. hot air heaters.
   b. hot water systems.
   c. steam heat systems.
   d. solar energy.

4 The heat supplied to the greenhouse is best controlled by:
   a. a sensitive hygrometer.
   b. an aspirated thermostat.
   c. a gate valve.
   d. a copper-constantan thermocouple.

5 Air distribution in the greenhouse is maintained by:
   a. convection currents.
   b. strategically located fans.
   c. perforated polyethylene tubes on fan-jets.
   d. all of the above.

6 The air leaving the holes in polyethylene tubing will be mixed with the greenhouse air when it is within:
   a. 10 to 20 diameters of the hole width.
   b. 20 to 30 diameters of the hole width.
   c. 30 to 40 diameters of the hole width.
   d. 40 to 50 diameters of the hole width.

7 In what season is it very important that outside air mix thoroughly with the inside air before it hits the crop?
   a. Spring
   b. Summer
   c. Fall
   d. Winter

8 When calculating the heat requirements of a greenhouse, the grower must know the:
   a. ground area of the greenhouse.
   b. total volume of the greenhouse.
   c. total exposed area of the greenhouse.
   d. Btu output of the heater.

9 The length of a roof bar may be calculated by using the formula for a right triangle. The correct formula is:
   a. $a^2 + b^2 = c$.
   b. $a + b^2 = c^2$.
   c. $a^2 + b^2 = c^2$.
   d. $a^2 + b = c^2$.

10 At an atmospheric pressure of 14.7 pounds per square inch (760 mm), water turns to steam at:
   a. 180°F (82°C).
   b. 212°F (100°C).
   c. 215°F (101°C).
   d. 222°F (105°C).

11 Gas heaters used in plastic greenhouses should be:
   a. properly grounded.
   b. burned only at night.
   c. properly vented to the outside.
   d. provided with an air intake to bring fresh air into the burner.

12 The primary purpose of a thermal blanket is to:
   a. reduce the light intensity by shading the crop.
   b. reduce the loss of heat through the greenhouse roof at night.
   c. cover the crop at the end of the day.
   d. provide short days to force chrysanthemums to flower.

13 Greenhouses are ventilated to:
   a. reduce air temperatures.
   b. let out the stale air.
   c. supply carbon dioxide to the plants.
   d. reduce relative humidity to help control diseases.

14 Heating air causes it to:
   a. become lighter and rise.
   b. become heavier and fall.
   c. expand with no change in weight.
   d. become more humid.

15 To calculate the fan size needed to cool a greenhouse, it is first necessary to calculate:
   a. the ground area of the greenhouse.
   b. the total exposed area of the greenhouse.
   c. the total volume of the greenhouse.
   d. the inside temperature minus the outside temperature.

16 The maximum effect of evaporative fan and pad cooling takes place when:
   a. the relative humidity of the outside air is very low.
   b. the relative humidity of the outside air is very high.
   c. both the relative humidity and temperature are very high.
   d. none of the above occur, as relative humidity has no effect on evaporative cooling.

17 In addition to fan and pad cooling, other methods of greenhouse cooling are:
   a. mechanical refrigeration units.
   b. shading the house with whitewash.
   c. shading the house with flowing water that is dyed green.
   d. desert coolers.

Effects of Environmental Factors on Plant Growth

CHAPTER

7

The Plant

OBJECTIVES:

After studying this chapter, the student should have learned:

■ parts of the typical plant cell.

■ the differences between meristematic and permanent tissues.

■ the various parts of a leaf.

■ the various parts of a complete flower.

Plants are the result of a continuing process of evolution that has brought them from single cells in the sea to their present form. Plants are unusual in that they have two very large absorption systems. One system is above ground and consists of leaves, stems, and branches. The other system is below ground and is formed by the roots of the plant. The aboveground system allows the plant to take carbon dioxide from the air. With the aid of light energy, the plant turns this gas into foods that are known as photosynthates. The belowground root system absorbs water and various nutrient salts. These materials then combine with the photosynthates produced by the leaves to nourish the plant.

Cells

Cells are the smallest units of which all living organisms are made. They are the building blocks in which the life processes function.

There are many types of cells in many forms. Some cells are square, some are rectangular, and others are long and narrow like pencils. Regardless of

FIGURE 7-1 Typical plant cell showing selected parts.

their shape, all cells have nearly the same interior structure. Figure 7-1 shows a typical plant cell with the most important structures indicated. The **primary wall** is the first part of the cell wall to develop. Where the walls of two cells meet, there is a layer between the walls, which is known as the middle lamella. A secondary wall is formed inside the primary wall of the cell. As the plant develops, the secondary wall becomes lignified, which includes the woody parts of the plant.

The cell walls contain a liquid known as the **cytoplasm**. The cytoplasm surrounds the various structures within the cell. Most of the life processes of the cell occur in the cytoplasm. One of the cell components is the **nucleus**. The nucleus may be thought of as the control center of the cell. The nucleus contains **chromosomes**, which are composed of **genes**, the carriers of the hereditary information that is passed from one generation to the next.

The cell cytoplasm also contains a large number of small but very complex bodies. Each cell has a number of **mitochondria**. These structures control many of the chemical reactions that occur in the cells. Small bodies known as **plastids** are also found in the cytoplasm. If these bodies contain the green pigment **chlorophyll**, they are **chloroplasts**. Other plastids may also contain various yellow and orange pigments called **carotenoids**.

Large, fluid-filled spaces surrounded by membranes within the cell are called **vacuoles**. When plants reach maturity, the vacuoles increase in size until they take up much of the volume of the cell. Vacuoles maintain the shape of the cell.

Tissues

Very primitive plants like algae consist of a single cell and thus are said to be **unicellular**. More complex (multicellular) plants have many different groups of cells. Large groups of cells that are all the same and do the same thing are called tissues. These tissues, in turn, form the organs and systems of the plant.

Tissues are divided into two major groups: meristematic tissues and permanent tissues. **Meristematic** tissues are made up of juvenile cells, which are embryonic. Active cell division takes place in meristematic tissue. Permanent tissues consist of more mature cells that are nondividing. Permanent cells may

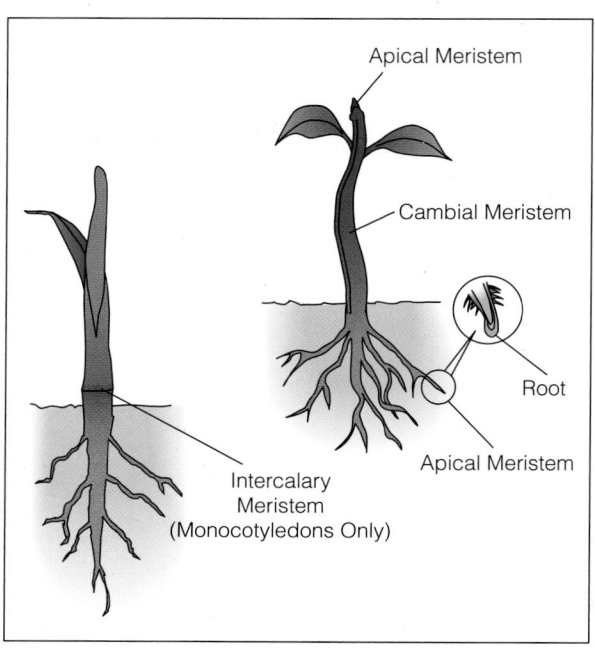

FIGURE 7-2 Areas of meristematic growth.

change back to meristematic cells under certain conditions, for example, when a plant is propagated by a cutting or when bark is wounded.

Meristematic Tissues

Active cell division occurs in meristematic tissue (Figure 7-2). The tips of roots and stems consist of apical meristems. The division that takes place in this tissue causes an increase in the length of the plant. A layer of growth tissue in the stem is known as **cambial meristem** or cambium. Cell division in this layer causes the plant to increase in size through its width, or girth. Monocotyledonous plants having a single seed leaf (cotyledon) have another area of meristematic growth. This area is called the **intercalary zone** and is located just above the node, or thickened stem portion, of the plant. Apical meristems, cambial meristems, and intercalary meristems are considered to be permanently embryonic.

Permanent Tissues

These tissues form the permanent nongrowing structure of the plant. Leaves, flowers, and fruits are examples of organs that arise from tissues that were once formed from lateral meristems. The outside of the plant is covered with a permanent tissue known as **epidermis**. The epidermis protects and supports the inner cells of the plant.

Underground Plant Structures

Roots

The tip of the apical meristem of the root is covered by loosely arranged cells called the **root cap** (Figure 7-3). As the root grows, the root cap cells protect the dividing cells.

As shown in Figure 7-3, there is an area of cell division just behind the root cap where new cells are formed. The next area of the root is a zone where the newly formed cells increase in size by taking up water. A lack of water in this zone causes a reduction in plant growth.

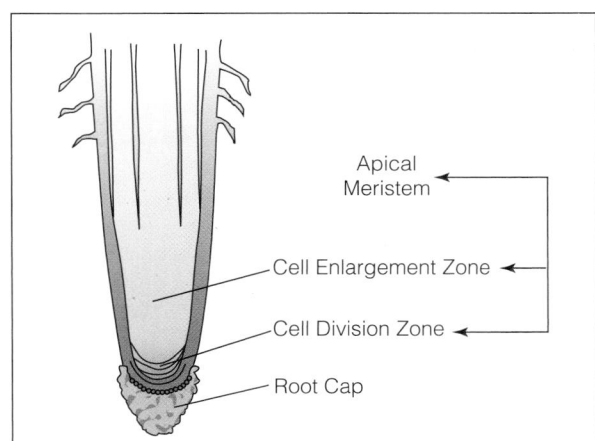

FIGURE 7-3 Simplified drawing of the root cap area. As roots grow through the soil, the first rows of the cells break off and slough away.

The zone of cell enlargement is followed by a zone of **maturation** or **differentiation**. In this zone, specialized cells develop to perform specific jobs. For example, special epidermal cells form root hairs that greatly increase the surface area of the roots. These root hairs grow rapidly, but they live for only a few days. As the root grows through the soil, new root hairs form continually. Most of the water and nutrients used by the plant are taken up by the root hairs.

When **seedlings** or even large plants are transplanted, many of the fragile root hairs are destroyed. This loss of root hairs greatly reduces the amount of water that can be taken up by the plants. To help the plants through this difficult time, they should be watered heavily and shaded. Less water loss occurs if the amount of transpiring area, where the plants give off moisture, is reduced. Pruning, or cutting away parts of the plant, reduces the transpiring area.

In addition to the root hair zone, other epidermal cells on the root take up water and nutrients dissolved in the soil solution. These materials then move into an inner plant tissue known as the **cortex**. The process by which these materials move is called simple diffusion. The water and dissolved plant nutrients pass through the cortex and move to the top of the plant by way of specialized tissue in the vascular system. It is in the root hair zone that the vascular tissue begins to develop.

Vascular System

The vascular system links the underground parts of the plant with the aboveground parts. The specialized tissues in this system carry food, water, and nutrients through the plant. One of these specialized tissues is **xylem** (Figure 7-4). Xylem tissue forms a continuous pathway from the roots through the stems into the leaves, delivering water and nutrients.

The second **vascular tissue** is the **phloem**. Phloem tissue carries plant foods made in the leaves down to the stems and the roots for storage, or to the growing points where they are used for new growth. Organic materials produced in the plant can move upward and downward in the phloem. In the xylem, however, water and nutrients are conducted upward only.

The movement of water, absorbed minerals, and manufactured food substances within the plant is called **translocation**. Translocation is carried on throughout the entire plant structure.

Apical Meristem

Leaf Structure

Stomate with Guard Cells

H_2O CO_2

Xylem Transports Water and Nutrients
Upward from Roots to the Leaves

Phloem Transports Synthesized Foods
from Leaves Upward to the Growing
Point and Downward to the Roots

Water Absorption Through Root Hairs
and Roots into Cortex and then into Xylem

FIGURE 7-4 Movement of water, dissolved nutrients, and manufactured food within the plant.

Aboveground Plant Structures

The aboveground parts of the plant are the stems, leaves, and flowers or reproductive parts.

Stem

The main purpose of the stem is to serve as a pipeline for water and plant nutrients moving from the roots to the top of the plant and for food moving from the top downward. The stem also supports the plant top. Cells in the stem develop thick walls that help hold the plant upright. Much of the stiffness of the plant is due to its water content. When all of the cells are filled with as much water as they can hold, the plant is said to be **turgid**. A plant is **flaccid** when it lacks water and wilts.

When a plant wilts, it suffers internal damage. The longer the plant is wilted, the greater the damage becomes. If a plant is watered soon enough, it usually recovers from the wilted condition and will continue to grow. If the plant is not watered in time, it reaches a stage called the **permanent wilting point**. At this point, the plant does not recover if there is more water loss and none is replaced. The permanent wilting point varies for different plant species. Plants that do not need much water, such as the cactus family, do not suffer as quickly from a lack of water as those plants that require a greater amount of water, such as corn.

Water Movement

The movement of water in plants normally takes place in one direction only. Water is taken up by the roots and is translocated through the xylem to the leaves where it is lost to the air in the form of vapor. **Transpiration** is the name given to the loss of water by leaves. Stems also lose water by transpiration, but the amount lost is much less than the water transpired through the leaves.

In the process of transpiration, carbon dioxide gas enters the leaf through small openings or pores known as stomata. The carbon dioxide is then dissolved into extremely thin films of water that surround cells within the leaf.

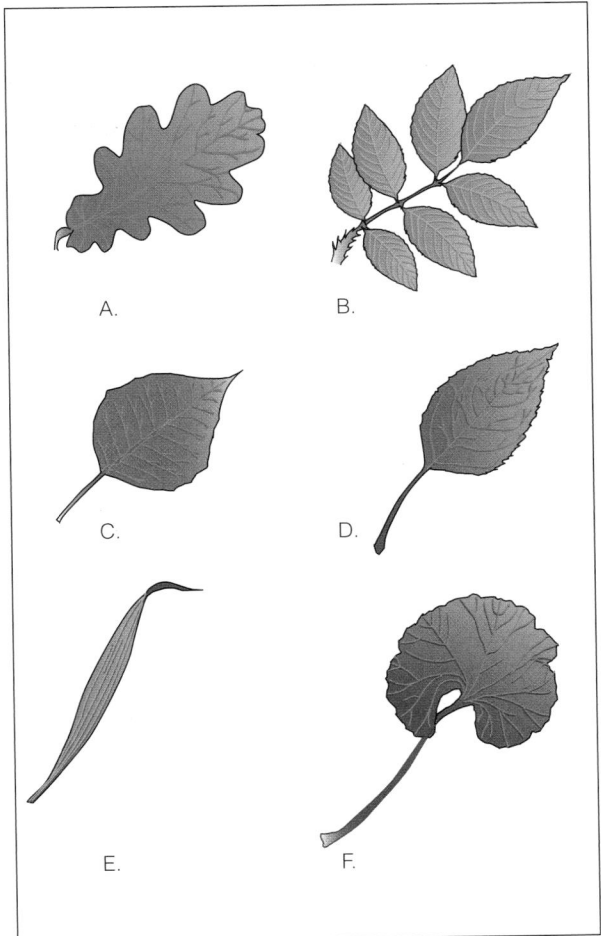

FIGURE 7-5 Leaf shapes: A, oak; B, compound rose leaf (note stipules at the base); C, poinsettia; D, apple; E, lily; F, geranium.

This water is lost to the air by evaporation. It must be replaced constantly from within the plant. More water is needed for transpiration than for any other process in the plant. The transpiration stream of water carries nutrients throughout the plant and effectively keeps the plant cool through evaporation.

Leaves

The food required by the plant for growth is manufactured in the leaves. There are many sizes and shapes of leaves (Figure 7-5). However, each leaf type has the same basic parts. These parts are the leaf blade; the **petiole.** or stem that attaches the leaf to the main stem; and the stipules, if present. **Stipules** are small, leaflike growths located at the base of the petiole. Many ornamental plants do not have stipules. Geraniums, roses, and some other crops have stipules.

Food is manufactured in the leaf blade by the process of photosynthesis. The organic products of photosynthesis move through the vascular system of the leaf and petiole to the phloem within the main stem. Water moves from the roots through the xylem into the petioles to the leaf blade.

Axillary buds are found at the base of the petiole in the angle (axil) between the petiole and the main stem. Axillary buds may be **vegetative** and produce new branches. Or, the buds may be flowering and produce flowers. The same stem may have both vegetative and flowering axillary buds.

Leaf Blade Structure

The leaf blade has both an upper and lower epidermis (Figure 7-6). A waxy substance called **cutin** is exuded by the epidermal cells. This substance forms a relatively waterproof layer covering the epidermis called the **cuticle**. Stomata (plural of **stoma**) are located in the epidermis.

The stomata are protected by specialized kidney-bean-shaped cells known as **guard cells**. The stomata make it possible for an exchange of gases to occur between the outside air and the intercellular spaces within the leaf. Oxygen for **respiration** and carbon dioxide for photosynthesis enter the leaf through the stomata.

When it is dark, the guard cells are flaccid and the stomata are closed. In the light, the guard cells become turgid (filled with water) and the stomata are opened. Photosynthesis can now take place. Any factor that affects the condition of the stomata has a direct effect on the rate of photosynthesis. For example, a plant that is wilting badly from a lack of water does not carry on photosynthesis, even though it may be daylight.

Internally, the leaf consists of a group of cells called the **mesophyll**. Typically there are two layers of cells in the mesophyll: the **palisade layer** and the spongy layer. The palisade layer contains a group of large cylindrical cells. The long axis of each cell is at a right angle to the epidermis. The palisade cells are usually filled with chloroplasts.

The spongy layer is located between the palisade layer and the lower epidermis. This layer consists of a mass of loosely joined, irregularly shaped cells. These cells also contain many chloroplasts although not as many as the palisade cells. The spongy layer has vascular bundles that consist of xylem and phloem tissues. In addition, there is a bundle sheath that gives more translocation capacity. The vascular bundles are the veins of the leaves. The veins carry the water and nutrients from the roots through the stem to the leaf blade. They also carry the organic foods manufactured in the leaf to the roots.

Leaf drop, which is normal for plants, is known as **abscission**. As a leaf ages and can no longer carry on photosynthesis, a special layer of cells is formed at the base of the leaf petiole. These cells are called the abscission zone. The leaf drops because the middle lamella dissolves or the cells in the abscission zone themselves dissolve and can no longer hold the leaf petiole to the stem.

Leaf abscission may result from injury to the plant or from the effects of other factors, such as exposure to gases. Roses, poinsettias, and azaleas often have a great amount of leaf drop. Excessive leaf drop is undesirable when it is

FIGURE 7-6 Cross section of a leaf showing typical structure, including chloroplasts in both the palisade layer and the spongy mesophyll cells.

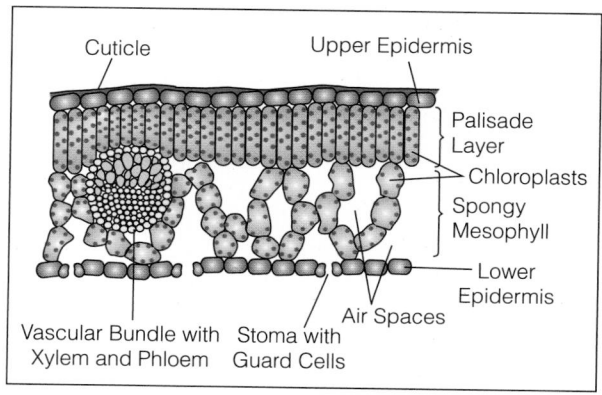

Cuticle

Upper Epidermis

Palisade Layer

Chloroplasts

Spongy Mesophyll

Lower Epidermis

Air Spaces

Vascular Bundle with Xylem and Phloem

Stoma with Guard Cells

due to factors other than natural old age. As the number of leaves that remain on a plant decreases, the amount of photosynthesis that is carried on also decreases. Greenhouse tomato growers often deleaf a plant below the lowest remaining fruiting cluster as these leaves not only carry on photosynthesis, but also drain energy from the remainder of the plant.

Flowers

Most floricultural crops are grown for their flowers. An exception to this statement is tropical green plants that are grown for their foliage. Although flowers take many different forms, they all have the same basic parts (Figure 7-7). The **receptacle** is the swollen portion of the stem at the base of a mature, complete flower that has grown from a bud. The flower parts are attached to the receptacle. The outermost parts of a closed flower are called the **sepals**. The sepals collectively are called the **calyx**. As the sepals unfold, they expose the colored parts of the flower, which are called the **petals**. The petals collectively are called the **corolla**.

The reproductive parts of the flower are the male part of the plant and consist of a slender stalk, or **filament** (Figure 7-8). One end of the filament has an enlarged part called the **anther**. Within the anther, the pollen grains are formed. Together the stalk or filament and the anther make up the stamen. The **stamens** are the same within any one type of plant. However, the number, size, and shape of stamens in a flower vary from one kind of plant to another. This variation is one method botanists use to identify plants. Collectively, the stamens are called the androecium.

The female part, or gynoecium, of the plant is called the **pistil** and is made up of a **stigma**, **style**, and **ovary**. The stigma receives the pollen grains deposited on it by insects or the wind. A pollen tube then grows through the style connecting the stigma to the ovary. In the ovary, the pollen fertilizes the ovule (egg). The fertilized ovule develops into the fruit of the flowering plant.

The colorful parts of plants are not always true flowers. For example, the petals that appear to be the flowers of the poinsettia and bougainvillea are actually modified leaves called bracts. The true flowers of these plants are small and often overlooked. The flower structures of a poinsettia are called cyathia.

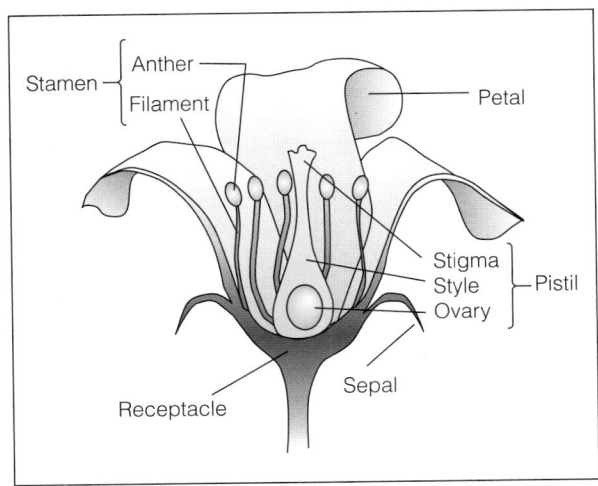

FIGURE 7-7 Longitudinal section of a complete flower.

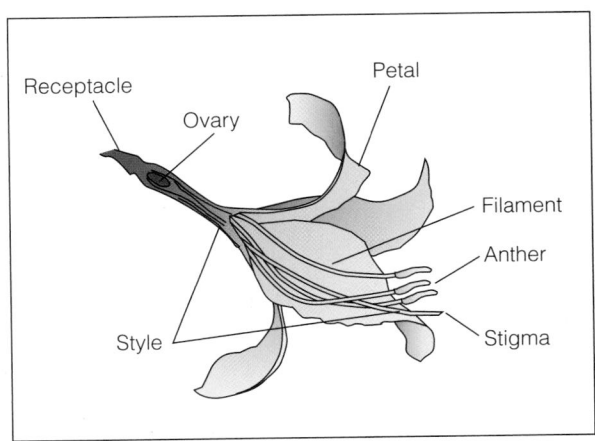

FIGURE 7-8 Longitudinal cutaway section of an Easter lily flower.

ACHIEVEMENT REVIEW

Complete each statement by filling in the blank with the correct word(s).

1 The aboveground absorbing system of plants is made up of _____, _____, and _____.

2 The basic units of which all plants and animals are made are _____.

3 The small bodies within the cell that contain chlorophyll are called _____.

4 Large groups of cells that act together to perform the same function are known as _____.

5 Apical meristems are responsible for an increase in the _____ of a stem.

6 The epidermis of the root provides _____ for the inner plant cells.

7 A group of cells at the tip of a root is called the root _____.

8 Most of the absorption of water and nutrients takes place through the _____.

9 Water moves from the roots to the leaves through the vascular tissue called the _____.

10 The other vascular tissue is called the _____.

11 The point at which a plant will no longer recover from a lack of water is known as the _____ point.

12 The process of photosynthesis occurs mainly in the leaf _____.

13 A bud that grows in the angle between the petiole and the main stem is called an _____ bud.

14 Pores or openings in the leaf where gas exchange takes place from the outer air to the intercellular leaf spaces are called _____.

15 The male parts of the flower are a slender stalk called the _____ and the enlarged part at the end called the _____ where pollen grains are formed.

16 The female parts of a flower are the _____, _____, and _____.

17 The showy, colorful parts of the poinsettia are called _____.

Light

OBJECTIVES:

After studying this chapter, the student should have learned:

■ the process of photosynthesis.

■ how light intensity affects growth.

■ the effect of day length on the flowering process.

■ the methods used to provide artificial long- and short-day conditions.

Photosynthesis

Photosynthesis is the process by which the light energy of the sun is changed by green plants into chemical energy. All life on this planet depends upon the photosynthesis carried on by green plants.

Chemical Reaction of Photosynthesis

The light energy of the sun acts on leaf cells containing the green pigment chlorophyll. A complex series of chemical processes combine the light energy and carbon dioxide and water in the chloroplasts to form simple sugars. The following simple chemical equation is generally used to express the complicated light reaction of photosynthesis.

$$6CO_2 + 12H_2O + \text{light} \quad \text{Chlorophyll} \quad 6O_2 + C_6H_{12}O_6 + 6H_2O$$

This expression says that carbon dioxide (CO_2) and water (H_2O) are combined in the presence of light through the action of chlorophyll. The end products of the reaction are oxygen (O_2) and the simple sugar ($C_6H_{12}O_6$) formed from a carbohydrate and water (H_2O). Gaseous oxygen and water vapor are given off through the stomata of the leaves. The simple sugar manufactured in the process is then used by the plant in a chain of chemical reactions. The energy released by the reactions affects plant growth.

Components of the Reaction

Carbon Dioxide

Carbon dioxide gas enters the leaf through the stomata. It is then dissolved in the water contained in the cells. The carbon dioxide content of air is approximately 0.033 percent, or 330 parts per million (ppm). The concentration of carbon dioxide around industrial areas may be well above this amount, often reaching 400 or even 500 ppm for short periods.

Water

The water used in the photosynthesis reaction is taken in by the roots. It is translocated through the vascular system to the leaves. The water supplies the hydrogen ions (H^+) that combine with carbon dioxide (CO_2) to form carbohydrate (CH_2O) compounds.

Light

Light is one of the most important environmental factors affecting plant growth. Light strikes only the aboveground parts of the plant and reaches only those tissues just beneath the epidermis. However, it affects every cell, tissue, organ, and physiological process of the plant. Light is unique in this respect, as compared to water, temperature, and nutrients that affect the entire plant directly.

Characteristics of Light

Light varies in intensity (brightness), duration (day length), and quality (color). Plants react in different ways to each of these factors. Light is commonly called sunlight. It is more correct to refer to light as solar energy or radiant heat.

The solar or radiant energy from the sun is in the form of waves. These waves have a similar shape to those of the ocean. Some light energy waves are short and others are long. The unit of measurement for these waves is the nanometer (nm). (One nanometer equals one billionth of a meter.) Short waves may measure little more than 100 nm from crest to crest. Long waves may be 60,000 nm between crests. Visible light has wavelengths from 390 to 700 nm. This range is less than 1 percent of the total energy spectrum. All of the wavelengths viewed together are seen as white light. When light is viewed through a thin film of moisture, the various wavelengths are separated to produce a spectrum of violet, blue, green, yellow, orange, and red. The spectrum of colors is visible as a rainbow after a summer shower. The spectral distribution of visible light is shown in Figure 8-1.

There are two areas of electromagnetic radiation that are just beyond the visible light range. These wavelengths have a great influence on plant growth. The ultraviolet region has a wavelength of less than 390 nm. The second region contains far-red and invisible **infrared radiation**. Far-red light is near the upper end of the visible range and has a wavelength of approximately 730 nm. Infrared energy has a wavelength greater than 750 nm.

The short, ultraviolet wavelengths cause chemical reactions, and start or speed up various plant responses often with undesirable results. Red and far-red light is known as photomorphogenic light. The long wavelength infrared radiation affects heat reactions in plants.

Light Intensity

There is a great variation in the light intensity, or the brightness of light, that falls on the surface of the earth. Differences in the brightness are due to the amount of cloud cover and the location on the earth's surface, such as at the

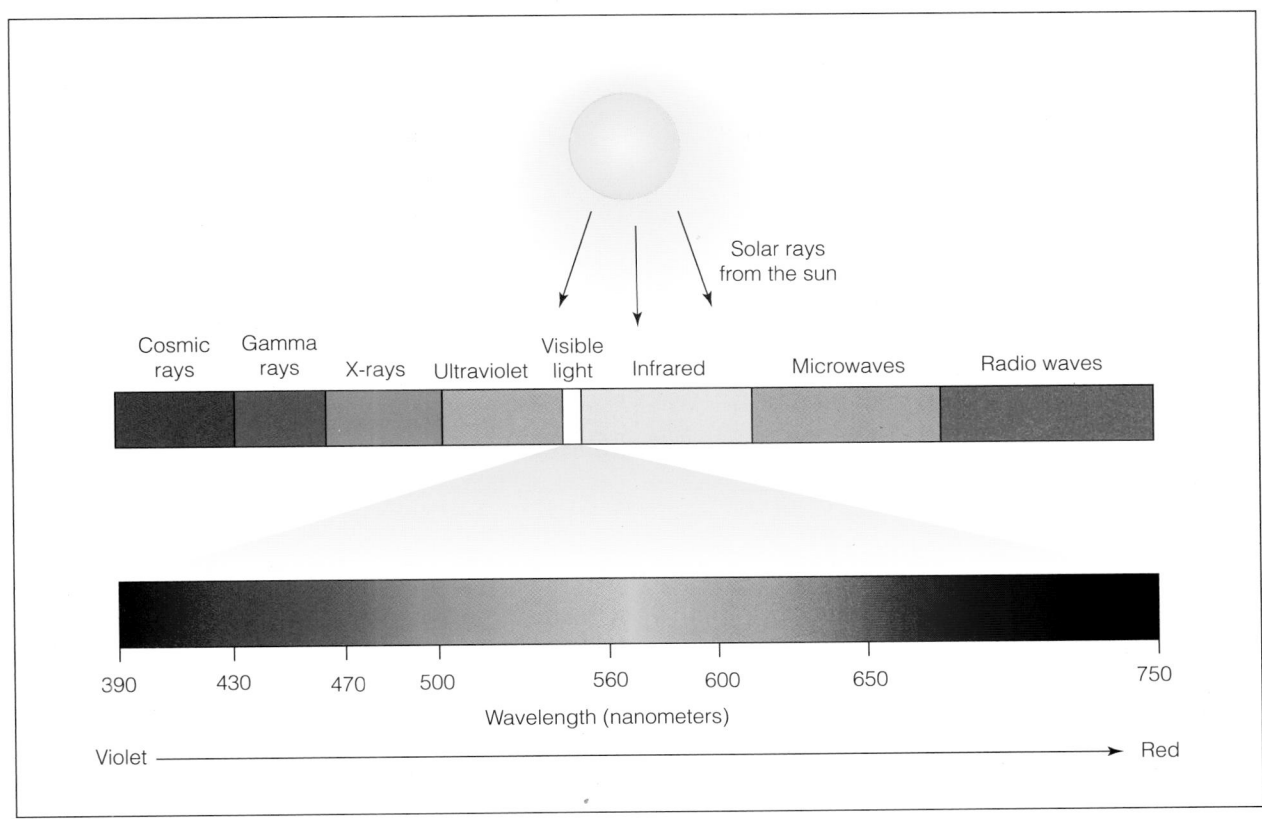

FIGURE 8-1 Radiant energy spectrum.

FIGURE 8-2 Light meter calibrated in footcandles.

equator or at the poles. The amount of dust and dirt in the air, the season of the year, and the time of day also affect the intensity of light.

Under ordinary conditions in the temperate areas of the earth, the light intensity varies from 0 footcandle when it is dark to 10,000 footcandles (107,700 lux) or more at noon on a clear, summer day. A footcandle (fc) is a measure of the intensity or brightness of light. One footcandle is the amount of light striking a surface at a distance of 1 foot from a standard wax candle. Other units of light measurement are **lumens,** lux, gram calories, and micro-einsteins per square centimeter per second ($\mu E \cdot cm^{-2} \cdot sec^{-1}$).

Light meters that measure intensity in footcandles can be used to determine the available visible light (Figure 8-2). Light meters that measure microeinsteins per square centimeter per second are called quantum sensors and measure photosynthetic active radiation. Light meters that measure gram calories are called pyranometers and measure total light energy. These sensors all measure light intensity at different wavelengths and are not uniformly convertible.

Providing the Proper Light Intensity by Artificial Means

On cloudy days, and especially in winter, light intensity levels may be very low. If the light level is low enough, photosynthesis does not take place.

Light is then said to be the limiting factor for growth. If natural light conditions are poor as during winter months, special **high intensity discharge (HID)** lamps help plant growth. Many growers commonly use HID lights to

FIGURE 8-3 High intensity discharge (HID) lights are used to supplement daylight during dark winter months. They may also be used to extend the daily light period for improved photosynthesis in the plants.

irradiate crops (Figure 8-3). These lamps provide 1,000 to 1,200 footcandles (10,750 to 12,910 lux) of light at an economical cost.

Lower light levels are required for plants such as tropical foliage plants, African violets, and similar types of houseplants. The proper light conditions can be obtained artificially with cool white fluorescent tubes. Warm white, natural white, and daylight fluorescent lamps are also satisfactory for growing plants without sunlight. Specially designed plant lamps are offered for sale by several companies. These lamps are expensive, but aid plant growth no more than standard fluorescent lamps.

When fluorescent tubes are new and the plants are placed close to the lamps, the light intensity may be as high as 1,000 footcandles (10,750 lux). Fluorescent tubes are cool in operation and plants can be placed close to them without fear of burning. For incandescent bulbs, only 10 to 12 percent of the energy used by the bulb is emitted as light. The remaining energy is present as heat. Plants placed close to **incandescent lamps** can be badly burned by the heat.

Plants requiring low light levels will grow very well if placed 6 to 12 inches (15 to 30 cm) from a fluorescent light source and are lighted for 14 to 18 hours a day. A two-tube lighting fixture is preferred (Figure 8-4). Single-tube strip fixtures are satisfactory if more than one is used and they are spaced about 3 inches apart on centers. Strip fixtures generally are 8 feet (2.4 m) long. They are not recommended for home hobby type installations.

Light Requirements of Plants

The light-intensity needs of various types of plants are not the same. Some plants require or can withstand very high light intensities. Other plants grow best at medium or low light-intensity levels.

Light Compensation Point

Photosynthesis is a food-making process. The food manufactured supplies the energy required for respiration. There is a point at which the amount of food produced by photosynthesis equals the amount used up in respiration. This point is called the **light compensation point**. The net photosynthesis is zero at this point.

FIGURE 8-4 Two-tube, 4-foot (1.2-m) long strip lighting fixtures installed over a collection of African violets.

FIGURE 8-5 Light saturation curve for plants grown under three light intensities.

The light compensation point varies for different plants of different species and genera. It is low for plants that usually grow in the shade and high for plants that normally need full sunlight.

Light Saturation Point

As the light intensity gets brighter during the day, there is a second point where growth is limited. The **light saturation point** is the intensity at which there is no further increase in photosynthesis with an increase in light intensity (Figure 8-5). A continuing increase in light intensity damages plants such as African violets, gloxinias, and Schefflera. Chlorophyll within the leaves is destroyed. If the light intensity is too high, leaves kept in this condition will develop a light, yellow-green appearance. If the plants are not shaded to reduce the light levels, or moved to an area of lower light intensity, they may die.

During the summer, the flowers of chrysanthemums, hydrangeas, and geraniums may be burned by very bright light. The burning is thought to be due to the extreme high temperature developed in the flower petals. The heat raises the rate of water loss from the petals to the point where enough water cannot be supplied from the roots. As a result, the tissue dies. High temperatures also bleach the flower colors, especially red, purple, and bronze.

Many flower crops require high light levels, including roses, carnations, chrysanthemums, poinsettias, lilies, and fuchsias. However, even these plants may require shading during the midsummer months to reduce the light intensity. Greenhouses are shaded mainly to reduce the temperature within the house. If too much shade is applied, the quality of the crops may be reduced.

Young, tender seedlings cannot survive a high light intensity. They must be protected by shading. As the plants grow, they are gradually exposed to more light until they are used to the full light levels required.

Tropical foliage plants, such as Schefflera (*Brassaia actinophylla*), grow in their natural surroundings in Australia under light intensities of 10,000 to 12,000 footcandles (107,700 to 129,200 lux). However, once these plants are conditioned to the low light levels normally found inside areas such as offices and shopping malls, the plants cannot be exposed again to high light intensities without suffering severe damage (Figure 8-6). Other foliage plants show the same response, which is called light acclimation. Tropical plants that have been grown under bright light conditions and not acclimated to lower interiorscape light levels more than likely will shed their leaves through abscission.

Effects of Low Light Levels

When the light intensity is low, the rate of photosynthesis and the rate of growth and development are also low. This means that the quality of the crops produced is low. On dark winter days, the light intensity may be so low that there is no photosynthesis. As mentioned previously, the required light for growth can be provided in winter by high intensity discharge lamps.

When the light intensity is too low, plants develop long stems. The distance between the leaves, the internode distance, stretches. The stems become weak and are unable to support the flower heads. This is especially true with carnations. As the leaves of foliage plants develop under reduced light levels, each one becomes smaller and smaller. Flowering is delayed or ceases completely. For example, gloxinias and African violets do not bloom when the light intensity drops below a certain level.

Plants are also affected by one-sided light exposure. The intensity of light in the winter is reduced because of the low angle of the sun in the sky. Under such conditions, for example, carnations bend toward the light. Houseplants grown

FIGURE 8-6 Leaf injury to *Ficus lyrata* (fiddleleaf fig) due to high light intensity.

near a window also act in this manner. The effect is known as **phototropism** (from *photo* = light and *tropism* = turning). Phototropism is due to the deposition of growth hormones (auxin) from the apical meristem on the side of the stem away from the light. The additional hormones cause the shaded side to grow faster than the lighted side. A curved stem is the result of this uneven growth. When plant shoots turn toward the light, the action is known as positive phototropism. Roots exhibit a negative phototropism and turn away from the light.

Photoperiod

Photoperiod is the term applied to the duration of time that light shines on a plant each day. Day length changes with the seasons of the year and with the latitude of the location. There are four days of the year when day length is especially important. (1) In the Northern Hemisphere on or about March 21 is the spring **equinox**, when the sun is directly over the equator. Both north and south of the equator, the days and nights are each 12 hours long. (2) On or about September 21 is the fall equinox, and again, day and night are equal in length. (3) On or about June 21 is the summer solstice. The sun reaches its most northern point above the equator, resulting in the longest day and shortest night for areas in the Northern Hemisphere, and the shortest day and longest night for locations south of the equator. (4) On or about December 21 is the winter solstice, when areas in the Northern Hemisphere have the shortest day and the longest night of the year. It is summer below the equator, and December 21 is the longest day of the year and the shortest night. Figure 8-7 shows the position of the sun on these four days for the Northern Hemisphere.

Effect of Day Length

The lengths of the light and dark periods have a definite effect on the growth of plants. Photosynthesis and respiration are greatly affected. The longer the light period, the more time there is for photosynthesis. Thus, growth takes place. Plants make their greatest amount of growth with a long day and a short night. Because less energy is used for respiration during the dark period than is made by photosynthesis in the light period, more energy is available for growth.

In many plants, the initiation (beginning) of flower buds depends upon the length of day the plants receive. This effect was first shown by two plant breeders

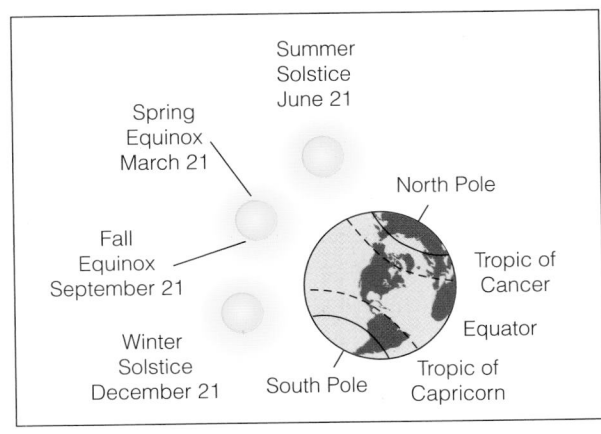

FIGURE 8-7 Position of the sun relative to earth for the four seasons in the northern hemisphere.

working for the U.S. Department of Agriculture, Garner and Allard. In the early 1920s, they were working on the flowering of Maryland Mammoth, a then new variety of tobacco, and soybeans. The tobacco failed to flower when grown outdoors in summer. When it was grown in the greenhouse in winter, it flowered and fruited very heavily. The soybeans were planted at different times during the spring and summer. However, they all came into bloom on the same date in late summer. The seeds sowed last grew the least in height.

The researchers tried to control flowering by varying several factors, including temperature, soil moisture, and fertilizers. None of these changes had any effect. Garner and Allard next controlled the length of day by placing the plants into dark chambers to give them a short day. They learned that the short day caused the plants to bloom. It was this exciting find that led them to investigate the effect of day length on many other plants. During the summer months, they shortened the natural day length artificially by covering the plants with an opaque, or lightproof, cover. In winter they increased the day length by using artificial light.

This plant response to day length is called photoperiodism. The research conducted by Garner and Allard enabled them to identify three classes of plants:

1. **Short-day plants (SDP)** flower when the day length is shorter than a critical number of hours. Chrysanthemums and poinsettias are both short-day plants.
2. **Long-day plants (LDP)** flower only if the day length is longer than a critical number of hours. Examples of long-day plants are tuberous rooted begonias, feverfew, and spinach.
3. **Day-neutral plants** are those plants for which the day length does not influence flowering directly. For these plants, the day length must be long enough for sufficient photosynthesis to occur to support growth. African violets and roses are examples of day-neutral plants.

There is a fourth group of day-length-controlled plants that was unknown to Garner and Allard. The plants in this group are known as quantitative photoperiodic responders. These plants have an improved flowering response for either long days or short days, but they will flower eventually even under unfavorable day-length conditions.

Quantitative long-day (QLD) plants initiate buds for any day-length conditions, but they flower in the shortest time when the days are long. The reverse is true for quantitative short-day (QSD) plants. In other words, they will initiate flower buds for any day length, but they flower in the shortest time with short days. The carnation (*Dianthus caryophyllus*) is a QLD plant. *Salvia splendens* is a QSD plant.

A fifth group of plants is known as obligate photoperiod responders. These plants flower only for a certain light period. Flowering does not occur if the day length is longer or shorter than a critical number of hours.

There are other variations in response to day length. Different **cultivars** (cultivated varieties) of the same species may require different day lengths for flowering. For example, cultivars of chrysanthemums that flower late in the fall require a shorter day length for bud initiation than do those cultivars that bloom in the early fall.

FIGURE 8-8 Another style of HID light. These units are also used to provide photoperiod control of flowering.

FIGURE 8-9 Incandescent lights are used to maintain long-day conditions on chrysanthemums grown under shade in southern Florida. These plants are maintained as stock blocks for vegetative cuttings.

Controlling Day Length

Photosynthesis requires high light intensities. In contrast, photoperiodism is controlled by red or far-red light at very low light intensities and short periods of lighting. Chrysanthemums will remain in a vegetative (nonflowering) condition when exposed to as little as 2 footcandles (22 lux) of light (Figures 8-8 and 8-9). Certain cultivars of poinsettias will remain vegetative if they are exposed to less than 1 footcandle (11 lux) of light. Flower growers must be very careful to prevent light from striking these plants and thus delaying bloom.

Extending Day Length

Incandescent light bulbs can be used to increase the day length. The light should be applied in the middle of the night. Chrysanthemums initiate flower buds when the day length is 14.5 hours and the dark period is 9.5 hours. Flower bud development takes place when the light period is 13.5 hours and the dark period is 10.5 hours. This decrease in day length occurs in the Northern Hemisphere in the fall as the days naturally are shorter. In the Southern Hemisphere it occurs in their fall, which is our spring.

Flower initiation in chrysanthemums and other SDP types can be stopped by using lights to break up the long dark period. Normally, the lights are turned on from 10 P.M. until 2 A.M.

Incandescent light is the most effective because it is high in red wavelengths. This energy breaks down the flowering stimulus that is built up during the dark period. This topic is discussed in more detail in the section, "Light Quality."

The light from cool white fluorescent lamps does not contain enough red wavelengths to prevent flowering. They should not be used to control day length. Another reason why fluorescent lights are less desirable is that their installation is more complicated than that of incandescent lights and the reflector and ballast cast large shadows.

Decreasing Day Length

The day length must be decreased or shortened whenever the grower wants short-day plants to flower out of their natural season. For chrysanthemums, the period from April 20–25 to August 20–25 at 42°N latitude has longer natural days than the critical day length needed for chrysanthemum flowering. To initiate bloom, the day length must be shortened.

The easiest way to decrease day length is to pull a lightproof cover over the plants. In Europe, black polyethylene is commonly used. In the United States, however, the average temperature is higher than in Europe. Plants grown under black polyethylene have more disease because of the high humidity that develops under the material. Black sateen cloth is used in the United States (Figures 8-10 and 8-11). Each day, the plants are covered with this material from 5 P.M. to 8 A.M. The plants are covered during this period because it is usually the same as the working hours.

Semiautomatic **black cloth** systems can cover entire crops in a few minutes (Figure 8-12). Such systems work well in clear span houses, but are not so easily adapted to houses with post supports.

FIGURE 8-10 Automatic black cloth shading of pot chrysanthemums.

FIGURE 8-11 An entire house can be blacked out at once for photoperiod control of flowering.

FIGURE 8-12 Semiautomatic black cloth system used to shade an entire house of chrysanthemums.

The semiautomatic system can be connected to a time clock. In this way, the cover can be drawn over the plants and removed automatically. Too often, however, mechanical devices fail to work properly. Thus, someone should be on hand to see that the system operates correctly. Black cloth systems also serve to conserve heat in the greenhouse during the winter months. Care must be taken to open the cover gradually in the morning. This slow exposure prevents the plants from being suddenly chilled by cold air that accumulates above the blanket during the night.

When black sateen cloth is new, it works very well in stopping light from reaching the plants. However, after several years of use, it becomes thin and faded, especially if the cloth is installed outside where it is exposed to rain and bright sunshine. As the cloth ages, it admits more light than is desirable. During the brightest periods of summer, unwanted delay in the blooming of chrysanthemums is due to excess light penetrating old cloth. To shade chrysanthemums during the high light periods of summer, only new black cloth should be used.

The short-day treatment is continued only as long as it is needed. When the days are naturally longer than the critical day length, the black cloth treatment is used until the flower buds are well developed and showing good color. If the treatment is ended before this point, flowering may be delayed. Once the proper stage of development is reached, the treatment is discontinued to save labor and avoid flower breakage.

During the natural short days of the year, black cloth is needed only if lighting from other sources can reach the plants and delay blooming. For example, long-day plants requiring extra lighting may be growing in the same greenhouse.

Some growers do not use the black cloth cover on weekends. Research has shown that for each day out of 7 days that the plants are not covered with black cloth, there is a delay of 1 day in the blooming date. If more than a day of shading each week is missed, there may be a serious delay in the blooming.

Other Day-Length Effects

The length of the day has other effects on plant growth and development in addition to flowering. For example, hemp plants develop more male (staminate) flowers in short photoperiods than long ones. The development of both staminate and pistillate (female) flowers of corn is affected by day length. For most strawberry varieties, short days cause flower bud initiation, but inhibit runner formation. Strawberry runners are produced during the long days of summer when flower bud initiation is inhibited.

Plants usually grow taller under long-day conditions. Accelerated growth of plant seedlings can be obtained by lighting them artificially for 18 or even 24 hours daily.

The formation of onion bulbs is a response to long day lengths. Dahlias, however, form thick storage roots under short-day conditions. Tuberous begonias form fine branching or fibrous roots with long-day conditions. With short days, they form thick, fleshy storage roots that resemble tubers. This is the reason why they are called "tuberous rooted begonias."

Detection of Day-Length Condition

All parts of the plant do not respond equally to the photoperiod. The change from the vegetative to the flowering state takes place at the growing point of the stem or the apical meristem. However, the signal for this change is not sensed at this location. It has been determined that chrysanthemum leaves respond to the day length to which they are exposed. The shoot tips of chrysanthemums do not respond to day-length conditions. Only the most recently matured leaves sense the length of day and send the signal to the growing tip. This response is described in more detail in the next section.

Light Quality

The quality or color of light is called the action spectrum. Visible light consists of different colors having wavelengths ranging from the very short wavelengths of ultraviolet to the very long wavelengths of infrared. Each color has a specific wavelength or band in the action spectrum. Each wavelength of light causes chemical and thermal (heat) responses in plants that influence various phases of growth.

White light is a blend of violet, indigo, blue, green, yellow, orange, and red. The wavelengths of these colors range from 400 to 700 nm. Light is measured in a unit called a nanometer. A nanometer is one-billionth of a meter (10^{-9}).

Photosynthetic light is that light absorbed by chlorophyll. There are two forms of chlorophyll, chlorophyll a and chlorophyll b. Chlorophyll a absorbs most of its light energy at 430 and 662 nm. Chlorophyll b absorbs most of its light energy at 453 and 642 nm. The wavelengths that are least absorbed by chlorophyll are green; thus, green light is reflected giving plants their characteristic green color.

Ultraviolet light is invisible. Its wavelength is from 200 to 400 nm. Infrared radiation is also invisible, with a wavelength greater than 750 nm. Within the plant, chemical reactions and the rates of various physiological processes are affected more by ultraviolet radiation than infrared radiation.

Ultraviolet radiation is divided into three subsets, UV-A, UV-B, and UV-C. UV-A (315 to 400 nm) is that radiation which creates smog and fades fabrics. UV-B (290 to 315 nm) is that radiation which affects plants and animals. It stimulates melanin in animals and is known to damage DNA and RNA, often resulting in cancer. UV-C (220 to 290 nm) is absorbed by the ozone layer in the atmosphere and does not reach the earth. UV-C radiation is lethal to most life as we know it. For example, high-intensity mercury vapor lamps that produce ultraviolet radiation may be used to provide light. If the outer glass of such a lamp is broken, the rays strike the plants directly and cause tissue death. The damage is known as sunburn or sun scald. Plants exposed to high levels of ultraviolet radiation, such as those growing at high altitudes, are stunted.

Infrared radiation has a greater heat effect on plants than radiation of shorter wavelengths. Plants grown under incandescent lights receive mostly red, infrared, and far-red energy. It is in this area of the action spectrum that the control of flowering by means of the photoperiod occurs.

This discovery was made by U.S. Department of Agriculture scientists who were working with Biloxi soybeans. They found that red light with a wavelength of 660 nm has the greatest effect in preventing the flowering response. Later they found that the inhibiting effect of red light of 660 nm can be reversed if the plants are exposed to far-red energy at 735 nm. The reversibility of these reactions means that plants can be made to flower if they are given red energy (660 nm) followed by far-red energy (735 nm). However, they do not flower if they are given far-red energy followed by red energy. The plants respond to the last form of energy received, regardless of how many times the energy is switched from one wavelength to the other (Figure 8-13).

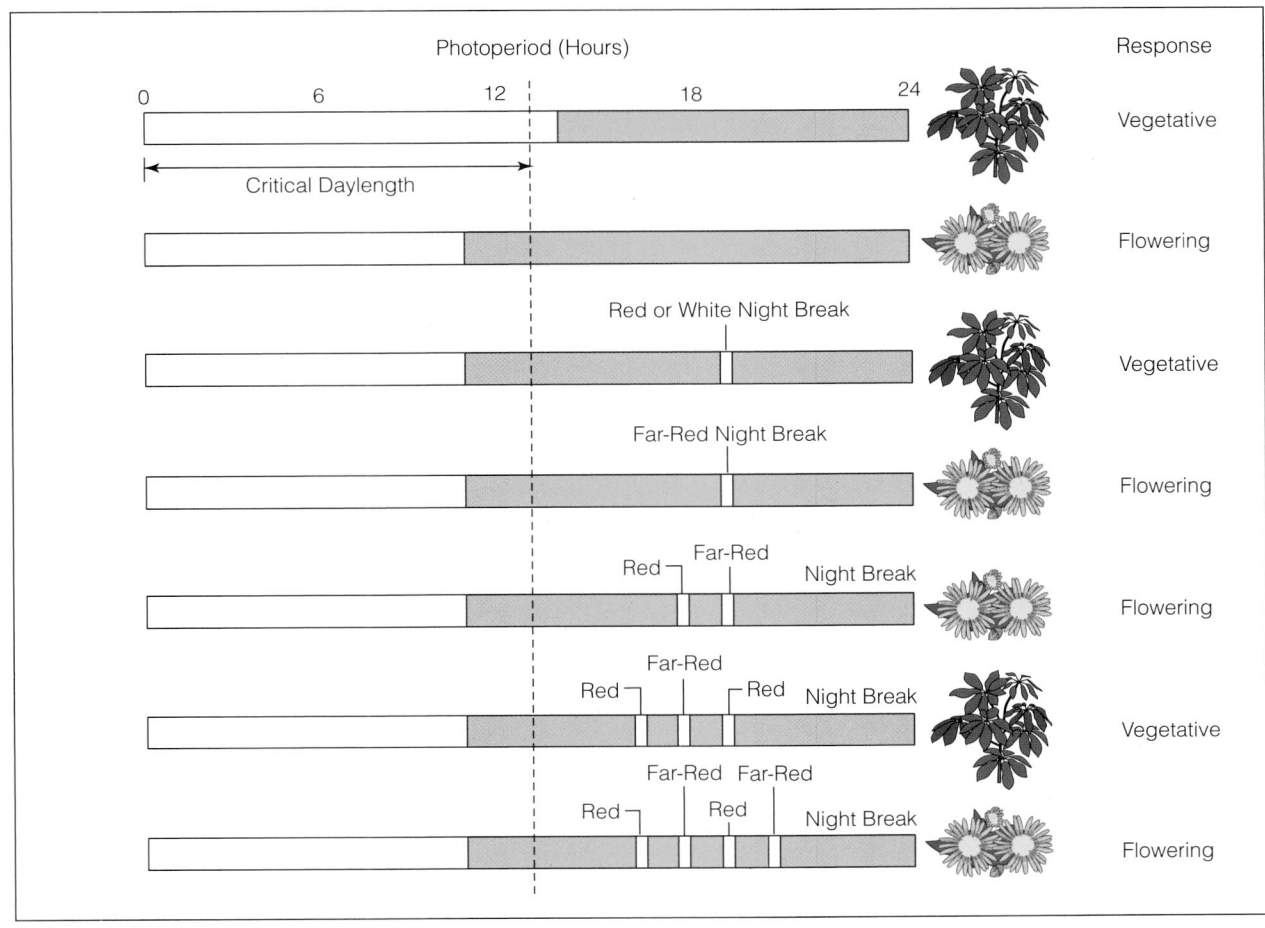

FIGURE 8-13 The response of short-day plants to various types of lighting at night.

The researchers determined that the light-sensitive substance in the plant is a soluble protein, which they named **phytochrome**. This protein is an enzyme. It is a biological catalyst that controls many kinds of plant responses in different plant tissues. For example, stem elongation and leaf growth seem to be controlled by phytochrome. Mature apples turn red in light having a wavelength of 660 nm. The light causes the production of the red pigment anthocyanin. The same red light that promotes the **germination** of some seeds also controls the production of anthocyanin in a number of seedling plants. Seed germination is inhibited by exposure to far-red light at a wavelength of 730 nm.

The germination response is also controlled by the radiation exposure given last. If the seeds are last exposed to red energy, germination occurs. If the energy is far-red, germination does not occur, regardless of the number of cycles of red and far-red energy.

Many discoveries have been made in recent years about the control of flowering and other plant responses to light. However, much information still remains to be discovered.

ACHIEVEMENT REVIEW

Complete each statement by filling in the blanks with the correct word(s).

1 The process by which green plants convert the light energy of the sun into chemical energy is called _____.

2 The green pigment in the leaf that helps convert the light energy is called _____.

3 Water used in this conversion process is absorbed by the roots. The water is the source of the _____ ions that combine with carbon dioxide to form carbohydrates.

4 Solar energy (light) is transmitted in the form of _____.

5 Two important radiation levels that are just beyond the visible range are the _____, which is less than 390 nm, and the _____, which is greater than 750 nm.

6 Light intensity is measured in units called _____.

7 When the light intensity is so low that photosynthesis cannot take place, the light is said to be the _____ for growth.

8 Fluorescent lamps are a better source of light than incandescent lamps because incandescent lamps give more _____ than light.

9 The light saturation point is reached when no further increase in _____ occurs with an increase in light intensity.

10 Excessively high light intensities in the summer may cause _____ of the flowers of chrysanthemums, hydrangeas, and geraniums.

11 House plants grown near a window in winter may lean toward the light in response to a reaction called _____.

12 The term used to describe the duration or length of the daily light exposure is called _____.

13 Those plants that flower when the day length is less than a critical number of hours are called _____-day plants.

14 Those plants that flower when the day length is longer than a critical number of hours are called _____-day plants.

15 Chrysanthemums exposed to as little as _____ footcandles of light remain vegetative (nonflowering).

16 The most recently matured _____ of the chrysanthemum respond to the length of day and signal the growing tip.

17 The light-sensitive substance within plants that responds to day length is a soluble protein called _____.

Temperature

OBJECTIVES:

After studying this chapter, the student should have learned:

- the minimum and maximum day and night air temperatures for ornamental crops.
- the effect of temperature on:
- photosynthesis.
- plant respiration.
- plant transpiration.
- flower bud initiation and development.
- how temperature affects flower color.

Temperature is another important environmental factor that influences the growth of plants. The response and growth of plants occur over a wide range of temperatures. This range may be defined by three basic levels: (1) The minimum temperature is the level below which growth does not take place; (2) the optimum temperature is the level at which growth is greatest; and (3) the maximum temperature is the level above which no growth occurs. The commercial grower tries to maintain the optimum temperature at all times for greenhouse crops. The temperatures are not allowed to drop below the minimum level in the winter months, except in the case of an accident with the heating plant.

In the summer, the temperature may go above the maximum, even when shading or a fan and pad cooling system is used. At such times, plant growth stops. If the temperature continues to rise, the plants will die. With increasing temperatures, plants cannot supply water fast enough to the tissues. This condition is known as heat scorch or **desiccation.**

Temperature affects each of the complex chemical reactions that form the basis for plant growth. The rate of these reactions is determined by the temperature to which the plant is exposed. Thus, the rate of plant growth is regulated by temperature.

Responses of Plants to Temperature at Various Stages of Growth

Plants do not respond in the same manner to temperatures at all stages of growth. For example, the seed may be alive after exposure to very high or very low temperatures for short periods of time. Such temperatures would kill the seedling. Germinating seeds are more sensitive to these same temperature extremes than dry seeds. Seedlings, newly rooted cuttings, and plants in early stages of growth may require warmer temperatures than mature plants. In addition, plant parts and portions of the plant processes may respond in different ways to temperature. Thus, root growth and stem growth may show different responses. The optimum temperature for vegetative growth may not be suitable for flower development. The grower must know how each crop responds to temperature to ensure the highest quality crop.

Night Temperature

Throughout this text, unless stated otherwise, the temperature noted is the night temperature (NT). Growers normally maintain day temperatures (DT) 10°F (5.56°C) higher than the night temperature on cloudy days and 15°F (8.33°C) or more on clear days. This assumes that the temperature at which the crop is grown can be controlled. Some growers increase the cloudy day temperature only 5°F (2.78°C) over the night temperature and 10°F (5.56°C) or more on clear days. In the winter, when the light intensity is poor, cool crops may be grown without any temperature variation. For example, carnations and snapdragons may be held at the same day and night temperature while cloudy conditions last.

The air temperatures listed are measured by an aspirated thermometer mounted in a shaded box, preferably at plant level. There is little benefit to the plants if the temperature at eye level is maintained at 60°F (15.5°C) NT and the plants are growing at ground level.

DIF Day–Night Temperature Relationships

For years the recommendations for flower crop production in commercial greenhouses has been to grow the plants cool at night. Day temperatures on cloudy days were raised to 10°F (5.56°F) above night temperatures. On clear, sunny days the day temperatures were allowed to increase 15°F (8.33°C) or more above the night temperature setting.

As an example of these recommendations, chrysanthemums were grown at 60°F (15.5°C) night temperature, 70°F (21°C) on cloudy days, and 75°F (24°C) on clear days. The theory behind such recommendations was that cool temperatures at night reduced respiration of the plants and thus conserved carbohydrates.

During the daylight hours, temperatures were increased to maximize the formation of photosynthates from the process of photosynthesis. As a result of following these temperature regimes, plants responded in the expected way. Active growth took place during the daylight hours with stem elongation (stretching) and flower development with little stem stretching at night.

Research at Michigan State University has resulted in recommendations for controlling plant growth that reverse the day–night temperatures used. The procedure is known as DIF, meaning the differential between day and night temperatures. The research found that plants that were grown at day temperatures that were cooler than night temperatures did not elongate or stretch as much as those grown under the normal low night, high day temperatures. The situation of warm night and cooler day temperatures is called negative DIF.

During the course of a 24-hour period, temperatures are usually at the lowest point just before sunrise. This fact of nature gives the grower a little better advantage to use DIF. The reason is that by giving 2 hours of cool temperatures immediately after the sunrise, the crops responded the same as if they were grown at that cool temperature all day long.

Cool temperatures can be obtained by delaying the changeover from the night to day temperature in winter. In spring and summer, ventilating the greenhouse and using fan and pad or fog cooling methods for these 2 hours is effective.

The use of negative DIF is an alternative to the use of chemical growth retardants to keep plants short. In addition, chemical growth retardants cannot be applied to vegetable transplants grown in the greenhouse and DIF is one of very few tools to maintain plant height.

The use of negative DIF is being widely practiced by many commercial flower growers. There do not seem to be any major bad side effects—no loss of quality nor delay in flowering. Some crops may show a temporary wilting or drooping of the leaves, which is called epinasty. This downward curling of the leaves is noticeable on the Easter lily cultivar 'Nellie White' (Figure 9-1). This

FIGURE 9-1 "Nellie White" lilies grown under positive DIF and negative DIF treatments. Notice the differences in height, leaf curling, and stage of flower development. *(Courtesy Dr. Allen Hammer, Purdue University.)*

curling occurs if the difference in temperature is very large, that is, if the +/− range of the DIF is greater than 8°F (4.4°C). If the curled leaves are young enough and still maturing, they will uncurl when the plant is grown on at normal temperatures.

Seedlings started as plugs, if subjected to extreme DIF conditions during the first week or two of growth, may develop chlorotic leaves. Permanent stunting of the plants may occur. If the DIF treatment is not greater than 2°F to 3°F (1.11°C to 1.67°C), the plants usually grow out of the condition.

Seed Germination

Temperature has the greatest effect on seed germination if there are no other limiting environmental factors. It is assumed that the seed has already passed through its after-ripening and dormancy period, if it has one. For each type of seed, there are upper and lower temperature limits beyond which the seed will not germinate. Some seeds germinate best if they are exposed to a **diurnal (day–night) cycle**. Diurnal changes take place as the temperature in the greenhouse is varied from day to night to day conditions.

The seeds of most ornamental crops germinate at an optimum air temperature range of 60°F to 70°F (15.5°C to 21°C). A few types of seeds germinate faster at slightly cooler temperatures. Another small group of seeds, mostly those of tropical foliage plants, germinate when the temperature range is 75°F to 80°F (24°C to 26.5°C).

Germination is hastened if **bottom heat** is applied by means of electric heating cables or other methods. The bottom heat temperature is usually maintained 5°F above the air temperature. When seeds are germinated under mist, bottom heat is especially important. The heat is necessary because the mist has an evaporative cooling effect that may reduce the media temperature below the desired level.

Vegetative Growth

Effect of Air Temperature on Photosynthesis

Temperature has a direct effect on the formation of carbohydrates in the photosynthetic process. If there are no other limiting factors, such as the amounts of water, carbon dioxide, or light, the rate of photosynthesis increases as the temperature increases. At 95°F (35°C) the rate of photosynthesis drops very quickly, and the process stops. Certain desert plants may withstand day temperatures of 122°F to 131°F (50°C to 55°C) for long periods because of their ability to store water in their tissues. However, little if any photosynthesis occurs at these high temperatures.

The lowest temperature at which photosynthesis can take place varies with the plant species. For example, photosynthesis in some arctic plants continues down to the freezing point of water, 32°F (0°C). Other plants, such as *Dieffenbachia amoena* (dumbcane) and *Saintpaulia ionantha* (African violet), are injured if the temperature drops below 50°F (10°C) for any extended period of time. Chlorophyll is destroyed in African violet leaves if they are exposed to subfreezing temperatures for as little as 2 to 3 minutes. As a result, the leaf tissue is marked.

Effect of Air Temperature on Respiration

Photosynthesis can take place only at certain levels of light intensity. In contrast, the process of respiration occurs in all living cells 24 hours a day. Food that is manufactured in daylight is used to carry on the various growth processes of the plants.

Temperature has a direct effect on respiration. The rate of respiration is greater at warm temperatures than at cool temperatures. The respiration rate increases as the temperature increases. However, the process can be slowed or interrupted if there is a lack of oxygen or if the carbon dioxide resulting from the process builds up to the point where it interferes with respiration.

If the night temperature is lower than the day temperature, the food manufactured by the plant during the day is conserved. There are times when the light is not strong enough and is the limiting factor for photosynthesis. For example, on dark winter days, the rate of daytime respiration may be faster than the rate of photosynthesis. As a result, there is a net loss of carbohydrates. Respiration is reduced if the greenhouse temperature is maintained at a cooler level on cloudy days.

The conservation of food reserves in plant tissues is an important factor in the lasting ability of cut flowers. Roses, carnations, and chrysanthemums can be stored after harvest without water for weeks or months at temperatures of 31°F to 33°F ($-0.5°C$ to $0.5°C$). At these temperatures, respiration is very slow, and food in the tissues is conserved.

Effect of Air Temperature on Transpiration

Transpiration is affected by temperature primarily at the leaves. As the leaf temperature rises, the rate of transpiration increases. This rate will continue to increase with the temperature as long as the plant can absorb water from the soil and the water can be transported to the leaves. Wilting occurs when the rate of transpiration increases to the point that water is used faster than it can be supplied to the leaves.

Wilting occurs frequently in winter after periods of dark, cloudy weather. Bright sunshine may raise the leaf temperature 5°F to 15°F more than the surrounding air temperatures. As a result, transpiration increases until the water supply can no longer keep up with the rate of water loss. This condition is noted particularly in snapdragons and carnations. Wilting is not corrected by adding more water to the soil. The plants must be shaded temporarily to reduce the **moisture stress**. Once they are shaded, the plants will recover slowly. After 2 to 3 days of sun, the plants are accustomed again to the brighter light levels and grow normally without wilting.

Air Currents

Warm or cold air currents or drafts also affect the temperature of a leaf. When exposed to air currents, the leaf will become the same temperature as the surrounding air. If the air temperature is below freezing (32°F or 0°C), the leaf may be killed. The growth of crops located near greenhouse doors may be severely retarded by cold winter air that flows in when the door is opened and closed several times a day. The use of HAF fans greatly reduces warm and cold air currents in the greenhouse, providing a more uniform temperature across the greenhouse.

Radiational Cooling

On clear, cold nights, plants growing close to greenhouse sidewalls will lose heat to the outside by radiation from the leaves. Although the greenhouse air temperature may be 55°F (13°C), the leaf temperature may be as much as 5°F colder when the outside temperature is 15°F to 20°F (−7.5°C to −10°C). This condition is known as **radiational cooling**. It may cause the development of the red pigment anthocyanin in geranium leaves and on marigolds. Radiational cooling is rapid on clear, cold nights. A cloud cover slows the rate of radiational cooling, and the leaves remain at or near the temperature of the greenhouse air.

Condensation

When the leaves are colder than the surrounding air, moisture in the air condenses on the leaf surface. This form of condensation is common in the spring and fall when the days are warm and bright and the nights are clear and cold. The moisture that collects on the leaves serves as a germination medium for certain disease-producing spores. For example, powdery mildew on roses is a problem under these conditions. If the very humid air is allowed to leave the greenhouse, the problem of condensation is reduced greatly. Rose growers often provide some heat in the greenhouse along with venting in the late afternoon to hasten the removal of the moist air.

Effect of Air Temperature on Other Metabolic Processes

In addition to photosynthesis, respiration, and transpiration, other growth processes that are affected by temperature include enzyme reactions, amino acid and protein synthesis, and carbohydrate and lipid (fat) conversions. Generally, it is believed that the temperature effect on these processes is much the same as for photosynthesis and respiration. As the temperature increases, the growth rate of the plant also increases. A law formulated by the Dutch chemist, Jacobus van't Hoff, states that "for every 10 degrees centigrade rise in temperature, the rate of reaction doubles." This is known as the Q_{10} rule.

Most plant processes stop at 95°F (35°C). At this temperature, protein becomes disorganized and is unable to remain functional. If the internal temperature of the plant rises above 95°F, the plant will soon die.

Effect of Soil Temperature on Soil Microorganisms

The activity of soil microorganisms and the rates of absorption of water, fertilizers, and other materials are greatly affected by the soil temperature. The greatest activity of soil microorganisms occurs when the soil temperature is in the range from just above freezing to slightly over 110°F (44°C). The activity is most rapid at the higher temperatures. Some organisms convert organic nitrogen fertilizers, such as urea, dried blood, and tankage, to more readily absorbed forms of nitrogen. These organisms are most active at temperatures above 60°F (15.5°C). The rate of nitrogen conversion below this temperature is slower. Thus, the nutrients are not available quickly to the crops. Organic

FIGURE 9-2 Bedding plant plugs should be grown on raised benches to prevent excessive chilling of the root zone. Benching systems need not be complicated as evident in these plug benches constructed out of heavy gage wire and stretched over frames that once supported roses grown for cut flowers.

FIGURE 9-3 Newly potted plants will benefit from warm root-zone temperatures provided by bottom heat circulating from hot water pipes suspended under benches.

fertilizers are not recommended for use on greenhouse crops from late October to late April because of the effect of the lower temperature.

Soil microorganisms also act on sources of ammoniacal nitrogen to convert them from ammonia to nitrites to nitrate forms that are readily absorbed by plants. The rate of these conversions is also slowed by cold soil temperatures. If plants are to be fertilized in dark weather and in winter, at least 50 percent of the nitrogen must be applied in the form of readily absorbed nitrates.

Spring bedding plant crops may have problems in absorbing nutrients from cold soils. Growers often place bedding plant packs directly on the cold ground. This means that the growing containers are cold as well. A nutrient deficiency known as **chlorosis** develops because the plant roots are unable to take up nutrients. To prevent this problem, the containers should be placed on an insulating layer or on raised benches (Figures 9-2 and 9-3).

North American growers in recent years are using root zone heating systems to control soil temperature. This is in addition to systems designed to increase the root-zone temperature for seed germination and vegetative propagation. European flower producers, particularly those in Denmark, Norway, and Sweden, heat the soil because it aids plant growth. They use small plastic pipes spaced 6 to 8 inches apart in the soil at the bottom of the bench. Modern benchtop heating systems use small-diameter tubing. Warm water is pumped through the pipes to keep the soil at the desired temperature (Figure 9-4). The minimum soil temperature is maintained at 60°F (15.5°C) under normal conditions.

In the United States, some rose growers place heating pipes beside ground beds to raise the soil temperature in the winter. However, this method does not result in uniform heat distribution across the bed.

The increasing cost of fuel makes it important to study the benefits of soil heating in more detail. It is possible that quality crops can be grown in cooler air temperatures if warmer soil temperatures are maintained.

FIGURE 9-4 Plastic pipes through which warm water is circulated will raise the temperature of the medium of these foliage plant transplants, which speeds growth.

Effect of Soil Temperature on Bulb Crops

Certain bulb crops require soil temperature controls. For example, Easter lilies, tulips, and narcissi must be exposed to cold temperatures for a definite period of time. The most common method of treating bulbs is to pot them and then place the pots in a bulb cellar, cold frame, or refrigerated room at the proper temperature. Many growers feel that the low temperature is needed to stimulate root growth. However, the cool temperature has the greatest effect on the initiation and development of the flowers.

Flowering Growth

Effects of Temperature on Floral Initiation

Floral initiation (formation of floral parts) and floral development are separate and distinct stages in the life of a plant. When a plant goes from vegetative growth to flowering growth, the resulting change in cell shape and structure is known as initiation. Generally, the pointed, terminal apex of the vegetative meristem changes in appearance to a rounded mushroom or flattened shape. The process of **floral bud initiation** usually occurs within the short period of a few days. Floral bud initiation can be controlled by temperature or by photoperiod as described in the previous chapter.

The next stage in plant growth is **floral bud development**. In this stage, the flowering tissues differentiate into the various floral parts. Much of the development stage is concerned with the increase in size and expansion of the newly initiated floral tissues. The development stage may last for several weeks.

The most important environmental factor that causes flowering in many species is the photoperiod. However, temperature also plays an important role in flower initiation for many other plants. Regardless of day length, *Cypripedium* orchids, Calceolarias, Cinerarias, *Matthiola incana* ('Stock'), and *Pelargonium domesticum* (regal geraniums) initiate flower buds when the temperature range is between 50°F and 60°F (10°C and 15.5°C). Chrysanthemums initiate flower buds at temperatures around 60°F (15.5°C). For some cultivars, the best flower bud development takes place at 60°F (15.5°C). Other cultivars require lower temperatures and still others require temperatures above 60°F

(15.5°C) for flower initiation. Growers select the cultivars that have the best flowering response to temperature at different times of the year.

Snapdragons are even more sensitive to temperature than chrysanthemums. Summer flowering snapdragons bloom best when the daily temperature is 80°F (26.5°C) or above. When these snapdragons are grown under cool conditions, 55°F (12.5°C), only vegetative growth occurs. The plants may reach a height of 6 feet or more and still will not flower unless they are exposed to the required temperature.

Floral initiation in bulb crops, such as iris, narcissi, and hyacinths, takes place rapidly at temperatures between 70°F and 80°F (21°C and 26.5°C). However, flower bud development in these crops is more rapid at cooler temperatures.

Temperatures above 80°F to 90°F (26.5°C to 32°C) are likely to delay the development of flower buds. Poinsettia flower development is greatly delayed if the night temperature is high. This is especially true if the light intensity is low as well. Because light intensity is so important to the flower development and growth of all crops, greenhouse glass should be cleaned in the fall before the natural low light days of winter arrive.

Flower Color and Lasting Life

Both high and low temperatures affect flower color and lasting life. Low temperatures tend to make pigment colors more intense, particularly red, pink, and bronze. If the nighttime temperature is lowered a few degrees, as poinsettias mature, the bracts become a darker red. Pot mums and other crops respond in the same manner. Lowering the temperature a few degrees slows growth, firms tissues, and deepens color. If the temperature is too low, the lower leaves of some plants, such as marigolds, geraniums, and chrysanthemums, turn red or bronze. This reaction to cold often occurs in the spring on crops grown outdoors. For geraniums, however, red lower leaves can also mean a nitrogen deficiency.

Very high temperatures cause flower colors to fade. The color pigments are destroyed after formation or they do not develop at all. Red flowers fade to pink, and bronze and yellow flowers fade to yellow and yellow-white. Bloom in chrysanthemums may be delayed a week or more if the temperature is allowed to exceed 85°C (29.4°C) for extended periods.

Flower crops that are grown at high temperatures prior to harvesting generally have a shorter lasting life than crops grown at cooler temperatures. One possible reason for this condition is that the higher temperature increases the rate of respiration. As a result, there is a reduction in the carbohydrate reserves within the plants. Fewer carbohydrates in the tissues mean that the lasting life is shorter.

ACHIEVEMENT REVIEW

Complete each statement by filling in the blanks with the correct word(s).

1 The three fundamental temperature levels that cover the range over which plants grow are the _____, above which no growth occurs; the _____, where the greatest growth occurs; and the _____, below which no growth takes place.

2 The optimum air temperature for most ornamental crop seeds is in the range between _____ °F and _____ °F.

3 In most plants, photosynthesis stops when the cell temperature exceeds _____ °F.

4 As the temperature rises, the rate of respiration in plants _____.

5 The rate of transpiration _____ as temperature increases.

6 Geraniums and marigold leaves develop a red color as a result of _____ air temperatures.

7 Soil microorganisms that convert organic nitrogen fertilizers are most active in temperatures above _____ °F.

8 Greenhouse temperatures greater than _____ °F to _____ °F will delay the development of poinsettia flower buds.

9 Low temperatures _____ the pigment colors of flowers.

10 Excessively high temperatures cause flower colors to _____.

Gases

OBJECTIVES:

After studying this chapter, the student should have learned:

- the gases in the normal atmosphere.
- why plants need oxygen.
- the effects of carbon dioxide on plants.
- the methods of adding carbon dioxide to the greenhouse atmosphere.
- the effects of various harmful gases on plants.

A large number of gases affect plants. Some gases, such as oxygen and carbon dioxide, are necessary for plant growth. Ethylene, ammonia, and fluorine are examples of gases that are harmful to plants. Substances used to control insects and plant diseases may be harmful to the plants they are supposed to protect. Severe damage to valuable crops, or even the death of the crop, may result from the improper handling of soil fumigants. Serious plant losses have occurred because of the uninformed or excessive use of wood preservatives and weed killers in the greenhouse. The effects on plants of beneficial and harmful gases are described in this chapter.

Oxygen

Atmospheric Oxygen

Humans, animals, and plants all need oxygen to sustain life. There are only a few organisms that can live without oxygen. Such organisms are said to be **anaerobic**; that is, they live in an environment with no free oxygen.

The normal atmosphere consists of many gases. The oxygen (O_2) content of the atmosphere is around 20.9 percent; the nitrogen (N) content is about 78 percent; and the carbon dioxide (CO_2) content is 0.033 percent. The remaining small percentage consists of hydrogen, neon, helium, argon, and other rare gases. The main building blocks of plant tissues are carbohydrates ($C_6H_{12}O_6$). Carbohydrate compounds are manufactured by plants from carbon and oxygen in the air. Hydrogen comes from water absorbed by the roots.

Oxygen is taken from the air by green plants and used in the process of respiration. Oxygen is produced by green plants as a by-product of photosynthesis. All plant cells require oxygen to carry on their life processes.

Although photosynthesis produces oxygen, the process requires oxygen as well. Generally, there is enough oxygen in the **greenhouse atmosphere** to support photosynthesis. However, there may be a lack of oxygen in the winter in a plastic-covered greenhouse where an open flame burner is in use with an unvented heater. In this case, the atmospheric oxygen in the greenhouse is quickly used up by the burning fuel. The plants may be damaged, but this is caused more by the poisonous products of combustion than a lack of oxygen.

Soil Oxygen and Aeration

It is more likely that there will be insufficient oxygen in the soil air. The oxygen content of the soil air is much lower than that of the greenhouse.

The proper amount of oxygen in the soil is influenced by the type of soil mixture used by the grower. The mixture should contain the proper ratios of organic matter, large inorganic particles, and pore space. Pore space is the name given to the empty spaces between soil particles. Water, oxygen, and other gases share these spaces. If there is too little pore space, a shortage of oxygen can develop quickly when too much water is applied. When the plant roots cannot respire (breathe), water uptake and other growth functions stop.

Oxygen is slightly soluble in water, but the rate of diffusion or movement of oxygen through water is very slow. The rate is too slow to meet the needs of plants when the soil is flooded.

Compaction of the soil may result in poor soil aeration. The soil can be compacted by applying water in a heavy stream from a hose directed on the soil. Over a period of time, the surface layers of the soil will pack together. Soil compaction can also occur as a result of stacking newly filled containers. This compaction limits the movement of both water and oxygen in the soil.

Pore space is also reduced by packing the growing medium when potting and repotting container plants. One of the advantages of a peat-lite mix for potted plants is the high pore space available for root development.

In areas where crops are field grown in soils with high clay content, puddling is a problem when the soil is extremely wet. That is, when wet clay soil is worked, the pore spaces containing air become filled with water. As a result, the

FIGURE 10-1 Snapdragon plugs grown for cut flowers are sensitive to low root-zone oxygen caused by soil compaction as a result of excessive steam sterilization or overwatering.

supply of oxygen in the soil is limited (Figure 10-1). Puddling is typically not a problem in greenhouses because growers use media containing large amounts of peat moss, perlite, vermiculite, and other materials. However, this may be an issue in greenhouses where crops are grown directly in ground beds.

Poor soil aeration can reduce the final yield of snapdragons if the seedlings are submerged in water for 24 hours. Both the rooting of cuttings and the germination of seeds can be delayed or prevented by poor soil aeration.

At one time research was conducted on the effects of forcing air through soils for impatiens. However, this practice is not used commercially for any crops.

Granulation

Some crops, such as chrysanthemums, are planted three or four times a year. For each new planting, the soil is prepared by tilling and steam pasteurization. Thus, there is rarely a problem with soil compaction and poor aeration. If steam pasteurization is used as a disease-control measure before planting, both soil drainage and aeration are improved. Steaming causes the soil particles to flocculate (stick together) and form small clumps of granular shapes. This granulation effect increases the pore space.

The use of calcium fertilizers also improves the granulation of soils. The chemical action of calcium binds the microscopic soil particles into small granules. On the other hand, the long-term use of sodium fertilizers, such as nitrate of soda, breaks down the soil particles into individual grains. Poor drainage and aeration result. The problems due to soil dispersion can be avoided by varying the types of fertilizers used on crops.

Carbon Dioxide

Carbon Dioxide Content of the Atmosphere

Carbon dioxide (CO_2) is the third most abundant gas in the atmosphere. For many years the concentration of 0.03 percent or 300 ppm (parts per million) was the average amount measured. As nations became more industrialized, the burning of fossil fuels along with the exhaust gases of automobiles, trucks, airplanes, and trains raised the average content of CO_2 to 330 to 340 ppm or more. Around some industrialized areas the CO_2 content of the air has been measured as high as 0.05 percent or 500 ppm. There is concern among many nations that the continuing increase of CO_2 levels may have harmful effects on all forms of life.

Carbon dioxide is the end product of respiration in both plants and animals. It is also given off when organic matter decays.

Plants use carbon dioxide in photosynthesis. A series of chemical reactions changes carbon dioxide, water, and oxygen into carbohydrates and other organic compounds. These compounds are the basic building blocks of plants.

Supply of Carbon Dioxide in the Greenhouse

In the past, commercial flower growers relied on the atmosphere of the greenhouse to supply the carbon dioxide required by plants for photosynthesis. The supply of carbon dioxide generally was adequate if the greenhouse was ventilated to bring in fresh air. During the winter, however, the ventilators would be closed for days or weeks. On sunny days, plants quickly used the carbon dioxide in the greenhouse for photosynthesis. A lack of CO_2 then became a limiting factor for growth as food production decreased. At the time, the growers did not know that there was too little CO_2 for optimum growth. They knew only that their plants were not growing as rapidly as they should be.

After the introduction of fan and pad cooling as a means of ventilating the greenhouse, some growers found that they could use the exhaust fan to exchange a very small amount of air on clear, cold winter days. The fans were run at a low speed to obtain a slow change of air in the greenhouse. Fresh air was drawn in through the laps of the glass. Whenever the fans were turned on during the day, even for only a few minutes, better crop growth resulted. It is now recognized that the renewal of the CO_2 in the greenhouse atmosphere caused the improved growth.

Artificial Methods of Increasing Carbon Dioxide Levels in Greenhouses

At the same time that growers were noting the effect of ventilation on plant growth, researchers began to experiment with increasing CO_2 levels in the greenhouse atmosphere by artificial means. The results of their experiments were quickly applied by commercial rose growers to increase production.

Dry Ice

One of the first methods of supplying CO_2 used large insulated tanks filled with dry ice. (Dry ice is carbon dioxide gas that has been chilled to a very low temperature and compressed.) As the dry ice **sublimed**, or changed from a solid to a gas, the CO_2 gas was piped into the greenhouse and mixed with the air by fans. During the daylight hours from late November to early April, the CO_2 concentrations were maintained at 1,000 to 1,200 ppm or more. For the rest of the year, CO_2 was not added because the ventilators were opened to reduce the heat levels and the CO_2 would have escaped to the outside air.

Rose growers and carnation growers were among the first in the industry to use carbon dioxide enrichment for their crops. They reported stronger and longer stems, larger flowers, and a larger overall yield with the addition of carbon dioxide.

FIGURE 10-2 This unit burns propane gas to produce carbon dioxide for improving plant growth.

FIGURE 10-3 Storage tank for liquefied carbon dioxide.

Propane Gas

The dry ice technique was expensive and bothersome. Thus, research continued into other methods of adding CO_2 to the greenhouse atmosphere. It was found that **propane gas** burned completely in the presence of oxygen to produce carbon dioxide, water, and heat. Each pound of propane gas burned gave off 3 pounds of carbon dioxide and 1.6 pounds of water.

Several types of burners are used. Some burners are installed outside the greenhouse, and the gas is brought inside through a stovepipe. One such burner, the Hy-Lo® unit (Figure 10-2), can be located throughout the greenhouse. The CO_2 output of each unit is known. Thus, it is possible to calculate the number of burners needed to maintain a CO_2 level of 1,000 to 1,200 ppm.

The practice of adding carbon dioxide to the greenhouse atmosphere is not commonly used today. The reasons for its decline are unknown. Several rose growers with large ranges are the only flower growers who still add CO_2 to the greenhouses in winter. Greenhouse vegetable growers regularly add CO_2 to increase productivity. These growers use either liquid CO_2 or CO_2 capture equipment installed onto their boiler exhaust stacks (Figure 10-3).

Carbon Dioxide Content of the Soil

The CO_2 content of the soil is much higher than that of the greenhouse atmosphere. Average soil concentrations range from 0.20 to 0.25 percent. However, much higher levels may occur under certain conditions. The carbon dioxide in the soil comes from two sources. The decay of organic matter provides the greatest amount of CO_2. Respiration by the plant roots contributes a small amount.

Carbon dioxide moves slowly through the soil by a process called **diffusion**. There is an exchange of gases at the soil surface. Carbon dioxide enters the air from the soil, and oxygen enters the soil from the air. The carbon dioxide given off to the air is used by the plant in photosynthesis.

Carbon dioxide in the soil may combine with the hydrogen ions (H^+) from the soil water to form carbonic acid (H_2CO_3). In addition, carbonates and bicarbonates of calcium, potassium, magnesium, and other minerals are formed as well. The role of carbonic acid in the process that makes nutrient elements available for absorption by plants is described later in this text.

If the level of carbon dioxide becomes too great, plant growth may be retarded. An excess of carbon dioxide in the soil usually causes the oxygen content of the soil to be lowered. This factor is the main reason for poor growth. Poor drainage and overwatering, leading to poor aeration, result in a high carbon dioxide level and a low oxygen level.

Harmful Gases

Plants may be damaged by harmful or toxic gases in the greenhouse.

Natural Gas

Natural gas is used mainly for heating. Plant damage can occur if the gas enters the greenhouse air because of improper burning of the fuel or leaks in the heating system. Leaks in the supply line outside the greenhouse may also allow natural gas to mix with the greenhouse air.

Poor combustion often is a problem in plastic-covered greenhouses. Because there is little ventilation, too little makeup air reaches the burner flame.

Plant damage from leaks in the supply system may occur in the winter or early spring. During this period, the ground is frozen and the greenhouse is not vented. Gas from a broken main can travel beneath the frozen crust of the soil and then surface in the unfrozen soil of the greenhouse. In some cases, natural gas has traveled several thousand feet in this manner.

Effect on Plants

Only a small amount of natural gas is needed to cause plant damage. One part of gas in 100,000 parts of air, or 10 ppm, is sufficient to damage certain crops. It is difficult to detect such a small amount of gas. However, a biological detector is one of the best methods. Both tomato and marigold plants are sensitive to gas in the air in concentrations of just 1 ppm. Plants exposed to natural gas suffer an injury called **epinasty**. This disfigured curling of the leaves results from the faster growth of the tissue on the upper side of the petioles. Castor beans are also used because they are sensitive to a gas concentration of 2.5 ppm.

Natural gas exposure causes carnation flowers to go to sleep; the flower petals turn upward at the ends. The flower buds of roses may open prematurely. High concentrations of gas cause roses to drop a large number of leaves.

Ethylene

Flowers, fruits, green plants, and vegetables all produce small amounts of ethylene gas. The growth process of fruit aging and ripening uses ethylene gas. Some fruits that must be picked green so they can be shipped long distances without damage, such as bananas, are ripened artificially by exposure to ethylene gas.

Snapdragons and calla lilies give off ethylene gas. Snapdragons react to the gas by dropping their florets. This response is called petal shatter. Calla lilies should not be packed for shipment in the same box with other flowers.

FIGURE 10-4A Carnation blooms exposed to increasing levels of ethylene gas. *(Courtesy Dr. Joe J. Hanan, Emeritus Professor of Floriculture, Colorado State University.)*

FIGURE 10-4B Carnation leaves with excessive downward curl or epinasty in response to ethylene gas. *(Courtesy of Dr. Joe J. Hanan, Emeritus Professor of Floriculture, Colorado State University.)*

Because they give off large amounts of ethylene gas, the concentration of gas can increase in a shipping box to the point of damaging other flowers packed in the same box.

Exposure to ethylene causes carnations to go to sleep. It has been noted that high concentrations of ethylene at airports can damage carnations in transit. To overcome this problem, ethylene absorbents have been placed in the shipping boxes. However, this remedy is only partially successful (Figures 10-4A and 10-4B). More effectively, 1-methylcyclopropene (1-MCP) sold as EthylBloc® is an ethylene action inhibitor. EthylBloc effectively stops the negative effects of ethylene which include premature wilting, leaf yellowing, premature opening, petal shatter, and premature death. It is easily applied as a gas to trucks or storage units.

Fruits produce large amounts of ethylene. Fruits and flowers should never be shipped together or stored in the same refrigerator.

Certain greens also produce large amounts of ethylene. For example, arborvitae, which is used in floral designs, should be refrigerated at a temperature below 40°F (4.5°C). These greens can also be kept in a separate cooler. At temperatures below 40°F (4.5°C), plant tissues give off little, if any, ethylene gas.

Controlled applications of ethylene are used on certain crops to hasten their flowering and fruiting. Pineapples are treated in the field so that all of the fruit ripens at the same time. In this way the fruit can all be harvested together. Bromeliads are forced to flower by exposure to small amounts of ethylene gas.

Florel® is used commercially to speed the development of ethylene in a plant. This product is applied to fruit and flower crops to stimulate their development. It is also used to stimulate branching. During geranium stock plant production, Florel® is applied to prevent flower development causing stronger vegetative growth.

Ammonia

Many years ago, the frequency of plant injury from ammonia gas was much higher than it is now. At that time, rose growers used manure to mulch their rose beds. Because decaying manure gave off large amounts of ammonia, the roses often suffered damage from the gas. In the modern greenhouse, ammonia is given off when soils containing large amounts of organic matter are steam sterilized at excessively high temperatures.

Mercury

Mercury is a heavy metal that once was used in many compounds applied to plants to control disease. It was also once a common component of paint. However, mercury has a number of harmful effects, and it lasts a long time in plant tissues. As a result, the Environmental Protection Agency (EPA) has banned the use of products containing mercury. The major sources of mercury in the modern greenhouse are equipment controls and thermometers. Automatic ventilator machines usually have mercury cutoff switches. If the container holding the mercury is broken, then mercury contaminates the greenhouse environment. High intensity discharge lamps containing mercury are another source of mercury that could damage crops in the greenhouse.

Roses are severely damaged by mercury when partly opened flowers do not continue to develop; the flower color darkens, and the leaves are scorched; and the peduncles (stems) of young buds turn yellow and later black. When mercury contamination occurs, as much mercury as possible must be removed. The area is then covered with a heavy layer of iron filings. The iron ties up the mercury and prevents further damage. The collected mercury must then be properly disposed through an approved hazardous waste management provider.

Wood Preservatives

Pentachlorophenol

Pentachlorophenol consists of chlorinated phenols dissolved in distilled petroleum solvents. This material usually contains about 5 percent by weight of chlorinated phenols. However, the amount may be higher or lower depending upon the solvent used.

Creosote

The distillation of various tar materials yields a number of creosote wood preservatives, including coal tar creosote, wood tar creosote, oil tar creosote, and watergas tar creosote. All of these preservatives are highly poisonous to plants and should not be used in greenhouses. The precautions given previously for pentachlorophenols also apply to creosote products.

If pentachlorophenol and creosote wood preservatives are used by mistake to treat greenhouse wood, a special paint can be applied over the preservative. The paint is known as B-I-N and can be purchased at any paint store. It is a flat white primer-sealer. Two coats must be applied to seal in the fumes from the preservative. The first coat must dry for at least 1 hour before the second coat is applied.

Once B-I-N is used on the treated wood, any paint, except polyurethanes, may be used to cover it. White is the recommended color for greenhouse wood.

Paints

It has been determined that some recent types of plant injuries are due to the use of certain paints in the greenhouse. For example, antirust paints are often used on steam pipes. When the steam is turned on for winter heating, the heat releases fumes from the chemicals in the paints. These volatile products have been identified as xylol and xylenes. These materials are poisonous to all plants, but are particularly damaging to roses.

For interior greenhouse use, it is recommended that the grower apply only those paints obtained from a reliable greenhouse equipment and product supplier.

Other Air Pollutants

General air pollution is now a serious problem throughout the world. Plants are one of the most sensitive living indicators of air pollution. Plants must be exposed to large volumes of air to obtain the proper amounts of carbon dioxide needed for photosynthesis. Thus, plants are especially sensitive to pollutants in the air.

There are four major **air pollutants** that affect the growth of plants: nitrogen dioxide, peroxyacetyl nitrate (PAN), ozone, and sulfur dioxide. Fluorine also affects plants to a lesser degree.

Nitrogen Dioxide

Nitrogen dioxide is a product of the burning of fuels. Generating plants supplying electrical energy produce large amounts of nitrogen dioxide. However, the source of the greatest amount of nitrogen dioxide in the atmosphere is automobile exhausts. Nearly 70 percent of the total amount of nitrogen dioxide added to the air is due to automobile exhaust fumes.

Plants respond to a nitrogen dioxide concentration of 10 to 200 ppm. Depending upon the type of plant, an exposure of 1 to 8 hours causes tissue collapse, bleaching, death of cells, and up to 100 percent leaf loss. The effects may be cumulative; that is, low concentrations of nitrogen dioxide that fail to cause injury over a period of 3 days may result in death on the fourth day.

Peroxyacetyl Nitrate

When sunlight acts on hydrocarbons and oxides of nitrogen, peroxyacetyl nitrate (PAN) is formed. Automobile exhausts are the main source of hydrocarbons and oxides of nitrogen. Exhausts also add carbon monoxide, carbon dioxide, and nitrogen dioxide to the air. All of these materials make up a mixture of pollutants called photochemical smog.

A PAN concentration of just 0.01 ppm causes reduced yield in sensitive plants such as tobacco and petunias. A concentration of 0.1 ppm destroys chloroplasts and causes lesions to develop. Another reaction to PAN at concentrations far below those that cause death is a weakening of the processes of respiration, photosynthesis, ion absorption, carbohydrate and protein synthesis, and enzyme activities (Figure 10-5).

FIGURE 10-5 Aster blooms damaged from peroxyacetyle nitrate.

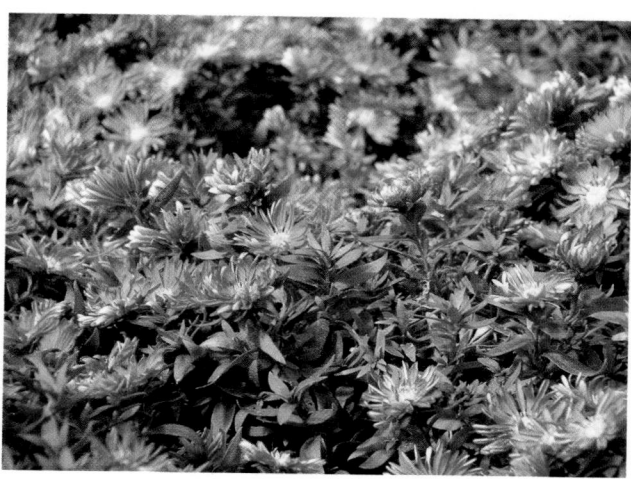

Ozone

Scientists are concerned about the harmful effects of aerosol sprays on the ozone layer in the upper atmosphere. However, little attention is being given to the amount of ozone in the air in and around cities. Ultraviolet light acts on nitrogen dioxide to produce ozone. The normal level of ozone in the air is 0.02 ppm. Around cities, however, ozone concentration is more than 0.05 ppm. Plant damage occurs at this level.

Ozone enters leaves through the stomata. Once inside the leaf, it damages cell membranes, increases permeability and respiration, reduces photosynthesis, and destroys enzymes. Even when the concentration of ozone is below the level at which tissue death occurs, photosynthesis may be reduced by 40 to 70 percent.

High ozone concentrations also occur where the level of automobile exhaust is high. This situation has affected some orchid growers. Because orchids are especially sensitive to ozone and other air pollutants, growers located near urban shopping centers have been forced to move.

Sulfur Dioxide

Oil and coal contain sulfur compounds. When these fuels are burned, sulfur dioxide (SO_2) is released into the atmosphere. Sulfur dioxide enters leaves through the stomata. It causes the breakdown of chloroplasts, plasmolysis, and the eventual death of mesophyll cells. The cells and tissues destroyed by SO_2 have a distinctive white color. An SO_2 concentration of 0.1 to 0.3 ppm reduces the rate of photosynthesis and retards general growth.

Sulfur dioxide in a humid atmosphere combines with water to form weak sulfuric acid. This acid corrodes metals and causes the erosion of marble, limestone, and other alkaline building materials.

Fluoride Toxicity

Fluorides are by-products of the aluminum industry and processes used in the manufacture of fertilizers, especially super phosphates. A fluoride concentration of just 5 parts per billion causes the leaf size to be 25 to 35 percent smaller than normal. The symptoms of fluoride exposure in most plants are the destruction of chlorophyll and smaller leaf size. Fluoride toxicity seems to be cumulative in that it builds up over a period of time. Thus, a plant will eventually die if exposed to low concentrations of fluoride for a long time.

Many cities and towns add fluoride to their water supplies to help prevent dental cavities (decay) in children's teeth. The usual amount added is 1 ppm. Members of the plant family *Liliaceae* (lily) are injured by concentrations of as little as 0.25 ppm of fluoride in the water supply. The injury is a necrosis (dying) of the tips of the leaves. *Chlorophytum comosum* (spider plant) and *Aspidistra elatior* (cast-iron plant) are two examples of plants sensitive to fluoride.

Soil Fumigants

Chapter 13 describes the common soil fumigants chloropicrin (tear gas), methyl bromide, Vapam, ethylene dibromide, and formaldehyde. These chemicals are poisonous to soil organisms. They are also poisonous to other animals, humans, and plants. Fumigants are valuable to the flower grower if they are used properly. However, if used incorrectly, they can cause severe damage to existing crops. The manufacturer's directions must be followed exactly when applying these materials.

Herbicides

Herbicides or weed killers are also a valuable tool to the flower producer when used properly. A limited number of weed killers are recommended for use inside the greenhouse. These materials can be used safely around growing crops if the directions are followed exactly.

Several other types of herbicides must not be used in or around a greenhouse if there is a possibility that the fumes may drift into the house. Severe plant damage is caused by 2, 4-D and similar volatile growth regulator types of weed killers.

Containers of 2, 4-D should not be stored in the head house or other area near the range. Metal containers quickly rust, and the fumes of the herbicide can leak through and drift into the production areas, causing extensive crop loss.

ACHIEVEMENT REVIEW

Complete each statement by filling in the blanks with the correct word(s).

1 The oxygen content of the atmosphere is about _____ percent.

2 Carbon dioxide is present in the air at an average concentration of _____ to _____ parts per million.

3 Soil _____ space is an area where oxygen and gases share a location with water in the soil.

4 One of the most important advantages of the peat-lite mixes is the large amount of _____ they provide for root development.

5 Overworking of heavy clay soils when they are excessively wet causes a condition called _____.

6 Carbon dioxide is one end product of _____ in both plants and animals.

7 Dry ice is greatly chilled and _____ carbon dioxide gas.

8 A disfiguring curl of the leaves as a result of exposure to natural gas fumes is called _____.

9 Ethylene gas causes snapdragons to drop their _____.

10 Carnations exposed to ethylene respond by going to _____.

11 Apples and oranges are _____ producers of ethylene gas.

12 Two wood preservatives that must not be used in greenhouse construction or for treating wooden flats or other wooden items in a greenhouse are _____ and _____.

13 About 70 percent of the total amount of the nitrogen dioxide in the atmosphere comes from _____.

14 Sulfur dioxide injury to leaves causes a distinctive _____ color to develop in the cells and tissues that are destroyed.

SECTION

4

Growing Media

Soils: Properties and Makeup

OBJECTIVES:

After studying this chapter, the student should have learned:

- the uses of soil.
- the proportions of solids, water, and gases in soils.
- the meaning of soil texture.
- the various types of organic matter added to soils.
- the inorganic materials added to soils.

Growing Media

Commercial greenhouse flower crops are not grown in soils alone. They are grown in mixtures of soil, **peat moss**, **perlite**, sand, **vermiculite**, sawdust, wood chips, and other products. Soil is only one part of the total mass that is called a **growing medium**. A **medium** (plural: *media* or *mediums*) is that substance in which the growing of a crop is accomplished.

The growing medium used by the greenhouse grower is different from the soil in the fields where the farmer grows crops. Soils cannot be removed from the field and placed directly in a pot or greenhouse bench. Other materials must be added to field soil to make a good growing medium.

In the field, the soil has certain advantages that are lost if the soil is removed from the field. For example, field soil has long, unbroken, tube-like passageways that reach from the upper layers of soil down to the water table. In some types of soils, these **capillary tubes** are 15 to 20 feet (4.57 to 6.10 meters) long.

These capillary tubes help to remove extra water from the upper layers of soil after a rain. When it is dry, the tubes bring water from deep in the soil to the upper layers where plant roots can take it up.

When soil is removed from the field to a pot or a greenhouse bench, the long capillary tubes are destroyed. Thus, the soil does not function in the same way it did in the field. There is no longer a tube system to remove the water from the upper soil layers. Unless the soil is loosened or opened up by the addition of sand, perlite, peat moss, or other matter, it stays too wet.

Perched Water Table

When field soil is placed in a pot, a condition develops that is known as a **perched water table**. This means that all of the air spaces at the bottom of the pot are filled with water because the drainage is so poor. With the soil tubes destroyed, there is no capillary action to pull the water from the soil to drain it.

The perched water table and the capillary tube action can be demonstrated using a common cellulose sponge. The sponge is soaked until it is full of water. It is not squeezed but is held carefully with its long axis horizontal. In this position, a certain amount of water will drain from the sponge due to gravity. The force pulling the water from the sponge is also called the suction force. Although the water will stop dripping from the sponge, there is still more water in it.

A close inspection of the sponge will show that there is an area at the bottom that is saturated with water. This area is about 0.25 inch (0.635 cm) deep and is known as the perched water table (Figure 11-1).

The sponge is then turned so that the long axis is straight up and down (vertical). More water drains from the sponge. The increased drainage is due to the longer capillary tubes formed when the sponge was turned. The longer tubes mean that there is an increased pull or suction force on the water.

The flower grower avoids the problem of poor drainage and a perched water table by adding materials to the soil to make it more loose and open. These materials are discussed in more detail later in this chapter.

Functions of a Growing Medium

One of the most important functions a medium performs is to hold the plant upright. Without the **mechanical support** the medium gives, the plant will fall over. For example, when plants are grown in a **hydroponic solution** (soilless liquid nutrient), they are held up by wires. If they were not supported in

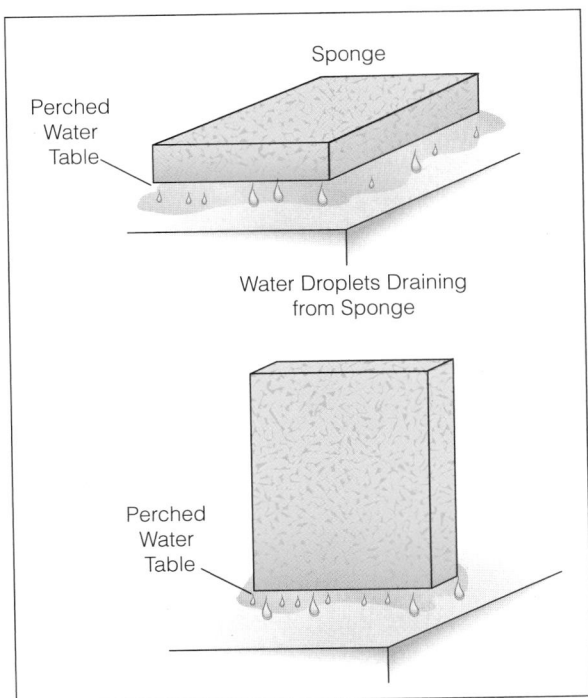

FIGURE 11-1 A cellulose sponge illustrates the condition known as the perched water table. When the wet sponge is flat, water drains to the bottom surface and from the sponge. When the sponge is turned on end, more water drains from it. The perched water table is at the bottom of the sponge in both positions.

this manner, the plants would fall over and either float or sink to the bottom of the tank.

The growing medium also stores minerals for plant nutrition and water for growth, and serves as a home for many living organisms. All of these factors have a constant changing relationship with each other. The environment of the medium is never static but is always changing.

Composition of the Growing Medium

The medium consists of three phases or parts.

1. The solid phase consists of all organic and inorganic materials. This phase is normally 50 percent of the total volume of the medium.
2. The liquid phase is made up of the soil solution. In an ideal medium, the liquid phase is 25 percent of the total volume.
3. Soil air is the remaining 25 percent of the volume. This phase contains all of the soil gases, primarily carbon dioxide, oxygen, and nitrogen. A number of other gases are present in such small amounts that they are neglected.

The relative amount of each of the phases in the medium greatly affects the plant growth. If the liquid phase increases too much as compared to the gas phase, the medium is said to be waterlogged. Poor drainage is a cause of water-logging. A growing medium that has a large amount of fine particles is easily overwatered.

When the gas phase is out of balance with the liquid phase, the medium is said to be very porous. It may be so porous that it does not hold enough water for good plant growth. A porous medium must be irrigated more often than is desired.

Greenhouse media are not at all like field soils. However, the student should be aware of the basic structure of true soils.

Soil Basics

Scientists who work with soils use terms such as *texture* and *structure* to describe the quality of the soil.

Soil Texture

Soil texture refers to the size of the various soil particles. These classes of texture, from the largest size to the smallest size, are gravel, sand, silt, and clay. There are variations in each class, which range from coarse to fine.

Clay particles are so fine that they cannot be seen unless greatly magnified. Soil researchers use electron microscopes and X-ray equipment to examine clay soil particles.

Most soils are mixtures of the various classes of particles. If a soil consists of at least two classes of particles, it is called a loam. It may be a gravelly loam, a sandy loam, a silt loam, or a clay loam. Its name depends upon the size of the most numerous particle. Figure 11-2 and Table 11-1 show how soils are classified by texture and named according to the most common particle.

Organic soils, such as peat soils and muck soils, contain a large amount of organic matter. To be classified as a peat soil, more than 50 percent of the

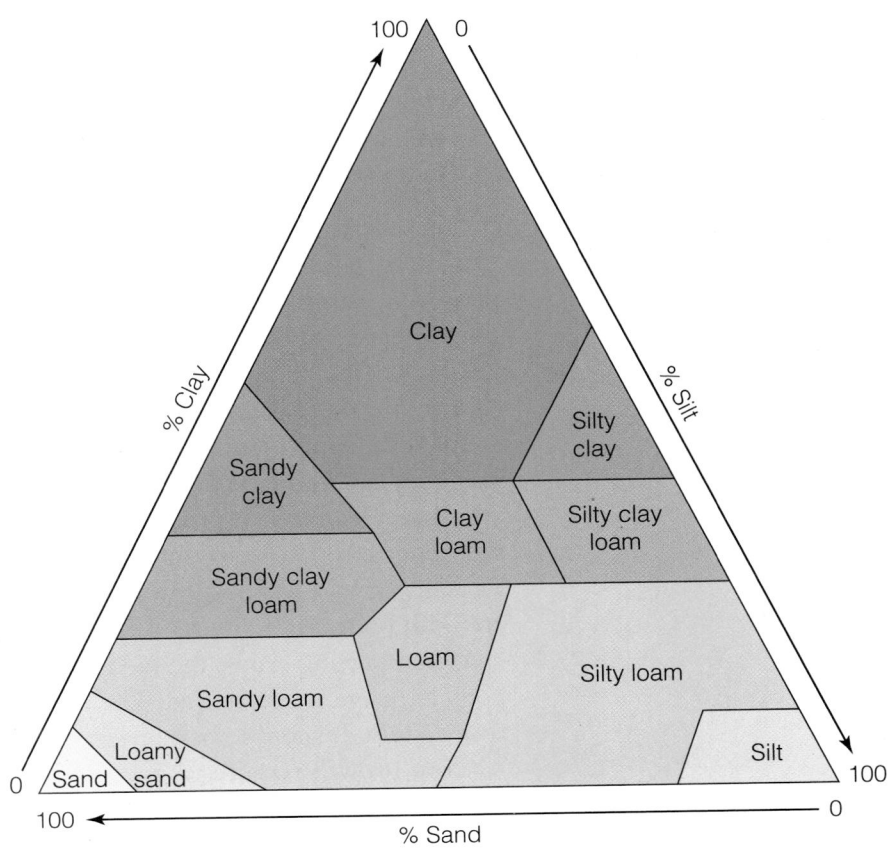

FIGURE 11-2 The soil triangle is used for classifying soils.

Table 11-1

Soil Textures	
Grade	**Type of Material**
Fine Texture	Clay, silty clay, clay sand, or sandy clay
Medium Texture	Loam, silt loam, and clay loam
Coarse Texture	Gravelly sand, coarse sand, sandy loam, coarse sandy loam, and others

total mass must be organic matter. A muck soil contains 20 to 50 percent organic matter, and the remainder is inorganic. The inorganic matter is usually a combination of clay, silt, sand, and some gravel.

Soil Structure

Soils are built up of units of various sizes and shapes. These units are to the soil what concrete blocks are to a wall made of such blocks. The term **soil structure** refers to the arrangement and grouping of the soil particles.

There are four main groups of structural shapes found in soils. These shapes are described as platelike, prismlike, blocklike, and spherical (round). There are also variations of each of these groups.

- Platelike units are similar to thin, flat plates stacked on top of each other.
- Prismlike units look like a series of rods or pillars that stand up straight. Some of these rods have flat tops, and others are rounded. Generally, the rods are six sided.

- Blocklike units look like blocks. They range in width from less than a fraction of an inch to several inches.
- Spherical or round soil particles are like grains of sugar or crumbs of bread. Usually, they are not more than 0.5 inch wide. Soils having a good crumb-like (granular) structure drain well and have good **aeration.**

These structural units can be changed greatly if the soil is plowed or worked when it is too wet. For example, if a fine-textured clay soil is too wet when it is plowed or rototilled, it becomes puddled. A **puddled soil** is one in which the liquid phase and the gas phase are unbalanced. That is, the soil is so tight and heavy that the aeration and drainage are poor and very little air space is left. As a result, there is less oxygen in the soil, and the plant growth is poor.

Even though greenhouse soils do not usually contain large amounts of fine clay, they can develop poor aeration. For example, heavy watering of crops with a hose can lead to this problem. The stream of water falling on the soil causes the upper layers to compact. That is, fine soil particles fill the pore spaces. As a result, air and water fail to penetrate the soil and the drainage and aeration are both poor.

When potting crops, the medium can be packed so much that aeration and good drainage are destroyed. The potting medium should be firmed around the root system. It should not be packed so hard that good drainage is impossible.

Crop growth requires a good soil structure. The soil structure affects the water penetration and drainage, the supply and ease of water uptake, the uptake of nutrient elements, aeration, and the way in which the plant roots grow.

A single-grain, open, porous soil structure permits free movement within the soil mass of air, water, and substances dissolved in the soil water. It is easy for plant roots to grow through such a soil. Certain crops need an open soil structure for best growth. Sands and sandy loams can be tilled when they are dry or very wet without damaging the structure of the soil.

Sandy soils are loose and porous and do not hold much water. They must be irrigated often. Natural fertilizers or plant nutrients added to sandy soils leach (wash) out easily. Because they are very low in nutrients, sandy soils must be fertilized regularly to grow good crops.

Organic Fraction of the Soil

The organic fraction of the soil is made up of plant roots, insects, bacteria, fungi, worms, and other small soil animals. When these soil organisms, insects, and animals die. Their decaying remains, as well as the leaves, stems, and branches that fall to the ground from nearby trees and vegetation, form most of the organic matter of the soil.

Humus is organic matter that is in an advanced state of decay. Humus particles are fine. When dry, humus is like coarse dust. Fibers of the parent material cannot be seen in true humus. The features of a soil can be greatly changed if large amounts of humus are added to it.

The organic fraction of the soil aids general plant growth by holding and storing large amounts of water, fertilizers, nutrients, and other matter dissolved in the soil water. As organic matter decays, it releases nitrogen, phosphorus, and sulfur, which are taken up by the plants. Carbon dioxide is also released. It combines with water to form weak acids. These soil acids break down the mineral parts of the soil to release other nutrients that are used by plants. These **plant nutrients** include calcium, potassium, magnesium, phosphorus, and certain trace elements. A large part of the nutrient needs of the plant are supplied by organic matter.

When organic matter is added to the mineral parts of the soil, the condition of the whole mass is improved. Organic matter loosens heavy soils and makes them more granular and porous. The properties of water absorption, aeration, underdrainage, and root penetration are all improved. Soil with organic matter resists being compacted.

In sandy soils, organic matter acts like glue to bind the mineral soil particles together. There is an improvement in the water- and nutrient-holding ability of the soil. An increase in the organic content of a sandy soil reduces the loss of nutrients through leaching.

The activity of soil microbes and other organisms benefits plant growth. Their activity is increased when organic matter is added to the soil.

Types of Organic Matter

The peat mosses, or peat, are the most important of the many types of organic matter that can be added to soils.

Peat Moss

Peat moss is the remains of dead plant materials that build up in layers for hundreds of years. It is estimated that a 1-millimeter depth of compressed peat moss (25.4 millimeters equal 1 inch) is equal to a year's growth of the *Sphagnum* moss species (Figure 11-3). Peat lands are found in all parts of the world, except deserts and arctic regions. Peat lands are mostly in the Northern Hemisphere, comprising 1 billion acres or 4.5 percent of the earth's land area. There are five important species of sphagnum mosses.

Sphagnum moss peat used in the United States is imported mostly from Canada (Figures 11-4A and 11-4B). The northern parts of some states bordering on Canada contain some bogs where sphagnum moss grows. However, the amount of peat obtained from these **peat bogs** is too little to meet the demands of American horticulture.

Sphagnum moss peat is also imported from Europe, where it is found in large amounts. Over one-third of the land area of Finland is covered with peat

FIGURE 11-3 Sphagnum moss—the main plant of peat bogs in the north. Sphagnum peat is acidic and an important part of horticultural potting mixes. *(Courtesy Premier Horticulture, Inc.)*

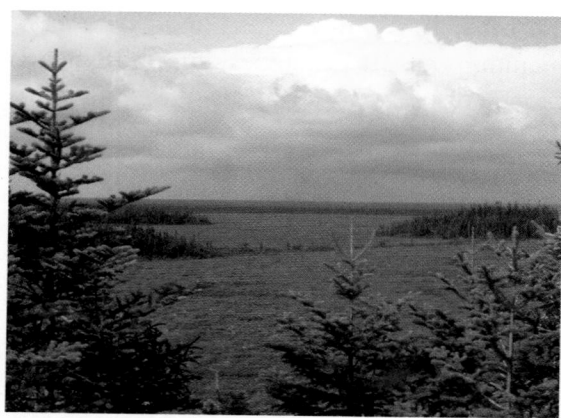

FIGURE 11-4A A Canadian peat bog before harvest. *(Courtesy Premier Horticulture, Inc.)*

FIGURE 11-4B A closer look at sphagnum moss growing in a Canadian peat bog. *(Courtesy Premier Horticulture, Inc.)*

FIGURE 11-5A Harvesting of sphagnum peat moss on a Canadian peat bog. *(Courtesy of Premier Horticulture, Inc.)*

FIGURE 11-5B Vacuum harvesting sphagnum peat moss on a Canadian peat bog. The surface is first loosened with a harrow. *(Courtesy of Premier Horticulture, Inc.)*

moss. About 14 percent of the land area of both Ireland and Sweden is also covered with peat moss. The Soviet Union has 64 percent of the world's supply of peat. The United States (not including Alaska) accounts for 5 percent.

Harvesting Sphagnum Peat Moss

In the past, all peat moss was harvested by hand. First, the bogs were trenched to drain the water. After a year of draining, workers cut the peat into blocks and stacked them to dry. The blocks were allowed to freeze during the winter to rupture the cells of the moss and thus increase its water-holding ability.

Labor is so costly today that peat harvesting is mechanized. The surface 3 to 4 inches of the bog is loosened through the use of a spike-tooth type harrow. After being allowed to air dry for 2 or 3 days, it is then vacuumed by huge machines (Figures 11-5A and 11-5B). The price of sphagnum peat moss is often controlled by the length of the summer harvesting period as well as the price of fuel for shipping from Canada.

In Ireland, sphagnum moss peat is harvested for horticultural export. It is also compressed into small bricks for local use as a fuel to heat homes. The peat bricks are used in place of coal in open fireplaces.

Properties of Sphagnum Moss Peat

Sphagnum moss peat has certain physical and chemical properties that make it important for use in horticulture. The age of the peat moss and its degree of decomposition have an important effect on its properties.

A method of measuring the degree of decomposition of peat moss was devised by a Swedish scientist named von Post. Using a scale of H1 to H10, von Post assigned a value of H1 to un-decomposed peat moss, or moss at the top of the bog. Completely decomposed peat moss, which contains no visible fibers, at the bottom of the bog was classed as H10. H10 peat moss has a wet, coallike appearance. It is often cut, dried, and used directly for fuel.

The best horticultural peat moss comes from layers H3, H4, and H5. These grades have good fiber content because they are not too decomposed. Peat moss from these layers has good physical qualities for horticultural use (Figures 11-6A through11-6E).

Dry peat moss is fairly light in weight. It weighs about 6.5 pounds per cubic foot (104 kg/m^3). When fully saturated with water, peat moss weighs between 97 and 130 pounds per cubic foot (1,556.8 and 2,086 kg/m^3). Fibrous peats can hold 15 to 20 times their own weight of water. This water-holding ability is one of the reasons peat moss is so valuable to horticulture. By mixing peat moss with soil, vermiculite, perlite, and other materials, the ability of the medium to hold water is greatly increased.

Peat moss also has a large amount of pore space. Since air and water are held in the pore space, a good growing medium must have a high porosity. The porosity of peat moss ranges from 85 to 98 percent of the total volume.

In a greenhouse medium, a total pore space of not less than 50 percent by volume is desired. If peat moss is a large part of the medium, then the required pore space is obtained.

The pH of most sphagnum moss peat used in horticulture is in the range of 4.0 to 4.5. (As described in Chapter 14, the pH is a measure of the alkalinity and acidity of the medium.) A pH of 4.0 is very acid. Some peat moss may have a pH less than 4.0 (more acid). Very few sphagnum mosses have a pH of less than 3.5.

Some sphagnum moss peats have a pH of 6.5 to 7.0 and even higher (alkaline). To determine the true pH, the peat moss must be tested.

Other Peats

Hypnum, reed, and sedge are three other major types of peat moss (Figure 11-7). These mosses are much more decomposed than sphagnum peat moss when offered for sale. Because of this, the benefits of using them are not as great as for sphagnum moss.

The pH of these peats ranges from a very acid 4.0 to an alkaline 7.5 or above. There are large differences in the nutrient and chemical content of these peat mosses as well. Although these peats are good for many horticultural uses, commercial growers prefer sphagnum peat moss.

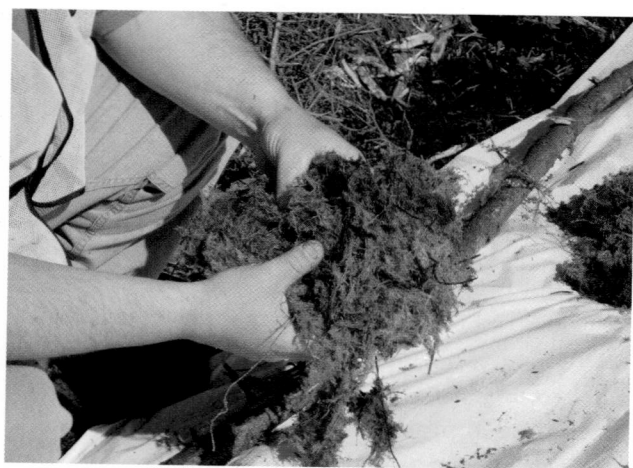

FIGURE 11-6A Sphagnum peat moss grade H1, which is undecomposed peat moss. *(Courtesy of Premier Horticulture, Inc.)*

FIGURE 11-6B Sphagnum peat moss grade H3, which is moderately decomposed peat moss. *(Courtesy of Premier Horticulture, Inc.)*

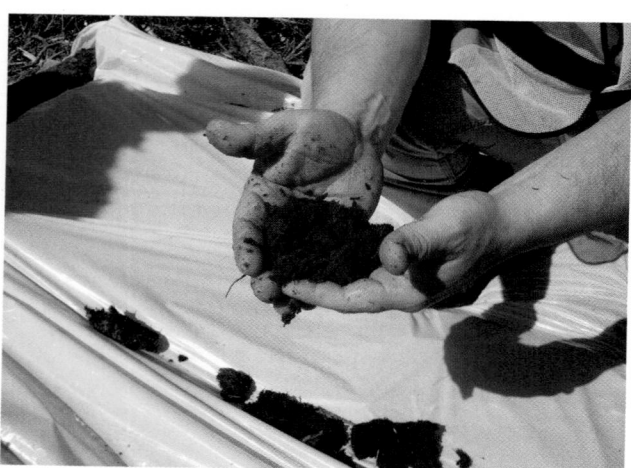

FIGURE 11-6C Sphagnum peat moss grade H4. As the H number increases, the level of decomposition increases. *(Courtesy of Premier Horticulture, Inc.)*

FIGURE 11-6D Sphagnum peat moss grade H5. As the H number increases, the level of decomposition increases. *(Courtesy of Premier Horticulture, Inc.)*

FIGURE 11-6E Sphagnum peat moss grade H6. As the H number increases, the level of decomposition increases. *(Courtesy of Premier Horticulture, Inc.)*

FIGURE 11-7 Reed sedge peat. *(Courtesy of Premier Horticulture, Inc.)*

Manures

Manures are the remains of animal wastes and contain a large amount of organic matter. Manures are often mixed with straw, wood chips, or other types of bedding materials. Manures are rarely used now in greenhouse crop production because they are not easily obtained except in regions with an established confined animal feeding industry. They have a low nutrient content for the amount that must be used. For example, 1 ton of good stable manure has the same fertilizer content as a 100-pound (45.4 kg) bag of 10-5-10 inorganic fertilizer.

Manure added to a flower crop medium may result in a large amount of soluble salts that can damage the crop (see Chapters 15 and 16).

If manure is used, it is never applied fresh to any greenhouse crop because it can burn the crop. Manure must be composted with soil for at least 1 year before it is safe for use on greenhouse crops.

Leaf Mold

Leaf mold is the rotted remains of leaves. It may be added to a greenhouse growing medium. The amount used is often determined by its availability. Some growers arrange with city street and parks departments to take all of the leaves that they collect. The leaves are piled and allowed to decompose. After the piles are turned over two or three times to hasten the breakdown of the leaves, the material is used.

The leaves may be contaminated by salts from street treatments to remove snow and ice or by weed killers used on lawns. As the leaves decompose, the chemicals are concentrated. When the leaves are used in the greenhouse, the chemicals may be strong enough to cause plant damage. Leaf mold may look like a cheap source of organic matter; however, it is very expensive if it damages the crop and is typically not worth the risk.

Wood By-Products

Sawdust may be added to growing media. In some areas of the country, it has replaced peat moss. However, there are a number of problems with sawdust. The following points should be considered before sawdust is used in a growing medium.

- Certain woods contain toxic (poisonous) substances that make the sawdust derived from them harmful to plants.
- Fresh, uncomposted redwood sawdust can cause root damage and discolored leaves in sensitive plant species. The toxins in redwood sawdust can be removed in two ways: (1) The sawdust can be left outside so that snow and rain can flush the toxins out naturally, or (2) the sawdust can be watered heavily for several hours so that the toxins are leached out of the material.
- Cedar and walnut sawdust should never be added to a growing medium. Both types contain toxins that cannot be removed by leaching.
- If used fresh from the sawmill, all sawdust causes a demand for nitrogen that is far above normal. The soil organisms that break down sawdust use nitrogen as their food source. Unless extra nitrogen is added, plants grown in a medium containing sawdust may show a lack of nitrogen.
- Sawdust should be nitrogen stabilized before it is added to a growing medium. For example, a fertilizer high in nitrogen can be mixed with the sawdust. If plants are growing in a medium with sawdust and they start to show signs of a lack of nitrogen (see Chapter 14), nitrogen fertilizer should be added.
- A small amount of nitrogen fertilizer, such as ammonium nitrate, may be added to the sawdust, which is then composted for 3 or 4 months. Composting starts the decaying process. The added nitrogen provides the energy needed by the organisms living in the sawdust. The amount of nitrogen needed depends upon the type of sawdust. Each cubic yard of sawdust may require from 0.33 to 3 pounds (0.4 to 2.4 kg/m^3) of nitrogen fertilizer.
- Some sawdust causes changes in the pH of the medium. The sawdust should be checked to see if it is acid or alkaline.

Woodchips have a larger particle size than sawdust. Usually, they are not added to flower growing mixes. The nursery industry uses them for woody plant growing media. They are often used as mulches around ornamental plants. Woodchips have the same need for nitrogen as sawdust.

Bark

Both hardwood and softwood bark materials are used in several commercial soilless media. **Hardwood barks** are those obtained from deciduous trees such as maple and oak. **Softwood barks** come from conifers such as fir, pine, or redwood. Both bark types can be quite variable in their physical and chemical properties.

Lignin is the main component of bark. As a result of the debarking process, usually some **cellulose** (wood) is included with the bark product. Lignin has a much slower rate of decomposition than cellulose. It is the cellulose fraction that requires the addition of extra nitrogen to assist decomposition.

The Mulch & Soil Council has established guidelines for classifying bark products. In order to be labeled a "hardwood bark" product, the material must contain at least 85 percent lignin. If there is less than 85 percent lignin, the product should be labeled a **"hardwood mulch."** A typical hardwood mulch contains 55 to 75 percent lignin. To be considered a **"softwood mulch"** base product, it must also have a lignin content of 85 percent. Typical softwood barks used in soilless media contain 95 percent lignin.

Hardwood mulches, because of their higher cellulose content, will decompose almost three times faster than softwood bark mulches. These differences have an important influence on the properties of aeration and water-holding capacity. They also affect the fertility of the growing media.

In chemical composition, hardwood bark contains more nitrogen, phosphorus, and potassium, but less calcium than softwood bark. A higher micronutrient content, particularly manganese, is found in softwood bark. Experience with the various bark mixes will help the grower to adjust the fertilization program as needed.

Both hardwood and softwood barks can contain substances that may be toxic to plants. Among the hardwoods, walnut, cherry, and silver maple contain growth inhibitors. These should never be used in preparing soilless mixes. The fresh bark of cedar, ponderosa pine, and white pine may also contain certain growth inhibitors. Bark from loblolly pine and slash pine does not. This is why the bark from these two species is widely used in soilless media.

Research by plant scientists at the Ohio Agricultural Research and Development Center at Wooster, Ohio, has shown that properly formulated bark mixes can suppress the development of *Pythium* (pith-e-um), *Phytophthora* (fi-topf-thor-a), and *Thielaviopsis* (theal-a-ve-op-sis) root rots. They also exert control over *Rhizoctonia* (rye-zock-tone-e-a) damping off, and *Fusarium* (few-sair-e-um) wilt and some nematodes. Nematodes are very tiny soil eel worms.

Although bark mixes can suppress some disease organisms, they cannot prevent plant diseases. If a grower does not follow good sanitation and cultural practices as well as use a good preventative program, diseases could become a problem.

Redwood and fir bark are used alone and combined with other media. Fir bark is often used as a growing medium by orchid growers.

Coir

Coir is the fibrous material from the mesocarp or middle layer of a coconut fruit (*Cocos nucifera*). Long coir fibers are used in the manufacture of brushes, automobile seat and mattress stuffing, drainage pipe filters, twine, and other products. The short fibers and dust were once discarded as a waste product. E. P. Hume in 1949 wrote of excellent plant growth obtained when this coir dust or, as he called it, "cocopeat," was used as the growing medium.

Coir dust is similar to peat in that it is light to dark brown in color and consists primarily of particles in the size range 0.2 to 2 millimeters (75 to 90 percent) with no sticks or other debris. Coir has a pore space at container capacity of 9 to 13 percent. Coir has a water-holding capacity higher than some peats and is as or more easily rewetted after drying with little shrinkage. These characteristics make coir an excellent growing medium for hanging baskets.

Coir has low levels of micronutrients, but significant levels of phosphorus and potassium. The pH of coir ranges from 5.5 to 6.5; therefore, no lime is needed for pH adjustment. Some concern has been addressed to high levels of chloride in coir, but recent advances in coir processing have mostly eliminated these issues. Coir is a common growing medium for greenhouse vegetable production (Figure 11-8).

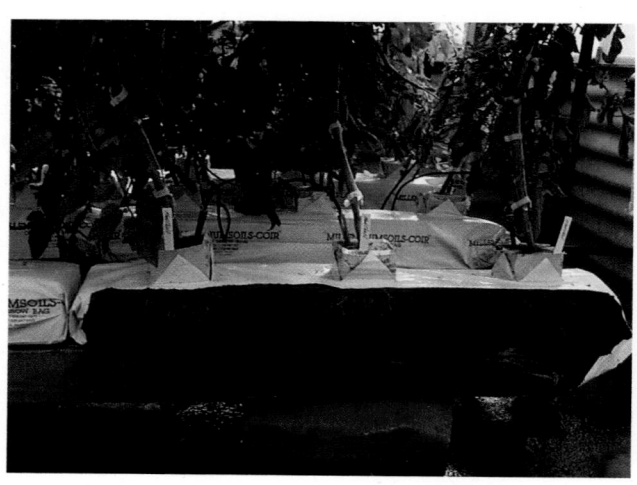

FIGURE 11-8 Greenhouse tomatoes grown in coir.

Other Organic Matter

Wheat and oat straw are added to soils to provide organic matter for cut flower crops. Straw must be chopped into small pieces from 2 to 4 inches (5.08 to 10.16 cm) long. Otherwise, it is very difficult to work the straw into the soil. Straw rots very quickly; thus, the benefits gained from adding straw to the soil last for only 3 to 4 months.

Soil with straw requires extra nitrogen for plant growth. Straw must be steam sterilized because of the large number of crop seeds remaining in the straw. Unsterilized straw can cause a weed problem in the bench. Straw may contain residual weed killers.

Other organic materials may be used to increase the organic content of the growing medium. The materials discussed are the ones commonly recommended and used.

Inorganic Materials Added to the Growing Medium

The addition of **organic material** alone to the growing medium does not always give the changes wanted. Inorganic materials must be added as well. An **inorganic material** contains no carbon. Sand, gravel, perlite, and vermiculite are examples of inorganic materials.

Sand

Sand is the most common addition to growing media. Sand is nothing more than finely ground stones. It has the same chemical content as the parent material. If the stone had high calcium content, the sand is also high in calcium. The pH of sand with a high calcium content is generally alkaline. Sands that come from materials having a high sodium content are also alkaline. Acid-loving plants, such as azaleas and rhododendrons, grow poorly in media containing alkaline sand.

Deep-mined, white mountain sands are mainly silica. They are usually free of disease organisms, insects, and weed seeds. These white sands are called sharp sands because the sand particles have flat sides. For mixing with other media, sharp sands are preferred. They do not pack down to reduce the pore space in the mix.

River bottom sands have rounded particles because the action of the water causes them to grind against each other. The same kind of wearing action takes place at the seashore when the surf washes up on the beach. Ocean sands are not used in horticulture because they have a very high salt content.

Gravel

Gravel can be added to a soil mix. The gravel used should be 0.125 inch (3 mm) in diameter or less. Gravel does not have the same effect as sand in changing the soil structure.

Both sand and gravel are heavy. A bushel (1.3 cubic feet) of moist sand weighs 100 pounds (45 kg). When sand is added to a mineral soil, the mix is heavy to handle for 8 or 10 hours a day. Sand is commonly added to cut flower soils. For container-grown plants, growers use two lightweight inorganic materials known as perlite and vermiculite. Sand is typically added to container media for tall plants where ballast is important to keep the plant upright.

Perlite

In its natural form, perlite is a volcanic rock. It is mined as an ore. After it is processed, it is heated to 1,800°F (982°C). At this temperature, each particle expands like a piece of popcorn. Expanded perlite is white and weighs only 6 to 9 pounds per cubic foot (95 to 145 kg/m^3). Its rough surface permits it to hold moisture well, but it also drains well.

Perlite is free of disease organisms, insects, weed seeds, and other living matter. It is sterile when packaged, but it can be contaminated. Careless handling methods may add unwanted diseases and insects.

Perlite is neutral with a pH of 7.0 to 7.5. It has no nutrients. Some poorer quality perlite sources have small amounts of sodium, aluminum, and fluoride that plants can take up.

Perlite is added to the growing medium to improve the aeration and drainage. It is a relatively hard material that does not break apart easily.

Perlite has two disadvantages: It is dusty, and it floats out of the medium. To overcome the dust problem, the perlite is moistened with water before it is used. There is no solution to the floating problem.

Vermiculite

Mica is a mineral found in large amounts in Montana and South Carolina. Vermiculite is a mica type mineral that is mined in both of these states. It is also mined in the state of Palobora in South Africa and is imported into the United States. The vermiculite mines in Montana have been closed due to asbestos risks.

Horticultural vermiculite is formed by expanding the flat, platelike pieces of mica ore at a temperature of 1,400°F (760°C). A piece of expanded vermiculite looks like an extended accordion bellows. It is a soft mineral. Thus, it has the disadvantage that it is easy to compress. When potting plants using a vermiculite medium, it should not be pressed down too hard.

Once vermiculite is expanded, it is size graded. Number 1 grade vermiculite is about the size of a garden pea; number 2 grade is 0.13 to 0.6 inch in diameter; and numbers 4 and 5 are even finer. Grades 2, 3, and 4 are used in horticulture.

Number 2 vermiculite is used for potted plants, number 3 is used for bedding annuals, and number 4 is used in germination media.

Dry vermiculite weighs from 6 to 9 pounds per cubic foot (95 to 145 kg/m³). When completely wetted (saturated), it holds about 500 percent water by weight. A cubic foot of saturated vermiculite weighs from 30 to 40 pounds (13.6 to 18 kg) and contains from 3 to 5 gallons (11 to 19 l) of water.

Two chemical properties of vermiculite are its **cation** (cat-eye-on) **exchange capacity** and its buffering capacity. The exchange capacity relates to nutrient uptake. The buffering capacity is the ability of the material to resist changes in pH.

Vermiculite contains the plant nutrients potassium, magnesium, and calcium in forms that plants can take in through their roots and use for growth. Vermiculite has enough potassium to satisfy the needs of some plants.

Vermiculite contains only a small amount of calcium. Because the amount is not great enough for plant growth, limestone must be added to the medium. The limestone supplies the required calcium and brings the pH into the correct range for nutrient uptake.

Horticultural grade vermiculite is used for plant growing. There is another grade of vermiculite that is called construction grade. This material is less costly, but it cannot be used in horticulture because it contains unwanted substances harmful to plants.

Rock Wool

The growing media known as rock wool is an inorganic product that is made by melting together at a very high temperature (2,700°F to 2,900°F [1,482°C to 1,600°C]) a mixture of basalt, coke, and limestone. In some cases blast furnace slag from iron production is substituted for the basalt. When blast furnace slag is used, the final product may have a high amount of sulfur in it.

When combined materials are melted, they are poured in a narrow stream onto four counter-rotating stainless steel wheels (Figure 11-9), from which the liquid, now in very small droplets, is thrown into a high-speed air blast. The droplets form strands of fibers, which are collected. From this point the product may be treated with a chemical that makes it resistant to water absorption. This product is used for insulation and acoustical purposes in homes and buildings. It is also fireproof.

Rock wool fibers used in agriculture may be treated with a wetting agent to improve water absorption. After treatment the fibers are collected and pressed into different shapes and forms. These shapes range in size from small cubes used for starting plants from seeds or cuttings to large slabs used for growing production plants (Figures 11-10A and 11-10B). Rock wool slabs are most often used in the production of greenhouse tomatoes and cucumbers.

In Europe and the United States, greenhouse roses are also grown in rock wool media (Figure 11-11).

Rock wool granules are also used. These are most often mixed with other media such as peat-lite type mixes to change the water-holding capacity and air-water balance.

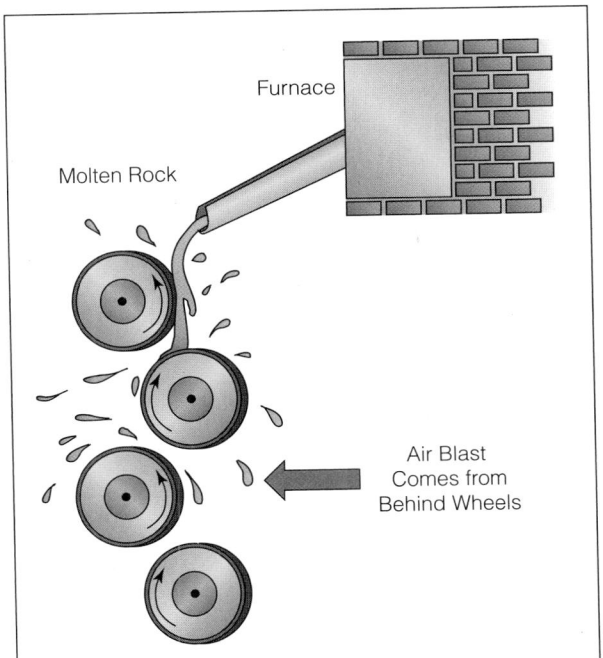

FIGURE 11-9 Molten rock drops onto four counter-rotating stainless steel wheels, which throw the droplets into the air blast where it cools and forms fibers. The fibers are collected, treated, and compressed into various shapes as rock wool media.

FIGURE 11-10A A sampling of rock wool cubes.

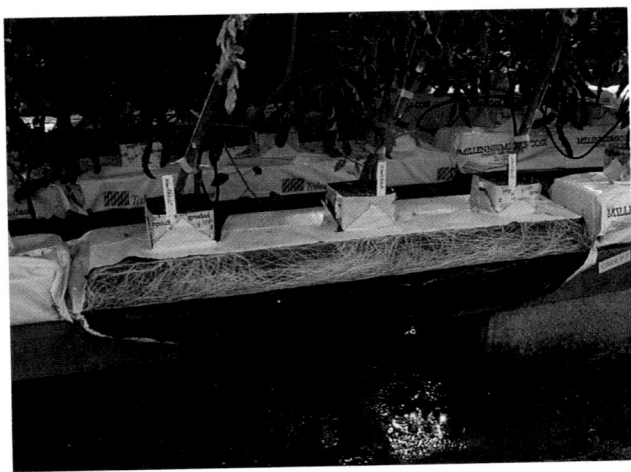

FIGURE 11-10B Greenhouse tomatoes grown in rock wool.

FIGURE 11-11 Roses being grown from cut flowers in rock wool media in Holland.

Rock wool fibers will hold a lot of water, because they contain only 3 to 4 percent solid matter and 96 to 97 percent pore space. When it is thoroughly wetted, rock wool media will drain 15 to 17 percent. This leaves a good air: water ratio for root growth.

Rock wool contains only a very small amount of plant nutrients such as calcium, magnesium, sulfur, iron, copper, and zinc. None of these is in enough quantity to supply the total nutrient needs of the plant.

The pH of rock wool media is alkaline, 7.5 to 8.5 or higher. However, the media is not highly buffered. This means that the pH will change to that of the nutrient solution used at the first irrigation. Rock wool media does not biodegrade, but it does slowly weather. Because it is not biodegradable, environmental protection agencies in several countries have banned its disposal in landfills.

ACHIEVEMENT REVIEW

Select the best answer or answers to complete each statement. List the appropriate letter(s).

1 Soils cannot be removed from the field and used directly for potting plants. Sand, perlite, peat moss, and similar materials must be added to:
a. provide nutrients.
b. increase the water-holding capacity.
c. increase the drainage.
d. make the soil lighter.

2 Two of the most important functions of soil are:
a. to provide a home for soil insects.
b. to give mechanical support to plants.
c. to increase the weight of plastic pots.
d. to act as a storehouse for nutrients and water.

3 The ideal percentages of solids, air, and water in a growing medium are:
a. 25 percent solid, 50 percent air, 25 percent water.
b. 33.3 percent solid, 33.4 percent air, 33.3 percent water.
c. 50 percent solid, 25 percent air, 25 percent water.
d. 25 percent solid, 25 percent air, 50 percent water.

4 The term *texture* refers to:
a. the hardness or softness of soils.
b. the sizes of the various soil particles.
c. how the soil is held together.
d. the smoothness or roughness of the soil.

5 A soil that has more clay particles than any other soil fraction may be called:
a. gravelly sand.
b. sandy clay.
c. clay loam.
d. silty loam.

6 Organic matter is added to a sandy soil to:
a. increase the pore space.
b. increase the nutrient content.
c. increase the water-holding capacity.
d. decrease the water-holding capacity.

7 The peat moss preferred for commercial horticulture is:
a. reed peat moss.
b. sedge peat moss.
c. Michigan peat.
d. sphagnum peat moss.

8 Horticultural peat moss is usually taken from a bog at the:
   a. H1 level (top of the bog).
   b. H3, H4, and H5 levels.
   c. H6 and H7 levels.
   d. H10 level (bottom of the bog).

9 Perlite is a volcanic rock that is expanded by heating it to:
   a. 500°F (260°C).
   b. 900°F (482°C).
   c. 1,800°F (982°C).
   d. 3,000°F (1,647°C).

10 Dry vermiculite and perlite both weigh:
   a. 3 to 4 pounds per cubic foot (48 to 64 kg/m^3).
   b. 6 to 9 pounds per cubic foot (95 to 145 kg/m^3).
   c. 8 to 12 pounds per cubic foot (127 to 190 kg/m^3).
   d. 10 to 14 pounds per cubic foot (159 to 225 kg/m^3).

11 The pH of rock wool media used in agriculture is:
   a. very acid, 4.5 to 5.0.
   b. slightly acid, 6.5 to 7.0.
   c. neutral, 7.0.
   d. alkaline, 7.0 to 8.5.

Student Project

Determine the percentage pore space and the water-holding capacity of three potting media.

A good potting medium supplies a plant with support and nutrition. It must also ensure that the plant has enough air and water to grow. Roots that do not get enough air will die. Different plants have different air and water needs.

Plants that need a highly aerated medium are: azaleas, ferns, and certain orchids. For these plants, the air space after drainage should be 20 percent of the total container volume. An air space of 10 to 20 percent is sufficient for African violets, begonias, foliage plants, and snapdragons. Plants requiring an air space of 5 to 10 percent (an intermediate amount) are camellias, chrysanthemums, hydrangeas, and lilies. Plants that will grow with a low percentage of air space (2 to 5 percent) are carnations, geraniums, ivies, and roses. Each of these groups contains many other plants. The plants listed are given as examples.

There are many published tables that give the water-holding capacity and pore (air) space of media. However, these charts may not be available when needed. In addition, a particular mix may not be included in the chart. Container size and shape also affect soil aeration. For these reasons, a grower should be able to determine the pore space and water-holding ability of the growing media.

Materials

Container used for growing (5-inch [13-cm] plastic pot)
Typical growing medium:
   a. peat-lite mix (**Jiffy-Mix®, Redi-Earth®, or Pro-Mix®**)
   b. heavy loam soil
   c. straight sand

Graduated cylinder marked in milliliters
Masking tape
Two plastic trays, 3 inches (7.5 cm) deep
Container for water
Pencil and paper

Procedure

1. Measure the volume of the pot.
2. Close the drainage holes by taping them. Place the tape on the outside of the pot on at least one hole as it must be removed after the pot is filled. The other holes can be taped inside the pot.
3. Fill the pot with water to within one-half inch of the top. Mark the fill line on the outside of the pot with a pencil.
4. Carefully pour the water into the graduated cylinder and record the amount in milliliters.
5. Dry the inside of the pot. Do not remove any of the tape.
6. Fill the pot with dry growing medium to the water line. Pack the medium as if potting a plant.
7. Using the graduated cylinder to measure the water, carefully pour water on the medium until it is thoroughly wet. Record the amount (volume) of water used. These values will be used in calculations later.
8. When the medium is saturated, there will be a thin film of water on top of the medium. Stop adding water and record the total number of milliliters used.

   Note: Some media are easier to wet than others. Dry peat moss is very hard to wet. Water uptake by peat moss media can be speeded up by using hot water or water to which a wetting agent is added (see Chapter 16). The water is added slowly so that it can be absorbed by the medium. Some media may need several hours to reach a totally wet condition. If a long wait is necessary, cover the top of the pot with saran wrap or aluminum foil to prevent evaporation. The total amount of water needed is equal to the total pore space of the medium. Both moisture and air are held in the pore space. By bringing the medium to a completely wet state, the air is driven out and the pores are filled with water.

9. The percentage porosity, or total pore space, can be calculated as follows:

$$\text{Percentage porosity (total pore space)} = \frac{\text{milliliters of water needed to wet the medium thoroughly}}{\text{total volume of the pot in milliliters}} \times 100$$

   Example: Total volume of a pot is 570 ml and 160 ml of water are required to wet the medium thoroughly.

$$\text{Percent porosity} = \frac{160 \text{ ml}}{750 \text{ ml}} \times 100 = 21.33\% \text{ porosity, or total pore space}$$

10. Once the medium is thoroughly wetted, remove the tape from one hole while holding the pot over the plastic pan. The water draining from the pot must be collected. DO NOT THROW THIS WATER AWAY.

11. After collecting all of the water that drains from the pot, measure it in the graduated cylinder. This water is called gravitational water. Its volume is equal to the pore space filled with air in the medium.
12. Calculate the percentage of air space.

$$\text{Percentage air space} = \frac{\text{volume (ml) of drained water}}{\text{total volume or pot (ml)}} \times 100$$

Example: 80 ml of water drains from the medium.

$$\text{Percentage pore space (air)} = \frac{80 \text{ ml}}{750 \text{ ml}} \times 100 = 10.67\%$$

13. Less water drains from the medium than is added to the pot. The difference between the amount added and the amount drained is the water-holding capacity of the soil. The answer can be obtained directly by subtracting the percentage of air pore space from the total porosity: 21.33% − 10.67% = 10.67%

In this example, the water-holding capacity is the same as the percentage of air pore space. This is not always true.

Activities

1. Determine the water-holding capacity of the three media.
2. Make up the other mixes and determine the pore space for each. This method allows the grower to prepare a growing medium with the exact amounts of air and pore space needed.

Artificial Soils

OBJECTIVES:

After studying this chapter, the student should have learned:

■ why artificial soils are needed.

■ the materials in a peat-lite mix.

■ several commercial peat-lite mixes.

Need for Artificial Soils

Artificial soils, or soilless mixes, were developed to satisfy several needs of the horticulture industry. One of the most important reasons is that good topsoil is becoming scarce. **Topsoil** always has been a large part of any plant-growing medium. The topsoil being marketed today has an unknown nutrient content. Disease organisms, insects, and weed seeds often are present. The quality of the drainage and aeration is unknown. The topsoil may contain chemical weed killers whose effect can last for 2 or 3 years. For these reasons, many growers no longer wish to use topsoil.

Standard Growing Media

A standard soilless growing medium has several advantages.

1. It consists of materials that have known properties.
2. It is prepared from a recipe and is the same from batch to batch, season to season.
3. The materials used are easy to get.
4. The materials can be handled and mixed using mechanical means.
5. The nutrient content is low, so fertilizers can be added in known amounts.
6. Vermiculite, one of the materials used, has some nutrients that are released slowly.
7. The drainage and aeration are designed to ensure good root growth.
8. Sterilization or pasteurization generally is not required. Thus, the costs of preparation are reduced, and energy is conserved.
9. The cost of preparing the artificial soil is often less than the cost of preparing soil mixes.
10. Soilless mixes are easy to use.

One of the most important facts about using a standard medium is that the grower soon learns how the crops grow in it. Knowing this, the grower has more control over production in each season of the year.

Cornell Peat-Lite Mixes

During the 1950s, Cornell University revolutionized the industry with a blend of sphagnum peat moss and vermiculite, called by most peat-lite. A half-century later, producers and manufacturers continue to modify and improve it. Soilless mixes are now the standard for growers. Few growers continue to use topsoil in their mix; however, good topsoil is scarce. Its weight and density prohibit sufficient water and air circulation and the prevalence of disease organisms also is an issue.

The Cornell University peat-lite mixes are standard, lightweight mixes that are widely used in the United States. These mixes were developed as growing media to start flower and vegetable transplants. Now, they are also used as growing media for other types of potted plants.

The peat-lite mixes consist of equal parts by volume of sphagnum peat moss and vermiculite (mix A), or peat moss and perlite (mix B). Table 12-1 shows the composition of mix A.

Other materials may be added to the mixes. Figures 12-1A and 12-1B show a few of these materials. **Calcined clays** are not used in the **Cornell peat-lite mixes.** Sand is used only to add weight as ballast for tall plants when it is needed.

Table 12-1

Cornell Peat-Lite Mix A for Seedlings, Bedding Plants, Pot Plants, and Greenhouse Tomatoes on Liquid Feeding		
Materials Used	Amount in 1 Cubic Yard (yd^3)/In 1 Cubic Meter (m^3)	
Sphagnum Peat Moss	0.5 yd^3 (13 bushels)[1]	0.5 m^3
Horticultural Grade Vermiculite, #2, #3, or #4 Size	0.5 yd^3 (13 bushels)	0.5 m^3
20 percent Superphosphate	1–2 lb	0.59–1.19 kg
or		
Treble Superphosphate	½–1 lb	0.29–0.59 kg
Ground Dolomitic Limestone	5–10 lb[2]	2.97–5.94 kg
Calcium or Potassium Nitrate	1 lb	0.59 kg
Trace Element Materials (*Use only one*):		
Fritted Trace Elements FTE 555	2 oz	74.25 gms
or ESMIGRAN	4 lb	2.38 kg
or PERK®	4 lb	2.38 kg
Wetting Agent:		
AquaGro® Liquid	3 oz	111.2 gms
or Surfside® Liquid 30	3 oz	111.2 gms
or AquaGro® Granular	1½ qt	1.85 l
or Surfside® Granular	1½ qt	1.85 l

1. A cubic yard equals 27 cubic feet, or approximately 22 bushels. A 15 to 20 percent shrink occurs in mixing. Therefore, an additional 5 ft^3 or 4 bushels are used to obtain a full cubic yard. A cubic meter equals 35.3 ft^3. Add 0.2 m^3 to obtain a full cubic meter.
2. Use the 10-lb (5.94-kg) rate for greenhouse tomatoes on liquid feeding.

FIGURE 12-1A Inorganic components of artificial media. From left to right, vermiculite, perlite, and sand.

FIGURE 12-1B Sphagnum peat moss. The left sample is a fine grind and the sample to the right is coarse.

FIGURE 12-2 Twister™ batch soil mixers. These units are designed to handle diverse types of media; peat moss, sand, soil, wood chips, as well as chemicals and fertilizers. The smaller unit will mix 1 cubic yard (0.8 m³); the unit on the right will mix 2 cubic yards (1.8 m³) at one time. *(Courtesy of Boulden & Lawson, McMinnville, TN 37110.)*

Preparation of Peat-Lite Mixes

The secret to success with any soilless medium is the thorough mixing of all materials without degrading the physical properties. Small batches can be turned over by hand. A small concrete mixer can prepare many yards of mix easily.

For larger volume amounts, specialized equipment is needed. Bouldin and Lawson of McMinnville, Tennessee, is one manufacturer of soil mixing equipment. Batches of 1 and 2 cubic yards can be prepared in their Twister I and Twister II mixers (Figure 12-2). Very large amounts of media are prepared by using continuous flow mixers (Figures 12-3 and 12-4). The various media and fertilizer components are metered into the continuously flowing product and blended to get a uniform mix. The Mix Maker 50 can prepare up to 50 cubic yards of media in 1 hour.

Regardless of the method of media preparation, overmixing, especially if too much water is added, can result in a poor final product. A good mix can be obtained in 3 to 5 minutes of turning.

Continuous mixers often discharge the media into semiautomatic flat fillers or pot fillers (Figures 12-5 and 12-6).

FIGURE 12-3 Mix Maker 10™, a reduced size version of Bouldin & Lawson's continuous media mixing system. The materials pass through a unique mixing head for thorough blending. Water is also added to moisten the media. *(Courtesy of Bouldin & Lawson, McMinnville, TN 37110.)*

FIGURE 12-4 Mix Maker 30™ is an intermediate size continuous media mixer. This unit will prepare 30 cubic yards (23 m³) of mix an hour. A still-larger version of this unit is the Mix Maker 50™, which can prepare 50 cubic yards (38 m³) of mix an hour. *(Courtesy of Bouldin & Lawson, McMinnville, TN 37110.)*

FIGURE 12-5 The new Mini Flat Filler™ can handle 12 plug trays a minute, bedding plant flats with inserts, and pots in flats up to 5 inches (12.5 cm) high. *(Courtesy of Bouldin & Lawson, McMinnville, TN 37110.)*

FIGURE 12-6 The Maxi Flat Filler™. This unit is larger and faster than the Mini Flat Filler. It can fill pots and trays up to 8.5 inches (21 cm) tall and 18 inches (45 cm) wide. *(Courtesy of Bouldin & Lawson, McMinnville, TN 37110.)*

The area where the mixes are prepared should be clean. Trash and dirt should not be allowed to accumulate in the mixing area. Debris may be the source of weed seed, insects, and disease that may contaminate the mix.

Dry peat moss can be difficult to wet. The use of a wetting agent (surfactant) is recommended (see Chapter 16).

Several variations of the original peat-lite mixes have been developed. Their formulas are given in Table 12-2.

A few brief comments follow about some of these other mixes.

Mix A—Alternate 1

This mix is used for seedlings, bedding plants, and general pot plants. A 10-10-10 fertilizer is typically substituted for the superphosphate and calcium or potassium nitrate at the rate of 5 pounds per cubic yard (2.97 kg/m^3). All other components are the same.

Mix A—Alternate 2

This mix is used for pot plants, except lilies, and bedding plants. A slow-release fertilizer is substituted for the calcium or potassium nitrate. Use 5 pounds (2.97 kg/m^3) of medium size **Magamp**® 7-40-6 plus 5 pounds (2.97 kg/m^3) of **Osmocote**® 14-14-14, or 18-9-9, or 19-6-12. Use only 1 pound (0.59 kg/m^3) of regular superphosphate or 0.5 pound (0.29 kg/m^3) of treble superphosphate.

Mix A—Alternate 3

This mix is used for greenhouse tomatoes that are fed with slow-release fertilizers. Liquid feed is not used. Add 5 pounds (2.97 kg/m^3) of Magamp® 7-40-6 plus 10 pounds (5.94 kg/m^3) of Osmocote® 14-14-14, or 18-9-9, or 19-6-12 to the other nutrients listed for liquid feed tomatoes.

Table 12-2

	Sphagnum Peat Moss (yd³)[1]	Horticulture Grade Vermiculite (yd³)	Horticulture Grade Perlite (yd³)	Ground Limestone (lb)	Treble Superphosphate (lb)	Trace Elements (Use only one) FTE 555 (oz)	ESMIGRAN (lb)	PERK® (lb)	Potassium or Calcium Nitrate (lb)	10-10-10 Fertilizer (lb)	Osmocote® 14-14-14 or 18-9-9 or 19-6-12 (lb)	Magamp® 7-40-6 (lb)	Wetting Agent Fl. oz	Wetting Agent Dry qt
Cornell Peat-Lite Mixes														
Mix A — For seedlings, bedding plants, pot plants (except lilies), and greenhouse tomatoes on liquid feed	0.5	0.5	—	5–10[2]	0.5–1	2	4	4	1	—	—	—	3	1.5
Mix A-1 — For the same crops as Mix A, except tomatoes	0.5	0.5	—	5–10	—	2	4	4	—	5	—	—	3	1.5
Mix A-2 — For pot plants (except lilies) and bedding plants (slow-release fertilizer)	0.5	0.5	—	5–10	0.5	2	4	4	—	—	5	5	3	1.5
Mix A-3 — For greenhouse tomatoes, no liquid feed	0.5	0.5	—	10	1	2	4	4	1.5	—	10	5	3	1.5
Mix B — For the same crops as Mix A, except tomatoes	0.5	—	0.5	5–10	0.5–1	2	4	4	1.5–2[3]	—	—	—	3	1.5
Mix B-1 — For the same crops as Mix B	0.5	—	0.5	5–10	—	2	4	4	—	8	—	—	3	1.5
Mix B-2 — For the same crops as Mix A-2	0.5	—	0.5	5–10	0.5	2	4	4	—	—	8	8	3	1.5

1. A cubic yard equals 27 ft³ or approximately 22 bushels. A 15 to 20 percent shrink occurs in mixing. Therefore, an additional 5 ft³ or 4 bushels are used to obtain a full cubic yard.
2. Use 10 lb for tomatoes.
3. Use potassium nitrate only.

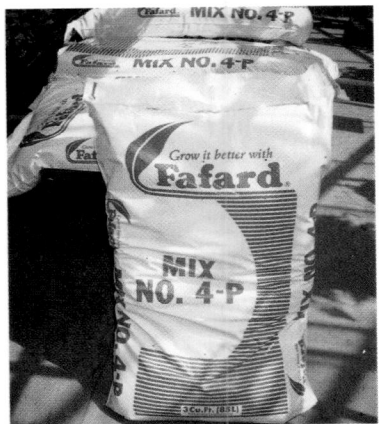

FIGURE 12-7 Fafard peat-like mix No. 4-P. This is only one of the many peat-lite type mixes that are available in the trade.

FIGURE 12-8 Fafard peat-lite mix No. 2. This is an additional peat-lite mix available for greenhouse growers. Bale type mixes are compressed for easier shipping and storage.

Mix B

This mix is used for the same crops as mix A except tomatoes. Horticultural grade perlite is substituted for vermiculite. Potassium nitrate is used in place of calcium nitrate at the rate of 1.5 to 2 pounds per cubic yard (0.88 to 1.19 kg/m3).

The great success of the original peat-lite mixes resulted in the development of many formula variations of other soilless media. Several peat moss suppliers added mixing equipment to their operations and started selling their mixes to the industry.

Fafard offers 13 mixes in 3- or 4-cubic-foot bags (Figure 12-7). They also have five growing mixes in 4-cubic-foot compressed bales. Compressed bales, when broken apart, usually have a fluff factor of 50 percent (Figure 12-8). This means that a 4-cubic-foot compressed bale will fluff to provide 6 cubic feet of media.

Sunshine® mixes are sold by Sun Gro Horticulture. These range from fine plug mixes for germinating seed to growing mixes, including ones that use bark. Sun Gro Horticulture acquired the Metro Mix® line from the Scott Company of Marysville, Ohio, which was originally owned by Grace-Sierra. These peat-lite mixes, in addition to the sphagnum peat moss, perlite, and vermiculite, contain granite sand for added weight and processed bark ash. Other formulations include the incorporation of coconut fiber or coir. Sun Gro Horticulture also markets the Redi-earth® products as well.

Michigan Peat Company has the Baccto® line of peat-lite mixes. Premier Horticulture offers several formulations of their mixes under the Pro-Mix® brand name. They have also introduced products with mycorrhizal formulations and products inoculated with bacteria serving as a biofungicide.

All of the suppliers offer mixes that contain no fertilizer to those with a small amount of starter fertilizer. Usually the starter fertilizer is enough to feed the plants for about 2 weeks. After that time period, the grower should begin a regular fertilization program for the specific crop being grown. Generally all the media suppliers also add a surfactant (wetting agent) to make them easier to wet. The difficulty of rewetting dry peat moss is well known. The addition of coconut fiber, called coir, may reduce this problem as coir is easier to rewet than peat.

Many growers use commercial mixes but add soil, sand, or other materials. This practice is not recommended because unknown and unwanted chemicals, diseases, organisms, or insects may be added to the medium. Growers that add soil should consider a pasteurization or sterilization step in their mix preparation. By altering the commercial mix, the grower gives up part of the control over the growth of the crop.

Other growers use local peats to make up their own mixes. However, depending on the local source, quality and consistency are often issues. After years of trial and error, they develop a mix that works for them. As long as they can obtain the basic ingredients, they will have no problem in blending their mix. However, if any of the ingredients must be changed, or if the source of supply is changed, they must again determine how their crops grow in the new medium. In the commercial peat-lite mixes, growers have a standardized, lightweight medium that they can depend upon for uniform results.

Industry Regulation

The names given earlier in this chapter do not include all the suppliers. The parent companies change with mergers, buyouts, and sellouts. Yet, many products remain the same only with different names.

With all the formula variations of the original Cornell Mix being made available, there was some concern voiced about whether the contents in the bag or bale were as listed on the label. Unlike the fertilizer industry, which is legislated in each state to guarantee that the product in the bag is what is claimed on the label, the soilless media industry is not regulated.

With the urging of some interested growers and agricultural extension agents in Georgia in 1993, a major campaign was undertaken to get the soilless media industry in the state legislated. The purpose of this effort was to get some standardization of the products being sold in the state. It was also aimed at obtaining a "truth in advertising" mandate, similar to that enforced for fertilizers. As a result of these efforts the Georgia Department of Agriculture passed the Georgia Horticultural Growing Media Act. Simply stated, this law requires that the contents of the bag/bale, or whatever the unit offered for sale, be listed on the label in a manner according to industry standards. Georgia is the first state to enact such legislation.

As a result of Georgia's actions there has been interest by other states in doing the same thing. In the hopes of preventing a hodgepodge of rules and regulations for each state, which would bring hardship to the suppliers of soilless media, the National Bark and Soil Production Association of Manassas, Virginia, has proposed a Model Legislative Bill for Horticultural Growing Media Labeling. The bill would amend the appropriate chapter in the Official Code of the state passing it that relates to commercial fertilizers, liming materials, and soil amendments.

The Mulch and Soil Council (MSC), an alliance of media manufacturers, has actively worked to influence legislation relating to soilless mixes. To ensure some stability in the industry, the association developed a national test procedure for volumetric inspection of mulch products.

ACHIEVEMENT REVIEW

Select the best answer or answers to complete each statement. List the appropriate letter(s).

1 Topsoil is no longer the choice of many plant growers. One of the main reasons is that:
   a. too many people are using it.
   b. there is no shortage of topsoil.
   c. it may still have active weed killers in it.
   d. the cost is too low.

2 Identify which one of the following is not a major advantage of a standard growing medium.
   a. It is prepared in the same way from batch to batch.
   b. The materials can be mixed by mechanical means.
   c. The nutrient content is high, and nothing must be added.
   d. It has excellent drainage and aeration.

3 Peat-lite media do not require steam sterilization. Thus, the grower can:
   a. save labor and conserve energy.
   b. plant into the media right away.
   c. forget about disease control.
   d. forget about the need for a steam boiler.

4 The most widely used peat-lite mixes were developed at:
   a. the University of California.
   b. Pennsylvania State University.
   c. Cornell University.
   d. Rutgers University.

5 Peat-lite mix A consists of:
   a. peat moss and perlite.
   b. peat moss and calcined clay.
   c. sand and peat moss.
   d. vermiculite and peat moss.

6 In addition to peat moss, vermiculite, and perlite, Metro Mix 200® contains:
   a. river bottom sand.
   b. granite sand.
   c. sharp mountain sand.
   d. none of the above.

7 When equal volumes of peat moss and perlite or vermiculite are mixed, some shrinkage in volume occurs. To overcome the shrinkage, how many bushels of media are added to a cubic yard of mix?
   a. 2
   b. 4
   c. 6
   d. 8

8 Magamp® is a slow-release fertilizer that has an analysis of:
   a. 6-40-7.
   b. 14-14-14.
   c. 19-6-12.
   d. 7-40-6.

9 Premier Brands sells a peat-lite mix called:
   a. Premier Mix®.
   b. Gro-Mix®.
   c. Pro-Gro®.
   d. Pro-Mix®.

10 The main objection to adding soil to a peat-lite mix is that it:
   a. adds weight to the mix.
   b. may add insects and disease organisms.
   c. compresses the vermiculite.
   d. decreases the pore space.

Sterilization (Pasteurization) of Growing Media

OBJECTIVES:

After studying this chapter, the student should have learned:

- several ways to sterilize (pasteurize) the growing media.
- the advantages of steam pasteurization.
- when chemicals can and cannot be used.
- the precautions to be used when fumigating with chemicals.

Sterilization is defined as "the act or process of killing all living cells, especially microorganisms." Microorganisms are bacteria, fungi, and other living organisms that can be seen only with the use of a microscope. These organisms may be pathogens that cause disease or helpful types that improve plant growth.

The treatment given to soils to free them of harmful insects, diseases, and weed seeds is not actually "soil sterilization." Although all harmful objects in the soil must be killed, all organisms that are useful to plants should be unaffected. Therefore, the process of pasteurization, not sterilization, is used to kill harmful organisms and most of the weed seeds. Beneficial organisms remain alive. The most commonly used methods are steam pasteurization, electrical pasteurization, or chemical fumigation. Figure 13-1 illustrates one situation in which a soil treatment may have prevented soil-borne disease.

Steam Pasteurization

Soil steaming is a method of applying heat to the soil to destroy harmful organisms. Although the hot steam must be handled carefully, it is not toxic and is the safest method of treating greenhouse soils and mixes. One bench can be steamed without causing damage to plants in an adjacent bench. Chemical soil treatment requires an empty greenhouse.

FIGURE 13-1 Overwatering and *Pythium*-infected potting medium as a result of poor sanitary conditions and lack of pasteurization resulted in the loss of this poinsettia cutting.

Steamed soils can be planted as soon as they reach a temperature of 80°F (27.5°C) or less. Chemically treated soils cannot be planted until they have aired for a period ranging from 24 hours to several weeks. The amount of time required for airing depends upon the chemical used.

Duration of Steaming

The previously recommended method of steaming the soil was to bring the temperature in the coldest part of the soil to 180°F (84°C). This temperature is maintained for at least 30 minutes. Some growers, however, oversteam the top soil layers by increasing the temperature to 212°F (100°C). They hold the temperature at this level for several hours. Oversteaming can cause a buildup of harmful substances in the soil. For example, manganese is a nutrient element normally used by plants in very small amounts. Oversteaming causes manganese to be released from the soil in quantities that are toxic (poisonous) to plants. Soil mixes with a high level of organic matter develop very high soluble salt levels as a result of oversteaming.

Fire molds are also common on greenhouse soils that have been oversteamed. These fungal organisms are often found on natural sites after a fire and are early colonizers of sterile sites. One species, *Pyronema*, is common on oversteamed soils. These organisms are not pathogenic to plants, but degrade organic matter. They may, however, form a fungal mat on the surface, which is difficult to wet (Figures 13-2A and 13-2B).

Aerated Steam

The preferred steaming method is the use of aerated steam at a temperature of 140°F (60°C) for 30 minutes.

Diseases that start in a soil recently steamed at 180°F (84°C) may spread very quickly because the heat also kills the natural enemies of the disease organisms. When the soil is treated with aerated steam at 140°F (60°C), these helpful organisms are unharmed, and thus slow the spread of disease.

Aerated steam is obtained by injecting air into the steam flow. The amount of air injected into the steam is controlled. As a result, the steam temperature can be held at a specific value as long as the system operates.

FIGURE 13-2A Fire molds are common to greenhouse media that have been excessively steamed. The snapdragon plugs planted into this site have been stunted due to the fungal mat interfering with irrigation and not infection.

FIGURE 13-2B Fire molds are common on greenhouse media that have been oversteamed. Fire molds such as *Pyronema* sp. typically are slime molds commonly found on the forest floor after a fire. They are saprophytic organisms that feed on organic matter.

FIGURE 13-3 Injection point for steam pasteurization of propagation bench media.

The use of aerated steam has some advantages: Less energy is needed for the process, the disease-causing organisms are destroyed, and the beneficial organisms are unharmed.

Preparing Soils for Steaming

Soils must be thoroughly mixed before they are steamed. The soil can be mixed by hand or rototilled by machine to break up lumps. Steam does not penetrate large lumps of soil to bring the temperature to the necessary level. Thus, disease organisms and weed seed may remain to cause problems later.

Fertilizers such as limestone and superphosphate are added before steaming. If they are added after steaming, the soil must be tilled a second time. This additional handling increases the likelihood of introducing diseases and also breaks down the soil structure. However, Osmocote® or other controlled-release fertilizers are never added before steaming (see Chapter 15).

The soil is covered with a tarpaulin, rubberized fabric, or some other type of material (Figure 13-3). The cover retains the heat once the proper temperature is reached. Any tools, planting row markers, or other equipment to be used in the soil after it is steamed should also be placed under the cover for treatment. Untreated tools and row markers may introduce unwanted disease organisms into a growing medium.

Steam treatment of soils improves the drainage and aeration of the soil. The high steam temperature causes substances in the soil to act like cement. This process of flocculation occurs when small soil particles stick together to form larger aggregates that improve the structure. If the soil is tilled after steaming, these aggregates are reduced again to small particles. The decreased pore space reduces drainage and aeration.

Buried Pipe Steaming

Steam is applied to soils in several ways. In one method, buried pipes with regularly spaced holes distribute the steam through the soil. One type of pipe used is 4-inch (10-cm) diameter aluminum pipe with 0.25-inch (0.6-cm) diameter holes spaced 1 foot (30.5 cm) apart. Downspout pipe can be used, but it wears out too quickly. The soil above the pipes should be twice the thickness of the soil below the pipes. The soil is usually pulled away from the sides of the bench and piled over the pipe (Figure 13-4).

For effective steaming, the soil should not be dry or too wet. Dry soil acts like an insulator, and a long time is required for the temperature to increase to the proper level. Very wet soil requires too much steam to heat the extra water. The soil moisture level desired for potting is the best level for steaming.

Surface Steaming

If surface steaming is to be done, the soil is prepared as for buried pipe steaming. However, the soil is graded so that it is even with the top of the bench. Perforated metal pipe is placed on top of the soil. A porous canvas hose is often used because it is easier to handle than pipe and works as well. Some growers do not use either pipe or hose. Instead, the steam is supplied under enough pressure to raise the cover.

Air is forced out of the soil ahead of the flowing steam. The air moves downward. Thus, there must be outlets at the bottom of the bench for the air to escape. Often, the air leaves through the drain holes in the bottom of raised benches.

FIGURE 13-4 Buried pipe method of steam sterilizing growing media in a greenhouse bench.

The soil at A is twice the depth of the soil at B. C is perforated pipe or downspout. D is a header connected to the steamhouse E. The cover F holds the steam in the growing media piled over and around the conductor pipe.

Surface steaming does not work well for ground beds. At the steam pressures used, the heat penetrates only 8 to 10 inches. For ground beds, the buried pipe method of steaming is preferred.

Chamber (Vault) Steaming

Chamber steaming is commonly used by propagators and some potted plant growers. Metal containers, flats, wooden boxes, and clay pots are filled with the growing medium. They are placed in a closed space or vault (Figures 13-5A through 13-5C). Steam is introduced until the proper temperature is reached. The chamber is held at this temperature until the media are sterilized. A remote reading thermometer is inserted in the medium in a container so that the grower can check the temperature. The vault cannot be opened to use a probe type thermometer (Figure 13-6).

After steaming is completed, the containers are removed and allowed to cool before planting. Sterilized media should be planted right away. If the media are not used, soluble salts build up in 2 or 3 weeks. Unplanted media that are left uncovered for long periods become contaminated with disease organisms and weed seed.

Many containers used for growing plants are made of plastic. Some types of plastic are not affected by the steam temperatures. Other plastics melt at these

FIGURE 13-5A Vault for steam pasteurizing growing media.

FIGURE 13-5B Bottom of a vault for steam pasteurizing growing media with a perforated pipe for distributing steam.

FIGURE 13-5C Vault for steam pasteurizing growing media in use with heavy tarpaulin to contain steam.

FIGURE 13-6 Probe type thermometer used to check soil (medium) temperature during steam pasteurization.

temperatures. Unfortunately, there is no way to tell beforehand if a plastic container will melt.

Chemical Soil Treatment

Plant growers who do not have a way of steaming their growing media must use chemicals to do the job. Although chemical fumigants are not as effective as steam sterilization, they are a satisfactory substitute.

Soil fumigation is effective only if the media are at the proper temperature. The gases do not diffuse easily (spread throughout the media) in cold soils.

None of the chemicals presently in use will kill all of the organisms in growing media. If high dosage rates of fumigant are needed to destroy a particular organism, there may be unwanted side effects. A few of the most commonly used fumigants are described in the following paragraphs.

Chloropicrin (tear gas) is sold under various trade names. If enough of the fumes are inhaled, it causes vomiting and the eyes to stream. It can be added to soils with a hand applicator or a small, tractor type fumigator. It is applied at the rate of 12 pounds (5.5 kg) per 1,000 square feet (92.8 m^2) of area. If the gas is to work properly, the soil can be no colder than 60°F (15.5°C). Heavy clay soils may require more of the chemical. The gas must be held in the treated area. A water seal is made by sprinkling the area immediately after treatment. A gastight cover can also be used. Even very large fields of several acres can be treated and then covered with polyethylene. The plastic is sealed to form a single large sheet.

The cover is left on for 24 hours. After the cover is removed, the soil must be aired for 10 to 21 days before it can be planted.

The fumes from chloropicrin are toxic to plants. If it is to be used in a greenhouse, all plants must be taken out of the building. Chloropicrin kills weed seeds, most fungi, nematodes (tiny soil eel-worms), and other soil insects. It also controls *Verticillium* wilt disease.

Vapam is a liquid carbamate soil fumigant that kills nematodes, most weeds, and fungi. It is effective at temperatures between 50°F (10°C) and 90°F (32°C). Vapam is applied at the rate of 1 quart (946 ml) per 100 square feet (9.28 m^2) of area. A sprinkling can or a proportioner attached to a hose can be used to apply the liquid. The fumes are toxic to plants. It should not be used in a greenhouse where plants are growing. Vapam must be watered thoroughly so that it will spread into the media. Vapam-treated media must be aired for 2 to 3 weeks before being used for planting. Although Vapam is not poisonous to humans, unnecessary exposure should be avoided.

Methyl bromide that is labeled for sterilization of potting mixes and soil is a mixture of 98 percent methyl bromide and 2 percent chloropicrin. Methyl bromide is colorless, odorless, and tasteless. Under ordinary conditions, it cannot be detected. Thus, chloropicrin is added to act as a warning of the presence of the gas. Methyl bromide controls weeds, insects, and nematodes when applied at the rate of 1 pound (0.45 kg) per 100 square feet (9.28 m^2) of area for 24 hours. When it is used at the rate of 4 pounds (1.8 kg) over the same area, and the area is sealed for 24 hours, it will kill bacteria and most fungi.

To treat bulk soils or other media, use 4 pounds (1.8 kg) of methyl bromide per 100 cubic feet (9.28 m^3), or about 1 pound (0.59 kg) per cubic yard (cubic meter) (Figure 13-7). The soil is covered for 24 hours. It must be aired for 3 to 7 days before it can be used. Carnations and salvia are sensitive to fumigants that contain bromine, such as methyl bromide. These crops must not be planted in soils treated with bromine. Although the fumes are slightly toxic to crop plants, the material can be used in a greenhouse if the house is ventilated.

The Montreal Protocol on Substances that Deplete the Ozone Layer was adopted on September 16, 1987, at the Headquarters of the International Civil Aviation Organization in Montreal. The Montreal Protocol is an international treaty designed to protect the ozone layer by phasing out the production of a number of substances believed to be responsible for ozone depletion, of which methyl bromide is included.

The U.S. EPA has declared methyl bromide a class I threat to the earth's protective layer of ozone. As of January 1, 2001, the product can no longer be manufactured, nor is it legal to use. As of 2005, methyl bromide is not to be used in the United States and other developed countries except where critical use exemptions have been granted. There are active research programs in place to develop replacements for this product, but none to date have been released as satisfactory alternatives.

Formalin is often used to treat soil for seed sowing. It contains 40 percent formaldehyde and is effective against fungi and bacteria. The soil to be treated should be warmer than 60°F (15.5°C). Use 2.5 tablespoons (37.5 ml) of formalin in 1 cup (240 ml) of water. The mixture is sprinkled thoroughly over each bushel of soil mix. The containers are then covered with a plastic sheet and left for 24 hours. Following seeding, the containers should be watered thoroughly.

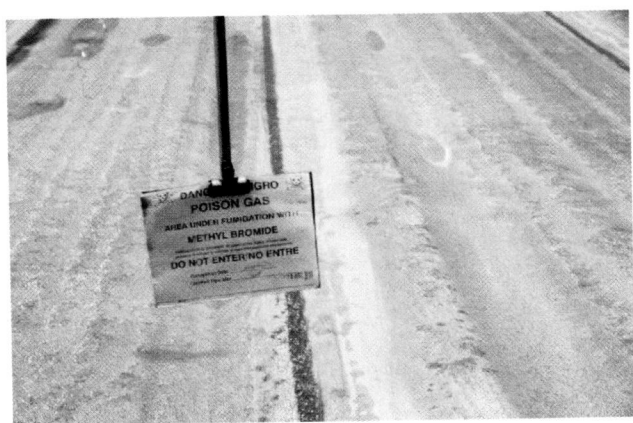

FIGURE 13-7 Methyl bromide gas is injected into covered soil (medium).

ACHIEVEMENT REVIEW

Select the best answer or answers to complete each statement. List the appropriate letter(s).

1 The process of killing all living cells, especially microorganisms, is known as:
   a. sterilization.
   b. fumigation.
   c. pasteurization.
   d. suberization.

2 Soil is steamed at a temperature of:
   a. 180°F (82°C) for 30 minutes.
   b. 212°F (100°C) for 30 minutes.
   c. 212°F (100°C) for 60 minutes.
   d. 140°F (60°C) for 30 minutes.

3 When steam is applied through buried perforated pipes, the depth of the soil on top of the pipes should be:
   a. one-half the depth of the soil below the pipes.
   b. the same depth as the soil below the pipes.
   c. twice the depth of the soil below the pipes.
   d. three times the depth of the soil below the pipes.

4 Surface steaming of soils requires thorough preparation of the area to be treated. This method of steaming is not as effective as the buried pipe method because the steam will penetrate to a depth of only:
   a. 4 to 6 inches (10 to 15 cm).
   b. 8 to 10 inches (20 to 25 cm).
   c. 12 to 15 inches (30 to 37.5 cm).
   d. 16 to 18 inches (40 to 45 cm).

5 Soil containers, such as clay pots, flats, or metal bushels, may be steamed in a confined space. This method is known as:
   a. pot steaming.
   b. surface steaming.
   c. chamber steaming.
   d. injection steaming.

6 Many chemicals used to treat soils are toxic to humans. When these chemicals are being applied, which of the following are required?
   a. Rubber gloves only
   b. A gas mask or respirator
   c. Protective clothing for the entire body
   d. All of the above

7 Chemicals in the form of gases diffuse throughout the soil mass. Thus, the soil must be at the proper temperature if the chemicals are to be effective. The minimum required soil temperature is:
   a. 40°F (4.4°C).
   b. 50°F (10°C).
   c. 60°F (15.5°C).
   d. 65°F (18.3°C).

8 Chloropicrin (tear gas) must be injected under a gas-proof cover or water seal. The soil must be aired for _____ days before it can be planted.
   a. 2 to 4
   b. 7 to 10
   c. 10 to 21
   d. 21 to 24

9 Methyl bromide is applied at the rate of 4 pounds (1.8 kg) per 100 square feet (9.28 m^2) for 24 hours. At this rate, it will kill:
   a. weeds only.
   b. weeds, insects, and nematodes.
   c. bacteria and most fungi.
   d. weeds, insects, nematodes, bacteria, and most fungi.

10 Methyl bromide is a very poisonous gas. Because it is odorless, a warning agent is added. The agent is:
   a. Vapam.
   b. chloropicrin.
   c. formalin.
   d. hydrogen sulfide.

11 Soils treated with methyl bromide should not be used for:
   a. tomato seedlings.
   b. petunias and peppers.
   c. carnations and salvia.
   d. marigolds and zinnias.

12 When soils to be used for seed sowing are treated with formalin, the soil must be covered with a plastic sheet for _____ hours for complete fumigation.
   a. 4
   b. 8
   c. 12
   d. 24

Nutrition
and Watering

CHAPTER 14

Essential Elements for Plant Growth

OBJECTIVES:

After studying this chapter, the student should have learned:

- the major and trace mineral elements needed for plant growth.

- the plant symptoms due to a deficiency of each element.

- the deficiency symptoms to determine if an element is mobile or immobile.

- how a nutrient element is tested to determine if it is essential for plant growth.

Types of Mineral Elements

Like people, plants require proper nutrition for optimum growth and development. The process of photosynthesis supplies plants with food in the form of carbohydrates. In addition to these substances, plants require specific mineral nutrient elements.

Plants are typically not selective in absorbing nutrient elements from the growing medium. This means that the presence in a plant of a particular element is not an indication that the element is necessary for the growth of the plant. If plants could distinguish between the materials they absorb from the soil, it is likely that they would not absorb chemical weed killers applied to the soil to destroy them.

Test to Determine if an Element Is Essential

A mineral is considered necessary or essential for plant nutrition if it passes a three-part test.

1. Does a lack of the element make it impossible for the plant to complete the vegetative or reproductive (flowering and fruiting) stage of life?
2. Can the **deficiency** symptom of a particular element be prevented or corrected only by supplying the missing element?
3. Is the element used directly in the nutrition of the plant, apart from any possible effect in correcting some microbiological or chemical condition in the soil or culture medium?

Essential Elements

Sixteen elements are considered to be **essential elements** for plant growth: carbon, hydrogen, oxygen, nitrogen, phosphorus, potassium, calcium, magnesium, sulfur, iron, manganese, copper, boron, zinc, molybdenum, and chlorine.

Macro (major) **elements** are needed by plants in large amounts. The first nine elements listed are considered to be macro elements. Trace elements, or micronutrients, are needed in extremely small quantities. The remaining seven elements in the list are micronutrients. There are vast differences between the quantities of macro and trace elements within plants. For example, a well-grown chrysanthemum plant contains 5 to 6 percent nitrogen (a macro element) on a dry weight basis. This amount is equal to 50,000 to 60,000 **ppm.** In contrast, the amount of copper (a trace element) required to maintain a plant may be as little as 0.006 to 0.008 percent, or 6 to 8 ppm. Although trace elements are required in very small amounts, they play a major role in the growth of plants. Insufficient amounts of any trace element within the plant may result in abnormal growth.

As stated previously, plants also take in elements other than those needed for growth. For example, plants often take in silicon and selenium, which are not necessary to plant nutrition. In some areas of the far western United States, plants take up so much selenium that they become toxic. Cattle feeding on these plants often die of selenium poisoning. Silicon, on the other hand, is taken up by many grass species and increases stem rigidity.

The fact that plants take up chemicals from the soil is used by growers as a means of controlling insects and diseases. Some insect poisons (**insecticides**) are applied to the soil. These chemicals are slowly taken up by a plant and are moved through the vascular system to all parts of the plant. Sucking insects such as red spider mites, white-flies, and aphids are killed by the poison as they nourish themselves on juices of the plant.

Distribution of Mineral Elements in Plants

Plants are made up of mineral compounds and nonmineral (organic and inorganic) compounds. (Organic compounds contain carbon.) The mineral elements of a plant can be obtained by removing the water from plant tissues by drying. (Water is 90 to 95 percent of the weight of a plant.) The remaining plant tissue is burned at 932°F (500°C) for 4 hours or longer. The resulting ashes are the mineral elements of the plant less carbon. These minerals may equal from 1 percent to more than 25 percent of the dry weight of the plant.

The amount of mineral elements in plants varies with the kind of plant, the organ or tissue of the plant, and its age. For example, seeds normally have a higher mineral content than fruits. Small roots have more minerals than large roots. Once the nutrient content of a plant is known, a chemical analysis can determine a deficiency of one or more elements.

Mineral Element Absorption

Plants use carbon, hydrogen, and oxygen from the air and soil water to make simple foods by the process of photosynthesis. These substances are used to make amino acids, proteins, and protoplasm. Carbon, hydrogen, and oxygen are not classified as mineral nutrient elements. However, they are essential to plant growth. By regulating the water supply and the amount of carbon dioxide in the greenhouse atmosphere, the grower has some control over the amounts of these elements available to plants.

Most mineral elements are taken up by plants through the roots. Moderate amounts are also absorbed through the leaves, stems, and twigs. Because plants can absorb nutrients through the leaves, the grower can apply needed trace elements in the form of foliar sprays. The foliar application of nutrients is becoming more common as plant scientists learn how the surroundings of the plant affect the uptake of these materials.

Absorption through the Root System

A plant with a mature, healthy root system will absorb all of the nutrients it needs for good growth. If the root system is damaged by disease, insects, low oxygen levels, or high levels of soluble salts in the soil, the nutrient uptake is reduced. As a result, the plant soon develops symptoms of nutrient deficiency. In many cases, the plant grower can tell what is lacking because certain elements work inside the plant in specific ways and produce recognizable symptoms. Unfortunately, by the time the problem is seen in the leaves, the damage to the plant may be so extensive that it is usually not reversible and the plant dies.

There are several ways in which mineral elements make their way to the root system. The flower grower may supply the elements in the form of **fertilizers** (Chapter 15). In addition, the breakdown of the soil and the organic matter in the soil releases mineral elements. Plant roots respire (breathe) carbon dioxide. This carbon dioxide combines with the soil water to form very weak carbonic acid, as well as other acids. These acids break down the organic matter in the soil, the soil particles, and fertilizers. The resulting organic salts or ions are then taken up by the roots.

Inorganic salts are broken apart by a process called dissociation. At any one time, both molecules and separate ions of the salt are present. A molecule consists of two or more ions. For example, potassium chloride (KCl) supplied as a fertilizer is dissociated in the soil solution into potassium (K^+) and chloride (Cl^-) ions. Ions with negative ($-$) charges are called anions. The ions are then absorbed by the roots through a semipermeable membrane, which surrounds each cell within the root. In exchange, the membrane will release a cation (H^+) or anion (OH^-) to maintain ionic balance.

Cation Exchange Capacity

The plant always takes in mineral elements in the form of ions. There is no difference between the ions supplied by an inorganic fertilizer and an organic fertilizer. Once dissociation occurs, the plant cannot distinguish between the same ions from different sources.

The quantity of nutrients in the plant roots is always greater than the concentration of nutrients in the surrounding soil solutions. Thus, energy is required if additional nutrients are to be absorbed. The energy is supplied by the carbohydrates manufactured in the leaves by photosynthesis. These sugars are transported to the roots where they are broken down by the process of respiration.

The actual process of nutrient uptake by plants is controlled by the cation exchange capacity (**CEC**), or base exchange capacity, of the soil. This action is associated with the clay particles (colloids) of a mineral soil. Organic materials, such as manures and peats, also have a cation exchange capacity. Generally, the CEC of organic materials is less than that of clays. The CEC of soils can be determined and is expressed as the "exchange capacity in milliequivalents per 100 grams of dry soil." Testing laboratories that specialize in greenhouse soils with peat often express CEC in milliequivalents per 100 ml.

The clay colloid has a negative (−) surface charge and attracts **cations** (+ charge). As explained previously, hydrogen ions are released when carbonic acid (H_2CO_3) is formed from the combination of the hydrogen from the soil water and the carbon dioxide resulting from root respiration. These hydrogen ions, which have a positive (+) charge, exchange their positions in the soil solution for positively charged cations held on the surface of the clay colloid. For example, two H^+ ions replace one Ca^{++} ion on the colloid, making the calcium available for absorption by the roots (Figure 14-1). A single H^+ ion replaces one K^+ ion because they each have one positive charge. It is for this reason that fertilizers high in ammoniacal nitrogen tend to reduce the pH of the soil whereas fertilizers high in nitrate nitrogen raise the pH of the soil.

Anion Exchange Capacity

Plants also need anions for good growth. Nitrates (NO_3^-), chlorides (Cl^-), and sulfates (SO_4^-) are examples of anions. Negatively charged anions are not attracted by the negative charge of the clay colloid. Thus, they are not held like cations. This repulsion effect can be compared to the reaction between two negatively charged magnets. When they are placed near each other, they repel each other. The negative charge of an anion keeps it in solution unless it is either absorbed by the plant or is lost through leaching. Because anions are not held firmly by soil colloids, they can be washed from soil solutions when large amounts of extra water are applied. This loss of anions is a problem with sandy type soils because they have a very low exchange

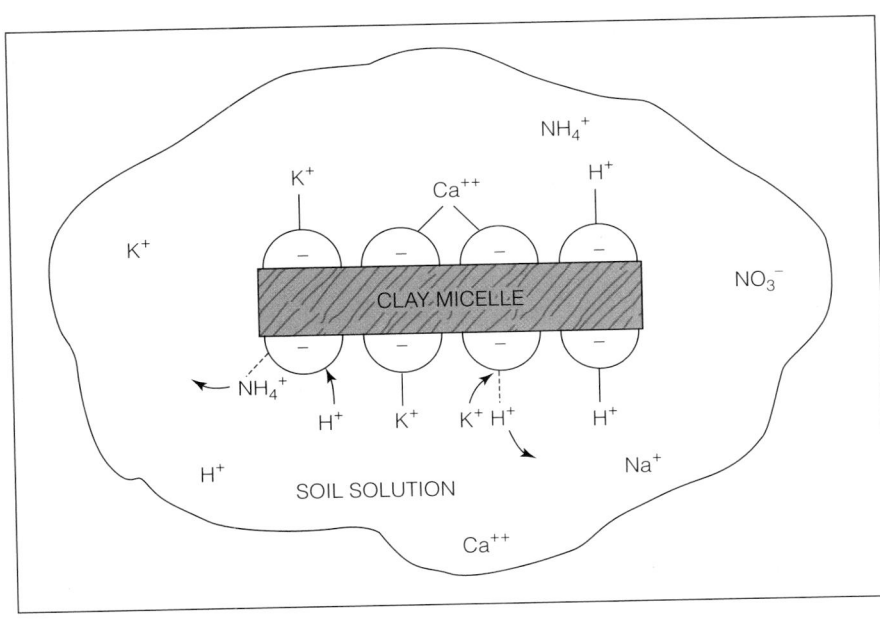

FIGURE 14-1 The colloid surface is negatively charged, so it absorbs cations. The cations are "exchangeable" in that if one leaves, another takes it place. (*Note: The semicircles used to represent exchange sites are for illustration only.*)

capacity. Sandy soils must be fertilized often to maintain the nutrient levels needed for good growth.

If there are too many anions in the soil solution, the total soluble salt level becomes too high. However, it is simple to remove the excess ions from the soil by leaching or applying large amounts of water. Leaching is a process that growers can use if their crops are overfertilized.

pH Effect on Nutrient Absorption

Nutrient uptake by plants is also affected by the pH of the soil. The term **pH** refers to the degree of acidity or alkalinity of the soil solution or growing medium. The pH scale ranges from 0 (acid) to 14 (alkaline). The neutral point occurs at 7.0 (Figure 14-2).

A neutral pH means that the number of H^+ ions (acid) is equal to the number of alkaline OH^- (hydroxyl) ions. If a soil is said to be acid, it has more hydrogen (H^+) ions than hydroxyl (OH^-) ions. The pH of an acid soil is less than 7.0. An alkaline soil contains more hydroxyl ions than hydrogen ions. An alkaline soil has a pH above 7.0.

Only a very small percentage of the water molecules in the soil solution are dissociated in H^+ and OH^- ions. This means that it is very awkward to express the concentrations of these ions in regular chemical terms. To solve this problem, the pH system was devised. Mathematically, pH is defined as the common logarithm of the reciprocal of the hydrogen ion concentration. This statement is written as follows:

$$pH = \text{logarithm } \frac{1}{(H^+)}$$

FIGURE 14-2 pH scale for solutions.

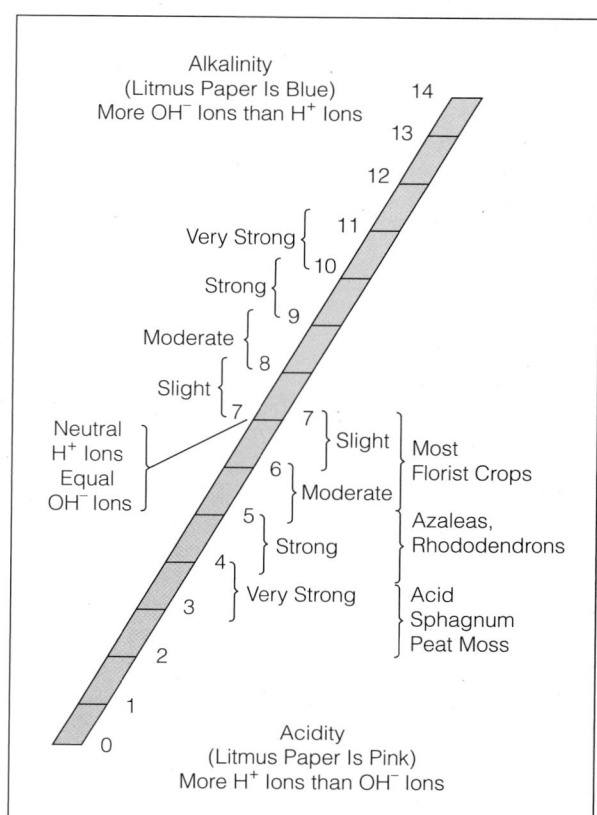

Because the pH scale is logarithmic, a soil with a pH of 6 is 10 times more acid than a soil with a pH of 7. Similarly, soil of pH 5 is 10 times more acid than one of pH 6. By the law of logarithms, a pH 5 soil is 100 (10 × 10) times more acid than a soil of pH 7. This factor of 100 is the reason it is more difficult to raise the pH of an acid soil from pH 5 to pH 7 than it is to raise the soil from pH 6 to pH 7. This means that if 10 pounds (4.5 kg) of ground limestone are required to raise the pH from 5 to 6, then 100 pounds (45.4 kg) are required to raise the pH from 5 to 7. In actual situations, other factors influence the ratios to change the values.

The pH of the soil or growing medium affects the availability of nutrients, particularly the trace elements. The pH range of most soils is from 4.0 to 8.5. The majority of agricultural plants grow best in a pH range of 5.5 to 7.0. Ericaceous plants are an exception. For example, azaleas and rhododendrons grow best at a pH of 5.0 to 5.5.

At the highest and lowest pH values, the nutrient elements are combined with other chemicals in the soil and are not available for plant growth. In very acid soils, the trace elements combine with aluminum and iron. Trace elements in very alkaline soils form insoluble compounds with carbonates, chlorides, and sulfates.

The availability of nutrients in organic field soils differs from that of soilless, peat-lite type growing media. Figures 14-3 and 14-4 show the pH ranges for optimum nutrient availability in an organic field soil and a soilless peat-lite media. For the field soil (Figure 14-3), the optimum pH range is 5.8 to 6.5; for peat-lite media (Figure 14-4), the pH range is 5.3 to 6.0. The soil pH does not

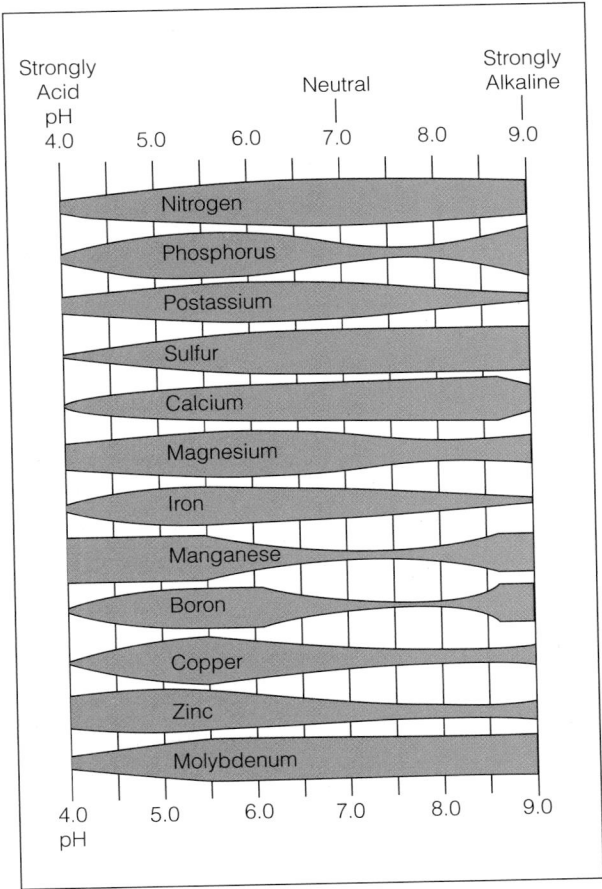

FIGURE 14-3 Effect of pH on nutrient availability in organic field soils.

FIGURE 14-4 The effect of pH on the availability of nutrients for plant growth in a soilless peat-lite-type growing media.

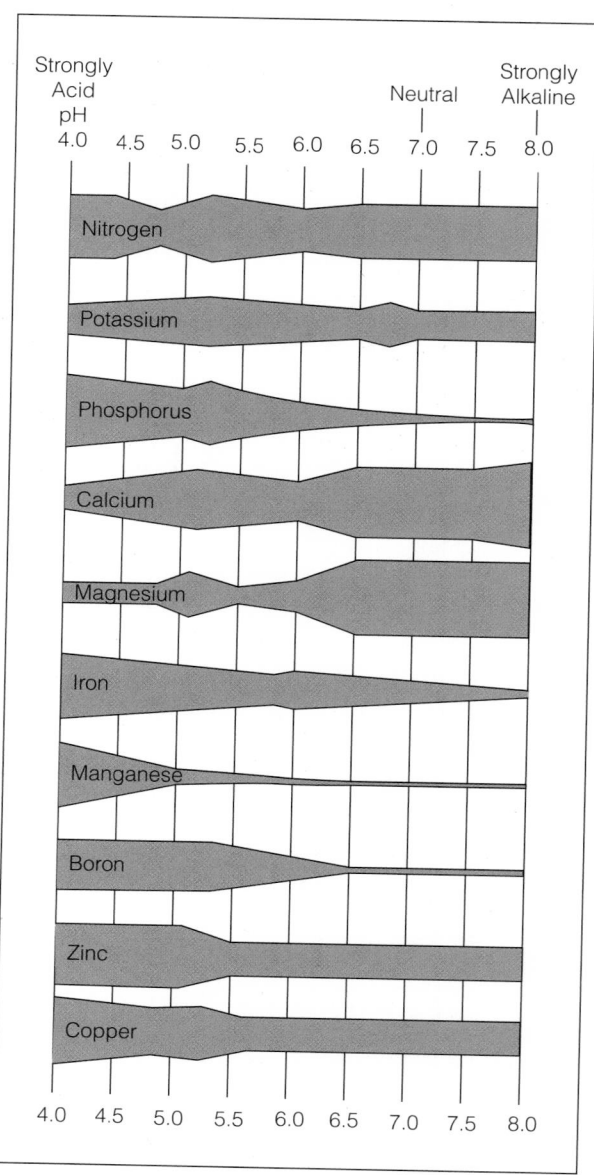

have the same effect on all elements. The macro elements react to pH in a different manner than do the trace or **micro elements.**

Essential Macro Elements

Nitrogen (N)

Nitrogen is very important to plant growth and is usually found in the largest amounts in plant tissues. On a dry weight basis, 5 to 6 percent of a healthy chrysanthemum is nitrogen.

Most plants cannot absorb the elemental (N) form of nitrogen. Legumes, such as beans and peas, are an exception to this rule. They can fix elemental nitrogen from the air directly into forms usable by the plant through a symbiotic relationship with a bacterium. Nonleguminous plants absorb nitrogen in the form of nitrates (NO_3^-) and ammonium (NH_4^+). Young plants in the early stages of growth seem to take up ammonium ions better than nitrate ions.

Some crops, such as rice, grow best when all of the nitrogen they require is in the ammonium form.

Nitrogen is present in the soil in both organic and inorganic compounds (Figure 14-5). Organic compounds are broken down by soil microbes and chemical reactions in the soil. Nitrogen is released for absorption by plants. Inorganic nitrogen in the ammonium form is changed by soil microorganisms through a process called nitrification. *Nitrosomonas* **bacteria** change the ammonium ions (NH_4^+) to nitrite ions (NO_2). The nitrite ions are poisonous to plants even in small amounts. Thus, they are quickly changed to nitrate ions (NO_3^-) by *Nitrobacter* **bacteria.**

Once nitrogen in any form is taken inside the plant, it is changed to an amide form. The amide then combines with certain acids to produce amino acids, the basic building blocks of protein.

Nitrogen is also a major ingredient of chlorophyll. The green color of plants is due mostly to the nitrogen content of chlorophyll.

Relationship between Supply of Nitrogen and Growth

Figure 14-6 shows a plant growth curve for various levels of nitrogen. This curve is typical for all of the essential elements. If the supply of nitrogen is low, the amount of growth is limited. When sufficient nitrogen is available, optimum growth occurs. The plant enters a stage of **luxury consumption** if additional nitrogen is made available. Finally, a toxic level of nitrogen is reached and the plant is damaged. Plant injury from an oversupply of trace elements is much less common than injury due to macro elements, except for boron.

An oversupply of nitrogen fertilizer increases the length of the growing period and delays the maturity of the crop. As a result, plants make rank growth, which is soft and succulent. Too much nitrogen often increases disease.

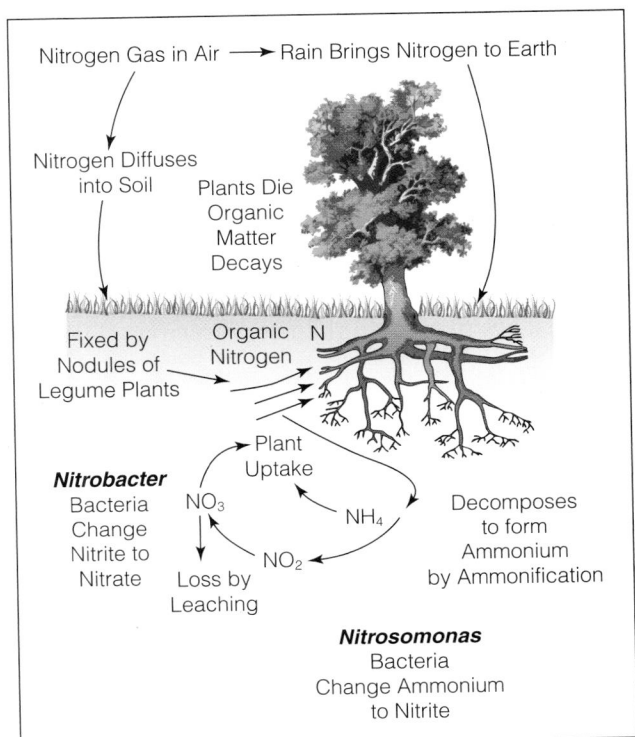

FIGURE 14-5 Nitrogen cycle and nitrification process in soils.

FIGURE 14-6 An idealizedgrowth curve for nitrogen.
A similar curve may be drawn
for any of the other nutrient
elements.

FIGURE 14-7A Lower leaves that turn completely yellow or
yellowish-green illustrate the typical symptoms of nitrogen
deficiency on this ageratum plant. Upon severe nitrogen
deficiency, the lower leaves will dry up and drop.
(Courtesy Dr. Brian Whipker, North Carolina State University.)

FIGURE 14-7B Nitrogen deficiency symptoms on cleome.
(Courtesy Dr. Brian Whipker, North Carolina State University.)

The proper amount of nitrogen produces vigorous, vegetative growth and deep green color. Flower crops generally need more nitrogen in their juvenile stages of growth than at maturity.

Deficiency

A nitrogen deficiency (lack) causes the leaves of plants to lose their dark green color. The resulting chlorosis (loss of green color) usually shows first on the lower, basal leaves of the plant. The upper leaves remain green. If the deficiency is not corrected, but increases, the lower leaves turn brown and die and the upper leaves turn light green in color (Figures 14-7A and 14-7B).

Nitrogen is said to be mobile within the plant. It is this **mobility** that causes the older leaves of a plant to lose their color while the young leaves remain green. If the roots are not able to take up enough nitrogen to meet the growing needs of the plant, a change occurs in the nitrogen compounds in the older basal leaves. The protein nitrogen is turned into a soluble form. This mobile form of nitrogen is translocated to the growing regions of the plant where it is used to make protoplasm. The basal leaves lack sufficient nitrogen to carry on life processes and soon die.

Not all nutrient elements are mobile within the plant. The condition of mobility helps the grower to diagnose a nutrient deficiency.

Phosphorus (P)

Phosphorus is the second of the three macro nutrients. The level of phosphorus in a plant is much less than that of either nitrogen or potassium. On a dry weight basis, the optimum level of phosphorus in the leaf tissue of a healthy chrysanthemum is 0.25 to 0.35 percent. Ornamental leaf tissues commonly have a ratio of 18 to 20 parts of nitrogen to one part of phosphorus.

Most of the phosphorus taken up by plants is in the form of the primary orthophosphate ion ($H_2PO_4^-$). Smaller amounts of secondary orthophosphate ions (HPO_4^-) and organic phosphorus compounds are also absorbed.

Phosphorus has several important functions. It must be available in sufficient quantities early in the life of the plant to assist in cell division and differentiation of the cells into the tissues of the reproductive parts of plants. It is also required for root growth and the formation of seeds. Both the respiratory and photosynthetic processes require phosphorus for high-energy phosphate bonds.

Deficiency

A phosphorous deficiency results in reduced plant size and a deep green color. The heightened color is due to an increase of nitrates in the plant tissue. A lack of phosphorus in tomatoes and marigolds causes the undersides of the leaves to turn purple. The purple color is easily noticed on seedlings. However, cold weather and overwatering can also cause the same reaction. Care should be taken in determining if a plant is reacting to a lack of phosphorus or some other factor.

Phosphorus is slightly mobile within plant tissues. Unlike nitrogen, however, a lack of phosphorus does not cause leaf symptoms (Figures 14-8A and 14-8B).

FIGURE 14-8A Phosphorus, when it is deficient, is demonstrated with lower leaf purple or reddish coloration and often appears when the plants are grown wet or cold. This geranium has typical lower leaf symptoms because of cold growing. (*Courtesy Dr. Brian Whipker, North Carolina State University.*)

FIGURE 14-8B This rosemary plant has been grown too wet and is phosphorus deficient. (*Courtesy Dr. Brian Whipker, North Carolina State University.*)

An excess of phosphorus in plant tissues rarely causes any special leaf symptoms. Phosphorus and iron levels in plants act in opposition to each other. At a high level of phosphorus, an iron deficiency may develop. Similarly, a high level of iron may cause a phosphorus deficiency.

When phosphorus is applied to a mineral soil, it is quickly fixed or held by the soil colloid. This means that the plant cannot absorb the phosphorus and it cannot be leached easily from the root zone. In artificial soils, such as the peat-lite mixes, very little phosphorus is held. Thus, much of the phosphorus applied as a preplanting dressing can be lost if the crops are overwatered.

Potassium (K)

Potassium is the third macro element required by plants. It is absorbed as the potassium ion (K^+). Soils normally have a high level of potassium, but only a small amount is available for plant growth in a form that can be absorbed easily. Nitrogen and phosphorus are converted into compounds that aid in plant growth. Potassium is found in plant tissues as a soluble, inorganic salt. It may be combined with the anions of certain organic acids, but it is not an integral part of plant tissue and is found primarily in the cytoplasm.

Potassium is essential for growth and usually is needed in large amounts. However, plants often show luxury consumption of potassium. That is, it is absorbed in amounts much greater than the levels actually needed for good growth.

Potassium cannot be replaced completely by other elements. Although sodium and lithium are closely related chemically to potassium, they can only partially make up for a lack of this element.

It is thought that potassium is needed for the production and translocation of carbohydrates. A lack of potassium may result in a temporary increase in carbohydrates followed by a rapid decrease. A potassium deficiency may occur while the plant tissues have a high level of reducing sugars.

Potassium also appears to be involved in nitrogen metabolism. Plants lacking potassium have a high water-soluble nitrogen level. A shortage of potassium seems to be made worse when the plants are fertilized with ammonium forms of nitrogen. If plants lacking potassium are supplied with ammonium nitrogen, they rapidly show severe symptoms of tissue injury. A buildup of ammonium ions apparently causes the injury. It appears that the reduced nitrogen is not changed into protein and thus increases until it becomes toxic.

Young, actively growing tissues often have much more potassium than older tissues. This condition is to be expected because potassium is highly mobile. The older leaves usually show a deficiency first with a **marginal necrosis,** or browning. As the deficiency becomes more severe, the leaf begins to lose some of its green color and turns yellow. The yellowing begins at the edge of the leaf and progresses inward toward the center. The symptom of the deficiency may begin as a series of yellowish or white spots near the margin of the leaf. The spots enlarge and soon join each other to form a mass of dead tissue (Figures 14-9 and 14-10).

Plants contain more potassium than any other monovalent (single charge) cation. The amount of potassium in the tissues frequently equals or is greater than the nitrogen concentration. On a dry weight basis, potassium may be 5 to 6 percent of a healthy chrysanthemum plant. Some carnations have shown 9 percent potassium. Roses usually average 2 to 3 percent.

FIGURE 14-9 Initially potassium deficient petunia plants are smaller in size when compared to the control plants receiving adequate levels of potassium. *(Courtesy Dr. Brian Whipker, North Carolina State University.)*

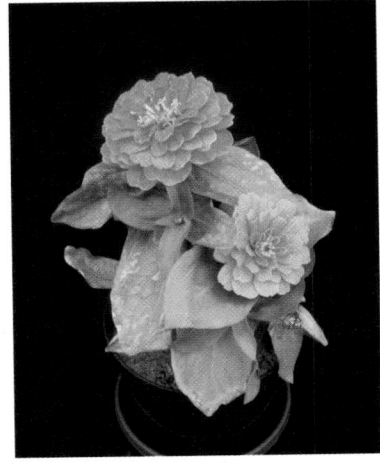

FIGURE 14-10 Under severe potassium deficiency conditions, the lower leaves of this zinnia plant have become necrotic. *(Courtesy Dr. Brian Whipker, North Carolina State University.)*

Calcium (Ca)

Calcium is absorbed in the ionic form (Ca^{++}). Most of the calcium in a plant is in the form of calcium pectate in the middle lamellae of the cell walls of the leaves. The calcium prevents the leaching of the mineral salts from the cells. Much akin to bone structure, the stiffness of a plant is due to calcium.

Calcium forms salts with organic acids. Some species contain calcium in the form of calcium oxalate. *Dieffenbachia* is known as "dumbcane" because of the effects of its high level of calcium oxalate crystals. If a leaf or stem of *Dieffenbachia* is chewed, the calcium oxalate crystals penetrate the tissues of the tongue and throat and cause instant swelling and possibly suffocation.

Calcium appears to influence the growth of the apical meristems of the plant. Calcium is immobile and is not translocated from older tissue to younger regions of growth. As a result, a calcium deficiency first appears at the youngest growing points of the plant, the shoot tips and the root tips (Figure 14-11).

In many plants, a calcium deficiency affects terminal bud development. Either the terminal bud may fail to develop at all or only partial development occurs. Carnations also show dieback with the leaf tips turning up at a 90 degree angle. Other plants, such as poinsettias, are affected by calcium deficiency through lack of leaf vein elongation, creating a puckered look because the interveinal tissue continues to grow (Figure 14-12).

On a dry weight basis, chrysanthemum leaves usually contain from 1 to 6 percent calcium. Calcium levels below 1 percent are thought to be too low for normal growth.

Magnesium (Mg)

Magnesium has a vital role as part of the chlorophyll molecule. With nitrogen, it is the only mineral element contained in chlorophyll. Magnesium appears to be related to phosphorus metabolism. A number of plant enzyme systems require magnesium to work properly.

Magnesium is mobile within plant tissues. Thus, symptoms of a lack of magnesium show up first on older leaves in the form of **interveinal** (between the

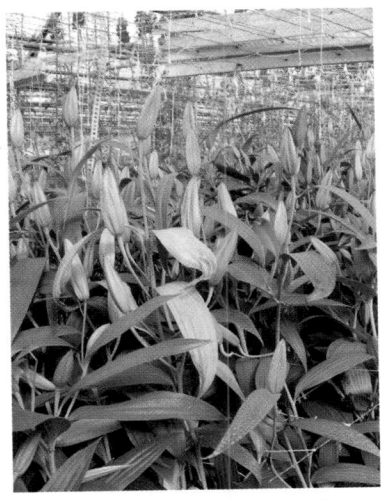

FIGURE 14-11 Recent research at Cornell University has revealed that upper leaf necrosis of lilies grown for cut flowers is related to calcium deficiency. These symptoms can be reduced by increasing air movement over the crop, thus increasing evapotranspiration carrying more calcium to the upper part of the plant.

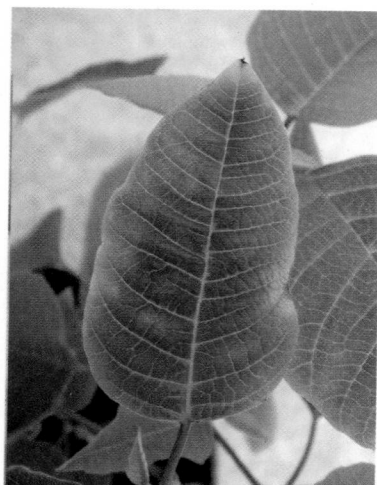

FIGURE 14-12 Calcium deficiency can occur on poinsettia leaves when prolonged cloudy weather occurs during times of rapid leaf growth. *(Courtesy Dr. Brian Whipker, North Carolina State University.)*

veins) **chlorosis**. The veins remain green as the interveinal tissue turns yellow. As the condition worsens, the leaves become uniformly pale yellow in color and then turn brown and die. A lack of magnesium in carnations appears as a group of irregular blotches on older leaves (Figures 14-13A and 14-13B). Magnesium deficiency is often easily corrected through the application of Epsom salts.

Magnesium deficiency in chrysanthemums may be made more severe by excessive applications of potassium and may be more pronounced in certain cultivars. The problem is not corrected by adding magnesium in the form of Epsom salts or magnesium sulfate. The amount of potassium supplied to the crop must be reduced. The plant can then absorb the magnesium needed for growth. Potassium prevents the proper absorption of magnesium and is said to be an antagonistic ion.

On a dry weight basis, the normal magnesium content of chrysanthemum leaves is from 0.3 to 1 percent. A level of less than 0.3 percent is considered to be too little for optimum growth.

Sulfur (S)

It is believed that sulfur is taken up from the soil in the form of the sulfate ion (SO_4). Small amounts of sulfur may be taken in through the leaves as sulfur dioxide (SO_2). If a large amount of sulfur dioxide is present, the leaf may suffer from air pollution injury (see Chapter 13). Sulfur seems to be involved in the formation of chlorophyll, but it is not a component of the chlorophyll molecule.

Growth is retarded if too little sulfur is available. A chlorosis develops that affects the whole plant, except for the topmost leaves. This chlorosis is similar to that caused by a nitrogen deficiency. Detection is only assured through a mineral test of the plant tissue.

Superphosphate contains sulfur. Thus, when it is used in the soil, sulfur is made available to plants. Gypsum, or calcium sulfate, also contains sulfur. A number of low-analysis fertilizers contain some sulfur as an impurity. Sulfur may also be included in the filler used to bring fertilizer packages up to the legal weight.

FIGURE 14-13A Lower leaf interveinal chlorosis is the initial symptom of magnesium deficiency on poinsettia. *(Courtesy Dr. Brian Whipker, North Carolina State University.)*

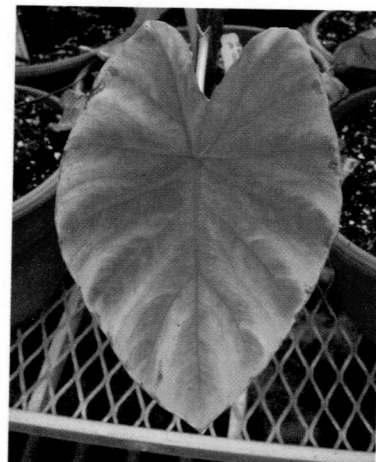

FIGURE 14-13B Leaf interveinal chlorosis from magnesium deficiency on *Colocasia*. *(Courtesy Dr. Brian Whipker, North Carolina State University.)*

Essential Trace or Micro Elements

The trace or micro elements are as important as the macro elements to normal, healthy plant growth. Micro elements are contained by plants in much smaller quantities than the macro elements. Iron, boron, manganese, copper, zinc, molybdenum, and chlorine are the essential trace mineral elements.

For many years, the importance of the trace elements was unknown. Part of the reason for this was the use of low-analysis fertilizers, such as 5-10-5 and 10-10-10. Large amounts of trace elements are contained as impurities in the fillers used in these fertilizers. As more chemically pure fertilizers were introduced, such as 20-20-20, it was found that crops needed more than nitrogen, phosphorus, and potassium for normal growth. As a result, modern high-analysis fertilizers, such as 15-5-15, are often fortified with trace element additives.

Iron (Fe)

Iron is sometimes considered to be a macro element because symptoms resulting from a lack of iron are so common. But iron is actually a trace mineral. Plants take up iron in the form of ferrous ions (Fe^{++}) or complex organic salts. Iron may also be absorbed in the ferric ion (Fe^{+++}) form. Plants may contain large amounts of ferric iron, but still show severe iron deficiency symptoms. Ferric iron is held within the plant tissue and is not available for plant use. The active growth form of iron is the ferrous iron.

Iron acts with certain enzyme systems that carry on respiration. It is also required in the formation of chlorophyll. Unlike magnesium, it is not a component in the chlorophyll molecule.

Iron is immobile. Thus, a deficiency of iron appears first in the youngest, developing tissues as an interveinal chlorosis. The large veins remain green while the small veins lose their color. If the deficiency is not corrected, the leaves turn light yellow and then almost completely white. Growth ceases and necrosis or death of the leaves follows soon after the leaves become white (Figures 14-14A and 14-14B).

FIGURE 14-14A Interveinal chlorosis of the upper foliage occurs when iron is deficient. In most cases it is caused by too high of a substrate pH as with azaleas, but can also occur with root rots, wet soils, or cold growing conditions.

FIGURE 14-14B In severe cases of iron deficiency, the leaves are bleached white as with this pansy plant. (*Courtesy Dr. Brian Whipker, North Carolina State University.*)

An iron deficiency may be caused by a high concentration of lime or calcium in the soil. By making the soil more acid (lowering the pH), more iron is made available for plant uptake. Cold soils can also cause iron deficiency in ericaceous species.

Iron deficiencies generally are corrected by adding iron chelates (key-lates) to the soil. A chelate is an organic-inorganic compound in which the metal is inactive. The word *chelate* is based on a Greek word meaning "claw." Thus, the iron ion in the chelate is held as if by a claw so that it cannot be tied up by the soil. As a result, it is available to the plant as a nutrient. Chelates of zinc, manganese, and copper can also be added to soils, but **chelated iron** is most commonly used in ornamental crop production.

An optimum level of iron in chrysanthemum leaf tissues is considered to be 100 to 400 ppm.

Boron (B)

Boron is taken up by plants in several ionic forms. The function of boron in plant growth is not known fully. It may be important in protein synthesis. It is thought to be related to calcium and potassium metabolism. A high level of calcium in carnations causes boron deficiency. In addition, it is thought to be involved in both water and sugar translocation in plants. Boron is immobile in most crops. The initial symptom of a lack of boron is the absence of growth in the terminal meristem regions of the shoots and roots (Figures 14-15A and 14-15B). Another symptom is the development of a large number of axillary side shoots at the terminal growing point. This growth, known as **witches' broom**, is typical in roses, chrysanthemums, and snapdragons. A boron deficiency in 'Giant Indianapolis White No. 4' chrysanthemums causes a halo effect within 48 hours just as the flowers become ready for harvest. The halo ruins the flowers for sale. This condition occurs only on the 'Indianapolis' cultivars. It has not been reported on any other chrysanthemums.

Chrysanthemum leaf tissue normally contains 25 to 100 ppm of boron. At concentrations above 100 ppm, boron causes marginal burn of leaves. Boron toxicity symptoms appear first on the older leaves and then progress up the stem.

FIGURE 14-15A Distorted new growth of *Cordyline* sp. that is boron deficient. *(Courtesy Laura Pottorff, Colorado Department of Agriculture.)*

FIGURE 14-15B Distorted new growth of *Dracaena* sp. that is boron deficient. *(Courtesy Laura Pottorff. Colorado State University Extension Regional Specialist.)*

Manganese (Mn)

Plants absorb manganese in the form of the manganous ion (Mn^{++}). Very small amounts are sufficient for normal growth. Manganese, like iron, may be taken in directly through the leaves. It is used in the active growing parts of the plant and in certain enzyme systems that oxidize other elements, such as iron. A manganese excess may cause iron deficiency symptoms. This situation can only be determined through a mineral test of the tissue.

Manganese is immobile. Thus, a deficiency appears first in the new growth. Manganese and iron deficiencies may be confused because the symptoms are similar. However, leaves that lack manganese have a very fine network of green veins. Even the smallest veins of the leaf remain green. The normal manganese content of chrysanthemum leaves is in the range between 250 and 500 ppm.

Copper (Cu)

Plants absorb copper in the form of the cupric ion (Cu^{++}). Ornamental plants require very small amounts of copper. Only molybdenum is used by plants in smaller amounts. If large amounts of copper are taken up by the plant, severe tissue damage often results.

Copper is needed for the proper function of enzyme systems. It stabilizes chlorophyll and delays its breakdown. Thus, copper helps to increase the effective life of leaves. A copper deficiency may be confused with a lack of boron because the symptoms are similar. The growing points of the plant are affected first. Eventually, terminal die-back occurs, and a witches' broom develops.

Highly organic humus or peat soils may be deficient in copper. Copper sulfate or chelated copper can be added to correct the problem. The normal foliar concentration of copper in chrysanthemums is in the range from 20 to 500 ppm. However, levels close to 10 to 20 ppm have been noted.

Zinc (Zn)

Roots absorb the zinc ion (Zn^{++}). Leaves also take in zinc from foliar sprays of zinc sulfate and other zinc compounds. It is also obtained from zinc-based compounds and from zinc-based fungicides sprayed on leaves to control diseases.

The role of zinc in plant nutrition is not known fully. A lack of zinc often causes stunted growth and the failure of seeds to form properly, sometimes known as rosetting. A type of chlorosis known as **mottle leaf** develops on peach and citrus trees due to a deficiency of zinc. Ornamental crops grown in greenhouses rarely show symptoms of a zinc deficiency, which is often confused with an aphid infestation, and also causes rosetting of terminal stems. The zinc content of chrysanthemum leaves is from 20 to 50 ppm.

Molybdenum (Mo)

Only very small amounts of molybdenum are required by plants. It appears that molybdenum is used in the nitrogen cycle in the formation of nitrogen compounds and the breakdown of nitrates. Often, the first signs of a molybdenum deficiency are the same as those of a simple nitrogen deficiency. The plant loses its good green color in the lower leaves. When a plant lacks molybdenum, nitrates are not absorbed properly, even when there are large amounts of nitrates in the soil.

At pH values below 6.0, molybdenum may be strongly tied up by soil colloids. This means that it is not available for growth. If molybdenum is not added to peat-lite mixes, poinsettia stock plants often develop serious deficiency symptoms. The problem can be avoided by adding the recommended trace elements and liming the medium to pH 6.0 or above.

Chlorine (Cl)

The use of chlorine by some plants for growth and development has only recently been established. Chlorine is probably absorbed in the ion form (Cl^-).

Symptoms of a chlorine deficiency are wilting followed by a mild chlorosis and bronzing of older leaves.

Reduced leaf size and slower growth are the only consistent general deficiency symptoms. Many plants show sharply reduced growth without any visible leaf symptoms.

Too much chloride in soils causes more problems than a lack of chloride. The excess chloride may be due to an overapplication of fertilizers such as potassium chloride. It may also be due to contamination of wells used for irrigation. Sodium or calcium chloride salts used to melt snow and ice in the winter may find their way into the wells through water runoff.

Excess chloride causes scorching or firing (necrosis) of leaf tips or margins, bronzing, premature yellowing, and abscission (dropping) of leaves. Generally, there is no chlorosis. See Table 14-1 for a quick reference key to nutrient deficiencies.

Other Elements

Several other mineral elements are found in plant tissues. Although these elements are classified as nonessential, plant growth is improved when some of these elements are present.

Fairly large amounts of sodium, aluminum, and silicon are found in plant tissue. Levels of sodium greater than 10,000 ppm (1 percent of the dry weight) in the leaf tissue of certain cultivars of carnations cause the buds to fail to open. In 'Sweetheart Supreme' azaleas, symptoms of leaf scorch and reduced growth occur from concentrations of 6,000 ppm sodium in the leaf tissues.

Aluminum forms complex salts in plant tissues. It is found in the greatest amounts in root tissue. Plants growing in acid soils contain more aluminum than plants growing in alkaline soils. Aluminum is more readily available at a low pH.

Hydrangea flower color is directly related to the amount of free aluminum in the soil. If there is enough aluminum, certain cultivars will produce blue flowers. If free aluminum is lacking, these same flowers will be pink or red. The flower grower uses phosphorus fertilizers to control the availability of free aluminum in soils. Phosphorus ties up the aluminum ions in a form that the plant will not absorb.

Cereal grain straws and similar types of plants contain large amounts of silicon. Except for oxygen, soils contain more silicon than any other element. To date, a specific function in the growth of plants has not been determined for silicon. Silicon has been shown in recent studies at the University of California at Davis to disrupt larval development of leaf miners in chrysanthemum leaves.

Table 14-1

Quick Reference Key to Nutrient Deficiency Symptoms	
Deficiency Symptom	**Deficient Nutrient**
I. Major symptom is yellow color (chlorosis) of the leaves.	
A. Entire leaf blade is chlorotic or partially chlorotic.	
1. Chlorosis is on the basal lower leaves. Leaves turn brown and die (necrosis) and fall.	Nitrogen
2. Chlorosis is uniform on all leaves.	Sulfur
B. The chlorosis is mainly between the veins of the leaves (interveinal)	
1. The older leaves show the symptoms first.	Magnesium
2. New growth, younger leaves are affected, appearance may be interveinal to almost white in severe cases.	Iron
a) Interveinal chlorosis is seen with even the smallest veins remaining green giving leaves a lacelike appearance; there may be gray or tan necrotic spots noted also.	Manganese
b) Young leaves have interveinal chlorosis, tips and lobes of leaves may remain green. In severe cases, veinal chlorosis develops followed by rapid necrosis of the leaves.	Copper
c) Newly developing leaves are very small, may be straplike, growing tip appears as a rosette.	Zinc
II. Leaf chlorosis is not the major symptom.	
A. Basal lower leaves are affected first.	
1. Leaves develop dark green color due to high nitrate accumulation, plants become stunted in growth; in young seedlings, tomatoes and marigolds especially, a purple pigment develops on the underside of the older leaves.	Phosphorus
2. Margins of older leaves become chlorotic and then necrotic. With some crops, small chlorotic spots also appear at the leaf margin. Continued deficiency results in interveinal chlorosis, necrosis, and finally leaf drop.	Potassium
B. Terminal new growth is affected first.	
1. Buds die, axillary vegetative shoots frequently develop giving the plant a witches' broom effect. Young leaves may become thick, older leaves become brittle and break easily. Purpling of new buds occurs on some crops as does a cracking of the stem, which becomes corky in appearance. Hollow stems of snapdragons are a sign of a deficiency.	Boron
2. New growth fails to develop and dies, buds that have formed abort, surrounding leaves turn necrotic. Leaf margins may curl under and also fail to form properly. Root growth is restricted. Roots appear short and stubby with blunt ends.	Calcium

Silicon does affect the cation exchange capacity of the soil. A large amount of silicon in a clay mineral causes the exchange capacity and the fertility level of that soil to be increased.

Fluorine levels are of concern to flower growers. Fluorine occurs widely in nature, but usually in very small amounts. However, the water in some parts of the country has a natural fluorine content of 3 to 4 ppm. Some cities add 1 ppm fluoride to their local water supplies to help prevent tooth decay.

Members of the *Liliaceae* (lily family) react to as little as 0.25 ppm fluoride in the water supply by showing leaf tipburn. This kind of fluorine injury can be avoided by liming the soil mixture to a pH of 6.0 to 6.5. At this pH, fluoride ions are held by the calcium ions and are unavailable for absorption by the plants.

Fluorine contamination in the air from industrial plants manufacturing superphosphate fertilizers can damage plants. For example, gladioli in Florida have been damaged in this manner although the fields were many miles from the source of the fluorine.

Fumes from glass cleaning compounds containing fluorine may cause marginal leaf burn on roses.

ACHIEVEMENT REVIEW

Select the best answer or answers to complete each statement. List the appropriate letter(s).

1 A nutrient element is considered to be essential for plant growth when a deficiency (lack) of the element:
   a. prevents the plant from completing both the vegetative and flowering-fruiting stages of life.
   b. can be prevented by supplying the element.
   c. may prevent a soil microorganism from working properly.
   d. has no effect on plant growth.

2 Essential elements that are supplied from the air and water are:
   a. carbon, nitrogen, phosphorus.
   b. carbon, hydrogen, sulfur.
   c. nitrogen, hydrogen, oxygen.
   d. carbon, hydrogen, oxygen.

3 The nutrient elements needed in the greatest amounts by plants are called:
   a. essential elements.
   b. trace elements.
   c. macro elements.
   d. minor elements.

4 A deficiency symptom that results from a deficiency of a nutrient element in the plant may be due to:
   a. a lack of the element in the soil.
   b. insect and disease injury to roots that prevents uptake.
   c. high soluble salt injury to the roots.
   d. none of the above.

5 Nutrient elements are absorbed from soils in the form of:
   a. fertilizers applied by the grower.
   b. individual ions resulting from molecule dissociation (breakdown).
   c. individual molecules.
   d. a combination of b and c.

6 The positive charge of the cation means that the soil particles will:
   a. hold it too tightly for plants to absorb.
   b. allow it to be exchanged through cation exchange capacity.
   c. repel it, after which it is lost by leaching.
   d. have no effect at all.

7 The pH of the soil or medium is a function of:
   a. the size of the individual soil particles.
   b. the balance of H^+ to OH^- ions.
   c. its water-holding capacity.
   d. the amount of pore space.

8 Plants such as rhododendrons and azaleas grow best when the pH is:
   a. below 5.0.
   b. 5.0 to 5.5.

c. 5.5 to 6.5.

d. at any level.

9 Nitrogen and potassium are two macro elements that are said to be mobile within the plant. As a result, a deficiency of either element appears first in:

a. the newly developing growth.

b. the entire plant.

c. the older basal leaves.

d. only the roots.

10 A purple color on the underside of the seedling leaves of marigolds and tomatoes is a sign that they:

a. lack potassium and have been grown warm.

b. lack phosphorus and have been grown warm.

c. lack nitrogen and have been grown cold.

d. lack phosphorus and have been grown cold.

11 Deficiencies of iron and manganese can be confused because they both appear first in new growth. A manganese deficiency can be distinguished from an iron deficiency because:

a. the leaves turn evenly yellow.

b. the large veins remain green.

c. all of the veins remain green, resulting in a lacelike appearance of the leaf.

d. the veins of the leaves turn white.

12 Three elements that affect the terminal growing regions (meristematic tissues) of the plant are:

a. nitrogen, calcium, potassium.

b. boron, calcium, copper.

c. boron, calcium, zinc.

d. iron, boron, zinc.

13 Only one mineral element is a component of the chlorophyll molecule. This element is:

a. nitrogen.

b. iron.

c. magnesium.

d. molybdenum.

14 If excessive boron is applied to growing plants, the resulting damage appears:

a. on the entire plant.

b. at the terminal growing regions first.

c. at the base of each separate leaf.

d. first as a necrosis of the leaf tips on the basal leaves.

15 Fluorine may damage certain foliage plants that are members of the *Liliaceae* (lily family). To prevent such damage:

a. the soil is limed to a pH of 6.0 to 6.5.

b. perlite is added to the soil to absorb the fluorine.

c. the soil is acidified to a pH of 5.0 to 5.5.

d. high levels of nitrogen fertility are maintained.

15

Fertilizers and Methods of Application

OBJECTIVES:

After studying this chapter, the student should have learned:

- the difference between organic and inorganic fertilizers.
- materials used in fertilizers that supply nitrogen.
- the soil reaction of nitrogen fertilizers (pH effect).
- materials used in fertilizers to supply phosphorus and potassium.
- sources of trace element nutrients.
- the difference between biweekly applications and fertilizing at every watering.
- the difference between the two basic methods of fertilizer proportioner operation.
- how to solve simple fertilizer arithmetic problems.
- facts regarding total soluble salts, including where they come from, how they are controlled, and how they are measured.

Commercial flower growers use soil mixtures that often contain little or no fertilizers. Their crops are grown in small volumes of soil, such as 6-inch (15-cm) deep benches, pots, and market packs. The limited amount of soil used is a poor storehouse for water and nutrients. Thus, growers must apply fertilizers to crops on a regular schedule.

The farmer puts fertilizer on the ground before planting and follows up with a later side dressing. However, the commercial flower grower may apply fertilizer each time the crop is watered. There is a wide assortment of fertilizers from which to choose.

Organic Fertilizers

Organic fertilizer materials are made from not only plant and animal residues, but also mined minerals. The Organic Materials Review Institute

(OMRI) classifies "crop fertilizers (CF) to contain one or more recognized plant nutrients. Used primarily for their plant nutrient content, they may be applied to the soil or to the foliage of plants. They include compost, animal manures, blended fertilizers, mined minerals, micronutrients, blood/bone meals, and plant extracts that make plant nutrient claims." Although urea is an organic formulation of nitrogen fertilizer in the chemical sense, it is not classified as a certifiable organic fertilizer.

Except for urea, commercial growers use few organic fertilizers. There are several reasons why they are not popular: They have a low nutrient analysis; nitrogen is primarily in the ammoniacal form leading to low pH in the soil after uptake and soft plant growth; the nutrient elements are released too slowly for good plant growth; the nutrient release rate is unknown; and too little nutrients are released. In addition, organic fertilizers must be used in large amounts to obtain the desired nutrition levels, these fertilizers are scarce, and they are usually expensive.

To equal the fertility supplied by 100 pounds (45.4 kg) of inorganic 10-5-10 ($N-P_2O_5-K_2O$), 1 ton of cow manure must be applied. The ease of applying 100 pounds (45.4 kg) of 10-5-10 fertilizer compared to that of 1 ton of manure is obvious.

The following organic materials are used on occasion by flower growers. The nutrient supplied is given in parentheses: dried blood (N), tankage from slaughter houses (N), cottonseed meal (N), guano (bird and bat droppings collected from large deposits on islands off the shore of South America) (N), and bone meal (both raw and steamed) (P_2O_5). Steaming makes the phosphorus in the bone meal easier for plants to take up. No other comments will be made on organic fertilizers.

Inorganic Fertilizers

Single-Element Materials

A large number of single-element fertilizer materials are available to growers. The nutrient requirements of the crop govern the selection of the fertilizer materials. The essential elements needed for plant growth were discussed in Chapter 14. The fertilizer materials that can supply these elements are described in the following sections.

Nitrogen

There are more nitrogen suppliers than carriers of other nutrient elements. The material used depends upon the crop being grown, the stage of crop growth, the season of the year, the method used to apply the fertilizer, and the soil reaction desired.

Crop Being Grown

Some crops have higher nitrogen needs than others. A well-grown chrysanthemum crop may contain 5 to 6 percent nitrogen on a dry weight basis. Chrysanthemums are said to be heavy feeders and must be fertilized accordingly. Roses, on the other hand, average only 3 to 4 percent nitrogen and high fertility rates waste fertilizer and may contaminate the environment.

Bulb crops such as tulips, narcissus, and hyacinths do not require additional nitrogen. The bulb typically contains all of the elements needed for flowering.

Stage of Crop Development

Young, tender seedlings or rooted cuttings must be fertilized with weak solutions of fertilizers. If full-strength fertilizers are added before the root systems are fully developed, the plant may be seriously damaged. Following planting, 7 to 10 days must be allowed before full-strength fertilizer is added on a regular schedule.

In the early stages of growth, most crops need more nitrogen than other nutrient elements. As the crop matures and stem strength increases, less nitrogen is needed. This is especially true of chrysanthemums and snapdragons. In fact, fertilizing chrysanthemums after the flower buds form can reduce the shelf life of the flower after sale.

Method of Fertilization

When fertilizer is applied with every watering, less material is used than when plants are fertilized every 2 weeks or monthly. The amount of fertilizer salts that are applied to a crop at one time is limited to prevent damage. This method of fertilization is discussed in more detail in the next section.

Soil Reaction Required

Nitrogen fertilizers can change the pH of the soil. If ammonium sulfate alone is used over a long period of time, the cation ammonium is taken up by the plant, releasing hydrogen (H^+) and causing the soil to slowly become more acid and the pH to decrease. Limestone must be added to correct the problem. Otherwise, the pH may reach such a low value that other nutrient elements, especially the trace elements, are made less available for plant uptake.

If calcium nitrate is used as the only source of nitrogen, the soil pH will slowly rise. The nitrate anion is taken up by the plant, releasing a hydroxyl (OH^-) and causing the soil to become more alkaline.

Some fertilizers have a balance of acid and alkaline ions. The continued use of such fertilizers does not cause a great change in pH. By selecting the proper material, a fertilizer program can be designed to produce a neutral soil reaction.

Nitrogen Sources—Acid Reaction

Ammonium Sulfate [$(NH_4)_2SO_4$]

This compound is a by-product of the process of manufacturing fuel coke. It is also produced by mixing anhydrous ammonia liquid with sulfuric acid under carefully controlled conditions. The nitrogen content of ammonium sulfate is 20 percent. The constant use of ammonium sulfate as a fertilizer lowers the pH of the soil.

Ammonium sulfate should not be applied to cold soils during the winter because the conversion of ammonium to the nitrate form is too slow. A buildup of **soluble salts** may occur. The maximum rate of applying ammonium sulfate is 1 pound to 100 square feet (0.45 kg to 9.28 m²) of area. Applying more than this amount at any one time may cause soluble salt injury.

Ammonium Nitrate (NH_4NO_3)

Pure ammonium nitrate has a total nitrogen content of 35 percent. This compound absorbs moisture from the air. Thus, conditioning materials are added to prevent caking and the formation of lumps. As a result, the nitrogen content is reduced to 33.5 percent. One-fourth of this amount is in the ammonium form and three-fourths is in the nitrate form. Nitrogen from ammonium nitrate is quickly available to plants. At any one time, the maximum amount that can be applied to growing crops is 0.75 pound to 100 square feet of area (0.34 kg to 9.28 m^2).

Ammonium nitrate fertilizer is a strong oxidizer and was combined with diesel fuel to create the explosives used in the World Trade Center bombing in 1993 and the Oklahoma City bombing in 1995. Therefore, its sale and distribution has been severely cut back by the Department of Homeland Security. Organizations such as The Fertilizer Institute are working with government officials to maintain ammonium nitrate as a viable fertilizer product for agriculture.

Ammonium Phosphate

Monoammonium phosphate ($NH_4H_2PO_4$) and diammonium phosphate [$(NH_4)_2HPO_4$] are the two important forms of ammonium phosphate. To obtain fertilizer grade material, phosphate rock is treated with sulfuric acid. It is then mixed with ammonia in the proper proportions. Monoammonium phosphate contains 11 percent nitrogen and 48 percent available phosphorus. Diammonium phosphate contains 21 percent nitrogen and 53 percent phosphorus. Because of the double ammonium ion, diammonium phosphate makes the soil more acid than monoammonium phosphate. Either form is applied at the rate of 1 to 2 pounds to 100 square feet (0.45 to 0.90 kg to 9.28 m^2) of area.

Nitrogen Sources—Alkaline Reaction

Sodium Nitrate (NaNO$_3$)

Sodium nitrate, or nitrate of soda, is the sodium salt of nitric acid. Chilean nitrate of soda results when the natural crude ore called caliche is leached with hot water. This ore also contains very small amounts of the trace elements boron, copper, iodine, magnesium, manganese, and several others.

Manufactured nitrate of soda is made from synthetic ammonia by chemically changing it to nitric acid and sodium carbonate. Both the Chilean and manufactured nitrates contain 16 percent nitrogen and 26 percent sodium. Sodium nitrate has a moderately high salt index. It should not be applied at a rate greater than 1 pound to 100 square feet (0.45 kg to 9.28 m^2) of area.

Calcium Nitrate [Ca(NO$_3$)$_2$]

The calcium salt of nitric acid is manufactured by reacting lime or limestone with nitric acid. Calcium nitrate is an excellent source of nitrate nitrogen and water-soluble calcium. The material is very hygroscopic (water absorbing) and should be stored in moisture proof bags or other containers.

Fertilizer grade calcium nitrate contains 15.5 percent nitrogen. Calcium nitrate has a lower salt index than sodium nitrate and can be applied safely at a higher rate to growing crops. However, at any one time, no more than 1.5 pounds (0.68 kg) should be used to fertilize 100 square feet (9.28 m^2) of area.

Neutral Reaction

Urea [$CO(NH_2)_2$]

Urea is a synthetically produced organic nitrogen source. Urea nitrogen is quickly absorbed by plants. It is manufactured by subjecting ammonia and carbon dioxide to a high pressure. Fertilizer grade urea contains 45 percent nitrogen. Urea is completely water soluble. It is commonly applied as a solution consisting of 0.5 to 0.75 pound (0.23 to 0.34 kg) of urea in 25 gallons (94.6 l) of water applied to 100 square feet (9.28 m^2) of area. Even though this compound contains carbon and is chemically classified as organic, it is not considered to be a naturally organic compound.

Ureaformaldehyde, a 38 percent nitrogen material, is obtained by reacting urea with formaldehyde. The resulting product is sold under the names Borden's 38®, Nitroform®, and Uramite®. The nitrogen in these compounds is made available to plants slowly. These materials may be used as long-lasting sources of nitrogen for both pot plants and cut flower crops. The release of nitrogen is controlled by warm temperatures. These materials should not be used on cool-grown crops such as carnations or snapdragons. The normal rate of application is 3 to 6 pounds (1.35 to 2.73 kg) to 100 square feet (9.28 m^2) of area. These materials are not broadcast on the surface of the soil. They must be mixed thoroughly with the soil. As with urea, this compound contains carbon and is chemically classified as organic, but it is not considered to be a naturally organic compound.

Potassium Nitrate (KNO_3)

Although potassium nitrate is used mainly for the potassium it supplies, it is also applied for its nitrate nitrogen content. Potassium nitrate is the potassium salt of nitric acid. It contains 14 percent nitrate nitrogen and 44 percent potash (K_2O). Although it is generally considered to be neutral in reaction, the long-term use of potassium nitrate does cause a slight rise in the pH of the media. Potassium nitrate can be used both in solution and as a dry fertilizer. It is applied to growing crops at the rate of not more than 0.75 pound (0.34 kg) to 100 square feet (9.28 m^2) of area.

Phosphorus Sources

Superphosphate

Superphosphate is the most commonly used source of phosphorus for plants. To obtain superphosphate, equal parts by weight of finely ground phosphate rock and sulfuric acid are mixed. The sulfuric acid reacts with the rock to make soluble phosphorus. In addition to the resulting superphosphate, calcium sulfate or gypsum is also formed. Superphosphate contains from 14 to 21 percent phosphorus (P_2O_5).

There are two grades of superphosphate with a higher analysis: double phosphate containing 32 percent P_2O_5, and treble superphosphate with 40 to 47 percent available P_2O_5. These materials are made by reacting single superphosphate with phosphoric acid. The higher analysis phosphorus products contain less gypsum than single superphosphate and are more readily available, thus making them a better choice for short-term crops.

The movement of phosphorus ions in the soil is very slow. As a result, it is important to distribute the fertilizer evenly. It is recommended that phosphate fertilizers be mixed with the soil before the crop is planted. The normal rate of applying 20 percent superphosphate is 5 pounds (2.28 kg) to 100 square feet (9.28 m^2) of bench area. For bulk soils, 2 pounds of 20 percent superphosphate are added to 1 cubic yard. One pound (0.45 kg) of treble superphosphate is added to each cubic yard (0.58 m^3) of soil.

Monocalcium Phosphate [Ca(H$_2$PO$_4$)$_2$]

Monocalcium phosphate is made by treating a high analysis rock phosphate with phosphoric acid (H$_3$PO$_4$). The available phosphorus in monocalcium phosphate is 50 to 55 percent. This material supplies phosphorus to plants more rapidly than superphosphates. It also supplies calcium. Monocalcium phosphate is applied at the rate of 1 pound (0.45 kg) to 100 square feet (9.28 m^2) of area.

Monoammonium and Diammonium Phosphate

These materials were described under the heading of nitrogen sources. However, both contain much more phosphorus than nitrogen.

Steamed Bone Meal

Bone meal is an organic source of 2 percent nitrogen and 25 percent phosphorus. It supplies these nutrients slowly to plants. Generally, it is used when potting bulb crops such as tulips, narcissus, and hyacinths in ground beds. There is a question of the value of using bone meal in this way because bulbs normally contain all of the nutrients they need for blooming and it is too slowly available for short production times.

Phosphate Rock

This material is a natural rock containing one or morecalcium phosphate minerals. It is the basic material from which the superphosphates are obtained. Finely ground rock phosphate releases phosphorus very slowly through natural weathering processes. The rate at which phosphorus is made available is so slow that this material is not recommended as a greenhouse fertilizer.

Potassium Sources

Potassium Nitrate (KNO$_3$)

Potassium nitrate is the most useful of the various potassium sources because both of its ions are needed for plant growth. Its nutrient content and rate of application were given in the section on nitrogen sources.

Potassium Sulfate (K$_2$SO$_4$)

Potassium sulfate is the potassium salt of sulfuric acid. It is also known as sulfate of potash. The potash (K$_2$O) content is 39 to 42 percent. The maximum rate of application of potassium sulfate on a growing crop is 0.75 pound (0.34 kg) to 100 square feet (9.28 m^2) of area.

Potassium Chloride (KCl)

Muriate of potash, or potassium chloride, is the cheapest form of potassium that can be applied to a crop. It provides the greatest amount of salts for each unit of nutrient applied. Potassium chloride is available with either 50 percent or 60 percent K_2O. It is applied at the rate of not more than 0.5 pound (0.23 kg) to 100 square feet (9.28 m^2) of area.

Complete Fertilizers

Dry Application

Complete fertilizers contain all three major nutrient elements for plant growth: nitrogen, phosphorus, and potassium. Bags of these materials list the percentage content of the major elements. The first number is the percentage of nitrogen (N), the second is the percentage of phosphorus (P_2O_5), and the third is the percentage of potash (K_2O) (Figure 15-1).

Fertilizers used for dry application are usually low in analysis, such as 5-10-10, 10-10-10, 6-8-6, or similar combinations. These materials typically are applied before planting. Few growers use low analysis complete fertilizers on a regular schedule. The nutrient content is low and is available at too slow a rate for quality crop growth. If these fertilizers are applied in large amounts, the resulting soluble salt levels are great enough to cause plant injury.

Liquid Application

Inorganic fertilizers applied as a solution have a higher total nutrient content than dry fertilizers. Typical analyses for these fertilizers are 15-30-15, 20-10-20, and 20-5-30 (Figure 15-2). Actually, these materials are completely soluble fertilizers used for liquid application. They are not liquids until they are dissolved in water. Soluble fertilizers are discussed in more detail later in this chapter.

Controlled-Release Fertilizers

All fertilizers have a degree of controlled nutrient release. However, only a few products are designed to release available nutrients slowly.

Urea Formaldehydes (UFN)

Nitroform®, Borden's 38®, and Uramite® were among the first of the controlled-release fertilizers to be developed. Each fertilizer is produced from the reaction of urea with formaldehyde.

The urea formaldehydes contain 38 percent nitrogen, which is nearly 85 percent available over a period of 6 months or longer. The remaining 15 percent is available so slowly that it cannot be relied upon as a nutrient. A warm temperature is the main factor that makes nitrogen available to the plants. It is recommended that these products be used during the spring, summer, and early fall months. If they are used during the winter months, especially on cool-grown crops such as snapdragons or carnations, it is likely that too little nitrogen will be made available. As a result, the crops will be deficient in this element.

Urea formaldehyde fertilizers must not be added to soils or soilless mixtures before steam sterilization. The high temperature used to steam soils causes all

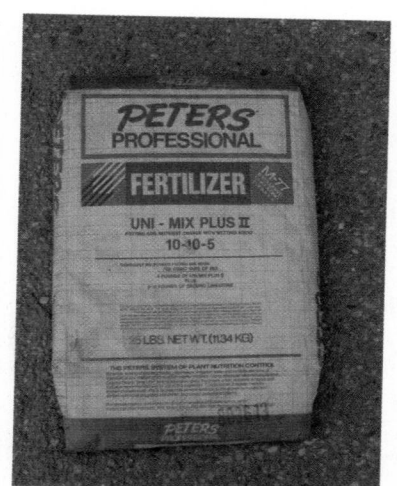

FIGURE 15-1 A bag of low analysis fertilizer incorporation into growing medium as a starter charge.

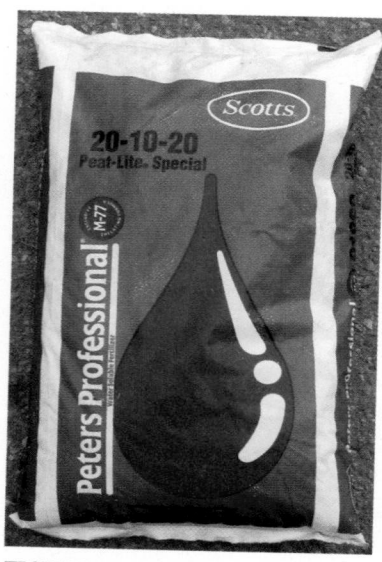

FIGURE 15-2 A high analysis 20-10-20 water soluble fertilizer. This particular formulation has a high nitrate to ammonium ratio.

of the nitrogen to be released at once. As a result, there is crop damage. These fertilizers are added to soils after steam sterilization and after the soil has cooled.

Magnesium Ammonium Phosphate (Magamp®)

Magamp® is a magnesium ammonium phosphate fertilizer developed by W.R. Grace Co., Inc. The analysis of this fertilizer is 7-40-6. It also contains 12 to 14 percent magnesium. Magamp® is supplied in two particle sizes: medium and coarse. The particle size is the method used to control the release of nutrients from the product. The smaller the particle, the more quickly available are the nutrients. That is, there are more units of fertilizer in a given amount of soil. Magamp® with medium size particles provides nutrients for as long as 3 months when it is used at the recommended rate. Magamp® with coarse particles provides nutrients for 6 months or longer.

Some crops grown in peat-lite or other soilless mixes fertilized only with Magamp® may develop a temporary lack of nitrogen during the first few weeks of growth. This lack occurs because all of the nitrogen is in the ammonium form. These mixes are said to have a slow rate of nitrification. In other words, the ammonium changes very slowly to nitrate forms of nitrogen.

Nitrate nitrogen fertilizers can be added to Magamp® to correct this problem. For example, 1 pound (0.59 kg) of potassium or calcium nitrate may be added to a cubic yard (cubic meter) of mix with Magamp®. The nitrate fertilizer supplies the required nitrogen until the ammonium can be converted to the nitrate form. Magamp® fertilizers are not recommended for use on cool temperature crops during the winter months.

Additional potassium may be required on long-term crops such as cut chrysanthemums. After 3 or 4 months of growth, 6 percent K_2O cannot maintain the required potassium level.

Soil moisture also affects the release of nutrients from Magamp® fertilizer. Moist soils usually have a more reliable nutrient supply than dry soils.

Additional Controlled-Release Fertilizers

Osmocote® is another popular brand of controlled-release fertilizer. High analysis, complete fertilizers such as Osmocote® are covered with a polymer coating. Fertilizers commonly used in this process are 14-14-14 and 19-6-12 products. The thickness of the coating and the type of coating material used control the length of time the nutrients are available for growth. The 14-14-14 product is prepared to last 120 days or 4 months. The 19-6-12 product is formulated to last 3 to 4 months or 6 to 9 months (Figures 15-3A and 15-3B).

As water slowly diffuses through the polymer shell surrounding the fertilizer, the shell expands slightly allowing a small amount of fertilizer solution to be released. This solution is then absorbed by the roots of plants. Some of the nutrients in the solution may be fixed by soil particles and made unavailable for plant growth. Other nutrients may be lost from the soil mass by the leaching action of the applied water. However, more nutrients per unit of CRF are taken up by plants than with standard liquid feed fertilizer formulations, thus justifying their use and expense.

The soil temperature has a great influence on the release of nutrients from CRFs. At a soil temperature of 40°F (4.5°C) or below, very little nutrient exchange takes place. At 80°F (26.5°C) and above, the rate of nutrient exchange may be

FIGURE 15-3A Osmocote® fertilizer particles of varying sizes.

FIGURE 15-3B A bag of Osmocote® 14-14-14 with a nutrient life up to 3 to 4 months.

greater than is desired for good plant growth. The levels of soluble salts may become too high.

CRF-treated soils must not be steam sterilized. The high temperature of steam sterilization melts the plastic particles and releases all of the fertilizer at once. Thus, a high soluble salt condition is created.

Moisture also affects the release of nutrients from CRF products. High moisture levels result in high fertility levels. Soils treated with CRF products before planting should be used immediately. Such soils must not be stored for more than a week. As long as treated soil stands, particularly moist soil, the soluble salt level increases.

Trace Elements

The use of trace elements or micro nutrients was described in Chapter 14. Extremely small amounts of the elements are required in plant nutrition. For example, a chrysanthemum may contain only 8 to 10 ppm of copper, compared to 50,000 to 60,000 ppm (5 to 6 percent) of nitrogen. However, a deficiency of a trace element will cause a plant to grow as poorly as if it lacked nitrogen.

Micro nutrients are available in many formulations, such as organic complexes or chelates, or inorganic sulfate forms. Micro nutrients may be added as individual elements through soil or leaf applications, or they may be mixed with other fertilizers and applied in a dry or liquid form.

Other commercial sources of trace elements commonly used in horticultural media are S.T.E.M. (soluble trace element mix), **FTE** (fritted trace elements) **555, ESMIGRAN**, and Micromax.

Frits

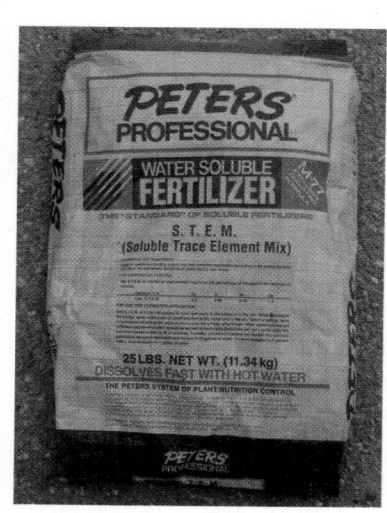

FIGURE 15-4 S.T.E.M. or Soluble Trace Element Mix includes a full complement of micronutrients required by plants in a readily available form that can be applied through a fertilizer injector.

Micro nutrients are available in a soluble form, S.T.E.M. (Figure 15-4), or in a form known as **frits**. Frits are made by cooling molten glass with a prepared solution of micro nutrient elements. Once the glass is cool, it is ground into a fine, dustlike powder. After frits are applied to the soil and are worked in, the weak acids that form in the soil slowly dissolve the soft glass particles. As the glass dissolves, the micro nutrients are released for plant growth.

Frits are safer to apply than soluble materials because there is less chance of applying too much. However, if frits are applied in excess, they can cause severe plant injury.

In general, frits are applied twice a year to growing crops such as roses. They are commonly applied in March and September. The amount used on growing crops depends upon the product purchased. Always follow the label recommendations.

Chelates

Flower growers generally use chelated forms of micro nutrient materials. Chelated micro nutrients are held in chemical bonds so that they are not taken up by soil particles and thus made unavailable. Chemical processes within plant roots act on the chelated micro nutrients and release them for uptake by the plant.

The most widely used chelates are compounds of iron. The iron (Fe) concentration may be 6 percent, 10 percent, or 12 percent. Manufacturers' recommendations for the amount to be applied should be followed. For bench grown crops, the recommended application is a 1-ounce (28.4-g) equivalent of material with a 12 percent iron content to 100 square feet (9.28 m^2) of area. If material with 6 percent iron is used, then 2 ounces (56.8 g) are applied to each 100 square feet (9.28 m^2) of area. Unless symptoms of iron deficiency are very severe, only one treatment is needed. A second application should not be applied prior to 3 weeks after the first treatment. The application of excess chelates can cause severe plant injury.

Zinc, manganese, and copper are also manufactured in chelated forms.

Boron and Molybdenum

The micro nutrients boron and molybdenum are applied as unchelated materials. Boron is commonly used in the form of borax (sodium tetraborate) or boric acid (H$_3$BO$_3$). Borax contains 11.8 percent actual boron (B). Boric acid contains 17.7 percent boron.

Borax is applied at the rate of not more than 1 ounce (28.4 g) to 100 square feet (9.28 m^2) of area once a year. Borax may be applied more often if the total application is not more than 1 ounce annually. When boron is applied with every watering, the amount used is 0.5 to 1 ppm. Overtreatment with boron can cause severe plant injury.

Salt Forms

PERK® is a micro nutrient fertilizer prepared from sulfate forms of the elements. An advantage of this product over FTE 555 is that it is used in relatively large amounts. Five pounds (2.95 kg) of PERK® are added to a cubic yard (cubic meter) of media (compared with 2 ounces of frit in the same soil volume). The larger the amount of material, the easier it is to ensure a better mix throughout the media mass.

ESMIGRAN is a recently introduced micro nutrient fertilizer. It is used at the rate of 4 to 6 pounds (2.4 to 3.5 kg) to a cubic yard (cubic meter) of media. Surface applications of this material on growing bench crops are made at the same rate on 100 square feet of bench area.

Methods of Fertilizer Application

Dry Applications

There are two basic methods of adding fertilizers to crops: dry or liquid. Dry type fertilizers usually supply nitrogen from an ammonium source such as ammonium sulfate. Phosphorus is supplied by superphosphate, and potassium is supplied by potassium chloride.

Complete fertilizers, or those that contain N-P_2O_5-K_2O, usually have a low analysis such as 5-10-5, 5-10-10, or 6-12-6 when they are prepared for dry application. A 5-10-10 fertilizer contains only 25 units of plant food. Thus, filler materials must be added to bring the weight to legal measure. However, 75 units of filler are not required. The raw materials that make up the 25 units of nutrients weigh more than 25 units. The nutrients may form as much as 75 to 80 percent of the total weight. This means that only an additional 20 to 25 percent must be added as filler.

Dolomitic limestone is often used as filler. This material adds weight and also supplies calcium and magnesium. It has an alkaline reaction that helps to offset the acidifying effect of the fertilizer.

Dry fertilizers are applied to the soil before the crop is planted. Digging distributes the material uniformly throughout the soil mass. Even distribution is especially important when superphosphate and liming materials are applied.

Osmocote® and Magamp® are classified as dry type fertilizers. They are applied and mixed into the soil before the crop is planted. On long-term crops, surface applications may be made from time to time. Surface applications of regular low analysis dry fertilizers are seldom used today. Liquid applications of nutrients are more accurate and efficient.

Liquid Application of Fertilizers

Solubility of Fertilizers

The fertilizer material must be soluble if it is to be used in liquid form. The fertilizer should be 100 percent soluble in cold water. However, this condition exists only when chemically pure materials are used. It is easier to dissolve many chemicals in hot water. If a solution is made up using 150°F (65.5°C) water, and the concentration is so strong that the saturation is close to 100 percent, some precipitation of the chemicals (formation of solids) takes place as the solution cools.

Table 15-1 lists the solubility of various chemically pure fertilizer materials. Commercial grade fertilizers are not chemically pure. Their solubility is slightly less than that of pure materials.

Salt Index

Fertilizers are salts. When they are added to soil, they increase the soluble salt level. The selection of the proper fertilizer material can help to keep the soluble salt concentration of a nutrient solution at a low level.

Table 15-2 lists the salt indexes of various commonly used fertilizers. The following example shows how the table is used. A soil test report shows that nitrogen must be added to the crop. One pound (0.45 kg) of nitrogen material (16 percent) is recommended for each 100 square feet (9.28 m²) of area. The grower may select from several materials. One pound (0.45 kg) of calcium

Table 15-1

Fertilizer Material	Ounces per Gallon (g/l)[1]			
	Cold Water		Hot Water	
Ammonia	13^{68}	(97.4^{20})	87^{140}	(651.6^{60})
Ammonium nitrate	158^{33}	$(1,183.4^{0.5})$	$1,163^{212}$	$(8,711.0^{100})$
Ammonium sulfate	94^{33}	$(704.1^{0.5})$	138^{212}	$(1,033.6^{100})$
Boric acid (H_3BO_3)	2.6^{33}	$(19.5^{0.5})$	52^{212}	(389.5^{100})
Calcium nitrate	355^{33}	$(2,658.9^{0.5})$	881^{86}	$(6,598.8^{30})$
Calcium sulfate (gypsum)	insoluble			
Diammonium phosphate	57^{32}	(426.9^{0})	142^{160}	$(1,063.6^{71})$
Dicalcium phosphate	insoluble			
Magnesium sulfate (Epsom salts)	94^{68}	(704.1^{20})	121^{104}	(906.3^{40})
Manganese sulfate	69^{40}	$(516.8^{4.4})$	93^{160}	(696.6^{71})
Monammonium phosphate	30^{33}	$(224.7^{0.5})$	231^{212}	$(1,730.2^{100})$
Monocalcium phosphate	insoluble			
Potassium chloride (Muriate of potash)	37^{33}	$(277.1^{0.5})$	75^{212}	(561.8^{100})
Potassium nitrate	18^{33}	$(134.8^{0.5})$	330^{212}	$(2,471.7^{100})$
Potassium sulfate	9^{33}	$(67.4^{0.5})$	32^{212}	(239.7^{100})
Urea	159^{76}	$(1,190.9^{24})$	—	

Source: Handbook of Chemistry and Physics, 88th Ed., 2008.
1. Superscript figures indicate the temperature of the water in degrees F (degrees C) used to dissolve the amounts given.
NOTE. Commercial grade materials are not chemically pure. Lesser amounts of these materials must be used to obtain a true solution.

Table 15-2

Fertilizer Material	Salt Index	Fertilizer Material	Salt Index
Nitrogen		**Potassium**	
Anhydrous ammonia, 82% N	47.1	Potassium chloride, 60% K_2O	116.2
Ammonium nitrate, 34% N	104.0	Potassium nitrate, 13% N, 44% K_2O	69.5
Ammonium sulfate, 21% N, 24% S	88.3	Potassium sulfate, 50% K_2O, 18% S	42.6
Urea, 46% N	74.4	Sulfate of potash-magnesia, 22% K_2O, 11% Mg, 22% S	43.4
Urea-ammonium nitrate solution:		Monopotassium phosphate, 52.2% P_2O_5, 34.6% K_2O	8.4
28% N (39% ammonium nitrate, 31% urea) 32% N	63.0	Potassium thiosulfate, 25% K_2O, 17% S	68.0
(44% ammonium nitrate, 35% urea)	71.1	**Sulfur**	
Calcium nitrate, 15.5% N	65.0	Ammonium thiosulfate, 12% N, 26% S	90.4
Sodium nitrate, 16.5% N	100.0	Ammonium polysulfide, 20% N, 40% S	59.2
Phosphorus		Gypsum, 23% Ca, 17% S	8.1
Ordinary superphosphate, 20% P_2O_5	7.8	Magnesium oxide, 60% Mg	1.7
Triple superphosphate, 45% P_2O_5	10.1	Magnesium sulfate, 10% Mg, 14% S	44.0
Monoammonium phosphate:		**Limestone**	
11% N, 52% P_2O_5	26.7	Calcium carbonate, lime, 35% Ca	4.7
10% N, 50% P_2O_5	24.3	Dolomite, 21.5% Ca, 11.5% Mg	0.8
Diammonium phosphate, 18% N, 46% P_2O_5	29.2		
Ammonium polyphosphate, 10% N, 34% P_2O_5	20.0		

Source: Western Fertilizer Handbook, 1985, The Interstate Publishers & Printers nitrate contains 15.5 percent nitrogen and has a **salt index** of 53. Sodium nitrate (16 percent nitrogen) has a salt index of 100. If 0.5 pound (0.23 kg) of ammonium nitrate (33.5 percent nitrogen) is used, the salt index is 52.5 (105 / 2 = 52.5). Four-fifths (1.76 kg) of a pound of ammonium sulfate (20 percent nitrogen) supplies a salt index of 55.2 (4/5 × 69 = 55.2). The choice of a material with a low salt index is preferred to the use of nitrate of soda, which has a salt index of 100. Similar comparisons for other fertilizers may be made using the salt index information.

Frequency of Application

Fertilizers in liquid form may be applied on a schedule. Some growers fertilize regularly every 2 weeks or every month. When the fertilizer is applied infrequently, the concentration of the solution must be high. When fertilizing once a month, the grower must apply 3 to 4 pounds (1.36 to 1.82 kg) of 20-20-20 fertilizer to 100 square feet (9.28 m²) of area. At the same time, however, the soluble salt level is becoming very high. As the crop uses the applied nutrients, the salt level will decrease. Each watering will cause a small amount of leaching to take place. This action reduces both the salt level and the nutrient content. Some time before the next scheduled application of fertilizer, the nutrient level of the soil cannot meet the needs of the crop. This condition of feast or famine as a result of fertilizing once a month creates excesses and deficiencies of nutrients.

These extremes can be reduced if the application schedule is reduced to once every 2 weeks or once weekly. A weaker fertilizer solution can be used if it is applied more often. Thus, the danger of injury from overfertilizing is reduced.

The preferred rate of applying liquid fertilizers is with every watering. An accurate mechanical proportioner is used to inject the small amount of concentrated fertilizer solution into the irrigation water. As a result, a weak fertilizer solution is applied to the crop each time it is watered.

Proportioning Equipment

There are two basic types of fluid injectors or proportioners. Inexpensive models are based on the Venturi principle. The Venturi principle is based on the pressure difference between the container of stock solution and the **irrigation** line. The pressure difference causes the stock solution to flow into the irrigation line. The most commonly used device is a siphon hose. One popular brand is the **HOZON**® proportioner (Figures 15-5 and 15-6).

When water moves through the faucet connection of a siphon hose proportioner, a low-pressure area develops at the top of the siphon tube. Atmospheric pressure on the surface of the stock solution forces the solution into the irrigation line.

Siphon hose proportioners are not very accurate because any change in the water pressure changes the amount of stock solution that is supplied to the irrigation water. The dilution ratios may range from 1:12 to 1:17 because of water pressure differences.

The second type of injector used is a positive displacement pump (Figure 15-7). It adds a known amount of concentrated stock solution to the irrigation line. Dosatron® is a popular, inexpensive example of this type of injector (Figure 15-8). Other types include a water pump, as in the Smith Measure-Mix®(Figure 15-9), or one metered electronically as with the H. E. Anderson injection system

FIGURE 15-5 In the HOZON®(Venturi-type) proportioner, atmospheric pressure forces concentrated stock solution up the tube from the reservoir. The low-pressure area also causes solution to be pulled into the tube.

FIGURE 15-6 A siphon-type proportioner. This device is attached in between the hose bib and the watering hose.

FIGURE 15-7 Positive displacement fertilizer injection pump uses water flow to drive pistons that draw concentrated stock solution, which is mixed with irrigation water.

(Figures 15-10A and 15-10B). These proportioners inject an accurate amount of stock solution into the water system. The displacement pump is generated by the flow of water through the system.

Dilution Ratios

Fertilizer proportioners are available with several dilution ratios. The most commonly used ratio is 1:100. This means that 1 unit of concentrated fertilizer

FIGURE 15-8 Dosotron® fertilizer injector. The injectors are shown installed in two pairs where each pair provides a different feed ratio. Fertilizer injectors often are installed in line to deliver two different fertilizer feeds into one solution to provide higher fertilizer concentrations or to deliver fertilizer products that may not be compatible when mixed in concentrated form.

FIGURE 15-9 A Smith Measure-Mix® injector. This unit has a double injection head to allow injection of two different fertilizers at the same time that may not be compatible in concentrated form.

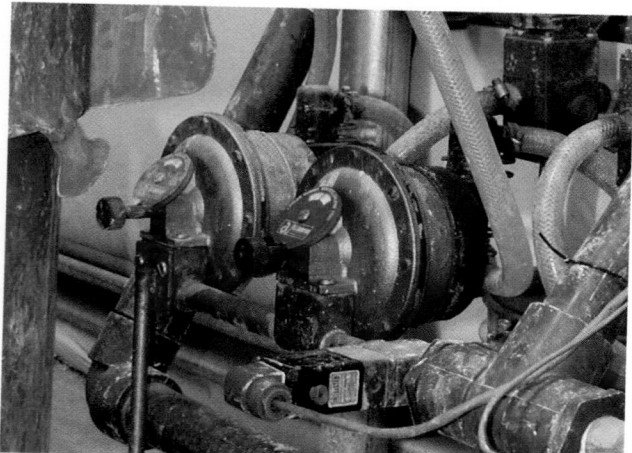

FIGURE 15-10A A pair of H.E. Anderson J Plus Ratio:Feeder® Dosing Pumps. H.E. Anderson injectors provide volume proportioning in that the injector measures the water flow to determine an exact volume proportion of fertilizer to inject.

FIGURE 15-10B An H.E. Anderson Ratio:Feeder® that measures the electrical conductivity (EC) of the irrigation water containing injected fertilizer. This unit maintains the precision of the injector.

stock solution is combined with 99 units of water. Thus, a fertilizer solution with a 1:100 dilution is added to the crop. The volume of the unit may be a pint, quart, gallon, or liter. The unit may be any volume as long as both the fertilizer and the water are measured in the same unit.

If the injector used has a dilution ratio of 1:200, then 1 unit of concentrated fertilizer stock solution is diluted with 199 units of water. Some fertilizer proportioners have dilution ratios of 1:1,000 and 1:2,000. Such proportioners are used on large outdoor ranges covering a number of acres.

A high dilution ratio is not always preferred. The solubility of fertilizer materials restricts their usefulness. For example, if a grower wants to apply fertilizer at the rate of 1 pound (0.45 kg) to each 100 gallons (378.5 l) of water, using an injector with a 1:1,000 dilution ratio, then 10 pounds of fertilizer must be

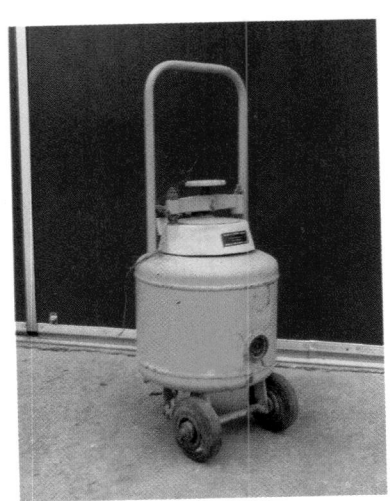

FIGURE 15-11 A small GEWA fertilizer proportioner used for greenhouse fertilizer injection. These injectors have no pumping device and rely on water pressure to push concentrated fertilizer from within a membrane inside the steel cylinder into the irrigation water flowing through a metering device.

dissolved in 1 gallon of stock solution. Not all fertilizers are soluble in this amount. The solubility differences for the various fertilizers are given in Table 15-1.

Some proportioners have controls to vary the dilution ratios (Figure 15-11). However, it is easier to make inaccurate injections of fertilizer stock solution using these machines because of incorrect operation of the controls. The calculation of the amount of fertilizer to be used is easier if it is being injected by a fixed ratio proportioner because the amount applied to the crop always has the same dilution ratio.

Fertilizer Arithmetic

Fertilizer arithmetic is the term applied to the calculation of the amount of fertilizer needed to provide the required solution strength. Whereas the farmer thinks in terms of tons of fertilizer applied to each acre, the greenhouse grower thinks in terms of pounds per 100 square feet, for dry applications, or parts per million (ppm), for liquid applications.

The term "parts per million" refers to the strength of the fertilizer in the solution being applied to the crop. In other words, there are so many parts of fertilizer per million parts of water applied.

Many growers are familiar with the Quick 75 method for calculating ppm. To calculate the amount of fertilizer required using the Quick 75, divide the desired ppm by 75 and then divide by the decimal fraction of the desired nutrient (such as nitrogen, potassium, or phosphorous) contained in the fertilizer. This results in the number of ounces of fertilizer to use in 100 gallons of water.

$$\frac{\text{desired ppm} \div 75}{\text{decimal fertilizer fraction}} = \text{oz fertilizer/100 gallons of water}$$

To use this equation, assume that the fertilizer recommendation calls for 200 ppm of nitrogen from ammonium nitrate (33 percent N). Using the above equation, divide 200 ppm by 75 resulting in 2.67, and then divide by 0.33, which is the percentage of the fertilizer converted to the decimal fraction. The answer is 8.09 ounces of ammonium nitrate which, when dissolved in 100 gallons of water, will yield 200 ppm nitrogen.

$$\frac{200 \div 75}{0.33} = 8.09 \text{ oz fertilizer/100 gallons of water}$$

The concept of ppm may be confusing to some. Examples of what 1 ppm represents under various conditions include: 1 crystal of salt in 5 pounds, 1 drop in 16 gallons, 1 inch in 158 miles, 1 minute in 1.9 years, 1 pound in 500 tons, and 1 cent in $10,000. Therefore, to understand the basis of the Quick 75 method to calculate ppm in 100 gallons of water, first multiply 100 gallons by 8.34 pounds per gallon which equals 834 pounds. Multiply 834 pounds by 16 ounces per pound which equals 13,344 ounces per 100 gallons. Therefore, 13,344 ounces per 13,344,000,000 ounces equals 1 part per million, or more simply 0.013344 ounces per 100 gallons of water equals 1 ppm.

    a. 100 gal * 8.34 lb/gal = 834 lb
    b. 834 lb * 16 oz/lb = 13,344 oz

    c. 13,344 oz/100 gal

    d. 13,344 oz/13,344,000,000 oz = 1 ppm

    e. 0.013344 oz/100 gal = 1 ppm or 1 oz/gal = 1 ppm

The next step is to multiply the desired ppm by 0.013344, which is 74.94. By rounding 74.94 to 75, it must be understood that the result will be neither entirely nor mathematically accurate, but perhaps is close enough for practical purposes. Within the units of the ratios commonly used in fertilizer solution, the error will be 0.02 or less per 100 gallons.

To use the Quick 75 method with a different fertilizer in a greenhouse situation, consider a standard rate of application of a complete fertilizer, such as 20-20-20 at 200 ppm each of nitrogen, phosphorus, and potash. The problem is to calculate the amount of fertilizer needed for the stock solution so that the strength of the solution applied to the crop is 200 ppm, using a 1:100 dilution ratio. The amount of each element required is obtained by dividing the desired ppm by 75 and then by taking that value and dividing by the percentage value of the fertilizer converted to a decimal form. The resulting number is the weight of fertilizer in ounces to be dissolved in 100 gallons of water.

For example, 20-20-20 fertilizer has a nitrogen content of 20 percent. Therefore, 20 percent N = 0.20 × 75 = 15 ppm.

$$\frac{200 \div 75}{0.20} = 13.3 \text{ oz fertilizer/100 gallons of water}$$

For each ounce of 20-20-20 fertilizer in 100 gallons of water there are 15 ppm of nitrogen. In this example, there are also 15 ppm of phosphorus (P_2O_5) and 15 ppm of potash (K_2O). If the fertilizer used is 20-5-30, then the amounts of phosphorus and potash will be different from the amount of nitrogen.

$$0.20 \times 75 = 15$$

If a proportioner with a dilution ratio of 1:100 is used, 13.3 ounces of 20-20-20 fertilizer must be dissolved in 1 gallon of stock solution. If the dilution ratio is 1:200, then twice the amount of fertilizer, or 26.667 oz, is needed for each gallon of stock solution. Similar calculations can be made for other dilution ratios and percentage fertilizer materials. Table 15-3 lists the amounts of fertilizer materials required to obtain the desired ppm strength at three different dilution ratios.

Checking the Proportioner

Fertilizer proportioners wear out and lose accuracy. If the operator is far from the unit, it is difficult to determine if the unit is working properly. Commercial fertilizers prepared for proportioners contain tracer or indicator dyes. The dye is visible in the water even when the system is supplying a weak concentration of nutrients. As long as the grower can see some color in the solution being applied to the crop, it is reasonably certain that the unit is working as intended.

Many greenhouse growers mix their own fertilizers. Individual fertilizer materials do not contain dyes, and some form of tracer is needed. Two materials

Table 15-3

	Amounts of Nitrogen Fertilizers Required to Make Stock Solutions for Dilution Ratios of 1:12 (HOZON®); 1:100, and 1:200 to Provide 100 ppm Nitrogen for Crop Fertilization		
Percent Nitrogen in the Fertilizer	Ounces per Gallon of Stock Solution		
	1:12 (HOZON®)	1:100	1:200
10	1.6	13.5	27.0
11	1.5	12.3	24.5
12	1.4	11.3	22.5
13	1.2	10.4	20.8
14	1.2	9.6	19.3
15	1.1	9.0	18.0
16	1.0	8.4	16.9
17	1.0	7.9	15.0
18	0.9	7.5	15.0
19	0.9	7.1	14.2
20	0.8	6.8	13.5
21	0.8	6.4	12.9
22	0.7	6.1	12.2
23	0.7	5.9	11.7
24	0.7	5.6	11.3
25	0.6	5.4	10.8
33 ½	0.5	4.0	8.1
44	0.4	3.1	6.2
45	0.4	3.0	6.0
46	0.4	2.9	5.9
48	0.3	2.8	5.6

used to color fertilizer solutions are Alphazurine FGND concentration 200 percent (blue) and uranine (fluorescent greenish-yellow). These dyes can be purchased from large chemical supply houses.

The grower can use dyes of different colors to color code the stock solutions. Color coding may prevent the application of the wrong solution to the crop.

Even though a dye is used and it appears in the water stream being applied to the crop, the proportioner may not be working properly. To check the accuracy of the unit, the solution is tested before it is applied to the crop. Every concentration of fertilizer, regardless of the material or the dilution ratio used, has a known soluble salt concentration. An instrument known as an electrical conductivity meter (Figure 15-12) is used to determine the salt concentration. The soluble salt concentrations and pH readings for three dilution ratios of several commonly used fertilizers are given in Table 15-4. If the salt concentration of the solution from a proportioner is not close to a value in the table, the machine is not working properly and should be checked.

FIGURE 15-12 Myron Meter used for measuring total soluble salts. This unit can also be used to measure pH.

Table 15-4

Nutrient Carrier	Total Soluble Salt Readings and pH Values of Nutrient Solutions Made from Selected Nutrient Carriers					
	Concentrations (ppm)					
	100		200		300	
	TSS[1]	pH[2]	TSS	pH	TSS	pH
NITROGEN						
Ammonium nitrate	54	6.3	100	6.4	146	6.3
Calcium nitrate	71	6.4	136	6.5	200	6.6
Potassium nitrate	92	6.1	178	6.0	263	5.7
Sodium nitrate	82	6.2	154	6.3	228	6.3
Ammonium sulfate	95	6.1	180	6.1	265	5.9
PHOSPHORUS						
Monoammonium phosphate	35	5.4	65	5.1	95	5.0
Diammonium phosphate	70	7.3	127	7.5	180	7.5
Monopotassium phosphate	35	5.3	66	5.1	96	5.0
POTASSIUM						
Potassium nitrate	40	6.1	73	6.2	105	6.1
Potassium chloride	42	6.1	80	6.2	110	6.1
Potassium sulfate	40	6.1	75	6.2	110	6.2
MIXED ANALYSIS						
Peters 15-0-15	73	6.0	139	5.7	205	5.0
Peters 15-20-25	70	5.6	131	5.4	196	5.3
Peters 20-20-20	46	5.6	89	5.4	130	5.3
Peters 20-5-30	53	6.1	102	6.0	152	60.0
Ammonium nitrate plus potassium nitrate (200 ppm each N and K)	–	–	130	6.1	–	–
Ammonium nitrate plus potassium chloride (200 ppm each N and K)	–	–	170	6.1	–	–
Ammonium nitrate plus calcium nitrate (200 ppm each N and K)	–	–	150	6.2	–	–

1. Total soluble salts determined on Model RDB 15 Solubridge®, using a 2:1 ratio volume of distilled or deionized (DI) water:media, results are given in mhos $\times 10^{-5}$.

2. pH readings determined on Fischer Accumet® pH meter.

Adjusting Soil Acidity and Alkalinity

High analysis water-soluble fertilizers used for liquid application rarely contain calcium. Two exceptions are calcium nitrate and monocalcium phosphate. Continued use of high analysis complete fertilizers leaves an acid residue that causes the acidity of the soil to increase. In other words, a low pH develops.

Increasing Soil Alkalinity

To correct acidity, some form of liming material is used to raise the pH and add calcium to the soil.

By keeping the soil pH in the proper range for plant growth, the most efficient use is made of the fertilizers applied. When the pH is too low or too high, certain nutrients are unavailable for plant growth.

Materials used for liming range from slow acting limestone to quick acting hydrated or burned lime. The slow acting materials are preferred because they are safer to use.

Ground Limestone

Calcium carbonate, or carbonate of lime ($CaCO_3$), is the most widely used liming material. It contains approximately 56 percent calcium oxide. The particle size affects the speed of the chemical reaction in the soil. The smaller the limestone particle, the faster the reaction will be. If limestone is ground to pass through a 200 mesh screen (200 openings per inch), it reacts quickly to raise the pH.

Dolomite limestone [$CaMg(CO_3)_2$] contains both calcium and magnesium. The percentage of magnesium depends upon the parent material used. Dolomitic limestone acts to neutralize soil acidity at about one-half the speed of the reaction of calcium limestone.

Limestone materials may be added to soils at the rate of 5 to 20 or more pounds in 100 square feet of area. The amount of limestone used depends upon the pH of the soil. If the soil is very acid, a large amount of limestone is used. The soil can be tested for pH to determine how much lime is needed to obtain a certain pH.

Calcium hydroxide, $Ca(OH)_2$, is prepared by adding water to burned lime. This material contains very large amounts of oxides. When exposed to moist air, it is converted readily to carbonates (limestone). Less calcium hydroxide is required, as compared to ground limestone, to obtain the same increase in pH.

Calcium oxide, CaO, is also known as burned lime or quick lime. This material is very caustic and causes severe burns when improperly handled. It should never be applied when the hands are wet. Protective gloves and goggles are recommended whenever calcium oxide is used.

Calcium oxide is produced by heating limestone to a very high temperature. Carbon dioxide and other gases are driven off, leaving impure calcium and magnesium oxides. Calcium oxide reacts much faster to change the soil pH than either limestone or calcium hydroxide. The rapid pH changes that occur can quickly damage growing crops. This material is seldom applied at a rate in excess of 5 pounds (2.27 kg) to 100 square feet (9.28 m^2) of bench area. The quantity normally applied to the soil is 1 to 2 pounds.

Calcium sulfate, $CaSO_4$, or gypsum (also known as land plaster), is not used to correct low pH conditions. It is applied to add needed calcium when the pH is to remain the same. The rate at which calcium sulfate is applied is usually

5 pounds (2.27 kg) to 100 square feet (9.28 m²) of area. If a greater amount of material is used, there may be a slight increase in the acidity of the soil. The rise in acidity is due to the combination of hydrogen ions from the soil water and sulfate ions to form sulfuric acid. This reaction is very slow.

The ability of a liming material to change the pH of a soil depends upon the type of soil and the buffering capacity of the soil. The buffering capacity refers to the ability of the soil to resist changes in pH. Heavy clay soils and soils containing large amounts of organic matter, such as peat soils, have a high buffering capacity. To change the pH of these soils, a large amount of liming materials must be applied.

Sandy soils have low buffer capacities. These soils contain very little organic matter. Only small amounts of liming materials are needed to cause large changes in the soil pH.

The ability of a soil to resist changes in pH through its buffering capacity guards against plant injury resulting from wide changes in the soil reaction.

Increasing Soil Acidity

In some instances, a lower soil pH or an increase in soil acidity is needed. Ericaceous crops, such as rhododendrons and azaleas, grow better in acid soils (pH 5.0 to 5.5) than in soils with a pH of 6.5 to 7.0. The soil may be too alkaline for good plant growth, and it is necessary to reduce the pH. If the irrigation water has too high a pH, the soil must be treated on a regular schedule to keep the pH in the proper range for plant growth.

Acidic Organic Matter

Acid organic matter can be added to lower the pH of soils being prepared for ornamental plants. For example, sphagnum peat moss has a pH of 4.0 to 4.2 (even lower at times). To obtain a growing medium with a low pH, sphagnum peat moss is combined with equal parts of soil. For some uses, one part of peat moss can be mixed with two parts of soil by volume to produce a medium with the desired pH.

Sphagnum peat moss is often used alone as a growing medium for azaleas and rhododendrons. However, it is too acidic for good growth, and limestone must be added to raise the pH slightly. The limestone also adds needed calcium.

Leaf mold, pine needles, pine bark, and some sawdusts are other organic materials that have an acid reaction. Each of these materials will lower the pH. Not all sawdusts have an acid reaction. The pH of the material should be checked before it is used.

Acidifying Chemicals

At times, it may not be possible to use organic matter to obtain the desired pH changes. This may be the case after the crop is planted and growing well. Potted azaleas or hydrangeas may develop iron chlorosis if the media pH increases. The need to apply a chemical becomes apparent.

Ferrous (Iron) Sulfate

Iron sulfate is often used to increase the acidity of soils. It is safe to handle, has a fairly rapid reaction rate, and supplies the iron needed by the plants. The acidifying action is short lived. Thus, frequent treatments are made at the rate of 1 pound (0.45 kg) to 100 square feet (9.28 m²) of area.

Sulfur

Three types of sulfur may be added to the soil: flowers of sulfur, dusting sulfur, and wettable sulfur. Each type is slow acting but has a long-term effect in the soil. Dusting sulfur has the finest particle size of the three types. It passes through a 325 mesh sieve (325 openings per inch [2.5 mm]). It also has the fastest reaction rate in the soil. Sulfur should be used when the soil temperature is 60°F (15.5°C) or above. Even though 2 or 3 months are required for the reaction to go to completion, the material has a long-term effect.

Aluminum Sulfate

If a soil lacks phosphorus, there may be a problem in the use of aluminum sulfate to acidify the soil. Without phosphorus, the free aluminum ions released with the dissociation of the molecule may become so concentrated that the plants are damaged.

An iron chlorosis is not corrected by applying aluminum sulfate to increase the acidity of the soil if there is a basic lack of iron in the soil.

There are several applications where aluminum sulfate works very well. For example, it is used to change the color of hydrangea flowers from red to blue.

Table 15-5 lists the amounts of liming materials needed to raise the pH of various soils. The amounts of acidifying materials needed to lower the pH are given in Table 15-6.

Table 15-5

Amounts of Limestone Needed to Raise the pH by 1/2 to 1 Unit			
	Pounds to 100 Square Feet of Area (kg to 9.28 m²)		
Material	Sandy Soil	Silt Soil	Clay Soil
Calcitic limestone	3.0 (1.36)	4.5 (2.04)	6.0 (2.72)
Dolomitic limestone[1]	3.0 (1.36)	4.5 (2.04)	6.0 (2.72)

1. The reaction of dolomitic limestone requires a longer time for a change in pH.

Table 15-6

Amounts of Material Required to Reduce the pH by 1/2 to 1 Unit			
	Pounds to 100 Square Feet of Area (kg to 9.28 m²)		
Material	Sandy Soil	Silt Soil	Clay Soil
Finely ground sulfur	3/8 (0.17)	1/2 (0.23)	3/4 (0.34)
Iron sulfate[1]	2.0 (0.90)	3.0 (1.36)	4.0 (1.82)
Aluminum sulfate[1]	2.0 (0.90)	3.0 (1.36)	4.0 (1.82)

1. Soluble in water and may be applied as a soil drench. Dissolve the amount given in 25 gallons (6.60 l) of water and apply to 100 square feet (9.28 m²) of bench area.

Determining Plant Nutrient Needs

There are two ways to determine the nutrient needs of plants: (1) Test the soil, and (2) test the plant tissue (foliar analysis).

Soil Testing

To test the soil, a small sample of soil is taken from several locations in a bench (for a cut flower crop) or from several different pots (for a pot plant crop). Collect the soil from a middle depth where root growth is active. Soil from the top or the bottom will not be representative. The samples are taken to a laboratory where they are mixed together, dried, screened, and put through a series of chemical tests to determine the nutrient contents. **Soil testing** determines the amounts of nitrate and ammonium nitrogen, phosphorus, potassium, calcium, and, at times, magnesium contained in the soil. Greenhouse growers also need to know the soil pH and the total soluble salts.

Several methods are used in commercial and state agricultural laboratories to test soils. The methods differ in the strength of the extracting solution used to treat the sample. In addition, the procedures used to analyze the extract may vary between laboratories. For this reason, the grower should not compare the results obtained from one laboratory with those of another laboratory for the same soil. The interpretation of the nutrient content of a soil sample and its relationship to the needs of the crop should be left to those who are experienced in this area.

The results of the soil test show only the nutrient content of the soil at the time the sample was taken. The test does not indicate if the plants have absorbed any of the nutrients.

The soil test may show adequate levels of nutrition. However, plants may be deficient in one or more essential nutrients because they cannot take up the nutrients. The problem of uptake may be due to diseased or insect-damaged roots, cold soil, poor aeration, overwatering, or other factors.

Foliar Analysis (Tissue Testing)

Foliar analysis (tissue testing) uses leaves or other parts of plants to determine the nutrient content. This procedure is more complicated than soil testing. Leaves from certain areas of the plant are sampled to obtain an accurate analysis of the nutritional status of the plant. The most recently matured and fully developed leaves are used.

Once the leaf sample is prepared and processed, it is analyzed. Both the macro and micro nutrient element content can be determined from the same sample. A typical report sheet will give values, as a percentage of dry weight, for nitrogen, phosphorus, potassium, calcium, and magnesium. Values in parts per million are given for sodium, manganese, zinc, iron, copper, boron, aluminum, and molybdenum.

The foliar analysis shows only the nutrients absorbed by the plant just before the sample was taken. This test does not indicate the nutrient reserves of the soil or the total soluble salt content. High levels of soluble salts and the other factors listed under soil testing prevent the uptake of nutrients.

There is one advantage of foliar analysis as compared to other methods of determining plant nutrient needs. Very small deficiencies of certain nutrient elements may be discovered several months before the effects appear on the plant. A low level of boron in carnations can be detected 2 months before it is visible as a deficiency symptom. This test allows the grower to add the deficient nutrient before the crop is visibly damaged and crop quality suffers.

To determine the nutrient requirements of flower crops, the best procedure is to use a combination of soil tests and foliar analysis. This procedure works best for long-term crops, such as carnations and roses, because it provides an opportunity to correct the problem before the crop is injured.

Total Soluble Salts

What Are Soluble Salts?

Soluble salts are chemical compounds that can be dissolved in the **soil solution**. They may be good or bad for plant growth. The effect obtained depends upon the strength of the solution for those materials considered to be good for plants.

The fertilizer ammonium sulfate supplies both ammonium and sulfate salts to the soil. The ammonium provides the nitrogen used by the plants. The sulfate provides some sulfur, which is needed in small amounts. Most of the sulfate remains in the soil solution unless it is leached away by repeated irrigations. Many applications of ammonium sulfate, or the application of too large an amount, can cause soluble salts to build up in the soil. Plant damage may result.

All inorganic fertilizers do not add the same amount of salts when they are used in equal quantities. The salt index is a measure of the effect of a fertilizer on the concentration of the soil solution. Ammonium nitrate has a salt index of 105, and calcium nitrate has a salt index of 53. Salt indexes for other fertilizers are given in Table 15-2.

Organic fertilizers also provide soluble salts. As organic materials decompose, they are broken down into inorganic units. The different fractions of these units all add salts to the soil.

The water used in the greenhouse may also add soluble salts to soils. In some areas, well water or pond water may have very high levels of soluble salts. In some locations, the use of salt to melt ice on the highways has increased the concentrations of sodium, chloride, and calcium ions in the ground water, thus contaminating wells. Underground sources of salts may also contaminate wells.

How Salts Accumulate

Salts accumulate in the soil basically because too much fertilizer is applied. This may happen at any one time or over a period of time.

Salt also builds up because too little water is used when an injector system applies fertilizer at each watering. To prevent salt buildup, it is recommended that 10 percent more solution be added. As a result, there is a small amount of leaching of salts each time the crop is irrigated.

A salt buildup may be due to poor drainage from a greenhouse bench or a hard pan in a ground bed. For example, the drainage cracks in wooden benches may swell until they are closed. In other types of benches, salts may accumulate and harden around the drain holes until they are blocked. The drainage openings should be checked regularly to ensure that they are open.

Steam or chemical soil sterilization may increase the soluble salt concentrations. The microorganisms living in the soil contain nitrogen tied up as protein. When the sterilization process kills these organisms, the nitrogen is changed to ammonium that is released into the soil. The ammonium adds to the salt content of the soil. Increased ammonium levels are a problem when large amounts of manure or spent mushroom compost are added to the soil.

Sterilization kills nitrifying bacteria more quickly than ammonifying bacteria. Nitrifying bacteria convert ammonia to nitrates and ammonifiers convert organic nitrogen to ammonia. For 2 to 3 weeks after sterilization, while the level of nitrifying bacteria builds up, the ammonifying bacteria cause the level of ammonia to increase to an excessive amount. To avoid this problem, a steam sterilized soil is planted as soon as it cools to room temperature. Similarly, a chemically treated soil is planted as soon as it is fully aerated. At the time of planting, both soils should be watered heavily so that some leaching occurs. If the crop cannot be planted immediately, it may be necessary to leach the soil heavily to lower the high salt levels. Between 3 to 5 gallons of water should be applied to each square foot of bench area. In 30 minutes, another heavy watering is applied. The first irrigation dissolves the salts, and the second irrigation flushes them from the soil.

The Action of Salt Injury

Salt injury usually occurs as osmosis. The strength of the salt solution in the soil is greater than that of the solution within the plant root cells. As a result, the water in the root cells is pulled out of the cells by a force exerted by the salt solution. This causes the cell contents to pull away from the wall and collapse. This condition is called plasmolysis. When plasmolysis takes place in a large number of cells, physiological drought sets in and the root suffers from excess water loss. Death follows shortly after. Examination of the root ball will show brown, dying, and dead roots.

The first symptoms of a high salt condition may be hard to see. If there are no unaffected plants, the slight retardation in overall growth may pass unnoticed.

As the salt level increases and the plasmolysis becomes worse, more visible symptoms develop. In some crops, such as snapdragons, a severe interveinal chlorosis occurs. The growth of the entire plant is affected. On other crops, there may be no chlorosis, but the basal leaves of the plant will turn brown. Depending upon how quickly the salts build up, the symptoms of plant damage can develop slowly or very quickly.

Young seedlings or cuttings are especially sensitive to high soluble salt concentrations. For this reason, the salt levels in the soil should be low when the crop is first planted.

High salt levels may result in poor seed germination. Old greenhouse soils that are used for potting almost always have a high soluble salt content.

Reducing High Salt Accumulations

To prevent the accumulation of high levels of salts, fertilizers must be applied properly in the correct concentrations. If the salt concentration has reached a level that is reducing or damaging the growth of the crop, the soil should be leached to remove the excess salts. Unfertilized irrigation water is applied at the rate of 3 to 5 gallons (122.8 to 203.8 l) to 1 square foot (1 m²) of area. In 30 minutes, a second, equally heavy application of water is made to flush the salts from the soil. If the salt level is extremely high, a third or a fourth watering may be necessary to leach the salts from the soil completely.

Good drainage is required if the leaching method is to be successful. It is nearly impossible to leach excess soluble salts from ground beds that lack good drainage.

High levels of soluble salts in ground beds can be reduced once the crop is harvested. First, the soil is allowed to dry. As the soil water evaporates, the salts move to the top layers of the soil. When the top 2 inches of soil are dust dry, this layer is skimmed off. It is replaced with field soil having a low salt content. A physical dilution takes place when the new soil is mixed thoroughly with the old soil. As a result, the total salt content is reduced. Additional applications of fertilizer should be controlled to avoid an immediate buildup of soluble salts.

Good soil drainage helps to control soluble salt concentrations. To improve the drainage by providing more pore space, coarse materials are added to the soil. These materials include peat moss, perlite, vermiculite, sand, and fine gravel.

In very heavy clay soils, some improvement in drainage may be obtained by adding gypsum to the soil. Gypsum, or calcium sulfate, also adds a small amount of salts. However, the calcium ion causes the soil particles to flocculate, or clump together to form larger aggregates. Thus, the soil is opened up for better water flow. The addition of calcium nitrate also improves drainage. On the other hand, the long-term use of sodium nitrate causes the soil colloids to break apart. This action destroys the drainage capacity of clay soils.

The addition of gypsum to greenhouse soils probably has little value because these soils have undergone many changes from their original character due to the addition of peat moss, perlite, vermiculite, and other materials.

In summary, soluble salt levels can be controlled by applying the correct concentrations with sufficient irrigation and by using growing media with good drainage and aeration.

Measuring Soluble Salt Content

The soluble salt content of soils, water, or other solutions is measured by an instrument called an electrical conductivity meter (Figure 15-13). This device is an excellent method of ensuring against crop losses. Modern meters are

FIGURE 15-13 Electrical conductivity (EC) meters and pH meters are available as *pen* type or larger. The larger, Hanna portable meter has a probe that measures both EC and pH.

available in a pocket form called pen meters. They are relatively inexpensive and a valuable tool for any greenhouse grower to carry while working.

There are four generally used methods for preparing solutions to be tested for total soluble salt content. The first method is the saturated-media extract (SME), sometimes called the saturated-paste extract. This method requires the use of a device to extract the solution from the media. Because a vacuum procedure is needed, this method is usually only used in a laboratory.

When using the SME method, a known volume of media is placed into a container that can have a suction force applied to it. Distilled or deionized water is added to the media until it is completely saturated. The surface of the media should just glisten with moisture. If too much water is added, this will dilute the soluble salts and an inaccurate reading is obtained. The media-water combination is allowed to react for 30 minutes or preferably 1 hour. At the end of this time period, a suction force is applied and the solution is collected. The total soluble salt measurement is made using a solubridge or similar instrument.

If distilled or deionized water is not available and tap water is used, the salt reading of the tap water must be taken and subtracted from the total soluble salt reading obtained on the mixture.

The second way of preparing the solution is to use one part by volume dry media. Mineral soils should be screened through a 10 mesh screen to remove stones. Two parts by volume water are added and the mixture stirred every 5 minutes for 30 minutes. After 30 minutes the soluble salt reading is made.

The third method is to use a 1:5 media:water ratio. The 30-minute reaction time is also used. A lower reading is obtained with the 1:5 dilution ratio than with the 1:2 dilution ratio.

The fourth method is a novel procedure developed by floriculture faculty at North Carolina State University. It is called the leachate pour-through method and is conducted on plants in containers without disturbing the plant. The first step is to saturate the soil with clear water and allow it to drain for 1 hour. Then set the container in a saucer and irrigate with enough water to collect 1.69 fluid ounces (50 ml), which is then tested for pH and electrical conductivity. Using too much water resulting in greater volumes of leachate will dilute the response. This method is growing in popularity with greenhouse growers as it is quick, easy, and nondestructive.

There are other methods of measuring the total soluble salt content of the media. Some of these use probes that are simply pushed into the media. Others use only a drop or two of extracted solution in a small depression to test for salts. The accuracy of readings taken with these instruments is less than that obtained by the four solution extraction methods described above.

Water alone may be tested by direct measurement. The interpretation of the usefulness of water for irrigation purposes is given in Table 15-7.

Electrical conductivity readings for mineral soils and peat-lite mixes are given in Table 15-8. Peat-lite mixes have greater cation exchange and water-holding capacities than mineral soils. For this reason, peat-lite mixes may have higher salt readings without causing plant injury.

Table 15-7

Solubridge® Readings for Water	
Solubridge® Readings (mho $\times 10^{-5}$)[1]	Usefulness for Irrigation
Below 25	Excellent
26–59	Good
60–149	Fair
150–200	Poor
Above 200	Excessively salty

1. The readings are given in a unit called a mho. This unit is the reciprocal of an ohm. An ohm is a measure of electrical resistance. A mho is a measure of electrical conductance. (The international Metric System of Measurement uses the unit "siemen" instead of mho.)

When the soluble salt levels are very high or excessive, the growing medium should be leached. If leaching cannot be done immediately, the medium should be kept moist. It must not be allowed to dry. If the soil does dry, the salt content increases in the remaining media solution, resulting in severe plant injury.

Table 15-8

Solubridge® Readings for Mineral Soils and Peat-Lite Mixes						
Solubridge® Reading				Saturated-Media Extract		
(mho $\times 10^{-5}$)[1]		(millimhos/cm)[2]				
Mineral Soils	Peat-Lite Mix	Mineral Soils	Peat-Lite Mix	Mineral Soils and Peat-Lite Mix	Rating	Interpretation
Above 200	Above 350	Above 2.00	Above 3.50	Over 8	Excessively high	Plants may be severely injured. The growing medium must be kept moist until leaching can be carried out.
176–200	226–350	1.76–2.00	2.26–3.50	6–8	Very high	Plants may grow, but the salt concentration is very near the danger zone. If the soil dries, injury may occur.
126–175	176–225	1.26–1.75	1.76–2.25	4–6	High	In general, this range is satisfactory for established growing plants. This range may be too high for seedlings or newly planted cuttings.
51–125	101–175	0.51–1.25	1.01–1.75	2–4	Medium	This range is satisfactory for general growth. It is the best range for plants fertilized at every watering.
0–50	0–100	0.00–0.50	0.00–1.00	0–2	Low	At this range, the plants are probably growing poorly due to a lack of nutrients.

1. The readings are given in units of mho $\times 10^{-5}$, a measure of electrical conductance. Some laboratories report readings in terms of millimho per centimeter (or millisiemens per centimeter). To change millimho per centimeter to mho $\times 10^{-5}$ (or to change millisiemens per centimeter to mS $\times 10^{-5}$), move the decimal point two places to the right. For example, 1.75 millimho per centimeter is equal to 175 $\times 10^{-5}$ mho (or 1.75 millisiemens per centimeter is equal to 175 $\times 10^{-5}$ mS).
2. mmho/cm or millisiemens/cm

ACHIEVEMENT REVIEW

Select the best answer or answers to complete each statement. List the appropriate letter(s).

1 A true organic fertilizer must contain _____ as an essential ingredient.
a. nitrogen
b. carbon
c. peat moss
d. humus

2 One ton of manure will supply as much nutrition as 100 pounds of:
a. 5-10-10.
b. 10-5-10.
c. 10-10-10.
d. 5-10-5.

3 Of the following materials, select the ones that are single-element fertilizers.
a. Ammonium nitrate
b. Calcium nitrate
c. Potassium nitrate
d. Potassium chloride

4 One of the following materials is not recommended for use as a greenhouse crop fertilizer during the months of December through February. The material is:
a. ammonium sulfate.
b. calcium nitrate.
c. potassium nitrate.
d. monocalcium phosphate.

5 The continued use of fertilizer can cause a change in the pH of the growing medium. Which of the following materials will make the pH more alkaline?
a. Ammonium sulfate
b. Calcium nitrate
c. Potassium chloride
d. Nitrate of soda

6 The three numbers on a bag of fertilizer refer to the percentages of:
a. N-P-K contained by the fertilizer.
b. NO_3-P-K contained.
c. N-P_2O_5-K_2O contained.
d. nutrients plus the trace elements contained.

7 Which of the following materials is not considered to be a controlled-release fertilizer?
a. Borden's 38®
b. 20 percent superphosphate
c. Osmocote® 14-14-14
d. Magamp® 7-40-6

8 The Osmocote® process controls the release of the fertilizer from the particle by the:
a. temperature at which the particle is made.
b. shape of the particle.

c. size of the particle.

d. thickness and type of plastic used to coat the fertilizer particles.

9 Moist soils containing Osmocote® should not be stored for more than 7 days before use because:

a. they will get moldy.

b. the soluble salt level may become extremely high.

c. all of the nutrients will dissolve.

d. they will dry out too much.

10 Trace elements or micro nutrients in the form of frits are normally applied to crops:

a. in liquid form.

b. at every irrigation.

c. once weekly.

d. twice yearly.

11 Fertilizer A has a salt index of 50. It has a nutrient content equal to that of fertilizer B, which has a salt index of 100. Fertilizer A is preferred for use because:

a. only half as much of A is needed to provide the same nutrient level as B.

b. twice as much nutrient element is available from A with the same salt content as B.

c. twice as much nutrient element is supplied from B when used at the same rate of application as A.

d. there is no difference in salt content when both are used at the same rate of application.

12 A fertilizer proportioner with a dilution ratio of 1:100 will dilute:

a. 1 pint of stock solution in 99 pints of water.

b. 1 pint of stock solution in 99 quarts of water.

c. 1 pint of stock solution in 99 gallons of water.

d. all of the above.

13 Which of the following materials is/are used to increase the pH of media?

a. Calcium hydroxide

b. Calcium oxide

c. Aluminum sulfate

d. Flowers of sulfur

14 Which of the following materials will require the greatest amount of limestone to change the pH?

a. Sandy loam

b. Loamy sand

c. Silty sand

d. Peat humus

15 A high total soluble salt content in a medium injures the plant roots because it:

a. increases the carbon dioxide around the roots.

b. decreases the drainage in the root zone.

c. pulls water from the root cells causing plasmolysis.

d. removes oxygen from the root cells causing suffocation.

CHAPTER

16

Water: Plant Requirements and Methods of Application

OBJECTIVES:

After studying this chapter, the student should have learned:

■ why a plant needs water.

■ how water is taken up and moved through the plant structures.

■ the effects of too much water on the growth of plants.

■ how moisture stress affects plants.

■ several ways in which water may be applied to crops.

Role of Water

Plant Requirements

The need of a plant for water is greater than its need for any other nutrient material. The water content of plant tissue may be as little as 10 percent, or the amount found in dry seeds. In leafy, succulent plants, such as lettuce and cabbage, water may equal more than 95 percent of the fresh weight. Fleshy fruits, young tender shoots, and roots all have a high moisture content.

The water of constitution is the amount that is bound within the plant. This amount, which makes up an integral part of the total dry matter, may be less than 1 percent. This quantity is a very small portion of the water needed to support vegetative growth and produce a good crop.

The production of 2.2 pounds (1 kilogram) of dry matter may require a plant to absorb 30 to 3,000 kilograms of water. The total amount of water absorbed depends upon the kind of plant and the conditions under which it is grown. As an average, a plant will absorb 500 units of water for each unit of dry weight produced. This value is commonly known as the water requirement of the plant.

Some plants require much more water than others. For example, pumpkins, beans, and cinerarias may absorb and transpire 10 times their own fresh weight in a 24-hour period. Most of this activity takes place during the daylight hours.

Water is very important to plant growth. As a solvent, water dissolves the nutrients, gases, minerals, and other materials needed by every cell of the plant. Plants can maintain turgor, or fully filled cells, because of their water content. When a plant lacks water, the cells shrink and become flaccid and the plant wilts.

Water Absorption (Uptake)

Absorption through the Roots

The greatest amount of water used by plants is taken in through the root system. Root hairs grow through soil particles and the spaces between them. As they grow, they absorb the soil solution and the dissolved nutrients in it. The life span of root hairs may range from a few hours to a few days. New root hairs are constantly being developed in the root hair zone. Recall that this area is located just behind the root cap. Each root has thousands of root hairs, which greatly increase its absorption area. It is estimated that a winter rye plant has 14 billion root hairs with a surface area of 1,310 square meters (1 square meter equals 1.2 square yards).

For a time, it was thought that the root hair zone was the main zone of water and nutrient **absorption**. More recent studies have shown that water is also taken in through the suberized (corky) covering of the older roots. Uptake through these areas is slower than absorption through the root hairs. However, the total area of suberized root tissue is large. Therefore, as much water may be taken in through these tissues as is absorbed through the root hair zone.

Water enters the root through a process known as diffusion. Simple diffusion is defined as the flow of ions from an area of high concentration to an area of low concentration. Energy is not required for simple diffusion to occur. This process is sometimes called passive diffusion as compared to active diffusion. A paper towel absorbs water by means of passive diffusion, that is, from wet to dry.

Diffusion against a resistance or gradient is a more complex process. It may be compared to the act of pushing a wagon uphill. Work or energy is needed to make the wagon (ions) move uphill (gradient). The plant supplies the necessary energy from the processes of photosynthesis and respiration. Most of the soil solution entering the root system does so by means of diffusion against a gradient.

After the water and the dissolved materials in it are absorbed through the epidermis, the fluid is known as the cell sap. The cell sap moves from the roots to the leaves by way of the vascular tissue known as the xylem. Recall that water moves upward from the roots through the xylem tissue.

The phloem is another part of the vascular bundle that extends from the roots to the leaves. Phloem tissue is the pathway by which metabolites, or manufactured food, move up and down the stem.

Effects of Submergence (Oversupply)

Plant growth can be adversely affected if the root system is submerged due to flooding or overwatering. Except for plants that grow in water (hydrophytes), the roots of ornamental plants cannot stand long periods under water. As living and growing tissues, roots need oxygen. There are no problems of root growth in a well-balanced growing medium. Such a medium has 50 percent solid matter, 25 percent moisture, and 25 percent soil gases. If the medium is very compacted or has poor drainage, or a combination of both problems, then many of the pore spaces of the medium may be collapsed. If those that remain are filled with water, too little oxygen is available to the roots. This condition soon leads to injury and eventual death of the roots.

An inadequate supply of oxygen for as little as 24 hours has a very great effect on the final growth of the plant. Seedlings are easily damaged by overwatering.

Absorption through Leaves and Stems

Most of the water used by the plant is taken up through the root system. However, some absorption also takes place through the leaves and stems. Absorption of water through the leaves is regulated by the type of leaf tissue involved and the frequency with which water strikes the leaf.

Leaves are covered with a waxy material called suberin. This material forms the cuticle, or the outside covering of the epidermis. Because the cuticle is waterproof, water rolls off smooth plant leaves easily. This fact also makes it difficult to wet the leaves when insect sprays or growth regulators are used.

The cuticle does not form a continuous, unbroken covering on the leaf. There are breaks in the cuticle that allow materials to enter the interior of the leaf. For example, foliar nutrient sprays reach the inside leaf tissues by means of these breaks.

Water vapor is also absorbed through the stomata. These kidney bean–shaped openings are located mainly on the undersides of leaves. Figure 16-1 shows some common leaf structures.

Water Losses

Transpiration

The process of transpiration accounts for most of the water used and lost by a plant. Transpiration occurs mainly through the leaves but can also occur through other organs.

Transpiration is one of the most vital processes occurring in the plant. It is not simply a loss of water. The cells of the spongy parenchyma of the leaves are surrounded by a microscopically thin film of water. Carbon dioxide in the air entering the stomata is exchanged in this film of water. Once the carbon

FIGURE 16-1 Electron microscopic views of leaf structure. Leaf structure views were taken with a scanning electron microscope (SEM) (A-D). Leaf cross sections were taken with a transmission electron microscope (TEM) (E and F). *(Courtesy of David Reed, Department of Horticulture, Texas A & M, College Station, Texas.)*

(A) (B) (C)

(D) (E) (F)

dioxide is dissolved, it passes through the cell walls by diffusion. It is then taken up by the contents of the cells and is used in photosynthesis. The carbon dioxide content is low, and the water film is continually evaporating. Therefore, both must be replaced. Water is conducted from the roots by the xylem tissue in the stem. Air containing carbon dioxide enters the stomata of the leaves. More water is needed for transpiration than for any other use in the plant. Several plant nutrients required for plant growth are translocated through the transpiration stream as well.

Most transpiration takes place during daylight. During this time, the largest potential gradient develops between the outside of the stomatal cells and the cell walls. The potential gradient refers to the force that acts to cause a loss of water. For example, if the relative humidity of the outside air is 100 percent, as is usually the case at night, there is little or no loss of moisture from the leaf. In this instance, the moisture potential of the stomatal cells is 100 percent.

As daylight begins and temperatures rise, the relative humidity of the atmosphere decreases, unless it is raining or foggy. As the relative humidity decreases, the potential gradient between the stomatal cells of the leaf and the outside air increases. As a result, there is a loss of water from the cells to the atmosphere.

When water is lost to the outside atmosphere, it creates a potential difference (suction force) between these cells and all other cells connected through the xylem to the roots. This suction force causes water to be pulled into the plants at the roots. This water is then moved through the plant to the stomata of the leaves. This process can be compared to the action of sipping soda through a straw. As long as a suction force is created at the top of the straw, soda flows into the bottom of the straw.

Several environmental factors affect the potential loss of water from the stomata. High air temperatures, low relative humidity, and moderate airflow over the leaves all increase the potential for water loss. Hot, dry winds can quickly lead to water stress in a plant.

Wilting occurs when the rate of transpiration water loss is greater than the rate of moisture uptake. The continuous columns of water in the xylem are broken. The cells no longer receive an adequate supply of water from the roots. The cells become flaccid, and the plant wilts. When wilting occurs, the guard cells of the stomata close. Carbon dioxide and oxygen are not exchanged, and photosynthesis stops temporarily. A wilted plant does not grow.

Guttation

There are other ways in which water is lost from the leaves of plants. Water vapor is lost by transpiration. **Guttation** is the term applied to the loss of fluid water from a plant.

Hydathodes are groups of cells located at the ends of the leaves of a number of plants. These small organs are found just inside the leaf tips. When the relative humidity is high, the water needs of the plant are satisfied by uptake through the roots. In this case, there is an excess of water within the plant. This water must be removed, but transpiration is at a minimum because of the high relative humidity. Thus, the water is forced out of the hydathodes and appears as small, hanging drops at the tip ends of the leaves. If the drops are undisturbed and evaporation occurs, the residue from the water of guttation shows on the tips of the leaves as a small amount of soluble salts. Only a very small amount of the total water supply of a plant is lost by guttation.

Moisture Stress

A lack of water is known as moisture stress. When water is deficient, photosynthesis is reduced because the stomata close and carbon dioxide no longer diffuses into the leaves. The loss of photosynthetic activity results in reduced dry weight and reduced protein synthesis.

The effect of water stress on the growth of the plant depends upon the stage of plant growth when the stress occurs. If seeds are planted and water is lacking, then the seeds will not germinate. If there is enough water for germination, but a shortage then occurs, the growth of the seedling will be affected and the resulting plant will be weakened.

Water stress has its greatest effect on the development of the cells. Immediately after cell division takes place, cells increase in size. To reach its maximum potential size, a cell must have an adequate supply of water. A cell fails to develop to its maximum size if it lacks water. Growth is permanently restricted as new cellular materials are laid down for the development of the cell wall.

In the root system, a lack of water at this critical stage of growth results in small, stunted roots. Such roots have a reduced ability to take up water and nutrients. In the stem, where the new leaves and flowers are being formed, the lack of water results in a permanent decrease in the size of the leaves. Smaller leaves mean that there is less total area for photosynthesis. Therefore, less growth occurs.

If the plant is at the critical stage of flowering, a lack of water greatly reduces the size of the flower that is formed. Flower spikes on snapdragons may be shortened and chrysanthemum flowers may be reduced in size, resulting in a downgrading from fancy to a lesser grade.

Causes of Moisture Stress

Moisture stress generally develops because the supply of moisture is limited. When the crop is dry, the amount of water applied should be great enough to wet the growing medium thoroughly. There should be enough extra water to drain from the pot or the bench. For a 6-inch (15-cm) deep bench, the normal volume of water applied is 1 gallon to 2 square feet of area (when the crop is fertilized at every watering). If the fertilizer is applied on a weekly or biweekly (2-week) basis, then the rate of watering is 1 gallon to 4 square feet (3.8 l to 0.4 m^2) of area. Crops are watered between fertilizer applications to cause some drainage from the bottom of the bench. This extra water prevents a buildup of the total soluble salts.

High Soluble Salts

Soluble salts result when fertilizers are applied to soils. When too much fertilizer is added, or too little water is applied to cause some leaching, the concentration of salts in the soil may increase until moisture stress occurs as a result of physiological drought.

In other words, though there may be enough moisture in the soil, the salt concentration (osmotic potential) of the soil solution is so great that the roots cannot take up water from the soil. In some instances, the force exerted by the soil solution is so strong that water is pulled from the root system (plasmo-lysis). This condition eventually leads to the death of the root tissue.

The concentration of salts may not be great enough to damage the roots. However, the concentration may be great enough that the root system must use additional energy to absorb the water it needs from the soil. The use of this extra energy causes a reduction in the total growth of the plant.

If there is a high level of soluble salts, leaching is the recommended method of reducing the level. If leaching is not possible, the soil should not be allowed to dry out. It should be kept moist at all times. The high moisture level keeps the salts diluted so that they cause no damage.

High Temperatures and Air Movement

High air temperatures are another factor leading to moisture stress. When the greenhouse temperature goes above 90°F (32°C), the plants are severely stressed. In bright sunshine, the leaf temperature may be as high as 120°F (49°C) due to the radiant energy of the sun. At this temperature, transpiration proceeds at a very rapid rate. If the temperature is not reduced by shading the plants, or by cooling the greenhouse through ventilation or fan and pad cool-ing, the plants will be severely checked in growth. Generally, the wilting that occurs is temporary. As the conditions that cause the wilting are relieved, the plants will recover turgidity.

Excessive air movement across plants can also cause water stress. In fan and pad cooled greenhouses, those plants located closest to the pads receive a high-speed blast of air as it enters the house. Although the air may be cool, the speed of its movement results in a very high transpirational moisture loss that is damaging.

Cold Soils Plus High Light Intensity

During the winter, there often are long periods of dark cloudy weather fol-lowed by one or two days of bright sunshine. Plants that have become accli-mated (accustomed) to the dark conditions cannot respond quickly to the light intensity of the clear days. As a result, too little water is taken up by the roots to replace the water lost through transpiration. As a result, wilting occurs. The grower may be misled into thinking that the plants need water. An inspection of the bench often shows that the soil is wet but cold. The cold is the reason that the roots are unable to take up water. The addition of more water only makes the problem worse. The stress on the plants is reduced by providing temporary shade. After the second day of bright light, the plants generally adapt to the light level, and shade is no longer needed. Root zone heat or bot-tom heat is another means to maintain better water relations.

Application of Water

When to Irrigate

Management of irrigation is a major factor in the success or failure of a crop. The greenhouse grower must acquire skill in scheduling irrigations. The mois-ture needs of crops vary with their stage of growth. That is, seedlings need less water than mature crops. The season of the year also affects the water require-ments of the crop. Water use is greater in the summer than in the winter. Other factors affecting irrigation include the type of growing medium used, the type of heating system installed, the type of pot used, and the crop itself. Some plants require much less water than others to produce the same amount of growth. It is a mistake to put an unskilled worker on the end of a water hose to manage a greenhouse crop.

There are few methods or devices available to the grower that can be used to determine when to apply water. Probably the best method is the judgment of the grower based on years of experience. The grower knows when the crop was last watered, how much water was applied, and how the crop is growing.

Soil particles hold water by exerting a certain amount of force, or tension. Thus, a mechanical means can be used to measure the tension. The **tensiometer** is the most widely used device for this purpose (Figure 16-2). The unit consists of a porous clay cup fastened to a tube that is connected directly to a vacuum gauge. The unit has a removable plug at the top so that it can be filled. It is filled with boiled, distilled water to which a small amount of dye is added. Boiling removes air from the water. Air causes an incorrect reading. Distilled water is used to retard the growth of algae inside the tube. The dye is added to make it easier to see the fluid level in the tube.

The unit is inserted into a hole in the soil. The hole has the same diameter as the porous cup. The unit should not be forced into the soil, because the soil is then compacted around the clay cup. As a result, the reading of the moisture content is incorrect.

As the soil dries, there is a suction force or pull on the moisture in the walls of the clay cup. This force, in turn, pulls on the water in the tube. A partial vacuum develops in the gauge. As the suction force increases, the needle of the vacuum gauge responds by moving to a higher number. The higher numbers reflect an increasing dryness of the soil.

Because of natural forces in soils, the tension of the moisture cannot be measured accurately beyond a range of 850 millibars, or a reading of 85 on the vacuum gauge. At this tension, the moisture column breaks, and the gauge no longer reflects the true moisture content of the soil. Electronic devices are available that can be connected to irrigation control computers.

FIGURE 16-2 Tensiometers are used to measure the moisture content of the soil. For each of the three different sizes shown, the dial measures moisture tension as it is sensed through the porous clay cup.

Bouyoucos Blocks and Granular Matrix Sensors

Bouyoucos blocks are made of plaster of Paris. Wires are embedded in the blocks for connection to a moisture meter. The blocks are buried in the soil at the depth where the moisture levels are to be measured. In field soils, the blocks may be buried at depths of 6, 12, or 18 or more inches. The blocks are connected to a moisture meter by means of the two wires. As the soil dries, a very small electrical resistance develops inside the block between the two wires. This resistance is measured and indicated on the meter as the level of wetness or dryness of the soil. The blocks and wires are left in place in the soil so that a meter can be connected quickly to obtain a reading.

Granular matrix sensors (GMS) are similar to bouyoucos blocks where they have electrodes embedded in a granular quartz material, which is surrounded by a membrane protected by a stainless steel mesh (Figure 16-3). Like the bouyoucos blocks, gypsum is used to buffer salts and the porous nature of the medium allows for measuring in wetter conditions, such as greenhouse media. As with bouyoucos blocks, the GMS probe is left in place in the soil so that a meter can be connected quickly to obtain a reading.

Moisture Probes

Small devices called moisture probes (Figure 16-4) were developed in recent years to measure soil moisture. These devices are used primarily by amateur growers. The end of the probe contains two unlike metals, separated by a plastic ring. The moisture in the soil causes a galvanic (electrical) reaction between the two metals. As a result, a very small electrical current is generated. The greater the moisture content of the soil, the stronger is the current. The current is increased by a small amplifier in the device. In this way, the signal can be read as a deflection (movement) of a needle on a meter at the top of the probe. The meter is marked with a scale of 1 to 10, or with a range from dry to wet. Instructions included by the manufacturer with each meter relate the meter reading to the desired moisture level for different crops.

A high soluble salt content in the soil causes a moisture probe to give a false reading. The meter will show the soil to be wet when it is actually on the dry side.

A long-standing practice used by growers is to lift pots to determine if the crop needs water. If the pot is light in weight, the medium is dry. This method

FIGURE 16-3 Granular matrix sensor manufactured by Watermark.

FIGURE 16-4 Inexpensive moisture probes.

of determining water needs is not too reliable, although it is satisfactory in many situations.

How Much Water Is Supplied to a Crop?

The amount of water required depends upon several factors: the type of crop, the stage of crop growth, the season of the year, the growing medium, and the method of fertilization being used. Fertilization is the most important factor. If the fertilizing schedule calls for an application every 7 to 14 days, the amount of water-fertilizer solution applied is just enough to wet the soil mass completely. There must be no excess water to drip through the bench or pot. This amount of solution is equal to nearly 1 quart for each square foot (10.2 l/m^2) of surface area. Plain tap water is applied between fertilizer applications at the same rate. Occasionally, the grower will apply a small amount of extra water so that some leaching occurs.

If fertilizer is applied at each watering, an extra amount of solution is applied. This amount is about 10 percent more than is needed to wet the soil mass. The excess solution causes some leaching at each irrigation. This procedure prevents a harmful buildup of soluble salts, yet may be considered excessive, risking nonpoint source pollution damage to groundwater. No leach irrigation is gaining acceptance by many even when plants are fertilized with each irrigation. The fertility rate is simply reduced by half, typically.

When the weather is cloudy, it may be necessary to reduce the frequency at which water is applied. Watering may be reduced greatly if the cloudy conditions continue for many days. Some growers feel that the crop should have some water at such times and will apply a small amount to the pot or the bench. This is a poor practice because the soil never dries out completely on the surface. Frequent applications of small volumes of water cause a serious reduction in soil aeration. As a result, root rot and other problems may develop. The grower should allow the soil or medium to become completely dry on the surface. Once this point is reached, a full volume of water can be applied. As the water moves down through the medium, it drives out the old air and pulls in new, fresh air behind it. This method of applying water reduces the incidence of an overly wet medium and a buildup of a high level of soluble salts.

FIGURE 16-5 A stiff piece of wire is easily transformed into a hose hanger to keep the end of the hose off of the ground to prevent contamination.

For pot plant crops, the amount of water applied should fill the reservoir at the top of the pot. The pot must not be overfilled with medium because it is then impossible to add the proper amounts of water and fertilizer to the pot.

Clay pots, once the mainstay of greenhouse production, must be irrigated more often than plastic pots. The walls of clay pots lose moisture. Plastic pots do not lose moisture in this way.

Methods of Application

There are many ways of applying water to crops. They range from a handheld hose to sophisticated semiautomatic systems that can be programmed to water an entire range of greenhouses.

Hose Watering

The use of a handheld hose is a time-consuming method of applying water to a crop. The present cost of labor makes this method too expensive for most large-scale use. When a hose is used for watering, the end of the hose must not be allowed to fall to the ground. The hose end should be hung from a small S-shaped hook or a broom clip (Figure 16-5).

Pythium is a very destructive disease that is found in every greenhouse. Allowing the end of the hose to lay on the ground between waterings is an excellent way to spread this disease to a crop.

If a hose is to be used to apply water, it should be 0.75 inch (1.9 cm) in diameter. A smaller size restricts the water flow, and more time is required to water the entire crop. Quarter-turn valves and quick couplers for connection make for quicker work and more efficient labor management (Figures 16-6A through 16-6D).

If a hose is placed on a walkway in full sunshine for several hours during the summer, the water in the hose is heated to a temperature that may be greater than 120°F (49°C). If water at this temperature is applied to plants, they will be severely damaged. Before the crop can be watered, the water must be turned on and allowed to run until it is comfortable to touch.

FIGURE 16-6A Dual hose bibs where the left provides clear water and is typically painted blue. The hose bib on the right that is painted red provides water where fertilizer has been injected.

FIGURE 16-6B Hanging water hoses on a cable system makes it convenient to move about while irrigating plants by hand.

FIGURE 16-6C Greenhouse workers require adequate water pressure when irrigating plants by hand.

FIGURE 16-6D Fog nozzles for misting newly planted seeds or cuttings can be mounted on multiple head devices and are typically home made.

Low-Volume Methods

Low-volume methods of watering have become popular within the last 10 years. Richard Chapin of Chapin Watermatics in Watertown, New York, pioneered the development of low-volume watering methods.

Tube Watering

Tube watering is used primarily for potted plant crops. Water at a pressure of 3 to 5 pounds per square inch is brought to a main pipe on the bench. Tubes lead off from the main to headers to which the individual pot trickle tubes are connected. A hollow core lead weight is attached at the end of each trickle tube to hold the tube in the pot. The weight also disperses the water so that it flows in a small trickle rather than a stream. The plastic tubes used have an outside diameter of 0.128 inch (3.25 mm) and an inside diameter ranging from 0.036 to 0.076 inch (0.9 to 1.9 mm).

Irrigation control devices for watering several benches or zones in rotation are also available (Figure 16-7). Figure 16-8 illustrates an installed trickle tube watering system for containers.

Ooze Tubes

Ooze tubes are used to water bench crops or crops grown in rows. A twin-wall plastic tube is rolled along the row of plants (Figures 16-9A and 16-9B). This tube is approximately 1 inch (2.5 cm) in diameter. The inner tube is the main water supply. At points along this tube, small holes permit water to bleed into a second tube surrounding the first. This water then seeps into the soil from tiny, uniformly spaced holes. These holes are made by sewing the plastic. The ooze tube has a low volume of delivery and is preferred where water supplies are limited or expensive. This method does conserve large amounts of water. Ooze tubes are available in a number of sizes.

Water Loops

Small loops of plastic may be wrapped around the stems of container grown plants. These water loops (Figure 16-10) apply water in much the same way as

FIGURE 16-7 Irrigation controllers designed for turf applications are easily adapted to greenhouse irrigation systems.

FIGURE 16-8 A bench of newly transplanted poinsettia cuttings with a trickle tube system installed.

FIGURE 16-9A Ooze tube installed on a bed of newly transplanted snapdragon plugs.

FIGURE 16-9B Ooze tubes spaced between rows of lisianthus planted for cut flowers. Note the wire mesh that will be lifted as the crops grows for plant support.

FIGURE 16-10 Close-up of a water loop.

ooze tubes. A header brings the water to the plants. The loop distributes the water around the base of the plant through the stitching of the plastic. Light-weight media, such as peat-lite mixes, are watered more evenly with the loops than with weighted tubes. Water tends to run in channels from weighted tubes. This means that the entire volume of medium is not thoroughly wetted.

Other types of low-volume trickle irrigation are also available to growers. Each uses the same basic principle of providing a small volume of water under low pressure directly to the growing plant.

Surface Irrigation Methods

Surface irrigation methods deliver water to the entire soil volume. These methods require a greater water supply and higher operating pressures than the trickle methods. Surface methods are used to water cut flower and row crops.

Gro-hose®

Gro-hose is 2-inch (5-cm) diameter plastic tubing that is pierced with tiny holes every 8 inches (20.3 cm) of its length. The tubing is made of black plastic to reduce the growth of algae within the hose. If allowed to flourish, algae will plug the holes. The tubing is connected to a header and is placed flat between the rows of plants (Figure 16-11). For a 4-foot (1.2-m) wide bench, four rows of tubes are adequate if they are evenly spaced. Water from the holes slowly wets the soil by the action of gravity. The lateral movement of the water is by capillary action.

Gro-hose systems can be manually operated or controlled by a semiautomatic remote system. The frequency of operation is determined by the greenhouse manager.

Gates Perimeter System

The Gates perimeter watering system uses a black plastic hose that is attached to the perimeter of the bench (Figure 16-12). Small nozzles with either 90 or 180 degree delivery are located at intervals in the hose. The nozzles produce a flat spray that wets the soil with little wetting of the foliage. By keeping the foliage dry, the chances for disease are greatly reduced.

Black Polyethylene "Lay Flat" Tubing

Water Emerging

Header for Water Distribution

FIGURE 16-11 Gro-hose® watering unit with black polyethylene tubes that lay flat when they are not filled with water.

FIGURE 16-12 Perimeter Gates watering system installed on ground beds for snapdragons grown for cut flowers.

The Gates hose should be removed from the bench when the soil is to be steam pasteurized between crop plantings. The high steaming temperatures damage the plastic.

Ohio State Skinner System

This system delivers a flat spray and consists of a pipe placed down the center of the bench. Flat spray nozzles with 360 degree delivery are spaced 3 feet apart. The system can apply water very quickly to a large growing area. The pipe is either copper or galvanized iron. Hard polyvinyl chloride (PVC) pipe can be used as well. The metal piping is preferred for long-term use, such as in rose houses.

Spray Stakes

Spray stakes are used to water pot plants, bedding plants, and other container grown crops (Figures 16-13). The flat spray is delivered over 360 degrees. The stakes are spaced so that there is some overlapping of the area being watered to prevent dry spots in the crop.

Boom Irrigation Systems

The advent of plug culture for starting plants required a whole new look at how to apply water and fertilizer solutions to the crop. In the small greenhouse, where only a few hundred plug trays were started, the handheld hose with a light spray nozzle worked well. However, where plug production numbers increased to the thousands and tens of thousands of trays, a better means of fertigation was needed.

Boom irrigation systems have been developed as custom made units for greenhouse production areas. This is because each greenhouse has a different shape and size. With as many greenhouse manufacturers as there are in the business and the different types of glazing materials used, it is impossible to think of a standard shape and size boom irrigation unit (Figure 16-14).

Boom irrigation systems are not designed for use with cut flower crops such as roses, carnations, or chrysanthemums. Other methods of fertigation are used for these crops.

FIGURE 16-13 Spray stakes installed among poinsettias.

FIGURE 16-14 Boom irrigation system. This unit can also be used to apply growth retardants and pesticides.

With plug trays, vegetatively propagated crops, bedding plants, and some potted plant crops, boom irrigation systems work very well. For maximum efficiency the growing area should be one that is clear span. That is, there are no support posts or other upright fixtures that will interfere with the travel of the boom.

Fertigation by means of a boom system has several advantages. Uniformity of solution application is one of the greatest assets. By proper selection of the nozzles used the solution can be applied as a gentle spray or a full deluge as dictated by the stage of the crop growth.

Labor saving is a great advantage of using a boom system of fertigation. The unit requires no hands-on operation except to prepare the fertilizer solution being used. After this is supplied to the system, the unit can be programmed by means of mechanical time clocks or computer controls. Some manufacturers claim that by using boom irrigation systems, there is a 40 percent savings in water and fertilizer when compared to hand watering.

The use of boom irrigation systems also improves the efficiency of space use in the greenhouse. Wide aisles formerly needed for hand watering operations are eliminated. The space saved is used for growing plants.

In addition to the application of fertilizer, boom systems can also be used to apply insecticides, fungicides, and plant growth regulators. Selection of the proper size and type nozzle is critical to success when applying these chemicals. Another benefit of boom systems when applying these chemicals is less worker exposure. Reentry waiting times do not change, so follow label restrictions as noted.

Appendix I contains a partial list of several manufacturers who supply boom irrigation systems.

Ebb and Flood Irrigation

The use of the ebb and flood technique of irrigation has become increasingly popular. Among the many reasons for this are the ever-restrictive rules of the EPA regarding water runoff and groundwater contamination.

The ebb and flood system, sometimes mistakenly referred to as ebb and flow system, was developed as a modern day refinement of the water injection system developed in the mid-1940s by researchers at Cornell University in Ithaca, New York. In that system, water or fertilizer solution was injected into the bench and allowed to remain long enough for the media in the pots to wet. The bench was then allowed to drain (Figure 16-15) beneath the bench.

In the modern ebb and flood system, the irrigation solution is pumped from a central storage tank into the waterproof bench (Figure 16-16). A series of channels helps to move the water quickly so that wetting of the media is done uniformly (Figure 16-17). The solution remains in the bench for a few minutes, after which it is drained back to the central storage tank for reuse. The entire system is monitored and controlled by computer.

Because the foliage never gets wet, which can promote the development of disease, plants can be watered-fertilized any time of the day or night (Figure 16-18). Normally this fertigation is done only once a day in winter. In summer or during high stress periods, the application of solution may be made two or three times daily.

Ebb and flood irrigation is used on raised greenhouse benches or through flooding of the entire floor area (Figure 16-19). Several companies have

FIGURE 16-15 Cross section of a water injection bench.

Water is added through the stand pipe to the gravel which distributes it across the bench. Pots may be placed on the shallow layer of sand (B). Water is then injected to the rim and allowed to drain through the tile. An alternate method is to plunge the pots into the sand (A) for more uniform moisture. Method "A" is often used for African Violets.

FIGURE 16-16 Chrysanthemums near flower bud set irrigated on an ebb and flood table.

FIGURE 16-17 A small ebb and flood bench.

FIGURE 16-18 African violets do not grow well when the leaves are wet. Ebb and flood irrigation works well with this crop.

FIGURE 16-19 Large containers of geraniums grown on an ebb and flood floor. Note the large number of hanging baskets overhead increasing the productivity of the greenhouse space.

developed unitized bench systems. These units enable growers to retrofit older greenhouses to ebb and flood techniques.

Where the entire floor is involved, it usually means that construction of a new greenhouse is needed. There are cases where existing greenhouses have been retrofitted for an ebb and flood system, but the job is much easier done in new construction.

To be successful in operation, the floor must be brought to within 0.5 inch of what the final grade will be. The rough graded floor must be absolutely level with no variation or slope. This precision construction is frequently done using a laser level mounted on a tripod. When this is done, the trenches for the main feed and return lines are dug. The pipe for this feed line must be capable of carrying large volumes of water quickly. As many as 600 to 800 or more gallons of water per minute may be pumped through the system. This large volume permits an entire bay to be filled and drained in 6 to 8 minutes. The need for rapid draining and filling is so that all pots are irrigated at essentially the same time so they absorb an equal volume of solution. Figures 16-20A through 16-20C demonstrate the flooding process of an ebb and flood greenhouse

FIGURE 16-20A Large containers of geraniums grown on an ebb and flood floor. Here the irrigation water is beginning to rise from the center spur line.

FIGURE 16-20B Large containers of geraniums grown on an ebb and flood floor. Here the irrigation water has nearly flooded the floor section.

FIGURE 16-20C Large containers of geraniums grown on an ebb and flood floor. Here the irrigation water has reached its flooded limit. Typically this level is held for 15 minutes or until the growing medium is saturated and then allowed to drain.

floor. Zones are easily separated by rubber strips installed in slots cut in the floor (Figure 16-21).

Branching off from the main lines are the spur lines that supply each bay. The number of spur lines needed for each bay depends on the width of the bay. After the spur lines are installed and backfilled, the ground is again leveled to within 0.5 inch of the final grade.

Reinforcing bars or wire mesh is laid down on top of where the floor heating is installed. Following the heating system installation, the concrete is poured. The concrete is a special blend of Portland and slag cements. This combination is less affected by fertilizers than if Portland cement alone is used. The slag cement also adds extra strength to the floor.

Because of the high precision that is needed throughout the entire installation, only professional contractors who have experience in doing ebb and flood floor construction should be used.

Capillary Mat

The **capillary mat** or pad method of watering is an outgrowth of the injection method (Figure 16-22). Originally, sand was used as the base upon which the pots were set. At the present, various types of capillary mats are used. As a result, the system is easy to set up and maintain.

The operation of the capillary mat system is based on the fact that water rises to a certain height in tubes with very narrow diameters through a process known as capillarity. This same force operates in field soils to bring water up to the top layers of dry soil from the lower layers of moist soil and the water table. In potted plants, the continuous pore spaces in the growing medium act as capillary tubes and lift the water from the mat to the roots inside the pot.

Capillary watering was widely used in Europe before American growers began to adapt this system to their production methods.

Any bench that is relatively level throughout its length can be adapted to capillary watering methods. The first step is to lay a piece of polyethylene, or

FIGURE 16-21 The rubber strip to the right of the containers allows for separation of irrigation zones. Rubber strips are used as they will fold down when carts or other wheeled equipment cross over.

FIGURE 16-22 Calceolaria or pocket book plants grown on a capillary mat. The fabric mat is underneath the black sheet shown. The black sheet shown is perforated to allow pot and mat contact while blocking light avoiding excessive algae growth on the mat. The circles and triangles indicate different potential crop spacing patterns.

other plastic, on the bench. The plastic distributes the water uniformly and prevents it from running through holes or cracks in the bench. The capillary mat material is then unrolled and smoothed over the plastic. The ends of the mat and plastic are not turned up at the sides. Water is allowed to run off the bench to prevent an accumulation of soluble salts in the mat.

Water is distributed on the mat by an ooze tube or by drip tube units. The mat should be wetted uniformly. Thus, care should be taken when placing the tubes. On benches wider than 3 feet (0.9 m), it may be necessary to place two ooze tubes along the length of the bench to ensure proper wetting.

Plastic pots are preferred to clay pots for use with the capillary system. Moisture loss through the porous side walls of the clay pots may be greater than the amount of water taken up from the mat. Under very high stress conditions, plants growing in clay pots may wilt on a wet mat.

The bottom of the plastic pot should have several holes to ensure good contact with the pad. Single-hole pots, pots with projecting "feet," and pots that have holes on the side wall are not recommended. Styrofoam pots and other thick-wall plastic pots should not be used because they do not make the proper contact with the mat.

After the pots are placed on the mat, they should be hose watered from the top. This watering forms the unbroken water columns that extend from the media to the mat. These columns are required so that the medium will remain wet. If the pots are disturbed in some way and the capillary action is destroyed, the pots must be rewatered from the top to reestablish the capillary action.

As the root system develops within the pot, some rooting will take place into the mat. This type of rooting is increased if the mat is allowed to dry out between irrigations. The pots should be lifted periodically from the mats to prevent excessive root growth into the mat. If too large a root system develops in the mat, the plant will depend on this system. This means that when the plant is removed for sale, excessive wilting occurs because half the root system is left in the mat.

Another problem with the capillary watering method is the movement of soluble salts from the bottom of the pot to the top of the soil ball. Moisture movement is always in one direction. Therefore, the forces of evaporation carry the salts to the surface of the medium. Problems can occur due to the high soluble salt levels after the plants are removed from the mats. To avoid these problems, it is recommended that the plants be watered from the top at least twice, and preferably three times, during the last week before they are removed from the greenhouse. These waterings will cause the excess salts to leach from the top of the pot and will distribute them more evenly throughout the growing medium.

Crops grown on a capillary mat are fertilized with a slowly available complete nutrient source such as Osmocote® 14-14-14. The fertilizer is mixed with the medium at the time of potting. Algae growth is encouraged if soluble fertilizers are added through the water system and the mats. Efforts to control algae growth using chemicals have not been successful. Although the chemicals kill the algae, they also damage the plants.

The only positive way of removing algae from capillary mats is to steam pasteurize the system between crops. However, this process shortens the life of the mats and increases the cost of the system. Some mat materials cannot withstand the high temperatures of the steaming procedure.

Capillary mats can supply too much water to plants at certain times of the year, particularly foliage plants in the winter. Poor growth results, and root diseases, often develop. Before an entire greenhouse range is converted to capillary watering, the grower should obtain some experience with the system on a small scale.

Water Quality

Water is evaporated from various bodies of water, particularly the oceans, into the atmosphere. It is carried over the land and falls to earth as precipitation in the form of rain or snow. Any moisture that plants do not use immediately drains into the soil where it is held in aquifers (underground storage areas). As these groundwater supplies are filled, seepage occurs to the surface. Streams and rivers are formed that eventually return this excess water to the oceans where the process begins all over again.

As the water moves from the ocean to land and back to the ocean, it is influenced by many factors. One of the most important factors is pollution from industrial wastes, untreated sewage, farmyard runoff, and other sources. Household detergents that are not biologically degradable, or capable of being broken down by soil microorganisms, have contaminated groundwater supplies miles from the point where they went into the soil.

The application of salt (sodium chloride) to highways in winter to melt snow and ice has become a serious factor in water contamination in the northeastern United States and in other areas. The use of fertilizers has overloaded the waters of many rivers in the Southwest to the point that they are nearly useless as sources of irrigation water.

The quality of water varies from region to region. Nature, as well as human beings, influences water quality.

Hard and Soft Water

Water is classified as hard or soft depending upon the amount of dissolved minerals it contains. Calcium and/or magnesium carbonates are the most common minerals that cause hardness. Although household laundry and dishwashing tasks are easier if the water is softened, plants will not grow as well if they are given water softened by chemical means.

Most water softeners use an ion exchange resin (Zeolite) and sodium chloride (salt) to exchange sodium ions with the calcium and magnesium ions of the hard water. When **softened water** is used to irrigate plants, the sodium ions rapidly build up in the soil. Sodium destroys the soil structure and results in poor drainage. It also builds up within the tissues of certain plants. For example, carnations with 10,000 ppm (1 percent) sodium in their tissues fail to flower although they have buds showing good color. The petals stick together and do not unfold properly.

Water-softening procedures may change the pH and add to the total soluble salt content of the water. Some changes that take place when water is softened are shown in Figure 16-23. The white bars represent readings for hard water, and the grey bars are the values for soft water.

Acidifying Water

In some regions, the pH of the water may be too high for use on plants. When the pH is above 7.5, water is considered to be too alkaline. Continued use of such water will result in a high pH in the soil. This high pH level may interfere

FIGURE 16-23 Comparison of hard water versus soft water.

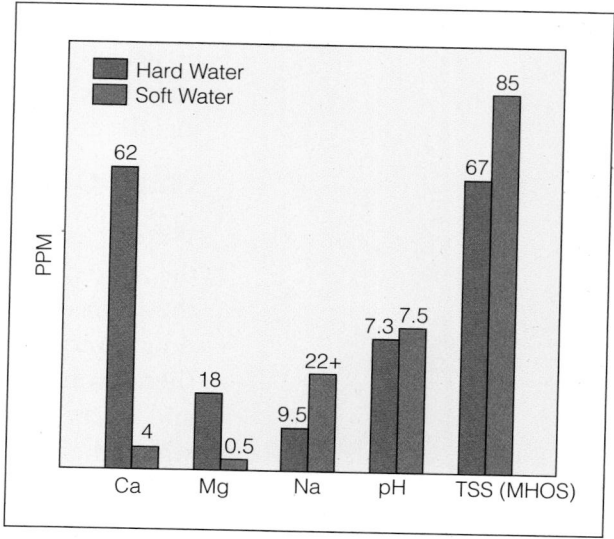

with plant uptake of certain trace mineral elements. This water must be made more acid by adding phosphoric acid (H_3PO_4).

The acid is added to the water in diluted form by an injector. However, even the diluted acid is very corrosive and may in time ruin metal pipes.

When phosphoric acid is used to correct a high pH condition, less phosphorus fertilizer is required. The phosphorus supplied by the acid is used by the plants.

If the pH is not too high, the alkalinity may be overcome by adding acid sphagnum peat moss to the medium. Fertilizers such as ammonium sulfate, ammonium nitrate, potassium sulfate, and most of the complete analysis products will have an acidifying effect when added to the soil. Regular applications of iron sulfate, aluminum sulfate, or sulfur also help to keep the soil pH in the proper range.

Chlorine

Chlorine is added to water to kill harmful bacteria and other organisms that may cause disease in humans. The chlorine added to the water readily escapes as a gas when the water is aerated. Although water may have a strong chlorine smell or taste, it will not harm plants as long as it is potable (suitable for humans to drink). The amount of chlorine needed to be harmful to plants would make any water treated at that rate unfit for humans.

Fluorine

Fluorine is another poison that occurs naturally or is added to water used as a source of drinking water. Some communities add fluorides to their water supplies in the amount of 1 ppm to help prevent tooth decay. In some parts of the country, the natural fluoride content of water may be as high as 2 or 3 ppm.

Research has shown that foliage plants of the lily family (*Liliaceae*) may be injured by as little as 0.25 ppm of fluoride in the water. Aspidistra, dracaena, chlorophytum, and cordyline all show **leaf tip burn** from excessive amounts of fluorides (Figure 16-24). Sphagnum peat moss, single superphosphate, perlite, and irrigation waters all contain fluorides.

The damage from fluorides can be avoided by adding ground limestone to the soil to raise the pH to 6.5. At this pH, calcium combines with the fluoride to form an insoluble compound.

FIGURE 16-24 Tip burn from fluoride accumulation.

FIGURE 16-25 Myron Meter used for measuring total soluble salts. This unit can also be used to measure pH.

Total Soluble Salts

The quality of water is affected by the total soluble salt content. Fertilizers and other dissolved minerals all add to the soluble salt content of the soil. If the concentration of salts in the soil is too high, water absorption is reduced, growth is retarded, and plant death can occur.

Because the soluble salt content of soils has such an important effect on plant growth and is so easily measured, every grower should check the salt content of the growing media, water supply, and fertilizer solution on a regular basis. An instrument known as a Myron Meter is available from the Myron L. Company, 2450 Impala Dr., Carlsbad, CA 92010 (Figure 16-25).

The water being tested should be compared to the following scale to interpret its quality.

Soluble Salts Reading Myron L Deluxe DS Meter (micromhos)	Quality Interpretation
Below 0.25	Excellent
0.26–0.59	Good
0.60–1.49	Fair
1.50–2.0	Poor
Above 2.00	Excessively salty

Water with a reading of 1.50 to 2.00 micromhos may be used to grow crops. However, there will be a buildup of salts in the soil. Extra water must be applied each time the crop is irrigated to obtain leaching. If the initial salt level is 1.50 to 2.00, the addition of any amount of fertilizer will cause a large increase in the salt concentration of the soil.

If the unfertilized water supply has a natural salt content of 2.00 or more, an alternate water supply should be obtained.

FIGURE 16-26 Aqua-Gro® L PsiMatric L. This is a surfactant (wetting agent) that is widely used in the horticulture industry.

Surfactants (Water Wetters)

Surfactants (surface acting agents) are chemical compounds that reduce the surface tension of water (Figure 16-26). Although water is considered to be an extremely mobile fluid, it has certain physical properties that give it strength and shape. For example, a drop of water on a leaf keeps its form because of the force of surface tension. Rain falls as drops because surface tension holds the water in a round shape.

When water is to be used as a vehicle for insecticides or fungicides, water wetters or surfactants are added to the spray tank to improve the wetness of the water. Household detergents used for washing dishes make water wetter and break down grease and oil on dishes.

Industry uses hundreds of surfactants in many different ways. Most of the surfactants made for industrial use are not safe for use on plants. Only a few surfactants are known to be safe for wetting plant growing soils. Most of these compounds are nonionic.

Neutral-charge **wetting agents** are made up of an alcohol and an ether or an ester. These chemicals act to make one part of the water molecule soluble but not the other. This condition weakens the adhesive or attractive forces of the water. As a result, the surface tension is reduced, and the water becomes more fluid. A few drops of a surfactant in a gallon of water are sufficient to cause the wetting effect.

If surfactant treated water is added to a hydrophobic (water resistant) medium, such as peat moss, the peat is wetted very quickly. The rapid wetting is followed by rapid drainage of any excess water. This drainage is important. If excessive water remains in the medium for a long time, a lack of oxygen can develop. Carbon dioxide and other gases then build up, reducing crop growth.

The use of surfactants also slows the loss of water from a growing medium due to evaporation. This means that water must be applied less often. Water added to the medium is used by the plants and is not lost to the atmosphere.

Surfactants are very useful when plants are transplanted. By reducing the energy needed to take up water from the growing medium, the plant has a better chance for survival.

ACHIEVEMENT REVIEW

Select the best answer or answers to complete each statement. List the appropriate letter(s).

1 Of the following nutrient materials, a plant has the greatest need for:
  a. oxygen.
  b. carbon dioxide.
  c. water.
  d. fertilizers.

2 The average number of units of water absorbed for each unit of dry weight produced is:
  a. 100.
  b. 250.
  c. 500.
  d. 750.

3 The absorption of water into the plant occurs mainly through the:
  a. leaves.
  b. stems.
  c. bark.
  d. roots.

4 Water enters a plant through a process known as simple diffusion. Diffusion is defined as a flow of ions from:
  a. areas of equal concentration.
  b. an area of low concentration to an area of high concentration.
  c. an area of high concentration to an area of low concentration.

5 Diffusion against a gradient is similar to the action required to:
  a. ski downhill.
  b. glide on ice skates.
  c. push a wagon uphill.
  d. ride a wagon downhill.

6 Cell sap (water) moves through a continuous pipeline of vascular tissues called:
  a. phloem.
  b. cambium.
  c. xylem.
  d. epidermis.

7 If roots are submerged for too long a time, the main effect is:
  a. a lack of carbon dioxide.
  b. a lack of oxygen.
  c. a lack of water.
  d. decreased nitrogen uptake.

8 Even though the cuticle is waterproof to some extent, water absorption takes place through the leaves because:
   a. the cuticle acts like a sponge.
   b. there are cracks in the cuticle covering.
   c. water has a positive charge.
   d. the cuticle has a negative charge.

9 The major pathway for water loss from plants is:
   a. respiration.
   b. leaching from leaves.
   c. transpiration.
   d. dry soils.

10 If the outside relative humidity is 100 percent, the potential for moisture loss from the plant is:
   a. 0 percent.
   b. 50 percent.
   c. 80 percent.
   d. 100 percent.

11 When a plant is wilted, photosynthesis:
   a. temporarily stops until turgidity is regained.
   b. is not affected by wilting.
   c. is speeded up.
   d. continues at half the normal rate.

12 High levels of soluble salts in the soil solution:
   a. speed up water absorption.
   b. have no effect on water absorption.
   c. slow down water absorption.
   d. can pull water from the root cells.

13 Snapdragons wilt on bright, sunny days in winter following 2 weeks of cloudy weather. This reaction may be reduced by:
   a. watering the soil.
   b. temporarily shading the plants.
   c. increasing the temperature of the greenhouse.
   d. decreasing the temperature of the greenhouse.

14 When fertilizer is applied with every irrigation, the amount of water-fertilizer solution used is:
   a. 1 quart to 1 square foot of area.
   b. just enough to wet the soil.
   c. about 10 percent more than is needed to wet the soil mass.
   d. 1 gallon to 1 square foot of area.

15 If the water supply is limited, the method that conserves the most water is:
   a. the Ohio State flat spray.
   b. the Gro-hose®.
   c. trickle irrigation.
   d. a capillary mat.

16 If pots on a capillary mat system become dry, they are rewetted by:
  a. keeping the mat constantly wet.
  b. applying water at the same rate.
  c. applying water to the top of the pot using a hose.
  d. flooding the bench by injection.

17 If a pot plant is being grown on a capillary mat, the fertilizer salts tend to:
  a. settle to the bottom of the pot.
  b. be unaffected by the system.
  c. migrate toward the surface of the medium due to evaporation.
  d. build up around the inside of the pot wall.

18 Water that is softened by a sodium ion exchange resin system:
  a. will improve the growth of plants like African violets.
  b. will accumulate within the tissues of certain crops.
  c. should never be used when watering plants.
  d. has a high percentage content of calcium and magnesium.

19 If it is suspected that fluorine has caused the tip burn of foliage plants of the lily family, the growing medium should be:
  a. mixed with perlite to improve the drainage.
  b. treated with aluminum sulfate to bring the pH to 4.5.
  c. maintained on the dry side.
  d. limed to bring the pH to a minimum of 6.5.

20 A surfactant acts on water:
  a. to increase the surface tension.
  b. to decrease the surface tension.
  c. by not affecting the surface tension.
  d. to increase the rate of evaporation from the soil.

Plant Propagation

Sexual (Seed) Propagation

OBJECTIVES:

After studying this chapter, the student should have learned:

■ how to draw and identify the major parts of a seed.

■ the functions of each of these seed parts.

■ how the dormant stage affects seed germination.

■ the influence of oxygen, temperature, and moisture on seed germination.

■ the materials and amounts needed to make a seed germination medium.

■ the procedure for the chemical fumigation of seed germination media.

■ how to prepare a flat for sowing seeds.

Human life is maintained by plant life. Plants must be reproduced continually to provide an uninterrupted food supply. New plants are obtained through a process called propagation. Propagation is defined as the increase of a plant species from one generation to the next. Plant propagation is both an art and a science. The science involves the knowledge of the parts of plants and how they work together to cause growth. The art is in knowing how and when to work with the plant parts to obtain the desired result.

There are two basic ways in which plants can be reproduced: sexual propagation and asexual propagation. **Sexual propagation** uses the seeds of plants. Seeds are produced when the male and female gametes, or tissues, unite within the flower. **Asexual (vegetative) propagation** occurs when various parts of a plant are placed in the proper environment until they root and eventually grow into a new plant. This chapter will describe methods of sexual propagation.

Life Cycle of Plants

Before describing the sexual propagation of plants, a brief review of the life cycle of plants is needed (Figure 17-1). After a seed is sown and germination takes place, the new plant, or seedling, begins to grow. This stage of vegetative or nonflowering growth is called the **juvenile stage of growth.** As the plant increases in size through the formation and growth of millions of new cells, the length of the stem and the cross section of the plant both increase. Leaves, stems, and other plant parts develop in this stage. A plant usually cannot be made to flower during the juvenile stage of growth.

After the proper length of time, the plant enters the **adult stage of growth.** This stage may vary from as short a period of time as a few weeks for herbaceous plants, such as the dandelion, to as long as several years for some trees. During the adult stage, the plant may be stimulated naturally to flower and produce fruit and/or seed. Not all plants produce fruit, but almost all plants produce seed. Some plants are said to be incomplete. In other words, they lack some flower parts. Such plants can serve only as the male parent and cannot produce seed.

Plants are classified according to the length of their life cycle. Three different types of life cycles are recognized.

Annual Plants

Annual plants pass through their entire life cycle in one season. For example, petunias are annuals. The seeds of an annual are planted in the spring. Once they germinate, the plants pass through the juvenile stage and reach maturity, bloom, and then form seeds. All of these stages occur in one season. The seeds will winter over in the ground and will germinate the following spring if they are not winter killed.

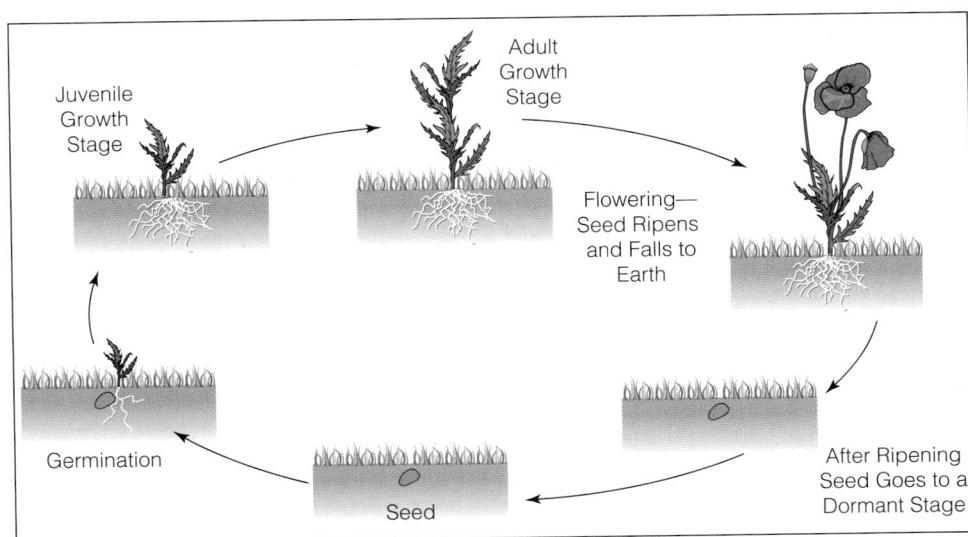

FIGURE 17-1 Life cycle of an annual flowering plant.

Biennial Plants

Biennial plants require 2 years to complete their life cycle. During the first year, the seeds germinate, and the plants are in the juvenile stage of growth. Biennial plants require a period of cooler temperatures before they will flower. During the winter months, they are dormant. Then, in the second growing season, the plants complete their growth and flower. After flowering, the plant sets seed and then dies.

Perennial Plants

Perennial plants live for more than 2 years. They have an annual vegetative-reproductive growth cycle. Their growth pattern may be either the warm–cold cycle of an annual plant or a wet–dry cycle. Plants with wet–dry growth cycles are usually found in semitropical or tropical areas where freezing temperatures do not occur.

Plants whose shoots die during the cold or dry period are known as **herbaceous perennials.** One example of such a plant is the peony which survives the winter as an underground fleshy root after the tops are killed back by frost.

Trees and shrubs are woody perennials, which increase in size each year through root, shoot tip, and cambial growth. The cambium is a specialized part of the plant that is discussed later in this chapter.

Seeds

Plant seeds vary greatly in size, shape, appearance, and internal structure (Figure 17-2). Because of these differences, it is possible to separate and identify the seeds from various plants. Seed size ranges from that of the coconut of the tropics to the microscopic seed of the orchid.

Parts of the Seed

The seed has three basic parts: the seed covering or coat, food storage tissues, and the embryo.

FIGURE 17-2 Differences in seed size are shown by *Aesculus glabra* or Ohio buckeye (left) and *Dianthus gigantea* or tall pinks (right).

Seed Coat

The seed coat has a hard surface that protects the interior parts. Thus, the seed can be handled, stored, and shipped without injury. The seed coat also plays an important part in germination. This function is described later in this chapter.

The seed coat is made up of an inner part and an outer part. The outer part is called the **testa**. It is hard, dry, and generally darker in color than the transparent, skinlike inner coat. The peanut is a good example of both parts of the seed coat. The wrinkled, roughened outer shell is the testa. The edible nut, or the seed, is surrounded by a thin, brownish inner coat.

Food Storage Tissue

The type of storage tissue a seed contains varies with the plant species. For some plants, the tissue is known as the **endosperm**. In other plants, the storage tissue is formed by the **cotyledons**. These tissues contain the carbohydrates, fats, and proteins needed to nourish the developing plant until it can begin to make its own food through photosynthesis.

Embryo

The **embryo** is a miniature plant that develops from the fertilized egg (gamete). In the mature resting seed, the embryo is always in a dormant condition. The embryo has four distinct parts in its most advanced stage: the plumule, hypocotyl, radicle, and cotyledons (Figure 17-3).

The **plumule** is the first terminal bud of the embryo. It develops into the first shoot that emerges from the seed.

The **hypocotyl** is the first true stem of the plant. In many seeds, the increase in the length of the hypocotyl causes the cotyledons and plumule to emerge from the seed during germination.

The **radicle** is located at the base of the hypocotyl. It is the first root of the plant. The tip of the radicle is always the first part of the embryo to emerge from the seed during germination.

The cotyledons are storage tissues that give rise to the first leaves that develop upon germination of the seed.

Seed Dormancy

It was stated that the seed is actually a tiny plant in a dormant condition. **Dormancy** is a protective condition that prevents the seed from germinating

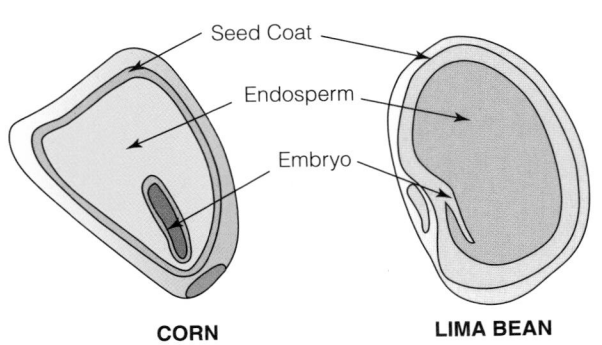

Seed Coat

Endosperm

Embryo

CORN   **LIMA BEAN**

FIGURE 17-3 Composition of a seed.

until all of the environmental factors required for optimum growth are present. These factors include moisture, warmth, and light. For example, an annual seed that matures in the fall must not germinate as soon as it reaches the ground. The developing plant will be killed by the first frost. The seed must remain dormant until spring when the improved conditions will help its growth. Dormancy is also a resting stage. The seed undergoes physiological and biochemical changes during the resting stage, or after-ripening period, which helps the seed to germinate under the proper conditions. This after-ripening period typically requires moist chilling or stratification of the seed from 35°F to 45°F (2°C to 7°C) for 30 to 90 days.

Not all ornamental flower seeds need an after-ripening period before germination will occur. Some seeds require a rest period that is not an after-ripening period and that is not as quiescence. Some tropical plant seeds must be planted as soon as they ripen or the probability of germination decreases rapidly.

Some types of seeds are dormant because of certain conditions of the seed coat. For example, the seed coat may be extremely hard, or it may contain a chemical inhibitor that must be diluted or washed away before germination can occur. The seeds of most flower and ornamental plants used by the commercial grower germinate readily without special treatment. Some seeds, however, must be treated in various ways before they will germinate.

Geranium seed has a very hard seed coat that must undergo **scarification** (be scraped or scratched) before it can absorb water. To penetrate the coat, the seed can be rubbed over coarse sandpaper or the ends can be chipped off, or a small nick can be filed in the seed coat. A more common method of treating the seed is to soak it for a few minutes in concentrated sulfuric acid.

After soaking in the acid, the seeds are washed several times with water to remove the acid. The seeds are then dried. The acid soak softens the seed coat so that water can enter the seed. Seed suppliers normally treat geranium seed before it is distributed for sale. In this way, florists and commercial growers are not required to spend valuable time treating their seed to ensure germination.

Seed Germination

Seed germination is a complex physiological and biochemical process. All of the parts of this process are not known completely. It is known that certain conditions must exist before seeds will germinate. The seeds must have water, oxygen, and a favorable temperature. Some seeds need light for germination; others require darkness.

Water Absorption

The process of germination begins when the seed absorbs water from the soil. It is better for the seed to be in a moist atmosphere than be covered by water. In this way, oxygen can be absorbed with the moisture. As the moisture enters the seed, enzymes are activated causing the embryo to produce a very small amount of a growth hormone called gibberellin. The hormone moves into a layer of cells called the aleurone, which surrounds the endosperm. These cells produce enzymes that cause the endosperm cells to begin the process of conversion from starches to sugar. This process releases other growth-regulating hormones called cytokinins and auxin. These compounds stimulate the growth of the embryo. As a result, its cells enlarge, and new cells are formed through the process of division (mitosis). As the germination process continues, the radicle, or the first root, emerges from the seed. It begins to absorb water and nutrients from the soil.

The plumule then appears. This first shoot of the developing plant is also called the coleoptile. If the shoot points down into the soil, auxin moves to the lower side of the shoot and causes it to grow faster. As a result, the growing point turns up toward the surface of the soil. After the seedling leaves appear, photosynthesis begins and the plant can make its own food. In many plants, the cotyledons continue to supply food to the developing seedling until it forms its first true leaves. The cotyledons then shrivel and dry up because they are no longer needed for life support.

Oxygen Supply

All living cells require oxygen. The seed is a compact mass of living cells and thus needs oxygen for respiration. When the seed is in the dormant stage, the cells need only a very small amount of oxygen. When the seed begins to germinate and grow, the demand for oxygen increases greatly. If too little air is available as a source of oxygen, the seed cannot complete the germination process.

Temperature

All seeds do not germinate at the same temperature. Some seeds require very warm temperatures, and others must have cool conditions. The temperature range in which seeds germinate extends from a minimum temperature (below which there is no germination), to an optimum temperature (which is the desired level for each species), to a maximum temperature (above which germination cannot occur). Most seeds fail to germinate below a temperature range of 32°F to 39°F (0°C to 5°C). Few seeds will germinate above 113°F to 120°F (45°C to 48°C). The optimum temperature for many seeds is 68°F to 86°F (20°C to 30°C).

Light

The effect of light and darkness on seed germination has been studied for many years. Light favorably affects the germination of the seeds of a large number of plant species (Figure 17-4). Such seeds should be covered lightly, if at all, after sowing. Other seeds germinate poorly when exposed to light. These seeds should be covered to improve their germination. The seeds of a small number of plant species do not respond to any light conditions and will germinate in light or darkness. Table 17-1 lists the light needs of the seeds of various ornamental plants.

FIGURE 17-4 Many seeds require light for germination. Specialty growers will use walk-in germination rooms that include light as well as high humidity. The seeds are sown into trays and placed on racks until the radicle emerges.

Table 17-1

	Seed Germination Guidelines for Selected Ornamental Annuals and Pot Plants[1]			
Common Name	Genus and Species	Optimum Temperature for Germination	Light or Dark*	Days to Emerge
Ageratum	*Ageratum houstonianum*	70	L	5
Alyssum	*Lobularia maritime* (L.) Desv	70	DL	5
Aster	*Callistephus chinensis* hybrids	70	DL	15
Begonia	*Begonia X semperflorens cultorum*, Link & Otto	70	L	15
Browallia	*Browallia viscosa* HBK *compacta*	70	L	15
Calceolaria	*Calceolaria X herbeo-hybrida*, Voss	70	L	15
Calendula	*Calendula officinalis*	70	D	10
Celosia	*Celosia argentea* L.	70	DL	10
Cineraria 'Maritima Diamond' (Dusty Miller)	*Senecio cineraria* Dc	75	L	10
Coleus	*Coleus X hybridus*	65	L	10
Cyclamen	*Cyclamen persicum* Mill.	60	D	50
Dahlia-Unwins dwarf mix	*Dahlia pinnata* Cav.	70	DL	5
Dianthus	*Dianthus chinensis* L.	70	DL	5
Geranium (seed)	*Pelargonium*	75	DL	5-10
Gloxinia	*Sinningia speciosa* Benth & Hook	65	L	15
Gypsophila	*Gypsophila elegans* Bieb	70	DL	10
Hollyhock	*Althaea rosea* (L.) Cav.	60	DL	10
Impatiens Holstii	*Impatiens holstii* Engler & Warb.	70	L	15
Kalanchoe	*Kalanchoe blossfeldiana*	70	L	10
Larkspur	*Delphinum ajacis* L.	55	D	20
Lobelia	*L. erinus, L. var. comp.* Nich.	70	DL	20
Marigold	*Tagetes erecta* L.	70	DL	5
Morning Glory	*Convolvulus* Sp.	65	DL	5
Nasturtium	*Tropaeolum majus* L.	65	D	8
Pansy	*Viola*	65	D	10
Petunia	*Petunia hybrida* Vilm	70	L	10
Phlox	*Phlox drummondii* Hook	65	D	10
Portulaca	*Portulaca grandiflora*	70	D	10
Primula malacoides	*Primula malacoides* Franch.	70	L	25
Rudbeckia single (Gloriosa Daisy)	*Rudbeckia lacinata* L.	70	DL	20
Sage, perennial	*Salvia officinalis* L.	70	D	15
Salvia	*Salvia splendens* Sello	70	D	15
Schizanthus	*Schizanthus pinnatus* Ruiz & Pav	60	D	20
Shamrock	*Trifolium dubium* Sibth	65	D	10
Snapdragon	*Antirrhinum majus* L.	65	L	10
Stock	*Matthiola incana* (L.) R/Br.	70	DL	10
Sweet pea	*Lathyrus odoratus*	55	D	15
Torenia	*Torenia fournieri* Lind.	70	DL	15
Verbena	*Verbena hybrida* Voss	65	D	20
Vinca, Periwinkle	*Catharanthus roseus*	70	D	15
Zinnia	*Zinnia elegans* Jacq.	70	DL	5

[1]Cathey, H. M., Guidelines for the germination of annual, pot plant and ornamental herb seeds. 161 (4174) 35-36, 1977.
*Key to light requirements:
D = exposure to continuous darkness during germination; L = exposure to continuous 300 fc (3.24K lux) cool white fluorescent light during germination; DL = presence or absence of light has no effect on germination.

Germination Media

Requirements

The moisture supply, oxygen availability, and temperature and light requirements are all influenced by the propagation (germination) medium used.

The medium must provide and maintain the proper levels of moisture and oxygen if it is to be considered the correct environment for seed germination. It must also provide physical support for the plants. The medium should not be used if the seedlings cannot stand up after germination. In addition, the medium must contain a small supply of nutrients to maintain growth until the seedlings are transplanted. It must be free of insects, disease organisms, and unwanted weed seeds.

Components

The germination medium usually consists of several materials mixed together. Sphagnum peat moss commonly forms the largest amount by volume of germination media. The advantages of peat moss are its large moisture-holding capacity and its good porosity (large amount of pore space). The porosity, in turn, ensures good aeration and excellent drainage.

Sphagnum peat moss is usually mixed with other materials. The most common addition is vermiculite (Figure 17-5), a micaceous mineral that is mined in South Carolina. To obtain vermiculite, the ore is placed in a furnace where it is heated to about 1,400°F (760°C). At this temperature, water inside the ore turns to steam. The particles of vermiculite ore are like a series of compressed plates. The heat of the steam causes them to expand, or exfoliate, into an accordion-like form. The properties of vermiculite were described in Chapter 12.

Perlite (Figure 17-6) is another mineral that reacts to a high temperature in the same way as popcorn. Both perlite and vermiculite are ideal propagation media. They are light in weight and have good moisture-holding capacity. As they leave the furnace, they are free from insects, diseases, and other harmful materials. In other words, they are sterile. However, they can be easily contaminated if they are not stored appropriately after the package is opened or if dirty tools are used.

FIGURE 17-5 Comparison of two sizes of vermiculite: No. 4 is on the right, and No. 2 is on the left.

FIGURE 17-6 Horticultural grade vermiculite, perlite, and coarse sand from left to right.

Peat moss and vermiculite, or peat moss and perlite, are excellent seed germination mixtures. To prepare two bushels of propagation media, the following ingredients are mixed thoroughly.

> Shredded sphagnum peat moss—1 bushel
> Horticultural vermiculite, grade 3—1 bushel
> Ground limestone—10 T
> 20 percent superphosphate, powdered—5 T
> Ammonium nitrate—4 T
> Chelated iron Sequestrene 300®—1 level teaspoon
> Note: T = a level tablespoon

When dry peat moss is used, warm water is added so that it will wet faster. If a mechanical mixer is used, the ingredients should be turned for 3 to 5 minutes. If mixing by hand, the materials should be turned until all ingredients are evenly distributed in the mix.

In some cases, equal volumes of sand and peat moss are mixed to form a germination medium. Nutrient materials should be added, including limestone, superphosphate, and nitrogen fertilizer. Sand and peat mixes must be checked more frequently for moisture content. Because sand does not hold water, mixes using sand dry out much more rapidly than either peat and perlite or peat and vermiculite.

Soil mixtures that do not contain peat moss, vermiculite, or perlite should not be used as **germination media.** Such soils have poor drainage and aeration, leading to root rot. In addition, seedlings that develop in soil mixes are difficult to separate without damaging their root systems. If soil is used, it must be steam pasteurized or chemically sterilized to kill disease organisms, insects, and weed seeds. Pasteurization and chemical soil sterilization are covered in more detail in Chapter 13.

Soil Pasteurization

Steam Treatment

When field soil is made part of any propagation or growing medium, the preferred method of pasteurizing the soil is by use of steam. It is also safer than using poisonous chemicals to disinfect soil. By maintaining a minimum temperature of 160°F (71°C) for 30 minutes in the coldest part of the volume of soil being steamed, most disease-causing organisms, soil insects, and weed seeds are killed. A thermometer (Figure 17-7) is used to determine that the proper temperature has been reached.

The widespread and predominant use of soilless, peat-lite mixes for starting seedlings and plugs has all but eliminated the use of steam pasteurization of seed starting media. Opinions differ as to the need for steaming peat-lite media. Sphagnum peat moss contains natural microorganisms that combat plant disease organisms. When peat-lite media are steam treated, these natural enemies of plant pathogens are killed, thereby eliminating their protective action. By killing the beneficial microorganisms, the media becomes susceptible to infestation by disease-causing organisms.

When plants are grown in outdoor field soils—as occurs in many locations where flower crops are grown—steam pasteurization of those soils is the preferred method of treatment.

FIGURE 17-7 Probe type thermometer used to check the soil (medium) temperature during steam pasteurization.

Chemical Fumigation

When it is not possible to steam soil, chemicals are used. These chemicals are very specific in their action. Many of them kill only nematodes (tiny soil eel-worms) and weed seeds, and not certain bacteria and fungi. Chloropicrin (tear gas), methyl bromide, Vorlex, mylone, and Vapam are other commonly used soil fumigants. Chemical treatment of soils requires the soil temperature to be warm. The minimum temperature for good results is 60°F (15.5°C). Following treatment, the soil must air for at least 3 weeks before it is safe for planting.

Before treatment, the soil must be homogeneous with lumps no larger than 1 inch (2.5 cm) in diameter. The fumes from the chemical used cannot penetrate large lumps of soil. Untreated soil may recontaminate the remainder of the treated soil.

The soil is piled to a depth of 12 inches (30 cm). The chemical is injected using the method recommended by the manufacturer. If the soil pile is deeper than 12 inches (30 cm), the fumigant will not be evenly distributed and the treatment will not be thorough.

A cover is used to hold the fumes in the soil for the required time period. Generally, a canvas or plastic cover is sufficient. Some chemicals can be held in the soil with a water seal, which can be applied with a hose. After the required waiting period is over, the soil is uncovered and aired thoroughly. If the pile freezes shortly after treatment and is undisturbed through the winter, it may still contain some gas in the spring. To prevent damage to the plants, the soil should be run through a shredder or turned over several times to release all of the gas.

Soils fumigated with methyl bromide (Bromomethane) or other bromine compounds cannot be used for seed germination or growing of alyssum, aster, calendula, carnations, other members of the *Dianthus* genus, celery, coleus, coreopsis, *dendranthema* (chrysanthemum), digitalis, garlic, godetia, matricaria, nierembergia, onions, salpiglossis, salvia, snapdragon, verbena, and viola. These plants are known to be damaged by bromide residues that remain in the soil after treatment with methyl bromide.

Fungicidal Drenches

If the growing medium cannot be steam pasteurized or chemically fumigated, a **fungicidal drench** can be used to destroy preemergent and postemergent damping off **disease organisms**.

Preemergent damping off fungal pathogens kill the young plant just as it develops from the germinated seed and before the seedling emerges from the medium. The disease organism *Pythium* is the most common cause of preemergent damping off fungal pathogen and is more prevalent at cooler temperatures.

Postemergent damping off generally is caused by the disease organism called *Rhizoctonia*. Stem rot at the soil line causes seedlings infested with *Rhizoctonia* to topple over. This form of damping off is commonly seen by growers and is more prevalent at warmer temperatures.

The fungi that cause **damping off** can be transferred to clean soil from infested soils, flats, tools, or the end of the watering hose. These organisms can be controlled by fungicidal drenches, if they are applied promptly according to the manufacturer's label recommendations. Yet, prevention through clean greenhouse practices is more effective.

Germination Environment

Aeration

The importance of oxygen in germination was pointed out earlier in this chapter. The developing plant requires increasing amounts of oxygen. To supply the required oxygen, the germination medium must be open and porous (Figure 17-8). The percentage of germination will be reduced in a hard, compacted medium because oxygen cannot diffuse to the seeds. In addition, a low oxygen level is created in a medium that is overwatered and lacks the proper drainage.

Temperature

Germination occurs most rapidly when the proper media and air temperatures are used. The use of bottom heat speeds germination (Figures 17-9 and 17-10). Bottom heat can be obtained from electric heating cables buried in a layer of sand, or from heat pipes mounted under the propagation bench. The bottom heat provided should be 5°F to 10°F (3°C to 7°C) warmer than the air temperature.

FIGURE 17-8 To supply the required oxygen, the germination medium must be open and porous. Vermiculite is used to prevent the medium from becoming hard and compacted because oxygen cannot diffuse to the seeds.

FIGURE 17-9 Bottom heat supplied by hot water pipes under the bench. Notice the cleanliness of the ground area.

FIGURE 17-10 Bio-energy system supplies bottom heat to the crop. Hot water is circulated in the tubes.

Table 17-1 lists the optimum air temperature for the germination of many of the flower crops started from seed.

After the seedlings emerge from the medium and the first true leaf appears, they should be moved to an area where the temperature is cooler. Under this condition, the seedlings become more firm. Seedlings that are too soft and succulent do not transplant well.

Moisture

Without moisture, the seed cannot germinate. A uniform supply of water is critical to the success of germination. Too much water may cause the seed to rot. If there is too little water, the young plants may die.

Intermittent low-pressure misting is a commonly used method of providing the proper moisture levels for germination. The mist is applied to the medium for a definite length of time in seconds every 2 or 3 minutes. As a result, the medium remains evenly moist.

The mist can be controlled by a time clock, a solar-activated counter, or by variations of an "electronic leaf." For each of these devices, the length of the on–off cycle can be adjusted. The solar counter responds to the intensity of the sunlight. In bright light, the counter turns at a fast rate, and the mist is applied frequently. For cloudy conditions, the counter turns slowly, and the mist is applied less often.

The electronic leaf control reacts to the amount of water on its surface. As the medium is misted, water builds up on the electronic leaf. When the amount of water reaches a preset value, an electrical circuit is closed and the mist stops. When the moisture evaporates, the circuit reopens, the mist starts, and the cycle is repeated. In areas of hard water, residue salts in the water may build up on the leaf. This increases the weight of the unit and results in an improper misting cycle. The leaf should be inspected daily and cleaned as needed.

Regardless of how the mist is controlled, it is applied only during daylight hours. Evaporative cooling occurs as a result of the mist. Thus, bottom heat must be provided to maintain the proper germination temperature of the medium.

If a misting system is not used to supply water, the grower uses a fine-spray nozzle to water the seed flats. A heavy stream of water is never used, because it can wash fine seeds out of the flat and can break off small seedlings. Overwatering should be avoided.

The proper moisture levels can also be maintained by **subirrigation.** The seed flats are soaked with water from the bottom and then are covered with opaque white plastic. Clear plastic is not recommended because heat can build up in the medium on bright sunny days. The heat cooks the seeds and kills them. The temperature does not build up to such high levels under white plastic.

As soon as the seeds germinate, the plastic is removed. Exposing the seedlings to the air reduces the possibility of disease and speeds the hardening of the seedlings.

Light Requirements

Many seeds require light for germination. For example, lettuce seed germinates poorly in darkness. With the proper amount of light, however, lettuce seed germinates promptly and uniformly over a wide range of conditions. The seed must be moistened before it will respond to light. Dry lettuce seed is insensitive to light. After the seed is moistened and exposed to light, it may be dried. The seed will react later to this stimulation when it is remoistened. It will then germinate in the dark.

The red wavelengths of the electromagnetic spectrum have the most influence on germination. Far-red light, or the wavelengths between the visible red and infrared wavelengths, inhibits seed germination. Seedlings grown in far-red light develop leggy (**etiolated**), weak stems. Such seedlings also lack chlorophyll. This same condition occurs when seedlings are grown in the dark.

Other types of seeds are inhibited by light and germinate only in the dark. Cyclamen seed require complete darkness before they will germinate. The light and temperature requirements for several ornamental crop seeds are given in Table 17-1.

Seed germination cabinets give growers excellent control of germination for large numbers of seeds. These cabinets provide the necessary temperatures and light intensities of 500 to 1,000 footcandles (5.3K to 10.75K lux) at plant level.

After the seeds are sown, the containers are put into plastic bags. They are then placed under fluorescent lights at a distance of 6 to 8 inches (15 to 20 cm). Several types of fluorescent lamps, including cool white, warm white, natural white, and daylight lamps, are satisfactory sources of light. Special plant-growing fluorescent lamps have not been shown to be better than standard lamps.

After the seedlings develop, the plastic bags are removed to expose the plants to the full intensity of the light. Within 3 weeks after the seeds are sown, the seedlings started under fluorescent lights are ready for transplanting. To prevent sunburn of the transplanted seedlings, they are placed in light shade for several days until they become accustomed to the natural light.

Seed Sowing

Pretreatment

Most ornamental seeds are not pretreated by seed suppliers (except for those seeds requiring scarification). Because there is the possibility that seed-borne diseases may damage the crop, a small amount of fungicide, just enough to cover the point of a knife blade, is placed in the seed packet. By shaking the packet, the seeds are coated with a thin layer of fungicide. No other form of pretreatment is required.

Pelleted Seed

Coated seed may be obtained from some seed producers. Clay, or a similar material, is used to form a protective coating around the seed. Such seed can be handled without damage in mechanical seed-sowing devices. Certain crops are not transplanted, but are sown directly in the growing packs or plug trays. Round seed, such as that of portulaca, petunias, and alyssum, have been used successfully in pelleted form. Seed with unusual shapes, such as that of marigolds and zinnias, are not coated.

Sowing Small Seed

Generally, seed is planted at a depth equal to twice the diameter of the seed. It is difficult to follow this rule when planting fine seeds such as those of petunias, snapdragons, and begonias. These seeds normally are sown on the surface of the medium. They are so fine that they fall into the spaces between the peat and vermiculite particles. Coarse seed, such as that of marigold, salvia, verbena, and tomatoes, are covered with medium to a depth of 0.13 to 0.25 inch (0.3 to 0.6 cm). If the seed is planted too deep, it will not receive the necessary amount of oxygen. As a result, germination may be delayed or prevented. Seeds that require darkness for germination must be covered. Those that germinate best in light should not be covered.

Sowing in Rows or Broadcasting

The grower may choose to sow seed in rows or spread seed randomly (broadcast it). There are several disadvantages to **broadcast sowing**. The seedlings are difficult to transplant without damaging the roots. If damping off begins, it usually spreads through the entire seed flat unless it is treated with a fungicide. Growers have noted that if damping off starts in seedlings growing in rows, the disease progresses to the end of the row and stops. Thus, only one row of plants is lost. Row sowing also makes it easier to handle the seedlings when they are transplanted. Figure 17-11 shows one type of seedling flat designed for row sowing.

Plug Seedling Production

The major change in the production of plants, particularly bedding plants, is the use of plug methods for starting seedlings. It is estimated that over 7 billion bedding plant transplants are produced each year in the United States. Of this number, at least three-fourths are started as plugs.

The use of plug culture has revolutionized the production of bedding plant crops. Where seed was often sown by hand, it is now sown mechanically by semiautomatic seed-sowing machines (Figures 17-12, 17-13, and 17-14). As a result, the ability to sow seeds mechanically has led to an explosive expansion of starting plants in plugs (Figure 17-15). Small growers, who would have sown seed using conventional methods and then laboriously transplanting them, can now buy plug started seedlings from specialist growers. Now instead of starting the seed sowing in December to have plants large enough to transplant in 4 to 6 weeks, depending on the crop grown, they can delay starting production and the need for heating the greenhouse in the coldest part of the year, until a few days before their scheduled deliveries are due. Another advantage of plug started seedlings over bare root plants is that plugs can be held for several days

FIGURE 17-11 A specially formed plastic flat used to sow seeds in row.

FIGURE 17-12 A Blackmore needle seeder used for sowing plug seedling trays.

FIGURE 17-13 A closer view of a Blackmore needle seeder used for sowing plug seedling trays. Note how each needle catches and holds individual seeds. This system works well for irregularly shaped seeds. (*Courtesy of Blackmore Company, Belleville, MI 48111.*)

before they need to be transplanted. Holding them too long may cause problems in that the plants become root bound in the trays and when planted are slow in starting into growth.

For large producers of bedding plants, the use of semiautomated transplanting machines has become common. Such units are very expensive, costing between $50,000 and $150,000 or more per unit (Figures 17-16 and 17-17). However, transplanting machines that were once only available to those

FIGURE 17-14 Hamilton plug tray seeder.

FIGURE 17-15 Well developed plug seedlings ready for transplanting.

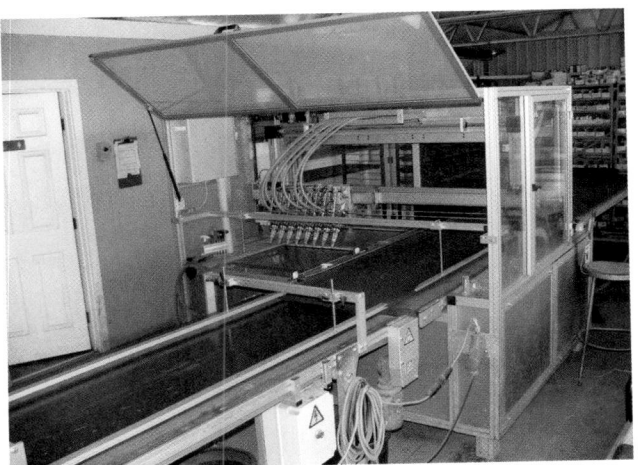

FIGURE 17-16 Visser plug transplanting machine.

FIGURE 17-17 Harrison Robotic plug transplanting machine. *(Courtesy of Blackmore Company, Belleville, MI.)*

bedding plant growers who produce several acres of plug started plants are now available at more affordable prices to a larger group of producers than before.

Because of the need to attempt to get 100 percent of the cavities in a plug tray growing a transplantable seedling, seed companies have resorted to specialized treatment of certain types of seed. The use of pelleted seed, particularly of the very small seeded crops sown, has been mentioned.

Large, unusual shaped seeds like zinnias and marigolds present special problems. The marigold has a fuzzy tail that interferes with the mechanical seed-sowing process. To overcome this problem, seed suppliers have resorted to "de-tailing" marigold seed. No seed supplier has developed a treatment for zinnia seed. It can only be handled by a few of the presently available seed-sowing machines.

Germination Chambers for Plug Seedling Production

Plug culture of seedlings requires sophisticated environmental control for success. Modern transplanting machines require near to or 100 percent seedlings in each tray. To ensure success, plug growers rely on germination chambers. These chambers can range in size from a growth cabinet with room to germinate a dozen or more trays of plugs to rooms several hundred square feet in size designed to accommodate multiple tall carts of plug trays.

Germination chambers or rooms require precise control of temperature, humidity, and light. If a seed requires light for germination, the chamber must have a light source. The relative humidity in a greenhouse chamber is typically maintained with a fog or misting system that provides enough moisture in the air to not allow the seeds or the seedling medium to either dry out or gain moisture.

Ideally, a germination temperature should be maintained at the optimum temperature for a specific species to germinate; however, a typical greenhouse grower is going to be germinating several species at any point in time. To accommodate this need, it is ideal for three temperatures to be maintained in three separate chambers based on the germination requirements of that species: low temperature 60°F to 65°F (15°C to 18°C); medium temperature 72°F to 75°F (22°C to 24°C); and high temperature 76°F to 82°F (24°C to 28°C). Of course, having three separate chambers is usually not possible, so many greenhouse growers opt to use only two separate chambers with the following temperature ranges as a compromise: 63°C to 68°F (17°C to 20°C) and 74°F to 78°F (23°C to 26°C).

Plug Seedling Culture

Growing seedlings with plug culture is best described in four or five stages.

1. Stage 1 is that period when the sown plug trays are moved into the germination chamber and concludes with the emergence of the radicle of the seedling. The main goal is rapid and uniform radicle emergence. The moisture level of the medium should not change during this period and no supplemental fertility is added.

2. Stage 2 is that period from radicle emergence to the formation of the first true leaves. Stages 1 and 2 often overlap due to variable germination rates. The conditions for stage 2 are identical for stage 1 except light is required. At this point, most growers move the plug flats from the germination chambers to a greenhouse area with HID lights and high humidity. The primary goal for stage 2 is root development; therefore, uniform moisture is critical and typically 50 to 100 ppm N is added from a N-NO$_3$ source.

3. Stage 3 is that period from the formation of the first true leaves to transplantable size. This is the longest period of plug production. The goal of the greenhouse grower is to maintain active, healthy root systems while growing acceptable foliage. When the foliage covers the tray, stage 3 is complete. A grower may opt to alternate wet and dry irrigation cycles and include 100 to 200 ppm N from a N-NO$_3$ source. Many growers also include plant growth regulators during this stage.

4. Stage 4 is that period from when the foliage just covers the tray to final transplant. Many greenhouse growers who grow their own plugs actually never schedule a stage 4, allowing stages 3 and 4 to overlap and move into the transplanting stage as demand requires. The moisture levels for stage 4 are similar to stage 3, but minimum use of fertilizers should be maintained. This prevents excessive growth and internode stretch. Natural light is adequate at this point, but the grower should be on alert monitoring *Botrytis* infections and use fungicides if deemed appropriate.

5. Not all greenhouse growers consider a stage 5 for growing plugs. Stage 5 is best described as a holding period. The condition that may trigger a stage 5 holding period is a backup of greenhouse space or access to transplanting facilities. Specialty plug growers are also often faced with a stage 5 when a disruption in shipping or sales occurs.

Stage 5 plug trays are typically stored at lower temperatures or refrigerated with supplemental light. The moisture levels should be maintained at the same levels as stages 3 and 4, irrigating in the morning so that the foliage is dry during the evening, and preventing *Botrytis* infection using a fungicide if necessary. Many growers use chemical growth retardants at this stage as well, but use care so as to not stunt or slow the plants excessively.

Fertilizer Requirements

For Germination

Most flower seeds contain enough stored food to support germination and the formation of the first sets of leaves. After photosynthesis begins, the plant must obtain its nutrients from the growing medium. High levels of fertilizer in the medium may kill tender seedlings. Thus, weak fertilizer applications are made to improve growth. The peat-lite recipe, given earlier in this chapter, lists the proper amounts of fertilizers to be added to the germination medium. Additional fertilizer is not required until the seedlings are transplanted to the growing containers.

If a peat-lite medium is not used, then ground limestone is added to the medium to bring the pH to 6.0 to 6.8. A fertilizer is mixed thoroughly with each bushel of medium at the rate of at least 1 ounce (28.35 g) of finely powdered 20 percent superphosphate, or 0.5 ounce (14.17 g) of treble phosphate. Phosphorus is very important to the development of strong root systems.

In Seed Flats

Following germination, a complete analysis fertilizer, such as 20-10-20 ($N-P_2O_5-K_2O$), can be applied at the rate of 1 ounce (28.35 g) to 3 gallons (11.4 1) of water. The fertilizer should not be applied more than once a week because the growth of the seedlings becomes too soft. This condition is undesirable as it makes the seedlings hard to transplant and their chances of survival after transplanting are decreased.

Transplanting

When to Transplant

A seedling should be transplanted when the first true leaf is fully developed. However, most seedlings are very small at this stage of their development, and it is difficult to transplant them. Many growers transplant seedlings at more advanced stages of growth simply because the plants grow faster than transplanting can be accomplished. If the seedlings waiting to be transplanted become too crowded, become too tall, and show signs of nutrient deficiencies, they should be discarded.

Seedlings grown as plugs are typically transplanted mechanically as previously described.

Precautions

Seedlings removed for transplanting should be planted immediately. Never remove more seedlings than can be planted in a few minutes. Exposure to the air quickly damages tender roots. Seedlings removed for transplanting will die quickly if placed in direct sunlight. As soon as they are transplanted, the seedlings are watered to prevent further drying.

Storage

It may not be possible to transplant all of the seedlings when they reach the proper size. To prevent them from growing any larger, they can be placed in cold storage. The seed flats should be well watered and then placed in a 40°F (5.5°C) refrigerator under cool white fluorescent lights. Three 40-watt fluorescent tubes, spaced 4 inches (10 cm) apart and 12 inches (30 cm) above the seed flats, will supply approximately 250 footcandles (2.7K lux) of light. The refrigerated plants are lighted for 14 hours a day. As a result, the plants remain in good condition with no additional growth. One day before they are scheduled for transplanting, they are removed from cold storage and allowed to warm up to the greenhouse temperature. The seedling flats must not be placed in direct sunlight as soon as they are removed from the refrigerator. After transplanting, these seedlings are handled the same as any other seedlings.

Seedlings grown in peat-lite mixes can be placed in unrefrigerated storage. The seed flats are placed in greenhouses covered with opaque white plastic. The air temperature in the greenhouse should be kept as cool as possible. The seedlings are watered only as needed. This type of storage is not as controlled as refrigerated storage, but it increases the useful life of the seedlings by as much as a week or two.

ACHIEVEMENT REVIEW

Select the best answer or answers to complete each statement. List the appropriate letter(s).

1 The life cycle of an annual plant:
   a. requires two growing seasons.
   b. continues from year to year.
   c. is completed in one growing season.

2 The first part of the embryo to emerge when a seed germinates is the:
   a. plumule.
   b. hypocotyl.
   c. radicle.
   d. cotyledons.

3 Hard seed coats may be scarified by:
   a. chipping off an end.
   b. filing a nick in the coat.
   c. treating the seed with sulfuric acid.
   d. using any of the above methods.

4 The first step in seed germination is the:
   a. absorption of oxygen.
   b. absorption of water.
   c. formation of gibberellic acid.
   d. development of cytokinins.

5 Which of the following is the best type of peat moss for mixing with perlite or vermiculite to prepare a germination medium?
   a. Michigan peat
   b. Reed peat
   c. Sedge peat
   d. Sphagnum moss peat

6 The main objection to using soil in a seed germinating mix is that it:
   a. contains disease organisms, insects, and weed seeds.
   b. is difficult to mix thoroughly with other ingredients.
   c. varies in nutrient content.
   d. has the wrong pH.

7 Before it can be used, soil fumigated with chemicals must be aired for:
   a. 24 hours.
   b. 10 days.
   c. 3 weeks.
   d. no specific amount of time as it may be used immediately.

8 Carnations or other members of the *Dianthus* genus should not be grown in soils treated with:
   a. vorlex.
   b. vapam.
   c. methyl bromide.
   d. tear gas.

9 Seeds that germinate best in light should receive:
   a. 10 fc (108 lux) of warm white light for 12 hours a day.
   b. 15 fc (130 lux) of Gro-lux lighting.
   c. 300 fc (3.2K lux) of continuous cool white fluorescent light.
   d. none of the above.

10 Seeds sown in rows are:
   a. not subject to damping off.
   b. always affected by damping off.
   c. sometimes affected by damping off, but the disease is usually restricted to a single row of plants.

11 Seedlings should be fertilized:
   a. daily with a strong solution.
   b. daily with a weak solution.
   c. once a week using 20-20-20 fertilizer applied at a rate of 3 ounces (85 g) to 1 gallon (3.785 l) of water.
   d. once a week using a 20-20-20 fertilizer applied at the rate of 1 ounce (28.35 g) to 3 gallons (11.4 l) of water.

12 Seedlings should be transplanted:
   a. when three sets of true leaves have developed.
   b. when the cotyledon leaves are fully expanded.
   c. when the first true leaf is fully developed.
   d. as soon as they can be handled.

13 When storing seedlings in refrigerated storage, the recommended temperature is:
   a. 32°F (0°C).
   b. 36°F (2°C).
   c. 40°F (4°C).
   d. 45.5°F (7.5°C).

14 Successful nonrefrigerated storage of seedlings can be done in a greenhouse covered with:
   a. glass.
   b. fiberglass.
   c. opaque white plastic.
   d. clear polyethylene.

STUDENT PROJECT: SOWING SEED

Objectives

This exercise teaches the student:

1. selection of the proper medium for seed germination.
2. the proper procedure for sowing seeds of various sizes.
3. the correct watering procedures.
4. how to identify the seeded flat.

Materials

1. Clean 144, 288, or 512 cavity plastic plug flat
2. Peat-lite medium, preferably the finer particle size used for starting plug seedlings
3. Seeds of bedding plant and/or vegetable crops
4. Plastic or wooden label
5. Soft lead pencil or permanent marking pen

Procedures

Note: Refer to Figures 17-18 through 17-24.

1. Sanitize the potting table with an approved disinfectant (Figure 17-18).
2. Assemble all materials (Figure 17-19).
3. Moisten the peat-lite medium using warm tap water. Do not fill a plug tray with dry medium and then try to wet it. It takes a long time to thoroughly saturate dry media in a plug tray.
4. Fill the tray with medium. Try to have the same amount of media density in each cavity (Figure 17-20).
5. Strike off the medium level with the top of the flat.
6. Fine seeds such as ageratum, alyssum, and petunias are sown on top of the medium and not covered.

FIGURE 17-18 Three disinfectant materials approved for sanitizing potting tables, flats, pots and bench surfaces, knives, walls, and floors. They are not to be used for disinfecting growing media.

FIGURE 17-19 Assemble materials needed: peat-lite plug mix, 288 tray, label and soft pencil or permanent marking pen, and seed packets.

FIGURE 17-20 Fill tray uniformly with peat-lite plug mix and strike off level.

7. Marigolds, zinnias, and other large seeds are lightly covered to maintain moisture around them. Some growers prefer to use coarse 0.13 inch (3 mm) vermiculite as a covering material.

8. Prepare the label by noting the date of sowing, the crop name, and the cultivar. Use a soft lead pencil or permanent marking pen. Do not use a ball point pen as water will wash away the information (Figure 17-21).

9. Water the flat using a sprinkling can having a fine diffuser or a FOGGIT® nozzle or place the flat under mist (Figures 17-22 and 17-23).

10. Place the flat in the germination environment with bottom heat of 70°F to 75°F (21°C to 24°C). If mist is not available, the tray can be placed into a large white plastic bag (Figure 17-24). Do not place the unit in direct sunlight, because the bag will retain heat and the seed will cook. As soon as the seed has germinated, remove the plastic bag and maintain normal cultural methods.

FIGURE 17-21 Sow seed in each cavity. Mark label with date, crop, and cultivar name. Attach to tray.

FIGURE 17-22 Watering planted tray with fine spray from watering can.

FIGURE 17-23 Wetting planted tray by placement under mist spray nozzle.

FIGURE 17-24 Insert tray into white plastic bag and place in germination environment.

Asexual (Vegetative) Propagation

OBJECTIVES:

After studying this chapter, the student should have learned:

■ why plants are propagated asexually.

■ the various types of asexual propagation methods.

■ why rooting hormones are needed.

■ how to draw a cross-sectional view of a propagation bench naming all of the parts.

Asexual, or vegetative, propagation of plants uses the leaves, buds, stems, and roots of plants to produce new plants. Each new plant has the same characteristics as the mother plant. Many plants grown from seed are not exact reproductions of the mother plant because they vary slightly in genetic makeup. These variations occur because the genetic message contained in the seed is heterogeneous, or highly mixed. Plants that reproduce exactly when grown from seed are said to be homogeneous, or the same, genetically.

In addition to producing true offspring, vegetative propagation methods are used because some plants fail to produce viable seed. In other words, seed that will germinate is said to be viable. Bananas, grapes, and figs may produce nonviable seeds. In many cases, it is faster to reproduce plants asexually than by sowing seed. For example, geraniums produced from cuttings will flower 3 to 4 weeks earlier than plants started from seed at the same time. Asexual (vegetative) propagation of some plants is less expensive than sexual (seed) propagation. For other plants, however, the opposite is true.

Basis for Vegetative Propagation

The scientific basis for vegetative propagation is the unique ability of each cell of a plant to divide and reproduce itself. Each cell contains all of the genetic information needed to make a new plant. This characteristic is known as **totipotency**, or more simply, total potency. For this discussion, potency is defined as the ability to reproduce.

Mitosis

The process of cell division is known as **mitosis**. All plant growth occurs because of mitosis.

A dicotyledonous plant, or one having a seed with two cotyledons, has three main areas where mitosis takes place. These areas are the stem tip, the root tip of primary and secondary roots, and the cambium layer. Each area is known as a meristematic region of growth.

Mitosis occurs in four phases (Figure 18-1). In the correct order of sequence, these phases are the prophase, the metaphase, the anaphase, and the telophase. Each phase of cell division smoothly follows the preceding phase. The progression of phases is continuous, just as the moon passes smoothly from the first quarter, to full moon, to last quarter, and finally to new moon.

The **prophase** is the first phase in the division of the nucleus. In this phase, the tightly curled chromosomes uncurl and form rodlike structures. The chromosomes carry the hereditary or genetic messages from the parent cells to the daughter cells and ensure the development of an identical plant. The prophase

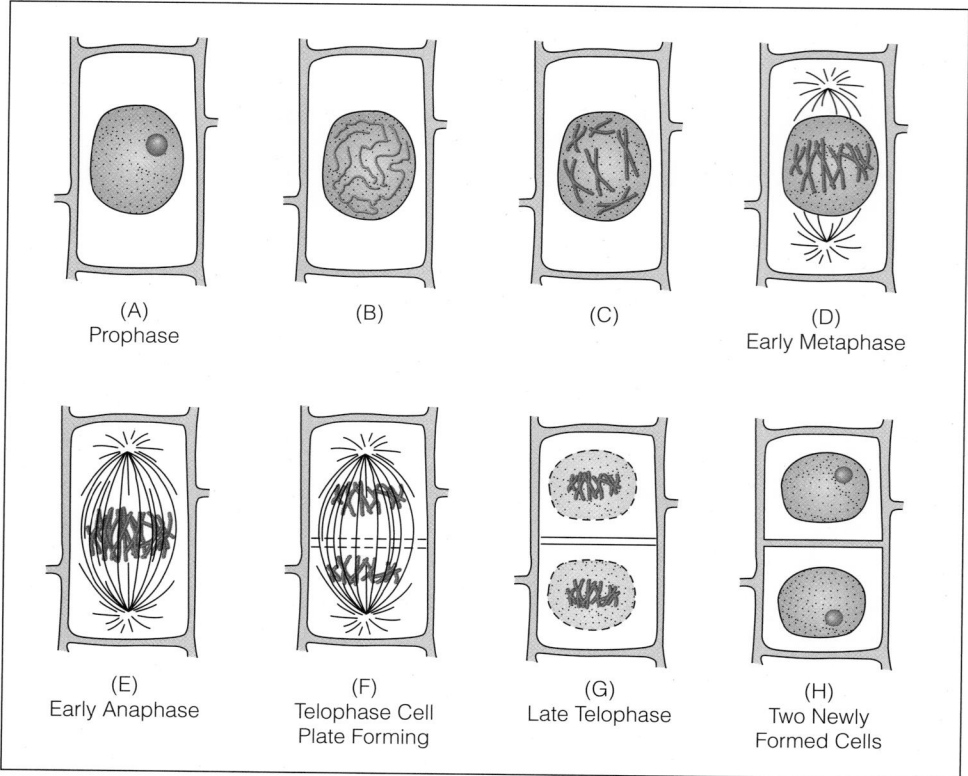

(A) Prophase

(B)

(C)

(D) Early Metaphase

(E) Early Anaphase

(F) Telophase Cell Plate Forming

(G) Late Telophase

(H) Two Newly Formed Cells

FIGURE 18-1 The phases in the process of mitosis in root tip cells are: (A), (B), (C), prophase; (D), early metaphase; (E), early anaphase; (F), (G), telophase; and (H), two new cells formed.

is followed by the metaphase, in which the chromosomes split lengthwise to form two sets of identical material.

In the **metaphase**, the two sets of chromosomes line up at the center, or equator, of the cell. Then they move quickly toward the poles of the cell. One-half of the chromosomes go to each pole. This movement of the chromosomes is called the **anaphase**.

As the chromosomes approach the polar regions of the cell, each group begins to form the new nucleus of a daughter cell. This stage is known as the **telophase**. New cell walls form to separate the two daughter cells. Each daughter cell has the same number of chromosomes as the parent cell. After a short period of growth, each daughter cell begins to divide. The continual process of cell division makes new tissue, resulting in growth. Monocotyledonous plants have seeds containing a single cotyledon. Corn is an example of a monocot. In these plants, there is a fourth area of cell division called the intercalary meristem. This region occurs just above each node. Growth in the intercalary meristem produces an increase in the height of the stem. This increase is in addition to the height obtained by growth at the top of the plant.

Mitosis also occurs when a plant part is wounded and callus tissue forms. Callus tissue is a group of cells that develops in response to a wound stimulus. In plants, callus formation is a healing mechanism. When two plant parts are grafted, the formation of callus tissue joins the parts. New roots that form on the callus tissues of cuttings are called adventitious roots.

Methods of Asexual Propagation

Stem Tip or Terminal Cuttings

A common method of propagating plants is to take stem tip or **terminal cuttings** (Figure 18-2). A wide range of plants can be propagated by stem tip cuttings, including herbaceous plants, such as chrysanthemums and carnations, soft wood plants, semihardwood, and hardwood plants. A sharp knife is used to remove 2- or 3-inch-long stem ends. The cut is made at the node or the internode (the space between the nodes). Plants with soft stems, such as chrysanthemums and carnations, can be broken with the fingers. The leaves are removed from the lower third of the stem. Thus, it is easier to stick the cutting into the propagation medium.

FIGURE 18-2 Terminal stem tip cutting of a mint selection.

Stem Section Cuttings

A stem section is cut just above a bud and its attached leaf. The cutting should include one-, two-, or three-eye (bud) sections of the stem (Figure 18-3). Two-eye cuttings consist of two buds and leaves. The end of the stem from the leaf to the next bud or cut is placed in the medium.

Stem section cuttings are made when many cuttings are needed and only a few plants are available. However, plants grown from stem section cuttings may require 2 to 3 weeks longer to reach a certain size as compared to plants grown from terminal cuttings.

Leaf Cuttings

New plants can be produced from the **leaf cuttings** of plants such as *Saintpaulia* (African violet), *Bryophyllum*, *Begonia rex*, and *Tolmiea* (piggyback plants) (Figure 18-4). The leaves can be placed on top of the rooting medium or can be inserted into the medium. New plants develop on the edges of the leaf where there is a leaf vein. After roots develop and the new plants can support themselves, the old leaf dies.

Saintpaulia (African violet), *Sansevieria* (snake plant), *Crassula* (Jade plant), and other plants produce new plantlets from meristematic cells located at the base of the leaf blade or petiole. African violets are usually propagated with a portion of the petiole (stem) attached to the leaves.

Leaf Bud Cuttings

Leaf bud cuttings are made when there is limited propagating material and it is desired to obtain as many cuttings as possible. This type of cutting consists of a leaf blade, the petiole, and a short piece of stem with an attached bud.

The cuttings are inserted into the rooting medium so that the bud is 0.5 to 1 inch (1.127 to 2.54 cm) below the surface. High humidity and bottom heat are required to stimulate rooting.

Rooting Hormones

Rooting hormones, or more correctly, plant growth regulators (PGRs), are chemical substances that promote the development of roots. These chemical

FIGURE 18-3 From left to right, terminal tip cutting, one-eye, and three-eye stem section cuttings from a *Bougainvillea* vine.

FIGURE 18-4 African violet leaf cuttings in individual cells.

substances are used to increase the percentage of cuttings that root. Rooting hormones are auxin type growth regulators that are applied to the bases of the cuttings. The cuttings of certain plants root easily and may not need hormone treatments. However, even these plants will make roots more quickly and uniformly when treated with a rooting hormone (Figure 18-5). Indolebutyric acid (IBA) and naphthaleneacetic acid (NAA) are two synthetic rooting hormones that work well in stimulating the growth of roots.

Different types of rooting hormones are made with varying concentrations of active ingredients. The weaker materials are used for easy-to-root cuttings. The stronger materials are applied to plants that are more difficult to root. The container label gives directions for the proper use of the rooting hormone on different crops.

Rooting hormones are manufactured in the form of powders or solutions. Powders generally are applied to cut stems with a puffer duster. Dusting is recommended for two reasons. First, the duster applies a small amount of material to the base of the stem. The use of too much hormone may cause the stem to rot and thus prevent root formation (Figure 18-6). Second, disease organisms are spread to healthy cuttings if they are dipped into the hormone powder after a diseased cutting. The use of the puffer duster prevents this type of infection (Figure 18-7).

FIGURE 18-5 Rooting hormone and puffer duster used to apply hormone to basal ends of cuttings.

FIGURE 18-6 Geranium cuttings that were overtreated with rooting hormone, probably by dipping method.

FIGURE 18-7 Puffer dusting basal ends of geranium cuttings.

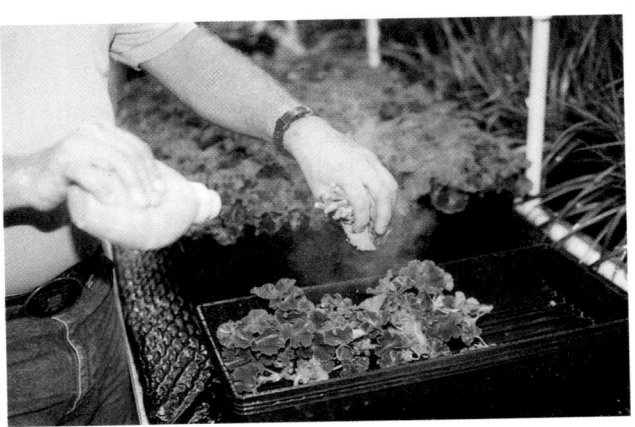

An older method of applying a rooting hormone is to soak the cuttings in a weak solution of the material. However, this method is no longer in common use. Cuttings may be dipped in a strong hormone solution. This method takes less time than soaking the cuttings in a weak solution, but again, all of the cuttings may be infected from one diseased cutting. A spray bottle can be used to apply liquid rooting hormones to cuttings (Figure 18-8).

Only fresh materials are used. New materials are obtained for each season. Always discard old or contaminated materials to prevent plant injury. To ensure that the contents of the main container are not contaminated, remove only the amount of material needed for the job. If a small amount of hormone remains after all of the cuttings are treated, throw it out. Do not return it to the main supply.

Mixing Rooting Hormones with Fungicides

Some growers mix a rooting hormone with a fungicide to protect the cuttings from fungus infections. Fungicides such as Captan may be mixed with rooting hormones. Generally, one part of Captan is mixed with nine parts by weight of hormone. Larger amounts of fungicides should not be used because rooting is prevented in some plant species and may be considered an off label treatment, which is an illegal use of that fungicide.

Propagation Environment

Cuttings generally are used to propagate floriculture crops. The conditions required by these plants for successful rooting differ from those required by semi-hardwood or woody plants. Leafy, herbaceous cuttings require an air temperature of 65°F to 75°F (18°C to 27°C). They also need sufficient moisture to prevent wilting, enough light to carry on photosynthesis, and a rooting medium that holds moisture but also drains freely, thus providing good aeration (Figure 18-9).

One method of supplying these conditions is the **propagation environment** (propagation house). This unit should be located near the workroom area of the head house. Thus, the grower is able to "stick" the cuttings into the propagation medium with very little delay.

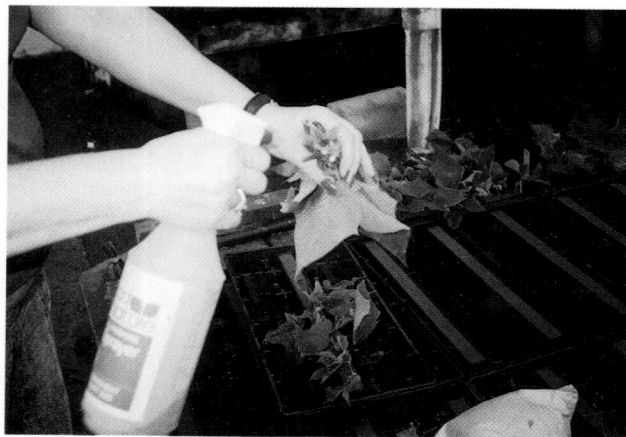

FIGURE 18-8 Applying liquid rooting hormone to the basal ends of cuttings by means of a spray bottle. This method is preferred to dipping the cuttings into the solution and possibly contaminating it with disease organisms.

FIGURE 18-9 Table with nozzles for providing intermittent mist. Note that this bench is covered with plastic to retain the humidity.

The **propagation benches** should be raised above the floor to reduce contamination. Benches made of concrete, plastic, and metal are preferred to those made of wood. As wooden benches age, it is more difficult to sterilize them between crops of cuttings. The propagation house should be located so that it receives all available light. If necessary, the light intensity can be reduced artificially by shading. The entire greenhouse must have excellent drainage because large amounts of water are used during propagation.

Moisture Requirements

Cuttings no longer have a root system for absorbing moisture. Thus, they cannot take up water as quickly as it is lost through transpiration. To prevent the cuttings from wilting, they can be wetted or misted frequently, or they can be shaded. Shading reduces the rate of photosynthesis.

Frequent wetting is the best method because the cuttings can be exposed to full light conditions, resulting in rapid rooting. One way of applying moisture frequently is the use of an intermittent misting system. The misting system used to water seed germination containers can also be used on vegetative cuttings.

Misting Schedule

If the weather is hot when the cuttings are first stuck, or inserted into the medium, it may be necessary to mist them almost continuously during the daylight hours to prevent wilting. The application of too much water can leach nutrients out of the leaves. Thus, it is important to apply only the amount of water needed to prevent wilting.

As the plants begin to develop roots, less mist is used. The plants gradually become more dependent on the water they take up through the new roots. This process is known as **hardening off**. For the last 3 or 4 days the cuttings remain in the propagation bed, mist is not used. They are watered by hose, traveling boom, or subirrigated on ebb and flood benches (Figures 18-10, 18-11, and 18-12). If plants are hardened off, they survive transplanting much better than do plants misted until just before transplanting.

FIGURE 18-10 Traveling boom irrigating newly directly stuck cuttings. A boom such as this can also be used to apply fungicides and growth retardants.

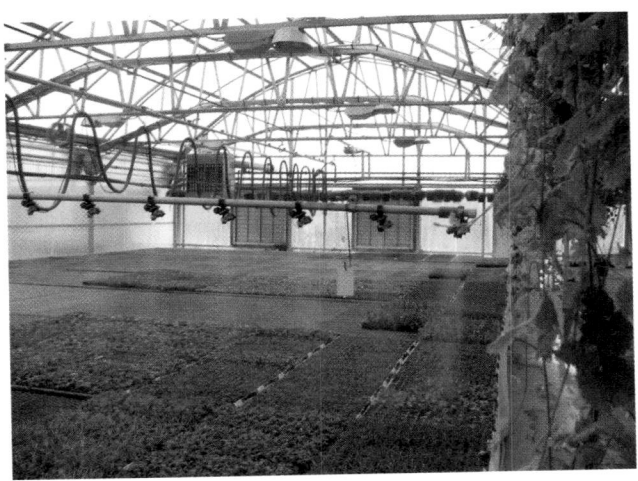

FIGURE 18-11 Traveling booms can be used for rooting cuttings newly stuck into flats and programmed to only mist bench areas with plants.

Each crop has its own schedule of misting. Most florist crops, except azaleas, root in 2 to 4 weeks when propagated under mist.

Nutrient Mist

Many crops benefit from small amounts of fertilizer applied by the mist system. Poinsettias and chrysanthemums respond well to a **nutrient mist. RAPID-GRO®** is a soluble fertilizer with an analysis of 23-19-17 (nitrogen, phosphorus, and potash). It can be used at the rate of 4 to 6 ounces (140 to 170 g) in 100 gallons (26.4 l) of water. The nutrient mist is turned on daily for 10 to 12 seconds every 2.5 minutes from 8:00 A.M. until 5:00 P.M.

Cuttings that are rooted under a nutrient mist must be potted or planted immediately in their final locations. The tissue is very soft and rots quickly if the cuttings are refrigerated for more than 24 hours.

Fog

Some growers changed from using mist to using fog as a means of keeping the relative humidity at or near to 100 percent in the propagation house (Figure 18-13). Water is pumped under very high pressure, 600 to 800 psi, through very fine orifices in nozzles to provide droplets of fog 6 to 8 microns in size. The fog keeps the plants from wilting. Through the process of evaporation of the fog droplets, the greenhouse temperature is also kept cooler than the surrounding air temperature. The frequency of fog application is controlled either by computer, mechanical timer, or humidistat.

Fog systems are expensive to install and require high-quality water and constant maintenance. However, with increased rooting success and lower disease rates, the expense can easily be justified.

Light Requirements

Cuttings can be rooted in full light intensity if mist is used. This means that the cuttings root more quickly and are kept in the propagation bench for a shorter length of time. However, at certain times of the year, the light intensity and the temperature may be too high even when mist is used. Under these conditions,

FIGURE 18-12 Ebb and flood bench being used for geranium propagation.

FIGURE 18-13 Fog system in propagation house.

shading is required to prevent water stress (Figure 18-14). Cheesecloth or saran shade cloth can be used to shade single benches. To shade an entire greenhouse, a shading compound can be applied to the glazing material. Do not place newspapers or other materials directly on the cuttings to shade them. Such a cover reduces the air movement around the cuttings and makes it easier for diseases to develop.

During the short days of winter, it is necessary to provide artificially long days for some plants. A lighting system is used to supply the additional hours required. Lighting fixtures should be installed above the mist lines to prevent water from striking the electrical device's center.

Temperature

The temperature of the air and the propagation medium must be carefully controlled. The air temperature should be in the range of 65°F to 75°F (18°C to 27°C). The bottom heat generally is maintained 5° to 10° higher than the air temperature. Bottom heat may be supplied by heating pipes installed under the benches (Figure 18-15), or by an electric heating cable with a thermostat control. To ensure that the heat is distributed evenly from a heating cable, it is covered with 0.5-inch mesh wire screening. The screen is then covered with the propagation medium. The screen prevents hot spots above the cable.

Media

Vermiculite and perlite are common propagation media. They may be used alone or in combination with sphagnum peat moss. These materials satisfy the requirements of good propagation media. They hold water and nutrients, are free of disease organisms, and provide good drainage and aeration. Cuttings propagated in such media can be removed for potting with little or no damage to the root systems.

Sand is a good rooting medium for some woody ornamentals. However, it is not used for most floriculture crops. Sand does not hold moisture well. As a result, any plants propagated in sand must be heavily shaded and hand watered.

Bulk media, as just described for rooting cuttings, have given way to preformed media. Preformed media are lightweight, easy to handle when setting

FIGURE 18-14 This automatic shading system is controlled by a photoelectric cell. On cloudy days the shade is retracted, but on sunny days it is pulled into position to protect the crop.

FIGURE 18-15 Bottom heat is supplied from heat pipes under the benches. Notice the cleanliness of the area, no weeds or litter.

up a propagation area, available in several shapes, and generally free of insects and disease organisms when received from suppliers.

The major suppliers of preformed media in the United States include Smithers Oasis, U.S.A. who manufactures Oasis Root Cubes® and Horticubes®. Jiffy Products of America supplies Jiffy® compressed peat pellets; Agro-Dynamics and Grodan, Inc. are two suppliers of rock wool media.

Oasis Root Cube® media are made from phenolic resin as the base product with six other chemicals. The resultant media contains only 2 percent solid matter. The rest of the volume is pore space filled with either air or water. Root Cube® media has 28 to 36 percent drainage. This means when thoroughly saturated and allowed to drain, 28 to 36 percent of the remaining 98 percent pore space is filled with air. The remaining pores are filled with water. Oasis Root Cube® contains a small amount of 1-1-1 analysis fertilizer.

Oasis Horticube® is used for seed starting of plants in hydroponic production such as lettuce and spinach. It is also excellent for starting herbs from seed. Other crops such as foliage that have high moisture and also high aeration requirements are propagated in Horticubes®. Horticubes® media has 55 to 65 percent drainage. This medium has no contained fertilizer. Both Horticubes® and Root Cubes® contain limestone, which supplies some calcium and magnesium. The pH of the media is 6.0, plus or minus 0.5 unit. Both media are available in several shapes (Figures 18-16 and 18-17).

Jiffy products are made using sphagnum peat moss and a small amount of wood pulp. The units are compressed under very high pressure to form the pellets. Jiffy products are also available in several sizes (Figure 18-18). These units must be hydrated by wetting to cause them to expand to usable size.

The manufacturing process used to make rock wool has been covered in Chapter 11. Rock wool is available in several sizes for vegetative propagation (Figure 18-19). Rock wool is also formed into slabs. These units are made in widths of 6 to 18 inches (15 to 45 cm), and lengths of 30 to 39 inches

FIGURE 18-16 Smithers Oasis Root Cube® and Wedge® propagation media. Top: left to right, 1.0 inch (2.5 cm), 1.25 inch (3.2 cm), and 1.50 inch (3.8 cm). Bottom: front to back, 5635, poinsettia wedge strip, 5625, geranium double wedge strip, and 5644, 10 × 20 inch (25 × 50 cm) wedge tray.

FIGURE 18-17 Carnation cuttings rooted in Smithers Oasis Wedge® media.

(75 to 100 cm). The thickness is usually 3 inches (7.5 cm). Rock wool slabs are widely used in Europe and Canada for greenhouse production of tomatoes, cucumbers, peppers, eggplants, and a few other crops.

Sanitation

Success in propagating cuttings also depends upon good sanitation. Clean knives must be used when taking cuttings. For some crops, several knives are used because they must be disinfected in alcohol, or a similar solution, after each cut.

New cuttings are placed in clean plastic bags for transfer to the rooting area. Do not lay cuttings on the ground or on benches where the stock plants are growing. The cuttings can pick up diseases through such careless handling. The cuttings can also be transferred to the rooting area in clean flats lined with newspapers.

The propagating media should be steam sterilized between each crop of cuttings (Figure 18-20). Even when steam is used, diseases may start in the propagation bench. If this happens, a fungicidal drench may be used at the strength recommended on the label. A fungicide drench cannot be used as a substitute for steam sterilization. In addition, the use of a **fungicide** should not take the place of good cultural practices. Excessive application of a fungicide may reduce or prevent rooting of some crops.

Fertilization

Unrooted cuttings should not be fertilized heavily if the fertilizer is not applied with a mist. The cuttings can take up very little fertilizer added to the rooting

FIGURE 18-18 Jiffy® compressed peat pellets: *left to right*, 0.7 inch (18 cm), 1.0 inch (2.5 cm), 1.2 inch (3.0 cm), 1.4 inch (3.6 cm), and 1.6 inch (4.2 cm). *Front*, compressed; *rear*, hydrated and expanded for use.

FIGURE 18-19 Rock wool propagation media, SBS system.

FIGURE 18-20 Steam pasteurization of propagation media and bench. Notice clamps needed to hold the cover in place.

medium until they develop root systems. Generally, a light application of fertilizer can be made 3 or 4 days before the rooted cuttings are removed from the bench. Then, a complete analysis fertilizer, such as 20-10-20, is applied at the rate of 1 pound (0.45 kg) to 100 gallons (378.5 l) of water.

If the pH of the propagating medium is low, ground limestone is added to bring the pH to a range of 6.0 to 6.5. Most cuttings root well at this pH range (with the exception of azaleas and other ericaceous types).

Handling After Rooting

The cuttings should be removed from the propagation bench as soon as a mass of roots develops. The root system should consist of several roots 1 to 2 inches (2.5 to 5 cm) long with many smaller roots.

There will also be some **secondary root** development. These new roots are very fragile. Great care must be used in removing the plants from the medium. Do not grasp the cuttings at the top and pull them from the medium. This treatment will tear many of the young roots. To remove the cutting, grasp the top with the left hand, slide the right hand under the cutting, and gently lift it from the medium. This method preserves the greatest number of roots. Some of the propagation medium may cling to the root system. Excessive amounts can be removed by gently shaking the cutting. The remaining material helps to retain the root ball and can be planted with the cutting.

The rooted cuttings should be planted no deeper than they were stuck in the propagation medium. If they are planted too deep, tender stem tissue is exposed to any disease organisms in the medium. As soon as they are planted, they should be watered thoroughly. If planting takes place on a hot, bright day, the newly potted plants should be heavily shaded or they should remain overnight in the head house. This means that the root systems can continue to grow without being required to provide more water than they can absorb. Only 1 or 2 days of shading are needed if the plants were properly hardened off in the propagation bench. The plants should be watched closely until they start active growth.

Direct Rooting in Finish Size Pots

Some crops, such as poinsettias and geraniums, are propagated directly in the finish pots in which they will be sold. Compared to the use of propagation benches, this method requires much more space. If the required area is available, this practice has the advantage of eliminating the potting step, thus reducing the labor involved.

When cuttings are to be rooted directly in finish pots, they must be graded carefully for length and stem thickness. Thin stemmed cuttings should not be planted with thick stemmed cuttings. The heavier cuttings may root more quickly and outgrow the thin plants. When several cuttings are placed in one pot and one or more of them fail to root, other cuttings must be substituted until all take root, which reduces the finished plant quality.

Plants rooted directly in finish pots are hardened off from the mist system in the same manner as cuttings in propagation benches. After the plants are removed from the propagation area, it may be necessary to mist them by hand. A FOGGIT® nozzle can be used several times a day until the plants are conditioned to the new, drier environment. Properly hardened plants need a minimum of hand fogging.

Storage After Rooting

There are times when the planting area may not be ready to receive the rooted cuttings. For example, the previous crop may have been delayed or other factors may require that the cuttings be held for a time before planting. If the cuttings continue to grow in the propagation bench, they will become hard. To prevent this, the cuttings are removed from the bench and placed in cold storage. Rooted chrysanthemums can be stored safely for 7 to 10 days, and carnations can be held for 3 weeks at 33°F to 35°F (0.5°C to 1.5°C). The cuttings are removed from the media and are placed in plastic bags. The open bags are then put into cardboard boxes in the refrigerator. Unrooted cuttings of chrysanthemums have been stored in sealed plastic bags at 31°F (0.5°C) for 3 to 4 weeks. Unrooted carnation cuttings have been stored at 31°F (0.5°C) for up to 6 months. After storage, both crops are rooted just as if they were fresh cuttings.

Poinsettias and geraniums are severely damaged if they are placed in cold storage for any length of time. Both crops should be planted as soon as they are rooted.

Other Vegetative Propagation Methods

Separations

Bulb crops, such as hyacinths, tulips, lilies, and narcissus, can be increased by a method known as **separation**. The clumps of bulbs are separated into individual bulbs. Small bulbs removed from a clump are planted in beds where they continue to grow for a year or more until they reach flowering size.

Crops grown from corms, such as gladiolus and crocus, are also propagated by separation. Young, newly developed cormels are removed from older plants (Figure 18-21).

New plants can be obtained from large corms by cutting the corms into pieces. Each piece must contain a bud or eye if a new plant is to develop.

Divisions

The method of plant **division** is similar to separation and is also used as a way of propagating new plants.

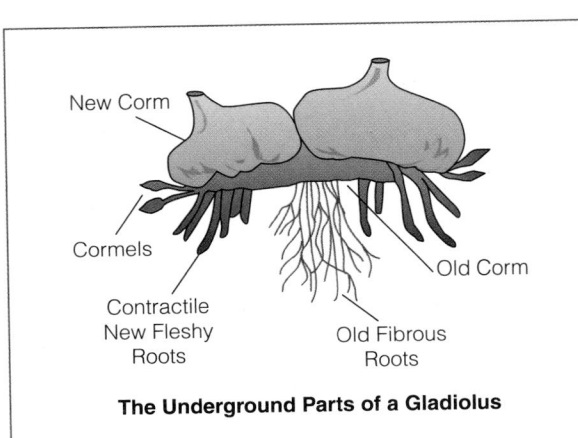

FIGURE 18-21 New gladiolus corms are formed from the old corm. Cormels will require from 2 to 3 years to develop to flowering size.

New Corm

Cormels

Contractile New Fleshy Roots

Old Corm

Old Fibrous Roots

The Underground Parts of a Gladiolus

Crowns

Crowns are formed when a number of plantlets grow in a close arrangement. African violets, *Philodendron selloum*, and other plants grow from crowns. Crowns are divided by breaking them apart (Figure 18-22).

Tuberous Roots and Stems

Some plants develop thickened roots or stems that act as storage tissues. For example, *Dahlias* form **tuberous roots** underground. Buds occur only at the stem end. When *Dahlias* are divided, each stem section must include a bud section if a new plant is to develop (Figure 18-23).

Rhizomes

Rhizomes are specialized stem structures. The main axis of the plant grows horizontally along or just below the soil surface. Iris and lily of the valley (*Convallaria majalis*) are flowers that form rhizomes.

Rhizomes are propagated by cutting them into sections. Each section must have an eye or bud where the new growth will develop.

Tubers

Tubers are specialized stem structures formed mainly from enlarged food storage tissues. The common white potato, *Solanum tuberosum*, is a tuber. Jerusalem artichoke (*Helianthus tuberosus*) and caladium are tuber crops that are propagated by cutting the tuber into sections. Each section must have an eye (axillary bud).

Cyclamen, gloxinias, and tuberous begonias produce tuberous stem tissues, which have the characteristics of stems. To produce new plants from tuberous stem sections, a bud must be attached to each section.

Offsets

Offsets are small plantlets that develop from the main plant. *Dendrobium* species orchids, pineapples, and bromeliads are propagated by means of offsets.

FIGURE 18-22 A multicrown African violet is separated to provide new plants.

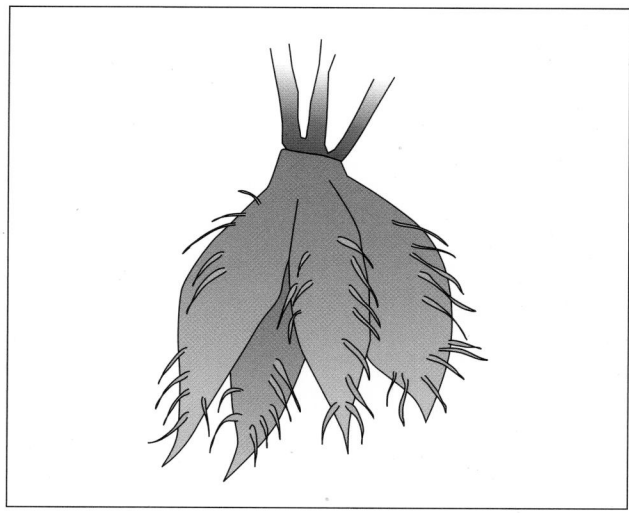

FIGURE 18-23 When the tuberous root system of the Dahlia is divided, a stem section with a bud must be included with the root or a new plant will not develop.

Pseudobulbs

Orchids produce specialized aboveground storage tissues called **pseudobulbs**. The appearance of these "false bulbs" varies with the orchid species. *Dendrobium* species orchids produce long, narrow pseudobulbs. The pseudobulbs of the *Cattleya* species orchids are more rounded. When orchids are propagated asexually, two or three pseudobulbs are separated as one unit from the rest of the plant.

Layering

Layering is a method of propagation that is used to produce a new plant that is still attached to the parent plant. In layering, the parent plant serves as a life-support system until the root development on the new plant is active enough to support its own growth. There are several methods of layering plants.

Air layering is commonly used for ornamental plants. Other methods of layering are mound, simple, tip, serpentine, and trench layering.

The development of roots in layered plants is governed by the same environmental factors that affect the rooting of cuttings. These factors include the proper temperature, adequate supplies of moisture and oxygen, the lack of light that stimulates the development of fruits, and the age of the wood.

Air Layering

Ficus, *Philodendron*, and similar species are normally propagated by air layering. The stem is girdled (cut around), and the cut is dusted with a rooting hormone. Growers generally prefer to make a girdling cut rather than a slit in the stem. The slit may grow together again unless it is held open by a toothpick or a wooden match. The cut is covered with moist sphagnum moss. The moss is then wrapped with a piece of aluminum foil to hold it in place. The aluminum reflects heat and keeps the temperature of the rooting medium at a reasonable level. The foil also prevents the moss from drying out.

As soon as the plant develops visible roots, it is cut away from the old stem and potted. Air layers, or mossings, as they are sometimes called, are often made by southern growers. When the plants are cut from the old stems, they are shipped to northern growers who then pot them and grow them for sale.

Mound Layering

Mound layering is a method used in the field. The stems of the plant are cut back during the dormant season, and soil is mounded around them (Figure 18-24). These stems form new shoots and later develop roots. After the root system develops to the point that it can support the new plant, the plant is removed from the mother plant. Roses are often propagated by mound layering.

Simple Layering

In a simple layering procedure, the stems of branches are bent to the ground where they are covered with soil or rooting medium. A shallow knife cut may be made in the part of the stem that is to be buried to stimulate the growth of new roots.

Young shoots less than a year old are recommended for layering because older stems do not develop roots easily. *Philodendron*, *Dieffenbachia*, and similar

FIGURE 18-24 When roses are mound layered, the plants are cut back during the dormant season and the first soil mound is placed. The second mound is made in the following spring.

Second Soil Mound New Roots Develop in this Mound

First Soil Mound

foliage plants are rooted by simple layering. Forsythia and climbing roses are also propagated by this method.

Tip Layering

Tip layering is similar to simple layering. In this procedure, the tips of branches are covered with soil or a rooting medium. As growth continues, the tip emerges from the soil. New roots develop on the part of the stem that is covered. In general, an entire growing season is required to obtain a plant from a tip layer. This method is used for plants having flexible stems, such as raspberries, blackberries, and blueberries.

Serpentine or Compound Layering

Serpentine or compound layering is similar to simple layering. This method is used to propagate vinelike plants. A long stem is placed on the ground. It is alternately covered with soil and exposed to air. The covered stem sections are nicked with a knife to promote rooting. The exposed stem parts should each have a visible bud where a new shoot will develop. After roots develop, the plant is cut into sections, which are then potted.

Trench Layering

In the trench layering method, the central portion of a stem is buried to a depth of 2 to 4 inches (5 to 10 cm). The stem is cut slightly at intervals to stimulate rooting. When the new shoots are large enough, the individual plants are separated and grown on. Roses, *Spirea*, and *Rhododendron* are often propagated by trench layering.

Grafting

Grafting is a procedure that joins two plant parts together so that they fuse or unite and continue growth as one plant (Figure 18-25). Grafting is widely used to produce new fruit trees. In ornamental horticulture, however, there is little need for grafting, with the exception of certain cultivars of azaleas, camellias, lilacs, and rhododendrons. Plant pathologists use grafting to determine if a plant is free of viruses. In this case, the procedure is known as virus indexing.

Roses are reproduced by a special grafting technique called **budding** (Figure 18-26). It is a highly specialized procedure that requires a great deal of

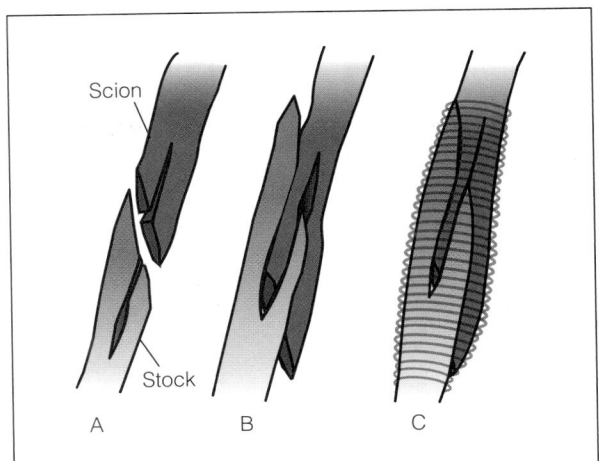

FIGURE 18-25 The steps involved in "whip" or "tongue" grafting are: A, prepared stock and scion, B, parts united, C, graft wrapped with waxed string.

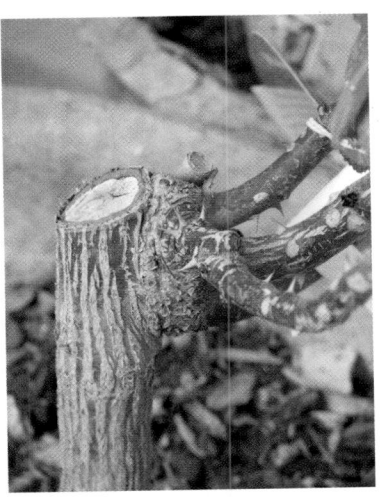

FIGURE 18-26 Budded rose plant showing the bud union, or the swollen area at the base of the plant.

skill and extensive knowledge of plant physiology. The commercial grower usually purchases roses from a specialist skilled in this procedure.

Micropropagation

The technique of **micropropagation** was once considered novel, but is now widely used by commercial growers. The procedure is also known as tissue culture. Tissue culture is a term that describes several specific methods of micropropagation (Figures 18-27, 18-28, and 18-29).

Micropropagation involves the use of extremely small pieces of plant tissue, usually less than 1 millimeter in length. Under aseptic (germ-free) conditions, this tissue is placed in a specially prepared growing medium. Root and shoot growth is controlled by regulating the amounts and kinds of chemicals and growth regulators that are used in the medium. When the plants are large enough to be handled, they are transferred to another growing medium. Finally, they are taken into the greenhouse where they complete their normal growth cycle.

Meristem Culture

Meristem, or shoot tip, culture uses the active growing points, or meristems, of plants to develop new plants. In meristem culture, the apical meristem and one or two leaf primordia (beginning leaves) below it are carefully cut from the plant. Using aseptic methods, the meristem is transferred to a sterile growing medium in a test tube (Figure 18-29). The culture medium must not be contaminated with molds and fungi because these organisms will kill the plant tissue.

The test tubes are then placed in an environment with a medium light intensity and a temperature of 72°F to 75°F (22°C to 24.5°C). The new plantlets develop in this environment until they reach the proper size and can be transferred to a soil type growing medium.

Tissue Culture

The term *tissue culture* is used to describe variations of meristem culture. Each plant cell has the ability to produce a complete plant from itself if it is given the proper environmental conditions. This ability is called totipotency, or total potency. Totipotency means the total potential a cell has for new growth. The nucleus of each cell contains the genetic message that enables the cell to grow into a new plant.

FIGURE 18-27 Laminar flow hood used for aseptic culture techniques.

FIGURE 18-28 Micro-tissue cultures are placed in solution medium on racks under artificial light.

FIGURE 18-29 Callus tissue culture transplanted to solid medium in test tubes.

FIGURE 18-30 Newly transplanted tissue culture plantlets under poly sheeting for hardening before being placed into the greenhouse.

Because of the totipotency of a cell, the tissue culture scientist can take a cell or group of cells from any part of a plant and produce new plants (using the proper techniques).

All of the plants produced by this method will be similar in vegetative appearance and blooming characteristics. Plants grown from a single tissue source are known as clones. Thus, orchid cultivars, or clones, produced by tissue culture methods will all bloom at the same time. This uniformity in the date of bloom helps the grower to harvest the crop and prepare it for market on schedule.

The earliest experiments with tissue culture techniques used orchids. At that time, scientists were attempting to overcome a serious virus disease that threatened the orchid industry in the United States. Since then, tissue culture techniques have been adapted to ornamental plants. Large numbers of ferns are now propagated commercially by tissue culture methods.

Tissue culture procedures require germ-free methods and surroundings. An increasing number of laboratories are presently working to supply the needs of commercial growers. This is not the type of procedure that a grower can set up in a section of a greenhouse.

ACHIEVEMENT REVIEW

Select the best answer or answers to complete each statement. List the appropriate letter(s).

1 Asexual or vegetative propagation uses:
   a. stems and roots.
   b. seeds.
   c. leaves and buds.
   d. all of the above plant parts.

2 Asexual propagation is required because:
   a. some plants do not produce seed.
   b. seed-grown plants are all the same.
   c. plants produced by this method flower sooner than those grown from seed.
   d. the costs of production are less than the costs of using seed.

3 Plants can be propagated vegetatively because of the process of cell division that is known as:
   a. symbiosis.
   b. mitosis.
   c. meiosis.
   d. hexoses.

4 Terminal or stem tip cuttings are prepared from:
   a. second year tissue.
   b. new growth.
   c. the bases of stems.
   d. any part of the plant.

5 Leaf cuttings are used to propagate:
   a. carnations.
   b. lilies.
   c. *Bryophyllum*.
   d. Rex begonias.

6 Leaf bud cuttings are used when:
   a. there is a lot of plant material available.
   b. plants are needed in a hurry.
   c. the propagating material is scarce.
   d. a maximum number of cuttings is wanted.

7 Rooting hormones will:
   a. increase the number of plants from a cutting.
   b. increase the percentage of cuttings that root.
   c. decrease the rooting time.
   d. increase the number of roots that develop on a plant.

8 The best method of applying rooting hormones to cuttings is to:
   a. dip the moistened ends of the cuttings in the powder.
   b. brush the ends of the cuttings with the powder.
   c. dust the powder on the ends of the cuttings using a puffer duster.
   d. apply a shallow coating of the powder to the rooting medium.

9 If too much hormone is applied to a cutting, the cutting may:
   a. root too quickly.
   b. form callus tissue only and not roots.
   c. rot.
   d. not be affected.

10 A good propagation medium will:
   a. last forever without needing replacement.
   b. provide cutting support, moisture, and aeration.
   c. always be free of diseases.
   d. transfer heat readily.

11 Rooting cuttings under mist:
   a. reduces the time of propagation.
   b. allows the use of full light intensity.
   c. leaches nutrients from the leaves.
   d. reduces the amount of labor required to water.

12 Mist is applied:
   a. continuously, day and night.
   b. to keep the cuttings moist without wilting.
   c. for 10 seconds every hour.
   d. three times a day.

13 The proper temperature for rooting cuttings is:
   a. 41°F (5°C) to conserve heat.
   b. 65°F to 75°F (18°C to 24°C) with bottom heat 10° higher.
   c. 60°F (16°C) air and medium temperature.
   d. 85°F (29°C) air temperature and 75°F (24°C) bottom heat.

14 Propagating directly in finish size pots:
   a. requires less space than rooting in benches.
   b. saves labor.
   c. can be used for all crops.
   d. requires very careful grading of the cuttings.

15 Bulb type crops are commonly propagated by:
   a. stem tip cuttings.
   b. root cuttings.
   c. separation or divisions.
   d. seed.

16 When dividing tubers, it is important that each section of tuber:
   a. is dusted with rooting hormone.
   b. is cut precisely at right angles.
   c. contains an eye, or axillary bud.
   d. includes roots.

17 The rubber plant, *Ficus elastica*, is propagated by:
   a. stem tip cuttings.
   b. air layering.
   c. stem tip layering.
   d. root division.

18 Air layered plants are removed from the parent plant:
    a. when the roots are 4 to 6 inches (10 to 15 cm) long.
    b. when the plants are sold.
    c. as soon as the roots are visible.
    d. immediately after layering.

19 Micropropagation is a relatively new technique that:
    a. is easy to do.
    b. requires germ-free (aseptic) conditions.
    c. guarantees the same kind of plants.
    d. uses meristematic tissue only.

20 Some of the earliest experiments in micropropagation used:
    a. carrots.
    b. asparagus.
    c. orchids.
    d. chrysanthemums.

STUDENT PROJECT: MAKING TERMINAL CUTTINGS

Objectives

This exercise teaches the student:
1. how to prepare the rooting medium.
2. how to prepare cuttings for propagation.
3. the procedure for sticking cuttings.
4. how different plant materials root at different speeds.

Materials

1. Peat moss and vermiculite, or peat moss and perlite
2. Six-inch (15-cm) diameter plastic pot (Figure 18-31)
3. Plant materials, such as coleus, chrysanthemum, geranium, and ivy
4. Sharp knife
5. Rooting hormone
6. Puffer duster
7. Ten-quart (9.46 l) plastic bag and tie
8. Plastic marking label and soft lead pencil or permanent ink pen

FIGURE 18-31 Assemble materials needed-plastic bag, label and marking pen, peat-lite media in pot, geranium cuttings, rooting hormone, and puffer duster.

Procedure

1. Assemble all materials.
2. Fill the pot with propagation medium. Firm the medium by pressing down moderately.
3. Water the medium thoroughly until water runs from the drain holes.
4. Make a terminal shoot tip cutting, 3 inches (7.5 cm) long, of the plant material being propagated. Strip away any bottom leaves so that they are not buried in the medium. There should be 1 inch (2.5 cm) of bare stem.
5. Dust the end of each cutting with rooting hormone (Figure 18-32).
6. Insert the cuttings 1 to 1.5 inches (2.5 to 3.7 cm) deep in the medium. Space the cuttings 1.5 to 2 inches (3.7 to 5 cm) apart (Figure 18-33).
7. Water the containers thoroughly and allow them to drain.
8. Place the pot in a plastic bag. Inflate the bag by blowing into it. Close the bag and secure the top with the tie. The plastic bag maintains a very humid atmosphere around the cuttings (Figure 18-34).

FIGURE 18-32 Puffer dusting rooting hormone on basal ends of cuttings. This is done to promote uniform rooting.

FIGURE 18-33 Cuttings inserted into media. Notice label with date and cultivar name.

FIGURE 18-34 Container enclosed in plastic bag. Pot is placed in an appropriate rooting environment.

9. Place the pot in the rooting environment at 70°F to 75°F (21°C to 24°C) with moderate light. Do not place the pot in direct sunlight because the plants will become overheated and die.

10. Examine the cuttings after 10 to 14 days. Carefully remove each cutting by prying it gently from the pot with a wooden label. Note how various plants root at different rates. Any cuttings that do not have roots 1 to 1.5 inches long are to be placed back in the pot.

11. When the roots are 1 to 1.5 inches long, transplant the cuttings to small individual pots.

Pest Management and Chemical Plant Growth Control

Insects, Mites, and Other Invertebrate Greenhouse Pests

OBJECTIVES:

After studying this chapter, the student should have learned:

■ the foundations and principles of integrated pest management (IPM).

■ appropriate pest scouting strategies.

■ effective pest management techniques.

■ the basics of biological pest control.

■ about the common insect, mite, and other invertebrate pests that commonly infest greenhouse plants.

The professional greenhouse manager's job is to plan, monitor, and control the activities in the greenhouse resulting in a high-quality and profitable crop. Planning how to manage pests in the greenhouse is a major part of this job. Planning provides more control over what responses need to be made in reaction to a pest infestation in the future. Planning provides a baseline in advance on what to do, how to do it, when to do it, and who should do it. It bridges the gap from where the organization is to where it wants to be.

Planning a pest control strategy uses a concept known as **integrated pest management (IPM)**. IPM is a practice that reduces pest damage on a crop with minimum impacts on human health, nontarget organisms, and the environment. Here, the term *pest* is used to identify many problems, including diseases, undesirable insects, mites and other invertebrates, and weeds.

Planning an IPM program includes pest prevention techniques, monitoring techniques, and management strategies. Pesticides should only be used when monitoring reveals that they are required. The use of pesticides as a first response to a pest infestation is typically due to poor planning.

Steve Dreistadt[1] of the University of California IPM Center has established that the basis of a successful IPM program should include five key components.

1. Problem prevention
2. Regular monitoring of crops and growing areas
3. Accurate diagnoses
4. Appropriate action thresholds
5. Effective management methods

To develop an effective IPM program, the manager must learn the problems to which the crop may be susceptible and take appropriate action to prevent problems from occurring. An ancient Chinese idiom says,

So it is said that if you know your enemies and know yourself, you will win a hundred times in a hundred battles. If you only know yourself, but not your opponent, you win one and lose the next. If you do not know yourself or your enemy, you will always lose.

We can use this saying in the greenhouse as well. By knowing the pests, knowing how those pests will affect the crop, knowing the conditions of the greenhouse, and knowing his or her own expertise, the manager will better maintain a clean and healthy crop. The following steps will help guarantee success.

Objectives of an IPM program. Objectives may include saving money, improving crop quality, reducing pesticide use, and switching from broad-spectrum toxic pesticides to environmentally *softer* pesticides.

Summarize the IPM program for each crop. Assembling the potential problems before they occur will make decision making much simpler, and quicker action can be taken. The summary should include what potential problems may occur, the scouting procedures to be followed, appropriate management strategies, acceptable pesticides to use, alternative pest control options, and the pest control thresholds.

Determine which pesticides and application rates are acceptable to your situation and crops. Maintain a list of pesticides that are both acceptable to the crop and user's level of experience, and are legally available in the state where they are to be used. Note that some pesticides are considered *restricted use pesticides* and require additional licensure or certification for handling and use.

Identify alternative treatment options. Often many growers tend to immediately use a chemical pesticide to control an outbreak. This is usually not the best first strategy. If the infestation is small, it may be simpler and cheaper to rogue out the infested plants than perform a full-scale pesticide application. Biological control may also be an option.

Determine appropriate action thresholds. The action threshold is very different from crop to crop. For instance, the action threshold for a whitefly population on a crop of poinsettias is very low compared to a crop of greenhouse tomatoes. The poinsettia plant itself is to be sold, whereas in the tomato, the fruit is to be sold. Any whitefly residual infestation on the fruit can be washed, but not so for the poinsettia plant.

---

[1]Dreistadt, S. H. *Integrated Pest Management for Floriculture and Nurseries.* Publication 3402 (Oakland: University of California, Agriculture and Natural Resources, Communication Services). 2001.

Maintain records. Keeping records of all pesticide applications is the law, but in addition, it is important to record what pest infestation occurs, when it occurs, the crops involved, environmental conditions, the action taken, and the relative success of the actions. Referring back to those records during the planning process helps in future growing success.

Pest Prevention

The adage, "An ounce of prevention is worth a pound of cure," aptly represents pest management in the greenhouse. Prevention in the greenhouse includes four critical practices.

1. Practicing good sanitation and pest exclusion
2. Proper management of the greenhouse environment and other cultural practices
3. Monitoring or scouting, which refers to regular, systematic inspection of crops and growing areas
4. Advanced planning of crop production and IPM programs

Sanitation

Sanitation involves clean practices in the greenhouse. Clean practices include eliminating weed infestations inside and outside the greenhouse. Weedy plants under the bench or around the perimeter of the greenhouse may harbor pests and diseases (Figure 19-1). Controlling pest infestations on crops will have no benefit if the weeds under the bench are not treated as well. Having to treat weeds for pest control is an unnecessary expense. Weeds outside the greenhouse are just as important as those under the bench. Keeping weeds clear of the exterior of the greenhouse reduces the potential population of pests that may migrate into the greenhouse. Mowing will reduce grass flowering, which in turn reduces thrips populations.

When choosing chemical weed control for the greenhouse, select only those herbicides that are labeled for greenhouse use. Many herbicides may volatize (become gaseous) in the greenhouse and will damage plants. Most herbicides are registered for greenhouse use only when the greenhouse is empty of plants.

FIGURE 19-1 Weeds in the greenhouse can harbor insect pests.

Clethodim (Vantage®) is a postemergence herbicide labeled for weedy grasses.

Diquat (Reward®) is a contact herbicide labeled for all weeds. When choosing diquat, it has a high mammalian toxicity rating.

Glufosinate (Finale®) is a broad-spectrum postemergence herbicide that is best used under benches, floors, and foundations.

Glyphosate (Roundup®) is a broad-spectrum contact herbicide labeled for use in an empty greenhouse.

Pelargonic acid (Scythe®) is a contact herbicide for use under greenhouse benches.

A nonchemical means to eliminate weeds in the greenhouse is the use of a propane torch. Using a torch easily eliminates weeds on gravel or concrete. However, it does not work well around polypropylene fabrics and must not be used near plastic greenhouse glazing. The propane torch also creates a great deal of smoke and uses oxygen in the greenhouse. When using a torch, make sure that there is adequate ventilation.

The best weed control in the greenhouse is prevention. The use of weed barrier fabrics works well, yet weeds may still become established. Make sure that underneath the fabric there is adequate drainage. Many growers are using quicklime or hydrated lime (calcium hydroxide) as a weed control measure. It is applied as a slurry through a sprayer, literally whitewashing the floor until it turns white. This layer of quicklime then serves as a chemical barrier to weed seed germination and keeps the greenhouse floor clean and white.

Algae control can also be included with weed control. Algae growth on the floors and benches creates a habitat that supports fungus gnats, shore flies, and drain flies. Although adult fliers are primarily a nuisance, fungus gnat larvae will damage plant roots. Algae, in addition to being unsightly, also provide a serious risk to worker safety in the greenhouse creating slick floors.

Algae should be killed using an appropriate **disinfectant**. Disinfectants that are currently registered for use in the greenhouse for algae as well as disease control include quaternary ammonium compounds, such as Green-Shield®, Physan 20®, and Triathlon®; hydrogen dioxide, such as ZeroTol® and Oxidate®; chlorine dioxide, such as Selectrocide™; and chlorine bleach.

Exclusion

Reducing pest entry through **exclusion** is a good management strategy. Checking and quarantining plants that are being brought into the greenhouse for pest infestations is important. It is best to isolate and quarantine any new plants to determine if an insect or disease is being transmitted. Of course, during busy seasons this is a difficult practice, yet all plants must be carefully inspected. Pet plants are an additional risk. Every greenhouse seems to have a few plants around that are very old or have some emotional tie to someone in the operation. These plants have a tendency to not receive the best care and often are infested with insects and diseases. Unless the business is to rejuvenate house or interiorscape plants for customers, these types should be kept out.

Screening is an effective means to prevent pest infestations in the greenhouse (Figure 19-2). Screens are placed over intake vents to prevent insect entry (Figure 19-3). Properly installed screens will restrict the entry of insects and

FIGURE 19-2 Screening vent openings can help to exclude many insect pests.

Pleated*

Lean-To

Top cut out view of a typical frame holding a pleated screen

Pleated screens provide a greater surface area than an unpleated screen, but takes up less space.

Box

End of Greenhouse

Flush

FIGURE 19-3 Some example greenhouse screen configurations.

pests while reducing worker exposure to pesticides. Screen openings are based on the size of the insect to be excluded (Table 19-1). If the greenhouse is shipping to an area that restricts the importation of plant products grown in states under Japanese beetle restrictions, fly screening may be sufficient. However, if the exclusion of western flower thrips is the goal, screens with very small holes will be required (Table 19-1).

The smaller the hole of the insect exclusion screen, the greater will be the resistance to airflow (Table 19-2). With increased resistance to airflow, the

Table 19-1

Insect Screen Sizing[1]		
Insect Pest (Common Name)	Size (μm/inches) Width of Thorax	Size (μm/inches) Width of Abdomen
Western Flower Thrips	215/0.0085	265/0.0104
Silverleaf Whitefly	239/0.0094	565/0.0222
Greenhouse Whitefly	288/0.0113	708/0.0279
Melon Aphid	355/0.0140	2,394/0.0549
Green Peach Aphid	434/0.0171	2,295/0.0904
Citrus Leafminer	435/0.0171	810/0.0319
Serpentine Leafminer	608/0.0239	850/0.0335

1. Bethke, J. A., and T. D. Paine. 1991 Screen hole size and barriers for exclusion of insect pests of glasshouse crops. *J. Entomol. Sci.* 26: 169–177.

Table 19-2

Insect Exclusion by Different Screen Styles			
Screen Style	Hole Size (μm)	Airflow Reduction (%)	Insect Pests Excluded
ECONET B[1]	1,000 × 4,000	5	Large insects such as bees and Japanese beetle
ECONET L[1]	600 × 600	25	Whitefly and larger
ECONET M[1]	400 × 450	30	Some thrips and larger
ECONET SF[1]	300 × 750	25	Some thrips and larger
ECONET T[1]	150 × 350	40	Thrips and larger
ECONET S[1]	140 × 140	45	Ultimate thrips exclusion
Anti-virus[2]	266 × 818		Leafminer and larger
No-Thrips[2]	150 × 150		Western flower thrips and larger

1. AB Ludvig Svensson Screens, Kinna Sweden.
2. Green-Tek, Edgerton, Wisconsin.

greater static pressure will drop, which will reduce air exchange, increase energy consumption by the exhaust fans, and increase wear on fan motors, ultimately increasing greenhouse temperatures. To prevent airflow problems, choose the correct screen and design the correct screen surface area. Do not forget to screen roof vents, doorways, and other openings as well.

An alternative to screening is the use of positive-pressure cooling. Conventional greenhouse cooling systems use exhaust fans that utilize negative pressure to pull air through a vent into the greenhouse from outside. Positive-pressure cooling systems pull air through a series of pads and filters providing clean air (Figure 19-4). These systems are well justified when used for research facilities and propagation greenhouses where insect exclusion is a priority.

Greenhouse Environment

Plants under stress are predisposed to pest infestations. Drastic temperature fluctuations as well as excessive wet and dry cycling will predispose the plants

FIGURE 19-4 Positive-pressure greenhouse cooling mechanism can be used to help exclude insect pests.

to damage. The use of proper environmental controls—heating, cooling, lighting, proper cultural practices, fertility, and irrigation—will favor high-quality crop production and discourage potential pest and disease problems. Choose growing media that has been properly pasteurized and has the proper aeration, drainage, and electrical conductivity for good plant growth.

Choose cultivars that are pest resistant when available. Regular rotation of crops will prevent many pest issues. Proper use of fans, heating systems, and lighting to control humidity also can reduce many pest and disease issues.

Monitoring

Monitoring the greenhouse crop will allow early detection of any pest or disease outbreaks at a time when they are easy to manage. Once a pest infestation becomes widespread, the crop may be severely damaged and the management options become limited. Many refer to monitoring as scouting. Scouting is the regular, systematic inspection of crops and growing areas. According to Steve Dreistadt[2] of the University of California IPM Center has established that the scouting will:

- prevent problems or reduce the damage and cost of control.
- determine the specific cause and severity of the problem.
- identify locations that require treatment.
- help to avoid treatment where control is not required.
- determine the most effective and economical solution.
- allow slower acting methods, which may be more environmentally sound and safe for workers.
- evaluate control efficacy.
- build a database of pest management history.
- improve crop quality and increase sales revenue.

The best way to begin a monitoring or scouting program is to follow some simple guidelines and expand as experience and needs grow. After gaining experience, the manager can modify the monitoring program based on a self-evaluation.

Before beginning a scouting program, the manager develops a clear and concise plan that determines who is responsible for the scouting and who is

[2]Dreistadt, S. H. *Integrated Pest Management for Floriculture and Nurseries*. Publication 3402 (Oakland: University of California, Agriculture and Natural Resources, Communication Services). 2001.

responsible for crop cultural practices, environmental conditions, and pest management. Each growing area should be evaluated to identify potential problems including but not limited to weeds, water leaks, crop debris, poor pest exclusion, standing water, and algae. The manager should note previous crops that were present that may lead to potential infestations not otherwise expected. Once the scouting plan is in place, it is important that the scouting procedure be followed the same way each time and at regular intervals, preferably each week.

The manager designates an individual or group of individuals to serve as **scouts**. An IPM scout is an individual who collects, records, and summarizes pest infestation data for the crop or greenhouse. The scout may be a professional consultant or grower who has had the task assigned. A grower or employee scout has the advantage of a more complete knowledge of the greenhouse operation than an outside consultant. However, employee scouts rarely have adequate pest management expertise or motivation, compared to a professional scout. An employee scout's abilities will increase with training and experience. Because employee scouts usually have many responsibilities, they should have a clearly defined time devoted entirely to monitoring as part of their regular duties.

Professional scouts and consultants are likely to be more effective at monitoring a crop. When employing a professional scout or consultant, the individual should have formal training, regular professional continuing education, and a complete knowledge of the pests specific to the crops being grown. Many states require licensure of pest control advisors.

A strong team relationship between the grower, scout, and applicator is maintained through good communication. Regular meetings between all those involved discussing monitoring results and management decisions must be a regular part of a successful IPM program. Scouts must provide clear and timely summaries along with suggested control actions and population thresholds. The scout should provide maps of areas to be treated and mark those areas with colored tape or flags. These markings can be used to assess treatment control efficacy.

A weekly monitoring program is typically adequate for most greenhouse situations. More frequent monitoring may be deemed necessary for high value crops or for pest-prone or pest-infested areas. Plugs should be inspected every 3 to 4 days. When choosing a monitoring strategy, the manager spends a predetermined time per area of growing space. When returning to monitor that space, the manager follows the same pattern (Figure 19-5).

Plants are inspected randomly, with a focus on areas near doorways, vents, and fans. The manager chooses one-third of the plants randomly and the remainder at targeted locations throughout the greenhouse. Sticky traps or cards are available for monitoring insect populations. Sticky cards are typically yellow and attract most flying insects (Figure 19-6). For small greenhouse sections, one card per 1,000 square feet is adequate and for large greenhouse sections, one card per 10,000 square feet will suffice. Blue sticky cards are used to attract and monitor thrips populations. Cards must be monitored and replaced weekly in order to effectively determine pest outbreaks. Remember that sticky cards only monitor flying insects.

With good record keeping, including the acceptability of the finished crop with comparison to pest scouting and control records, a series of thresholds can be established for each crop. Thresholds can be both quantitative and numerical. Thresholds can be established for the number of pests per trap each week

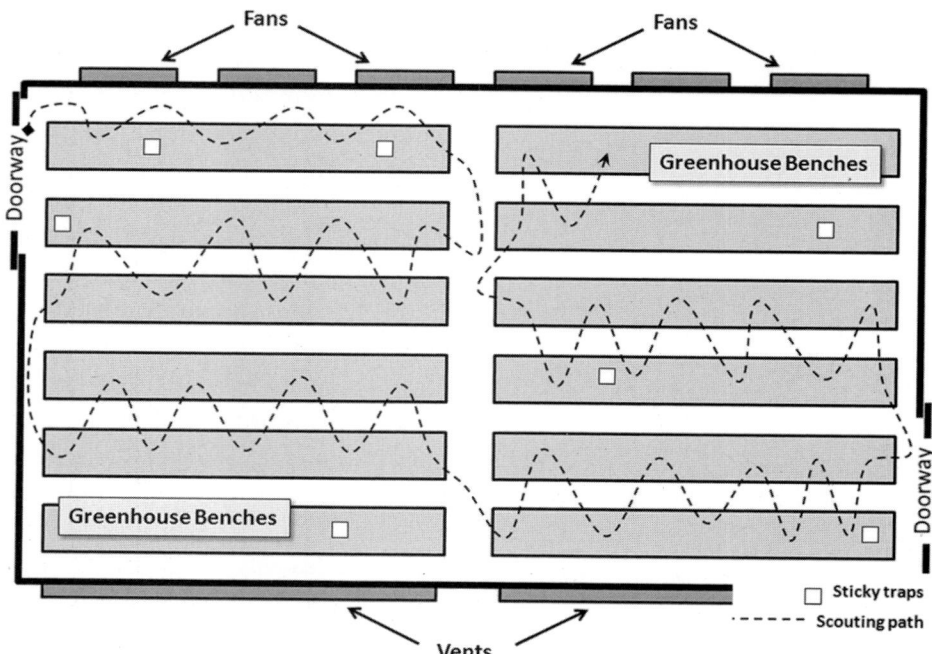

FIGURE 19-5 Typical scouting pattern for a greenhouse section.

FIGURE 19-6 Yellow sticky cards can be used to help monitor insect populations.

or the percentage of plants determined to be infested or damaged. Quantitative thresholds are established through monitoring methods that determine that conditions are conducive to pest and disease development. Thresholds will be different for different crops and different pests. For instance, the action threshold for whiteflies on greenhouse tomatoes may be quite high compared to whiteflies on a poinsettia crop. Ornamental plant consumers expect that their blooming plants be insect and disease free for a proper display in their home; the tomato consumer, however, does not care if whiteflies dwell on the tomato vine in the greenhouse, but the fruit must be clean and blemish free.

Management

An effective IPM program incorporates mechanical, environmental, cultural, biological, and chemical control measures. Typically, using two or more methods results in more effective control than using a single method. The preferred techniques use cultural practices and environmental conditions to prevent pest development rather than the use of chemical pesticides. When pesticides are required, the least toxic approach is preferred to minimize the impact on other pests, beneficial organisms, and IPM methods.

Mechanical Control

Mechanical control implies the use of labor and equipment to reduce pest populations directly. Mechanical control may be as simple as removing infested plants. Other typical mechanical control activities include hand-pulling and mulch application. Installation of screens and sticky barriers are other examples of mechanical control.

Environmental Control

Modification of the greenhouse environment by manipulation of temperature, light, and humidity is an effective means to control pests and prevent damage. Heating greenhouse media through solarization, steam pasteurization, or proper composting will control most soil-borne pests. Altering humidity and temperature by increasing ventilation, increasing heat, or modifying irrigation practices, for example, switching from sprinkler to drip, can reduce many foliar pathogens. Maintaining higher humidity can reduce spider mite populations. Eliminating standing water will reduce algae growth and subsequent fungus gnat populations.

Cultural Control

Cultural control measures to prevent pest outbreaks include choosing crop species or cultivars that are less susceptible to infestations than others, rotating crops from susceptible to nonsusceptible, altering planting times, and adjusting the duration and frequency of irrigation intervals. Also, cultural control includes fertility. Too much fertility or incorrect fertilizer choice may result in succulent growth that favors high populations of insect pests, such as aphids.

Biological Control

There are many beneficial organisms that can be used in a greenhouse to reduce pest populations. Biological control methods to be effective must be integrated with other methods, such as exclusion and sanitation. Pesticides are rarely used unless they are compatible with the pest's natural enemies. A low-level pest population must be present for a successful biological control program. Living organisms that can be used to reduce pest populations include predators, parasites, and pathogens.

A **predator** is typically an insect or insect-related animal that attacks or feeds on a host or individual pest. This attack is typically part of the predator's natural feeding cycle. Some predators also feed on pollen, nectar, and honeydew. Many feed on prey that is abundant including a variety of insects and mites as well as other invertebrates. Typical examples include ladybugs and predatory mites.

A **parasite** or parasitoid is an organism that attacks and kills a single host. Typically the parasite completes all or part of its life cycle in or on this individual pest. Many parasitoids feed on the host in their immature stage and live free as an adult. *Encarsia formosa* is an example of a parasitoid that oviposits its eggs into an immature stage of a whitefly. The nymph subsequently then consumes the whitefly. Another example of a parasitoid would be the parasitic nematode *Steinernema feltiae*. This nematode enters an insect larva and develops within the larvae, eventually killing the insect larvae.

A **pathogen** is typically a microorganism that brings about a disease on a host animal or pest. *Bacillus thuringiensis* is a good example of a bacterial organism that infects worm larvae, and *Beauveria bassiana* is a good example of a fungal organism that infects the exoskeleton of many insect pests.

Chemical Control

Chemical control refers to the use of pesticides. Whether your pesticide is considered to be organic or conventional, it is a product designed to control, repel, or mitigate a pest problem. The correct pesticide choice can prevent crop damage and temporarily control certain pests if applied at the right time and rate. Incorrect pesticide use, such as choosing the wrong pesticide or using it at the wrong rate, often does more harm than good. Incorrect pesticide use can result in damage to beneficial organisms, crop damage, applicator injury, and an increase in pesticide resistance.

Before using any pesticide, it is important to understand its relative toxicity, mode of action, rate of application, mixing and application guidelines, reentry intervals, persistence, and environmental impact. One can learn most or all of these issues by reading the most current label. This is required for effective, safe, and legal use of any pesticide.

Sanitation and Exclusion

One of the best methods of pest control is to control what pests have access to in the greenhouse. Exclusion methods include screening of vents, doorways, and other openings; inspecting newly introduced plants or plant shipments; using pest-free stock; controlling weeds; removing crop debris; promptly removing infested plants or plant parts; and maintaining as pest-free a growing area as possible.

Keeping the greenhouse pest free before planting is the most effective and least expensive pest control strategy available. Control after planting is more difficult because growers may be limited to what measures or strategies they can use. The use of sanitizers, weed control, and eradication in an empty greenhouse poses little risk to plants and often is easier to accomplish.

Pests can be transported in or on propagation materials. To avoid contamination, use high-quality propagation material from reputable suppliers. If using on-site stock plants, implement stringent controls as well as testing and scouting to maintain the plants as pest free as possible. Many growers will acquire culture-indexed plants from specialty growers and build stock blocks from those plants. The initial investment of these more expensive plants, which can reproduce many generations, is well worth the investment with reduced pest and disease problems.

Resistant Cultivars and Alternative Crops

Prior to planting, learn whether a species or cultivar is less susceptible to pest infestations. Some species or cultivars can tolerate a certain level of infestation without any damage. Maintaining good records that describe susceptibilities across multiple cultivars may allow the grower to make more informed pest management decisions.

Rotating crops with different species and cultivars in the greenhouse can break many pest infestation cycles. Choosing an alternative crop that is less susceptible to pest infestations or individual pest populations can be used to disrupt insect or mite life cycles.

Pesticide Selection

Chemical control implies the use of a pesticide. The Environmental Protection Agency defines a pesticide as any substance or mixture of substances intended for preventing, destroying, repelling, or mitigating any pest. Pesticides are often misunderstood to refer only to insecticides, but the term also applies to herbicides, fungicides, and various other substances used to control pests. Some pesticides are considered to be organic. Organic pesticides are those that come from natural sources. These natural sources can be plants, fungi, bacteria, or similar by-products. Examples include pyrethrum, spinosad, and minerals such as boric acid, cryolite, or diatomaceous earth. Remember, however, that even if a product is labeled as organic, it is still a pesticide and may be toxic to nontarget organisms.

Correct pesticide use can prevent crop damage and temporarily control certain pests without harm to the environment or human risk. Incorrect pesticide use often does more harm than good. Incorrect pesticide use includes application of the wrong pesticide, application at the wrong rate, or application in the wrong manner.

Pesticides traditionally used for flowering crops are typically broad spectrum and persistent. That is, they control a broad range of pests and secondary pests, which may include many natural enemies. They also have a tendency to pose hazards to workers and the environment. Many pesticides with a long persistence can increase pesticide resistance among numerous organisms. A successful IPM program that relies on biological, pest-specific, and reduced-risk pesticides can reduce many of these risks.

Pesticide Safety

Reduced-risk pesticides that are IPM compatible are typically targeted to specific pests, pose a reduced risk to nontarget organisms, pose less risk to the environment, have lower application rates, have lower potential for development of pest resistance, provide reduced potential for groundwater contamination, and offer better worker safety.

Pesticides are categorized into three general groups according to the target organism: insecticides to control insects, herbicides to control weeds, and fungicides to control fungal diseases. Also, more specific pesticides include miticides to control mites and bactericides to control bacterial diseases. Further classification of pesticides includes the source of the material: botanicals or plant extracts such as neem oil; inorganic pesticides such as sulfur and copper; synthetic organic pesticides from petroleum distillation including organophosphates and chlorinated hydrocarbons; and biological extracts or fermentation products. Table 19-3 includes insecticides registered for control of greenhouse insects and mites.

Table 19-3

Pesticides Labeled for Controlling Insects and Other Pests Common to the Greenhouse						
(This information is provided for general information only. It is your responsibility to use a pesticide to read and follow the label that has been approved for the particular state or locality in which the pesticide is to be used, and to comply with all federal, state, and local laws and regulations relating to the use of pesticides.)						
Class	MOA*	Common Name	Trade Name	REI	Manufacturer	Pests Controlled (Lists from Current Manufacturer Labels)
2, 4 – Diphenyloxzoline Derivatives	10B	Etoxazole	TetraSan™	12	Valent USA Corp.	Spider Mites Including: Citrus Red Mite, European Red Mite, Lewis Spider Mite, Pacific Spider Mite, Southern Red Mite, Spruce Spider Mite, Two-spotted Spider Mite
Benzoyl Urea Insect Growth Regulators	15	Diflubenzuron	Adept®	12	OHP, Inc.	Drench application: Fungus gnats, Shore Flies Foliar application: Armyworms, Leafminers, Whiteflies
	15	Novaluron	Pedestal™	12	OHP, Inc.	Thrips: Western Flower Trips, Flower Thrips, Gladiolus Thrips, Citrus Thrips, Whiteflies: Green Whitefly, Silverleaf Whitefly, Sweetpotato Whitefly, Armyworm: Beet Armyworm, Lawn Armyworm, Yellow Striped Armyworm, Fall Armyworm, Southern Armyworm, Leafminers (Suppression): Serpentine Leafminer, Citrus Leafminer
	16	Buprofezin	Talus®	12	SePRO Corp.	Leafhoppers, Mealybugs, Scales, Whiteflies. *Also labeled for greenhouse tomatoes*
Biopesticides	11	*Bacillus thuringiensis* Kurstaki	DiPel® Pro DF *For organic production*	12	Valent USA Corp.	Flowers, Bedding Plants, and Ornamentals: Looper, *Heliothis*, Tobacco Budworm, Omnivorous Looper, Omnivorous Leafroller, Diamondback Moth, Armyworm, Ello Moth, Lomoth, Oleander Moth, Azalea Caterpillar Herbs, Spices and Mints: Looper, Saltmarsh Caterpillar, Armyworm
	11		Deliver® *For organic production*	12	Certis USA, LLC	Leafrollers, Codling Moth, Cutworms, Imported Cabbageworm, Diamondback Moth, Beet Armyworm, Green Cloverworm, Velvetbean Caterpillar, Fruitworm, Hornworms, Tobacco Budworm, Cotton Bollworm, Gypsy Moth, Grapeleaf Skeletonizer, European Corn Borer
	11	*Bacillus thuringiensis* Israelensis	Gnatrol®	12	Valent USA Corp.	Fungus Gnats
	M	*Beauveria bassiana*	Naturalis L® *For organic production*	4	Troy Biosciences	Ants, Aphids, Armyworms, Bermuda Grass Mites, Billbugs, Budworm and Bollworm, Cutworms, Chinch Bugs, Citrus Blackfly, Colorado Potato Beetle, Corn Borers, Leafhoppers, Beetles- Flea Cucumber, Elm Leaf, Fungus Gnats, Grasshoppers, Japanese Beetles, Elm Leaf Leafhoppers, Leaf-feeding caterpillars, Leafrollers, Loopers, Lygus Bug, Mealybugs, Millipedes, Mole Crickets, Tetranychid Mites, Pear Psylla, Root Weevils, Psyllids, Sowbugs, Spittlebugs, Sod Webworms, Shore flies, Tarnished Plant Bug, Thrips, Tomato Fruitworm, Weevils and Whiteflies, Eggs of Lepidopteran Pests, Larvae of the following pests: White bugs, Annual White Grub, Black Turfgrass Ataenius,

Class	MOA*	Common Name	Trade Name	REI	Manufacturer	Pests Controlled (Lists from Current Manufacturer Labels)
						European Chafer, Green June Beetle, Japanese Beetle, Northern Masked Chafer, European Cranefly Larvae
	M		BotaniGard®	4	BioWorks, Inc.	Adult and larval stages of Whitefly, Aphids, Thrips, Psyllids, Weevils, and Mealybugs
Biopesticide Insect Growth Regulators	18	Azadirachtin	Azatin® XL	4	OHP, Inc.	Aphids, Armyworms, Beetles, Cabbage Looper, Caterpillars, Cutworms, Fungus Gnats (as a drench), Moths, Leafhopper, Leafminers, Leafrollers, Shore-Flies, Thrips, and Whiteflies.
			Ornazin®	12	SePRO Corp.	Aphids, Beetles, Caterpillars, Fungus Gnats, Shore flies, Lacewings, Leafhoppers, Leafminers, Mealybugs, Nematodes, Psylids, Sawflies, Soft scales, Thrips, Weevils, Whiteflies
	7	s-Kinporene	Enstar®	4	Wellmark International	Whiteflies, Thrips, Aphids, Soft-bodied and Armored Scales, Fungus Gnat, Root (ground) Mealybug
Boric Acid	M	Sodium Tetraborohydrate Decahydrate	TriCon™	12	BioWorks, Inc.	Aphids, Mealybugs, Mites, Soft Scales, Whiteflies
Botanicals	3	Pyrethrins	Pyrethrum® TR	12	Whitmire Micro-Gen	Aphids, Fungus Gnats, Mites, and Whiteflies
	3		Pyreth-It™	12	Whitmire Micro-Gen	Aphids, Armyworms, Asparagus Beetle, Blister Beetles, Cabbage Looper, Caterpillars, 12-spotted Cucumber Beetle, Colorado Potato Beetles, Corn Earworm, Crickets, Cross-striped Cabbageworm, Cucumber Beetles, Deer Fly, Diamondback Larvae, Fireworms, Flea Beetles, Fruit Flies, Fruittree Leafroller, Grape Leafhopper, Green Peach Aphids, Greenhouse Thrips, Gypsy Moth (adults & larvae), Harlequin Bug, *Heliothis* sp., Imported Cabbageworm, Leafhopper, Leafrollers, Leaftiers, Mexican Bean Beetle, Potato Leafhopper, Psyllids, Silverfish, Skippers, Stink Bugs, Tabanidae, Thrips, Webworms, Whiteflies
	4B	Nicotine	Fulex Nicotine Fumigator	*	Fuller System, Inc.	Aphids, Whiteflies, Mealybugs, Spider Mites, and Thrips
			*Depends on greenhouse ventilation			
Carbamates	1B	Carbaryl	Sevin®	12	Bayer Environmental Science	Aphids, Beetles, Borers, Caterpillars, Gall Insects, Grasshoppers, Leafhoppers, Leafminers, Mites, Scale insects, Weevils
	1B	Methiocarb	Mesurol®	24	Gowan Company	Aphids, Mites, Western Flower Thrips, Snails, Slugs
Carbamate Insect Growth Regulators	7	Fenoxycarb	Preclude® TR	12	Whitmire Micro-Gen	Thrips, Whiteflies, Scale, and Aphids
Carbazates	25	Bifenazate	Floramite®	4	OHP, Inc.	Variety of species of mites and all life stages of Tetranychid Spider Mites
Cyclodiene organochlorines	2	Endosulfan	Fulex Thiodan insecticidal Smoke	*	Fuller System, Inc.	Aphids and Whiteflies (resistant adults)
			*Depends on greenhouse ventilation			

(continues)

Table 19-3 *(Continued)*

Class	MOA*	Common Name	Trade Name	REI	Manufacturer	Pests Controlled (Lists from Current Manufacturer Labels)
Glycosides	6	Abamectin	Avid®	12	Syngenta	Leafminers, Mites, Aphids (suppression), Whiteflies, and Thrips
	6	Milbemectin	Ultiflora™	12	Gowan Company	Spider Mites (two-spotted and carmine), Red Mites (citrus and European)
Napthoquinone derivatives	20	Acequinocyl	Shuttle™	12	Arysta Life Sciences	Two-spotted Spider Mite and Spruce Spider Mite
Neonicotinoids	4A	Acetamiprid	TriStar®	12	Cleary Chemical Corp.	Adelgids, Aphid, Beetles, Caterpillars, Crane Fly Larvae, Fungus Gnat Larvae, Leafhoppers, Mealybugs, Plant Bugs, Psyllids, Sawfly, Scale (hard & soft), Swede Midge, Tentiform Leafminer, Thrips, Weevils, Whiteflies
	4A	Clothianidin	Clero™	12	Arysta Life Sciences	Aphids, Whiteflies, and Mealybugs
	4A	Dinotefuran	Safari™	12	Valent USA Corp.	Foliar spray: Aphids (suppression), Asian Cycad Scale, Florida Wax Scale, Japanese Beetles (adults), Leafminers (serpentine), Mealybugs, Scale (armored and soft), Tea Scale, Thrips (suppression), Whiteflies: Giant Whitefly, Greenhouse Whitefly, Silverleaf Whitefly Drench: Aphids, Asian Cycad Scale, Florida Wax Scale, Fungus Gnats, Leafminers, Mealybugs, Scale (armored and soft), Tea Scale, Thrips, Whiteflies: Giant Whitefly, Greenhouse Whitefly, Silverleaf Whitefly
	4A	Imidacloprid	Marathon®	12	OHP, Inc.	Aldegids, Aphids, Thrips, Whiteflies, Armored Scales, Elm Leaf Beetles, Japanese Beetles, Lacebugs, Mealybugs, Psyllids, Root Mealybugs, Root Weevil, Soft Scale, White Grub Larvae
	4A	Thiamethoxam	Flagship™	12	Syngenta	Aphids, Lacebugs, Whiteflies, Mealybugs, Soft Scales, Sawflies, Leafhoppers
MOils	M	Clarified hydrophobic extract of neem oil	Triact® 70	4	OHP, Inc.	A broad-spectrum of pest including Aphids, Whitefly, Thrips, and Mites
	M	Paraffinic oil	Ultra-Fine Spray Oil	4	Whitmire Micro-Gen	Adelgids, Aphids, Caterpillars, Lacebugs, Leafhoppers, Leafminers (larvae), Mealybugs, Mites, Plant Bugs, Psyllids, Sawfly larvae, Scale, Thrips, Whiteflies, and eggs of aphids, mites, and certain caterpillars
Organophosphates	1A	Acephate	Orthene® TT&O	24	Valent USA Corp.	Aphids, Leafrollers, Mealybugs, Scale (crawlers), Thrips, Whiteflies (greenhouse and silverleaf)
			Orthene® TR	24	Whitmire Micro-Gen	Aphids, Cabbage Loopers, Cankerworms, Fall Webworms, Fungus Gnats, Lacebugs, Leafminers, Leafrollers, Lygus Bugs, Mealybugs, Scale, Spider Mites, Tent Caterpillars, Thrips, and Whiteflies

Class	MOA*	Common Name	Trade Name	REI	Manufacturer	Pests Controlled (Lists from Current Manufacturer Labels)
	1A	Chlorpyrofos	DuraGuard™	24	Whitmire Micro-Gen	Aphids, Beetles, Caterpillars (Cabbage Worms, Rose Slugs, Hornworms), Cockroaches, Colorado Potato Beetles, Crickets, Cucumber Beetles, Flea Beetles, Firebrats, Flies, Fungus Gnats, Fungus Gnat Larvae, Gnats, Japanese Beetles, Leafhoppers, Leafminer, Mealybugs, Root Mealybugs, Root Weevils, Rose Aphids, Rosechafer, Scale, Scorpions, Shore Flies, Spider Mites, Stink Bugs, Tarnished Plant Bugs, Thrips and Watermelon Beetles
	1A	Dichlorvos	Fulex DDVP	*	Fuller System, Inc.	Spider Mites, Aphids, Whitefly, and Mealybugs
			*Depends on greenhouse ventilation			
	1A	Dimethoate	Cygon 2E	48	Value Garden Supply	Aphids, Thrips, Mites, Bactra Moth Larvae, Leafhoppers, Leafminers, Mealybugs, Scale
	1A	Malathion	Gowan Malathion 8F	12	Gowan Company	Aphids, Cucumber Beetle, Japanese Beetle, Potato Leafhopper, Mexican Bean Beetle, Nitidulid Beetle, Cutworm, Spider Mites, Thrips
Organotins	12	Fenbutatin-oxide	ProMite™	48	SePRO Corp.	Two-spotted Spider Mites, *Oligonychus* sp., Clover Mites, Southern Red Mites
Phenoxypyrazoles	21	Fenpyroximate	Akari™	12	SePRO, Corp.	Spider Mites
Pyrethroids	3	Bifenthrin	Talstar® F	12	FMC Corp.	Aphids, Bagworms, Brown Soft Scales, Broad Mites, Budworms, California Red Scale (crawlers), Cutworms, Fungus Gnats, Leaf Feeding Caterpillars, Whiteflies, Spider Mites, European Red Mite, Thrips, Leafminers
			Attain® TR	12	Whitmire Micro-Gen	Aphids, Armyworms, Bagworms, Caterpillars, Fungus Gnats, Mealybugs, Omnivorous Leafrollers, Plant Bugs, Scale, Thrips, Mites, and Whiteflies
	3	Cyfluthrin	Decathlon®	12	OHP, Inc.	Aphids, Armyworms, Bagworms, Beetles, Caterpillar, Cutworm, Fungus Gnats, Gypsy Moths, Lacebugs, Leafhopper, Leafroller, Mealybugs, Mites, Scales, Thrips, Webworm, and Whiteflies.
	3	Fenpropathrin	Tame®	24	Valent USA Corp.	Aphids, Beet Armyworm, Black Vine Weevil (adults), Japanese Beetle, Lacebugs, Leafhoppers, Mealybugs, Spider Mites: European Red Mite, Southern Red Mite, Two-spotted Mite, Spotted Tentiform Leafminer, Whiteflies: Greenhouse and Silverleaf
	3	Fluvalinate	Mavrik®	12	Wellmark International	Aphids, Thrips, Mites, Clover Mites, Two-spotted Spider Mites, Whiteflies, Mealybugs, Leaf-Feeding Caterpillars (i.e., Tent, Orlando, Gypsy Moth, Bagworm, Canker Worm, etc.), Ticks, Brown Dog Ticks, American Dog Ticks, Deer Ticks, Fleas, Ants, Fire Ants, Crickets, Armyworms, Cutworms, Springtails (Collembola), Adelgids,

(continues)

Table 19-3 *(Continued)*

Class	MOA*	Common Name	Trade Name	REI	Manufacturer	Pests Controlled (Lists from Current Manufacturer Labels)
						Mosquitoes, Chiggers, Earwigs, Lygus Plant Bugs, Beetles (Flea, Cucumber, Elm Leaf) Leafhoppers, Weevils (Strawberry Root, Obscure Root, Blackvine Root), Psyllids, Millipedes, American Cockroaches, Smokey Brown Cockroaches, Oriental Cockroaches, Boxelder Bugs, Spiders, Silverfish, Firebrats, Chinch Bugs, Asian Lady Beetles, Sowbugs, Glassy-Wing Sharpshooters
	3	Lamda-Cyhalothrin	Scimitar® GC	24	Syngenta	Armyworms, Azalea Caterpillars, Aphids, Bagworms, Black Vine weevils (adult), Budworms, Cutworms, European Sawflies, Gypsy Moth Larvae, Japanese beetles (adults), Lacebugs, Leaf-feeding Caterpillars, Leafhoppers, Leafminers (adults), Leafrollers, Sawflies, Scale insects (crawlers), Thrips, Broadmites and Spider Mites, Brown Softscale, California Redscale (crawler), Clover Mites, Mealybugs, Whiteflies
	3	Permethrin	Astro®	12	FMC Corp.	Aphids, Bagworm, Beet Armyworm, Loopers, Thrips, Fungus Gnat, Leaf-feeding Caterpillars, Leafminers, Leafhoppers, Mealybugs, Whiteflies.
	3		Perm-Up**	12	United Phosphorus	Beet Armyworm, Cabbage Looper, Omnivorous Leafroller
						**Greenhouse roses only*
	3		Permethrin 3.2 EC**	12	Helena Chemical Co.	Beet Armyworm, Cabbage Looper, Omnivorous Leafroller
						**Greenhouse roses only*
Pyridazinones	21	Pyridaben	Sanmite®	12	Gowan Company	Broad Mite, European Red Mite, Southern Red Mite, Tumid Mite, Two-spotted Mite, Whiteflies
Pyridine azomethines	9	Pymetrozine	Endeavor™	12	Syngenta	Whiteflies and Aphids
Pyridine carboxamides	9	Flonicamid	Aria™	12	FMC Corp.	Aphids, Thrips, Whiteflies, Mealybugs
Pyridine Insect Growth Regulators	7	Pyripoxyfen	Distance®	12	Valent USA Corp.	Aphids (suppression), Fungus Gnats, Mealybugs Scale: Black Scale, Euonymus Scale, Florida Wax Scale, San Jose Scale and Snow Scale, Shore Flies, Whiteflies: Greenhouse Whitefly, Silverleaf Whitefly
Pyrroles	13	Chlorfenapyr	Pylon®	12	OHP, Inc.	Caterpillar, Foliar Nematodes and Fungus Gnats
Soaps	M	Potassium salts of fatty acids	Insecticidal Soap 40%	12	Value Garden Supply	Leafminers, Aphids, Whiteflies, Thrips, Plant Bugs, Spider Mites, Russet Mite, Leafhoppers
Spinosyns	5	Spinosad	Conserve®	4	Dow AgroSciences LLC	Thrips, Lepidopterous Larvae, Foliar Feeding Worms, Shore flies, Spider Mites

Class	MOA*	Common Name	Trade Name	REI	Manufacturer	Pests Controlled (lists from current manufacturer labels)
Spinosyns	5	Spinosad	Conserve®	4	Dow AgroSciences LLC	Thrips, Lepidopterous Larvae, Foliar Feeding Worms, Shore flies, Spider Mites
Tetrazines	10A	Clofentezine	Ovation™	12	Scotts Company	Two-spotted Spider Mite, European Red Mite, Pacific Spider Mite, Yellow Spider Mite, McDaniel Spider Mite
Tetronic acids	23	Spiromesifen	Judo™	12	OHP, Inc.	Mites and Whiteflies
Thiazolidinones	10A	Hexythizox	Hexygon® DF	12	Gowan Company	Spider Mites (ovacide/miticde) *Not for use on orchids*
Triazine Insect Growth Regulators	17	Cyromazine	Citation®	12	Syngenta	Dipterous Leafminers, Fungus Gnats, and Shore Flies
Combination Products						
Organophosphate + Pyrethroid	1+3	Acephate + Fenpropathrin	Tame/Orthene™ TR	24	Whitmire Micro-Gen	Aphids, Cabbage Loopers, Cankerworms, Fall Webworms, Fungus Gnats, Lacebugs, Leafminers, Leafrollers, Lygus Bugs, Mealybugs, Scale, Spider Mites, Tent Caterpillars, Thrips, and Whiteflies
	1+3	Chlorpyrifos + Cyfluthrin	DuraPlex® TR	24	Whitmire Micro-Gen	Aphids, Caterpillars, Fungus Gnats, Mealybugs, Mites, Scale, Shore Flies, Thrips, and Whiteflies
Pyrethroid + Neonicotinoid	3+4A	Cyfluthrin + Imadacloprid	Discus®	12	OHP, Inc.	Aphids, Leafhoppers, Leafminers, Mealybugs, Scale, Strawberry Root Weevil, Whiteflies, White Grub Larvae, and many other insects

*Insecticides / Miticides Modes of Action

1. Acetyl cholinesterase inhibitors. Inhibition of the enzyme acetylcholinesterase, interrupting the transmission of nerve impulses
2. GABA-gated chloride channel antagonists: Interferes with GABA receptors of insect neurons, leading to repetitive nervous discharges
3. Sodium channel modulators: Acts as an axonic poison by interfering with the sodium channels of both the peripheral and central nervous system stimulating repetitive nervous discharges, leading to paralysis
4. Acetylcholine receptor agonists/antagonists: Binds to nicotinic acetylcholine receptor disrupting nerve transmission
5. Nicotine acetylcholine receptor agonists (not group 4)
6. Chloride Channel Activators: Interferes with the GABA nerve receptor of insects
7. Juvenile hormone mimics (Insect growth regulator): Mimic juvenile hormones, which prevent molting from the larval to the adult stage
9. Compounds of unknown or nonspecific mode of action (selective feeding blockers)
10. Compounds of unknown or nonspecific target site of action (mite growth inhibitors)
11. Microbial disruptors of insect midgut membranes
12. Inhibition of oxidative phosphorylation at the site of dinitrophenol uncoupling (Disrupt ATP formation)
13. Uncoupler of oxidative phosphorylation (disrupt H proton gradient formation)
15. Inhibit chitin biosynthesis – type 0, Lepidopteran
16. Inhibit chitin biosynthesis – type 1, Homopteran
17. Molting disruptor, Dipteran
18. Ecdysone agonists/molting disruptors
20. Coupling site II electron transport inhibitors (Complex III)
21. Site I electron transport inhibitor
23. Inhibitors of lipid biosynthesis
25. Neuronal inhibitors (unknown mode of action)
 M Miscellaneous

This list was developed by the Insecticide Resistance Action Committee (IRAC) in cooperation with the Environmental Protection Agency.

Table 19-4

Steps for Safe Use of Pesticides
1. Read all labels before purchasing or using any pesticide.
2. Choose the correct pesticide for your pest and location.
3. Learn and follow all state and federal regulations.
4. Train all employees that mix, apply, handle, or work around pesticides.
5. Transport pesticides safely.
6. Wear appropriate personal protection equipment according to the label.
7. Mix pesticides properly and under safe conditions.
8. Use appropriate pesticide application equipment according to the label instructions.
9. Apply pesticides safely and according to the label instructions.
10. Plan for accidents and conduct accident practice drills.
11. Clean any pesticide spills immediately. Dispose of waste product appropriately.
12. Dispose of pesticide containers properly.
13. Keep good records and maintain employee access to those records.

The skull and crossbones is the universal symbol for a poison.

FIGURE 19-7 The skull and crossbones is the universal symbol for a poison.

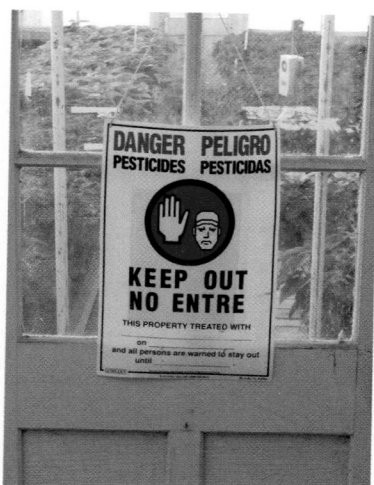

FIGURE 19-8 Signs must be posted on greenhouse entrances indicating when it is safe to enter the greenhouse.

Always remember that pesticides are toxic. Wise use of pesticides includes avoiding exposure to people and nontarget organisms. This can be done by wisely choosing when, how, and what pesticides are applied. Always consider the least hazardous pesticide for the situation (Table 19-4).

Adverse effects of a pesticide range from slight to extreme. All pesticides sold in the United States must include a product label with a signal word indicating the potential hazard of immediate or acute injury. The signal words are CAUTION, WARNING, and DANGER (Table 19-5). Pesticides that require the signal word DANGER are the most hazardous and require a skull and crossbones symbol (Figure 19-7). These products are typically considered to be restricted use and are available only to certified applicators who have completed special training.

Long-term or chronic health effects are not required on a pesticide label. This information is included on a Materials Safety Data Sheet (MSDS). The MSDS details any potential hazards that are known. It is important to study the MSDS as well as the pesticide label.

Employers are required to ensure that all workers who apply, handle, or mix pesticides have been properly trained. Federal Worker Protection Standards require that all employees working in fields, greenhouses, and nurseries that have been treated with a pesticide receive pesticide safety training, regardless if they work directly with pesticides. Restricted entry intervals (REI) are the required number of hours or days after application during which entry into treated areas is prohibited (Figure 19-8). The purpose of the REI regulations is to minimize employee exposure to pesticides. This information is available on the pesticide label.

Pesticide Phytotoxicity

Many pesticides have the potential to be phytotoxic. Certain active ingredients, formulations, and application methods have higher risks of phytotoxicity than

Table 19-5

Pesticide Signal Words and Toxicity Categories			
Signal Word	Toxicity Category	Study	Common Label Statements
Danger	I	Acute oral	Fatal if swallowed. Wash thoroughly with soap and water after handling and before eating, drinking, chewing gum, or using tobacco.
		Acute dermal	Fatal if absorbed through skin. Do not get in eyes, on skin, or on clothing. Wash thoroughly with soap and water after handling and before eating, drinking, chewing gum, or using tobacco. Wear (appropriate protective clothing listed here). Remove and wash contaminated clothing before reuse.
		Acute inhalation (4-hour)	Fatal if inhaled. Do not breathe (dust, vapor, or spray mist listed here). Wear (appropriate respiratory protection listed here). Remove and wash contaminated clothing before reuse.
		Primary eye irritation	Corrosive. Causes irreversible eye damage. Do not get in eyes or on clothing. Wear (appropriate protective eyewear such as goggles, face shield, or safety glasses listed here). Wash thoroughly with soap and water after handling and before eating, drinking, chewing gum, or using tobacco. Remove and wash contaminated clothing before reuse.
		Primary skin irritation	Corrosive. Causes skin burns. Do not get in eyes, on skin, or on clothing. Wear (appropriate protective clothing and gloves listed here). Wash thoroughly with soap and water after handling and before eating, drinking, chewing gum, or using tobacco. Remove and wash contaminated clothing before reuse.
Warning	II	Acute oral	May be fatal if swallowed. Wash thoroughly with soap and water after handling and before eating, drinking, chewing gum, or using tobacco.
		Acute dermal	May be fatal if absorbed through skin. Do not get in eyes, on skin, or on clothing. Wash thoroughly with soap and water after handling and before eating, drinking, chewing gum, or using tobacco. Wear (appropriate protective clothing listed here). Remove and wash contaminated clothing before reuse.
		Acute inhalation (4-hour)	May be fatal if inhaled. Do not breathe (dust, vapor, or spray mist listed here). Wear (appropriate respiratory protection listed here). Remove and wash contaminated clothing before reuse.
		Primary eye irritation	Causes substantial but temporary eye injury. Do not get in eyes or on clothing. Wear (appropriate protective eyewear such as goggles, face shield, or safety glasses listed here). Wash thoroughly with soap and water after handling and before eating, drinking, chewing gum, or using tobacco. Remove and wash contaminated clothing before reuse.
		Primary skin irritation	Causes skin irritation. Do not get on skin or on clothing. Wear (appropriate protective clothing and gloves listed here). Wash thoroughly with soap and water after handling and before eating, drinking, chewing gum, or using tobacco. Remove and wash contaminated clothing before reuse.
Caution	III	Acute oral	Harmful if swallowed. Wash thoroughly with soap and water after handling and before eating, drinking, chewing gum, or using tobacco.
		Acute dermal	Harmful if absorbed through skin. Avoid contact with skin, eyes, or clothing. Wash thoroughly with soap and water after handling and before eating, drinking, chewing gum, or using tobacco. Wear (appropriate protective clothing listed here). Remove and wash contaminated clothing before reuse.
		Acute inhalation (4-hour)	Harmful if inhaled. Avoid breathing (dust, vapor, or spray mist listed here). Remove and wash contaminated clothing before reuse.
		Primary eye irritation	Causes moderate eye irritation. Avoid contact with eyes or clothing. Wear (specify protective eyewear, if appropriate, here). Wash thoroughly with soap and water after handling and before eating, drinking, chewing gum, or using tobacco.
		Primary skin irritation	Avoid contact with skin or clothing. Wash thoroughly with soap and water after handling and before eating, drinking, chewing gum, or using tobacco. Wear (appropriate protective clothing and gloves listed here).

others. These risks can be compounded through overapplication, incorrect mixing, careless application, drift, failure to follow label recommendations, or using a sprayer previously used for herbicides.

Phytotoxicity can result from improper tank mixing of more than one pesticide product. For instance, many emulsifiable concentrates (often labeled as EC) have petroleum distillates to keep the active ingredient in suspension or dissolved. When mixing two or more products that are both considered emulsifiable concentrates, the petroleum distillate concentrate is doubled, thus putting the plants at risk for damage.

Pesticide Resistance

Pesticide resistance is a growing concern among pesticide manufacturers and pesticide applicators. Many insects and pathogens are developing resistance to certain pesticides. The use of broad-spectrum toxic pesticides in the past was not as likely to allow pesticide resistance, but modern, softer pesticides in use today that are often targeted to a limited number of pests allow the development of pesticide resistance. For this reason it is important for the applicator to rotate pesticide families and pesticide modes of action for maximum management of pest populations.

A common pesticide rotation practice involves the application of a pesticide product two or three times followed by a switch in pesticide family or pesticide mode of action. The mixing of two or more pesticide families is controversial and may result in the development of multiple pesticide resistance. This practice is not recommended.

Pesticide Application

Proper use of pesticides includes proper application or delivery of the pesticide to its intended target. To accomplish this, the characteristics of the pest must be known as well as the characteristics of the pesticide. Prior identification of the pest, location of the pest, and the most susceptible stage of the pest to the pesticide is critical for proper control. Many pesticides are systemic, whereas others are contact only. Systemic pesticides are most effective on insect pests that are phloem feeders, such as aphids and whiteflies, while contact pesticides are best on insects that feed on foliage, such as caterpillars.

Methods of pesticide application include sprays (high-volume, low-volume, and aerosol), thermal fog, smoke, and root zone. Which application technique is chosen is based upon label requirements or recommendations, existing equipment, weather conditions, the type or stage of development of the crop or pest, susceptibility of the crop to phytotoxicity, and the economics of the method in question.

High-volume spray application is the most common method of applying pesticides in a greenhouse (Figure 19-9). The equipment available is inexpensive and most pesticides are formulated for this form of application. Spray droplets are large, 100 or more microns in diameter, and spray volumes are typically applied at 50 to 100 gallons per acre. High-volume sprayers range in size from 1- to 4-gallon hand units up to powered sprayers from 10 to 200 gallons.

FIGURE 19-9 High-volume sprayer.

FIGURE 19-10 Low-volume electrostatic spray system.

Larger sprayers are mounted on wheels and can be easily maneuvered through the main aisles of a greenhouse.

Low-volume sprayers use high pressure to reduce the droplet size of the pesticide formulation (Figure 19-10). Low-volume sprays are more effective in reaching the internal foliage as it is carried around the outer leaves via turbulence of the air stream. Low-volume applications include the use of less pesticide at a more concentrated rate. This can reduce application time while increasing application uniformity. However, care must be taken in that low-volume pesticide application formulation is not as visible as with high-volume sprays and there is a risk of overapplication.

There are several forms of low-volume pesticide application equipment. Some use a spinning disk upon which the pesticide flows under low pressure. The pesticide is then moved across the spinning disk via centrifugal force and sheared off with a fan. Other low-volume application equipment uses pressures from 1,000 to 3,000 psi. These pressures shear the droplet size down to 30 microns. A modification of the high-pressure applicator is an electrostatic sprayer. Electrostatic sprayers use air assist to gain the pressure and the sheared droplet size while passing the droplet through an electric current. This electric current charges the droplet, increasing its tendency to attach to the foliage or target pests while avoiding other droplets with similar charges.

Aerosol application is a form of low-volume application where the pesticide is purchased in an aerosol can or cylinder under pressure. Aerosol applicators are available in two primary forms, total release and conventional applicators. The active ingredient or pesticide in an aerosol device is premixed with a propellant under pressure. This eliminates much of the risk to the applicator in that the product is premixed. When the applicator is actuated, the pesticide and the propellant expand rapidly as a gas moving at a high velocity. The pesticide is delivered in small droplets, which drift throughout the greenhouse and finally settle on the plants and target pests.

Total release aerosols can be set into a greenhouse, actuated, and allowed to fill the greenhouse. Critical calculations of greenhouse volume are important in order to ensure proper application rates. Greenhouse vents must also be closed and fans turned off; therefore, timing is also critical so that application occurs during the cool of the day. Conventional aerosols are convenient for spot spray treatments of pest outbreaks. Often a small pest outbreak can be efficiently and economically controlled with an aerosol application of a pesticide more conveniently than through loading and mixing a tank of pesticide spray.

Thermal fog application is another form of low-volume application where the pesticide is diluted into a specialized carrier and heated by a device that delivers the aerosol in the form of a white fog. Foggers typically burn gasoline or propane that heats a pipe upon which the pesticide formulation is injected. These units are carried on a backpack or pulled through a greenhouse on a cart.

Smoke application of pesticides is one of the simplest and easiest techniques available to a greenhouse. Smokes are only available to a few pesticides, which can withstand intense heat. Smoke applicators are formulated with the active ingredient and a combustible product, which is ignited with a sparkler similar to those used to celebrate the Fourth of July.

Smoke application requires that the greenhouse be tightly sealed to contain the pesticide and that the volume of the greenhouse be carefully calculated to ensure proper application rates. Typically, smoke applications are less phytotoxic, but care should be taken to ensure that plants are well watered and the foliage is dry prior to application. Smoke applications are ideal for greenhouses with a great deal of foliage, such as tomato, roses, and carnations. Smoke applications ensure total contact of all surfaces in a greenhouse, which conventional sprays will not reach.

Root substrate application of pesticides is used to eradicate soil-borne pests, but is used to deliver systemic pesticides as well. Many pesticides are systemic in nature and can be taken up by the root systems and delivered throughout the plant. Root substrate application of systemic insecticides is an effective and convenient means to kill many phloem-feeding insects. For example, poinsettia growers often apply imidacloprid early in the season and enjoy a whitefly-free crop with only spot treatments toward the end of the season. Pesticides formulated for root substrate applications are typically granular in form for easy delivery. When moistened, the pesticide is released into the soil solution and taken up by the plant.

Insects, Mites, and Other Invertebrate Pests

Aphids

Aphids are small, soft-bodied insects that feed on plant sap through piercing and sucking mouthparts. The most common species in the greenhouse is the green peach aphid (*Myzus persicae*) (Figure 19-11). Aphids feed in groups on stems and leaves, usually closely associated near phloem tissue along leaf veins. Some aphids transmit certain viruses that cause diseases in flowering crops. Aphids that feed on young leaves and buds will cause chlorotic pinpoints and distorted foliage. They also excrete honeydew, a sticky sugary sap. Honeydew, if left unchecked, will develop into a black sooty mold rendering ornamental plants unsaleable.

Aphids reproduce quickly under warm conditions, often completing a generation in 6 to 7 days. Most aphids are female and reproduce without mating,

FIGURE 19-11 Green peach aphids with a winged form. *(Courtesy of Whitney Cranshaw, Colorado State University, Bugwood.org.)*

giving birth to live young. A single aphid can give birth to 60 to 100 nymphs over a period of 20 to 30 days. Aphids are typically wingless in the greenhouse, but upon overcrowding, winged females will appear for migration. Male and female aphids appear out of doors as winter approaches. It is at this time that they will mate and lay eggs for overwintering.

Aphid control includes physical, cultural, and biological methods. Aphids are typically found on new succulent tissue. Avoiding shearing or pinching that stimulates succulent growth during times of aphid outbreaks is advised. The insecticides that are effective on aphids include reduced-risk products such as azadirachtin, *Beauveria bassiana*, fenoxycarb, imidacloprid, kinoprene, and pyridaben (Table 19-3).

Fungus Gnats, Shore Flies, and Moth Flies

Fungus gnats, shore flies, and moth flies or drain flies are often confused in that they are similar in appearance and appear under wet, overwatered conditions. They are easily distinguished when at rest. A fungus gnat (*Bradysia* sp. and *Sciara* sp.) (Figure 19-12A) at rest holds its wings slightly spread when viewed from above. Shore flies (family Ephydridae, commonly *Scatella stagnalis*) (Figure 19-12B) are a bit more robust in appearance with bristle-like antennae, which are shorter than the head. Moth flies or drain flies (family Psychodidae) are grayish in color with many fine hairs.

Adult fungus gnats are commonly found on soil and other surfaces with standing water. They mostly feed on dead and decaying tissue, being primarily a nuisance. The larvae typically feed on soil fungi and decaying plant matter as well, but in high populations will attack roots and other soil-borne plant organs such as tubers and corms. Fungus gnat larvae can also damage roots, allowing secondary pathogen infections. Adult fungus gnats have been reported to spread fungal pathogens such as *Botrytis, Fusarium, Phoma, Pythium,* and *Verticillium* species on their feet as they fly from infected plants to healthy plants.

The easiest control strategy for fungus gnats is management through cultural control, sanitation, and exclusion. Overwatering is the number one cause of fungus gnat problems. Insecticides applied as a drench are the most effective for controlling fungus gnats. Insecticides known as insect growth regulators (IGRs) are very effective in controlling the larvae as well as *Bacillus thuringiensis*, a biological insecticide that is only toxic to fly larvae.

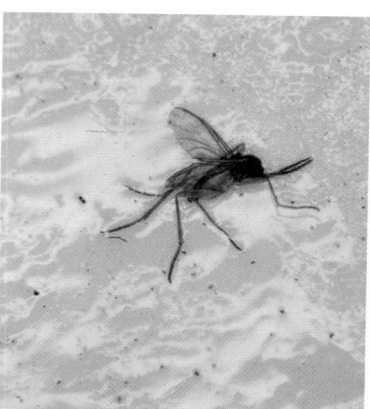

FIGURE 19-12A Adult fungus gnat on a yellow sticky card. *(Courtesy of Whitney Cranshaw, Colorado State University, Bugwood.org.)*

FIGURE 19-12B Shore fly infestation on a bedding plant flat. Note the fly specks. *(Courtesy of Whitney Cranshaw, Colorado State University, Bugwood.org.)*

Parasitic nematodes (*Steinernema feltiaei*) are effective in suppressing fungus gnat populations.

Shore flies typically feed on algae and not plant tissue. They do, however, appear under the same conditions as fungus gnats and are often confused. Adult shore flies are larger and more robust than fungus gnats. The adults are primarily a nuisance. Use the same cultural control strategies for shore flies as fungus gnats.

Moth flies also thrive under moist and overwatered conditions. Excess fertility creating algal scum in and around drains and bench supports tend to attract drain flies. Moth flies have been reported to feed on plants, but little documentation is available. Like shore flies, moth flies are primarily a nuisance and cultural management is the best control.

Leaf Miners

Moths and flies from several families have larvae that cause damage by tunneling through leaves (Figure 19-13). The serpentine leaf miner (*Liriomyzia trifolii*) and the chrysanthemum leaf miner (*Phytomyza atricornis*) are the two most prevalent species in the greenhouse. The adult female punctures the leaf using an ovipositor, inserting her eggs. The larvae hatch within the leaf and tunnel through the leaf for about 2 weeks. Afterwards, they drop from the leaf and pupate in the soil.

It is difficult to control leaf miner larvae as they are protected within the leaf. Systemic insecticides are not very effective in that translocation of most systemic insecticides is limited to vascular tissue and cells in close proximity. There is some evidence from research at the University of California at Davis on reducing leaf miner tunneling through using fertilizers with potassium silicate. Apparently, the silicon in leaf tissue disrupts tunneling.

The most effective control of leaf miners is to exclude the adults. Exclusion with insect screening works well. Adult leaf miners are also attracted to greenhouse lighting.

Mealybugs

The citrus mealybug (*Planococcus*) (Figure 19-14) is the most common mealybug found on greenhouse crops, but the longtailed mealybug (*Pseudococcus*

FIGURE 19-13 Leaf miner damage to columbine foliage. *(Courtesy of Whitney Cranshaw, Colorado State University, Bugwood.org.)*

FIGURE 19-14 Citrus mealybug. *(Courtesy of Whitney Cranshaw, Colorado State University, Bugwood.org.)*

longispinus) is found as well. Adult females are grayish, soft-bodied, wingless insects that are elongated, segmented, and coated with a whitish cottony wax. It is the wax coating that protects the mealybug and therefore makes contact insecticides less effective. Contact sprays are most effective at the first-instar nymphal stage. Surfactants can improve insecticide penetration through the waxy protection.

Thorough coverage with a contact spray is important for control of mealybugs at the crawler stage, especially on inner and under plant parts where they tend to congregate. Systemic insecticides, such as imidacloprid, are effective against mealybugs.

Mealybugs excrete honeydew as do aphids and whiteflies. Observing ant columns is one means to track down infestations. Management of mealybugs is simplest with early detection. Sanitation and inspection are the best defense. Plants with heavy infestations visualized as white cottony masses are best destroyed and eliminated from the greenhouse.

Mites

Mites are arachnids and not insects. Arachnids include tics and spiders. They do not have segmented bodies, wings, or antennae. Mites common to the greenhouse include the spider mites and red mites: carmine mite (*Tetranychus cinnabarinus*), privet mite (*Brevipalpus obovatus*), two-spotted mite (*Tetranychus urticae*), broad mite (*Polyphagotarsonemus latus*), cyclamen mite (*Phytonemus pallidus*), bulb scale mite (*Steneotarsonemus laticeps*), bulb mite (*Rhizoglyphus* spp.), and bud mites such as the carnation bud mite (*Aceria paradianthi*). Of these mites listed, the two-spotted mite is probably the most common; however, others are becoming more prevalent with crop diversity (Figure 19-15).

Foliage with mite damage appears stippled or flecked where the mites have been feeding. Severe infestations will stunt plant growth and cause leaves to drop. *Tetranychus* species will produce large amounts of fine webbing, hence the name spider mite. Yet with close monitoring, spider mite infestations should not reach this level.

Hot and dry greenhouse conditions favor mite development. Drought stressed plants provide an ideal condition for mite reproduction as the plants are less able to tolerate feeding damage. Keeping the greenhouse cooler and well

FIGURE 19-15 Two-spotted spider mite infestation. *(Courtesy of Whitney Cranshaw, Colorado State University, Bugwood.org.)*

watered will suppress mite populations in general. Spider mites require about 1 week at 80°F (26.7°C) to 3 weeks at 63°F (17.2°C) to complete one generation.

Inspection of new plant material coming into the greenhouse requires a 10× to 15× hand lens. This will prevent some outbreaks. Mites are easily spread by wind or on infested plants, equipment, and workers. They feed on many different plants, so weed control in and outside of the greenhouse is imperative.

Chemical control of mites includes the use of insecticides known as miticides or acaricides. Some broad-spectrum insecticides are effective as well. However, insecticides during hot and dry periods that might kill mite predators as well can often cause large mite outbreaks in only a few days.

Mites have the biological capacity to become resistant to many acaricides if used over a long period of time. Resistance can be controlled by rotating among three classes of acaricides switching about every two or three mite generations or every month.

Scale Insects

There are many genera of scale insects common to greenhouse crops, all from the order Homoptera. They feed on plant tissue by piercing plant tissue and sucking sap. They also inject toxins that stunt growth as well as spot and streak the foliage. Some species produce honeydew such as aphids, mealybugs, and whiteflies.

There are two groups of scale common to the greenhouse, armored and soft. Armored scale (family Diaspididae) has a platelike covering that protects the soft insect body underneath. Armored scales do not secrete honeydew. The surface of soft scales (family Coccidae) is smooth and brown. Soft scales are typically larger than armored scales.

Inspect all new plants coming into the greenhouse. The most effective time to apply a contact insecticide to control scale is when the crawlers are active. Yet diligent sanitation and exclusion is a more effective management technique. Plants heavily infested with scale are best discarded.

Insecticides that are registered for control of scale on greenhouse crops include buprofezin, imidacloprid, kinoprene, and pyridaben. Imidacloprid is especially effective in that it is systemically applied to the growing media. Insecticidal soaps and horticultural oils are also effective.

Slugs and Snails

Slugs and snails are not insects, but *mollusks*. Slugs are differentiated from snails in that they lack a shell. Slugs and snails feed at night, and during daylight they hide in cracks and crevices on the bench, beneath pots, or in plant litter on the ground. They prefer dark and moist hiding places. Slugs and snails are most active when the conditions are cool and moist.

Slugs and snails feed on seedlings and low growing herbaceous crops. Their chewing mouthparts have thousands of tiny teeth. They chew irregular holes with smooth edges. They are easily tracked during the day as they leave slime that dries into silvery trails.

Monitor new plants by inspecting not only the plant, but the underside of the containers as well. Inspect the underside of benches and other hiding areas. Keeping the greenhouse floor and bench areas dry and free of debris will eliminate or reduce snail and slug problems. Switching to trickle tube drip irrigation from sprinkler irrigation can reduce snail and slug populations where practical. Molluscicides must be combined with a bait in order to attract and kill snails and slugs. The two most common products are metaldehyde and methiocarb, and are formulated with bran bait.

Thrips

Thrips (order Thysanoptera) commonly attack many greenhouse crops. The western flower thrips (*Frankliniella occidentalis*) pose the most serious thrips problem in greenhouses (Figure 19-16). Thrips damage is often not apparent until sometime after feeding. They feed on flowers, buds, terminals, bulbs, and corms. Thrips hide deep in these tissues and the damage is not visible until after the tissue elongates.

Adult and larval stages of thrips feed on plant tissue by puncturing cells and feeding on the oozing sap. The tissue, as it grows, becomes silvery, stippled, blotched, streaked, or deformed. Western flower thrips also feed on pollen and some mites.

Western flower thrips are the vector for two economically important viral strains, impatiens necrotic spot virus (INSV) and tomato spotted wilt virus (TSWV). These viruses are the most significant result of western flower thrips

FIGURE 19-16 Western flower thrips. (*Courtesy Jack T. Reed, Mississippi State University, Bugwood.org.*)

in the greenhouse as they are incurable once the plant is infected and must be destroyed. Virus control relies on control of western flower thrips.

Both INSV and TSWV are the only plant-infecting members of the Bunyaviridae virus family, which includes some nasty mammalian viruses such as blue tongue and hantavirus. Thrips acquire the viruses only as larvae when they feed on infected plants. Then, once thrips mature into adults they can transmit the disease to healthy plants. The resulting damage is not reversible, thus early detection is crucial.

Thrips are difficult to manage. It takes only a small population to render a greenhouse crop unsaleable. Screening a greenhouse helps minimize populations to a great degree. Yet any new plants entering the greenhouse must be inspected. When inspecting plants, take time to carefully observe all plant parts where thrips may hide. One technique to find thrips is to tap a flower on a white note card. The thrips will drop onto the card and will be more easily visible.

Thrips are protected during much of their life cycle from insecticide sprays. Newly hatched larvae and young adults are the most susceptible to insecticide applications. To control high populations, use three applications 5 days apart and increase the interval thereafter. The use of ULV or fog can better penetrate protected plant parts and are more effective in thrips control. Frequent application over long periods of many insecticides can result in increased insecticide resistance.

Whiteflies

Whiteflies are in the order Homoptera and are not true flies, which is the order Diptera. The primary whiteflies that infest greenhouse plants are the greenhouse whitefly (*Trialeurodes vaporariorum*) (Figure 19-17A) and the silverleaf whitefly (*Bemisia argentifolii*) (Figure 19-17B).

Whitefly nymphs and adults feed on plants with piercing and sucking mouthparts and are associated with phloem tissue. Whitefly feeding can stunt plant growth and kill young plants. Whiteflies produce honeydew much like aphids, mealybugs, and scale.

Whiteflies typically lay eggs on the underside of leaves, from which crawlers emerge and wander before settling on the leaf surface to feed. Once they settle

FIGURE 19-17A Greenhouse whitefly. (*Courtesy of Whitney Cranshaw, Colorado State University, Bugwood.org.*)

FIGURE 19-17B Silverleaf whitefly. (*Courtesy Scott Bauer, USDA Agricultural Research Service, Bugwood.org.*)

to feed, whiteflies are virtually immobile during this period through the fourth instar. The semitransparent nymphs are flattened and oval. A single generation of whiteflies varies from one to several months depending on the host plants and the temperature. Whiteflies differ from most Homoptera in that their fourth instars are flattened and, as they develop, they swell a bit and species filaments develop or become visible. The final stage in the life cycle of a whitefly is the pupa. Whiteflies do not have complete metamorphosis; therefore, there is no true pupal stage.

The greenhouse whitefly has no markings on the wings and is larger than the silverleaf whitefly. It holds its wings flat and parallel to its yellowish body. The greenhouse whitefly pupae have long submarginal waxy filaments, and short marginal waxy fringe that is elevated with its edges perpendicular to the leaf surface.

The silverleaf whitefly has no markings, similar to the greenhouse whitefly, but it holds its wings tilted at about a 45-degree angle. The yellowish body of the silverleaf is visible between its wings. The convex or rounded pupae do not have a waxy fringe, but short filaments arise from the top.

Visually inspect all plants that come into the greenhouse for the immature stages of whiteflies. Escaping adults are easily spotted when opening up a box of cuttings or plugs that are infested. Monitoring whiteflies in the greenhouse includes yellow sticky cards, which serve as traps. However, trapping does not eliminate the need for regular inspection of foliage.

Sanitation, exclusion, and cultural practices are the most effective means for preventing infestations of whiteflies. Soft, succulent growth is susceptible to whitefly infestations; therefore, avoiding excess irrigation and fertility can reduce some infestations.

Eggs and pupae are not susceptible to most insecticides, whereas the early stage nymphs and adults are more susceptible to insecticides. Making it more difficult to use pesticides, whiteflies typically have overlapping life stages and have developed resistance to many broad-spectrum insecticides. Resistance can be managed by rotating insecticide mode of action every two or three generations.

ACHIEVEMENT REVIEW

Select the best answer or answers to complete each statement. List the appropriate letter(s).

1 Integrated pest management is a pest control strategy that includes:
   a. broad-spectrum chemical pesticide application to completely eradicate all pests.
   b. pest prevention techniques that include monitoring and management strategies.
   c. the establishment of appropriate action thresholds that trigger pest management action.
   d. pest exclusion rather than pest eradication.

2 Which of the following are appropriate data to include when developing a pest control data recording system?
   a. Weather and greenhouse environment
   b. Plants infested
   c. Pesticide applied and efficacy
   d. Cost of pesticide applied

3 The use of screening on greenhouse openings is known as:
   a. sanitation.
   b. exclusion.
   c. monitoring.
   d. eradication.

4 Herbicides are never used in the greenhouse.
   a. True
   b. False

5 Weed control in the greenhouse is best accomplished by:
   a. herbicides.
   b. burning with a propane torch.
   c. prevention by using weed barrier fabric.
   d. application of a quicklime slurry.

6 Algae control is:
   a. not an issue in the greenhouse due to high humidity.
   b. an important part of managing fungus gnats, shore flies, and drain flies.
   c. best controlled with a fungicide.
   d. an issue with worker safety.

7 Issues to consider when insect screening is used on a greenhouse include:
   a. exhaust cooling fan size.
   b. size of insect to be screened.
   c. screen surface area.
   d. worker safety.

8 Which of the following has the potential to lead to pest infestations?
a. Excessive wet and dry fluctuations
b. Conditions that favor high-quality crop production
c. Pasteurized growing media
d. Regular rotation of crops and not maintaining species uniformity

9 A greenhouse scout is:
a. always a professional consultant who works in the greenhouse independent of the management.
b. responsible for identifying potential problems, such as weeds, water leaks, crop debris, poor pest exclusion, and standing water that may lead to algae.
c. responsible for applying any pesticides.
d. responsible for determining action thresholds with management.

10 The removal of infested plants is an example of:
a. mechanical pest control.
b. chemical pest control.
c. environmental pest control.
d. biological pest control.

11 Predatory insects that attack or feed on an insect host is an example of:
a. mechanical pest control.
b. cultural pest control.
c. environmental pest control.
d. biological pest control.

12 Chemical control is best conducted immediately once a pest outbreak is detected to prevent any further spread.
a. True
b. False

13 Organic pesticides are always a safer choice when selecting a pesticide.
a. True
b. False

14 Pesticides that have long environmental persistence are often considered to be:
a. broad spectrum and beneficial in the greenhouse, killing a broader range of insects.
b. a potential risk to the greenhouse by promoting an increase in the pesticide resistance of the target organism.
c. biological control products.
d. organic.

15 Pesticides that are best suited for integrated pest management programs are:
a. broad spectrum.
b. organic.
c. pest specific.
d. effective on a broad range of pests.

16 Which of the following alert a pesticide applicator to the relative toxicity of a pesticide?
a. The restricted entry interval
b. The signal words
c. The application rate
d. The EPA registration number

17 When rotating pesticides to manage insect resistance, which of the following is important to consider?
a. Pesticide group or mode of action
b. The signal words
c. The chemical name of the pesticide
d. The EPA registration number

18 Which of the following is not a signal word illustrating the toxicity of a pesticide?
a. Caution
b. Warning
c. Danger
d. Restricted use

19 Which of the following application methods will best deliver a pesticide not only to pests on the plants, but also to the cracks and crevices of the greenhouse?
a. High-volume spray
b. Low-volume spray
c. Electrostatic spray
d. Smoke

20 Which of the following is best for spot spraying a small pest outbreak?
a. Aerosol
b. Fog
c. Smoke
d. Total release

21 Which of the following insects are considered to be phloem feeders?
a. Aphids
b. Whiteflies
c. Mealybugs
d. Fungus gnats

22 Which of the following are in reality nothing more than a nuisance in the greenhouse?
a. Aphids
b. Whiteflies
c. Mealybugs
d. Fungus gnats

23 Mealybugs are best controlled:
a. as crawlers.
b. as mature adults.
c. as eggs with an ovacide.
d. when they are stationary in a white cottony mass.

24 Spider mite development is favored by which of the following conditions?
a. Hot and dry greenhouse conditions
b. Hot and humid greenhouse conditions
c. Cool and damp greenhouse conditions
d. None of the above, as spider mite development is not influenced by greenhouse conditions

25 Western flower thrips are responsible for:
a. vectoring tospoviruses such as impatiens necrotic spot virus and tomato spotted wilt virus.
b. vectoring tobacco mosaic virus.
c. slivery, stippled, blotched, streaked, or deformed new growth of leaves and flowers.
d. parasitizing whiteflies.

Greenhouse Diseases

OBJECTIVES:

After studying this chapter, the student should have learned:

■ how to apply integrated pest management (IPM) techniques to disease control of greenhouse crops.

■ how the environment, plant, and pathogen interact to cause disease in the greenhouse.

■ what organisms cause plant disease in the greenhouse.

■ diseases common to greenhouse plants.

Diseases of greenhouse crops can be caused by pathogenic bacteria, fungi, viruses, and phytoplasmas. Many abiotic or noninfectious conditions mimic diseases or lead to plant disease. It is important to be able to distinguish between the two. If a disease is suspected and a pesticide is applied to remedy the situation caused by an abiotic stress, then undue expense, labor, and risk have occurred.

Chapter 19 focused primarily on insects and introduced integrated pest management (IPM), which is appropriate for this chapter as well. Planning an IPM program for disease control includes sanitation techniques, monitoring techniques, and management strategies. Pesticides should only be used when monitoring reveals that they are required. The use of pesticides as a first response to a disease outbreak is typically due to poor planning. Steve Dreistadt[1] of the University of California IPM Center has established that the basis of a successful IPM program should include five key components.

■ Problem prevention
■ Regular monitoring of crops and growing areas

---

[1]Dreistadt, S. H. *Integrated Pest Management for Floriculture and Nurseries*. Publication 3402 (Oakland: University of California, Agriculture and Natural Resources, Communication Services). 2001.

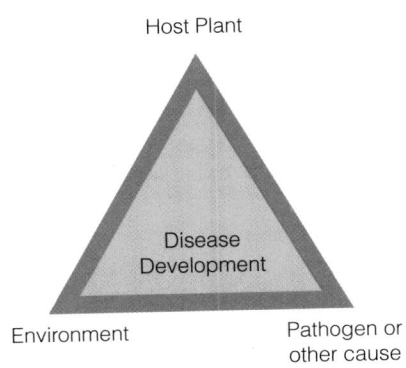

FIGURE 20-1 The disease triangle

■ Accurate diagnoses
■ Appropriate action thresholds
■ Effective management methods

The occurrence of plant disease is a result of a complex interaction of a susceptible host plant, the presence of a pathogenic causal organism, and the environment (Figure 20-1). For disease symptoms and damage to occur, the host plant's genetic characteristics, growth stage, and growth vigor must allow the plant to be susceptible to infection from the causal organism. If the environmental conditions are poor for pathogen development and plant vigor is high, many pathogens have little or no effect on their hosts. Yet, if the plants are stressed and abiotic factors such as temperature and humidity favor pathogen development, then plant disease can be expected.

Plant Disease

Despite the four basic forms of plant disease causal organisms that commonly infect greenhouse crops (fungi, bacteria, viruses, and phytoplasmas), many microorganisms are beneficial. Beneficial microorganisms control pests, build soil structure, and decompose organic material as well as eliminate toxins. Elimination of all microorganisms results in a lack of natural balance causing disruption and may encourage devastating outbreaks of disease-causing organisms. To determine if a microorganism is the causal agent of a disease, pathologists use four criteria, known as Koch's postulates.

Robert Koch was a German physician who first isolated anthrax (*Bacillus anthracis*), tuberculosis (*Tuberculosis bacillus*), and cholera (*Cholera vibri*). Koch and Louis Pasteur are considered to be the founders of modern bacteriology. Koch developed a series of four postulates that establish guidelines for establishing that a microorganism is the causal agent of disease. These postulates were based on Koch's work with anthrax and tuberculosis in 1884. Koch's postulates are as follows:

1. The microorganism must be found in the organism when the disease occurs.
2. The microorganism must be isolated from the diseased organism and grown in pure culture.
3. The cultured microorganism must cause disease when inoculated into a healthy organism.
4. The microorganism must be reisolated from the inoculated, diseased experimental organism and identified as being identical to the original specific causative agent.

Fungi

Fungi are difficult to describe because they are organisms found differing in many forms, behaviors, and life cycles. These organisms are characterized by a chitinous cell wall and filamentous growth called hyphae, which forms mycelium. Some fungi grow as single cells. Sexual and asexual reproduction is by way of spores. They are heterotrophic, lacking photosynthesis, which means that they exist either as a saprophyte or as a parasite. Saprophytes share a role in the decay of plant and animal remains into simpler forms that can be later absorbed by future generations of plants and animals. Fungi are grouped into four categories.

1. *Obligate saprophytes.* Live only on dead and decaying organic matter.
2. *Obligate parasites.* Attack and live only on living tissues.
3. *Facultative saprophytes.* Parasitic and attacks living tissue, but can survive on dead and decaying organic matter.
4. *Facultative parasites.* Normally saprophytic, but can exist as a parasite under the proper conditions.

Botrytis Blight

Botrytis blight or gray mold will infect most greenhouse crops and is one of the most important pathogens of aboveground plant parts typically caused by *Botrytis cinerea*. The symptoms include brown, water-soaked spots or decay on leaves or petals (Figure 20-2). A well-developed infection will include fuzzy gray, brown, or tan spores on rotted tissue. *B. cinerea* will not infect healthy tissue directly, but only tender young seedlings or petals, weakened tissue, and damaged or old plant parts. Once a plant is infected, the disease will spread. High humidity greater than 90 percent, splashing, condensation, and exudation all result in free water on foliage. Six or more hours in contact with water will promote *Botrytis* spore germination.

Proper management of *Botrytis* relies on good cultural practices and maintaining a lower relative humidity. Strict sanitation guidelines include removal of any infected plants, removal of dead or dying flowers, disbuds, prunings, and fallen leaves, and hand washing. Fungicides to control *Botrytis* are only used when greenhouse practices and conditions cannot be avoided (Table 20-1). Fungicides available for *Botrytis* control are preventative only and must be applied prior to infection.

Downy Mildew

Often confused with powdery mildew, downy mildew almost always is found in patches on the underside of foliage as soft and fluffy gray, purplish, or light brown sporulation. It is sometimes found on stems or buds. *Peronospora* and *Plasmopara* species are the primary causal fungal agents on flowering greenhouse crops.

In certain species, snapdragons for instance, downy mildews can become systemic when young plants are infected, resulting in stunted, malformed, yellowish growth (Figures 20-3A and 20-3B). Downy mildew infection is known to be carried in seeds or bulbs.

FIGURE 20-2 *Botrytis* infection on poinsettia cuttings under intermittent mist.

Table 20-1

Pesticides Labeled for Controlling Diseases Common to the Greenhouse						

(This information is provided for general information only. It is your responsibility when using a pesticide to read and follow the label that has been approved for the particular state or locality in which the pesticide is to be used, and to comply with all federal, state, and local laws and regulations relating to the use of pesticides.)

Class	MOA*	Common Name	Trade Name	REI	Manufacturer	Pests Controlled (Lists from Current Manufacturer Labels)
Acylamine	4	Mefenoxam	Subdue® MAXX	0	Syngenta	For control of damping-off, and root and stem rot diseases caused by *Pythium* and *Phytophthora*.
Aromatic Hydrocarbon	14	PCNB	Terraclor®	12	OHP, Inc.	Anthracnose, Botrytis, Black Spot, Leaf Spots, Powdery Mildew, Rhizoctonia, and many other diseases
Benzamides	7	Flutolanil	ProStar®	12	Bayer Environmental Science	*Rhizoctonia* and *Sclerotium, rolfsii*
Bicarbonate	NC	Potassium bicarbonate	Amicarb 100	4	Helena Chemical Co.	*Alternaria, Anthracnose, Botrytis*, Powdery Mildew *Monilinia, Septoria*
			Milstop®	1	BioWorks, Inc.	*Alternaria* Leaf Spot, *Anthracnose, Botrytis*, Downy Mildew, *Fusarium, Phoma, Phytophthora*, Powdery Mildew
Biopesticide	NC	*Trichoderma harzianum* T22	PlantShield® HC	0	BioWorks, Inc.	*Pythium, Fusarium,* and *Rhizoctonia*
		Trichoderma virens GL21	SoilGard®	4	Certis USA, LLC OHP, Inc.	Damping-off and root rots
		Bacillus subtilis GB03	Companion®	4	Growth Products	*Rhizoctonia, Anthracnose, Magnaporthe* Poae, *Pythium* Blight, and Microdochium Nivale
		Bacillus subtilis QST713	Rhapsody® AS	4	AgrQuest, Inc.	Suppresses diseases caused by *Rhizoctonia, Pythium, Fusarium,* and *Phytophthora*
		Streptomyces lydicus WYEC 108	Actinovate®	4	Natural Industries, Inc.	*Pythium, Phytophthora, Fusarium, Rhizoctonia, Verticillium,* and other root decay fungi
Boric Acid	NC	Sodium tetraborohydrate decahydrate	TriCon™	12	BioWorks, Inc.	Downy Mildew and Powdery Mildew
Botanical Extract	NC	*Macleaya* extract	Qwel®	12	Camas Technologies	Powdery Mildew
		Raynoutria sachalinesis	Milsana	24	Western Farm Service, Inc.	Powdery Mildew, some control of *Botrytis*
Carbamate	28	Propamocarb	Banol®	24	Bayer Environmental Science	*Pythium* sp. and *Phytopthora* sp.
Chloronitriles	M5	Chlorothalonil	Daconile® Ultrez®	12	Syngenta	*Anthracnose* sp., *Botrytis* sp., Leaf Spots, *Phytophthora*, Powdery Mildew
			AllPro® ExoTherm Termil	*	Value Garden Supply	Leaf and Flower Spots, *Phytophthora* Leaf Blight, Powdery Mildew, Rust
				* Depends on greenhouse ventilation		
Cinnamic Acid Amides	40	Dimethomorph	Stature®	12	BASF	Downy Mildew and *Phytophthora* Root, Crown, and Stem Diseases

(continues)

Table 20-1 *(Continued)*

Class	MOA*	Common Name	Trade Name	REI	Manufacturer	Pests Controlled (Lists from Current Manufacturer Labels)
Copper, Complex	M1	Copper Sulfate	Camelot®	12	SePRO Corp.	Bacterial Pathogens - Bacterial Blight, Bacterial Leaf Spot, *Erwinia*, Fire Blight, *Pseudomonas*, *Xanthomonas* Fungal Pathogens - *Alternaria*, *Anthracnose*, *Botrytis*, *Cercospora* Leaf Spot, *Diplodia* Tip Blight, *Dothistroma* Needle Blight, Needle Casts, Downy Mildew, Leaf Gall, Leaf Spot, *Phomopsis*, *Phytophthora* Dieback, Powdery Mildew, Rust, Scab, Twig Blight
			Phyton 27®	24	Phyton Corp.	*Anthracnose*, *Botrytis*, Powdery Mildew, *Erwinia*, *Pseudomonas*, *Xanthomonas*, *Cylindrocladium*
Copper, Fixed	M1	Copper hydroxide	CuPro™	24	SePRO Corp.	*Alternaria* Blight, *Anthracnose*, *Phomopsis*, Bacterial Leaf Spot, Gray Leaf Mold, *Septoria* Leaf Spot, *Xanthomonas*, *Cercospora*, *Pseudomonas*, *Erwinia*, *Alternaria*, *Cercospora* Leaf Blight, *Botrytis* Blight, *Phytophthora* Dieback, Powdery Mildew, *Pestalotia* Leaf Spot, *Entomosporium Maculata*, *Colletotrichum* sp., Algal Leaf Spot, *Botrytis* Leaf Rot, Flower Rot, *Volutella* Leaf Blight, Fire Blight, Fungal Leaf Spot, Scab, *Exoporium* Leaf Spot, *Phytophthora* Bud Rot, Black Spot, and Downy Mildew
Dicoarboximides	2	Iprodione	OHP 26 GT®	12	OHP, Inc.	*Alternaria*, *Botrytis*, *Rhizoctonia*, *Fusarium*, and *Cylindrocladium*.
			OHP Chipco® 26019 N/G	12	OHP, Inc.	*Anthracnose*, *Botrytis*, Black Spot, Leaf Spots, Powdery Mildew, *Rhizoctonia*, and many other diseases
			Sextant™	12	OHP, Inc.	Blights, *Botrytis*, *Fusarium* Leaf Spot, Ink Spot, and *Rhizoctonia* Stem and Root Rot
Dithiocarbamates	M3	Mancozeb	Dithane®	24	Dow AgroSciences LLC	*Alternaria*, *Anthracnose*, *Botrytis*, Downy Mildew, *Rhizoctonia*
			Fore®	24	Dow AgroSciences LLC	*Alternaria*, *Anthracnose*, *Botrytis*, *Entomosporium*, *Rhizoctonia*
			Junction™	24	SePRO Corp.	*Alternaria* Blight, *Anthracnose*, Fungal Leaf Spot, *Cercospora*, *Pseudomonas* Leaf Spot, *Botrytis* Blight, *Septoria* Leaf Spot
			Pehtathlon™	24	SePRO Corp.	*Botrytis*, *Cercospora*, Rust, Leaf Spot, Downy Mildew, *Phomopsis*, *Septoria*
		Manganese + Zinc	Protect®	24	Cleary Chemical Corp.	*Alternaria* - Leaf Spots and Flower Blights; *Anthracnose* Blights - *Colletotrichum*, *Discula*, *Elsinoe*, *Gloeodes*, *Gloeosporium*, *Glomerella*, *Gnomonia*, *Guignardia* *Aschochyta* Blights; Blackspot of Rose - *Diplocarpon*; *Botryosphaeria* Dieback; *Botrytis* Blights, Gray Mold; Brown Rot Blight - *Monolinia*, *Sclerotinia*, *Whetzelinia*; *Cephalosporium* Leaf Spot; *Cylindrocladium* Leaf Blight; Downy Mildews - *Plasmopara*, *Peronospora*; *Exobasidium* Blights; Flower Blights - *Aschochyta*, *Botrytis*, *Monolinia*, *Sclerotinia*, *Whetzelinia*; *Fusarium* Leaf Spot and Blight; Gray Mold – *Botrytis*; Ink Spot - *Ciborinia*, *Sclerotinia*, *Whetzelinia*; *Kabatina* Dieback; Leaf

Class	MOA*	Common Name	Trade Name	REI	Manufacturer	Pests Controlled (Lists from Current Manufacturer Labels)
						Spots - *Alternaria, Bipolaris, Cephalosporium, Cercospora, Cylindrosporium, Dactylaria, Didymellina, Diplocarpon, Dreschleria, Exerohilum, Helminthosporium, Entomosporium, Fabraea, Leptosphaeria, Marssonina, Mycosphaerella, Pestalotia, Phyllosticta, Rhytisma, Septoria, Stigmina; Ovulinia* Blight; *Rhizoctonia* Aerial or Web Blight; Rusts - *Cronartium, Endocronartium, Gymnosporangium, Melampsora, Melampsoridium, Phragmidium, Puccinia, Uromyces;* Scab - *Fusicladium, Venturia, Sphaceloma, Cladosporium;* Shothole Leafspot - *Blumeriella, Coccomyces Stemphyllium* Blight; Tip Blights - *Phomopsis, Diaporthe*
Ethyl Phosphonates	33	Fosetyl-Al	Aliette®	12	OHP, Inc.	Downy Mildew, *Pythium, Phytophthora*
Hydrogen dioxide	NC		ZeroTol™	0	BioSafe Systems	Algae, Bacteria, Fungi, and Viral Plant Pathogens on contact
Hydroxyanilide	17	Fenhexamid	Decree®	4	SePRO Corp.	*Botrytis cinerea*
Imidazole	3	Triflumizole	Terraguard®	12	OHP, Inc.	*Rhizoctonia* Root Rot, *Thielaviopsis, Cylindrocladium,* Powdery Mildew, and Leaf Spots
Oil	NC	Clarified hydrophobic extract of neem oil	Triact® 70	4	OHP, Inc.	Triact® 70 coats fungal spores and then dehydrates them to stop the disease cycle.
Phenylpyrrole	12	Fludioxonil	Medallion®	12	Syngenta	Foliar diseases caused by *Rhizoctonia* spp., *Botrytis* spp., *Cercospora* spp. and *Myrothecium* spp. Stem and root diseases caused by *Rhizoctonia* spp. and *Fusarium* spp., *Cylindrocladium* spp., *Sclerotium* spp., and *Thielaviopsis* spp.
Phosphite	33	Phosphorus acid	Alude®	4	Cleary Chemical Corp.	Downy Mildew, *Phytophthora* spp., *Pythium* spp.
Phthalimides	M4	Captan	CuPro™ 2005 Captan 50W	96	Micro Flo Company LLC	*Alternaria* Leaf Spot, *Botrytis* Flower Blight, *Septoria* Leaf Spot, Damping-Off Diseases
Piperadine	5	Piperalin	Pipron®	12	SePRO Corp.	Powdery Mildew eradicant
Strobilurins	11	Azoxystrobin®	Heritage®	4	Syngenta	*Anthracnose, Rhizoctonia,* Rust, *Phythium*
		Kresoxim-methyl®	Cygnus®	12	BASF	Protective and curative activity against powdery mildew, and protective activity against certain leaf spots, rusts, and downy mildews.
		Trifloxystrobi	Compass® O	12	OHP, Inc.	Powdery Mildew, *Botrytis, Anthracnose, Septoria* Leaf Spot, *Myrothecium,* Rust, Scab, Blackspot, Downy Mildew, *Rhizoctonia* Root Rot
		Pyraclostrobin	Insignia®	12	BASF	*Anthracnose* spp., *Rhizoctonia solani, Phythium,*
		Fenamidone	FenStop™	12	OHP, Inc.	Downy Mildew, Foliar *Phytophthora, Phytophthora, Pythium*
Thiadizole	14	Etridiazole	Truban®	12	Scotts Company	*Pythium, Phytophthora*
			Terrazole®	12	OHP, Inc.	*Pythium* and *Phytophthora*

(continues)

Table 20-1 *(Continued)*

Class	MOA*	Common Name	Trade Name	REI	Manufacturer	Pests Controlled (Lists from Current Manufacturer Labels)
Thiophanates	1	Thiophanate-methyl	3336™	12	Cleary Chemical Corp.	*Botrytis, Cylindrocladium, Fusarium, Gliocladium, Myrothecium, Penicillium, Ramularia, Rhizoctonia, Sclerotinia,* and *Thielaviopsis*
			OHP 6672™	12	OHP, Inc.	*Anthracnose,* Black Spot, Blights, *Botrytis, Fusarium,* Leaf Spot, Powdery Mildews, *Rhizoctonia,* and Scabs
			AllBan®	12	Scotts Company	Foliar Diseases - *Anthracnose Colletotrichum;* Black Spot of Rose *Diplocarpon rosae;* Leaf Spots and Blights- *Ascochyta, Blumeriella, Botrytis, Cercospora,Coccomyces, Corynespora, Curvularia, Idymellina, Entomosporium, Fabraea, Fusarium, Ramularia, Rhizoctonia, Marssonina, Mycosphaerella, Myrothecium, Phoma, Physalospora, Schizothyrium, Septoria, Sphaceloma;* Ovulinia Blight; Powdery Mildews - *Erysiphe, Microsphaera, Phyllactinia, Podosphaera, Oidium, Sphaerotheca;* Rust Diseases caused by: *Puccinia, Gymnosporangium, Uromyces*
Triazole	3	Triadimefon	Strike®	12	OHP, Inc.	Leaf Blight, Leaf Spot, Powdery Mildew, Rust, and Sclerotinia
		Mycolbutanil	Eagle®	24	Dow AgroSciences LLC	*Anthracnose,* Powdery Mildew, Rusts, *Septoria* Leaf Spot
Combined Products						
Acylalanine + Phenylpyrrole	4+12	Mefanoxam + Fludioxonil	Hurricane™	48	Syngenta	*Rhizoctonia* spp., *Phytophthora* spp., and *Pythium* spp.
Thiophanate + Chloronitrile	1+M5	Thiophanate-methyl + chlorothalonil	Spectro®	12	Cleary Chemical Corp.	Leaf Spots/Foliar Blights - *Actinopelte* Leaf Spot, *Alternaria* Leaf Spot, Leaf Blight, *Anthracnose Ascochyta* Blight, *Bipolaris (Helminthosporium),* Leaf Spot Black Spot of Rose, *Blumeriella* spp., *Botrytis* Leaf Spot, Leaf Blight, *Cephalosporium* Leaf Spot, *Cercospora* Leaf Spot, *Cercosporidium* Leaf Spot, *Coccomyces* spp., *Colletotrichum* Leaf Blotch, Leaf Spot, Blights, *Corynespora* Leaf Spot, *Coryneum* Blight, Shothole *Curvularia* Leaf Spot, Tan Leaf Spot, *Cylindrosporium* Leaf Spot, *Dactylaria* Leaf Spot, *Didymellina* Leaf Spot *Diplodia* spp., *Drechslera* Leaf Spot, Ink Spot, *Exobasidium* Leaf Blister, *Entomosporium,* (Fabraea) Leaf Spot, *Fusarium* Leaf Spot, *Gloeosporium* Black Leaf Spot, *Marssonina* Leaf Spot, *Monilinia* Blossom Blight, Twig blight, *Mycosphaerella* Ray Blight, *Myrothecium* Leaf Spot, Brown Rot, *Nematostoma* Leaf Blight, *Phoma* spp., *Phyllsoticta* Leaf Spot, *Physalospora* spp., *Phytophthora* Aerial Blight, *Ramularia* Leaf Spot, Rhizoctonia Web Blight, *Schizothyrium* spp., *Septoria* Leaf Spot, *Sphaceloma* spp., *Sphaeropsis* Leaf Spot, *Stagonospora* Leaf Scorch, *Taphrina* Leaf Blister, *Volutella* Leaf Blight Flower Spots/Blights - *Botrytis* Flower Spot, Flower Blight, *Curvularia* Flower Spot, *Monilinia* Blossom Blight, *Ovulinia* Flower Blight, *Rhizopus* Blossom Blight, *Sclerotinia* Flower Blight, Powdery Mildews - *Erysiphe*

Class	MOA*	Common Name	Trade Name	REI	Manufacturer	Pests Controlled (Lists from Current Manufacturer Labels)
						spp., *Microsphaera* spp., *Phyllactinia* spp., *Podosphaera* spp., *Oidium* spp. *Sphaerotheca* spp., Rusts - *Gymnosporangium* spp., *Puccinia* spp., *Uromyces* spp. Scabs - *Venturia* spp. Stem Rots/Crown Rots – *Bipolaris*, *(Helminthosporium)* spp., *Botrytis* spp., *Cylindrocladium* stem canker, *Fusarium* spp., *Gliocladium* spp., *Myrothecium* spp., *Ramularia* spp., *Rhizoctonia* spp., *Sclerotinia* spp.
Thiophanate + Dicrobxamide	1+2	Thiophanate + iprodione	26/36™	12	Cleary Chemical Corp.	*Anthracnose, Ascochyta* Leaf Blight
Thiophanate + Dithiocarbamate	1+M3	Thiophanate-methyl + Mancozeb	Zyban®	24	Scotts Company	Scab, Rust, Leaf Spot, Blight, *Anthracnose*, Downy and Powdery Mildew
Thiphanate + Thiadiazole	1 +14	Thiophanate-methyl + etridiazole	Banrot®	12	Scotts Company	*Fusarium, Phytophthora, Pythium, Rhizoctonia, Thielaviopsis*

*Fungicides Modes of Action
 1. Inhibition of tubulin formation in mitosis
 2. Affect cell division, DNA and RNA synthesis and metabolism
 3. DMI (Demethylation Inhibitor): Inhibition of sterol synthesis
 4. Phenylamides-Affect RNA synthesis
 5. Inhibition of an isomerase in sterol biosynthesis- Piperadines, Morpholines
 7. Affect mitochondrial transport chain
11. Quinone outside inhibitors (QOI)
12. MAP protein kinase in osmotic signal transduction
14. Lipid peroxidation (proposed)
17. 3-keto reductase during C4 demethylation in sterol biosynthesis
19. Chitin synthase inhibition in cell wall development
28. Affect cell membrane permeability (proposed)
33. Mode of action unknown. The mode of action cannot be placed within any other grouping.
40. Phospholipid biosynthesis and cell wall deposition (proposed)
 M Multi-site activity. Chemicals that act at several sites, which may differ among the group members
NC Not classified

This list is from the U.S. Environmental Protection Agency, in cooperation with the Fungicide Resistance Action Committee (FRAC). FRAC is a technical working group within the Global Crop Protection Federation (GCPF)

FIGURE 20-3A Downy mildew infection on snapdragons grown for cut flowers.

FIGURE 20-3B Underside of snapdragon leaf showing sporulation of downy mildew.

Infection is favored by cool greenhouse temperatures and high humidity. Free moisture is required for spore germination. For this reason, downy mildew is becoming more prevalent as fuel prices increase and growers are maintaining lower greenhouse temperatures.

Downy mildew is best controlled by maintaining low greenhouse humidity and air circulation with proper plant spacing and HAF fans. Immediate removal of infected plants will reduce disease inoculum. Several fungicides are registered for preventative and a few for eradicant control of downy mildew (Table 20-1).

Powdery Mildews

Powdery mildews are a group of fungi including the genera *Erysiphe*, *Leveillula*, and *Sphaerotheca* that produce gray or white powdery growth on leaves (Figure 20-4). Many species are host species specific. For instance, *Sphaerotheca pannosa* var. *rosae* only infects members of the Rosaceae family.

Powdery mildews are obligate parasites existing only on living tissue and do not require plant stress or injury to infect host plants. Warm greenhouse temperatures during the day and cool temperatures at night favor powdery mildew infection. Free water is not required for spore germination.

Management of powdery mildews relies on early detection, sanitation, cultural practices, environmental control, and fungicides. Prune or rogue infected plants while wet to prevent spread of spores. Bagging infected leaves works as a control measure as well. Syringing infected tissues with a spray of water is an old cut rose grower technique that continues to have merit.

Maintaining high light and low humidity can control powdery mildew to a degree, even though the spores do not require free water for germination. Air circulation with HAF fans can reduce infections as well. Both eradicant and protectant fungicides are registered for powdery mildew controls (Table 20-1). Protectants prevent powdery mildew from developing, whereas eradicants stop pathogen development from any further damage. Therefore, apply protectants when the conditions are conducive to powdery mildew outbreaks; yet apply eradicants only when the pathogen is present. Otherwise the effort and expense are wasted.

Soil-Borne Diseases

Soil-borne diseases include the species *Pythium*, *Phytophthora*, *Rhizoctonia*, and *Thielaviopsis*, which collectively include the majority of the fungi that infect

FIGURE 20-4 Heavy powdery mildew infection on cut rose plant.

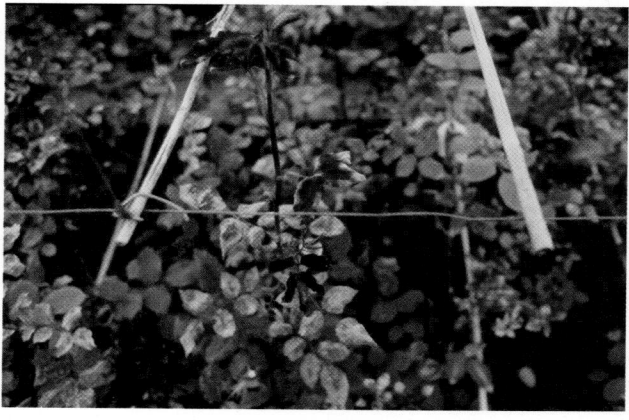

roots and crowns of plants. Most greenhouse crops are susceptible to one or more of these causal organisms.

Damping-off is a term that generally refers to sudden plant death in the seedling stage due to the attack of soil-borne fungi. The foliage becomes dull or wilted followed by yellowing of aboveground plant parts. Established plants infected by damping-off fungi show symptoms of wilting and poor growth. It is usually the new developing rootlets that are infected, resulting in root rot. Pasteurization of media prior to planting can prevent many problems with damping-off fungi.

Pythium infection is more common in wet and poorly drained soils (Figure 20-5). It infects young plant tissues, typically root tips, and spreads into primary roots. It causes a soft dark brown to black wet rot that disintegrates root cortical tissues, which will sluff leaving only the vascular stele.

There are many species of *Pythium*, but only a few that are pathogenic. *P. aphanidermatum* and a few other species are only active at warm temperatures; whereas *P. ultimum* is more active at cooler temperatures. *P. ultimum* has a broad host range infecting most greenhouse species.

Management of *Pythium* is simplest by using pathogen-free media, plants, water, and equipment. Providing good drainage and maintaining healthy plants will help to prevent *Pythium* outbreaks. Fungicide applied as a drench can help to prevent *Pythium* outbreaks (Table 20-1).

Rhizoctonia solani is the most common species to cause root rots of greenhouse plants (Figure 20-6). *Rhizoctonia* infection typically results in moist, brown lesions or cankers on crowns and roots. Good sanitation and media pasteurization is effective in preventing *Rhizoctonia* infection. Fungicide applied as a drench can control *Rhizoctonia* (Table 20-1). Research at Ohio State University has shown that the addition of bark or compost to the root zone media will suppress *Rhizoctonia* development.

Thielaviopsis basicola infected roots will become very black and thickened, hence the name black root rot. The lesions will be dry compared to those found with *Rhizoctonia* infection. Poor drainage, high pH (>5.5), and excessive fertility promotes *Thielaviopsis* development. Pasteurization, good drainage, adjusted pH (4.5), and thiophanate fungicides applied as a drench are all common means to control *Thielaviopsis*.

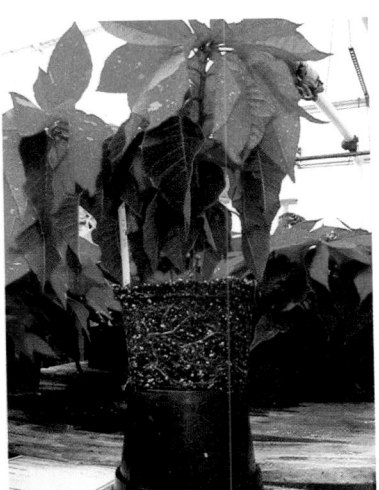

FIGURE 20-5 *Pythium* infected poinsettia roots. Note that the plant is under severe drought stress.

FIGURE 20-6 *Rhizoctonia* infection has killed this poinsettia cutting.

Bacteria

Bacteria are single-celled microorganisms. Unlike cells of plants, bacterial cells do not contain a nucleus. Bacteria can be found in every habitat on earth, soil, hot springs, radioactive waste, seawater, and deep within the earth's crust.

Bacterial diseases found that infect greenhouse plants are difficult to control other than through prevention, sanitation, and removal of infected plants. Disinfectants can be used to sanitize greenhouse tools, benches, and pots to provide some protection. Bacterial species that cause greenhouse plant diseases include bacterial wilt of carnation (*Pseudomonas caryophylli*), bacterial blight of geranium (*Xanthomonas pelargonii*), soft rot (*Erwinia chrysanthemi*), and crown gall on rose, chrysanthemum, and geranium (*Agrobacterium tumefaciens*).

Many organisms are subject to quarantine to prevent them from becoming established in the United States. One such species is *Ralstonia solanacearum* race 3 biovar 2, a bacterial pathogen not known to occur in this country. It causes a wilt disease in several important agricultural crops such as potatoes, tomatoes, peppers, and eggplant. Inadvertent introductions of this biovar occurred from shipments of geranium cuttings in several states in 2003, introduced from Kenya, and 2004, introduced from Guatemala. These introductions were contained and the plants were subsequently eradicated. *R. solanacearum* race 1 is endemic to the South and it is known as southern wilt, bacterial wilt, and brown rot of potato. *R. solanacearum* race 3 biovar 2 does not exist in North America, yet pathologists suspect that it has the potential to become pandemic throughout the region causing significant economic damage to agriculture. There are no pesticides that control this disease.

Bacterial blight of geranium (*X. pelargonii*) causes a systemic infection resulting in leaf spots. The spots can be found in a yellow wedge-shaped area of the leaf followed by wilting and eventual dieback of growing points. During hot and humid periods, bacteria can be spread from infected leaves into the stem becoming systemic.

Control of *Xanthomonas* in geraniums is maintained by growing culture-indexed cuttings and stock plants under strict sanitation guidelines. Strict hand washing and sanitation practices are to be enforced where geranium stock plants are grown (Figures 20-7A and 20-7B).

Viruses

Viruses can be defined as noncellular organisms. These organisms consist of nucleic acids, RNA and DNA surrounded by protein, which obligately reproduce inside host cells using the host's metabolic machinery and ribosomes. This reproductive process forms products called virions, which protect the virus and are transferred to other cells. Viruses were once considered to be nonliving in that they do not respire, they are not irritable, and they neither move nor grow; however, they do most certainly reproduce, and they regularly adapt to new hosts. They are too small to be seen under a light microscope, but they can be observed using an electron microscope.

FIGURE 20-7A Hand sanitizers strategically placed in the greenhouse can slow the transmission of fungal and bacterial diseases.

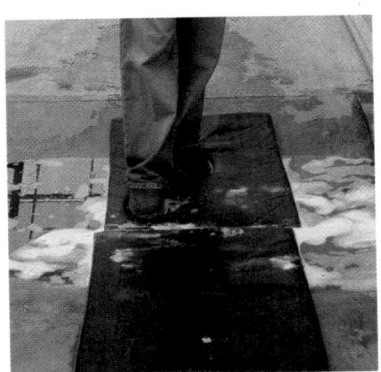

FIGURE 20-7B Floor mats soaked with disinfectant strategically placed between greenhouse sections can slow the spread of fungal and bacterial infections.

Plant viruses are not as well understood as their animal counterparts in spite of the fact that the first virus discovered was tobacco mosaic virus. There are 73 genera and 49 families of plant viruses. Some representative viruses common to greenhouse crops are listed in Table 20-2. Plant viruses are transmitted directly through sap by contact through plant wounds with contaminated tools, hands, or by animals, mainly aphids, leafhoppers, and thrips feeding on the plant. The virus then spreads systemically throughout the plant via the vascular system.

Viruses rarely kill their host plants, but they can alter their appearance, thus reducing their value and saleability. Yet some viruses are cultivated for their aesthetic value. The pink and white petal blotching of a tulip flower results from infection by the tulip breaking virus. Viruses retard plant growth; alter the growth of foliage (distort or stunt), flowers, and fruits; and discolor foliage (e.g., spot, streak, or mottle). Flowers may be dwarfed, deformed, streaked, or faded. They may even remain green.

Identification of viruses has traditionally been through observation of damage symptoms, hosts lists, grower knowledge, and local disease history. Visual virus diagnosis is common, but many of the symptoms are easily confused with nutritional disorders, pesticide damage, and other nonviral pathogens. Modern virus diagnosis relies on serological assays.

Modern molecular procedures can be used to identify nucleic acid structures or antigens that are unique to each virus or family of viruses. Antigens are pathogen proteins that induce an immune response in their host. Antigens are identified by a procedure called enzyme-linked immunosorbent assay, or ELISA.

ELISA procedures are common to most plant disease diagnostic clinics, but kits are available for onsite testing of viral pathogens. Plant tissue that is suspected to be infected with a virus is collected, ground, and suspended in a liquid. A small volume of this liquid is then incubated with an indicator solution, which contains products that change color in the presence of antigens specific to the suspected virus. For even easier use, some kits use dip sticks that change color when dipped into the plant extract.

Table 20-2

Common Viruses that Infect Greenhouse Crops. This is Only a Short Listing of the Many Host Plants That are Infected by Viruses. There are More Than 400 Plant Species That are Infected by Cucumber Mosaic and More Than 600 Plant Species That are Infected by Tomato Spotted Wilt Virus. And the Lists are Growing.			
Virus and Family	Vector	Greenhouse Crops	Weeds
Bean yellow mosaic (potyvirus)	Aphids	Freesia, fuchsia, gladiolus, sweet pea, violet	*Chenopodium* spp., clovers, sweet clover
Beet curly top (rhabdovirus)	Leafhoppers	Celosia, coreopsis, cosmos, geranium, Marguerite daisy, nasturtium, pansy, petunia, strawflower, stock, sweet William, zinnia	*Atriplex* spp., *Chenopodium* spp., clovers, fennel, *Polygonum* spp., *Rumex* spp., Russian thistle, shepherd's-purse
Cauliflower mosaic (caulimovirus)	Aphids	Honesty, lunaria, stock	Mustard, *Raphanus* spp., shepherd's-purse
Cucumber mosaic	Aphids	Aster, begonia, buddleia, calendula, columbine, dahlia, daphne, delphinium, freesia, fuschia, geranium, gladiolus, iris, larkspur, ligustrum, lily, lobelia, nasturtium, passion vine, penstemon, periwinkle, primula, snapdragon, vinca, zinnia	*Commelina*, lambsquarter, lupine, milkweed, pigweed, pokeweed
Impatiens necrotic spot virus (tospovirus)	Thrips	African violet, begonia, browallia, calla, chrysanthemum, cineraria, cyclamen, dahlia, exacum, gloxinia, impatiens, kalanchoe, lisianthus, lyceum, odontoglossum orchids, phalaenopsis orchids, pouch flower, primrose, snapdragon, stephanotis, streptocarpus	Bindweed, bittercress, chickweed, *Emilia* spp., jimsonweed, knotweed, lupine, malva, *Mesembryanthemum* spp., miner's lettuce, nightshade, oxalis, physalis, pigweed, shepherd's-purse
Tobacco mosaic (tobamovirus)	Mechanical, carried in seed, soilborne	Delphinium, flowering tobacco, petunia, phlox, wisteria	*Emilia* spp.
Tomato spotted wilt virus (tospovirus)	Thrips	African violet, ageratum alstroemeria, amaryllis, anemone, aralia, aster, begonia, calendula, calla, carnation, Christmas cactus, chrysanthemum, cineraria, columbine, coreopsis, cosmos, dahlia, dieffenbachia, everlasting, exacum, forget-me-not, fuchsia, geranium, gerbera, gladiolus, gloxinia, godetia, gypsophilia, hydrangea, impatiens, larkspur, lily, lisianthus, marigold, nasturtium, peony, periwinkle, petunia, phlox, poppy, pouch flower, primrose, ranunculus, sage, saponaria, scabiosa, schefflera, snapdragon, statice, stephanotis, stock, sweet pea, verbena, zinnia	Bindweed, bittercress, chickweed, *Emilia* spp., jimsonweed, knotweed, lupine, malva, *Mesembryanthemum* spp., miner's lettuce, nightshade, oxalis, physalis, pigweed, shepherd's-purse
Turnip mosaic (potyvirus)	Aphids	Anemone, begonia, nasturtium, petunia, safflower, statice, stock, sweet William, wallflower, zinnia	Cruciferous weeds

ELISA test kits are reasonably priced and allow for rapid virus detection; however, they are usually only accurate to the virus family. For instance, the ELISA kit for impatiens necrotic spot virus actually tests only for tospoviruses. Therefore, it tests for tomato spotted wilt virus as well. For greenhouse use,

FIGURE 20-8 Screening of greenhouse vents can help to exclude insects that vector viruses.

this diagnosis is enough as the two viruses are both transmitted by thrips and they are equally damaging.

There are no pesticides that can cure a virus infection of a plant. Viruses are managed through sanitation, monitoring, and exclusion. Prevention is the most effective method of virus control. Keep the greenhouse, inside and out, free of weeds that serve as virus reservoirs. Break virus cycles by using a plant-free period if possible. Use virus-free propagation materials by purchasing culture-indexed or virus-indexed stock.

Excluding and managing insect vectors is the primary means of controlling tomato spotted wilt virus and impatiens necrotic spot virus. These viruses are only transmitted by thrips. Excluding thrips from the greenhouse using insect screens will maintain a virus-free greenhouse (Figure 20-8).

Phytoplasmas

Phytoplasmas are considered to be specialized bacteria that are obligate plant parasites found in the phloem. Phytoplasmas were originally thought to be plant viruses, but ultrathin sections of phloem tissues revealed the presence of mycoplasma-like organisms that could not be cultured in cell-free media. Aster yellows is the most significant phytoplasma to infect greenhouse plants and is vectored by leafhoppers. It is not spread by seed, tools, cuttings, or any other insects. Control of aster yellows is through the use of pathogen-free plants and control of the insect vector. Aster yellows is characterized by green, stunted growth.

Another characteristic of phytoplasma infections is excessive branching. Since 1923, the presence of an endophytic biological agent has been suspected to occur on poinsettias to produce free-branching morphotypes. This conclusion was surmised when it was discovered that the ability of free-branching poinsettia cultivars to induce axillary shoots was lost after the plants were subjected to procedures to eliminate poinsettia mosaic virus (PnMV). The free-branching characteristic could be restored when the treated plants were approach-grafted onto free-branching stock plants. The phytoplasma associated with free-branching poinsettia cultivars has been determined to belong to the peach X-disease phytoplasma group. Not all plant diseases are bad.

ACHIEVEMENT REVIEW

Select the best answer or answers to complete each statement. List the appropriate letter(s).

1 Which of the following causes disease in greenhouse plants?
a. Bacteria
b. Fungi
c. Viruses
d. Phytoplasmas

2 Which of the following lead to disease in greenhouse plants?
a. Presence of a pathogenic causal organism
b. Susceptible plants
c. Presence of saprophytic organisms
d. Koch's postulates

3 An obligate parasite lives only on dead and decaying organic matter.
a. True
b. False

4 Facultative saprophytes can exist as a:
a. parasite.
b. saprophyte.
c. pathogen.
d. phytoplasma.

5 Fungi differ from bacteria in that they have:
a. a chitinous cell wall.
b. no nucleus.
c. multicellular filamentous growth.
d. photosynthetic capacity.

6 A foliar disease that infects mostly aboveground plant parts with fuzzy gray, brown, or tan spores on rotted tissue is most likely:
a. *Botrytis.*
b. powdery mildew.
c. *Rhizoctonia.*
d. *Pythium.*

7 *Pythium* infections differ from *Rhizoctonia* infections in that:
a. *Pythium* is a bacterial organism and *Rhizoctonia* is a fungal organism.
b. *Pythium* is found in wet and poorly drained soils and *Rhizoctonia* is found in soils with a high pH.
c. *Pythium* typically first infects root tips and young plant tissues, whereas *Rhizoctonia* causes moist, brown lesions or cankers on crowns and roots.
d. *Pythium* can be controlled by blending compost or bark in the root zone media.

8 Downy mildew is often confused with:
   a. powdery mildew.
   b. *Botrytis* blight.
   c. *Phytophthora.*
   d. *Xanthomonas.*

9 Bacterial infections are best controlled through:
   a. fungicide applications.
   b. sanitation practices.
   c. federal quarantine.
   d. exclusion of western flower thrips.

10 Bacteria include organisms:
   a. that consist only of nucleic acids.
   b. called phytoplasmas.
   c. with chitinous cell walls.
   d. with cells that do not contain nuclei.

11 The bacterial pathogen *Ralstonia solanacearum* race 3 biovar 2 is not found in the United States; therefore, it is of no risk to greenhouse crops.
   a. True
   b. False

12 Viruses are nonliving.
   a. True
   b. False

13 Plant viruses may:
   a. create aesthetic value by improving flower color.
   b. be transmitted by insects.
   c. be considered multicellular organisms.
   d. be considered to be saprophytes.

14 Which of the following has the potential to spread a virus in the greenhouse?
   a. Tools
   b. Employees
   c. Insects
   d. Nematodes

15 Impatiens necrotic spot virus is best controlled by:
   a. application of chemical fungicide sprays.
   b. exclusion and/or chemical control of western flower thrips.
   c. sanitation of pruning tools and hand washing.
   d. using disinfectants on bench tops.

Regulating Plant Growth

OBJECTIVES:

After studying this chapter, the student should have learned:

■ the need for growth regulation in the greenhouse.
■ the common chemical growth regulators used in the greenhouse.
■ the concept of mechanical conditioning by brushing.

A well-managed greenhouse should provide an environment that it is optimum for plant growth. A greenhouse that is kept too warm may encourage greenhouse crops to stretch, and a greenhouse that is kept too cool may create short plants with delayed flowering. It is the goal of the greenhouse grower to integrate light intensity, photoperiod, temperature, and the relationship of night to day temperature to control plant growth. Plant growth is described as an increase in plant size or weight. This occurs through stem elongation, thickening of leaves, and the addition of branches. Typical greenhouse conditions often create an environment where plants are too tall due to excessive stem elongation. Some conditions that promote stem elongation include:

■ shady or dark greenhouse conditions.
■ crowded greenhouse bench spacing.
■ long production schedules (i.e., more time for growth).
■ long-day length.
■ cool night and warm day temperatures.

Plants that are excessively tall will tend to topple on the bench, be prone to stem breakage, be difficult to ship, and not be marketable. Market conditions typically require that attention be paid to the height and shape of the plant. Customers prefer plants that range from 2.6 to 3.5 times the pot diameter (e.g., 16 to 21 inches tall for a 6-inch pot). Commercial greenhouse growers will use many cultural and chemical tools to finish plants at the correct marketable size and form.

Plant height is typically controlled in the greenhouse using temperature, light, chemical growth regulators, and mechanical conditioning. Using temperature to control plant height employs a concept known as **DIF** and is more fully discussed in Chapter 7. Light and photoperiod are used to control plant growth and flowering and are more fully discussed in Chapter 8.

Chemical Plant Growth Regulators

A **chemical plant growth regulator (PGR)** is a natural or synthetic substance that in very small quantities regulates or controls some aspect of plant growth. These include stem length, rooting, flowering, leaf abscission, fruiting, and cold hardiness. This chapter will focus primarily on those chemical PGRs that regulate stem length, rooting, and flowering. These PGRs are known as chemical growth retardants.

Most PGRs that are chemical growth retardants are considered to be antigibberellins. **Gibberellins** are naturally occurring plant hormones, which are responsible for the promotion of growth and cell wall elongation. When greenhouse conditions are dark, plants tend to produce gibberellins in larger quantities as a natural response, triggering plants to grow taller to outcompete their neighbor. Growth retardants applied at the proper time can be used to manage plant height (Figure 21-1). The amount of growth retardant to apply depends on the chemical used, the height reduction desired, the species and cultivar, crop size at time of application, method of application, root zone medium, and cultural conditions.

Chemical Plant Growth Retardants

Growth retardant application and timing of application is as much art and skill as science. Early application, before the need for growth retardation is not yet apparent, will provide moderate control of elongation and improve plant form.

FIGURE 21-1
The control plant on the left received no growth retardants while the one on the right was sprayed with retardants which resulted in a fuller and shorter plant. Courtesy Delmar/Cengage Learning, Photo by Jack Ingels.

Application after growth begins is more popular, however, because the need for retardation is more obvious. Experience and growth records determine which growth retardant to be used and when to apply. Recording application methods, rates, and results of growth retardant treatments will promote consistent results and establish rates and volumes appropriate to the greenhouse conditions. The common growth retardants used in greenhouse production include **daminozide, chlormequat chloride, ancymidol, flurprimodol, paclobutrazol**, and **uniconazole**. All chemical PGRs are considered by the Environmental Pesticide Agency to be pesticides and all laws and regulations pertaining to such must be followed. Examine the label of any PGR prior to its use. No chemical plant growth retardant is labeled for application to greenhouse grown vegetables, vegetable transplants, and culinary herbs.

Daminozide

Daminozide is marketed under the trade name B-Nine®. It reduces ornamental plant size by reducing internode length. This internode shortening is the result of smaller, more compact cells. In addition to compact plants the leaves are also a deeper green color. This compactness provides greater stem strength with less stem breakage during shipping and handling. Daminozide will also reduce apical dominance. Treated shoots form terminal buds early and will have more branches. This yields plants with a compact habit of growth and more flower buds.

Daminozide is systemic and, to be effective, it must be absorbed through the leaves. It is to be dissolved in water and can be applied to the foliage with standard spray equipment. Using a fine spray that will remain on the leaf surface longer and dry more slowly is the most effective. Apply the spray to the point just prior to run-off. Avoid spraying plants during the heat of the day and under low humidity, which will result in rapid drying of the spray reducing effectiveness and will leave white deposits on leaves. Spraying early in the morning or in late afternoon or evening is preferable. The plants to be treated should be well watered and the foliage should be dry at the time of application. It is not active when applied as a drench.

Daminozide should be applied at a 0.25 to 0.50 percent (2,500 to 5,000 ppm) concentration. This concentration is generally effective in controlling excessive stem elongation of most bedding, perennial, and other flowering plant species. Daminozide applications should begin when 1.5 to 2 inches of new growth has occurred after transplant or pinch, typically after 2 weeks. The applications can be repeated at 7- to 10-day intervals. For poinsettias, do not apply daminozide after the onset of short days. This will result in smaller bracts.

Daminozide is often tank mixed with paclobutrazol or chlormequat chloride for greater activity on hard to control species or for extended residual activity. The recommended rates are 2,500 ppm and 15 ppm for daminozide and paclobutrazol, respectively. The recommended rates for daminozide and chlormequat chloride are found in Table 21-1.

Chlormequat Chloride

Chlormequat chloride is marketed under the trade name Cycocel®. As with daminozide, treated plants are more compact with shorter internodes, stronger stems, and greener leaves. Spray application rates of chlormequat chloride typically range from 1,000 to 2,000 ppm, but can be applied up to 4,000 ppm in

Table 21-1

Recommended Rates of Daminozide and Paclobutrazol When Applied as a Tank Mix for Optimum Height Control Under Various Growing Conditions and Species		
Desired Activity Level	Daminozide (ppm)	Paclobutrazol (ppm)
Very High	5,000	1,500
High	2,500	1,500
Medium	1,250	1,250
Low	800	1,000

some cases. Chlormequat chloride is often tanked mixed with daminozide to control growth of plants that are not very sensitive to chlormequat chloride or when two applications may be required. When used together, these products are synergistic, producing greater results than either chemical used alone. The optimum rates are shown in Table 21-1.

Chlormequat chloride will typically control internode elongation for about 1 to 3 weeks after application. This depends on crop culture, environmental conditions, and plant growth habit. Many growers will use multiple applications, but chlormequat chloride has its greatest effect on final plant height when applied at the beginning of rapid stem elongation and will have little or no effect if applied when shoots are not elongating or at the end of the elongation phase.

Chlormequat chloride is absorbed through young expanding leaves, mature leaves, and stems while the spray solution stays wet. Therefore, chlormequat chloride sprays are best applied under conditions that support slow drying of spray solutions on the foliage.

Chlormequat chloride can be applied as a drench and taken up by the plant through the roots where it is translocated to the stem tips where it is active. Drenches should be applied when the plants are well watered so that the root zone medium is evenly saturated with the product. Drench treatments range from 2,000 to 4,000 ppm of chlormequat chloride.

Foliar spray applications of chlormequat chloride will often be yellowed slightly near the leaf margins or at the tip of leaves. Mature leaves at time of spray and leaves formed after application are not affected. This discoloration typically appears 3 to 5 days after the spray treatment. Discolored areas usually regain most or all green color by the end of the crop cycle. Drench applications do not cause leaf yellowing.

Ancymidol

Ancymidol is marketed under the trade name A-Rest® and is absorbed by roots, leaves, and stems. Ancymidol inhibits gibberellin synthesis and activity, resulting in shortened stem internodes and a more compact plant. Secondary effects that result from ancymidol application include lower plant water requirements, better plant tone, and darker green leaf color. Ancymidol is absorbed through both foliage and roots and may be applied as a spray or drench to control plant height. Some growers will use split or sequential applications allowing greater treatment flexibility. Suggested application rates are listed in Table 21-2.

Ancymidol is an effective growth retardant when applied to bedding plant plugs. It is typically applied as what is called a *sprench* when plants have reached the one to two true leaf stage. A sprench is a common application method used

Table 21-2

Recommended Spray and Drench Application Rates for Ancymidol by Plant Type		
Plant Type	Rate Range (ppm)	Drench Rate Range (ppm)
Bedding Plants	6-66	1-2
Bedding Plant Plugs	3-35	0.5-1
Flowering/Foliage (annual or perennial) Herbaceous Species Woody Species	20-50 50	1-2 2
Bulb Crops	25-50	2

in bedding plant plugs where the chemical is applied to both the foliage and the root zone medium as a spray and a drench.

Ancymidol is effective for controlling plant height of most bulb crops. It is most effective when applied as a drench rather than a spray. The recommended starting rate for drenching bulbs is 1 to 4 ppm and 50 ppm for sprays and applied when the plants are 2 to 6 inches tall. A second application can be made 2 weeks after the first.

The efficacy of ancymidol when applied as a drench to plants grown in root zone media containing bark may be slightly reduced. The bark does not inhibit the activity of the growth retardant; it simply is slightly bound to the bark reducing its effect. When considering a shift to media containing bark, its impact must be evaluated on growth retardant application rates.

Flurprimodol

Flurprimodol is marketed under the trade name Topflor®. Flurprimodol reduces internodal elongation through the inhibition of gibberellic acid biosynthesis. It has been shown to increase plant quality, without growth reduction. These characteristics include enhanced leaf and bloom color, darker foliage, increased leaf thickness, stronger stems, and decreased water use.

Flurprimodol is absorbed by plants through foliage, stems, and roots. It may be applied as a spray or drench using standard application equipment. Multiple or split applications provide greater treatment flexibility, more uniform growth regulation, and may prevent overapplication of the product.

Flurprimodol is active on a wide range of species at a wide range of application rates. These rates range from 0.5 to 80 ppm. Refer to the product label for specific guidelines for species and rates for plugs and finished bedding plants. Flurprimodol has been widely used in Europe for many years and is fairly new to the U.S. market. Its use will increase as growers gain experience and trust with the product.

Paclobutrazol and Uniconazole

Paclobutrazol and uniconazole are plant growth retardants that are in a chemical class called triazoles. Many other triazole compounds are fungicides. Triazole compounds are very active chemicals and control growth of most species at low rates.

Triazole growth retardants can be applied either as a spray or drench. However, translocation of triazoles is limited to the xylem. If applied to the foliage, the growth retardant will not move into the phloem, but if applied to roots or stems, the chemical will migrate to the xylem for translocation throughout the plant. When applied as a drench, triazoles have little effect on flower size and bud opening.

Paclobutrazol and uniconazole are very similar chemicals. Yet they are quite different in their activity. Uniconazole tends to be two to four times more active than paclobutrazol to achieve the same height control. Paclobutrazol is applied at concentrations of 2 to 90 ppm for different crops and uniconazole is applied at concentrations of 1 to 50 ppm. Note that these concentrations are considerably lower than any of the other growth retardants discussed to this point and great care must be taken when measuring and mixing spray solutions. Recommended rates for paclobutrazol are found in Table 21-3; and for uniconazole, recommended rates are found in Table 21-4. As with ancymidol, bark in the root zone medium will impact paclobutrazol and uniconazole when applied as a drench.

Table 21-3

Recommended Rates for Applying Paclobutrazol by General Plant Type			
Plant Type	Spray (ppm)	Drench (ppm)	Bulb Soak (ppm)
Bedding Plants	30	1	N/A
Bedding Plant Plugs	5	N/R	N/A
Flowering/Foliage Plants			
Herbaceous	30	1	N/A
Woody	50	2	N/A
Woody Landscape Plants	100	4	N/A
Bulb Crops	100	10	20 (@15 minutes)

N/R = Use is not recommended.
N/A = Use is not applicable.

Table 21-4

Recommended Rates for Applying Uniconazole by General Plant Type				
Plant Type	Spray (ppm)	Soil Surface (ppm)	Drench (ppm)	Bulb Soak (ppm)
Bedding Plants	1–50	0.5–20	0.1–2	N/A
Bedding Plant Plugs	0.5–10	0.5–10	N/R	N/A
Flowering/Foliage Plants				
Herbaceous	5–40	0.5–10	0.5–1	N/A
Woody	20–50	1.5–20	0.5–2	N/A
Woody Landscape Plants	10–50	1–20	1–2	N/A
Bulb Crops	2.5–20	0.5–10	1–3	1–10

N/R = Use is not recommended.
N/A = Use is not applicable.

Other Chemical Plant Growth Regulators

Auxins

Auxins control many plant growth functions including cell enlargement, apical dominance, and stimulation of roots. **Indoleacetic acid (IAA)** is a naturally occurring auxin, but is not used commercially as it breaks down rapidly in light. **Indolebutyric acid (IBA)** and **naphthalene acetic acid (NAA)** are synthetic auxins that are used commercially. IAA and NAA are regularly used to stimulate and speed adventitious rooting for many species (see Chapter 18).

Ethylene

Throughout most of this book, ethylene gas has been discussed as a greenhouse pollutant causing flower senescence and was to be avoided. Yet, there are many beneficial effects from ethylene that can be used during greenhouse production. However, ethylene is a gas and difficult to apply. **Ethephon**, which is sold under the trade names Florel®, Pistill®, and Ethrel®, is a compound carried in liquid form as 2-chloroethylphosphonic acid. As an acid, the ethephon is absorbed by the plant, and within the plant cell, which is basic, the acid is broken down into a phosphate and ethylene gas. It is within the cell where ethylene activity is manifested.

Ethephon is used to prevent flowering, decrease internode length, and increase lateral branching for several crops. It is used regularly for stock plant production of geraniums, New Guinea impatiens, and poinsettias. By preventing flowering of geranium stock plants, more plant energy is devoted toward vegetative growth. When used to finish hanging baskets to promote branching and delay flowering, fuller more well-developed plants are the result. A general rule of thumb is to apply ethephon for a 6-week delay in flowering.

It is important to make sure that ethephon is applied in a spray solution of pH 4 or below or a prerelease of ethylene gas will occur. Once applied, ethephon is not translocated, so good coverage is essential. Along with a surfactant, ethephon sprays are applied at 100 to 500 ppm.

Gibberellins

Gibberellins are involved in many different plant processes. They are produced in leaf tips and small leaves. Gibberellins stimulate cell division and cell enlargement. Thus, plants treated with gibberellins have larger leaves and larger flowers. As an emergency treatment, gibberellins can be applied to offset a misapplication of a growth retardant, but success is erratic.

Gibberellins have a role in seed germination. Application of gibberellins to dormant seeds can promote germination. This is an important tool for seed propagation of species with little or no cultural information that prove to be difficult to germinate. Gibberellins have also been used to shorten the chilling time required to break flower dormancy of azaleas.

Cytokinins

Cytokinins are naturally occurring hormones produced in plant roots. They serve as a means for the root and shoots to communicate. Cytokinins promote lateral bud development. Applications of cytokinins have been used to promote lateral branching on flower crops, but not predictably. Cytokinins have also been demonstrated to delay leaf senescence. It is being used commercially as a blend with gibberellic acid, called Fascination®, to reduce leaf yellowing.

Fascination® is a blend of gibberellic acids (GA_{4+7}) and 6-benzyladenine (6-BA), a cytokinin that prevents leaves from yellowing, increases flower size, improves flower number, enhances the life of a flower, and advances overall plant vigor and growth.

Fascination® is commonly applied to Easter lilies and hybrid lilies to decrease yellowing and necrosis of the lower leaves and to delay flower senescence. Gradual leaf yellowing is common and typically begins prior to or near the visible bud stage. Postharvest leaf yellowing is most dramatic during cold storage of finished plants. Fascination® will inhibit both forms of leaf yellowing when used preventatively. Fascination® is applied as a spray and only the leaves covered by the spray will be protected from chlorosis.

Fascination® is to be applied at 10/10 ppm (GA_{4+7}/6-BA) to Easter lilies 7 to 10 days prior to the visible bud stage for early season protection, which lasts up to 3 weeks. Midseason application at 100/100 ppm (GA_{4+7}/6-BA) should occur 7 to 10 days after visible bud, taking care not to spray immature leaves that may lead to upward curling. Again, the effect should last up to 3 weeks. Late season application of Fascination® at 100/100 ppm (GA_{4+7}/6-BA) to all foliage and flower buds will prevent leaf yellowing and prolong flowering of Easter lilies during shipping and in the postharvest environment. Do not apply more than 14 days prior to placing in the cooler or shipping.

Mechanical Conditioning

Mechanical conditioning is a practice that is used to control plant height without the use of any chemicals. This practice, along with DIF, is one of the only means of controlling plant height of greenhouse grown vegetable transplants, which includes culinary herbs. Mechanical treatments that can be used to reduce plant height include rubbing the stems, brushing shoots, shaking potted plants or whole flats, vibrating the canopy, or disturbing the plants with wind or water. Most of these treatments are not practical on a large scale in the greenhouse without serious engineering investments. Wind application using a leaf blower has some merit, but may cause excessive drying.

Research conducted by Dr. Joyce Latimer at Virginia Polytechnic Institute and State University has demonstrated that brushing plants is the most effective and efficient method of mechanical conditioning when used on high density plantings, such as bedding plants. Treatments include brushing plant shoots with a dusting brush, suspended dowels or PVC pipes, a steel rod suspended in a cloth sling, or multiple layers of burlap. The brushing material must be sufficiently strong as to physically move the shoot when passed overhead. The brushing device must be mounted using a rack in such a way to pass over the plants uniformly. Care must be taken to not tear the foliage. Plants in bloom are generally not treated as the brushing will cause bruising of petals.

Mechanical conditioning using brushing has the potential to reduce plant height by 20 to 50 percent compared to untreated plants. For maximum effectiveness, the plants must be brushed twice daily with a minimum contact time of 1.5 to 2 minutes. Plants treated in this manner appear darker green with thicker leaves, and a denser canopy, much like those treated with chemical growth retardants. Yet this practice is not widespread due to engineering constraints.

ACHIEVEMENT REVIEW

Select the best answer or answers to complete each statement. List the appropriate letter(s).

1 Which of the following will lead to excessively tall greenhouse plants?
   a. Shady or dark greenhouse conditions
   b. Short-day photoperiods
   c. Cool day and warm night temperatures
   d. Crowded greenhouse bench spacing

2 Most customers prefer plants that are:
   a. 16 to 21 inches tall when grown in a 6-inch pot.
   b. 10 to 16 inches tall when grown in a 6-inch pot.
   c. 22 to 28 inches tall when grown in a 6-inch pot.
   d. 27 to 35 inches tall when grown in a 10-inch pot.

3 Plant growth regulators are controlled by the US-EPA as pesticides.
   a. True
   b. False

4 Which of the following plant growth regulators is considered to be a growth retardant?
   a. Gibberellins
   b. Auxins
   c. Daminozide
   d. Ethylene

5 Plant growth retardants impact plant growth by:
   a. controlling gibberellin metabolism.
   b. controlling auxin metabolism.
   c. stimulating ethylene production.
   d. reducing cytokinin metabolism.

6 The commercial product called Fascination® is a blend of gibberellic acid and a synthetic cytokinin and is used to:
   a. control plant height.
   b. control seed dormancy.
   c. prevent leaf yellowing.
   d. increase flower longevity.

7 Darker, more green foliage can be a result of:
   a. brushing.
   b. daminozide application.
   c. triazole application.
   d. cytokinin application.

8 Which of the following products will result in flower senescence?
   a. Chlormequat chloride
   b. Ethephon
   c. Fascination® (GA_{4+7}/6-BA)
   d. Flurprimodol

9 The activity of ethephon is increased when applied as a:
   a. spray solution at a pH of 4 or less.
   b. spray solution at a pH of 5 or greater.
   c. gas fumigant.
   d. granular product.

10 Mechanical conditioning is best accomplished by:
   a. the use of wind generated by a leaf blower.
   b. shaking or vibrating benches.
   c. brushing.
   d. withholding water.

Container Grown Crops

OBJECTIVES:

After studying this chapter, the student should have learned:

- advantages and disadvantages of clay pots.
- advantages and disadvantages of plastic pots.
- different styles of pots used to grow plants.
- characteristics of a good medium for potted plants.
- the correct method of potting plants.
- how plants are prepared for market.

More than 80 percent of the floriculture crops sold today are grown in containers. Round clay pots in many sizes were once the main type of containers in use. Now plastic pots in different sizes and shapes are primarily used for bench grown crops and for crops grown in hanging baskets. Multicell plastic containers are used to raise annual spring flowering bedding plants. The largest containers used in the industry at present are pots for large foliage plants.

Various types of crops are grown in containers, including flowering potted plants sold for holidays, such as poinsettias for Christmas and lilies for Easter. *Dendranthema* (chrysanthemums) are beautiful at any time of the year. Tropical foliage plants are grown for their attractive shapes and colors. Spring bedding plants are the last group of plants grown in containers.

Regardless of the type of container used, they all have certain common features. Compared to the volume of soil available to a plant growing in the field, a potted plant has only a small volume of medium in which to grow. This means that the supply of nutrients is limited. In addition, the medium can hold only a limited amount of water. Thus, both water and fertilizer must be applied regularly. Less frequent feedings are needed if a long-lasting, **controlled-release fertilizer** is used.

Clay Pots

Clay pots have been used to grow ornamental plants for centuries. The Romans and Egyptians used clay containers to display potted plants and flowers. For years, clay pots were used as containers in the florist industry. They ranged in size from small thumb pots to pots 24 inches (61 cm) or more in diameter (Figure 22-1). With the development of plastic pots the general use of clay pots in commercial greenhouses rapidly declined.

There are a few places in the United States where clay pots are favored by flower growers. In the Cincinnati, Ohio, region and also around St. Louis, Missouri, some growers produce plants in clay pots.

Pots and other items made from clay are greatly in favor in retail garden centers and other horticultural outlets (Figure 22-2). Many customers like the products for their functional beauty. They attribute an aura of prestige to clay containers as opposed to plastic. Yet many larger containers are now available with the look of clay, but with the light weight and durability of plastic.

Clay pots and other decorations manufactured from clay should not be stored outside where temperatures go below freezing, as they will absorb moisture. Alternate freeze and thaw cycles will break them into pieces.

Plastic Pots

Many growers use molded polystyrene pots, whereas other growers use hard plastic, polypropylene pots for greenhouse crops. Polypropylene plastic pots range in size from 2 inches in diameter to 12 inches in diameter or more (Figure 22-3). Plastic pots weigh much less than clay pots.

Plastic pots are nonporous. This means that they do not lose moisture through their side walls and they cannot exchange gases. As a result, plants grown in plastic pots must be watered with great care. When the drying

FIGURE 22-1 Assorted clay pots in common use. The large pot to the left is half as tall as wide and known as a bulb pan. The two pots to the far right are azalea pots, which are slightly less tall than wide.

FIGURE 22-2 Clay containers and figurines or containers and figurines designed to appear as clay offered for sale by a retail-grower florist.

FIGURE 22-3 Hard plastic pots in various sizes and shapes.

FIGURE 22-4 Three disinfectants approved for greenhouse use. These are not to be used for treating growing media.

conditions are poor, as in winter because of reduced sunshine, it is easy to over-water crops grown in plastic pots.

The slow rate of drying that results when plastic pots are used is an advantage in the summer when pot crops often dry too quickly. **Leaf scorch** due to water stress is a common plant injury in summer.

In addition to round plastic pots, square pots are also manufactured. Square pots make better use of the growing area when they are lined up on the green-house bench. However, there is less air movement around square pots than there is when round pots are used. This reduced air movement can lead to an increase in leaf disease. The largest square pot is rarely more than 5 inches (12.5 cm) in width.

Plastic pots cannot withstand the high temperatures of steam pasteurization. They lose their shape at these temperatures. The pots can be cleaned by means of a chemical treatment. They are soaked for 10 minutes in a solution of Green-Shield®, Floralife® D.C.D., or Triathlon® (Figure 22-4). The pots are then rinsed in clear water and allowed to air thoroughly before use. This method does not achieve the same amount of sterilization as steam pasteurization. Before the pots are soaked, they must be washed until all dirt particles are removed. If dirt particles are allowed to remain on the pots, disease organisms within the particles may not be reached by the disinfectant solution during the soak. These organisms can then infect any crop planted later in the pots.

Some growers have reported that root development is poor on plants growing in white plastic pots because light passes through the walls of the pots and reaches the roots. Research has not shown that light has a negative effect on root growth. Several types of glass containers were used in recent studies to grow foliage plants. All plants showed excellent growth, both when light reached the roots and when it did not penetrate the walls of the pots.

Peat Pots and Multicell Paks

Peat pots are made of compressed sphagnum peat moss and newspaper fiber. **Multicell paks** are molded from various types of plastics. These containers are often used to start plants, particularly spring flowering annuals. The features and uses of these containers are covered in Chapter 28.

Pot Sizes and Shapes

Clay pots and plastic pots are manufactured in a range of sizes. The standard pot is as wide at the top as it is high. For example, a 6-inch (15-cm) **standard pot** is 6 inches (15 cm) wide at the top inside measurement and 6 inches (15 cm) deep. A 4-inch (10-cm) standard pot is 4 inches (10 cm) wide and 4 inches (10 cm) deep.

Another type of pot is known as an **azalea pot.** The height of this pot is three-fourths of the inside diameter measurement at the top of the pot. Azalea pots have a wider base than standard pots and are less likely to tip over.

Pans are often used for poinsettias and bulb crops. These containers have a height equal to one-half of the diameter. Pans are used less frequently now than in past years. Growers now prefer to use three-quarter azalea pots for poinsettias because of the greater soil volume.

Hanging Baskets

Hanging baskets are specialized containers used for foliage plants and bedding plants (Figure 22-5). Many hanging baskets are sold with a saucer to catch drainage water. It is important that this saucer be removed during greenhouse production to prevent waterlogging of the medium. It can be snapped back into place prior to shipping or sale.

Growing Media

Regardless of the type of container used, the growing medium placed in the container must have all of the features listed in previous chapters for a proper medium. It must drain freely but hold enough water for good plant growth. It should be loose and well aerated. It must contain organic matter. It must be free of insects and disease organisms. It should serve as a storehouse for applied nutrients.

Field soil is not a good growing medium when it is removed from the field. Field soils in place have certain physical features relating to drainage that are destroyed when they are removed and placed in shallow containers. Undisturbed field soil has long capillary tubes that conduct water through the soil. When the soil is removed, these tubes are broken. As a result, the downward

FIGURE 22-5 Hanging basket of ivy geraniums ready for sale.

movement of water by the combined action of gravity and capillarity is greatly reduced. When the soil is placed in a container, the condition known as a perched water table develops at the bottom of the container. This means that 100 percent of the pore spaces in the soil are filled with water and there is no room for air. The excessive water content causes the roots to die.

Drainage can be improved in soils removed from the field by adding materials to make the soil more porous. Sphagnum peat moss, perlite, rock wool, vermiculite, and sand are commonly used for this purpose.

The Cornell peat-lite mix A, consisting of 50 percent by volume of sphagnum peat moss and 50 percent horticultural vermiculite, is used by many growers for poinsettias, potted chrysanthemums, foliage plants, and bedding plants. This mix meets all of the requirements of a good growing medium.

Another good medium for potted plants consists of equal parts by volume of sphagnum peat moss, loam soil, and either perlite or sand. This mix is considerably heavier than the Cornell mix A. The 1:1:1 mixture has the good drainage and aeration needed by pot plant crops. Although this mix has recently fallen out of favor with many growers, other greenhouse growers use loam soil in their mix to increase the cation exchange capacity creating an additional buffer for nutrient retention.

When plants are first potted, the fertility of the medium should be at a low level. If the fertilizer content is too high, the level of soluble salts may be high enough to damage the roots of young transplants or rooted cuttings.

A controlled-release fertilizer, such as Osmocote® or Magamp®, may be added to the potting medium before planting. If such a fertilizer is not used, then only ground limestone and superphosphate are added, if necessary. The medium should be tested before the crop is planted to determine if these materials are needed. If the pH is in the proper range for the crop to be grown, the limestone may be omitted. Similarly, if the phosphorus levels are above the recommended minimum value, then superphosphate is not required.

The medium should be tested every 3 or 4 weeks following the planting of the crop to check the nutrient levels. It is recommended that fertilizer be applied at every watering rather than by weekly or biweekly applications.

Transplanting, Potting, and Repotting

To make the potting operation more efficient, all required materials should be on hand and in their proper places before work is started. The medium must have the proper moisture content for potting. To test for moisture, squeeze a handful of the medium. If the handful is then dropped and it crumbles easily, the moisture content is in the right range. If the medium breaks into two or three large pieces, it is probably too wet. A very wet medium can be packed too hard, resulting in poor aeration and drainage. If the medium breaks into many pieces, it is probably too dry. A very dry medium is difficult to use. In addition, it may take so much water away from the plant roots that the plants are permanently damaged.

Plants being potted or repotted are placed in new containers at a depth equal to their depth in the propagation medium or a smaller pot. Most plants cannot be shortened by placing the roots at the bottom of the pot. The plant will soon die from a lack of oxygen.

If the plants were rooted in peat pots, or in other types of **fiber pots** (Figure 22-6), the top of the peat pot must be covered with the medium when the plant is repotted. If the rim of the pot extends above the soil line, wicking occurs. That is, moisture is absorbed from the medium by the peat material and evaporates to the atmosphere. This action causes the medium inside the peat pot to dry faster than it can take up moisture from the surrounding potting medium, even when the surrounding medium is thoroughly moist. If there is moisture stress because of high temperatures, the root ball may not receive enough water from the potting medium. As a result, the plant wilts. In newly panned poinsettias, this problem causes sun scorch of the leaves. Another solution is to peel off the top edge of the peat pot, but this requires a lot of labor and time if there are many plants to transplant.

The plants should be placed at the same depth in each pot. Because there is the same amount of space at the top of each pot, each plant will receive the same amount of water and fertilizer at each application. The space allowed at the top of the pot is usually one-half the depth of the pot collar (Figure 22-7).

The potting medium is firmed only until the plant is held in place. Too much pressure on the medium may break the roots (Figure 22-8). Do not press along the sides of the stem as root damage may result. As the pot is filled, it is rotated and pressure is applied on the medium with the thumbs toward the perimeter of the pot.

Small plants should not be potted in large containers. An oversized pot has too large a volume of soil. This means that the soil remains too wet for small plants, resulting in poor growth and eventual death of the plant. Each plant should be placed in a pot of the proper size. The plant can be repotted into larger pots as it grows.

As plants grow and fill the medium with roots, they must be repotted. Plant growth is stunted if the plant is not repotted to a larger pot at the proper time.

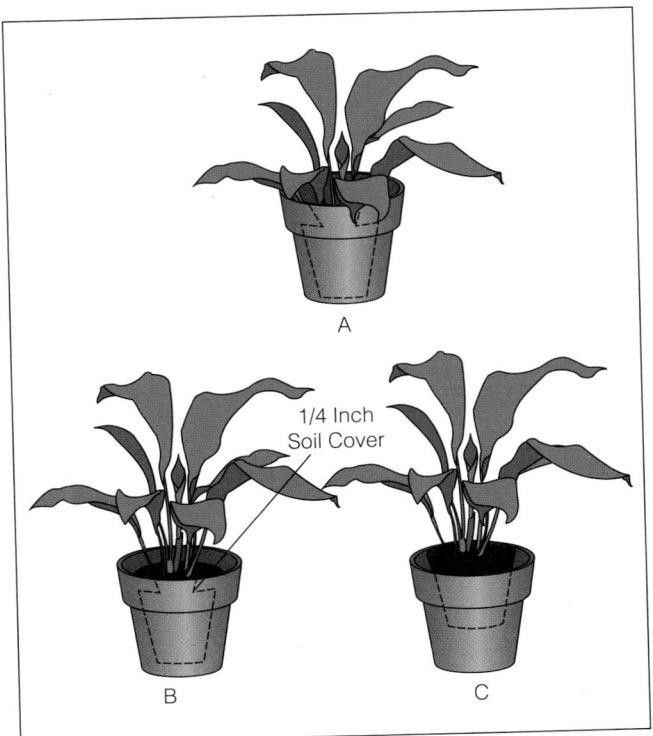

A

1/4 Inch
Soil Cover

B   C

FIGURE 22-6 The proper planting depth for plants started in peat pots is shown at B. The plant at A is set too deep in the medium. The peat pot at C is too shallow, resulting in "wicking" of moisture from the top of the peat pot.

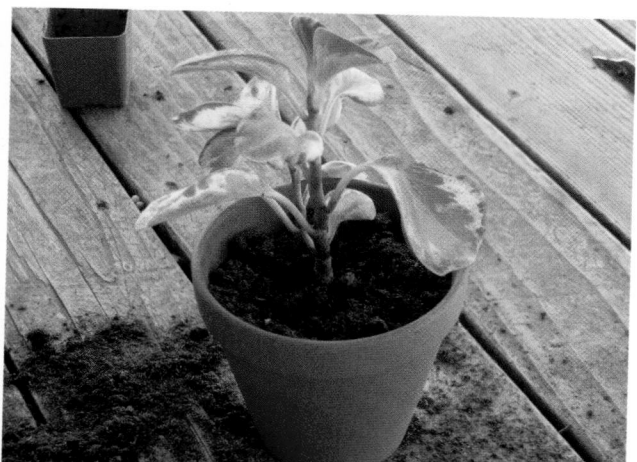

FIGURE 22-7 This reservoir at the top of the pot is at the proper depth for applied water and fertilizer solutions.

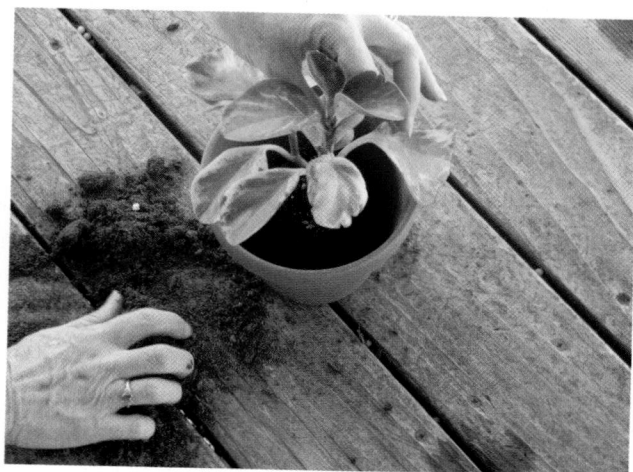

FIGURE 22-8 Proper placement of hands and fingers is important when firming the medium around the plant as the pot is turned. If the medium is pressed too close to the stem, roots may be broken.

FIGURE 22-9 This plant is ready to be transplanted. The roots at the bottom of the pot are beginning to circle and form a mat of roots.

The root system of a potbound plant may completely cover the growing medium (Figure 22-9). A potbound condition develops more rapidly in clay pots than in plastic pots. Because the clay pots are porous, air reaches the growing medium through the pot wall. Roots grow toward the outside of the medium ball. They then circle inside the pot wall and wind around the medium ball to form a tightly interwoven mass. The roots do not turn inward and grow back into the ball.

In plastic pots, the roots grow more evenly throughout the medium. They are not concentrated around the outside of the medium as in clay pots.

When a heavily potbound plant is to be repotted, it is recommended that the root ball be given a hard squeeze. This action breaks some of the roots, and new root growth occurs at the points of injury after repotting.

Preparation for Marketing

Potted plants must be prepared for marketing. That is, they are hardened off to increase the carbohydrate levels in the plant tissues. Hardening off consists of lowering the temperature a few degrees below the temperature at which the crop is normally grown. During the last few weeks of production, poinsettias are often grown at a **night temperature** of 55°F (13°C) to deepen the color of the bracts. Lilies and potted chrysanthemums also benefit from a reduction in the night temperature prior to sale.

The grower should not withhold fertilizer or water from plants because they are going to market. The practice places too much stress on the plants, resulting in premature leaf drop when the plant reaches the poorer growing environment of the customer's home. Yet some species, such as poinsettias and chrysanthemums, benefit from stopping fertilization 2 weeks before market. Stopping fertilization at that time will intensify the color on poinsettia bracts and extend the life of the chrysanthemum in the home.

The grower should clean the pots until they are free of algae before shipping them to the retailer. Very often, however, the cleaning is left for the retailer. Plastic pots are less subject to algae growth than clay pots. Extremely dirty clay pots must be washed before they are sold. It is a disservice to the customer to sell a plant in a dirty pot.

Potted plants are watered thoroughly before they are shipped. Plants shipped with a dry medium may be damaged in transit if they are moved long distances. There are several hydrophilic water-holding products in the marketplace. When mixed into the potting medium, they are supposed to extend the time period between needed applications of water. Upon being wetted, the particles expand many times, one claims up to 400 times, their original size in the form of a gelatin-like substance. Several researchers have found that these products do not perform as claimed. The grower should experiment with them on a small scale before making a large application of these materials in a crop production program.

During the late fall, winter, and early spring, flowers must be protected from cold temperatures. Paper or polyethylene sleeves can be fitted over single pots to give protection from the cold. Plants that are to be moved long distances in subfreezing weather generally are placed into sleeves. They are then packed into heavy cardboard boxes.

The bracts of poinsettias are easily bruised. Care should be taken to not damage the bracts when the plant is placed in a sleeve. They should be wrapped no more than 24 hours before shipment. If they are enclosed longer than this time, ethylene gas builds up and causes the petioles of the bracts to curl. This condition is called epinasty.

Potted plants should be loaded into heated trucks from heated shipping areas. They should not be exposed to temperatures below 40°F (4.5°C) for extended periods of time; severe damage may result. Plants like African violets and other gesneriads are damaged at temperatures below 50°F (10°C).

The grower's responsibility for producing high-quality, flowering potted plants does not end until they are delivered safely to market.

Select the best answer or answers to complete each statement. List the appropriate letter(s).

1 Of the floriculture crops sold today, what percentage of these crops is grown in pots or other types of containers?
a. 35 percent
b. 43 percent
c. 50 percent
d. More than 80 percent

2 Root systems of plants grown in clay pots:
a. grow entirely inside the potting medium.
b. quickly grow to the edge of the medium and curl around the inside pot wall.
c. may become potbound very quickly.
d. always have enough room.

3 Plants grown in plastic pots may be:
a. overwatered in the spring and fall.
b. overwatered in the summer.
c. overwatered in the winter.
d. watered the same year-round.

4 Plastic pots that are reused for potting plants should be:
a. fumigated with methyl bromide.
b. steam sterilized.
c. washed thoroughly and then soaked for 10 minutes in a solution of approved disinfectant.
d. given no special treatment.

5 The phrase "perched water table" describes a condition in a plant container where:
a. the container is sitting in water.
b. all of the pore spaces in the soil at the bottom of the container are filled with water.
c. one-half of the pore spaces in the soil at the bottom of the container are filled with air.
d. all of the pore spaces in the medium are filled with water.

6 When newly rooted cuttings or seedlings are potted, the soluble salt content of the media should be:
a. zero.
b. on the low side.
c. medium high.
d. very high.

7 Potted plants properly prepared for market should be:
a. hardened off by lowering the night temperature.
b. grown without fertilizer for 3 weeks before sales.
c. sent out unwrapped in winter.
d. watered well before being shipped.

Flowering Plants I

OBJECTIVES:

After studying this chapter, the student should have learned the cultural details for the flowering potted plants Abutilon to Calceolaria. For each crop, the student should be able to describe:

■ methods of propagation.

■ soil or growing medium requirements.

■ fertilizer recommendations.

■ special methods used to control growth.

■ disease and insects that affect the plants.

The production of flowering potted plants may be only a small part of a retail grower's business. For a large wholesale grower, however, the sale of flowering potted plants may be a major part of the business. Few growers produce just one type of flowering potted plant. Normally, 10 to 20 or more different types of plants are grown. Many of these plants are seasonal or are grown for certain holidays.

The cultural information for each crop given in this chapter and following chapters should be viewed by the student as a foundation for future study. The information presented here should not be regarded as the only possible approach to plant production. The student should build a fund of knowledge by referring to one-crop manuals, trade papers, state flower grower organization bulletins, cooperative extension service bulletins, and other published material. Additional knowledge is obtained by attending growers' schools and meetings, state florists' conventions, and similar meetings. Finally, experience will teach the student the cultural requirements of crops under specific conditions. The true student of horticulture maintains the search for knowledge throughout his or her lifetime.

Format of Cultural Information

The information presented for all of the crops listed in this section follows the same format. The common name of the crop is given first. This name is followed by the scientific genus, species, and the **family** to which it belongs. According to the convention used in horticultural science, the genus and species names are given in italics.

⊠ **Abutilon**

⊠ **Achimenes**

⊠ **Azalea**

⊠ **Begonia**

⊠ **Bromeliads**

⊠ **Bougainvillea**

⊠ **Caladium**

⊠ **Calceolaria**

A brief description of the crop is given when possible. Then, the specific requirements of each crop are listed for the following factors: propagation, growing media, watering, fertilizers, and temperature. Any special needs or handling details are presented. The discussion for each crop ends with a brief review of the cultural problems, if any, and a description of the diseases and insects that attack the crop.

Exact disease and insect control recommendations generally are not given. The Environmental Protection Agency (EPA) continually issues changes and restrictions on the use of pesticides. A substance that is registered for use today may have that registration withdrawn next week. There is also a great variation in the pesticide regulations of individual states. Thus, it is unwise to even attempt to give detailed control recommendations.

ABUTILON (Flowering Maple) ⊠ *Abutilon species, Malvaceae*

Abutilon has bell-shaped flowers and is attractive as a potted plant. Abutilon hybrid cultivars usually are grown for spring sales. The seed can be sown in October. The plants are grown at a temperature of 60°F (15.5°C). The plants will flower in March and April in the northeastern United States. Less time is needed to reach the flowering stage in areas where the light intensity is higher during the winter.

Terminal cuttings propagated in January or February will be ready for sale in May as 4-inch (10-cm) pot plants. Full sunlight is needed for best growth.

The peat-lite mixes are excellent as both propagation and growing media for Abutilon cultivars. The plants are fertilized at every watering using a solution of 20-10-20 analysis fertilizer at a strength of 150 ppm.

Problems

Aphids, mealybugs, thrips, and spider mites attack this crop.

ACHIMENES (Magic Flower) ⊠ *Achimenes hybrids, Gesneriaceae*

Achimenes are trailing plants with tubular white, blue, purple, pink, and reddish flowers. These herbaceous perennials are one of the best plants for hanging baskets. They grow from small underground, pinecone-like rhizomes. The plants may be erect or drooping in growth habit.

Achimenes are similar to African violets in their cultural needs. They grow best when the night temperature is warm or 65°F (18.5°C). A porous, peat-lite growing medium is used.

Achimenes are propagated by dividing the rhizomes in March or April. The pieces of rhizome are covered with a 1-inch (2.5-cm) layer of medium. The propagation flat must be kept evenly moist. A low-pressure mist can be used to maintain the proper moisture level.

Achimenes are fertilized at every watering with a solution of 20-10-20 fertilizer at a strength of 100 ppm.

The light intensity should be no greater than 2,000 fc (2.5K lux). Light that is too bright causes leaf burn. Leaf spotting occurs if cold water is dropped on the leaves. This is the same kind of injury that is seen on African violets.

After flowering, the plants should be allowed to dry gradually by withholding water. The rhizomes are then removed from the medium. They are shaken or brushed carefully until they are free of old soil. They are then stored in dry sand or vermiculite at 45°F (7°C) until February when they are repotted.

Problems

Cultural Disorders

A high light intensity produces yellowing of the leaves. When water at a temperature less than 60°F (15.5°C) strikes the leaves, chlorophyll at these points dies and leaf spotting occurs.

Diseases

Overwatering causes crown rot.

Insects

The plants are attacked by aphids, mealybugs, and thrips.

AZALEA ✖ *Rhododendron species, Ericaceae*

Azaleas are one of the most popular of the flowering potted plants (Figure 23-1). They are grown in almost all parts of the United States. However, in most of the United States, azaleas must be protected from severe winter weather.

There are several azalea production areas for florist azaleas in the United States. Each of them, due to differences in climate and environmental conditions, produces the crop in a different way.

On Long Island, New York, Kurt Weiss Greenhouses, owned by Russell Weiss and Son, are the principal azalea producers. They also have production facilities in Florida, South Carolina, and Connecticut. Their production is natural season, bringing plants on line as early as September for very early forcing. Over winter the plants are held in polyethylene covered greenhouses heated just enough to keep them from freezing. These plants are then forced into bloom for Christmas and Valentine's Day sales. As spring progresses, the plants are moved into refrigerated rooms where they are held until brought out to force for Easter and Mother's Day, the last major holiday of spring for azalea sales.

Weiss propagates his own material. Sales are primarily to food stores and mass market outlets. Sales are also made through brokers to retail florists. Kurt Weiss Greenhouses produce several million azaleas annually.

In Florida there are three major producers. Yoder Brothers of Fort Myers is the largest. The other two growers are natural season producers. Due to Florida's warmer temperatures the natural season is shorter than in other areas. Their first flowering crops are not available until November and December. These growers service grocery stores, garden centers, and landscape nurserymen.

Yoder Brothers produces azaleas on a 52-week, or year-round, schedule. They have built their business on production of dormant azaleas. These are plants that have been propagated and grown on with several pinches to develop multiple branches. After reaching bud stage, the plants are sold to wholesale growers for precooling, either naturally or artificially in refrigerated cool rooms. They also

FIGURE 23-1 Azalea (*Rhododendron*). *(Courtesy of the Professional Landcare Network, formerly the Associated Landscape Contractors of America.)*

sell azaleas that are in the candle stage buds just beginning to show color. These plants are primarily sold to wholesalers who supply retail florists.

Yoder Brothers from Florida supplies florists with azaleas from Florida as far north as southern Canada and west to the Rocky Mountain states. They supply product 52 weeks of the year.

A third product form is also available. This is a "ready to force" azalea. This is a plant that has been grown outdoors in Florida under Saran cloth that provides partial shade. It has also received the proper precooling treatment to break dormancy of the buds. Because they are given the precooling treatment in temperature-controlled cold rooms, they have received a precise number of hours of chilling. This ensures that the forcing time needed to bring them into flower will be the more consistent regardless of the season the plants are forced.

The last production area is in the Pacific Northwest. Six major producers are located in Oregon. There is another grower in Washington.

These growers sell blooming plants to wholesale and retail florists. Depending on the grower, they also sell precooled plants to wholesale forcers.

They start shipping naturally precooled plants to their customers in August and September. The amount of chilling these plants receive depends on ambient weather conditions. If summers are unusually warm, the plants may not be fully precooled and will require a longer period to force to bloom than if they had received the proper amount of chilling. Flower bud initiation occurs during the warm weather of June and July.

Among the very early azalea cultivars that can be forced into flower are the Vogels. These plants originated in Germany. The original cultivar is called 'Helmut Vogel.' It is a red flower cultivar that mutates very readily. There are more than 50 different color sports (mutations) of the original Vogel cultivar in the trade. These are mostly sold in Europe. This group of azaleas requires very little cold temperature to break dormancy. They can be readily forced for Thanksgiving and Christmas sales as well as for Valentine's Day.

Most Oregon growers plant a crop for the season. Only a few attempt to produce plants on a 52-week, year-round schedule. They do this by first planting those cultivars that can be forced early. Among these, the Vogels are considered to be very early. These are followed by plantings of midseason flowering types and then the late flowering cultivars.

Oregon growers begin shipping plants in August and September. They continue shipping until February and March, at which time they are done for the year.

Oregon growers service wholesalers, traditional retail florists, mass market outlets, and grocery stores through grower wholesalers.

Origin

Azaleas are members of the genus *Rhododendron*. This genus is very large and is divided into series and subseries. The azalea is one of 43 series described by horticulturists working within the genus *Rhododendron*. However, these series do not represent clearly defined and separable groups and are not often considered legitimate by traditional taxonomists.

Most azaleas available today are the result of crossbreeding plants from Japan or China. The majority of the plants used for breeding originated in Japan. *Rhododendron indicum* was found in southern Japan. It is one of the earliest cultivars used by the Belgians for breeding purposes. *Rhododendron indicum*, or more correctly *R. indica*, is a late forcing cultivar with large flowers.

The *Kurume* azaleas were bred from the Japanese *R. kurume*. The genetic makeup of modern *Kurume* cultivars is greatly diluted because of extensive crossbreeding with other azalea cultivars. *Kurume* azaleas have small, single flowers that range in color from white to red. There are large numbers of flowers on each plant. When these azaleas are planted outdoors, they are winter hardy down the eastern coast from Connecticut to Georgia.

The *pericat* azaleas were developed in the late 1920s by Alphonse Pericat of Collingswood, Pennsylvania. Mr. Pericat did not keep accurate records of the parent plants used in the breeding program. However, it is thought that he used Indian (Belgian) and *Kurume* hybrids. The pericats continue to be found in the trade from specialty growers.

The first hybrid azaleas patented in the United States were Rutherford azaleas developed by L. C. Bobbink of East Rutherford, New Jersey. They were introduced about 1935. Again, there is some question regarding the parentage of this azalea. It is thought by some authorities that *Kurume* and *indica* hybrids were used.

The U.S. Department of Agriculture conducted two large breeding programs that produced the Glenn Dale azaleas. The resulting cultivars have large flowers and can withstand colder temperatures in many parts of the United States.

Charles Sanders crossed *R. obtusum* with Indian hybrids to produce the Sanders azaleas.

Another U.S. Department of Agriculture breeding program in the 1940s and 1950s produced the Yerkes-Pryor azaleas. Guy E. Yerkes and then R. L. Pryor tested and evaluated these new *Kurume* hybrids.

Propagation

Azaleas may be propagated by seed, leaf bud cuttings, terminal cuttings, grafting, or layering. However, most growers propagate azaleas by terminal cuttings.

Grafting

The use of grafting as a method of azalea propagation is almost extinct. Less than 1 percent of the azaleas grown in the United States today are grafted.

Cuttings

Terminal cuttings are mainly taken from production plants as they are being shaped and sheared. The only time that stock plants are grown specifically for propagation purposes is when a new cultivar or mutation is being increased for production purposes. The time of year when cuttings are taken depends on the type of cropping being done. Where production is maintained on a 52-week schedule, cuttings are taken every week. In other production methods, cuttings may be taken from late spring to early fall.

Rooting hormone is not often used in propagating the plants. Growers have had problems with plant injury when hormones have been used.

Terminal cuttings generally are 2 to 4 inches (5 to 10 cm) long. The cut should be made with a sharp knife to minimize damage to the basal end of the cutting. The cut may be made at an angle or flat across the stem. It is easier to place a cutting with a slanted end into the propagating medium.

The rooting medium may range from 100 percent sphagnum peat moss (100 percent by volume) to mixtures of equal parts of perlite and vermiculite and peat moss. The mixture should not contain less than 50 percent by volume of peat moss because it will dry out too quickly. A low-pressure mist can be used to maintain the necessary moisture levels around the cuttings. If the night temperature is less than 60°F (15.5°C), bottom heat of 65°F (18.5°C) should be used.

Full light intensity is seldom used in propagation. Diffused light is preferred and is obtained by using various density shade cloths over the plants. If there are flower buds present on the cuttings, they are removed.

It takes 10 to 12 weeks or more for an azalea cutting to develop a good root system. Some cultivars may need more time in the propagation bench until the cuttings have developed sufficient roots.

The rooting medium should be steam pasteurized or chemically treated between batches of cuttings to ensure that it is free of diseases. If the medium cannot be steamed, it should be replaced after every batch of cuttings. The medium should not be drenched with fungicides because these are known to reduce the percentage of rooted cuttings.

Some growers are propagating in cavity trays using peat-lite type media. In this situation the media is potted with the plants, so new propagation media is used each time cuttings are rooted. This avoids the problem of disease carry-over in the propagation media. Some growers have had success with preformed media such as the Smithers-Oasis Wedge® product. The Horticube® media, which has 55 to 65 percent drainage, is the preferred product.

Unrooted Cutting Storage

Unrooted *Kurume* azalea cuttings can be stored for 4 weeks at storage temperatures of 32°F, 36°F, and 39°F (0°C, 2°C, and 4°C). **Cold storage** is useful when the grower has a limited amount of propagation space and a large number of cuttings to root.

Production

After the cuttings are rooted, they are planted directly to the finish size pot. Other plants that are to be grown on as liners are planted out to the field. These plants are later potted up for forcing.

Azaleas make their best growth in an acid medium ranging from a pH of 5.0 to 5.5. The medium may be 100 percent sphagnum peat moss or various mixtures of soil and/or sand and acid peat moss.

Sphagnum peat moss is preferred to other types of peats. It has the greatest water-holding capacity and the best porosity for azaleas. The pH of the peat should be checked before the medium is used. Some sphagnum peats tested have had an alkaline pH. Others have been too acidic for azaleas. In this case, limestone must be added to increase the pH. Calcium may be added as well. If the pH is below 5.0, limestone must be added. Normally, 1 to 2 pounds (0.60 to 1.2 kg) of limestone is added to each cubic yard (cubic meter) of peat and then is mixed in thoroughly.

Very dry peat moss is difficult to wet. This problem can be overcome by using a surfactant.

Fertilization

Azaleas have very fine root systems that are easily damaged by excessive fertilization, resulting in high soluble salt levels. Azaleas do not need high levels of fertility; however, they do need large amounts of iron and manganese. It is easier for azalea roots to take up these mineral elements when the pH of the medium is acid. The control of the pH level is an important factor in preventing iron chlorosis.

When the medium is prepared, treble super phosphate 40 to 47 percent P_2O_5 is mixed in thoroughly at the rate of 1 pound (0.6 kg) to a cubic yard (cubic meter).

Nitrogen and potassium are applied in liquid form at concentrations of 100 ppm each with every watering. Iron can also be applied at each watering. The amount of iron used depends upon the percentage of the element in the material applied. Chelated iron sources containing 12 percent active iron (Fe) are applied at the rate of 5 ppm or less at each irrigation. If a single treatment is to be made to correct iron deficiency symptoms, a product containing 12 percent iron is applied at the rate of 1 ounce (28.4 g) in 25 gallons (94.6 l) of water. One-half pint (237 ml) of solution is applied to each 6-inch (15-cm) tall plant. This treatment should not be repeated for at least 3 weeks to prevent serious plant injury. A product containing 6 percent active iron is applied at the rate of 2 ounces (56.8 g) in 25 gallons (94.6 l) of water.

Computer-controlled sensors may be used to accurately determine when the crop needs watering. Leachate from the pots is tested frequently to determine when fertilization is needed also. To successfully use these sensors, accurate records must be maintained in order to determine trends during the crop cycle and between cropping cycles.

Pruning and Shearing

During the growing period, the plants are pruned and sheared to control and increase head growth. Long, straggly growth of single branches makes a plant unattractive for sale. When a plant is sheared, only the tips of the branches are removed. This treatment causes more shoots to develop. Each shoot tip is the potential location for a flower bud. Thus, the more shoots a plant has, the greater the number of buds that may develop.

Chemical pruning agents have been used with erratic success. Often the pinching patterns resulting from a chemical pruning agent are not uniform. When using any chemical on a plant, always follow the directions on the label exactly.

Flower Bud Formation

The azalea flower cluster is made up of several buds. The number of buds, which may range from three to seven or more is determined genetically. Some cultivars have more buds than others. It is also determined by the number of days that elapse during the flower initiation period from the time of the last pinch. A natural season crop is usually overbudded. The reason for this is the extended period from the last pinch in spring until the cooler weather of fall that stops initiation and starts bud development. Some azalea cultivars respond to a photoperiod. That is, the bud initiation occurs sooner when the plants are placed under short-day conditions. Most azalea cultivars produce fully developed flower buds when exposed to a temperature of not less than 65°F (18.5°C) for 8 to 10 weeks after the last pinch. In the natural state, this condition occurs in the summer and early fall. The flower buds that form are ready for precooling by mid to late September.

Flower bud initiation can be made to occur sooner by applying growth retardants such as Cycocel® and B-NINE SP®. These materials are applied as foliar sprays at the rate of 2,000 ppm for Cycocel® and 1,500 ppm for B-NINE SP®. The plants are sprayed when shoots are 1 to 1.5 inches (2.5 to 4 cm) long following the last pinch.

The time gained by the earlier bud set is lost during the forcing period. As compared to an untreated plant, a longer time is required to force a plant into bloom after it is treated with a growth retardant.

As the plant matures, fertilizer is applied regularly. Extra iron is applied if chlorosis develops. The fertilizer used is a nitrogen type that has an acid reaction in the soil.

If the pH of the water is alkaline, regular treatments with iron are required. Some midwestern growers must use water with a high pH. They acidify the water by adding an acid. Sulfuric and phosphoric acid are two of the more common acids in use. They are corrosive chemicals and must be used carefully. The amount of acid needed to acidify a water supply to the desired pH depends upon three conditions: the acid used, the original pH of the water, and the buffering capacity of the water, which is determined by measuring the alkalinity. Each problem with pH is corrected on an individual basis.

Precooling Treatment

After the flower buds have set and developed to a certain stage, they then need a cool temperature treatment. This precooling treatment, as it is known, causes the buds to continue development. It also overcomes the dormancy of the buds, which allows the grower to force the plants so that all the flowers bloom at the same time. This is the quality plant the consumer prefers to purchase.

There are three methods of precooling azaleas: natural cold treatment and mechanically refrigerated treatments where plants are kept in either absolute dark or lighted cold rooms.

Natural cooling is obtained by exposing the plants to the increasingly cooler temperatures of fall; in areas where winter temperatures regularly go below freezing, the plants must be protected. This is normally done by placing them in greenhouses and supplying enough heat to maintain a minimum temperature of 35°F (1.7°C). If a plastic covered greenhouse is used, the greenhouse is covered with white rather than clear plastic. White plastic keeps the winter sunshine on clear days from overheating the greenhouses.

Before the plants are placed in the greenhouse, they are thoroughly irrigated. It is important that the foliage be kept dry while in storage to help minimize disease development.

Dark storage in mechanically refrigerated cold rooms requires precise temperature and relative humidity conditions. The thermostat must be accurate to within plus or minus 0.5°. The temperature required for the 28-day dark storage period is 36°F (2.2°C) plus or minus 0.5°. The **relative humidity** must be kept between 85 and 90 percent. The plants must not be allowed to dry out. If these conditions are maintained exactly, there should be little or no foliage turning yellow or leaves dropping. If this temperature cannot be met and the storage period is going to be longer than 28 days, lights must be used. The air movement in the cold room must be very gentle. A low air velocity chiller is recommended. If this is not available, baffles must be used to keep drafts from hitting the plants. If plants are exposed to rapid air movement, they quickly dry out, wither, and drop leaves.

There must be no fruit or vegetables stored in the cold room with azaleas that are being precooled. Even at low temperatures fruits and vegetables produce ethylene gas, which causes leaf drop.

While in cold storage the foliage should never be wet. If the plants require irrigation more than once during the 4-week precooling storage, then proper conditions are not being maintained. The best way to hold the 85 to 90 percent relative humidity is to keep the floor moist. This can be done by installation of a drip watering system, or a fog system controlled by a humidistat.

Lighted Storage

The conditions for precooling azaleas in lighted storage require a temperature of 38°F to 42°F (3.3°C to 4.4°C). Light intensity of 30 fc (324 lux) is provided by two 40-watt, cool white fluorescent lamps spaced 22 inches (55 cm) apart on 48-inch (120-cm) wide shelves. The vertical spacing between the shelves is 22 inches (55 cm). This will put the lamps about 6 inches (15 cm) above the plants. Reflectors are placed above the lights to direct the light downward and to protect the units from dripping water. Unless ballasts are designed for low temperature conditions, the ballasts for the lights are mounted remotely, outside of the cold room to prevent excessive cold from affecting the ballast's temperature. The lights are kept on for 12 hours daily. The lights are usually turned on at night when outside temperature is lower than day temperature. In some areas a lower cost for electricity may be obtained at night. The relative humidity is maintained at 80 to 90 percent.

In lighted storage, plants can be held for a maximum of 6 weeks. After this they are removed for forcing purposes.

Prior to placement in storage the plants are thoroughly irrigated. The foliage must be completely dry before the plants go into storage, or leaf injury may occur. Leaf diseases may also get started. 'Dorothy Gish' cultivars are especially sensitive to wetting of the leaves. If this occurs the first 10 days the plants are in storage, the leave may turn black and drop.

If leaf drop occurs and the leaves are green, with no loss of color, ethylene gas may be the expected cause. Leaves that turn yellow and/or brown and fall are a result of the plant drying, disease starting, or a combination of both.

Forcing for Holidays

After the required period of precooling is completed, the plants can be forced into flower. Because certain cultivars have less of a chilling requirement than others, these plants can be forced for early holidays. The Vogel cultivars are considered to be very early for forcing purposes. Table 23-1 gives the details of treatments needed for selected cultivars for various holiday sales.

Some azaleas are available for Thanksgiving and Christmas. Larger numbers become available for Valentine's Day, Easter, and Mother's Day. Easter and Mother's Day are the most important holidays in terms of sales volume.

Table 23-1

A Partial Listing of Recommended Azalea Cultivators for Holiday Forcing		
Holiday	Cultivars Classification	Treatment Required
Thanksgiving	'Helmut Vogel'—red 'Inga'—pink/white 'Jacinth'—pink 'Marcelle'—white 'Marianne Harig'—red/orange 'Nicolette Keessen'—pink/white 'Nordlicht'—red 'Paloma'—pink blush opens to white 'Edmond Troch'—white	Very early flowering. These cultivars will force best for Thanksgiving, Christmas, and Valentine's Day sales. They need the least amount of precooling to force uniformly. They are not recommended for Easter or Mother's Day sales from natural season production because of the difficulty in holding them.

Christmas	All of the cultivars forced for Thanksgiving can be forced for Christmas. Additional cultivars are:
	   'Alaska'—white
	   Bertina'—pink
	   'Coral Bell'—pink
	   'Coral Dogwood'—coral
	   'Dogwood'—white
	   'Erie'—salmon/white
	   'Julia'—pink/white
	   'Nancy Marie'—pink/white
	   'Oregon Alaska'—white
	   'Pink Dream'—pink
	   'Prize'—red
	   'Sarah Ann'—pink
	   'Terra Nova'—light pink
	   'Variegated Dogwood—variegated
		Early flowering. This group requires more budding time and precooling than the early flowering cultivars. They are recommended for Christmas sales as well as later holidays.

Valentine's Day	All of the cultivars forced for Thanksgiving and Christmas sales can be forced Valentine's Day. Additional cultivars are:
	   'Adonia'—purple
	   'Del. Valley White'—white
	   'Dorothy Gish'—Salmon
	   'Gish White'—white
	   'Dr. Koster'—red
	   'Friedhelm Scherrer'—deep red
	   'George Struppek'—pink
	   'Gloria'—pink/white
	   'Hershey Red'—red
	   'Hinodegiri'—pink/red
	   'Madame Kint Glasser'—white/pink
	   'Red Ruffles'—red
	   'Road Runner'—deep pink
	   'Rosalie'—pink
	   'Satellite'—rose/white variegated
	   'Snow'—white
	   'Sweetheart Supreme'—light pink
	   'Tradition'—pink
	   'Valentin'—pink
	   'Violacea'—purple
		Midseason flowering. These cultivators will flower well in natural season treatments for Valentine's Day and later holidays. They can also be held for Easter and Mother's Day sales in cold rooms. They should not be forced to bloom before January 1st. Inadequate cooling will cause bud blasting and bypass growth.

Easter and Mother's Day	Any of the cultivars forced for Christmas of Valentine's Day can be forced for Easter and Mother's Day. Medium to late blooming cultivars are preferred. Early forcing cultivars are too difficult to hold back. Other suitable cultivars are:
	   'Barbara Gali'—pink
	   'De Weales Favorite'—pink/white
	   'Knute Erwin'—red
	   'Laura'—rose pink
	   'Pink Ruffles'—pink
	   'Pink Supreme'—medium pink
	   'Stella Maris'—white/pink
		Late flowering. These cultivars are best used for Easter and Mother's Day sales. Refrigerated cold rooms are needed to hold them, especially if spring weather is warm. These plants should not be forced before January 15th. Bud blast and bypass growth will occur from inadequate cooling.

Year-Round Flowering

Yoder Brothers with production facilities in Fort Myers, Florida, is a major supplier of azaleas year-round. These plants require special handling procedures to maximize quality flowering products.

When year-round dormant azaleas are received in the months of May through October, the following are Yoder's recommended handling procedures. When the boxes arrive, unpack them immediately. This allows any condensation that may have built up in the box to be released. If they cannot be unpacked immediately, place them in a 35°F to 40°F (1.1°C to 4.4°C) cold room. Do not allow them to sit in the sun or be exposed to freezing temperatures.

After unpacking, place the plants in the greenhouse for 7 days at 3,000 to 4,000 fc (32.3K to 42.3K lux) of light. This permits the rebuilding of carbohydrate reserves that were depleted during shipping. It also prevents ammonia leaf burn from chemical buildup in the leaves during the cold storage period.

The plants should not be held in low light or high temperature conditions such as a foliage plant house or a packing shed. Maintain a 65°F (18.3°C) night temperature when possible. **Day temperatures** should not go above 85°F (29.4°C). As soon as the plants are placed in the greenhouse, they must be irrigated twice. This restores moisture to the media that was lost during shipping and also helps to warm the root ball. Do not fertilize the plants.

Year-round azaleas that are received during the months of November to April should be unpacked immediately and thoroughly irrigated two times. The second irrigation immediately follows the first.

As soon as the foliage is dry, the plants are placed in the cold room. Do not place them in the greenhouse, because this is not necessary and may cause uneven flowering when forced.

Maintain the proper temperatures, relative humidity, and light conditions as previously given. The environmental conditions during the precooling period influence the number of days needed to force the plants into bloom, the uniformity of flowering, and the retention of green foliage.

Forcing to Flower

After the proper precooling treatment, the plants are removed from the cold room and placed in the greenhouse. If maximum greenhouse temperatures can be held below 75°F (24°C), the plants can go directly into the greenhouse. They are irrigated immediately to help warm the root ball. Overhead misting or syringing several times a day helps to acclimate the plants to the greenhouse environment. The desired light intensity is 2,000 to 4,000 fc (21.5K to 43.1K lux). If needed, to reduce light intensity and keep temperatures down, the greenhouse can be shaded. Light intensities of 3,000 to 4,000 fc (32.3K to 43.1K lux) may be too high. These light levels may cause leaf burn and/or flower color fading.

In low natural light areas in winter, azaleas can benefit from the use of high intensity discharge (HID) lights to provide 350 fc (3.8K lux) at plant level for 12 to 18 hours daily. Lighting the plants helps to reduce forcing time, improves the uniformity of flowering, and most importantly increases flower count. Maintain a night temperature of 65°F (18.3°C) and a maximum day temperature of 85°F (29.4°C). If temperature goes above this, forcing time is reduced but flower color is faded as important carbohydrates are used quickly.

In northern greenhouses, bottom heat of 65°F (18.3°C) is beneficial in warming the root zone and reactivating the root system.

Gibberellic Acid

Research has shown that the use of **gibberellic acid** GA_3 growth hormone will help to improve the uniformity of flowering. GA_3 is prepared as a 250 to 500 ppm concentration. The solution is sprayed on the foliage three times: the third, tenth, and seventeenth day after removal from the cold storage room.

Space the plants adequately to maximize quality growth. There should be no overlap of foliage. Keep the media moist. Do not allow the media to dry out because buds may abort. No fertilizer is needed during the forcing period. Prune extraneous vegetative shoots, bypass shoots, as they develop. Do not let them grow excessively because they detract from the quality of the plant. Prevent petal blight by keeping the foliage dry in the cold room and also during the forcing period in the greenhouse. It is especially important to keep the opening flowers dry going into the night. Sanitation in the forcing area is critical to maintaining healthy plants. Clean up all fallen leaves and plant debris periodically and remove it from the greenhouse. Plant debris is a prime source of inocula for leaf diseases and petal blight.

Follow a good insect preventative control program. Periodic spraying will help to prevent thrips injury to the developing flowers.

The time that is needed to force the plants into flower varies with cultivars and the season of the year. Some cultivars will require 3 to 5 weeks to force for Christmas and Valentine's Day. Other cultivars will need 4 to 6 weeks forcing for the same holidays. These same cultivars will force more quickly for Easter and Mother's Day sales due to warmer temperatures and higher light intensities at that time of year.

A number of schedules have been proposed for forcing azalea blooms. All of the schedules include the following steps: (1) Produce enough vegetative growth; (2) provide a period for flower bud initiation; (3) provide the necessary cold temperature treatment; and (4) force the plants into bloom. Table 23-2 gives two sample schedules.

The following cultivars have been used in year-round flowering programs.

'Alaska'	'Kingfisher'*
'Barbara Gail'	'Red Macaw"
'Dogwood'	'Red Wing'
'Dorothy Gish'	'Roadrunner'*
'Gloria'	'Skylark'*
'Hershey's Red'	'White Christmas'*
'Hexe'	'White Gish'

Note: Cultivars shown with an asterisk (*) should be treated with a growth retardant. Treatment with B-NINE SP® reduces flower size and delays blooming. The use of Cycocel® delays flowering and plant size but does not reduce flower size.

In the summer, the light intensity should be in the range of 3,000 to 4,000 fc (37.3K to 43.2K lux). Full light exposure is required in the fall, winter, and spring. During periods of naturally short days, night lights are used from 10 P.M.

Table 23-2

Two Schedules for Year-Round Forcing of Azaleas

A. North Carolina State program for high light intensity areas. The time interval between pinching and blooming is 20 to 22 weeks.

Pot

6 weeks of long days at 65°F (18.3°C) → ← 6 weeks of short days for flower bud initiation and development → ← 6 weeks cooling at 48°F (17.8°C) with 20 fc for 12 hr. daily → ← Force (2-4 weeks) to bloom at 60°F-65°F (15.5°C-18.3°C) →

Pinch

NOTE: The North Carolina State program for the low winter light areas adds two weeks for short days.

B. Yoder Brothers, Barberton, Ohio. The time interval between pinching and blooming is 29 weeks.

2,500 ppm B-9, 2nd application of B-9 SP® (if needed)

← Pinch in 2 weeks 12 weeks of long days after final pinch → ← 6 weeks of short days at 65°F (18.3°C) for bud set and development → ← 4 weeks dark cooling at 34°F-36°F (1.1°C-2.2°C) or 6 weeks cooling at 38°F-40°F (3.3°C-4.4°C) with 7 fc for 12 hr. daily → ← 4-5 weeks force to bloom at 65°F (18.3°C) Depends on Season →

0   1   2   3   4   5   6   8   10   12   14   16   18   21   24   27   30-33

Week Number

to 2 A.M. daily. Incandescent bulbs rated at 100 watts are spaced 4 feet apart, 2 feet above the plants.

Holding and Preparation for Sales

When the plants have developed several open flowers and a large number of buds are showing good color, they can be placed in a cold room. These plants can be held for 1 week maximum at 35°F (1.7°C) without affecting quality. If held longer than this or at higher temperatures, the lasting life is affected and flower color is reduced.

Use proper sleeving material when packing the plants. Paper sleeves were once preferred over plastic sleeves, which retain moisture that may lead to the development of petal blight. Modern sleeves are a combination of a nonwoven polyester fabric and plastic glued together. This product will allow for gas and moisture exchange. The marketable stage of the plants depends on the local market requirements. If buds are just showing color, many of them will not open. Others that do open will be small and open slowly.

Problems

Cultural Disorders

Cultural problems may lead to wilting, chlorosis, leaf drop, and finally death. Other cultural disorders are related to poor flower formation, uneven and delayed flowering, and bypassing of buds when nonflowering shoots are forced.

Wilting

Wilting occurs for the following reasons: root damage due to very high soluble salt levels, keeping the root ball too dry, and keeping the root ball too cold.

In addition, very high light levels cause the leaves to lose moisture faster than it can be replaced by root uptake.

Chlorosis

Chlorosis appears as a yellowing of the leaves between the veins. Generally, chlorosis is due to root damage, which prevents the uptake of nutrients, especially iron. A deficiency of iron can also result from a lack of iron in the medium; too high a pH level; a very high soluble salt level; a high level of bicarbonates in the water supply, which results in high medium pH; overwatering or underwatering of the plants; or soil insects or root diseases.

Leaf Drop

Leaf drop occurs when the plants are too dry; the soluble salt levels are too high; the plants are not handled properly in storage; the temperature is too high; there is too little light at temperatures above 40°F (4.5°C); the relative humidity is low; ethylene gas accumulates; or root insects or diseases attack the plants.

Poor Flower Formation and Development

Problems with flower formation and development are due to the wrong day length and/or temperature conditions during the flower formation period. Another reason is that the plants may be placed in cold storage before the flower buds are far enough developed to respond to the low temperature treatment.

Diseases

Cylindrocladium scoparium and *Rhizoctonia solani* cause a cutting to rot and a graft union to decay. This problem may occur even after the plants are removed from the rooting area. *Exobasidium vaccinii* is a leaf gall that causes the leaves to become thick and fleshy. They also turn pale green to whitish in color. During forcing, brown spots on the flowers are usually due to *Botrytis cinerea*. *Phytophthora* and *Pythium* cause root rot. *Septoria* leaf spot is caused by a fungus specific to azaleas. *Ovulinia* petal spot is another fungus specific to azaleas. Only the petal tissue is attacked by this fungus. Moist, humid conditions increase the probability of fungus infections in azaleas.

Insects

The most troublesome soil pests are the black vine weevil and **nematodes** that attack the roots of azalea plants. Aphids, mealybugs, and spider mites are a problem during the forcing period primarily.

An insect and pest management program should be in place, which may or may not include an insecticide spray program. The program should be completed before the plants are brought into storage to minimize the insect problem.

BEGONIA ▓ *Begonia hybrids, Begoniaceae*

Commercial flower growers use several types of begonia. The most important florist crop is the semituberous *Begonia socotrana*, also known as the Christmas, Melior, or Norwegian begonia. *B. tuberhybrida* is a commonly used spring and summer flowering tuberous-rooted begonia. Rhizomatous begonias such as

B Rex, *B. heracleifolia*, and *B. ricinifolia* are grown for their colorful foliage (Figure 23-2). *B. × semperflorens-cultorum*, which is also known as the wax or fibrous-rooted begonia, is an everblooming type (Figure 23-3). It is used as a large potted specimen or for bedding purposes.

In recent years, the Rieger elatior begonias were introduced in this country. These hybrids were developed by the firm of Otto Rieger of Nurtingen, Germany. They quickly became popular with consumers because of their excellent colors and keeping quality.

Begonia Socotrana

B. socotrana is sold as a potted plant at Christmas. A year or more is required to produce a well-grown, blooming plant. Propagation is by leaf petiole cuttings made in November and early December. Rooting occurs in 4 weeks when a moist medium is used with bottom heat applied at 70°F (21°C).

The cuttings should remain in the propagation medium until new shoots develop, usually in 8 to 10 weeks. The cuttings are then potted to 2.25-inch (5.7-cm) pots and are grown on at 60°F to 65°F (15.5°C to 18.5°C).

Growth can be hastened by giving the plants artificial long-day conditions. (**Note:** These are the same conditions maintained for chrysanthemums.) By April, the natural days are long enough and lights are not required to keep the plants in the vegetative stage. From April through mid-September, the plants are shaded to reduce the light intensity. If the light intensity is too high, the chlorophyll in the leaves fades and they become chlorotic.

B. socotrana can also be propagated later in the winter. If the plants are lighted, they will grow as large as those propagated in November. The reduced growing period means that less energy is needed to produce saleable plants.

Plants propagated in December are moved to larger pots in early May or June. Six-inch (15-cm) pots are preferred if greenhouse space is available. If space is a problem, 4- or 5-inch (10- or 13-cm) pots may be used. The plants are transplanted to 6-inch (15-cm) pots no later than early August so that they will be ready for Christmas sales.

FIGURE 23-2 A small Rex Begonia grown as a bedding plant.

FIGURE 23-3 Wax or fibrous-rooted begonias grown for bedding purposes. The red flowered cultivar is a dark-leafed selection and the white flowered cultivar is a green-leafed selection.

The shoots are pinched in June or July to produce branching. If enough of the stem is removed, these tops can be rooted and grown on. Small plants will be ready for Christmas, or they may be kept as stock plants. If the plants are to be used as stock, they must be grown under lights starting October 1. These long-day conditions will keep them growing vegetatively during the short days of winter.

The normal date of flower bud formation at latitude 42° north is between October 10 and October 20. If the plants are to flower before Christmas, they must have short days. The black cloth treatment is given from 5 P.M. until 8 A.M. daily. The plants will bloom 8 weeks after starting short days.

B. socotrana can be made to flower for Easter and Mother's Day as well. Plants for Easter are propagated in October as terminal cuttings or leaf petiole cuttings. They are given 4 hours of artificial light from 10 P.M. to 2 A.M. until mid-January. Then the plants are grown under natural day length until they flower, usually around April 1.

Plants intended for Mother's Day sales are given long days until mid-February. They then continue growing under natural day length until April 1. To guarantee bloom, they are given the short-day treatment using black cloth from April 1 to May 1. Five-inch (13-cm) plants of good quality should be ready for Mother's Day following this schedule.

Rieger Elatior Hybrids

Rieger elatior are winter flowering begonias and are the result of crossing *B. socotrana* with various Andean tuberose species, or with *B.* × *tuberhybrida* cultivars. The winter flowering habit of *B. socotrana* is combined with the large, colorful flowers of the Andean tuberose begonia (Figure 23-4A and B). The flower color ranges from white to pink, red, orange, or yellow. The flowers may be single or double.

There are two main types of Rieger elatior begonias: the Schwabenland series and the Aphrodite series. The Schwabenland begonias are propagated by leaf petiole cuttings. The Aphrodite series is propagated by terminal and/or stem cuttings. As compared to the Aphrodite series, the Schwabenland begonias have stronger, heavier, and more numerous stems.

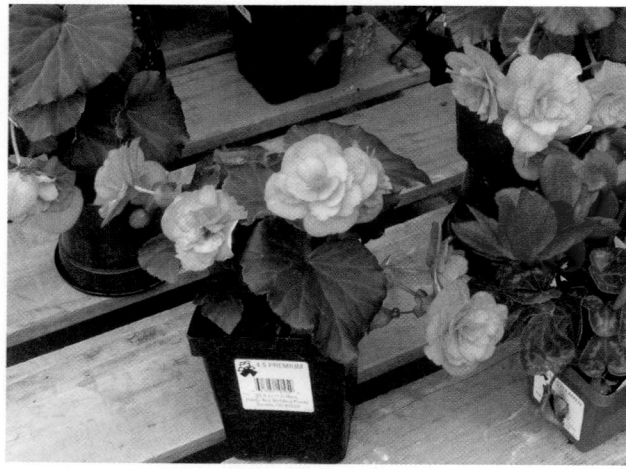

FIGURE 23-4A A small *Rieger elatior* begonia.

FIGURE 23-4B A mixed container of *Rieger elatior* begonias.

Schwabenland

Ten to 12 weeks are required for cuttings of this series to root and grow to small plants. After a good root system develops, the plants are placed in 2-inch (5-cm) pots. As the root system continues to develop and before the plants become pot bound, they are shifted to 4-, 5- or 6-inch (10-, 13-, 15-cm) pots. A single plant of the Schwabenland series is placed in each pot because of its good branching habit.

The preferred growing medium is a highly organic, well-aerated medium such as a peat-lite mix. Young plants are grown at 70°F to 72°F (21°C to 22°C) between pottings.

The plants are given additional light from September 1 to April 15. These plants do not respond to the length of day. Rather, the additional light is given to promote photosynthesis to improve vegetative growth. The light intensity should be at least 20 fc (216 lux). Preferably, the intensity should be 50 fc (540 lux). The lights are used to extend the daylight hours. They are not used in the middle of the night as they are to control flowering in *B. socotrana*. The lighting schedule is given in Table 23-3.

For the rest of the year, the black cloth treatment is given from 5 P.M. to 8 A.M. daily to hasten the development of the plants and promote more even flowering. A small fan is used to circulate the air under the black cloth to prevent a buildup of humidity. A high humidity level promotes disease conditions. Humidity can also be lowered by raising the cloth after dark and lowering it before daybreak. If the black cloth installation is semiautomatic, a time clock can be used to control the pulling of the cloth.

As the plants grow, they should be spaced to prevent stretching. Strong growing shoots are pruned and pinched to produce a compact, well-proportioned plant.

When the plants are spaced for the first time, the temperature is reduced to 62°F to 65°F (16.5°C to 18.5°C) to provide the best conditions for flower formation. The crop should finish its growth at 60°F to 62°F (15.5°C to 16.5°C) to improve the quality. The plant may be held at 56°F to 58°F (13.5°C to 14.5°C) for several weeks after the flowers bloom if the market conditions are not the best for sale.

Rieger begonias grow best with low levels of fertility. The use of too much fertilizer damages the roots, makes the foliage more brittle, and reduces the number of flowers. A solution of 20-10-20 fertilizer at a concentration of 50 to 75 ppm applied to the crop at every irrigation will produce top-quality plants.

Table 23-3

Number of Hours That 20–50 fc (216–540 lux) of Artificial Light Are Used in Rieger elatior Begonias to Promote Vegetative Growth During Low-Light Periods	
Lighting Period	Hours of Light Applied
Sept. 1–15, April 1–15	1
Sept. 16–30, March 16–31	2
Oct. 1–15, March 1–15	3
Oct. 16–31, Feb. 16–28	4
Nov. 1–15, Feb. 1–15	5
Nov. 16–Jan. 31	6

Osmocote® 14-14-14, Magamp® 7-40-6, and similar materials are used at one-third the amounts recommended by the manufacturer.

Do not wet the foliage when watering the plants. The moisture on the foliage supports the development of fungus diseases. A trickle irrigation system or ebb and flood is preferred to hose watering. Good air circulation is maintained around the plants by removing the basal leaves as they grow large.

Aphrodite

This series of Rieger elatior begonias is not as strong a grower as the Schwabenland series. These begonias are commonly used for hanging baskets because of their soft growth.

Terminal tip cuttings are used for propagation. The plants must be pinched regularly so that a bushy plant will develop. However, **pinching** increases the amount of time required before the plants bloom. The time for cropping may be shortened a little by planting several cuttings in a pot. Otherwise, the cultural practices for the Aphrodite series are the same as those for the Schwabenland series.

Problems

Diseases

The most serious disease that affects Rieger elatior begonias is *Xanthomonas begoniae*, a bacterial disease. Meristem tissue culture is the only propagation method that ensures disease-free plants. Once the plants are infected, there is no chemical means of controlling *Xanthomonas begoniae*.

Mildew and botrytis also affect begonias. The proper heating and ventilation practices will reduce the potential for these diseases.

Insects

Aphids, foliar and root nematodes, mealybugs, mites, scale, and thrips are all problems on begonias. Preventive control includes regular use of smoke fumigation, aerosol treatments, or spraying. Spraying is not recommended because the moisture on the leaves helps to spread diseases.

Tuberous-Rooted Begonias

B. tuberhybrida is the showiest of the various types of begonias and has the largest flowers. These begonias are used as potted specimens and as bedding plants. They flower naturally from March to September because they are long-day, photoperiod responsive plants.

The seed of *B. tuberhybrida* is very fine. Sowing consists of scattering the seed on top of a peat-lite mix. The seed is not covered. The seeds are sown from November through January to obtain spring flowering plants. The germination temperatures should be in the range of 65°F to 70°F (18.5°C to 21°C).

As soon as the seedlings are large enough to transplant, they are placed in nursery flats or are planted in individual 2.25-inch pots. Later they are transplanted to grow to finish size in 5- and 6-inch pots.

Plants subject to cool night temperatures of 55°F to 60°F (13°C to 15.5°C) produce the best flowers. The light intensity should be reduced to one-third of the full intensity because these plants do best in shade. The fertilization requirements are the same as those for the Christmas begonia.

As the day length decreases in the fall, the plants stop growing. The tubers are allowed to ripen by withholding water. When the medium is completely dry, the tubers are removed and cleaned. They are stored in dry peat moss at 45°F to 50°F (7°C to 10°C) until they are needed for forcing in the spring.

Growth of the tubers is started by burying them in flats of moist peat moss. The temperature is held at 75°F to 80°F (24°C to 27.5°C). After several leaves have developed, the plants are potted in a peat-lite medium in 5- or 6-inch pots. Then they are grown on at cool temperatures. The plants must not be placed too deep in the medium because stem rot may occur.

Begonia Semperflorens

The ever-flowering or wax begonia is the most commonly grown type of begonia. It is used as a bedding plant and in combination plantings in the spring. It is also grown as an inexpensive specimen plant at any season. The best growth is obtained under high light intensity.

The seed is very fine and is lightly pressed into the surface of a peat-lite germinating medium. To shorten the growing period, the plants can be propagated by stem-tip cuttings.

Seed sown in December or January provides plants for use as bedding plants and spring specimen plants in 4-inch pots. Seed sown in June will provide 3-inch plants at Christmas and 4-inch plants by Valentine's Day. The plants are grown at 60°F (15.5°C). If the temperature goes above 70°F (21°C), especially under short days, the plants remain vegetative.

The plants do well in either full sunlight or shade. Flower and foliage color is brighter in full sunlight.

Problems

Diseases

Begonias are affected by damping off (young seedlings), powdery mildew, botrytis stem rot, and bacterial leaf spot caused by *Xanthomonas begoniae*.

Insects

Aphids, mealybugs, spider mites, and whiteflies infest the plants. Root knot nematodes are a serious problem. Steam pasteurization practices are recommended to control the nematodes.

BROMELIADS ✄ *Bromeliaceae*

Bromeliads are epiphytes, or air-loving plants, that are slowly becoming more popular. They are grown mainly for their attractive foliage colors. When they bloom, their unusual flowers are borne on spikes (Figure 23-5). The foliage is arranged in such a way that a cuplike structure is formed at the base. Bromeliads are also known as vase plants because of this growth habit.

At one time, most of the bromeliads grown were used in dish gardens. Now, more of the plants are being used as specimen plants. The most common genera are *Acme, Billbergia, Cryptanthus,* and *Vriesia.*

Although the plants may be started from seeds, the most common method of propagation is to divide large plants. In general, bromeliads are grown in subtropical areas where the temperature and humidity are high. The plants do

FIGURE 23-5 Bromeliad *Aechmea*. (*Courtesy of the Professional Landcare Network, formerly the Associated Landscape Contractors of America*).

best in moderate light intensity. In their natural surroundings, they are found growing in the upper branches of trees where they get filtered sunlight. In northern greenhouses, some shade is required to reduce the light intensity from April 15 to October 1.

Because they are epiphytes, bromeliads do best in a loose, porous medium. A standard peat-lite mix with one-third coarse perlite or fir bark chips added provides the proper conditions for rooting and growth. Other suitable media are *Osmunda* fiber and sphagnum moss.

The minimum temperature for growth is 60°F (15.5°C). The preferred temperature is 65°F (18.5°C). Light fertilization is required. The cup at the base of the plants must contain water at all times. For this reason, overhead watering is preferred.

Ethrel® is used to stimulate the plants to flower. The manufacturer's directions on the label are to be followed in applying this material. Bromeliads have few cultural problems.

BOUGAINVILLEA ⌗ *Bougainvillea glabra, Nyctaginaceae*

Bougainvillea is a climbing plant native to Brazil. It is a showy plant with color provided by pink and white bracts rather than the tiny flowers (Figures 23-6A and 23-6B). It is a common plant in tropical and subtropical regions where it provides a cascade of color.

Plants may be grown from seed or from 5- to 6-inch (13- to 15-cm) long semihardwood cuttings. The cuttings may be rooted from March to May. Bottom heat is needed to ensure rooting.

The plants need full light intensity. However, they should be grown as cool as possible during the summer. During the winter months, the recommended night temperature is 50°F (10°C).

The once preferred growing medium consisted of equal parts by volume of loam soil, sphagnum peat moss, and perlite; however, many growers now use a traditional peat-lite medium. Super phosphate is added at the rate of 2 pounds (1.2 kg) to a cubic yard (cubic meter) when using a medium containing loam soil.

FIGURE 23-6A *Bougainvillea* grown in hanging baskets. The selection to the left is a variegated cultivar.

FIGURE 23-6B A close view of the *Bougainvillea* inflorescence with colorful bracts.

The trailing plants should be supported on some type of wire frame. The vinelike growth is wrapped to the wire. The long stems are not pruned because this practice delays blooming.

In spring, the plants are grown at 60°F (15.5°C) so that they will bloom for Easter. Fertilizer is applied sparingly in the winter. In the spring, the amount of fertilizer used is increased as active growth takes place. Constant fertilization with 100 ppm of full analysis N-P_2O_5-K_2O fertilizer in winter is increased to 200 ppm in the spring and throughout the summer.

Problems

The plants have few cultural problems.

CALADIUM (Fancy-Leaved Caladium)
Caladium hortulanum, Araceae

Fancy-leaved caladiums are sold as potted plants for Easter and Mother's Day, as well as during the summer and fall (Figure 23-7). Caladiums are native to Brazil, Peru, and the West Indies. The plants are grown from field produced tubers in Florida and California.

The propagation of caladiums is similar to that of potatoes. The caladium tuber is cut into pieces to produce new plants. Each piece must contain an eye or small bud, where the new plant will develop. The tubers are planted in the field in March and April. The plants are grown until September or later. At this point, the foliage ripens and eventually dies. After the foliage dies, the tubers are removed from the soil and are stored in a well-ventilated area in shallow trays. The minimum storage temperature is 60°F (15.5°C). Florists force these tubers to obtain plants for spring sales.

Tubers obtained in December or later are started in moist peat moss or sand. The medium temperature is in the range of 70°F to 80°F (21°C to 27.5°C). In 10 to 14 days, the tubers should have large enough roots to permit potting. Great care should be used in removing the tubers from the peat moss as the roots are easily broken.

The preferred medium must have a high organic content. A single tuber or cluster is planted to a 5- or 6-inch (13- or 15-cm) pot. Two or more tubers are placed in pots larger than 6 inches (15 cm). The tubers are forced at a minimum temperature of 65°F (18.5°C). The plants must be shaded from bright sunlight to avoid leaf burn. High light intensity may be used if the plants are kept moist at all times. However, less leaf burn occurs with shading than when an attempt is made to keep the plants evenly moist.

Caladiums are fertilized with 100 ppm of full analysis fertilizer at each watering. They do not require high levels of nutrition. The roots are easily damaged if the soluble salt content of the soil is too high.

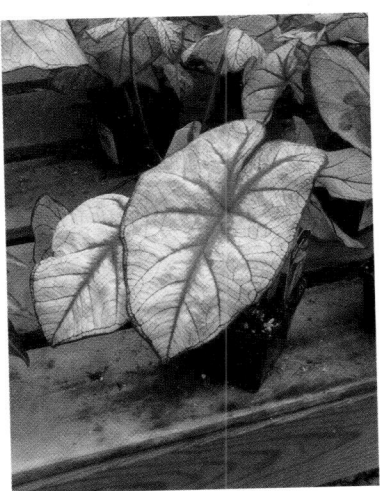

FIGURE 23-7 Fancy leafed *Caladium.*

Problems

Cultural Disorders

Foliage damage results if the plants remain too dry under high light conditions. Failure to start growth may be due to chilling of the tubers, a very low temperature in the starting medium, or a lack of water.

Diseases

Exposure of the tubers to temperatures below 50°F (10°C) causes rot. Caladium tubers must not be stored in a refrigerator.

Insects

Aphids and thrips are the most common pests.

CALCEOLARIA (Pocketbook Plant) ▨
Calceolaria species, Scrophulariaceae

Calceolarias are spring flowering potted plants that are grown mainly for Easter and Mother's Day sales. However, they are also grown as bedding plants along the west coast of the United States.

The sales of these plants are limited. Growers who sell 1,500 plants profitably may lose money if they overproduce by just 100 plants.

Calceolaria herbeohybrida is a herbaceous plant with large flowers. Typical flower colors are yellow, bronze, and spotted purple and brown. *C. integrifolia* is a shrub type plant with slightly smaller flowers in pink, red, or yellow.

Seed of *C. herbeohybrida* is sown in July or August. The seed is extremely small and is sprinkled over the surface of the medium. It is not necessary to cover the seed. The preferred medium is a peat-lite mix using number four vermiculite. The medium is kept moist and at a temperature of 65°F (18.5°C). If a low-pressure mist is used, bottom heat is required because the evaporative cooling effect of the water will cause the medium temperature to become too low.

The seedlings are heavily shaded from the sun. As soon as they are large enough to be handled, they are transplanted to a community flat or to small pots. However, small pots dry out quickly and need almost constant attention. When the plants reach the proper size, they are moved to finish size 5- or 6-inch (13- to 15-cm) pots.

C. integrifolia is propagated from terminal cuttings from July through October. The stock plants must be grown as cool as possible during the summer months under heavy shade. Fan and pad cooling is ideal for this crop. The cuttings root in 3 or 4 weeks. They are then potted using a well-drained, open medium such as a peat-lite mix.

Young plants may be grown at temperatures of 60°F (15.5°C) for the first few months. They form flower buds at temperatures below 60°F (15.5°C). As the natural day length increases after January, flower development is hastened by supplying artificial light from 10 P.M. until 2 A.M. daily. Incandescent 40- to 60-watt bulbs are spaced 4 feet (1.2 m) apart, 2 feet (0.6 m) above the plants. A light intensity of 5 fc (54 lux) is needed. To obtain flowering plants for Valentine's Day, lights are turned on beginning November 15. For early March bloom, lighting is started December 20. If flowering in early April is required, lights are turned on beginning January 20. Natural flowering takes place in May.

The plants are grown at a cool temperature, 50°F (10°C), after they have made 3 to 4 inches (7.5 to 10 cm) of growth. As soon as buds are visible, the temperature is raised to 60°F (15.5°C). Day temperatures should not exceed 65°F (18.5°C).

The full intensity of light is used during the winter. Light shade is applied beginning around March 1. Too much light burns the foliage and flowers.

Fertilization

The plants are sensitive to high soluble salt levels. A 20-10-20 fertilizer is applied at each watering at concentrations of 100 ppm each of $N-P_2O-K_2O$. Some growers prefer to apply fertilizer with every third watering using 250 ppm each of N and K.

Super phosphate is mixed into the medium before transplanting. Overfertilization causes excessive foliage growth.

Moisture

It is difficult to provide the proper amount of moisture for calceolarias. They are sensitive to overwatering but also need sufficient water for maximum growth. Excess moisture on the leaves causes rot. A capillary watering system ebb and flood is an ideal solution to the problem of watering calceolarias. If this type of system is not available, then foliage must be kept dry and the plants must not be overwatered.

Supporting the Plants

Unless grown with chemical growth retardants, the plants must be staked and tied to support the heavy weight of the top growth. Green bamboo stakes are placed in the pot, and the stems are tied together with green florist string. The flowers are especially sensitive to sun scorch. This means that heavy shading is required in regions with bright winter sun.

Problems

Cultural Disorders

The major cause of stem rot is placing the plants too deep in the medium in the pots. Stem rot may also occur because of overwatering in heavy media. Ethylene gas causes the flowers to drop.

Diseases

A virus disease that is carried by aphids may cause the development of light green leaf sections. This problem is reduced by controlling the aphids.

Insects

Calceolarias may be infested by aphids, leaf rollers, mealybugs, and thrips. Some growers use calceolarias as an indicator plant as aphids will infest this species often before any others.

ACHIEVEMENT REVIEW

Select the best answer or answers to complete each statement. List the appropriate letter(s).

1 Achimenes are propagated by:
  a. seeds.
  b. terminal cuttings.
  c. division of rhizomes.
  d. division of tubers.

2 Azaleas originated in:
  a. the eastern United States.
  b. the western United States.
  c. China and Japan.
  d. West Germany.

3 Azaleas are commonly propagated by:
  a. leaf bud cuttings.
  b. terminal cuttings.
  c. grafting.
  d. layering.

4 Azalea flowers are made up of:
  a. single blossoms.
  b. clusters of three to five buds.
  c. clusters of two to four buds.
  d. multiple clusters.

5 Azaleas initiate (set) flower buds:
  a. when the days are longer than 12 hours.
  b. when the temperature remains above 65°F (18.5°C) for 8 to 10 weeks.
  c. when the temperature remains below 65°F (18.5°C) for 8 to 10 weeks.
  d. when the temperature is held constant at 65°F (18.5°C) for 8 to 10 weeks.

6 After flower initiation, azaleas need an additional period of cool temperatures for buds to develop normally. They are stored:
  a. in complete darkness below 32°F (0°C).
  b. between 35.5°F and 36.5°F (1.9°C and 2.2°C) for 4 weeks.
  c. between 35.5°F and 36.5°F (1.9°C and 2.2°C) for 8 weeks.
  d. between 39°F and 41°F (4°C and 5°C) for 6 weeks.

7 The time needed to force azaleas into flower for spring holidays after cold storage:
  a. remains the same regardless of the month.
  b. becomes longer as spring progresses.
  c. becomes shorter as spring progresses.
  d. is twice as long in May as in February.

8 The interveinal yellowing (chlorosis) that develops in the newly formed leaves of azaleas is due to:
a. lack of boron.
b. lack of nitrogen.
c. lack of potassium.
d. root damage and a lack of iron uptake.

9 The growth of Christmas begonia, *B. socotrana*, can be hastened by:
a. keeping the plants under short days.
b. giving the plants artificial long days during winter.
c. spraying the foliage with gibberellic acid.
d. spraying the foliage with B-NINE®.

10 *Rieger elatior* begonias are lighted to improve vegetative growth using:
a. 2 to 5 fc (22 to 54 lux).
b. 20 to 50 fc (216 to 540 lux).
c. 200 to 500 fc (2.15K to 5.38K lux).
d. 2,000 to 5,000 fc (21.5K to 53.8K lux).

11 Lighting for *R. elatior* begonias is provided:
a. from 10 P.M. until 2 A.M.
b. from 11 P.M. until 1 A.M.
c. from midnight until 4 A.M.
d. at the end of the natural day.

12 Bromeliads belong to the group of plants called:
a. hydrophytes.
b. xerophytes.
c. epiphytes.
d. halophytes.

13 Propagation of caladiums is the same as that of:
a. beans.
b. tomatoes.
c. white potatoes.
d. corn.

14 Calceolarias require night temperatures that are held at _____ for flower bud formation.
a. 40°F (4.5°C)
b. 45°F (7°C)
c. 50°F (10°C)
d. 60°F (15.5°C)

15 Plants that are sensitive to high soluble salt levels should not be fertilized at every watering with solutions that have more than:
a. 100 ppm each $N-P_2O_5-K_2O$.
b. 200 ppm each $N-P_2O_5-K_2O$.
c. 250 ppm each $N-P_2O_5-K_2O$.
d. 300 ppm each $N-P_2O_5-K_2O$.

Flowering Plants II

❈ Christmas Cactus

❈ Christmas Cherry

❈ Cineraria

❈ Clerodendrum

❈ Cyclamen

❈ *Dendranthema*
  (Chrysanthemum)

❈ Exacum

❈ Fuchsia

❈ Gardenia

❈ Genista

OBJECTIVES:

After studying this chapter, the student should have learned the cultural details for the flowering potted plants, "Christmas" cactus to Genista. For each crop, the student should be able to describe:

■ methods of propagation.

■ the soil or growing medium requirements.

■ the fertilizer recommendations.

■ methods for controlling growth, diseases, and insects.

CHRISTMAS CACTUS ❈ *Schlumbergera bridgesii, Cactaceae*

The epiphytic (air-loving) plant is a native of Brazil. In its natural surroundings, it grows on the branches and bark of trees. Modern breeding techniques have produced many variations of the original plants collected from their natural surroundings. The "Christmas" cactus and "Thanksgiving" cactus are members of the same genus, but different species. The "Christmas cactus is *Schlumbergera bridgesii* and the "Thanksgiving" cactus is *Schlumbergera truncata*; the "Easter" cactus is *Rhipsalidopsis gaertneri* and is another member of these holiday cacti.

The plants can be distinguished from each other by slight differences in the appearance of the margins of the stem sections (Figure 24-1). For example, the stem segments of *S. bridgesii* have rounded points on the margins. In *S. truncata*, the segments have saw-toothed points with the teeth pointing forward.

The plants are propagated vegetatively using three-segment or joint cuttings. The propagation medium may be sphagnum moss, equal parts by volume of peat moss and perlite, or equal parts by volume of peat moss and sand. The cuttings should be watered well when they are stuck in the medium. Overwatering must be avoided thereafter. A low-pressure mist is used with a short on–long off cycle.

Generally, propagation is by simple cleft grafting. The desired scion is grafted to *Selenicereus* stock.

The potting medium must have good porosity and drainage. The plants grow very well in the peat-lite mixes. If such mixes are not used, the soil

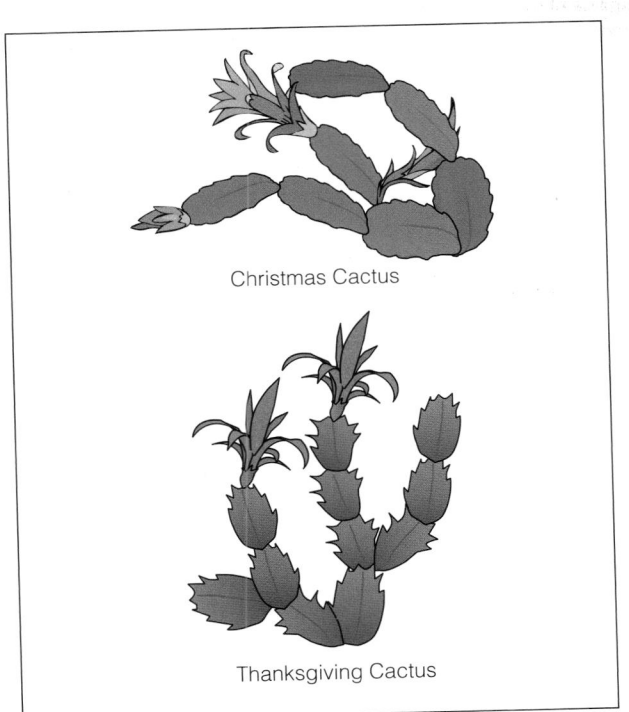

FIGURE 24-1 The "Christmas" cactus (top) has joints with rounded margins and the "Thanksgiving" cactus (bottom) has sections with pointed or toothed margins.

FIGURE 24-2 "Christmas" cactus plants can live for many years in the home. This specimen is more than 45 years old.

mix should contain at least one-third peat moss by volume. Unlike other cactus types, the flowering "Christmas" cactus is not grown on the dry side.

The plants flower in response to day length and temperature. The "Thanksgiving" cactus initiates flowers sooner than the "Christmas" cactus. Buds on the "Thanksgiving" cactus become visible around September 20 at 42° north latitude. This means that as short-day plants, their critical day length is slightly more than 12 hours. "Christmas" cactus flower buds are first visible around October 25 at 42° north latitude (Figure 24-2).

Day-length control for *S. bridgesii* begins on September 15 using a 9-hour day until October 10. A minimum night temperature of 55°F (13°C) is maintained. Blooming occurs 2.5 to 3 months after the start of short days.

To obtain even flowering on the whole plant, the ends of the leaves are trimmed just before the start of the short days.

During the short-day period, the plants are kept on the dry side, resulting in more even flowering. However, if they are kept too dry, blooming may be delayed and the flower size may be reduced.

S. bridgesii requires some shading from bright sunlight in the summer. Best growth in the late fall and winter occurs in full light intensity.

Problems

Diseases

Generally, *S. bridgesii* is free of disease when grown using clean cultural practices.

Insects

The most troublesome insects are mealybugs and scale.

CHRISTMAS CHERRY ❋ *Solanum pseudocapsicum, Solananaceae*

Solanum pseudocapsicum is a dwarf shrub. It is commonly known as the Jerusalem cherry or the Cleveland cherry. The seeds are sown from January through March so that the potted plants will be ready for sale at Christmas. After germination, the seedlings are shifted at intervals from 3-inch plant bands or pots to 5- or 6-inch finish size pots. The final **transplanting** should be completed in late May. The plants are placed outdoors during the summer in cold frames. They may also be planted directly to beds after there is no further danger of frost. The plants are removed from the beds and potted in September.

Flowering and fruit set occur in the summer. The use of too much nitrogen fertilizer may cause poor fruiting. The most striking feature of the plant is its scarlet and yellow berries. Therefore, poor fruiting usually has a negative effect on sales. It is recommended that a balanced fertilizer, such as 20-10-20, be applied with each watering at a concentration of 200 ppm N.

The plants can be shaped by pinching long shoots and pruning when the plants are small. Additional trimming is not required.

The plants are grown in a 50°F to 55°F (10°C to 13°C) greenhouse until Christmas. Full light intensity is used.

Problems

Cultural Disorders

The lack of fruit set may be due to poor light intensity, overwatering, or temperatures that are too low.

Leaf and fruit drop are caused by low humidity, inadequate moisture, overwatering to the point that the soil is waterlogged, temperatures that are too low, or gas fumes of various types.

Insects

Aphids, thrips, whiteflies, and mealybugs are the most common pests.

CINERARIA ❋ *Senecio cruentus, Asteraceae*

Common florist's cineraria, *Senecio cruentus*, is one of the most colorful commercial pot plants. A large mass of small, daisylike flowers opens at one time. The colors range the spectrum from red to blue.

The plant is grown as an annual, but it responds to temperature as a true biennial. That is, it will set flower buds only if it has 6 weeks at a temperature below 60°F (15.5°C). The plants are often grown to bloom for sale during the period from Easter to Mother's Day.

Cineraria seed is very fine, and care must be used in choosing the germinating medium. The peat-lite mixes are excellent for these plants. Sowings are made from June through October. During the warm days of summer, the plants grow vegetatively. Early sowings will produce blooming plants for January and February. Seed sown in September and early October will produce 5- or 6-inch (13- or 15-cm) pot size plants for Easter.

Cool temperatures are needed for flower bud formation and development. The temperature should not exceed 60°F (15.5°C).

At a night temperature of 45°F to 50°F (7°C to 10°C), bud set and formation will require 6 weeks. An additional 8 weeks is needed for full flower development.

To speed production, the plants are started and grown at a night temperature of 60°F (15.5°C) until they reach a height of 3 to 4 inches (7.5 to 10 cm). Then they are grown at the cooler temperature needed for bud set.

Earlier flowering can be obtained by providing artificial long days. The seedlings respond to the light intensity provided by 40-watt cool white fluorescent lamps. The lamps are lighted for 18 to 24 hours daily. After transplanting, the plants can be lighted using incandescent bulbs until color first shows in the buds. The incandescent lighting must not be continued beyond bud color. The red energy from this light source causes the stems to stretch too much. A growth retardant spray may be used to prevent the stretch.

The young plants must not become potbound. This condition checks their growth, and they never recover. To prevent this problem, the plants are shifted a little earlier than usual to the next larger size pot or to the finish size pot.

Compact plants have the greatest sales appeal. To achieve this form, the plants should be well spaced and grown in full light intensity. Cinerarias require large amounts of water when they are in bloom. A light shade on the greenhouse roof will help prevent wilting due to water stress.

The use of too much nitrogen fertilizer will result in soft growth. The recommended amount of fertilizer is 75 to 100 ppm each of N-P_2O_5-K_2O applied at every watering.

Problems

Cultural Disorders

Sudden wilting usually is caused by *Verticillium* wilt disease.

Diseases

Cineraria are subject to *Verticillium* wilt disease. To minimize the disease, all materials used for each new crop, such as growing media, flats, and pots, should either be steam sterilized or new. Wilting usually appears when the plants are ready for sale.

Mosaic is a virus disease carried by aphids. This disease causes irregular light and dark green mottling of the leaves. The plant tends to remain dwarfed.

Streak is a virus disease that is carried by thrips. It causes the leaves to curl. In addition, leaves have a rough appearance with reddish-brown areas. The discoloration also occurs in the stalks and petioles of the plants. The leaves die from the base of the plant to the top.

Stem rot is caused by *Phytophthora*, *Fusarium*, and other fungi.

Powdery mildew is due to an airborne fungus of the *Erysiphe* species.

Botrytis cinerea causes the common gray mold. To control both mildew and botrytis, the proper temperature and ventilation must be provided.

Insects

Many insects attack cinerarias, including aphids, cabbage loopers, cyclamen mites, leaf tier, mealybugs, spider mites, thrips, and whitefly.

CLERODENDRUM ❋ *Clerodendrum thomsonae, Verbenaceae*

This plant is an import from Europe, where it has been grown as a pot plant for many years. The plant has a vining or trailing type of growth unless it is treated with a growth retardant. White bracts enclose very bright red flowers. The bracts remain on the plants long after the red flowers have dropped. Both day length and treatments with a growth retardant promote flowering.

The plant is propagated by asexual means. Cuttings with a single eye are taken in September and October. These cuttings root quickly in 10 to 12 days. The air temperature is maintained at 70°F (21°C) with bottom heat 5° higher. The use of a low-pressure mist improves the rooting of the cuttings. The preferred propagation medium has a pH of 5.0 to 6.0 and consists of equal parts of sphagnum peat moss and perlite or vermiculite.

The rooted cuttings are potted using a light, porous medium. *Clerodendrum* grows very well in peat-lite mix A, but with only one-half the recommended amount of limestone. One or two cuttings are planted in a 4-inch (10-cm) pot. Three cuttings planted in a 5-inch (13-cm) pot produce a fuller looking plant.

When only a single cutting is placed in a pot, the plant must be pinched to cause branching. Pinching is done by removing the shoot tips only when the new growth is 1 to 2 inches (2.5 to 5 cm) long.

Fertilizer is applied at every watering at the rate of 150 ppm each of N-P_2O_5-K_2O using a 20-10-20 analysis product.

Ten-hour short days are needed to bring the plant into flower. The short-day treatment is started when the plants are pinched. The plants will bloom in 10 to 12 weeks if they are grown on at a constant temperature of 70°F (21°C), day and night.

The plants are very sensitive to excessive temperatures. The black cloth treatment during the summer may cause very high temperatures. This condition promotes shoot growth at the expense of flowers. Below 70°F (21°C), the growth of the plants is very slow.

The growth retardant A-REST® can be applied as a soil drench to hasten flower development. Follow the label recommendations for the correct amounts.

Problems

Cultural Disorders

The plants may fail to flower because of a temperature that is either above or below 70°F (21°C). A low light intensity also inhibits flower development.

Flower drop occurs when the plants are kept in the dark for longer than 24 hours in shipping or delivery. Another cause of flower drop is a temperature above 70°F (21°C). Insecticide fumigation smokes may also cause flower drop.

Diseases

A strain of tobacco mosaic virus may infest the plant. This virus destroys the chlorophyll in the plant, resulting in retarded growth. Because there is no prevention for this disease, infected plants must be destroyed. When propagating, use only those plants that appear healthy.

Septoria phalaenoides and *Cercospora* species cause a leaf spot. The spores of these diseases are spread by splashing water on the foliage. To reduce the amount of disease, keep the foliage dry, improve the air circulation and ventilation, and spray with an approved fungicide. Remove and destroy infected leaves and stems.

CYCLAMEN ▒ *Cyclamen persicum, Primulaceae*

The cyclamen is a member of the primrose family. It is native to the Mediterranean region and extends to central Europe.

The cyclamen is a herbaceous plant that has a rounded, flattened corm from which the long, petioled, leathery leaves arise. The leaves have light green or grayish patches on their upper surface. A single flower is borne on the end of a long petiole. The flowers will last up to 4 weeks after cutting. In the United States, cyclamen are grown mainly as pot plants and as bedding plants for the Gulf States. The cut flowers are not yet as popular as they are in Europe, where they are highly prized for use in arrangements.

The cyclamen is a traditional Christmas crop. At present, it is facing increased competition from several poinsettia cultivars. One disadvantage of the cyclamen as a crop is the long growing period from seed to flower. A standard growth program may require from 12 to 15 months. More recently, growers have produced flowering crops of the new F_1 hybrids in 7 to 8 months after sowing seed. This shorter growth period means that the greenhouse space can be used more productively. Both the 12- to 15-month growth program and the 7- to 8-month program are described in this section.

Standard Culture

Sowing Seed

Cyclamen seed is imported from Denmark, Holland, Germany, and Switzerland. Some American specialists produce their own seed. The methods used to germinate cyclamen seed are very exact. Success with this crop depends completely on the proper handling of the seed.

The seed coat of cyclamen contains a chemical **inhibitor** that slows germination. Although the exact nature of this inhibitor is unknown, some of it can be removed by washing the seed under a hard stream of water for 20 minutes.

To obtain a Christmas crop the following year, the seed is sown during September and October. The germination medium may be sphagnum peat moss with limestone added to bring the pH to 6.0, or it may be a mixture of equal parts of loamy soil, sand or perlite, and sphagnum peat moss. Cyclamen seed is large and can be sown individually. The germination and development of the seed to transplanting size may take as long as 2 months. Thus, the spacing of the seed in the seed flat is 1 inch by 1 inch (2.5 cm by 2.5 cm). The seed must be completely covered by peat moss because light prevents germination. The medium is kept uniformly moist. A temperature of 64°F to 68°F (18°C to 20°C) is recommended. Higher temperatures delay germination.

The germination of cyclamen seed is unlike that of other plants. Most plants develop a root, which is followed shortly by a stem. The cyclamen seed, however, develops a root about 2 weeks after sowing. There is then a period of growth in which a small corm is formed. After an additional 4 to 6 weeks, the first cotyledon leaf develops from the corm. The tiny corm is easily damaged by environmental conditions such as too much heat or too little water.

A second cotyledon inside the embryo begins to grow if the first cotyledon is injured or dies. If the first cotyledon remains healthy, the second cotyledon does not begin to grow.

Transplanting

As soon as the seedlings begin to crowd each other, they are transplanted to flats where they are spaced on a 2-inch by 2-inch (5-cm by 5-cm) or a 3-inch by 3-inch (7.5-cm by 7.5-cm) grid. Some growers use small pots or peat pellets (Jiffy-7®) rather than flats. However, in separate containers, the plants tend to dry out too quickly.

The corm must not be buried in the medium when it is transplanted. Half of the corm should show above the medium each time the plants are moved. This practice greatly reduces the problem of crown rot.

The wider spacing of the plants means that they can be grown for a longer time before it is necessary to shift them to finish size pots. The same type of medium may be used for germination and growing the plants in the flats. The medium must be sterilized to avoid problems with soil-borne insects and diseases, such as cyclamen grubs, cyclamen stunt disease, and root-knot nematodes.

Fertilization

Cyclamen do not grow well with high levels of fertilization. The preferred fertilization program begins with the addition of 1 pound (0.26 kg) of 20 percent superphosphate to a cubic yard (cubic meter) of medium, or the addition of 0.5 pound (0.3 kg) of treble superphosphate. Limestone is also added to bring the pH to 6.0. Then the plants are fertilized at each watering using a 20-10-20 analysis fertilizer at the rate of 100 ppm each of N and K_2O. Excess fertilizer damages the tender roots of cyclamen. High soluble salt levels from other sources should not be allowed to accumulate.

Final Shift

When the plants become crowded in the flats, they are moved to 4-inch (10-cm) or 5-inch (13-cm) three-quarter size finish pots. This final transplanting is made in late August or early September. This schedule permits the plants to develop good root systems for flowering. It was once thought that cyclamen plants would develop flowers only if they were potbound. This theory has never been scientifically documented.

Light and Temperature

During the spring and summer, cyclamen grow rapidly if they are given full light intensity. However, the temperature must remain cool at the same time, less than 65°F (18.5°C). As soon as the day temperature rises above 65°F (18.5°C), foliage burn occurs. It is possible to grow cyclamen under the full summer light intensity using a low-pressure intermittent mist to cool the leaves. This method of culture produces strong plants but does not affect later flowering. However, the misting must be done frequently, and the schedule is too exact for general crop use.

To keep cyclamen cool in the summer, the greenhouse is usually shaded. Fan and pad cooling may be used as well to produce excellent plants.

The greenhouse shade is removed with the cooler temperatures of fall to give the plants the maximum light intensity. During the fall flowering period, the best growth with the greatest number of flowers is obtained when both the day and night temperatures are maintained at 60°F (15.5°C). If the temperature during the day exceeds 60°F (15.5°C), the greenhouse must be cooled to the required night temperature, 60°F (15.5°C).

F_1 Hybrids

It is only recently that F_1 hybrids have begun to replace open-pollinated strains that were the mainstay of the industry. One authority states that more than one-half the cultivars grown today are F_1 hybrids. These are being offered by seed companies such as DeRuiter, Goldsmith, Pan American, S and G, Takii, and others.

The hybrids have become prominent because they take less time to produce a crop from potting to finish than the open-pollinated strains. Miniature and intermediate hybrids can be finished in 3 to 3.5 months. The large flowered hybrids take a little longer, 3.5 to 4 months.

Part of the reduced production time is gained by growing the plants in small pots. The miniatures and intermediate plants are being grown in 2.75-inch (7-cm) and 3-inch (7.5-cm) size pots. Standard cultivars are finished in 3.5-inch (9-cm), 4-inch (10-cm), and 4.5-inch (11-cm) size pots. Very large plants may be grown on to flowering size in 5.5-inch (14-cm) pots (Figure 24-3A and B).

The desirable attributes that producers of this crop look for are good flower shape, a free-flowering habit, strong stems, both foliage and flowers, and a well-proportioned but compact growth habit. Overall uniformity, particularly in flowering and maturity, is also important.

The genetic makeup of cyclamen restricts the range of colors available. Predominant are the pinks, through salmon, red, fuchsia, violet, and purple. There are also white cultivars as well as bicolors. These are usually classified as 'White with Eye' or 'Soft Pink with Eye.' There are no yellow, orange, or similar shades.

FIGURE 24-3A Assorted colors of small Cyclamen plants.

FIGURE 24-3B A closer view of a small Cyclamen plant.

In the standard series the following cultivars are offered by suppliers:

Sierra Series (Goldsmith)

'Bright Fuchsia'
'Deep Red'
'Deep Rose'
'Deep Salmon'
'Light Pink'
'Light Pink with Eye'
'Light Purple'
'Lilac'
'Purple'
'Salmon'
'Salmon with Eye'
'Scarlet'
'Sierra White'
'White with Eye'

Concerto Series (S and G)

'Apollo'—white
'Boheme'—deep wine red
'Caruso'—scarlet
'Esmeralda'- deep salmon
'Estrella'—deep salmon
'Fidelia'—salmon pink
'Finlandia'—white
'Giselle'—salmon
'Louisa'—rose pink
'Manon'—deep salmon
'Norma'—white with purple eye
'Ophelia'—light salmon rose with red eye
'Othello'—scarlet
'Sylvia'—dark lilac

Zodiac Series (Pan American)

'Cherry Red'
'Light Pink with Eye'
'Lilac'
'Pink'
'Purple'
'Red'
'Salmon Pink'
'Salmon Pink with Eye'
'Scarlet'
'Zodiac White'

Colorado Series (DeRuiter)

'Bright Pink'
'Colorado White'
'Deep Salmon'
'Lilac'
'Purple'
'Salmon'
'White with Purple Eye'
'Wine Red'

A separate cultivar, 'Pastellete,' has several pastel color variations that range from purple through white with eye.

For the intermediate or midi series the following cultivars are offered:

Laser Series (Goldsmith)

'Laser Scarlet'
'Rose'
'Salmon'
'Salmon with Eye'
'White'
'White/lilac with Eye'

Novella Series (Pan American)

'Novella Cherry Red'
'Pink with Eye'
'Rose Pink'
'Salmon'
'Violet'
'White'

Intermezzo Series (DeRuiter)

'Bright Pink'
'Flamed'
'Intermezzo Red'
'Light Pink with Eye'
'Pink'
'Purple'
'Salmon Pink'
'White'
'White with Eye'

Takii Cultivars

'Arashiyama'
'Daimonji'
'Kemo'
'Saga'

Miniatures (mini hybrids)

Dixie Series (S and G)

'Burgundy'
'Dixie Scarlet'
'Light Lilac'
'Pink'
'Salmon with Eye'
'Violet'
'White'
'Wine Red'

Marvel Series (Pan American)

'Dark Fuchsia'
'Marvel Cherry Red'
'Marvel Red'
'Pink Shades'
'Salmon Pink Shades'
'Soft Pink with Eye'
'Violet Shades'
'White with Eye'

Minimate Series (DeRuiter)

'Bright Pink'
'Minimate Red'
'Purple'
'Salmon Pink'
'White'
'White with Eye'

Miracle Series (Goldsmith)

'Miracle Scarlet'
'Rose'
'Salmon'
'White'
'White with Eye'

Seed Sowing

The preferred germinating medium for short season cultivars is sphagnum peat moss plus fertilizers. To each **bushel** of sphagnum peat moss (1.25 cubic feet [0.035 m^3] in volume), the following materials are added.

- 200 grams of agricultural limestone
- 20 grams of magnesium sulfate (Epsom salts)
- 7 grams of potassium nitrate (KNO_3)
- 12 grams of 20 percent superphosphate, or 6 grams of treble superphosphate (46 percent P_2O_5)
- 16 grams of Osmocote® 14-14-14
- 1 gram of fitted trace element (FTE 555)

The addition of fertilizer to the medium means that it will not require more fertilizer for 2 months after the seed is sowed. After this period, nitrogen is applied at the rate of 100 ppm at each watering using a 30-10-20 analysis fertilizer.

Growth inhibitors are contained in the seed coat. To overcome this problem, the seed is soaked for 12 hours in 75°F (24°C) water. The seed is then dipped for 20 to 60 seconds in a surface sterilizing solution made up of 1 part by volume of calcium or sodium hypochlorite (household chlorine bleach) diluted with 19 parts of water. The seeds must not be dipped for more than 60 seconds in this solution.

After the seed is placed in the medium, it is covered with 0.13 to 0.25 inch (3 to 6 mm) of sphagnum peat moss. The media temperature must not be less than 65°F (18.5°C). The recommended temperature is 68°F (20°C). At a temperature above 72°F (22°C), the seed will not germinate. Because complete darkness is needed, the seed flats may be covered with black polyethylene. This cover must be lifted every other day to allow an exchange of air to take place. The plastic must be removed as soon as the cotyledon leaves are visible.

Light and Temperature

The young plants are grown in a moderate shade where the light intensity is 3,500 to 4,500 fc (35.6K to 48.4K lux). A constant temperature of 60°F (15.5°C) is maintained. When the seedlings become crowded in the germination flat, the plants are moved to 5-inch pots. This shift will be required in August for seed sowed in late March at 42° north latitude. The seedling is placed in the pot so that only half of the corm is covered with the growing medium. The pots may touch at first, but as soon as the plants are large enough for the leaves to touch each other, the pots are placed in their final 15-inch by 15-inch (38-cm by 38-cm) spacing on centers.

In October, the plants are given full light intensity. Good ventilation must be provided to obtain a quality cyclamen crop.

Fertilization

A regular program of fertilization is started 1 month after the plants are shifted to their final pots. A 30-10-20 analysis fertilizer is applied at the rate of 100 ppm of nitrogen at each watering. If this fertilizer is not available, a 20-10-20 analysis fertilizer may be used at the same rate.

Problems

Cultural Disorders

A number of cultural problems may affect cyclamen.

1. The production of few flowers is due to overfertilization, especially nitrogen, or a lack of phosphorus.
2. Excessive pedicel length is due to a lack of phosphorus, or irregular and improper irrigation.
3. Flowering may occur below the foliage on short stems; this condition is due to too little water or cyclamen stunt disease.
4. A delay in bloom occurs if there is a deficiency in phosphorus and potassium or if the pH of the medium is too high.
5. Dried and withered leaf edges are due to scorch caused by water stress and/or too much light. If older leaves are injured, it may be the result of a lack of potassium.
6. Flower bud abortion is due to weak growth.
7. Light colored leaves are the result of too little light.

Diseases

The most serious disease is crown rot caused by *Botrytis*, which may also cause petal spot. This disease can be reduced by improving aeration and ventilation. One method of improving the circulation of air around the pots is to place each plant on an inverted pot.

Stunt is caused by a seed-borne fungus, *Ramularia cyclaminicola*. Strict sanitation and pasteurization procedures are required to control this disease. All infested plants must be discarded.

Root-knot nematodes cause small swellings, or nodules, on the roots. The sterilization of the growing medium is the only method of controlling nematodes. Infested plants must be discarded, and all materials used in their production must be steam sterilized.

Erwinia caratovora causes a soft rot of the flower pedicels and leaf petioles. Again, infected plants are discarded. To control this disease, the growing medium is steam sterilized before the crop is planted. Strict sanitation procedures are required in all phases of growth.

Insects

Cyclamen grubs infesting the roots cause a sudden wilting of the leaves. Aphids and cyclamen mites cause distorted leaves and flowers. Mites delay flower opening. Thrips cause specks or streaks to appear on the flowers.

CHRYSANTHEMUM ▩ *Dendranthema* × *Grandiflora, Asteracea*

Dendranthema × *grandiflora* is the current genus and species name for what was previously known as *Chrysanthemum morifolium*. Despite the name change, this group of plants will be known as chrysanthemums for many years to come.

Chrysanthemums were once second to poinsettias in terms of dollar value for flowering potted plants. They are now third, being supplanted by potted orchids, which are now second to poinsettias. Among the many reasons for the popularity of chrysanthemums is their availability in a broad range of flower types and colors. Plant breeders' efforts have resulted in dramatic cultivar changes every year. Of the 20 most popular cultivars in 1987, only 7 were in the top 20 for 1992. Seventy percent of today's potted mums are cultivars that were introduced over the last 5 years. In January 1995, Yoder Brothers Inc. listed 24 new cultivars that were going to be offered to the trade, and in 2006, Yoder Brothers Inc. listed 11 more. This annual presentation of a new look has helped to maintain the popularity of the crop. Another important factor is their longevity. A quality plant should last 3 to 4 weeks or more in the conditions found in the average home.

Potted chrysanthemums are also one of the easiest flowering crops to grow on a production line basis. There is more information known about the cultural needs of this crop than any other. The versatility of potted mums allows the grower to produce them as mini-plants in 2-inch (5-cm) pots up to large tubs.

General Culture

The production of a quality crop starts with the cuttings. Growers can purchase cuttings from commercial suppliers who root them as a specialty crop. Two of the largest producers of cuttings are Yoder Brothers in Barberton, Ohio, and in Florida. In addition, other producers supply smaller amounts of **rooted cuttings.**

It is recommended that rooted cuttings be purchased by growers whose businesses can be classed as small or medium in size. These growers should not attempt propagation because the space needed for stock plants is an expense they cannot afford. The large specialty propagator can employ an expert staff to ensure that the cuttings produced from the stock plants are disease free and of the highest quality. This guarantee of a superior product is well worth the price of the plants.

Space Needed

Chrysanthemum crops normally are produced on a year-round basis. The cultural details for this type of production are well known. A growing area of nearly 380 square feet is required to produce a minimum of 50 flowering plants in 6-inch pots every 2 weeks. It is not recommended that less than 50 pots be grown on a regular program. More plants can be grown in the same area if smaller pots are used (less than 6 inches).

Receiving Cuttings

The plants are shipped from the specialty grower as unrooted or rooted cuttings. Most growers order rooted cuttings so that they do not have the additional expense of rooting them. When the boxes arrive, they open them immediately and check the condition of the plants. If any shipping or freezing damage is noted, the growers file a claim with the transportation carrier. Frozen chrysanthemum cuttings rarely survive after thawing.

Because the plants may have been in transit for some time, they may be dry. In general, they are still moist because of the special packing methods and shipping boxes used. If potting is not done immediately, the plants are placed in a refrigerator at 35°F to 40°F (1.5°C to 4.5°C). They must not be stored for more than a few days. The plants have already been without light for several days during shipping. They should be planted as soon as possible.

Growing Media

A well-prepared soil mix or soilless medium is used. A typical mixture once consisted of measured proportions of 1-1-1 by volume loam soil, sphagnum peat moss, and perlite. This mix has good drainage and aeration, is not too heavy, and can be pasteurized easily. Most peat-lite formulation will work, but any medium used for growing mums must be chemically or steam pasteurized to kill disease organisms.

Most chrysanthemum plants are now grown in peat-lite mixes. However, plants growing in a peat-lite mix often become top heavy and fall over unless sand is added to the mix (20 percent by volume) to provide the necessary weight, or the plants are grown in clay pots.

The pH of the soil mix should be in the range of 6.0 to 6.5. Two pounds of 20 percent superphosphate is added to each cubic yard of mix.

Potting

The rooted cuttings are potted directly in three-quarter size finish pots (Figure 24-4). The wider base of this type of pot provides stability so that the pot is not so likely to tip over as the plant grows larger. The shorter pot also looks better because it balances the top growth of the plant.

Cuttings stored in a refrigerator should be warmed to room temperature before they are potted, allowing 3 to 4 hours for warmup. The plants are injured if they are potted directly from the refrigerator and then placed in bright sunlight. That is, they transpire water faster than the roots can absorb it.

To avoid a "stovepipe" type of growth, the cuttings are planted at a 45-degree angle (Figures 24-5 and 24-6). As a result, the finished plant is more attractive. Five cuttings are planted around the edge of a 6-inch (15-cm) pot with one in the center.

FIGURE 24-4 Chrysanthemum cuttings are planted at a 45° angle to prevent a "chimney effect."

FIGURE 24-5 Chrysanthemum cuttings planted in a 6-inch pot where the cuttings are slanted toward the outside of the pot at a 45° angle.

If the weather is warm and sunny, it may be necessary to apply a mist over the tops of the plants. The nozzle used should provide a very fine mist. The mist reduces moisture stress until the plant roots can take up the required amount of water. For a day or two after the cuttings are potted, some growers provide a heavy shade to protect them from direct sunlight. Any shade used is removed as soon as the plants are established and turgid. The plants are normally grown in full sunlight.

Fertilizers

There are several ways to fertilize chrysanthemums to produce quality plants. Commonly used methods range from constant fertilization using an injector-type proportioner to single preplant applications of controlled-release fertilizers.

Constant Fertilization

A regular program of liquid fertilization is started as soon as the plants are potted. A 20-10-20 analysis fertilizer is put on at each watering at a concentration of 250 ppm of each of N-P_2O_5-K_2O. The solution is applied until leaching occurs. Some growers apply plain tap water once every 2 weeks to flush away accumulated salts. Superphosphate may be added before planting. In this case, concentrations of 250 ppm each of nitrogen and potassium are added by applying equal amounts of ammonium nitrate and potassium nitrate rather than 20-10-20 fertilizer.

To obtain improved flower and plant quality, the plants are finished using potassium nitrate at a concentration of 250 ppm K. This treatment is started when the buds show good color. However, many growers stop all applications of fertilizers 2 weeks prior to shipping to enhance keeping quality for the consumer.

Controlled-Release Fertilizers

For pot mums, the two most commonly used controlled-release fertilizers are Osmocote® and Magamp®. Osmocote® 14-14-14 is used at the rate of 10 to 12 pounds (5.9 to 7.1 kg) in a cubic yard (cubic meter) of medium. Before planting, 1 pound (0.45 kg) of superphosphate is also added to ensure the supply of phosphorus during the critical early stages of growth. Additional fertilizer is not required during the remaining growth of the crop.

Magamp® 7-40-6 with a medium particle size is used at the rate of 10 to 12 pounds (5.9 to 7.1 kg) in a cubic yard (cubic meter) of medium. The high phosphorus content of this fertilizer means that superphosphate is not required. One pound (0.45 kg) of potassium nitrate is added to provide nitrogen. The plants will use the nitrogen from the potassium nitrate until the ammonium nitrogen in Magamp® is changed to the nitrate form of nitrogen. The extra potassium also helps the crop during the early stages of growth. No other fertilizers are needed during the remaining growth of the crop.

If less than the recommended amounts (10 to 12 pounds [5.9 to 7.1 kg] to a cubic yard [cubic meter]) of medium of Osmocote® or Magamp® are used, fertilizer must be added as the crop grows. Some growers use half of the recommended amount of controlled-release fertilizer. They then make up the difference by adding **liquid fertilizer** solution. These growers say that this method of fertilizing gives them more control of crop growth.

FIGURE 24-6 A typical rooted chrysanthemum cutting ready for planting in the pot.

Water

Potted chrysanthemums need large amounts of water if they are to reach maximum growth. The plants should never be allowed to wilt for lack of water. Fortunately, it is easy to set up a semiautomatic watering system for pot mums.

Many growers also use the capillary mat method of watering. Review the section on capillary mat watering in Chapter 16 for the precautions about watering from the top.

Spacing

When the plants are first potted, the pots are placed so that they touch each other. As soon as the plants become crowded, they are given their final spacing. Six-inch (15-cm) diameter pots are spaced 14.5 inches by 14.5 inches (37 cm by 37 cm) on centers. Thus, each pot has 1.5 square feet (0.14 m²) of area. Smaller pots are given less space. Plant quality is reduced if the pots are crowded too close together. The lower foliage often turns yellow because of a lack of light due to crowding. In addition, air movement around the plants is poor, giving disease a chance to start.

Pinching

Plants are pinched to control the growth and to produce more flowers (Figure 24-7). A **soft pinch** is made by removing 0.75 to 1 inch (1.9 to 2.5 cm) of the growing tip of the stem. A **hard pinch** is not recommended. In a hard pinch, enough stem is removed to make a cutting. The axillary side shoots, where new branches grow, are slow to develop from the hard stem tissue that remains after a hard pinch.

There are a number of schedules for pinching. The schedules are often affected by environmental factors. Time is needed for new growth to take place. Good-quality flowering plants must have enough vegetative tissue. If the light levels are low, growth is slow. For example, during the late fall and winter, light levels are low. Slow growth at this time means that the plants are pinched after they are potted for 2 or 3 weeks. With good growing conditions in late spring and summer, plants are often pinched the day they are potted.

Plant height is controlled by the timing of the pinch and the start of short days. Chrysanthemum cultivars are classified into groups according to their

FIGURE 24-7 A plant tip is held directly above the chrysanthemum from which it was pinched. The tip measures about ¹/₂-inch in height. This results in a fuller plant with more side shoots. *(Courtesy H. Edward Reiley.)*

response time. This period is the number of weeks required after the start of short days to bring the plants into flower. There are 7-, 8-, 9-, 10-, 11-, 12-, 13-, and 14-week **response groups**. Generally, garden chrysanthemums are in the 7-week and 8-week response groups. Years ago, most winter grown chrysanthemums were in the 12-, 13-, and 14-week response groups. Modern growers use plants in the 9-, 10-, and 11-week response groups. This means that they have a faster turnover of the crops.

Exact schedules for bringing a crop of potted mums into bloom are printed in grower information bulletins and catalogs. Based upon when blooms are needed, these schedules tell when to pot and pinch the plants and when to use artificial long and short days. Although a grower may follow a schedule exactly, some cultivars are always too short or too tall. Frequently, these problems are due to local growing conditions.

For example, if the plants of a certain cultivar are always too short, the grower pinches them 7 to 10 days ahead of schedule. This extra time allows more growth to take place. Plants that grow too tall are pinched 7 to 10 days after the scheduled date. Normally, only a single pinch is made.

More than one pinch adds too much growing time and increases the cost of the crop.

Some crops, such as 'Indianapolis' cultivars, are grown single stem and are not pinched. These cultivars produce a very large flower that is most attractive. Single plants in 4-inch (10-cm) pots are always popular. As a rule of thumb, for the number of cuttings to use in large pots add one or two to the diameter of the pot. In a 7-inch (17.5-cm) diameter pot, eight or nine cuttings are evenly distributed.

Some crops that grow too tall are planted on the date when the pinch is scheduled. At times, however, these plants as well as pinched plants grow too tall, regardless of the adjustments made in the pinching program. To overcome this problem, growth retardants are used.

Growth Retardants

Several chemicals can be used to control the growth of flower crops. However, not all growth retardants have the same effect on all crops. The retardants registered for use on potted chrysanthemums are B-NINE SP® and A-REST® ancymidol. B-NINE SP and A-REST are concentrated and must be diluted with water. A solution of 0.25 percent B-NINE SP is made by mixing 5 teaspoons of powder in 1 gallon (3.785 l) of water. This solution is sprayed on the growing tips of the plants about 2 weeks after pinching. Only one treatment is needed.

A-REST is a very strong growth retardant. If too much of this material is used, plant growth may be severely stunted. Always follow the manufacturer's recommendations printed on the label when using this product.

Disbudding

Cultivars grown as pot mums produce more than one flower on each stem. To obtain the largest bloom possible, all buds below the terminal bud must be removed. This process is called **disbudding** (Figures 24-8A and 24-8B). The side buds are removed when they are about 0.13 to 0.25 inch (3 to 6 mm) in width. The easiest method of removing these buds is to rub them off with a

FIGURE 24-8A One stem of a chrysanthemum plant showing many buds at tip and along the stem. *(Courtesy H. Edward Reiley).*

FIGURE 24-8B The same stem after all buds except the terminal bud are removed. The removal of the buds results in larger, single flowers on each stem. *(Courtesy H. Edward Reiley).*

sideways motion of the thumb. Care must be used so that the main flower bud is not removed.

If disbudding is delayed until the buds are quite large, the task is more difficult, time consuming, and costly. In addition, energy that was used in the growth of the side buds should have gone into the central flower bud.

Temperature

Accurate temperature control is necessary to obtain quality plants. Chrysanthemums are highly sensitive to the effect of temperature on flower formation and development. The best flower quality results if the mums are grown at a night temperature of not less than 60°F (15.5°C). The day temperature should be 70°F (21°C) for cloudy conditions and 75°F (24°C) when it is clear. Very high temperatures on summer days delay flower formation and development. Fan and pad cooling is used in the summer to produce quality plants. Shading is not recommended to reduce the greenhouse temperature because the reduced light causes the plants to stretch.

To deepen the color of the blooms and improve their quality, the night temperature is reduced to 55°F (13°C) for the last 7 to 10 days before the plants are ready for sale.

Year-Round Flowering

Chrysanthemums are classified as short-day plants. Most cultivars initiate set buds when the day length is 14.5 hours or less. Flower development takes place when the day length decreases to 13.5 hours or less. Whenever the day length is longer than 14.5 hours, plant growth is vegetative.

The photoperiod response in chrysanthemums is controlled by giving the plants artificial long days or short days, depending upon the season of the year.

Long Days

Artificial long days are given from August 1 to May 15 of the following year at 42° north latitude. The dates change above or below this latitude. The flowering response is turned on or off by red-far-red radiant energy. Thus, the best light is given by incandescent bulbs because they emit large amounts of red energy.

Chrysanthemums have the same response to as little as 2 fc (22 lux) of light as they do to full light. When light is given in the middle of the dark period, only vegetative growth occurs. To ensure that flowering does not occur, it is recommended that minimum lighting of 10 fc intensity be used in the darkest part of the bench. The proper intensity is provided by 60-watt incandescent bulbs, mounted 2 feet (61 cm) above the plants and spaced 4 feet (122 cm) apart. Or, 100-watt bulbs can be spaced 6 feet (183 cm) apart, 3 feet (91 cm) above the plants. Reflectors are used on the bulbs to prevent the loss of light. High-intensity flood lights may be used to light an entire house. A light meter should be used to ensure that the plants receive a minimum of 10 fc (108 lux of light).

At 42° north latitude, light is applied at night according to the following schedule.

August, September, April, and May 1 to May 15	*2 hours*
October and March	*3*
November and February	*4*
December and January	*5*

The light break should be given close to the middle of the dark period. For example, if the sun sets at 6 P.M. and rises at 6 A.M., a 4-hour light break is given from 10 P.M. until 2 A.M. In large greenhouse ranges, various factors may make it possible to use only half of the lights needed. If this is the case, lighting is scheduled so that the dark period is no longer than 7 hours.

Short Days

When the natural day length is long enough to cause vegetative growth, the plants must have artificial short days if they are to flower. At 42° north latitude, black cloth is used to give short days from March 15 until October 20. There is some overlap in the schedule of short days and the lighting schedule. This procedure ensures that the day length is short enough to cause flowering.

The short-day treatment is given by pulling black sateen cloth or black polyethylene over the plants. Generally, the black cloth is pulled each day at 5 P.M. and 7 A.M. (These times are the normal working hours of greenhouse personnel.) When daylight savings time is in effect, the cloth is pulled when it is only 4 P.M. suntime. The sun is still high in the sky at this hour. As a result there is a heat buildup under the black cloth. The temperature may get as high as 129°F (54°C) or more. This high temperature can cause a serious delay in flowering. If possible, the cloth should be pulled at 6:30 or 7 P.M., when the outside temperature has decreased. The plants need only 12 hours of darkness to flower.

On the weekend, the black cloth generally is not pulled on Saturday night. For each day the cloth is not pulled, the grower can expect a delay of 1 day in the time for flowering.

If the plants are grown on a year-round schedule, it may be necessary to use black cloth shading, even when the days are naturally short. Lights that are used to keep plants vegetative must not be allowed to shine on plants that are scheduled for flowering. Flowering will be prevented if any stray light reaching the plants is more than 2 fc in intensity. As black cloth ages, it becomes less opaque. Cloth that is 2 or 3 years old may allow a small amount of light to pass through to the plants at the beginning and end of the shading period on bright summer days. The amount of light that reaches the plants may be enough to delay flower development. The grower should use only the newest and best grade of black cloth (or black polyethylene).

Problems

Cultural Disorders

Generally, the failure to form flower buds is due to an improper temperature at the time of bud set. The minimum temperature required is 60°F (15.5°C).

The failure of bud set may also be due to too much light. The black cloth treatment should be used to ensure short days.

Another cause for a delay in flowering is too much heat in the summer. The greenhouse must be cooled properly. Also, the black cloth should not be pulled until after the sun sets or the temperature has started to decrease.

Plants crowded together grow too tall. Another factor in tall growth is an inadequate light intensity. High-quality plants require full light intensity. It is recommended that the greenhouse glass be kept clean so as not to reduce the light intensity.

After a plant is pinched, too few branches, or breaks, may develop. The main reason for this condition is too little light. During the darker days of winter, some growers use HID lamps to supply from 400 to 800 fc (4.32K to 6.48K lux) each day from 6 A.M. to 6 P.M. The lights are turned on when the plants are pinched. They are not used again after 2 weeks. Other growers provide additional light for 30 days before the scheduled date of flowering. Additional factors that are important to the development of branches after the pinch are proper fertilization and watering.

Flower petal burn is caused by bright sunlight. Some cultivars need a light shade to prevent sun scorch when they come into bloom.

Disease causes the flower to rot as the plants bloom.

Diseases

Leaf spot is caused by two fungi, *Septoria obesa* and *S. chrysanthemella*. Plants should be sprayed with a fungicide. The leaf spot problem is reduced by the proper spacing of the plants and adequate ventilation.

The vascular system of the plant may be affected by bacterial stem blight, caused by *Erwinia chrysanthemi*, and *Verticillium* wilt, caused by a soil fungus. The only control for these diseases is the use of resistant cultivars and meristem cultured cuttings. Once these diseases enter a plant, the crop is lost. If the diseases develop in a planting, the plants must be destroyed by burning. All materials used to produce the crop, including pots, benches, flats, and tools, must be steam sterilized.

Powdery mildew and *Botrytis* petal spot are controlled with fungicide sprays.

As a protection against root and stem rot, the plants should be drenched with a fungicide as soon as they are potted.

Insects

Pot mums are attacked by aphids, leafrollers, mealybugs, midges, leaf and root-knot nematodes, spider mites, symphyllids, thrips, and whitefly. Adult leafminers are attracted to photoperiod lighting and may be an issue during the vegetative stage. Other troublesome pests include cabbage loopers, grasshoppers, leafminers, slugs, snails, sow bugs, and tarnished plant bug. A regular insect control program is needed to keep the plants free of insects.

If systemic insecticides are used, they should be applied only when the soil is moist. If the soil is dry, serious plant injury can result.

EXACUM (Arabian Violet) ⊠ *Exacum affine, Gentianaceae*

FIGURE 24-9 *Exacum affine.*
(Courtesy of the Professional Landcare Network, formerly the Associated Landscape Contractors of America.)

Exacum is a native of the Socotra Islands. *Exacum affine* is the most widely grown and popular cultivar. This plant is a small, shrublike annual having light green leaves. The mature plant develops many fragrant, star-shaped white, blue, or lavender blossoms (Figure 24-9).

Growers have found that these plants sell well from late spring into summer. The plants begin to flower while they are still small. Thus, they can be offered for sale while they are still in 2.5- or 3-inch (6- or 7.5-cm) pots. Most exacum are sold as 5- or 6-inch (13- or 15-cm) plants.

Seed is sown in December through March to produce flowering plants from April through July. The germination medium is a peat-lite mix. The recommended night temperature is 60°F to 65°F (15.5°C to 18.5°C) with the temperature during the day 10° warmer. Temperatures higher than the given values cause flower drop.

The plants should be shaded from direct sunlight. They are fertilized at every watering at the rate of 150 ppm of $N-P_2O_5-KO$ using a balanced 20-10-20 analysis fertilizer.

The plants are generally free of problems.

FUCHSIA ⊠ *Fuchsia hybrida, Onagraceae*

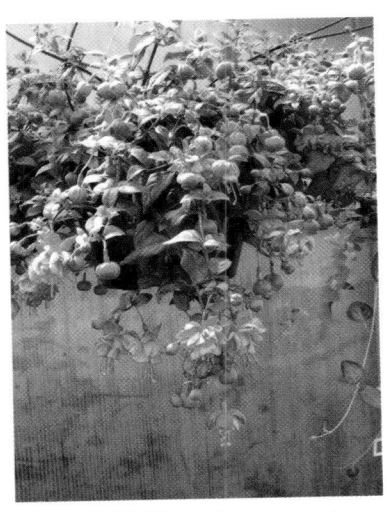

FIGURE 24-10 Fuchsia grown in a hanging basket.

The large number of fuchsia (pronounced *foo-she-a*) hybrids now available to consumers accounts for the increasing popularity of these plants. They are native to tropical America and New Zealand.

Some fuchsias are propagated from seed, but most are grown from terminal cuttings. The plants do very well in highly porous peat-lite mixes. Rooting of the cuttings occurs rapidly at an air temperature of 65°F (18.5°C).

Fuchsias need a high light intensity. However, to obtain maximum flower development, they must be shaded from bright sunlight with its high temperatures. A night temperature of 55°F (13°C) is needed. Bushy plants are produced by pinching the plants until about 8 weeks before they are offered for sale. The plants are fertilized at each watering at the rate of 150 ppm each of $N-P_2O_5-K_2O$.

Fuchsias grown in hanging baskets do extremely well for springtime sales (Figure 24-10). Cultivars that are especially suited for growing in hanging baskets are 'Dark Eyes,' 'Dollar Princess,' 'Marinka,' 'Southgate,' 'Swingtime,' and 'Winston Churchill.' Fuchsias with an upright growth habit are used as pot plants.

'Florabell' is a new introduction that can be grown from seed to produce an excellent 4-inch (10-cm) pot. The small flowers are purple with bright red sepals.

Problems

Insects

Fuchsias may be infested by aphids, cyclamen mite, mealybugs, scale, thrips, and whitefly. The most troublesome insect is the whitefly.

GARDENIA *Gardenia jasminoides, Rubiaceae*

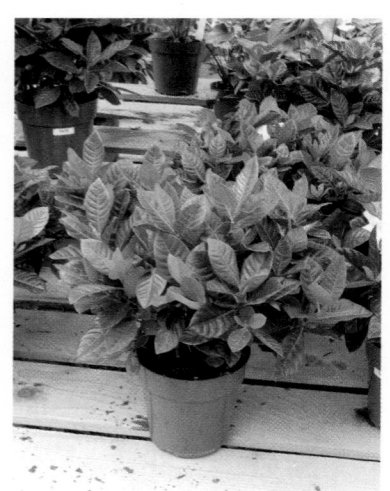

FIGURE 24-11 Potted Gardenia plant.

The fragrant flowers of the gardenia were once very popular for corsages. Now, gardenias are grown as pot plants for Easter and Mother's Day sales (Figure 24-11). A native of China, the cultivar Vietchii is preferred for pot plant culture because it produces many smaller blossoms.

To propagate gardenias, 4- to 6-inch (10- to 15-cm) long terminal cuttings are taken from November through March. The cut may be made at a node or an internode. Basal foliage is removed so that it will not be buried in the medium. Sand with a pH below 6.0 is one possible rooting medium. Combinations of sand and peat moss, and perlite and peat moss, are also used. The cuttings may be placed directly in small pots and thus reduce the number of steps in the production program.

The plants are very sensitive to a lack of moisture. A low-pressure mist should be used to prevent the plants from becoming hard in the propagating bench. A hard plant is one whose growth is checked by some environmental stress.

If a misting system is not available, the cuttings can be rooted under plastic or in a closed grafting case where high humidity can be maintained. Rapid rooting requires an air temperature of 75°F (24°C). The cuttings should be shaded from direct sunlight. A rooting hormone can be used to speed the development of roots. A good root system should be formed in 4 to 6 weeks.

Soils and Fertilizers

Gardenias need a growing medium with a pH of 5.5 to 6.0 (acidic). Once the cuttings are rooted, they are shifted to small pots. The medium is a peat-lite mix to which half the recommended amount of limestone is added. A mixture of equal parts of soil and sphagnum peat moss may be used also. Iron is added to the medium by mixing 1 pound (0.6 kg) of iron sulfate with a cubic yard (cubic meter) of the mixture. The plants are very sensitive to a lack of iron. If soil is used as a part of the mix, it must be steam pasteurized to kill disease organisms and soil nematodes.

Only those fertilizers having an acid reaction with the soil should be used on gardenias. For example, a complete analysis fertilizer (20-10-20) has an acid reaction. It is applied at the rate of 150 ppm at each watering. If the soil starts to become less acid, ammonium sulfate may be applied two or three times at the same rate to lower the pH. If the water supply has a high pH, specially formulated fertilizers may be used.

In gardenias, an iron deficiency is indicated by a chlorosis or yellowing of the new growth. To prevent this problem, a chelated iron product is applied regularly at the rate recommended on the label. Iron sulfate can also be used and is applied at the rate of 1 ounce in 3 gallons (28 g in 11.4 l) of water every 2 or 3 weeks.

Light and Temperature

Gardenias are grown at a temperature of 60°F (15.5°C) in full light intensity. It may be necessary to use light shade in the summer to control the temperature in the greenhouse. This crop grows best in a very humid environment. Therefore, the foliage should be syringed several times a day to maintain a humid condition.

Young plants are shifted gradually through a series of pots of increasing size. As soon as roots are well developed at the surface of the growing medium, the plants are shifted to a larger pot. If the plants become potbound, growth and flower bud development are retarded.

Some growers place two or three small plants in a 5- or 6-inch (13- or 15-cm) diameter pot. This practice yields a plant of marketable size more quickly than a single plant in a pot.

Regular pinching of long shoots promotes branching. The last pinch should be made no later than July 21 for plants that are to flower for Christmas. If the plants are to flower after Christmas, the last pinch is made no later than September 1.

Flower Bud Formation

A night temperature of 60°F to 62°F (15.5°C to 16.5°C) aids flower bud formation and development. Lower temperatures reduce plant growth and higher temperatures delay flower bud development. The day temperature must be higher than the night temperature to ensure good bud development.

To obtain flowers for Christmas, a short-day treatment may be used. For example, black cloth may be applied from 5 P.M. until 8 A.M. daily from July 21 to August 13. This treatment seems to delay the development of flower buds already formed. It does not affect the formation of buds.

Problems

Cultural Disorders

A lack of iron causes yellowing to occur between the veins of the leaves. Iron chelates or iron sulfate applied at regular intervals will prevent this problem. To increase the uptake of iron, the soil temperature should be kept above 60°F (15.5°C).

Diseases

Canker is caused by the fungus *Phomopsis gardeniae*. The stem of an infected plant will be orange-yellow in color when it is cut off at the soil level. To prevent infection, a fungicide dust may be used on the base of the cuttings.

Bacterial leaf spot is caused by a bacterium, *Pseudomonas gardeniae*. To prevent the spread of this disease, keep moisture off the tops of the plants.

The root-knot nematode is controlled by steam pasteurization of the growing medium.

Insects

Aphids, mealybugs, red spider mites, and thrips may infest the plants.

GENISTA � *Cytisus canariensis, Fabaceae*

As the name indicates, *Cytisus canariensis* is native to the Canary Islands. This plant is known as Genista and is often grown as a potted plant for sale in the spring.

To propagate Genista, terminal cuttings are taken from stock plants or from unsold plants that are held over to the next sale season. Cuttings may be taken from April through October. The earliest cuttings produce the largest plants. The best propagation medium for this crop is sand. Because Genista is a member of the Legume family, the propagation and growing medium must have a pH close to 7.0. A mixture of peat moss and perlite is too acid and must be heavily limed.

Rapid rooting occurs if bottom heat is provided at 65°F to 70°F (18.5°C to 21°C) and the air temperature is 60°F (15.5°C).

Once healthy root systems are established, the plants are moved from the propagation medium to 2.5- or 3-inch (6.3- or 7.5-cm) pots containing a mixture of soil and leaf mold or well-rotted manure. If peat moss is used, extra limestone must be added to raise the pH.

As a legume, the plants produce their own nitrogen once the roots are inoculated with nodule forming bacteria. A 10-20-20 analysis fertilizer is recommended for this crop's low nitrogen and high phosphorus and potassium.

During the summer and early fall, the plants are grown at a night temperature of 60°F (15.5°C). A compact shape is obtained by **shearing** the plants. Shearing should be stopped 6 weeks before cool temperatures are maintained. As soon as the weather outside permits, the plants are grown at a temperature of 50°F (10°C). At a temperature above 60°F (15.5°C), the plants will not flower. When the plants are coming into flower, full light intensity is preferred. After the plants bloom, shade is provided to reduce the temperature.

The plants will bloom about 4.5 months after the temperature can be held constant at 50°F (10°C).

Problems

Cultural Disorders

The plants will not form flower buds if the temperature is more than 60°F (15.5°C). High temperatures also cause the flower buds and blooms to drop.

Poor or slow rooting is a problem when the temperature is less than 60°F (15.5°C).

Leaf drop is the result of both too little moisture and too much moisture.

Insects

Aphids, mealybugs, red spider mites, and thrips may infest the plants.

ACHIEVEMENT REVIEW

Select the best answer or answers to complete each statement. List the appropriate letter(s).

1 "Christmas" cactus is propagated by:
   a. the division of rhizomes.
   b. terminal cuttings.
   c. meristem tissue culture.
   d. three-joint cuttings.

2 Christmas cherry fruit set is delayed by:
   a. overfertilization with nitrogen.
   b. poor light intensity.
   c. high summer temperature.
   d. overfertilization with potassium.

3 Chrysanthemum cuttings received in a frozen condition should be:
   a. planted immediately.
   b. held at 31°F (−0.5°C) in a refrigerator.
   c. held at 41°F (5.0°C) until thawed.
   d. discarded because they rarely recover from freezing.

4 An objection to the use of peat-lite mixes for growing pot mums is the:
   a. high cost.
   b. light weight of the mixes.
   c. nutrient content.
   d. drainage capacity.

5 The amount of 20 percent superphosphate added to each cubic yard (cubic meter) of medium for pot mums is:
   a. 1 pound (0.6 kg).
   b. 2 pounds (1.2 kg).
   c. 4 pounds (2.4 kg).
   d. 5 pounds (3 kg).

6 Rooted chrysanthemum cuttings are planted at a 45-degree angle to:
   a. support the stems on the pot edge.
   b. be able to grow more plants in a pot.
   c. avoid a "stovepipe" effect on growth.
   d. create a "stovepipe" effect on growth.

7 Osmocote® 14-14-14 fertilizer is used as a one-shot controlled nutrient program for pot mums at the rate of:
   a. 4 pounds (2.4 kg) to a cubic yard (cubic meter) of medium.
   b. 6 to 8 pounds (3.6 to 4.8 kg) to a cubic yard (cubic meter) of medium.
   c. 10 to 12 pounds (6 to 7.2 kg) to a cubic yard (cubic meter) of medium.
   d. 5 to 15 pounds (3 to 9 kg) to a cubic yard (cubic meter) of medium.

8 Pot mums are pinched to:
a. time the crop.
b. delay flowering.
c. increase the number of stems that will produce flowers.
d. increase the number of cuttings that can be taken.

9 The growth retardant B-NINE SP® is applied to pot mums:
a. as a soil drench.
b. along with the fertilizer.
c. by mixing it with the medium before planting.
d. as a foliar spray 2 weeks after pinching.

10 When chrysanthemums are ready for sale, the night temperature is reduced to:
a. conserve greenhouse heat.
b. reduce the need for water.
c. intensify colors and improve quality.
d. delay flowering.

11 The least amount of light energy that will trigger the photoperiod response in chrysanthemums is:
a. 0.5 fc (5.5 lux).
b. 2 fc (22 lux).
c. 10 fc (l08 lux).
d. 50 fc (540 lux).

12 To set flower buds on cineraria requires 6 weeks of temperatures in the range of:
a. 70°F to 75°F (21°C to 24°C).
b. 65°F to 70°F (18.5°C to 21°C).
c. 60°F to 65°F (15.5°C to 18.5°C).
d. below 60°F (15.5°C).

13 Clerodendrum grows best when the media pH is:
a. 4.0 to 5.0.
b. 4.5 to 5.5.
c. 5.0 to 6.0.
d. 5.5 to 7.0.

14 Cyclamen seed require washing under a stream of water or soaking before sowing to:
a. remove the hard seed coat.
b. get water into the seed.
c. wash away a germination inhibitor.
d. produce a high carbon dioxide environment.

15 When transplanting cyclamen, one-half of the corm remains above the medium to:
a. improve flower bud initiation.
b. increase the number of leaves that develop.
c. decrease the number of leaves that develop.
d. reduce the possibility of crown rot.

16 The usual method of keeping cyclamen cool in the summer is to:
   a. grow them under heavy shade.
   b. use low night temperatures.
   c. grow them under a low-pressure mist.
   d. submerge the pots in wet sand.

17 Flowering below the foliage on short stems is caused by:
   a. a lack of nitrogen.
   b. overwatering.
   c. a potassium deficiency.
   d. cyclamen stunt disease.

18 If a peat-lite mix is used for potting gardenias, it should be made using:
   a. no limestone.
   b. one-half the recommended amount of limestone.
   c. the recommended amount of limestone.
   d. twice the recommended amount of limestone.

19 Genista is a member of the *Leguminosae* family. Thus, it produces its own:
   a. nitrogen.
   c. potassium.
   b. phosphorus.
   d. iron.

CHAPTER

25

Flowering Plants III

- �належ African Violets
- ✷ Geraniums
- ✷ Gloxinia
- ✷ Hydrangea
- ✷ Kalanchoe
- ✷ Poinsettia
- ✷ Primrose
- ✷ Roses (Potted)

OBJECTIVES:

After studying this chapter, the student should have learned the cultural details for the flowering potted plants, African Violets to Roses. For each crop, the student should be able to describe:

- ■ methods of propagation.
- ■ the soil or growing medium requirements.
- ■ fertilizer recommendations.
- ■ methods of growth, disease, and insect control.

AFRICAN VIOLET ✷ *Saintpaulia ionantha, Gesneriaceae*

The African violet is probably the most popular and widely grown flowering potted plant in the United States (Figure 25-1). Its popularity is due to the fact that it thrives under the warm, dry, low-light conditions of the average home. As the name implies, it is native to the Tanzania's East Usambara mountains of tropical eastern Africa.

Propagation

The plants are propagated vegetatively by means of leaf petiole cuttings taken from mature plants. A 0.5-inch (1.2-cm) or 1-inch (2.5-cm) long

FIGURE 25-1 An assortment of African violets for retail sale.

470

petiole must remain attached to the leaf. Longer petioles tend to slow the development of the young plants.

The base of the petiole is dusted with a number one rooting hormone and inserted in the propagation medium. African violets root easily in many materials. A popular medium is a mixture of equal parts by volume of peat moss and vermiculite or perlite.

The leaf cuttings are spaced so that they do not touch each other in the row or between rows. The cuttings are inserted in the medium until the base of the leaf is just above the medium level. The base of the leaf should not touch the medium, as rot may develop.

The cuttings are rooted directly in the bench or in flats filled with medium. Rooting occurs sooner if bottom heat is used and the air temperature is maintained at 65°F (18.5°C). The small plants are large enough to be potted in 2.25-inch (5.7-cm) pots about 2 months after the cuttings are stuck in the medium. Before the plants can be sold as 3-inch (7.5-cm) plants with flowers, another 4 to 6 months of growth is needed.

African violet leaves are easily spotted when the plants are irrigated with cold water. The spots occur because the cold water chills the leaf, causing the chlorophyll to die in those areas. Once the leaf is spotted it remains that way until it dies or falls off.

At the time of propagation and during the growth of the plants, any water applied from overhead should be heated to a minimum temperature of 60°F (15.5°C). Watering from the bottom by the injection method or a capillary mat eliminates the need to heat the water.

Soils and Fertilizer

African violets have very fine root systems. They grow best in a highly porous medium with good drainage and aeration. The plants do not grow well in heavy loam or clay soils. Among amateur growers, there are as many soil mixtures for African violets as there are growers.

The peat-lite mixes are widely used for African violets. However, some growers prefer to make their own mixes using equal parts of loam soil, sand, peat moss, perlite, and leaf mold if it is available. Growers rarely use manure as part of the soil mix for several reasons: It is not readily available; it may contain nematodes if it is not pasteurized; and it can cause high soluble salt levels. The nematodes and high salt levels are very damaging to African violets.

When the cuttings are well rooted and small plants have developed, they are transplanted to 2.25- or 2.5-inch (5.7- or 6.25-cm) pots. At times, a plant with multiple crowns is produced. These plants are grown on until they reach saleable size and flower. If a grower is in the mail order business, the plants are kept in 2.25- or 2.5-inch (5.7- or 6.25-cm) pots until they are well established, usually 6 to 8 weeks after potting. At this point, the plants are ready to be shipped. It is less expensive to ship small, nonflowering plants. In addition, such plants suffer less damage during shipping than do larger plants.

One week after the plants are transplanted from the propagation medium, a regular fertilization program is started. A complete analysis fertilizer, such as 15-15-15 or 20-10-20, is applied at the rate of 100 ppm each of N and K_2O at each watering. Every seventh watering should be made with plain tap water to wash out salt buildup.

Temperature

African violets make their best growth at warm temperatures. The recommended temperatures are a minimum of 65°F (18.5°C) at night and 70°F to 75°F (21°C to 24°C) during the day. At cooler temperatures, the plants grow very slowly and often develop powdery mildew.

Light

Any discussion on the light needs of this crop must be divided into two areas: natural light and artificial light. There are probably more African violets being grown under artificial lights than any other flower crop in the country.

Natural Light

When plants are grown in the greenhouse under natural light conditions, the intensity of the light must be controlled. High-quality African violets can be obtained at moderate light levels. During the brightest part of the day the light intensity should be no more than 1,200 fc (12.9K lux). At higher levels the plants usually develop bleached foliage and short petioles. The reduced length of the petioles results in compact plants (Figures 25-2 and 25-3). If the light levels are too low, the petioles stretch and the plants become very leggy.

In the northeastern United States, shade is applied to the greenhouse in March and is continued until mid-October. More shading is needed during the bright days of midsummer than is required in the spring and fall. No shade is used on the glass between November and March because the light intensity is usually very low.

In areas where the light levels are high year-round, the greenhouse may be shaded permanently. In addition, some growers place Saran cloth, muslin, or cheesecloth over the benches to assist the roof shading (Figure 25-4).

FIGURE 25-2 An African violet with sun scorch.

FIGURE 25-3 An African violet with shortened petioles from excessive light. Note that the larger leaves have shipping damage.

FIGURE 25-4 A nice crop of African violets grown on ebb and flood tables under hanging baskets for shade ready for shipment.

FIGURE 25-5 African violets grow well under continuously on fluorescent lamps.

Artificial Light

Plants have been produced under artificial light for many years. Growers who wish to use the greenhouse space more effectively combine natural and artificial light to produce the crop. The benches are constructed at two or three levels. Plants on the top level receive natural light only. Plants on the other levels use whatever natural light reaches them. Fluorescent tubes installed on the lower benches provide most of the light these plants need (Figure 25-5).

A few growers use artificial light only to produce the crop. Excellent, high-quality plants are grown under these conditions.

There has been a great deal of debate about the use of special plant-growing lamps marketed by several companies. To date, no scientific studies have shown that these lamps are any better than cool white or warm white fluorescent lamps for growing plants. One Canadian study reported that the flowering of African violets is improved when a few incandescent bulbs are added to an area lighted with a popular plant-growing lamp installation. The added incandescent light was 10 percent of the total wattage.

To grow violets under lights, the intensity supplied by the cool white fluorescent lamps must not be less than 600 fc (6.48K lux). The lights are kept on for 14 to 18 hours daily. Lighting the plants for more than 18 hours is not harmful to them, but it does waste electrical energy because the added light does not benefit the growth.

Two, 2-tube, 8-foot (2.4-m) long strip fixtures spaced 12 inches (30.5 cm) apart, and mounted 12 inches (30.5 cm) above the plants will give a light intensity of 600 fc (6.48K lux). When new tubes are used, the intensity will be greater than 600 fc (6.48K lux). As the tubes get older, the light output decreases and they grow dimmer. A light meter is used to measure the actual intensity.

Plants grown under fluorescent lamps often are of better quality than those grown in the greenhouse. The more controlled growing conditions under artificial lights probably account for the improved quality. That is, the light level is constant, the relative humidity does not vary as much as in the greenhouse, and it is easier to hold the temperature at an even level.

Plants under lights grow faster and generally need slightly more fertilizer and water than do those being grown in the greenhouse. A complete analysis fertilizer, such as 20-10-20 or 15-15-15, is applied at the rate of 100 ppm each of N and K_2O at each watering. Because the plants must be irrigated more frequently than greenhouse plants, they receive more nutrients.

Problems

Cultural Disorders

Cold water on the leaves causes **ring spot.** If the light level is too low, the plants will not flower. This is not usually a problem in the greenhouse, but amateur growers often ask why their plants are not flowering, especially during the winter.

Diseases

Powdery mildew is caused by spores of *Oidium* species, which are carried in the air. Wet leaves spread the disease. Spraying with the proper chemicals controls the disease. Do not use cold water because the leaves will become spotted.

Both root-knot and leaf nematodes are especially troublesome. To prevent nematodes from infesting the crop, sterilize all media, pots, tools, benches, and beds. Generally, good control of plants infested with nematodes is not obtained with chemical treatment. If nematodes are suspected, the plants must not be subirrigated because this practice spreads the organisms.

Insects

Cyclamen mite is the most serious pest. It infests the crown of the plant and is very difficult to see with the naked eye. Its presence is noted by the distortion of the newly formed leaves. Cyclamen mites can be controlled by spraying with systemic miticides.

Mealybugs may be a problem, but they are easily controlled.

GERANIUMS 🔲 *Pelargonium* × *hortorum, Geraniaceae*

Geraniums are one of the most widely grown of all flowering pot plants. They are sold in the spring as flowering plants, and the bright red colors are becoming more popular for Christmas sales. Geraniums are also in demand as spring bedding plants.

Geraniums may be propagated by taking cuttings from stock plants, or they may be started from seed.

Production from Cuttings

Stock Plants

Geranium cuttings may be purchased from specialty growers as unrooted, callused, or rooted cutting, or the grower may maintain stock plants and take cuttings from them.

When stock plants are to be used, the grower must ensure that the plants from which the stock plants are grown are as disease free as possible. It is recommended that the grower buy rooted cuttings from specialty growers who use tissue culture or shoot-tip reproduction methods. Growers usually guarantee that these plants are free of all diseases.

The longtime leader in the field of geranium propagation is Oglevee, in Connellsville, Pennsylvania. Mr. Robert Oglevee, owner and founder of the company, saw the need for improving the geraniums that were being offered to the industry. In 1953, he began a cooperative study with Dr. James Tammen, then a plant pathologist at Pennsylvania State University, to develop the culture indexed geranium. This procedure ensured that the cuttings being produced were free from bacterial and fungal disease organisms. In the later years the protocol for also producing virus-free indexed geraniums was developed. Today, all of the major geranium propagators offer CVI certified, cultured virus indexed plants.

Cultivars

The ever increasing popularity of geraniums has resulted in a large number of cultivars that range in color from red, through pink to white. There are no blue geraniums.

The grower selects cultivars based on the demands of the local market. The environmental conditions of the part of the country in which the business is located also influence the selection of colors and types of geraniums to grow. More red geraniums are sold than any other color. One source recommends the following division of a crop: 60 to 70 percent red, 20 percent pink, and 10 to 20 percent white and other colors.

Oglevee offers 28 cultivars of the zonal types. Zonal geraniums are usually grown as potted plants from vegetative cuttings and as bedding plants started from seeds. They also have cultivars in a new group called floribundas. Ivy types and regal types, which are most often used in hanging baskets, make up the other classes. They offer 18 ivy and 24 regal type geraniums. The following cultivars are a partial list of their offerings.

Zonal Types

'Angel'—white
'Aurora'—lavender/magenta
'Ben Franklin'—brocade
'Isabell'—red/scarlet
'Kim'—red/scarlet
'Medallion Dark Red'—red/scarlet
'Pink Expectations'—pink
'Red Hots'—red/scarlet
'Sincerity'—red/scarlet
'Snowwhite'—white
'Veronica'—lavender/magenta
'Wendy Ann'—pink
'Yours Truly'—red/scarlet

Stardom-floribundas

'Angela'—light pink
'Elizabeth'—red

'Grace'—light pink
'Gypsy'—lavender
'Judy'—coral/white eye
'Julia'—coral
'Katherine'—white
'Lucille'—coral
'Marilyn'—candy pink

Ivy Types

'Balcon Royale'—red
'Beauty of Eastbourne'—red
'Cornell'—purple
'Double Lilac White'—white
'Monique'—pink
'Peppermint Candy'—striped
'Salmon Queen'—pink
'White Nicole'—white
'Yvette'—purple

Regal Types

'Erin'—salmon pink/red
'Lavender Grand Slam'— lavender/pink
'Sandra'—light pink/scarlet

Elegance Series

'Allure'—pink
'Crystal'—white
'Fantasy'—burgundy/white
'Splendor'—rose/red

Royalty Series

'Baron'—dark purple
'Baroness'—raspberry/red
'Duke'—light rose/dark burgundy
'Duchess'—burgundy pink/white

'Prince'—light rose purple/burgundy
'Princess'—lavender/burgundy purple

Vavra Series

'Dolly'—coral red
'Honey'—rose
'Josy'—light candy/rose
'Mary'—white
'Rosemary'—pink magenta
'Shirley'—bright scarlet

The Ball Horticultural Company, West Chicago, Illinois, has recently begun to offer a line of certified cultured and virus indexed geraniums. These cultivars are released through a subsidiary company, called Ball FloraPlant. They have many cultivars in the Designer series. Seven cultivars are in the Showcase series.

Representative Designer Series

'Bright Lilac'—lilac with white eye
'Bright Red'—carmine red
'Coral'—coral
'Hot Pink'—deep pink
'Light Pink'—soft pink

'Pink Parfait'—deep pink with dark spot
'Pink Pearl'—soft lavender pink
'Rose'—deep fuchsia rose
'Salmon'—salmon
'Scarlet'—scarlet red

Representative Showcase Series

'Bright Coral'—coral
'Dark Salmon'—dark salmon
'Light Scarlet'—light scarlet
'Pink'—pink with small white eyes
'Pink Heart'—pink with dark pink eyes
'Salmon'—soft salmon
'White'—pure white

Fischer USA, Inc., in Boulder, Colorado, supplies several trademark geranium lines. All of their cultivars are patented or have patents pending.

Representative Zonal Types

'Alba'—white
'Atlantis'—red/purple
'Blues'—pink
'Boogy'—red
'Brazil'—fluorescent lilac/pink
'Champion'—orange

'Dolce Vita'—light salmon pink
'El Dorado'—pink
'Explosive'—dark pink
'Grand Prix'—orange red
'Magic'—purple
'Montevideo'—salmon
'Rokoko'—salmon
'Tango'—dark red
'Tiffany'—hot pink

Cascade Types

'Acapulco'—bicolor/pink with white eye
'Blue-Blizzard'—light lavender
'Bright Cascade'—fluorescent red
'Red-Mini-Cascade'—red
'Sofie-Cascade'—light pink
'White Blizzard'—white

Representative Ivy Types

'Barock'—dark red
'Belladonna'—pink
'Chic'—fluorescent lilac pink
'Comedy'—dark violet
'GTI'—dark rose
'Luna'—white
'Molina'—light pink
'Ragtime'—dark fuchsia
'Shive'—dark red
'Sybil Holmes'—pink
'Wico'—light pink

Soils and Soilless Media

Soilless media are commonly used to produce a crop of geraniums. An objection to the use of soilless media for growing stock plants is that they are too light in weight. When stock plants are grown in separate containers for 10 to 12 months, they produce heavy top growth. As a result, the plants will fall over unless they are supported in some manner. The plants do not tip over so readily when they are grown in benches. Even in this case, however, some support must be used to keep the plants upright.

A 1:1:1 mixture of equal parts by volume of loam soil, sphagnum peat moss, and perlite was once a commonly used growing medium. To reduce disease, the growing medium, and all containers, benches, and tools, must be steam pasteurized. If steam is not available, the soil mix can be treated chemically. The containers are disinfected by soaking them for 10 minutes in a solution of Physan 20® or Green-Shield®. Physan 20 and Green-Shield solutions are prepared at label recommended rates. These treatments are to be used only when steam sterilization or other chemical treatments are not possible.

The geranium cuttings may be free of disease when they are received from the specialty grower, but they are not resistant to diseases. For this reason, a mother-block system of culture is used. That is, each plant is potted into its own pot. Six-, 8-, and 10-inch (15-, 20-, and 25-cm) pots have all been used for growing stock plants (Figure 25-6). Using the larger size containers permits larger plants to be grown and results in more cuttings during the propagation period. By growing plants in separate containers, it is easy to remove plants that show signs of disease. Plants that show yellow leaves, leaf spots, stem dieback, or other signs of poor, abnormal growth should be discarded completely. This does not mean moving the waste plant material to a compost pile for future reuse, but moved off site to a municipal landfill or burned.

The potting medium should have a pH of 6.0 to 6.5. Ground limestone may be added to bring the pH to the proper level. In addition, 2 pounds of 20 percent superphosphate or 1 pound of treble superphosphate is added to each cubic yard of mix. The total soluble salt content should be in the low to medium range. No other fertilizers are used before planting. One week after the cuttings are potted, a liquid feeding program is started using a 20-10-20 analysis or a 20-5-30 analysis fertilizer at the rate of 200 ppm of nitrogen at each watering. Soil tests are made every 3 or 4 weeks to monitor the nutrient levels. A heavy watering with plain tap water once a month is beneficial.

Methods of Growing

After the rooted cuttings are planted, the plants are grown on for 2 weeks. This time period permits the root system to become established and produces enough new growth so that the terminals can be pinched. This pinch stimulates lateral shoot growth. When these shoots reach a length of 2.5 to 3 inches (6.3 to 7.5 cm), they can be propagated. These cuttings are rooted, potted when rooted, and grown on as increase stock. This method provides a small number of cuttings the first few months. At the time the major number of cuttings is needed, which is January, February, and March, there will be large numbers of cuttings available from these increase plants.

If the stock plants are started in June, the maximum number of cuttings can be made. Each month of delay in starting the stock plants reduces the number of cuttings available by about 10 percent. The data in Table 25-1 show the effect of planting dates on potential cutting yield.

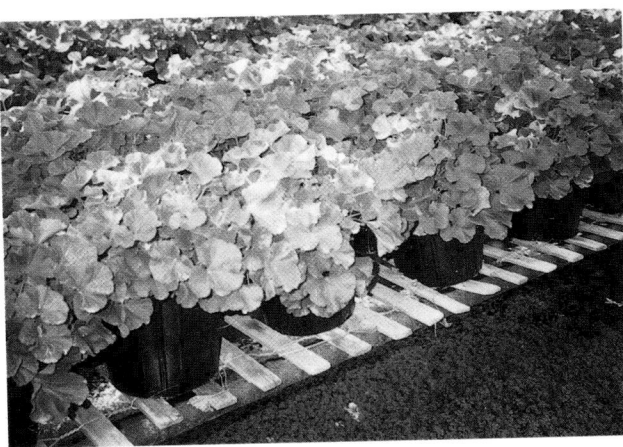

FIGURE 25-6 Geranium stock plants growing in 10-inch (25-cm) pots.

Table 25-1

Effect of Planting Date on Potential Cutting Yield for Geraniums Grown at 40 to 50 Degrees in N. Latitude	
Month Planted	Yield
June	100
July	90
August	80
September	65*
October	50
November	35
December	20
January	15

*Numbers get fewer as days get shorter and light intensity decreases.

Propagation

Vegetative cuttings, usually taken as terminals, are propagated whenever shoots are long enough to provide cuttings 2.25 to 2.5 inches (5.6 to 6.3 cm) long. It is recommended that the person taking the cuttings perform thorough handwashing with soap and water to prevent transmission of disease. Some producers require cutters to also disinfect their hands in a disinfectant solution.

At least two and preferably three knives are used in making cuttings. These are also disinfected. While one knife is being used to remove cuttings from the stock plant, the other two are in the disinfectant solution. Be sure to alternate use of the knives between stock plants. It is a good idea to briefly rinse the knife in plain water after the disinfectant solution. This prevents chemical injury to the base of the cutting.

Some growers will take cuttings by breaking them rather than using a knife. This results in a ragged end at the base of the cutting, which makes them difficult to insert into some media. Usually a flat cut is preferred to a slant cut. It makes no difference whether the cut is made at a node or between the nodes. Rooting is the same.

Cuttings should be taken in clean containers. Do not use old, dirty flats or other containers that might harbor disease organisms. Cleanliness and sanitation in propagation is of extreme importance to successful rooting of geraniums.

The cuttings are prepared for insertion into the propagation media by removal of the stipules from the base of the stem. Remove any leaves that would have their bases inserted in the rooting medium (Figure 25-7). Also remove all flower buds because these quickly die and become vectors for *Botrytis* disease to start.

To obtain uniform rooting and thus be able to clear the propagation bench at one time, a rooting hormone is recommended. Hormodin No. 1 and Rootone are two commercial preparations that are used. Powders are preferred to liquid hormones because powders can be dusted onto the bases of the cuttings (Figure 25-8). If cuttings are dipped into the powder or a liquid

FIGURE 25-7 Geranium cuttings being prepared for insertion into media. Stipules and basal leaf removed.

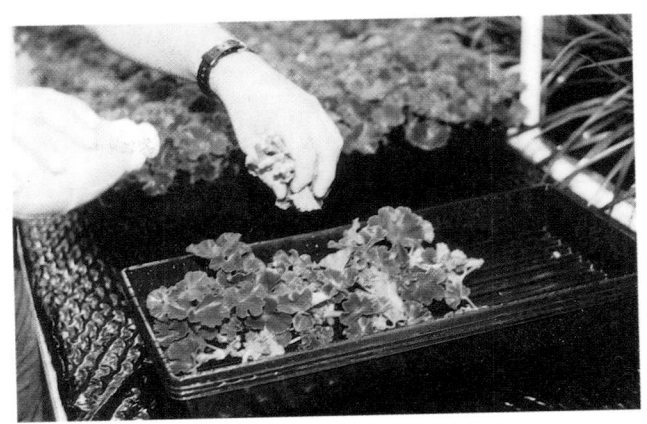

FIGURE 25-8 Dusting basal ends of cuttings with rooting hormone.

solution and there is disease present on any one of them, the rooting hormone will become contaminated. Every cutting thereafter will be inoculated with the disease. Another disadvantage of dipping cuttings into powders is the possibility of getting too much material on the stems. This can inhibit the development of roots.

The rooting medium used must have good aeration and drainage. It must also be free of disease organisms. Over the past 15 years preformed media have become very popular for rooting geraniums. Smithers Oasis Company of Kent, Ohio, produces a rooting medium made from phenolic foam called Root Cube®. The product is purchased as cubes or wedges (Figure 25-9). The wedge shape media has several advantages. Among these is the rapid development of root hairs on the new roots. This soil-type root transplants very readily with a minimum amount of delay in the establishment of the new plant. The wedge shape also makes it easy to transplant the rooted cutting. No large hole has to be made in the potting medium. If the potting medium is at the proper moisture level, the wedge rooted cutting can just be easily pushed into the media (Figure 25-10).

Other preformed media of different shapes and sizes are also used. Jiffy Products of America, Inc., in Norwalk, Ohio, has several different sizes of compressed peat moss pellets (Figure 25-11). These must be hydrated by soaking them in water or placing them under mist on the propagation bench.

Agro Dynamics, Inc. and Grodan, Inc., both of Coppell, Texas, were two of the early proponents of rock wool media. These media are also formed in several sizes for propagation of different crops (Figure 25-12).

Other media have also been used for rooting geranium cuttings. Various combinations of peat moss and vermiculite, perlite, and sand have been presented as bulk media or preformed. All of these media do an excellent job of rooting geranium cuttings when properly handled.

The cuttings are inserted 0.5 to 0.75 inch (1.3 to 1.9 cm) deep. When bulk media are used in cavity trays, the cuttings may be inserted to a 1-inch (2.5-cm) depth.

Preformed media such as Smithers Oasis Wedge® and rock wool have the spacing already established. In loose, bulk media the usual spacing of cuttings is 2 inches (5 cm) apart in rows 4 inches (10 cm) apart. The size of the cutting taken and the amount of foliage dictates the spacing needed. It is important not

FIGURE 25-9 Geranium cuttings propagated in Smither Oasis Wedge® media. They are at the perfect stage for transplanting.

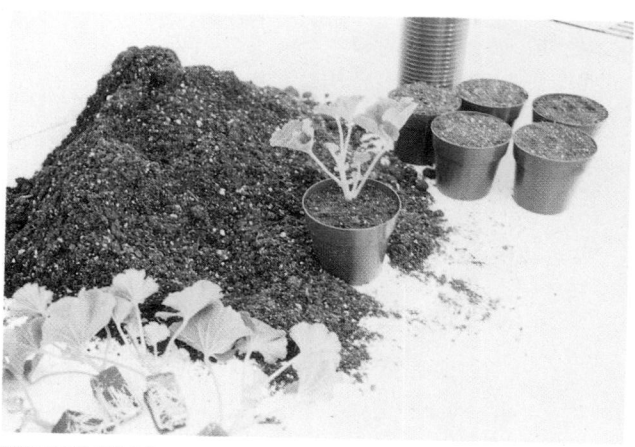

FIGURE 25-10 Potting Wedge® rooted geraniums in peat-lite media in 4-inch (10-cm) pots.

FIGURE 25-11 New Jiffy® compressed peat pellets. *Left to right:* 0.7 in (18 mm), 1.0 in. (25 mm), 1.2 in. (30 mm), 1.4 in. (36 mm), 1.6 in. (42 mm).

FIGURE 25-12 Rock wool propagation media, single block system (SBS).

to crowd them too closely because this is an open invitation for *Botrytis* leaf rot and *Rhizoctonia* stem rot to get started.

Immediately after being stuck, the cuttings should be watered in well. The first few days of propagation, low-pressure mist may be used to keep them from wilting. This should be applied infrequently. The mist timing cycle will depend on the environmental conditions at that time. If the weather is hot and dry, more mist will be needed than if it is cool and rainy. Mist applications to the foliage of geraniums should be kept to the minimum to avoid development of botrytis and other leaf and stem diseases.

The media temperature is maintained at 75°F (24°C) with the air temperature at 65°F (18.5°C).

In the propagation area, light levels must be reduced by shading the greenhouse and/or using shade cloth over the propagation benches. Zonal geraniums require light intensities of 3,500 to 4,500 fc (38.5K to 49.5K lux). The recommended light intensity levels for ivy and regal geraniums are 2,500 to 3,500 fc (27.5K to 38.5K lux).

Although some media, such as Oasis Root Cube® Wedge® products and Jiffy® products, contain a small amount of fertilizer, there is not enough to maintain healthy cuttings throughout the propagation period. Eight days after sticking the cuttings, callus formation occurs. At this time, a light application of fertilizer, 100 to 120 ppm of 20-10-20 or 15-15-15 analysis, should be made. This can be applied every other day until the cuttings are removed for potting.

Depending on the season of the year and the cultivar, geranium cuttings can be ready for potting in as little as 12 days after sticking. Usually it takes 18 to 24 days to have a cutting rooted well enough that it can withstand the rigors of potting.

Some growers do not use any propagation media for starting their new crops. They will direct stick cuttings into 4- or 4.5-inch (10- or 11-cm) pots. This system works well where light intensity is high and days are getting long, as found in spring in the northern United States. Direct sticking requires a large amount of greenhouse space compared to that needed for vegetative propagation in media. There is some saving in labor because the cuttings are only stuck once and do not have to be repotted to be grown on to sale size.

Production

Once the cuttings are rooted, the plants are potted in finish size 4-inch (10-cm) or 5-inch (12.5-cm) pots. The potting medium may be a mixture of equal parts by volume of loam soil, peat moss, and perlite. A mixture of two parts soil, one part peat moss, and one part perlite is used by some growers for the extra weight the soil provides. Many geraniums are grown in peat-lite mixes. Regardless of the type of potting medium used, a preplant application of phosphorus is required. A proportioner can be used to fertilize plants during the growing period. A 20-10-20 analysis fertilizer is applied at the rate of 200 ppm each of N and K_2O at each watering. The plants should be kept moist, but the water must not be applied to the foliage. A lack of water delays bloom and reduces both the plant and flower size. Keeping the plants dry does not force them to flower as some growers believe.

Plants grown in 4-inch (10-cm) pots are spaced in the bench with 1 inch (2.5 cm) between pots. No further spacing is needed as they grow to marketable size. The pots should not touch each other because the plants will be crowded, causing tall and leggy growth and undesirably long internodes.

With the proper spacing, it will not be necessary to pinch the plants to cause them to branch. Pinching delays bloom for as much as 2 or 3 weeks. If the plants are to be pinched, plants grown for Memorial Day sales should not be given a hard pinch later than mid-February. A soft pinch, where just the tip is removed, can be made until late March and the plants will still be ready for late May sales.

The minimum air temperature required at night is 62°F (16.5°C) and the daytime range is 65°F to 70°F (18.5°C to 21°C). Geraniums do not grow well when the night temperature is below 60°F (15.5°C). The plants are grown under full light conditions. At the time of flowering, some light shade on the greenhouse roof may help to prevent petal burn, especially on the white cultivars.

A light misting of gibberellic acid can be applied to the Memorial Day crop to increase the size of the blooms and keep the flowers from dropping. A solution of 5 or 10 ppm of gibberellic acid in water is sprayed on flower buds that are just beginning to show color. The plants must not be sprayed before May 1, because the life of a geranium bloom is about 4 weeks.

Problems

Cultural Disorders

Edema is a nondisease condition in which tiny, water-soaked blisters are formed on the underside of the leaves. The blisters break open, and the tissue becomes corklike in texture and brown in color. Edema commonly affects plants in the spring when the humidity is high and the days are dark and cloudy. An abnormally high moisture content in a plant is thought to give rise to this condition. The recommended control is to avoid overwatering and to provide good air circulation and ventilation. Plants with edema usually flower normally and can be sold.

Diseases

A strict sanitation program is the best way to prevent problems with the many diseases that affect geraniums. Such a program includes maintaining clean growing conditions, steam sterilizations of the media and equipment, the use of chemical sprays or drenches as needed, and the use of disease-free cuttings.

Mosaic, crinkles, and chlorotic leaf spot are viruses whose symptoms are easiest to see when the temperature is cool. Plants suspected of being infected with these viruses should be discarded.

Pythium is a disease that causes a shiny, coal black area on the stem of a plant. The disease can be serious in the cutting bench and among young plants. *Pythium* may attack the roots of the plants as well as the stems. Any infected plants should be discarded. There are some chemical substances that can be applied to the plants to help control the disease, but they are not always successful.

Botrytis is a gray mold that develops when the air circulation is poor and the humidity is high. Under these conditions it is difficult to protect the plants because the *Botrytis* spores are always in the air. The amount of botrytis can be reduced by removing old flowers and dead leaves. The air circulation can be improved by using fans and tube ventilating systems. Fungicides can be applied following label recommendations.

Thielaviopsis root rot attacks the roots of plants, causing black rotted areas. This disease is not a major problem in geraniums except in rooting cuttings. A preventive drench of fungicide may be an effective control.

Rhizoctonia stem rot may infest geraniums. The treatment for this disease is the same as for *Thielaviopsis*.

Xanthomonas pelargonii causes bacterial blight. The symptoms of this disease include leaf spot, stem rot, dieback, and blackleg. The bacteria are spread by the people working with the plants, water splashing on the leaves, and the knives used to make the cuttings. The only control is to use disease-free cuttings propagated by culture indexing or shoot-tip culture. Any plants showing symptoms of this disease should be burned. If the disease appears in the crop, all of the materials used in the propagation of the crop must be discarded or steam sterilized.

Insects

For geraniums, a general spray program using recommended chemicals can control aphids, mealybugs, plume moths, spider mites, beet army worms, and whitefly. The geranium plume moth is an especially difficult insect to control. Cuttings received from specialty growers may contain the tiny eggs of the plume moth. If an examination of the plants does not disclose the eggs, they soon hatch into larvae that actively feed on the geranium leaves. Another heavy feeder is the beet army worm.

Geraniums from Seed

Geraniums produced from seed are commonly sold as spring bedding plants. Their culture is covered in Chapter 28

GLOXINIA *Sinningia speciosa, Gesneriaceae*

Gloxinias formerly were grown only for spring and summer sales. However, new colorful hybrids have made the plants popular year-round. Natives of Brazil, gloxinias may be started from either seed or tubers. Under greenhouse conditions, a flowering plant can be produced from seed in 6 to 8 months, depending upon the season. Plants growing in the lower light intensity of winter take longer to flower.

Growing Under Lights

Flowering plants for Christmas sales can be grown under lights from a seed sowing made in August. Cool white fluorescent lamps are used. The lamps are placed 8 inches above the plants to provide an intensity of 850 to 1,000 fc (9.2K to 10.75K lux) for 16 hours daily. The temperature is kept constant at 72°F to 75°F (22°C to 24°C). A peat-lite mix is used. The plants are fertilized with a 20-10-20 analysis fertilizer at 150 ppm each of N and K_2O at each watering.

Greenhouse Growing

Starting Seeds

If the plants are to be grown in a greenhouse, seed may be sown at any time. The seed is very small, and it is sown on top of a peat-lite germinating medium. To prevent the seed from being washed too deeply into the mix, the crop is watered from below. A higher percentage of germination occurs when the day and night temperature is maintained at 70°F (21°C). The seed flat should be shaded from bright sunlight, using moderate shade.

Transplanting

The young plants are first transplanted to flats where they are spaced 1.5 inches (3.75 cm) apart in each direction. They are easily damaged at this stage, so care must be used in handling them. A peat-lite mix is used because of its excellent drainage and aeration characteristics. The plants should be kept moist but not overly wet. The light intensity at the brightest part of the day should be no more than 1,500 fc (16.1K lux).

When the plants become crowded, they are shifted to 3- or 4-inch (7.5- or 10-cm) pots. If the plants are being grown for sale to other florists, they are placed in 2.5-inch (6.25-cm) pots. A peat-lite mix is used for all growing stages.

Shifting Up

Repotting or shifting up is necessary when the plants begin to become pot bound. At this time, they are shifted to larger pots. Three-quarter pots or azalea pots are used because they are better suited to the size of the plants.

As the plants grow, they can tolerate higher light levels. At the brightest part of the day, the intensity should be no greater than 2,400 fc (25.9K lux).

Gloxinias develop crown rot very easily. Therefore, water must not be allowed to remain in the crown overnight. Many growers will transplant the plant slightly higher allowing the crown to be above the medium surface. Capillary mat irrigation is an excellent way of watering gloxinias.

The plants should not be grown at less than 65°F (18.5°C). Below this temperature growth is very slow.

Growing from Tubers

A quick crop of gloxinias is obtained by using tubers. For example, tubers are started in January for early spring flowering plants. To obtain plants for late spring and summer sales, tubers are started in February and March. The tubers are placed just touching each other in a flat of moist peat moss. They are kept at a temperature of 70°F (21°C). As soon as leaves develop, each plant is transplanted to a finish size pot. The plants are then grown on under the same cultural conditions used for plants started from seed.

Problems

Cultural Disorders

Lack of flower buds may be due to an insufficient light intensity, especially in winter. Low light levels also cause the leaves to curl under at the margins and the petioles of the flowers to stretch.

When the light intensity is too great, chlorophyll in the leaf cells dies and the leaves turn yellow.

Diseases

Powdery mildew is a problem when the plants are grown in a very humid atmosphere with poor ventilation.

Bud rot is caused by the fungus *Botrytis cinerea*. Improved air circulation and ventilation reduce the disease. To prevent sites where botrytis spore germination can occur, all dead leaves must be picked up. Crown and stem rot may also affect the plants. *Botrytis* may develop if plants are not spaced in time to prevent leaves from touching. When gloxinia leaves are allowed to touch, water may condense and start to rot.

Insects

Cyclamen mites, mealybugs, spider mites, and thrips may infest the plants.

HYDRANGEA (Snow Ball Plant)
Hydrangea macrophylla, Saxifragaceae

Hydrangeas are popular plants for Easter and Mother's Day sales (Figure 25-13). Originally discovered in China and Japan, the modern cultivars used by growers were developed by specialists in Europe and the United States.

Hydrangeas are grown in greenhouses only in those parts of the country where the plants can be given a cold temperature treatment to initiate flower bud formation. After the buds have developed, they can be forced to flower in any climate. Hydrangeas are prepared in the north and then are shipped in large numbers to Florida and other southern states where they are flowered for the spring holidays.

Propagation

The plants are propagated by cuttings taken from February through May. The type of cutting used depends upon the number of stock plants available. If the

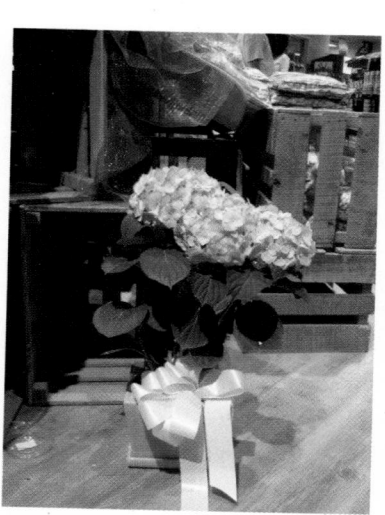

FIGURE 25-13 A hydrangea plant for retail sale for Mother's Day.

grower has a large number of plants, 3- to 4-inch (7.5- to 10-cm) long terminal cuttings are taken. In addition, blind wood from flowering shoots can be used for cuttings. Contrary to rumor, plants that develop from blind wood will flower.

If there is a limited number of stock plants, the entire plant is cut into as many pieces as possible. Leaf-bud cuttings are made by cutting through the stem about 0.25 inch (6 mm) above the node and 1 inch (2.5 cm) below the node. This cutting is then split in half. A leaf remains attached to each half of the stem. As compared to terminal cuttings, an additional 3 to 4 weeks are required for leaf-bud cuttings to grow to flowering size.

The cuttings root well in sand, sand and sphagnum peat moss, or perlite and sphagnum peat moss. A rooting hormone is recommended. Hydrangeas are very sensitive to low oxygen levels in the propagation medium. In other words, good aeration is required if the cuttings are to root. Low-pressure mist is used to maintain the high moisture levels required by these plants. The best rooting conditions include bottom heat at 65°F (18.5°C) and a minimum air temperature of 60°F (15.5°C).

If mist is not used, the cuttings will wilt if they are not heavily shaded with cheesecloth for a few days immediately after they are placed in the medium. The plants should root and be ready for potting in 3 to 4 weeks.

Cuttings propagated in June and July will be small plants ready for sale in the following spring.

After the plants are rooted, each one is transferred to a 3-inch (7.5-cm) pot. The potting medium is a well-drained, highly porous soil mixture of equal parts by volume of loam soil, sphagnum peat moss, and perlite. If the plants are to be grown for blue flowers, no phosphorus is added to the soil. Plants grown for pink and white flowers are fertilized with 20 percent superphosphate added at the rate of 2 pounds (1.2 kg) to each cubic yard (cubic meter) of potting medium.

Controlling Flower Color

Plants with blue flowers can be obtained from specific red or pink cultivars. The grower should select these cultivars from a reliable dealer. If the color is not controlled, these plants normally have pink or red flowers. Hydrangeas with white flowers cannot be forced into a color change. The flowers will always be white.

The factors having the greatest effect on flower color are the pH of the medium and the amount of phosphorus in the medium. To achieve blue flowers, both of these variables must be at the proper levels at the time the plants are taken from the propagation medium and are potted.

Blue Flowers

Blue flowers are obtained when the pH of the medium is acid, 4.5 to 5.0, and there are large amounts of free aluminum ions in the soil. Aluminum is more readily available in media with an acid pH and low levels of phosphorus. High phosphorus levels tie up the aluminum on the soil colloid. As a result, the plants cannot absorb the aluminum ions. It is for this reason that phosphorus is not added to media being prepared for hydrangeas that are to be grown for blue flowers.

Plants being grown for blue flowers are fertilized using acid nitrogen fertilizers, such as ammonium sulfate and ammonium nitrate. Potassium is also supplied from potassium chloride or potassium sulfate. Potassium nitrate should be used sparingly because it does have a slight alkalizing effect on the soil. The fertilizer combination used must supply 200 ppm each of nitrogen and potassium at each watering.

If the flowers are not clear blue when the proper pH and fertilizers are provided, then aluminum sulfate is added. This material acidifies the soil and also supplies free aluminum ions that the plants can absorb. Three or four applications are made at 7-day intervals using 1 pound (0.45 kg) of aluminum sulfate dissolved in 5 gallons (19 l) of water. The treatment is started after the plants are brought into the greenhouse for forcing and show some growth. The soil should be slightly moist when this solution is applied.

Pink Flowers

The best pink colors develop when the pH is between 6.2 and 6.5. In this case, phosphorus is added to the medium to tie up free aluminum ions. The flowers are a muddy pink when small amounts of aluminum are absorbed by the plants.

For pink and white flowers, the fertilizers used are high in phosphorus and leave an alkaline residue in the soil. A 15-30-15 analysis fertilizer is applied at the rate of 200 ppm N at each irrigation. Applications of calcium and sodium nitrate may be alternated with the 15-30-15 fertilizer. Each material is used at the rate of 200 ppm N at each irrigation.

Flower Bud Formation

Flower buds form when the night temperature is below 65°F (18.5°C) for not less than 6 weeks. In the northern part of the United States at 42° north latitude, the temperatures begin to drop below this level about September 1. At this time the plants are still in the cold frames or open field. In colder locations, it may be necessary to keep the plants in the greenhouse because night temperatures that are too cold prevent proper bud formation.

Field-grown plants should be protected from frosts that may occur during the 6-week period needed for flower bud formation.

After the buds are well formed, the plants are exposed to temperatures of 33°F to 40°F (0.5°C to 4.5°C) for an additional 6 weeks so that they develop to the stage where they can be forced to flower evenly.

During this second 6-week period, the plants remain in a darkened storage area where the temperature does not go below freezing. Because it is believed that the leaves contain a material that slows flowering, the leaves are removed. In addition, mold or *Botrytis* may develop on the plants while they are in dark storage. Removing the leaves and spraying with fungicides as needed will help to control these diseases. To speed the removal of leaves, the plants are gassed with ethylene or sprayed with Florel®. Apples that are just beginning to rot are a good source of ethylene gas. One bushel of apples placed in the storage area for each 200 cubic feet (5.7 m³) of space causes the leaves to drop over a period of 3 to 4 days. The temperature in the storage area at this time should be no less than 65°F (18.5°C). After the leaves fall, they are removed from the storage area at one time. The few that may remain on the plants can be hand picked.

Forcing into Bloom

Following the storage period, the plants can be brought into the greenhouse where they are **forced into bloom.** The temperature at which they are forced may range from 55°F to 80°F (13°C to 29.5°C). The temperature selected depends upon how quickly the flowering plants are needed. At 60°F (15.5°C), about 3 months are required to bring the plants into flower. As the forcing temperature is increased, less time is needed.

When the plants are first brought into the greenhouse, they should be misted from overhead several times a day. This high humidity helps to soften the buds and make it easier for them to start growth. The plants should be given full sunlight until after the flowers start to open. They are then shaded to prevent sun scorch. It is important to provide the proper spacing between the plants to prevent crowding because this condition causes the stems to stretch.

During the forcing period, only nitrogen fertilizer is needed. An alkaline reaction fertilizer, such as calcium nitrate, is used on plants being grown for pink and white flowers. Ammonium nitrate or ammonium sulfate is used on plants being treated for blue flower color. During forcing, the fertilizer is applied every 2 or 3 weeks at the rate of 2 pounds (0.9 kg) of calcium nitrate, or 1.5 pounds (0.7 kg) of ammonium sulfate in 100 gallons (378.5 l) of water. Ammonium nitrate is applied at the rate of 1 pound in 100 gallons (0.45 kg) in 378.5 l.

If the aluminum sulfate treatment for blue flower color is not used before the plants are placed in storage in the fall, the material can be applied early in the forcing period. The amount used is 1 pound in 5 gallons (0.45 kg in 19 l) of water. To ensure the proper color, this solution is used six or seven times, with 7 days between each application.

A lack of iron often occurs during the forcing period, causing a yellowing of new growth. To correct this deficiency, a chelated iron material is applied following the manufacturer's recommended rate of use. A 12 percent iron (Fe) chelate is used at the rate of 1 ounce in 25 gallons (28.4 g in 95 l) of water. This solution is given to the plants as a regular watering. The iron treatments should be spaced at least 10 days apart. Generally, after one treatment the plants are green again. The use of too much chelated iron causes marginal leaf scorch.

Timing Forcing

Easter occurs at different times from year to year. Therefore, there is no specific date on which the plants are taken from storage. Growers usually count back 12 weeks from the date of Easter and start the plants at that time, using a 60°F (15.5°C) forcing temperature. Eight weeks before Easter, the flower buds should be the size of a small pea. Six weeks before Easter, the buds should be the size of a nickel. At 4 weeks, they should be silver dollar size. Two to 3 weeks before Easter the buds should begin to show color. If the buds are not developing on schedule, the temperature can be raised to hasten bloom or lowered to slow it. As full blooms develop, shade is provided to prevent sun scorch of the flower petals.

After the plants reach a saleable size (Figures 25-14 and 25-15), they are hardened by lowering the night temperature to 55°F (13°C). The plants are in better condition for the homeowner if the grower holds them at a cool temperature.

FIGURE 25-14 Hydrangea plants displayed for Easter sales.

FIGURE 25-15 Hydrangeas ready for shipment. *(Courtesy of Douglas C. Needham, Longwood Gardens.)*

Problems

Cultural Disorders

Blindness, or failure to develop buds, is caused by several factors: a night temperature above 65°F (18.5°C); leaf drop due to early frost damage; placing the plants in dark storage before the flower buds are fully developed; too little light while the plants are growing; a severe disease condition, such as mildew, which injures the buds; plants growing too close together; and plants pruned or pinched too late in the fall, resulting in too little leaf area to provide the food needed for good bud development.

Yellowing of new leaves (chlorosis) is due to a lack of iron.

Brown leaf margins may be caused by high soluble salt levels, lack of water, or overtreatment with leaf sprays or soil drenches of iron chelates and excess boron.

Diseases

Powdery mildew and leaf spot are common problems in the fall and spring when the humidity is high. Improved ventilation, proper spacing of the plants, and fungicide sprays are effective controls for these diseases.

Botrytis gray mold is a major problem while the plants are in storage. Control measures include cleaning up all fallen leaves, using fans to improve air circulation, and applying the proper sprays.

Insects

Aphids, spider mites, and thrips are common pests on hydrangeas.

KALANCHOE *Kalanchoe blossfeldiana, Crassulaceae*

The original species of this plant is native to Madagascar. It normally flowers under short-day conditions. Natural bloom occurs in January. The plants can be made to flower at any time of the year by using artificial short days (black cloth treatment) for about 6 weeks. The rate of flower bud development is controlled by temperature.

Mikkelsens Inc., of Ashtabula, Ohio, was once the foremost hybridizer of kalanchoes in the United States. The following is a list of the "Bonanza" hybrid cultivars they offered.

'Attraction'—orange-red
'Avanti'—red
'Bali'—deep orange
'Bingo'—pink
'Cherry Jubilee'—rose-red
'Cinnabar'—bright orange
'Citation'—light pink
'Dignity'—light pink

'Eternity'—salmon
'Fascination'—light purple
'Flamboyant'—light orange
'Fortyniner'—yellow
'Garnet'—red
'Goldstrike'—yellow
'Inspiration'—dark red
'Pollux'—red

'Royalty'—rose-red
'Satisfaction'—lilac
'Sensation'—pink
'Seraya'—salmon-pink
'Singapore'—dark rose
'Tropicana'—orange

Kalanchoes are now available through cutting propagators from The Netherlands and others in the United States, such as Oglevee, LLC. These are available as unrooted cuttings, rooted cuttings in 72 packs, prefinished and finished in 48 packs, or prefinished and finished 4-inch (10-cm) and 6-inch (15-cm) pots.

The plants may be started from terminal tip or leaf petiole cuttings, but those grown from seed make the best plants. A large plant can be grown from seed in about 11 months (Figure 25-16). Smaller plants can be grown in 5 to 6 months. An attractive plant for sale is obtained by placing four or five plants in a 6-inch (15-cm) pan.

The seeds are extremely small and are sown on the surface of the medium. All germinating and growing media must be sterilized because the plant is easily killed by stem rot. The seeds germinate in about 2 weeks at a temperature of 65°F to 70°F (18.5°C to 21°C). The seed flats should be watered from the bottom to reduce disease problems.

As soon as they are large enough, the small plants are transplanted to 2.25- or 2.5-inch (5.7- or 6.3-cm) pots. Two plants are normally placed in each pot. A growing medium with excellent drainage is needed because excess moisture often leads to stem rot. When the plants reach the proper size, they are potted in a finish size pot. Three small plants placed in one pot will make a high-quality plant for sale at Christmas.

The plants are fertilized at each irrigation with 150 ppm each of N and K_2O. For best results, the foliage should be kept dry. An excellent method of watering

FIGURE 25-16 Kalanchoe ready for retail sale.

FIGURE 25-17 Kalanchoe grown on ebb and flood benches and are ready for shipment.

kalanchoe is the Chapin Automatic Watering system, which uses individual tubes for each pot. Capillary mat watering is not recommended because the pots are kept too moist, especially in late fall and winter, yet ebb and flood irrigation works well (Figure 25-17).

Plants started in January may be pinched in June or July. These top cuttings can be used to start other plants that will flower for Christmas.

Kalanchoe are grown in full light intensity except in the middle of summer. At this time, the light intensity can be reduced to 3,000 to 3,500 fc (32.3K to 37.6K lux) and still produce good growth.

Kalanchoe is responsive to photoperiod. The plants are given short-day conditions by using the black cloth treatment daily from 5 P.M. until 8 A.M. The short-day treatment is continued for 6 weeks after which the plants will go on to flower.

When grown at a night temperature of 60°F (15.5°C) at 42° north latitude, plants given short days from July 15 to September 15 will bloom October 15. Plants shaded from August 15 to October 1 will bloom in the period December 1 to 10. If the plants are shaded from September 1 to October 20, they will bloom December 25.

As soon as the flower buds begin to show color, the night temperature is lowered to 50°F (10°C). Flowering is not delayed by the cooler temperature, but the intensity of the red color is greatly increased.

If the plants are to flower in the summer, artificial short days must be given between March 12 and October 5. At this time of the year, the natural days are long enough to keep the plants growing vegetatively.

Problems

Cultural Disorders

Cupping and yellowing of the foliage occurs under short-day conditions. Stopping the black cloth treatment after 6 weeks helps to prevent the problem.

Diseases

Crown rot and basal stem rot are due to soil-borne fungi. To reduce these problems, steam pasteurize the growing medium and use clay pots and benches. The diseases may be made worse by keeping the soil too wet.

Powdery mildew may be a serious problem in the spring, summer, and fall. To reduce this problem, keep the foliage dry, increase the air movement around the plants, and improve the ventilation. A biweekly spray with a fungicide is recommended. Some control is provided by applying a sulfur paste to the heating pipes once weekly, or as needed, but the sulfur fumes may injure the leaves and bleach the flowers.

Insects

Aphids, broad mites, cyclamen mites, mealybugs, spider mites, and thrips may infest the plants.

POINSETTIA *Euphorbia pulcherrima, Euphorbiaceae*

The poinsettia was first brought to the United States from Mexico by Joel Roberts Poinsett in the 1820s while he was the U.S. Minister to Mexico under President James Monroe. For years the poinsettia has been the most important pot plant grown for Christmas. Plant breeding efforts over the last 20 years have produced a group of outstanding cultivars. Many of these cultivars have long-lasting qualities, exciting colors, and interesting flower forms, which remain attractive 6 months after Christmas.

Paul Ecke of Encinitas, California, probably produces more poinsettia cuttings than any other grower in the world; yet, several companies are increasing the level of competition. Tens of thousands of cuttings are sold every year from his ranch. These cuttings serve as stock plants that will produce the flowering plants sold at Christmas. The 'Ecke' cultivars are well known in the trade.

Cultivars

Many, but not all, of the commonly grown poinsettia cultivars are given in the following list. These cultivars are classified according to the originator.

Annette Hegg™ Series

'Annette Hegg Brilliant Diamond'—red
'Annette Hegg Dark Red'—dark red
'Annette Hegg Diva Starlight'—brick red
'Annette Hegg Hot Pink'—pink
'Annette Hegg Marble'—bicolor white and pink
'Annette Hegg Top White'—white

Eckespoint® Celebrate Series

'Celebrate Red'—red
'Celebrate 2'—pink
'Celebrate Pink'—pink
Eckespoint 'Freedom'—dark red

Eckespoint 'Jingle Bells 3'—bicolor, pink flecks on red bracts
Eckespoint 'Lemon Drop'—yellow

Eckespoint 'Lilo' Series

'Lilo Red'—red
'Lilo Pink'—pink
'Lilo White'—white
'Lilo Marble'—bicolor, white and pink
Eckespoint 'Monet'—cranberry red/rose/cream
Eckespoint 'Red Sails'—dark red
Eckespoint 'Success'—orange red

Fisher Pelfi® Star Series

'Bonita'—red
'Cortez'—red

'Dark Puebla'—bicolor pink and white
'Flirt'—pink
'Picacho'—red
'Puebla'—bicolor pink and white
Beckmann's Atrosa® 'Maren'—pink
PLA® 'Noblestar'—coral
GrossTM 'SUPJIBI red'—red
Gross 'SUPJIBI pink'—pink

Gutbier™ 'V-10' Series

'V-10 Amy'—light red
'V-10 Marble'—bicolor, white and pink
'V-10 Pink'—flesh pink
'V-10 White'—white

Gutbier™ 'V-14' Series

'V-14 Glory'—red
'V-14 Marble'—bicolor, soft white
  and pink
'V-14 Pink'—pink
'V-14 White'—white

Gutbier™ 'V-17 Angelika' Series

'V-17 Angelika Marble'—bicolor,
  white and pink
'V-17 Angelika Pink'—pink
'V-17 Arigelika Red'—light red
'V-17 Angelika White'—white

Oglevee®, Brightpoints

'Dynasty Red'—dark red
'Nutcracker' Series
'Nutcracker Red'—dark red
'Nutcracker Pink'—bold true pink
'Nutcracker White'—
  creamy white

Stock Plants

Growers maintain stock plants to reduce the cost per plant and to ensure that cuttings are on hand when needed.

Stock plants are started from rooted cuttings that are received from specialty growers in March, April, and May. At this time of year, the day length is still short enough to cause the development of flower buds. Therefore, the plants must be given artificial long days so that they will continue vegetative growth. A minimum of 10 fc (108 lux) of light is supplied from incandescent bulbs from 10 P.M. until 2 A.M. each night. Sixty-watt bulbs with reflectors can be spaced 4 feet (1.2 m) and 3 feet (0.9 m) above the plants to provide the required intensity. If 100-watt bulbs are used, they are spaced 6 feet (1.8 m) apart and 4 feet (1.2 m) above the plants. A light meter can be used to ensure that the right intensity is obtained. The lights must be used until May 15. After this date, the natural days are long enough to maintain vegetative growth.

As soon as the rooted cuttings or small plants are received, they are potted in a 6-inch (15-cm) diameter pot minimum size, or an 8-inch (20-cm) diameter pot preferred size (Figure 25-18). The larger pot provides more room for root growth, which means larger top growth and thus more cuttings from each plant.

Previously, the medium is a mixture of equal parts by volume of loam soil, sphagnum peat moss, and perlite or sand. Modern growers now use peat-lite media. If using loam soil and to prevent disease problems, the medium should be steam pasteurized or chemically treated. A peat-lite mix can be used. Some growers obtain extra weight by adding 10 to 20 percent pasteurized sand to the peat-lite medium.

The pH of the medium should be in the range 6.0 to 6.8. Ground limestone can be added if needed. Before the plants are potted, 20 percent superphosphate

FIGURE 25-18 Poinsettias actively growing vegetatively.

is added to the medium at the rate of 2 pounds (1.18 kg) to each cubic yard (cubic meter) of soil mix, or 1 pound (0.6 kg) of treble superphosphate is added. Poinsettias grow best when the fertilizer level is medium high. This level is obtained by applying 300 ppm of nitrogen from 20-10-20 or 20-5-30 analysis fertilizer at each irrigation. The maximum number of high-quality cuttings is obtained by using a regular fertilizer program. A molybdenum supplement may be necessary for stock plant production.

The plants are grown at a night temperature of 65°F (18.5°C) and a day temperature of 80°F (27.5°C). The most rapid growth is made at these high temperatures. The daytime temperature must not exceed 90°F (32°C) because the plants will stall. If fan and pad cooling is not used, a light shade used on the greenhouse glass will keep the temperature down. Do not apply a heavy shade because this causes soft growth that does not root easily.

The greatest number of cuttings is obtained when good growing conditions are maintained and pinches are properly timed. The stock plants are pinched on a regular schedule. At least 1 inch (2.5 cm) of the end of the stem is removed. A soft pinch (only the tip is removed) is not made because this method means that more time is needed for the side branches to develop than if a hard pinch is made.

The first pinch is made about 7 to 10 days after potting. From this point on, the plants should be inspected weekly. A shoot is ready for pinching as soon as a stem has put on enough growth that three or four fully matured leaves will remain on the plant after the pinch is made. Unless cuttings are to be taken, stems should not be allowed to grow longer than 6 inches (15 cm). It takes about 1 month for a shoot to reach cutting size after a pinch. If cuttings are to be made from August 6 to 10, the plants should not be pinched after July 10. Growth made in the shorter days at the end of the summer is not as strong as the growth produced earlier in the summer.

Cuttings can be made until the recommended end date for a locality. In the northeastern part of the United States, the last date for taking cuttings to be used for single stem plants is about September 5 to 10. Cuttings taken after this date are not likely to make enough growth to be ready for sale at Christmas.

The stock plant containers must be well spaced so that strong growth is obtained. Crowding the containers in the bench causes the stems to stretch, resulting in weak cuttings that root poorly.

Water is applied as needed. If the plants become too dry, the leaves scorch easily. Semiautomatic watering methods, such as an automatic watering system, works well with poinsettias. Consider a weekly spray with a fungicide as good preventive medicine if foliar diseases have been a problem in the past.

Propagation

Cuttings should be taken in the early morning when the temperature is cool and the stock plants are turgid. The cutting is made 3 or 4 inches (7.5 or 10 cm) long, depending upon the cultivar. For most of the Annette Hegg group, 3-inch (7.5-cm) cuttings are taken.

The cuttings should be removed using a sharp knife. A clean cut is better for rooting than the ragged edge caused by breaking the shoot with the fingers.

The cut may be made at a node or internode. Any basal leaves on the part of the stem to be buried in the propagation medium are removed so that they will not rot and become disease sites. It is important to leave as many leaves as possible on the stem for photosynthesis.

When taking cuttings, take only the number that can be immediately stuck in the propagation bench. Poinsettia cuttings should not be allowed to wilt before they are stuck in the propagation medium because they rarely recover to make good plants. The cuttings should be collected in clean plastic bags. Do not use old, dirty flats or boxes because they may contain disease organisms. It is very important to keep strict conditions of cleanliness during propagation.

The cuttings may be rooted in a propagation bench, small pots, peat pellets, other types of propagation units, or directly in finish size pans. The use of finish size pans requires more propagation space than the other containers listed.

The propagation area should be thoroughly sterilized before any cuttings are placed in the medium. In addition, the area is sterilized between each batch of cuttings. All watering hoses must be placed on hooks off the floor. Do not drench the rooting medium with chemicals or fungicides because this treatment will reduce the percentage of rooting.

The cuttings are usually spaced about 2 inches (5 cm) apart in rows 4 inches (10 cm) apart. *Botrytis* gray mold and *Rhizoctonia* stem rot are problems in the propagation bench. Good air movement through the plants helps to control these diseases (Figure 25-19).

Poinsettias root best when they are propagated under a low-pressure mist. The mist is programmed to keep the cuttings moist without becoming excessively wet. Too heavy a mist leaches nutrients from the leaves. As roots form on the cuttings, the mist is gradually reduced. To harden the plants, no mist is used for 3 or 4 days before the plants are potted. This hardening process helps the plants to withstand the shock of potting and the lower humidity of their new environment.

A weak solution of fertilizer can be applied through the mist lines. Two ounces (56.7 g) of potassium nitrate and 3 ounces (85.1 g) of calcium nitrate are mixed with each 100 gallons (378.5 l) of water applied. Mist-type fertilization follows the same schedule as plain tap water applications.

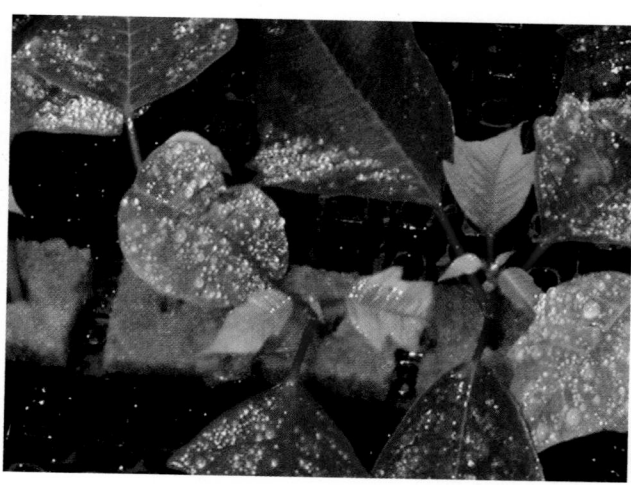

FIGURE 25-19 Poinsettia cuttings rooting into rock wool cubes under intermittent mist. The dark lesion on the leaf is Erwinia, a bacterial infection common when poinsettia cuttings are kept too wet.

FIGURE 25-20 Poinsettia bract color is intensified by lower temperatures late in the season.

If fertilizer is not applied with the mist, 2 pounds (0.9 kg) of 20-10-20 analysis fertilizer in 100 gallons (378.5 l) of water is applied 5 days after the plants are stuck in the medium. Thereafter, the cuttings are fertilized every 3 days until they are potted.

An air temperature of 65°F (18.5°C) and bottom heat of 72°F to 75°F (22°C to 24°C) are maintained. A light shade is used on the greenhouse roof to keep the temperature down. Poinsettia propagation is improved when fan and pad cooling is used on extremely hot days. Colors are intensified by using lower temperatures late in the season (Figure 25-20).

Direct Rooting in Finish Pans

Many growers root poinsettia cuttings directly in finish size pans. Less labor is needed because one potting operation is omitted. However, more propagation space is needed because the finish size pans are larger than other types of propagation units. Each grower must decide if the reduced labor cost is offset by the need for more space.

Guidelines for Direct Rooting in Finish Size Pans

The following guidelines are offered for those who wish to try this method. It is recommended that the grower start on a small scale first to get some experience before the entire crop is grown this way.

1. New cultivars only are to be used as they are more readily adapted to this method.
2. Use a well-drained, well-aerated, and pasteurized medium, such as the peat-lite mixes. New pots are placed on the bench so they touch.
3. All of the cuttings in one pan must be of equal length and thickness. The success of this method depends upon the uniformity of the cuttings. It is important that the cuttings root and grow at the same rate. Even growth will ensure that the plants will flower at the same time.
4. Stick the cuttings directly into preformed holes in the medium. Do not force the cuttings into the medium because this action damages the basal end of the stem. Stick all of the cuttings to the same depth. Do not let the stems touch the pot rim. Use a rooting hormone to hasten rooting.

5. Use a low-pressure mist, following the directions given for regular propagation. Be sure that the pots on the outside rows receive as much moisture as those on the inside rows.

6. The following temperatures are to be maintained: a night air temperature of 72°F (22°C), and a day air temperature of 80°F (27.5°C). Bottom heat is provided to give a minimum temperature of 75°F (24°C) in the pot. A probe-type thermometer is inserted directly in the pot to check the bottom heat.

7. Medium shade is used over the plants to reduce the stress due to high light intensity and high temperatures.

8. Reduce the mist gradually as the cuttings root. Grower experience is a factor at this point in determining when the cuttings have rooted. Under normal conditions, a rooted poinsettia cutting holds its leaves firmly erect in a fully turgid condition.

9. Apply potassium nitrate fertilizer at the rate of 0.5 pound in 100 gallons (0.2 kg in 378.5 l) of water 7 days after the cuttings are stuck in the medium.

10. A regular fertilization program is started about 2 weeks after the cuttings are stuck. The first application should be a **one-half strength** solution (150 ppm N).

11. The plants are given their final spacing when they become crowded.

12. Apply growth retardants according to label recommendations.

Production, Regular Program

Panning

Plants are placed in pans (**panning**) as soon as they are well rooted. The plants become hard if they are left in the propagation bench or in small pots too long. As a result, the quality of the finish plants is reduced.

The plants must not be placed any deeper in the pans than they were in the propagation medium. Plants started in peat pellets, Oasis Root Cubes®, or similar propagation units must be planted so that not more than 0.25 inch of potting medium covers the root ball of the small plant. If the top of the ball or peat pot is exposed to air, wicking occurs. As a result, the root ball dries faster than it can take up water from the potting medium. If the air temperature is high and the light intensity is high, this moisture stress results in leaf scorch. The leaves never recover from this damage.

If the root ball of the young plant is buried too deep in the potting medium, the plants may die from a lack of oxygen. In addition, the soft, tender stem tissue may be infected with *Rhizoctonia* stem rot.

The final panning of all plants should be completed by late October. Earlier panning allows the grower to treat the plants with growth retardants with the least amount of phytotoxicity (plant injury).

Pinching

Most of the newer cultivars available to growers branch freely and are grown as pinched plants. For very large specimen plants, such as those used in banks and churches, some growers prefer to produce single stem plants from non-branching cultivars.

Table 25-2

Number of Plants per Size of Pan		
	Number of Plants	
Size of Pan Inches (cm)	Pinched	Single Stem
4 (10.0)	1	1
5 (12.5)	1	1 or 3[1]
6 (15.0)	2	3 or 4
7 (17.5)	3	4 or 5
8 (20.0)	3 or 4	6 or 7
9 (22.5)	4 or 5	8 or 9

1. If cultivars with large flowers are used, the smaller number of plants is preferred.

The number of cuttings planted in a pan depends upon the size of the pan. Table 25-2 shows the number of pinched and single stem plants that can be placed in pans of different sizes.

Media

The medium used for poinsettias must have the following properties: excellent aeration, very good drainage, freedom from insects and diseases, and good nutrient- and water-holding capacity. A mix of equal parts by volume of loam soil, peat moss, and perlite, and the peat-lite mixes are commonly used for growing poinsettias. Some growers add soil or sand to the peat-lite mix for weight. Such additions must be steam pasteurized so that the mix is not contaminated.

Growers who blend mixes from the various materials available prepare only the amount needed at any one time. Fertilizer amounts are weighed accurately using a scale. A small postage scale is suitable for weighing the trace elements if FTE 555 is used. Too much fertilizer, particularly the trace elements, can cause serious plant injury.

The fertilizer materials to be added to a mix of equal parts by volume of soil, peat, and perlite are as follows:

	Amount per Cubic Yard	(kg/m^3)
Dolomitic limestone	5 pounds	2.98
20 percent superphosphate	8 pounds[1]	4.70
Potassium nitrate	0.5 pound	0.30
Chelated iron (Sequestrene 330®)	2 level tablespoons	39 cc
Fritted trace elements (FTE 555)	4 ounces[2]	0.15
	or	
ESMIGRAN	4 pounds	2.35
	or	
PERK®	4 pounds	2.35

1. Mikkelsen cultivars produce a larger bract when 10–14 pounds (5.88–8.32 kg/m^3) of superphosphate are used. Use one-half this amount if treble phosphate is used.
2. Use only one of the trace elements given.

Peat-lite Mix A for poinsettias consists of the following materials.

	Amount per Cubic Yard	(kg/m^3)
Shredded sphagnum peat moss	11 bushels	15 bu
No. 2 horticultural vermiculite is preferred, but No. 3 may be used.	11 bushels	15 bu
Dolomitic limestone	5 pounds	2.98
20 percent superphosphate	8 pounds[1]	4.70
Potassium nitrate	1 pound	0.60
Chelated iron (Sequestrene 330®)	2 level tablespoons	39 cc
Fritted trace elements (FTE 555)	4 ounces[2]	0.15
ESMIGRAN	4 pounds	2.35
or		
PERK®	4 pounds	2.35
AquaGro® or Surfside® wetting agent	3 ounces liquid	117 cc
or		
	1^1/$_2$ pounds granular	0.89

1. Mikkelsen cultivars produce a larger bract when 10–14 pounds (5.88–8.32 kg/m^3) of superphosphate are used. Use one-half this amount if treble phosphate is used.
2. Use only one of the trace elements given.

All soils and soilless mixes are prepared on a clean surface. The materials are thoroughly mixed to reduce cultural problems arising from poor distribution.

The mix tends to settle when it is watered. A small amount of extra medium should be added to all containers to allow for settling. If the mix is compacted too much during potting, the aeration and drainage properties are destroyed. There must be some space at the top of the pot to serve as a reservoir for water and the fertilizer solution.

Fertilization

If the grower is to achieve top-quality poinsettia crops, large amounts of fertilizer are needed. The proper fertilization practices are started when the plants are in the propagation bench and are continued until the plants are sold.

Fertilization at Every Watering

The best control of the nutrition of the crop is obtained by applying fertilizer at every watering. (The procedures for this method of fertilizing are given in Chapter 15.) Superphosphate must be applied before planting. Nitrate forms of fertilizer are used on poinsettias. This is especially true if peat-lite mixes are used as the growing medium. Ammonium sources of nitrogen damage poinsettias. Commercially prepared 20-20-20 and 20-5-30 fertilizers, as well as other similar complete analysis fertilizers, may contain high levels of ammonium. These fertilizers can be used on stock plants only and in August and September. Many prefer 20-10-20, which has no more than 40 percent of its nitrogen in the ammonium form.

This method of applying fertilizer provides the crop with a small quantity of nutrients and salts at every watering. To prevent a buildup of the total soluble salts, excess solution is added at each irrigation so that some leaching occurs. A solution containing 292 ppm N and 205 ppm K is made by dissolving 9 ounces (255 g) of potassium nitrate and 17 ounces (482 g) of calcium nitrate in 100 gallons (378.5 l) of water. If a 1:100 proportioner is used, the fertilizer is dissolved in 1 gallon (3.785 l) of water to make a concentrated stock solution.

For plants being grown in the peat moss and vermiculite mix A with added potassium nitrate, only nitrogen fertilizer is required during the growing period. Calcium nitrate is applied at the rate of 25 ounces (1.6 kg) in 100 gallons (378.5 l) of water to supply 292 ppm N. The vermiculite itself supplies enough potassium to ensure a high-quality crop. This method is used only for plants growing in a mixture of equal parts by volume of sphagnum peat moss and vermiculite.

Fertilization Every Week

When the plants are to be fertilized once a week, a stronger fertilizer solution is needed. Twelve ounces (340 g) of potassium nitrate and 23 ounces (1.4 kg) of calcium nitrate are dissolved in 100 gallons (378.5 l) of water to make a solution containing 393 ppm N and 336 ppm K. Tap water irrigations are made as needed to keep the medium moist between applications of fertilizer solution.

Special Fertilization Programs

If fritted trace elements are not added to the medium when it is prepared, soluble trace element fertilizers must be used during the growing period. Follow the label recommendations for the product so that excess trace elements are not applied to the plants.

Poinsettias growing in peat-lite mixes may show symptoms of too little molybdenum, even when this trace element is added to the medium as a preplant treatment. The common symptoms are a marginal scorch and interveinal yellowing of the older leaves. The student should be aware that the symptoms of ammonium injury are very similar to those of a lack of molybdenum.

To overcome the deficiency, a molybdenum solution is used regularly during the growing season. The quantity of molybdenum required is very small. Therefore, a concentrated stock solution is made up, and further dilutions of this stock are made as needed.

Sixteen ounces (0.45 kg) of either ammonium or sodium molybdate are dissolved in 5 gallons (19 l) of water. This special stock solution should be kept separate from the regular fertilizer stock solution.

A proportioner may be used to inject fertilizer at every irrigation. To obtain a 1:200 dilution ratio, add 15 fluid ounces (450 ml) of the molybdenum stock solution to 50 gallons (189 l) of concentrated fertilizer stock solution. If a 1:100 dilution ratio is desired, add 7.5 ounces (225 ml) to 50 gallons (189 l) of concentrated fertilizer stock solution.

If a fertilizer proportioner is used but the crop is fertilized less frequently (perhaps every second or third irrigation), mix 30 fluid ounces (900 ml) of the molybdenum stock solution with 50 gallons (189 l) of fertilizer stock solution. This solution is then applied through a 1:200 dilution ratio injector. If a 1:100

dilution ratio injector is used, add 15 fluid ounces (450 ml) of the molybdenum stock solution to 50 gallons (189 l) of fertilizer stock solution.

When a single treatment of molybdenum is required, 1.33 ounces (40 ml) of the molybdenum stock solution is added to each 100 gallons (378.5 l) of water being applied to the crop. When this solution is applied as a soil drench, even as late as the third week in November, good control of the problem is achieved. Of course, any plant that is so severely injured that leaves have died is beyond saving.

Environmental Control Practices

Temperature

The proper control of temperature is one of the most important factors in producing top-quality poinsettias. Temperature influences the speed of maturity, the sturdiness of the plants, stem strength, and leaf and bract color.

Poinsettias are so sensitive to temperature that attempts to reduce fuel costs by growing the crops at too low a temperature result in very-low-quality plants. If the greenhouse temperature must be lowered, it is safe to do so only after the bracts are well developed both in size and color.

Generally, the temperatures to be maintained are 65°F (18.5°C) at night and 70°F to 75°F (21°C to 24°C) during the day. Plants propagated late must be grown at a warmer temperature if the crop is to be ready for Christmas. The night temperatures must not rise above 80°F (27.5°C) because the crop will be seriously delayed or flowering may not take place at all.

Many of the new cultivars produce high-quality plants when they are finished at a cool temperature. After mid-November, red and pink 'Hegg' cultivars are grown at a night temperature of 65°F to 67°F (18°C to 19°C) and a day temperature of 70°F to 80°F (21°C to 26°C). 'Eckespoint Celebrate Series' cultivars must be finished at a night temperature of 60°F to 62°F (15°C to 17°C) during late November and into December. The best flower bud initiation in these cultivars is achieved when the night temperature is in the range of 65°F to 70°F (18.5°C to 21°C) during the last week of September and the first 2 weeks of October.

'Mikkel Rochford' cultivars grow well at the standard recommended night temperatures. White cultivars cannot be finished at a low temperature because the bracts develop a greenish color. Warm temperatures are needed to produce pure white bracts. Therefore, the minimum finishing night temperature for white cultivars is 60°F to 62°F (15°C to 17°C).

The thermometer(s) or thermostat(s) used to check the temperature should be installed at plant level, not eye level, to obtain accurate readings.

Photoperiod

Poinsettias are short-day plants. The critical light period for flower initiation (bud set) is from 12 hours to 12 hours and 15 minutes. The plants will remain in vegetative growth if the day length is longer than 12 hours. Rather than extending the day, it is recommended that the plants be given a light break in the middle of the night. In other words, the plants are given 2 hours of light from 11 P.M. until 1 A.M. daily. The minimum amount of light recommended is 10 fc from incandescent bulbs. One source states that poinsettias respond to

1/100 fc (0.01 fc) of light just as if it were full daylight. However, the recommended light intensity is 10 fc (108 lux).

The 'Hegg' cultivars will flower at Christmas if artificial lights are used from September until October 10. Black sateen cloth is then used for 2 weeks from 5 P.M. until 8 A.M. daily to ensure that bud initiation occurs.

The blooming of poinsettias being grown for Christmas may be delayed because of light from sources outside the greenhouse, such as shopping plazas, highway traffic with headlights shining into the greenhouse, and other night lights. If necessary, the plants can be screened from such lights by the use of black cloth curtains.

Any person checking greenhouse temperatures at night should use a flashlight instead of turning on the overhead greenhouse lights. Even this short period of exposure every hour may delay flowering.

Light Intensity

Some of the new poinsettia cultivars show better growth under reduced light intensity, especially during the early part of the season. These cultivars tend to become slightly yellowish-green if they are grown under full light levels. To prevent this problem, a light shade is applied to the greenhouse glass from mid-August to mid-October. The light intensity decreases with the shorter fall days. Therefore, the greenhouse glass must be cleaned thoroughly after mid-October so that the maximum amount of light will reach the crop.

Spacing

The quality of the plants is reduced if they are crowded on the bench. The space allowed for each plant depends upon the number of blooms present when the plant is panned. At least 0.5 square foot is allowed for each flower. Therefore, a four-bloom plant in a 6-inch (15-cm) pot will need 2 square feet (930 cm^2) of bench area. A more generous spacing may improve the plant quality slightly, but it will mean additional expense to the grower.

Water

Poinsettias need adequate water for best growth. If the pots are allowed to dry out, the resulting moisture stress rapidly causes leaf injury. This problem is common on weekends in the early part of the season when the outdoor temperature is still high. If the reduced greenhouse staff is not able to water the plants quickly enough, moisture stress can occur. Semiautomatic watering systems can help prevent this kind of damage.

Although the plants require a large amount of water, they should not be overwatered. The grower must be especially careful in the fall when the unpredictable weather causes long periods of cloudy conditions. The plants are very sensitive to root and stem rot. Both diseases develop more quickly when the growing medium is too wet.

Growth Retardants

There are five chemical plant growth regulator (PGR) retardants that are registered for use on poinsettias: AREST®, B-NINE®, Bonzi®, Cycocel®, and Sumagic®. There is a difference in chemical activity of these materials producing a retarding effect on growth. A-REST®, Bonzi®, and Sumagic® are applied at lower rates than B-NINE® and Cycocel®.

Whether to use a PGR or not is a choice the grower must make early in the production stages of the crop. PGRs are usually applied between pinching and the initiation of flowers. At this time the height problem may not seem to be apparent. If the weather turns bright and warm during the following weeks when the crop is maturing, a grower may wish that a PGR had been applied because the plants will stretch in growth quite rapidly under these conditions. If a PGR is applied and the weather turns dark and cold, then the growth of the plants may be naturally restricted. If this happens, then a treatment made with a PGR might not have been needed. Whether to treat or not to treat is a gamble no matter what the grower does.

Poinsettias grown in warm climates usually need higher concentrations of chemical than those grown in cool climates. In cool climate regions it is not recommended to use B-NINE®, Bonzi®, or Sumagic® after flower initiation because both bract size and bracket development may be adversely affected.

In making applications of PGRs, it is important to follow the label directions exactly as they are written. Otherwise the final results may be a disaster both plantwise and financially.

A-REST® concentrate is a 0.0264 percent liquid that has 250 mg of active ingredient ancymidol per quart (946.25 ml) of fluid. A-REST® is recommended to be applied as a media drench. Do not apply more than two drenches. Applications made too late in the production cycle may result in reduced final bract size. The recommended treatment is to prepare a solution using 1 fluid ounce (30 ml) of A-REST® in 1 gallon (3.785 l) of water. Apply 4 ounces (120 ml) to a 6-inch (15-cm) pot.

B-NINE® is a powder that contains 85 percent daminozide powder active ingredient in a water soluble form. B-NINE® is applied as a spray. It is not effective as a media drench. Application at rates higher than 2,500 ppm may cause plant injury in warm climates. A B-NINE®–Cycocel combination prepared at 1,000 to 2,500 ppm B-NINE® and 1,500 ppm Cycocel in 1 gallon (3.785 l) applied as a foliar spray to 200 square feet (18.6 m²) of bench area is effective. This treatment is best made in the early morning before air temperature reaches 80°F (26.6°C). No spreader/sticker is needed for the solution. This spray should be applied when new shoots following a pinch are 1.5 to 2 inches (3.8 to 5 cm) long.

Do not apply after flower initiation has occurred because bract size may be reduced and development delayed.

Bonzi® is available as a 0.4 percent liquid that has 3,784 g paclobutrazol (*pak-lo-bu-tra-zol*), the active ingredient per quart of liquid. This PGR may be used as a media drench or foliar spray application. Unlike other chemicals that are sprayed on the foliage, this chemical is better absorbed through the plant stems and leaf petioles. One gallon (3.785 l) is used to cover 200 square feet (18.6 m²) of bench area. Spray applications are best made when newly developing shoots are 2 inches (5 cm) long.

Spray application rates range from 10 to 50 ppm concentration depending on the cultivar treated and climatic region where applied. Higher rates are needed in the warmer climates. Bonzi® is less effective as a spray than as a media drench treatment.

Cycocel® is also a liquid, containing 11.8 percent chlormequat (*clor-me-kwat*), the active ingredient. This is the most widely used PGR on poinsettia crops. It is effective as a foliar spray on all colors of poinsettias. As a drench treatment it is only labeled for red poinsettias. Effective drench rates have not been determined for pink, white, or bicolor poinsettias.

Foliar applications of this PGR should only be made when the plants are fully turgid. The plants should be watered the day before treatments are made. Apply the product early in the morning when temperatures are cool. Make applications when side shoots (lateral branches) are 0.5 to 3 inches (1.3 to 7.5 cm) long. Early season applications are made at 3,000 ppm concentration. Later season applications are made at 1,000 ppm concentration. Under good growing conditions more than one application may be needed. Apply the chemical as a fine mist and avoid runoff. Excessive application may cause leaf injury in the form of yellow leaf margins (edges). Avoid spraying later than 6 weeks before plants are wanted for sales. Flowering delay and reduced bract size may occur otherwise.

Drench applications are made using a solution prepared at a 1:40 dilution rate. One quart (946.25 ml) of Cycocel® is diluted to 10 gallons (37.85 l) of water. Apply 1 fluid ounce (30 ml) for each inch (2.54 cm) of container diameter size (Table 25-3).

Sumagic® prepared as a 0.05 percent liquid contains 0.064 ounces (1.81 g) of Uniconazole-P (*uni-con-a-zol*), the active ingredient per gallon (3.785 l). This PGR may also be used as a foliar spray or media drench application. Sumagic® is a very potent chemical, so do not overapply the material. Spray applications are made at 2.5 to 10 ppm concentrations. Apply 1 gallon (3.785 l) to cover 200 square feet (18.6 m²) of bench area.

Recommended drench applications are made at the rate of 0.1 mg per pot. Apply 4 ounces (120 ml) of solution to each 6-inch (15-cm) pot. Make applications when lateral side branches are 1.5 to 2 inches (3.75 to 5 cm) in length. Do not make applications of Sumagic® after flower initiation has occurred.

A technique for monitoring plant height that helps the grower make wise choices for height control through computer technology is called graphical tracking. The concept of graphical tracking was developed by Dr. Paul Fisher of the University of New Hampshire during his doctoral work at Michigan State University with Dr. Royal Heins. Graphical tracking is a simple tool that uses a computer spreadsheet and was developed as a grower management tool where actual plant height is compared with desired plant height over time.

Table 25-3

Media Drench Applications of Cycocel® for Red Poinsettias	
Pot Diameter Inches (cm)	Dilute Solution Per Pot Fluid Oz (ml)
2¼ to 3 (5.7 to 7.5)	2 (60)
4 (10)	3 (90)
5 (12.5)	4 (120)
6 (15.0)	6 (180)
8 (20)	8 (240)

FIGURE 25-21 Screen shot of the University of New Hampshire Graphical Tracking for Poinsettia spreadsheet developed by Paul Fisher, Ph.D.

This allows the grower to make management decisions based on real-time data obtained from the greenhouse.

Graphical tracking is practiced by plotting plant height on a graph once or twice each week and make height-control decisions based on actual and target heights. Computer software using a spreadsheet program can be obtained from the University of New Hampshire (Figure 25-21).

Preparation for Market

Normally, poinsettias are sold during the Christmas season. Because the plants are very sensitive to cold, they must be well protected when they are delivered to market. An unprotected plant may be injured by the cold in the short time that it is carried from the greenhouse shop to the customer's automobile.

Some protection for the bracts and leaves is provided by use of plant sleeves applied at the greenhouse. Some growers also wrap the individual bracts in tissue paper for added protection. Epinasty (curling) of the petioles of the bracts occurs when they are wrapped for several days. The twisted bracts do return to normal, but the plants are unsightly and cannot be sold until the condition is corrected.

When the bracts are broken in shipment, some retailers use a red floral spray to conceal the damage. Severely injured or damaged plants should not be offered for sale.

When the winter weather is severe, the plant should be loaded into trucks inside heated buildings. Sleeve-wrapped plants may be placed in cardboard boxes for added protection.

The plants must be protected even though the production costs are increased because of the additional labor required and the cost of the covers. The costs are added to the sale price of the plant.

Problems

Diseases

Poinsettias are subject to several diseases. The best disease-control program is one of prevention. Maintaining clean growing conditions at every stage of production will minimize disease problems.

The greenhouse employees can be encouraged to develop an antidisease attitude if the grower posts information on the life cycles of disease-causing organisms and methods of controlling these organisms. It is much less expensive to follow a program of disease prevention than it is to apply chemicals to control diseases after they infect the crop.

Appendix III contains a list of University Plant Disease laboratories in the United States and Canada. If a disease condition is suspected and you cannot identify it, it is recommended that you use the services of the laboratory nearest you. There is a nominal fee for each diagnosis.

Propagation Control Measures

The previous section on propagation describes the control measures to be followed.

Production Control Measures

All potting benches, carts, and production benches where poinsettias are handled must be steam sterilized if possible or drenched with a chemical solution. The potting bench and the carts used to move plants can be drenched with an approved disinfectant used as directed on the label. Carts are treated outdoors, followed by an airing for several hours before they are returned to the greenhouse.

Use new pots only to grow poinsettias. Even though old pots are washed and chemically treated, they may still carry disease organisms that can infect the plants. The small savings realized by using old pots is offset by the risk of losing the crop to disease.

When greenhouse benches cannot be steam treated, they are thoroughly scrubbed and washed first. Then they are drenched with an approved disinfectant solution.

The most common diseases of poinsettias are three types of root rot due to *Rhizoctonia solani*, *Pythium*, and ***Thielaviopsis basicola***.

Rhizoctonia solani is a fungus that causes a stem rot at the soil line and root rot. *Rhizoctonia* is most damaging when moisture levels are low and the soil temperature is high. These conditions are common early in the growing season. Therefore, *Rhizoctonia* is most severe at this time.

Pythium is a root rot caused by *Pythium* species. This fungus is most active when the moisture level of the medium is high and the temperature is cool. Symptoms of the disease are wilting, stunted growth, and yellowing and loss of the lower leaves. Biological control is obtained by keeping the growing medium on the dry side and raising the soil temperature in an attempt to force the root system to grow faster than the disease. Some control may be achieved by improving root aeration. That is, the plant is knocked out of the pot, rotated one quarter turn, and replaced gently in the pot.

The *Thielaviopsis basicola* organism causes leaf yellowing and wilting. These symptoms usually appear early in December when the crop is ready for market. At this time it is often too late to save the plants. Prevention is the best control. Some control is obtained by growing the plants in a medium with an acid pH of 4.5 to 5.0.

These three organisms can be controlled in varying degrees by chemical drenches. The effectiveness of these drenches depends upon the timing of the treatments. Information on the latest chemical controls can be obtained from local cooperative extension specialists.

Botrytis cinerea produces a gray mold when the moisture levels are high and the temperature is cool. Fallen leaves are easily infested with *botrytis* spores.

Therefore, the chances of the disease spreading unchecked through the crop are greatly reduced when the fallen leaves are picked up frequently.

Increased air circulation helps to keep the foliage dry, thus reducing the disease. If the disease is not too widespread, the improved air circulation may even eliminate it. For added protection, the crop may be sprayed and smoked with appropriate chemicals at the recommended rates.

Erwinia caratovora is a bacterium that causes soft rot. It commonly attacks the cuttings in the propagation bench or at the time of panning. It may develop when the plants are being grown at a cool temperature just before they are placed on sale. The symptoms are a severe wilting followed by a soft, watery rot.

There is no control for this disease. Plants suspected of having the disease should be taken immediately from the greenhouse and discarded. The only method of prevention is to start with disease-free cuttings and then maintain very clean conditions during all stages of crop growth.

Crud is a yellowish mass that is the result of latex oozing from the plant. This condition is often seen in the very early stages of bract development. In addition, it can be seen as small drops on the stems. Severe cases of crud cause abnormal bract formation, which reduces the quality of the plants.

It is not known why crud occurs. Some growers believe that it is more common when the humidity is high and drying is very slow. These are the same conditions that cause edema in geraniums. However, there is no known relationship between these two physiological diseases. The practices that are helpful in reducing edema are also effective on crud. In other words, withhold water and improve the air circulation and ventilation.

Insects

It is far better to prevent insects from infesting plants than it is to apply control measures after they appear on the plants. Always remove or destroy weeds around the greenhouse, under the benches, and in the aisles. Weeds serve as hosts for whiteflies, red spider mites, and other insects that feed on poinsettias.

The greenhouse whitefly (*Trialeurodes vaporariorum*) and the sweet potato whitefly (*Bemisia argentifolia*) are the two most important whitefly pests on poinsettias as well as other greenhouse crops. A third species, the banded winged whitefly (*Trialeurodes abutilonia*), has become more common of late but has not reached epidemic levels on greenhouse plants. The whitefly has a complex life cycle, making it very difficult to control. During its growth from egg to adult, the whitefly goes through five distinct stages (Figure 25-22). At each stage, there is a limited number of insecticides that are effective in killing the insect at that stage of its life cycle.

Only a few insecticides are effective on eggs laid on the undersides of the leaves. When the eggs hatch, the insects appear as crawlers that move on the undersides of the leaves searching for a feeding place. The insects do not feed at this stage, however, and can be killed by contact insecticides only.

The insect then becomes a nonmoving, scalelike nymph, which is a highly efficient feeding stage. At this point, it is resistant to contact insecticides but can be killed by systemic insecticides. The nymph continues to feed for a time before the insect enters the pupal stage. The scale thickens, and the insect passes into an immobile, nonfeeding state that is resistant to insecticides.

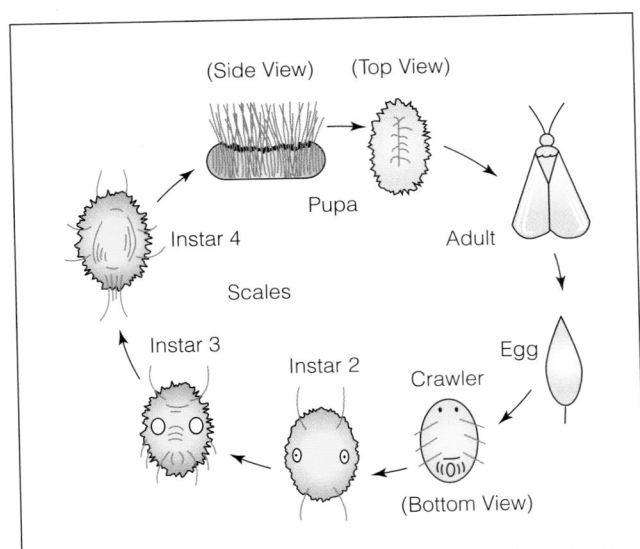

FIGURE 25-22 Stages of the life cycle of the whitefly.

The adult flies hatch 2 to 3 days after the nymph enters the pupal stage. The adults fly from plant to plant looking for places to lay eggs. The adults are easily controlled with a smoke type or aerosol fumigant. Some contact insecticides are also effective against adults.

All five stages of the life cycle can exist at the same time. This means that a single application of insecticide controls only one or two stages of the life cycle. The stages not affected continue to develop and soon reproduce.

For most insecticides, applications must be scheduled regularly to control whitefly. However, there are some systemic insecticides that are effective on all of the feeding stages. Generally, a single application is adequate. When fumigation smokes or aerosols are used, they must be applied every 3 or 4 days for 4 to 6 weeks. If one of the scheduled treatments is missed, the control will not be effective.

Red spider mites and aphids also attack poinsettias, but they are not considered to be serious pests.

Other Problems

Bedding plant growers often put in poinsettias as a second crop to give them a year-round income. When working with bedding plants, growers usually apply herbicides during the summer. These chemicals present no problem in the greenhouse when it is empty. When the greenhouse is closed for the heating season, however, fumes from the volatile chemicals may become concentrated and cause plant damage. This means that growers should not use the types of herbicides that are known to cause such damage. Chemicals approved for greenhouse use generally have a statement on the label to this effect.

Lumber treated with creosote and pentachlorophenol can damage poinsettias seriously. The grower must never use these materials in the greenhouse (see Chapter 4).

PRIMROSE *Primula species, Primulaceae*

Primula malacoides (small flowers) and *Primula obconica* (large flowers) are the two types of primroses grown as pot plants. Both species originated in China.

P. obconica is rarely grown now because many people develop a severe skin irritation when they come into contact with it.

The propagation of primroses is timed so that they are in flower for Christmas and during the spring. Primroses are cool temperature crops that grow poorly during the heat of summer. They are more popular in Europe than in the United States, probably because of the cooler summers there.

When primroses are started from seed in June, 5- or 6-inch (13- or 15-cm) pot plants will be ready by the following January. The seed does not germinate well in the heat of July and August. Therefore, additional sowings are delayed until September, resulting in flowering plants by the spring. The seed is very fine and is sown on the surface of a peat-lite mix. The seed is not covered because light improves the germination. Care should be taken to avoid damping-off problems. Maximum germination occurs at a temperature of 70°F (21°C). As soon as the small plants can be handled, they are transplanted to flats and spaced in a 1 inch by 1 inch (2.5 cm by 2.5 cm) pattern. Some growers prefer to use peat pellets or 2.25-inch (5.7-cm) pots. The plants are then kept as cool as possible. Light shade is used if the light intensity is too bright. The recommended growing temperature is 45°F to 50°F (7°C to 10°C).

When moving the plants to finish size pots, they are set with the crown just above the new soil line. If the plant is set too low in the soil, crown rot can develop. If the plant is set too high, it will topple over.

Regular applications of fertilizer are needed to keep the plant actively growing. The continued use of a complete analysis fertilizer, such as 20-10-20, may lead to an acid soil condition resulting in yellow leaves. This type of chlorosis can also be caused by overwatering and a lack of nitrogen and potassium. To correct the problem, chelated iron is added. To prevent the problem, calcium nitrate is alternated with the 20-10-20 fertilizer, and each is applied at the rate of 150 ppm N at every irrigation.

The best cultivars presently available are the *P. acaulis* F_1 hybrids. The color choice includes white, blue, yellow, orange, scarlet, pink, rose, and mixed colors. Customers commonly ask for white and light pink flowering plants.

Problems

Diseases

Crown rot can be prevented by setting the crown above the soil line when transplanting. In addition, the plants should not be overwatered.

Insects

Aphids, mealybugs, red spider mites, and whitefly are the most troublesome pests on primroses.

ROSES *Rosa minima, 'Chinensis', Rosaceae*

Miniature roses are grown in pots for spring sales. Such roses are excellent items to be sold with the usual bedding plants available at this time of year. Roses are also forced into bloom as specimen plants for sale at Easter and Mother's Day.

Forcing for Easter

Floribunda and hybrid tea type roses that are to be grown and forced as flowering specimen plants are ordered to arrive in late December. The preferred plants are dormant, 2-year, XXX grade budded stocks. The XXX grade has at least four strong canes and produces the best specimen plants. The lesser XX and X grades should be sold as nonstarted plants.

The plants used for forcing are grown in open fields in Arizona, California, and Texas. When the plants are removed from the field, both the tops and roots are pruned so that the plants are easier to handle. It is a common practice to dip the tops of the plants in an antitranspirant material. This dip leaves a wax-like coating on the stems that reduces moisture loss while the plants are in storage and as they move through the marketing chain from the producer to the flower grower. More rose plants are lost through drying than any other cause.

Once the plants are dipped, they are placed in cold storage where they are held at temperatures of 31°F to 35°F (−1°C to 3°C) until they are to be shipped. When an order is received, the roses are counted and packed in large cardboard boxes. The interior of each box has a special plastic coating to resist moisture. A small amount of wet excelsior is packed with the plants to help maintain high humidity in the box.

Handling upon Receipt

The plants are shipped in the winter, and they may freeze en route. The container should be opened as soon as it is received to check the condition of the plants. Any freezing or shipping damage should be noted and a claim filed with the carrier so that the grower can collect for damages.

If the plants are frozen, there may be ice on the stems or buds. If there is no visible ice but there is reason to believe the plants are frozen, the box is closed and stored in an area where the temperature remains in the range of 35°F to 40°F (3°C to 7°C). After 3 or 4 days, the plants are thawed and can be unpacked. The slow thawing process may bring them through without loss. However, if they are unpacked immediately and thawed quickly, they will probably be damaged.

Preplant Handling

As soon as the plants are unpacked, they are placed in water. Some suppliers place a package of root-activating hormone in the box with the plants. They suggest that this material be added to the water as soon as the plants are placed in water. There is some question as to the usefulness of this procedure.

As the plant roots soak in water, the tops of the plants should be covered with moist burlap or plastic to maintain a high level of humidity around them. To prevent damage to the unpotted plants, they must not be exposed to direct sunlight or high temperatures.

The plants need to be soaked for a few hours only. Then, any long, large, or broken roots are pruned. Sharp pruning shears should be used to make clean cuts. Ragged root edges are more easily infected by disease organisms than are clean edges. Very large roots should be cut back because they make it more difficult to pot the plant. The grower is more interested in the development of fine white roots.

The tops are pruned back to leave a stem 6 to 8 inches (15 to 20 cm) long as measured from the bud union. (The bud union is the swollen part of the stem where the canes emerge.)

Potting Soil

The soil mixture used to pot roses should be porous enough to provide good aeration and drainage, and it must be free from insects and diseases using steam pasteurization. In addition, the mixture must contain enough organic matter to ensure good moisture- and nutrient-holding capacity. Most peat-lite media will satisfy the requirements. Some growers prefer a slightly heavier mixture consisting of a 2-1-1 combination of loam soil, peat moss, and perlite or sand.

The plants are dormant when they are potted. Therefore, the soil must have low fertility and soluble salt levels: Only superphosphate is added at the rate of 2 pounds (1.18 kg) in each cubic yard (m³) of medium. The pH should be 6.0 to 6.5. Ground limestone may be added to correct the pH if necessary.

The pots used must be at least 6 to 7 inches (15 to 18 cm) in diameter. The larger size pot is preferred because it provides more room for the roots to grow. An attractive container should be used for specimen blooming rose plants. It will detract from the image of a quality plant if tarpaper pots are used.

To pot the plant, some soil is placed in the bottom of the pot. The root system is then carefully spread out so that it is not crowded. Soil is added until the graft (bud) union is just at the soil surface. The soil is then firmed around the roots to eliminate air pockets.

After the plants are potted, they are placed in a cool location and watered thoroughly. They may be placed in a shaded greenhouse where the temperature can be held at 50°F (10°C) or in a cold frame. The greenhouse is preferred because the temperature must be increased later to force the plants to flower. The plants should not be exposed to the full light intensity at this time.

After 2 to 3 weeks at 50°F (10°C), the roots will have started to develop and there will be just a small amount of top growth. If the plants are grown at a warmer temperature, the top growth is faster than the root growth. In this case, the plant may be severely checked if high moisture stress conditions develop. An underdeveloped root system is not able to supply all of the moisture needed by the excessive top growth, and severe wilting may occur. If the plant does not recover from the wilting, the plant will die.

When the tops of the plants are kept in a very humid atmosphere, root growth is stimulated. The plants are fogged or syringed several times a day. Another method of providing high humidity is to place opaque white plastic sheeting over the tops of the plants. Clear plastic is not used because sun shining on this material causes a high temperature to build up under the plastic. Straw placed around and over the plants will also help retain the humidity. However, extra labor is required to remove the straw, and tender shoots may be broken when the straw is removed from the plants. Also, attention to rodent control may be necessary.

Forcing the Blooms

As soon as the new shoots start to grow, the covering materials are taken off and the plants are exposed to the full light intensity. Any shade on the glass of the greenhouse or cold frame is removed. The temperature is increased to a minimum of 60°F (15.5°C) at night and 70°F (21°C) during the day. When growth begins, fertilizer is applied regularly. The first two applications are made with a half-strength solution of 0.75 to 1 pound (0.34 to 0.45 kg) of 20-10-20 or 20-5-30 fertilizer in 100 gallons (378.5 l) of water. Later applications are made with a solution of 1.5 to 2 pounds (0.68 to 0.91 kg) of fertilizers in 100 gallons (378.5 l) of water. The fertilizer is applied every 7 to 10 days. Three applications at full strength are usually made during the forcing period.

The crop can be timed for the holidays by controlling the temperature and pinching the growing tip. If Easter falls on April 1, then 7 to 8 weeks are required from the time the pinch is made until the newly developed shoot is in full bloom (Figure 25-23). In other words, the crop should be potted January 1 and pinched no later than February 1.

For a Mother's Day falling on May 13, the plants will bloom about a week earlier because of the warmer temperatures and the better light intensity of the late spring. Therefore, the Mother's Day crop must be pinched about March 21.

If the temperature is increased by a few degrees, the crop will flower several days earlier. A night temperature of 55°F (13°C) slows the growth and causes a delay in flowering of 4 or 5 days. It is preferable to maintain a uniform night temperature of 58°F to 60°F (14.5°C to 15.5°C). Increasing or decreasing the temperature to speed up or slow growth reduces the quality of the plant.

Plants not sold for Easter may be pruned back to leave two, five-leaflet leaves on a stem. These plants will flower again in about 6 weeks and will be ready for sale later in the spring.

Any plants forced for Easter can be planted in the garden as soon as there is no further danger of frost.

Garden Roses

Potted roses can also be grown for sale in the spring along with flowering annuals. These roses are started later than roses forced for Easter. Garden roses may not be offered for sale to customers until there is no more danger of frost. To suggest to customers that these plants can safely withstand a frost is unethical.

The dormant plants are ordered to arrive about 2 months before they will be placed on sale. If the plants must be delivered before this time, they are held in a refrigerator until they are started into growth.

FIGURE 25-23 Miniature rose with buds at various stages of opening.

The procedures for unpacking, inspection, and handling are the same as those given for forcing roses.

Garden roses need not be started in hard plastic pots, but they can be placed in papier-mâché pots or other containers made of similar materials. Such pots are less costly and are easy to remove when setting the plants in the garden. The pots should be at least 6 inches (15 cm) wide and 8 inches (20 cm) deep.

Soil

The soil mixture used for garden plants is the same as that used for forcing roses. After the plants are potted, the containers are watered thoroughly. The containers are placed on a surface having good drainage. The tops of the plants must be syringed or misted several times a day. To hold moisture around the plants and to protect them from drying winds, an opaque, white plastic cover can be installed. Once growth is well started, the plants are exposed gradually to less humid conditions by lifting the sides of the plastic cover for several hours every day. After 4 or 5 days, the plastic can be removed.

Roses started outdoors are already conditioned to the garden environment. Therefore, the customer should have little difficulty in establishing the plants in the garden after purchasing them from the grower or garden center. The fertilizer program used is the same one described for Easter plants.

Problems

Diseases

Powdery mildew and black spot develop when it is very humid and there is poor air circulation. These diseases appear quickly during wet and cloudy spring weather. Sprays of recommended chemicals are used to control both diseases.

Insects

Nothing discourages a customer more quickly than purchasing plants from a nursery and later finding them covered with aphids. The green peach aphid is the most common insect pest on roses. Plants held over from the spring for sale in the summer may become infested with red spider mites. Both mites and aphids are controlled through timely applications of specific insecticides.

Cultivars

Several factors determine the choice of cultivars to be grown. The area of the country where the plants are to be sold and climatic and environmental conditions are important considerations. It is particularly important to know the lowest temperature normally reached during the winter. Roses cannot survive extreme winter cold even when they are thought to have adequate protection. Suppliers can recommend cultivars suitable for specific localities.

In order of decreasing popularity, preferred rose colors are red, pink, yellow, white, and multicolored.

ACHIEVEMENT REVIEW

Select the best answer or answers to complete each statement. List the appropriate letter(s).

1 Shoot tip or meristem cultured geraniums are usually free of all diseases except:
   a. bacteria.
   b. fungi.
   c. water molds.
   d. viruses.

2 Geranium cuttings are ready for potting as soon as they:
   a. are callused.
   b. have roots 0.25 inch long.
   c. have roots 1 inch long.
   d. have roots 2 inches long.

3 If geraniums are pinched, flowering is:
   a. not affected.
   b. hastened by 1 week.
   c. delayed 2 to 3 weeks.
   d. delayed for 1 month.

4 The time needed to produce a flowering size gloxinia from seed is:
   a. 3 months.
   b. 4 to 6 months.
   c. 6 to 8 months.
   d. 1 year.

5 Gloxinia plants are usually grown in the greenhouse under a maximum light intensity of:
   a. 850 fc (9.18K lux).
   b. 1,500 fc (16.1K lux).
   c. 2,400 fc (25.92K lux).
   d. 3,200 fc (34.56K lux).

6 Production of hydrangeas to the stage where they can be forced into bloom is limited to areas where:
   a. there is high light intensity.
   b. there is low light intensity.
   c. it is warm year-round.
   d. the plants can be given a cold treatment.

7 Hydrangeas propagated from blind (nonflowering) shoots will produce:
   a. only blind plants.
   b. plants that will flower normally.
   c. pink flowers only.
   d. blue flowers only.

8 Hydrangeas in the propagation bench are very sensitive to a deficiency of:
a. nitrogen.
b. phosphorus.
c. carbon dioxide.
d. oxygen.

9 To force blue flowered plants, the soil should have a:
a. low pH and a high level of phosphorus.
b. low pH and no phosphorus.
c. high pH and no phosphorus.
d. high pH and a high level of phosphorus.

10 Which of the following are good fertilizers for blue flowered plants?
a. Ammonium sulfate
b. Ammonium nitrate
c. Calcium nitrate
d. Sodium nitrate

11 When a good blue color is wanted, the grower should apply:
a. ammonium nitrate.
b. aluminum nitrate.
c. aluminum sulfate.
d. potassium sulfate.

12 Generally, poinsettia stock plants are received as rooted cuttings in:
a. late winter.
b. early spring.
c. summer.
d. at any time of the year.

13 When propagating using a mist type fertilizer, the solution is applied:
a. more often than tap water misting.
b. on the same schedule as tap water misting.
c. less often than tap water misting.
d. half as often as tap water misting.

14 Direct rooting cuttings in finish size pans:
a. requires less space in a propagation bench.
b. is a labor-saving method.
c. makes use of new cultivars.
d. requires the application of retardants.

15 When poinsettias in 2.25-inch (5.7-cm) peat pots are panned, the pot should be placed:
a. in the bottom of the pan.
b. so that the top of the peat pot is above the level of the medium.
c. so that the top of the peat pot is at the level of the medium.
d. so that the top of the peat pot is 0.25 inch (6 mm) below the surface of the medium.

16 FTE 555 is added to 1 cubic yard (m³) of peat-lite mix at the rate of:
   a. 1.5 ounces (55.3 g).
   b. 3 ounces (136.5 g).
   c. 2 pounds (1.2 kg).
   d. 4 pounds (2.4 kg).

17 The symptoms of a deficiency of molybdenum are very similar to those of a lack of:
   a. iron.
   b. boron.
   c. potassium.
   d. nitrogen.

18 The critical day length for poinsettias is:
   a. 11 hours 30 minutes.
   b. 12 hours 15 minutes.
   c. 13 hours.
   d. 14 hours 15 minutes.

19 Poinsettia bracts wrapped in tissue paper may develop a curling of the petioles that is known as:
   a. phototropism.
   b. geotropism.
   c. epinasty.
   d. cold injury.

20 When African violets are grown under artificial lights alone, the recommended intensity is:
   a. 200 fc (2.16K lux).
   b. 400 fc (4.32K lux).
   c. 600 fc (6.48K lux).
   d. 1,200 fc (12.96K lux).

Flowers from Bulbous Species

OBJECTIVES:

After studying this chapter, the student should have learned:

- the need for a cold temperature treatment for certain bulbous species.

- how the cold temperature treatment is given.

- the process of potting bulb crops.

- how to identify disease organisms that affect bulb crops.

- how certain crops must be prepared for market.

A large number of the cut flowers and potted plants produced today are grown from bulbs, corms, tubers, and tuberous roots. The unusual life cycles of these plants dictate that many of them flower only as spring crops. It is fortunate that three of the holidays when florists make their greatest sales are in the spring. Valentine's Day, Easter, and Mother's Day come at ideal times for the bulb forcer.

A summer flowering species is gladiolus. These plants are grown from corms during the summer growing season in northern latitudes. In southern areas, gladioli are produced in the winter.

Bulbs for Forcing

The general term *bulb* is used to refer to the various corms, tubers, and tuberous roots that can be forced into flowering plants. Most bulbs originated in an area that lies between 23° to 45° north and south latitude. This area includes the Mediterranean Sea, Asia Minor, and China. Others originated in South America or South Africa. The weather pattern here consists of cool, moist winters followed by warm, dry summers. This alternate warm–cool cycle is required both to grow good bulbs and to force them into flowers.

Some types of bulbs are grown in the United States. Large numbers of *Narcissus* (daffodils), iris, and lily bulbs are grown in Washington State and Oregon. Some tulips are grown in Michigan, but the number is too small to have much impact on commercial growing. The majority of the bulbs used in the United States by flower growers are imported from the Netherlands.

Before continuing with specific details of other bulb crops, some comments about ordering bulbs should be made.

Ordering Bulbs

There are several factors to be considered before a grower can plan an order for bulbs. The first question concerns the market. It is foolish to invest time and money in a crop and then find that no one is willing to buy it. It may be necessary to develop the market. For example, growers normally place six or seven tulip bulbs in a large pan. However, three bulbs placed in a 4-inch (10-cm) pot may have more sales appeal. In addition, new species of bulbs may have better sales than the old standbys.

The grower should review records of previous seasons. If there are no records, other bulb users in the area may be consulted for information about forcing bulbs. Good records are essential because they serve as the best source of information for future crops. A bound, ledger type record book for each crop is preferred to a loose-leaf notebook because the loose pages are too easy to lose.

A close inspection should be made of the structures available for forcing bulbs. If cold frames are to be used, do they need repair? Cold frames can do the job, but modern bulb forcers use a rooting room. This is a protected area where the bulbs are stored for the **precooling** needed. It is worth the cost of construction if the rooting room has easy access for the removal of bulbs in the winter.

Placing the Order

Orders should be placed early to ensure that the cultivars and species wanted are available in the quantities needed. The seller must know if the bulbs are to be "prepared" for early forcing or are to receive standard handling for regular forcing. Are the flowers to be sold as cut flowers or are the crops to be grown as potted crops?

The bulb supplier normally informs the grower when the shipment is scheduled to arrive. The bulbs must be picked up as soon as they reach port. Any delays that take place in shipment can be harmful because bulbs are never dormant. That is, they are always in some phase of development. Temperature is very important to the success of forcing bulbs. If there are delays in shipment, the bulbs can be damaged if they are held at the wrong temperatures.

When the bulbs arrive, they are prepared for any additional required treatments. First of all, they are ventilated. The cases are separated, and air is circulated around them. The bulbs have been held for 2 to 4 weeks or more in sealed shipping containers. Ethylene gas buildup can cause severe problems with tulips. *Fusarium* disease on tulip bulbs produces ethylene gas. Ethylene damage may not be immediately apparent, but may result in flower abortion ruining the potted plant and potential sale.

As soon as the cases are open, the bulbs are inspected for damage or disease. Diseased and damaged bulbs are discarded. If there is any damage due to the carrier, a claim is filed immediately.

Depending upon the species, some bulbs will show floral parts by the time they are received. To determine the exact stage of development of floral parts, a cut is made lengthwise through the middle of the bulb.

The bulbs can be placed in storage at the proper temperatures. Following these preparatory steps, the storage temperature is critical to the success of forcing later.

Precooled versus Standard Forcing

Bulbs that are precooled are given a special cooling treatment in the shipping case or after they are planted. Precooling is a way to hasten the required changes inside the bulb that make it possible to force blooms. These same changes take place naturally during the winter in bulbs planted in the ground.

Standard forcing treatments use regular cooling methods. The topic of precooling is covered for each crop.

General Information

Containers

The containers used to force bulbs range from pots for holiday potted plants to boxes and greenhouse benches for cut flower crops. The most common type of pot used is called a bulb pan. This pan is one-half the depth of a standard pot. Some growers prefer three-quarter size pots because they provide a slightly greater volume for the medium. Because of their height, which makes them fall over easily, lilies are grown normally in three-quarter pots rather than bulb pans.

Most potted bulb crops are forced in hard plastic pots. If clay pots are used, they must be presoaked to prevent them from taking moisture from the medium surrounding the bulbs. A lack of moisture at this point can cause severe damage. If the bulbs are being forced for cut flowers, they are placed in flats that are not less than 4 inches (10 cm) deep.

Media

The medium serves two particular functions: (1) The roots of the bulb anchor the plant in the medium, and (2) the medium acts as a storehouse of moisture. In general, high fertility levels are not required for most bulb crops. If there is a specific nutrient deficiency to be overcome, a particular fertilizer can be added.

In the past, mixtures of loam soil, sand, peat moss, perlite, vermiculite, bark, and other materials were in common use for bulbs. These materials are mixed in various amounts by volume. The standard 1:1:1 mixture by volume of soil, sand, and peat moss can be used for all bulb crops. Modern bulb growers use peat-lite mix A or a commercially prepared peat-lite mix. Sand may be added at the rate of 10 percent by volume to give the medium more weight.

Excellent drainage and aeration are required in a bulb medium. A bulb crop can be ruined quickly by excess water in the root zone. The preferred pH level is 6.0 to 7.0. The total soluble salt content should be less than 100 mhos for a 1:2 soil:water dilution. High soluble salt levels prevent the growth of a good root system.

Bulb Storage Areas

Cold Frames

A bulb storage cellar, rooting room, or similar area is preferred to cold frames for storing bulbs. In terms of construction, cold frames are inexpensive structures in which bulbs can receive the cold treatment they need. However, they are the most expensive in terms of labor cost.

For example, extremely cold winter weather may cause the soil covering the pots to freeze so hard that an air-powered jackhammer must be used to free the pots. If the soil is not frozen, then the grower must work in heavy, wet, muddy conditions. The grower will realize little profit on a pot of tulips if this type of effort is needed.

When outdoor storage alone is used, the area must be selected carefully to ensure that excess water will drain away from the pots. The bulbs will die if they are submerged in a waterlogged medium.

The storage area is dug out to a depth of 12 to 15 inches (30 to 37.5 cm). The bottom is leveled, and the well-watered flats or pots are set in place. If mice are a problem in the area, naphthalene moth flakes can be sprinkled lightly over the tops of the pots.

Naphthalene keeps the mice away, but does not harm the bulbs.

A 1- to 2-inch (2.5- to 5-cm) layer of sand or perlite is spread on top of the moth flakes. This layer acts as a buffer between the medium in the pots and the soil used for frost protection. The pots are then covered with 6 to 8 inches (15 to 20 cm) of the soil removed to make the bed. In very cold climates, added frost protection is obtained by placing 8 to 10 inches (20 to 25 cm) of salt hay over the soil.

Ground polypropylene or shredded Styrofoam plastic provide excellent insulation. These materials can be poured directly on top of the pots. A buffer layer of sand or other material is not needed. These lightweight plastics can be blown away by a strong wind, exposing the pots to freezing temperatures. However, the plastic can be held in place by covering the area with a piece of Saran shade plastic and weighting it down with snow fence. The snow fence allows rain to enter the area and permits air exchange around the bulbs. The pots must not be covered with a solid plastic cover because the bulbs will rot.

It is helpful to make a map of the locations of the different crops. Because not all of the crops are brought in for forcing at the same time, the grower must be able to locate a specific crop without digging up the entire field.

The bulbs are usually panned and buried outside in late October. At this time, the soil temperature is in the range that promotes good root growth, 50°F to 55°F (10°C to 13°C). The plants must have well developed root systems before the temperature drops and root growth ceases. If prematurely cold weather is expected, the bed area must be given extra protection in the form of straw or salt hay. Snow is an ideal protective cover against frost.

It is common practice to dig cold frame bulbs and bring them into the greenhouse when the temperature is too low. By this time, the stems have grown several inches into the medium above. The plants must be handled very carefully because it is easy to break the stems. If the plants are frozen, they are moved into the greenhouse to thaw gradually at a temperature of 41°F to 45°F (5°C to 7°C). They must not be exposed to direct sunshine at this time. The radiant energy from the sun can damage the tissue. The bulbs may be placed under the greenhouse benches for 3 or 4 days until they are acclimated to their new surroundings. The pots can then be placed on top of the benches.

Any plants showing signs of severe disease problems are discarded promptly. A light mist spray of fungicide can be applied to the tops of the plants at this time.

CAUTION

Do not use moth flakes made from **paradichlorobenzene** *because this material is toxic to the bulbs.*

Bulb Storage Cellar-Rooting Room

A bulb storage cellar, or a rooting room, has several advantages over outdoor storage.

1. The flowering season can be extended.
2. Mass production of high-quality plants is possible year after year.
3. Regular inspections of the plants can be made to remove diseased or damaged plants and thus ensure a high-quality crop.
4. Mechanization can be introduced so that the labor needed is reduced.

The rooting room should be near the headhouse working area where the bulbs are potted. It should also be close to the greenhouse where the forcing takes place. It is expensive in terms of time and labor to move pots long distances, especially when hand equipment alone is used.

Ethylene gas can seriously damage bulb crops. Fruits and vegetables that produce ethylene gas must not be stored in the same room with the bulbs. In addition, the rooting room should be protected from exhaust gases from engines.

A ventilation system must be provided that will ensure one complete air change every 24 hours. A water line with regularly spaced connections is needed. A line with quick-connectors is preferred to one with screw-on hose connections because less time is required to attach the individual lines. The easiest system to use consists of drop lines from overhead mains.

Light must not enter the rooting room because light causes the stems of the crop to stretch. Ventilation ducts are baffled so that light cannot enter. Some artificial light is required by the workers who water and inspect the crop. This light is provided by 25- or 40-watt incandescent bulbs.

The room must be cleaned thoroughly before the potted bulb crop is brought inside for storage. A forceful hose washing is followed by a heavy fumigation. Then the entire room is sprayed thoroughly with Green-Shield® or a similar disinfectant.

The relative humidity of the room should be in the range of 90 to 95 percent. Excessive moisture can lead to disease. Floor drainage is required to carry away the water used to irrigate the pots. When the water does not drain freely from a pot after it is irrigated, a water seal occurs. To prevent this, expanded metal shelving is used. Number 9 weight galvanized metal will last for years in the conditions found in the rooting room.

Air movement around the pots minimizes disease problems and helps maintain a uniform temperature throughout the storage area. To promote good air movement, neither shelves nor the pots can be closer than 2 inches (5 cm) to any of the outside walls. In very large storage rooms, small oscillating fans can be installed in opposite corners to keep the air moving.

During extremely cold weather, air must not be allowed to enter the room at the outside temperature because the bulbs or developing plants may freeze. Baffles can be installed to allow the air to reach the storage room temperature before striking the plants.

Need for Cold Period

Bulbs need a cold period during which physical and chemical changes take place inside the tissues. These changes must take place if the bulb is to be forced to flower. The temperature during this period must be controlled accurately. The following sections in this chapter on specific crops discuss the temperature requirements.

The changes within the bulbs require a definite amount of time. The grower cannot hasten or decrease the speed of these changes. If the bulbs require a period of 4 weeks at 50°F (10°C), reducing the time to 3 weeks results in flowers of reduced quality. If the bulbs are forced too soon, the flowers may have short stems and be extremely small.

The sections that follow discuss the methods of forcing the most common bulb crops.

ANEMONES ▨ *Anemone coronaria, Ranunculaceae*

Anemones are native to the central European countries around the Mediterranean Sea. Although there are several species of anemones, growers generally use *A. coronaria* for forcing. Anemones need cool temperatures before they will flower. Thus, their flowering period is from late fall to late spring.

Most anemones are propagated from seed because the roots usually are diseased. Seed is sown in the peat-lite mix in April or May. As soon as the seedlings are large enough, the small plants are transplanted directly to the production benches or ground beds.

Media and Fertilizers

The growing medium must have very good drainage and aeration because anemones are prone to crown rot and *Botrytis*. In addition, the medium must be pasteurized. The preferred medium was once a 1:1:1 mixture of loam soil, peat moss, and perlite with a pH of 6.0 to 6.5. Except for when grown in ground beds, most growers now use a peat-lite medium. Anemones do not grow well when they are heavily fertilized. Regular 20 percent superphosphate is added before planting at the rate of 5 pounds (2.28 kg) to 100 square feet (9.28 m²) of area. As the crop grows, fertilizer is added at the rate of 100 ppm each of N and K at every irrigation. The plants are irrigated with clear water every third week, if needed, to leach out any buildup of soluble salts. When the plants start to flower, the quality of the blooms can be improved by increasing the amount of potassium applied to 150 ppm.

When anemones are planted, the crown of the plant is set above the soil surface to prevent crown rot. Some growers prefer to set the seedlings in small ridges. This practice helps to ensure good air movement around the plants. The plants are usually spaced in a grid of 5 inches by 10 inches (12.5 by 25.4 cm) in the bench.

To keep the foliage dry, an automatic watering system with ooze tubes is preferred. Watering from the bottom reduces the development of both crown rot and *Botrytis*.

Light and Temperature

Anemones are grown under high light intensity (2,500 to 5,000 fc [26.9K to 53.8K lux]). A high-quality flower is obtained if the plants are grown at a cool temperature, 41°F to 45°F (5°C to 7°C). A high temperature at night may cause the plants to go dormant.

Growers may be tempted to turn off the greenhouse heat too early in the spring because of the high cost of heating and the requirement for low temperatures. It should be realized that turning off the heat at this time threatens the entire crop. As the outside temperature decreases at night, the relative humidity inside the greenhouse soon reaches 100 percent. As a result, moisture condenses in the crown of leaves. These conditions are ideal for the development of botrytis. The excess moisture can be removed by using some heat and opening the ventilators slightly.

Trimming Leaves

The plants produce large numbers of fleshy leaves. These leaves should not be trimmed or thinned out unless they become so crowded that air circulation is reduced. It is important to keep as many leaves as possible because they supply food to the plant from photosynthesis. Trimming leaves does not improve the quality of bloom.

Harvesting the Crop

The flowers are harvested for market when they are three-quarters open. They are gathered in a bunch of a dozen blooms and wrapped in tissue paper. The flowers are hardened in a 35°F to 41°F (2°C to 5°C) refrigerator before they are sent to market. Do not store them in water. They should be stored dry.

Problems

Cultural Disorders

The main cultural problems are the diseases that affect the crown or leaves to cause short stems and small flowers. Overfertilization with nitrogen may have the same effect.

Diseases

The diseases affecting anemones are crown rot and botrytis petal blight.

Insects

Aphids, mealybugs, red spider mites, and leafrollers attack the plants.

CALLA (Calla Lily) ▓ *Zantedeschia species, Araceae*

Calla lilies are native to south central Africa. They also grow wild in New Zealand. They grow in normally wet areas that are subject to periods of drought. This natural wet–dry cycle is an important factor in making them bloom.

There are several species that are used for cut flowers. Israel, the Netherlands, New Zealand, and the United States are principal suppliers of calla rhizomes that are forced into bloom in the United States and Canada.

Zantedeschia aetheopica and *Z. albomaculata* have white flowers. *Z. rehmanni* has pink flowers and is generally forced as a potted plant. White calla lilies are very popular for Easter as church decorations and favored by many people for weddings.

The plants normally are grown from rhizomes that can be purchased from specialty growers. Some European growers produce rhizomes that are consistently disease free. Pink calla lilies can be grown from seed as well. The cut flowers have a high market value. Because of the value of the crop, some calla growers are very protective of their plants. They permit only the grower in charge to enter the greenhouse. This restriction is intended to reduce the chances of people accidentally bringing disease into the calla growing area.

Media and Fertilizers

A 1:1:1 mixture by volume of loam soil, sphagnum peat moss, and perlite or coarse sand was once the preferred medium because it must have good drainage and aeration, but many growers now use a peat-lite medium. The media must be steam pasteurized or treated with chemicals. Regular 20 percent superphosphate is added at the rate of 1 pound (0.56 kg) in a cubic yard (cubic meter). One-half pound (0.28 kg) of treble superphosphate is added to a cubic yard (cubic meter) of medium.

The rhizomes are planted in ground beds or in large separate containers. An 8- to 10-inch (20- or 25-cm) diameter pot provides adequate room for good root development. Planting in separate containers is a better practice than planting in benches or beds. If a potted plant is attacked by disease, the entire container including the medium and the plant can be discarded. In a planting made in a bed or bench, the diseased plant and some of the medium can be removed. However, there will always be some doubt about disease organisms remaining in the bed.

During the active growth period from September to June, the plants must be well fertilized. Liquid fertilizer can be applied at every watering at a rate of 150 ppm N using a 20-10-20 analysis fertilizer. Twice a month, the plants may be irrigated with plain tap water to leach any built-up salts from the medium. The total soluble salt content of the medium should be less than 150 mhos for a 1:2 soil:water dilution. Regular tests of the medium will ensure that the nutrient levels are in the correct range.

Light and Temperature

From September to June, the plants are grown under full light intensity. During the summer, shade is provided to reduce the intensity to 5,000 fc (53.8K lux), or one-half the normal light. White callas grow best at a night temperature of 55°F (13°C). Pink callas need a minimum night temperature of 60°F (15.5°C). These temperatures are difficult to maintain in the summer unless fan and pad cooling is used.

Water

Callas need large amounts of water to produce high-quality flowers. The Chapin watering system is ideal for this crop.

If the growing medium is allowed to dry at any time, the plant may go into its dormant stage. In its natural habitat, this stage is a part of the normal life cycle.

Some growers stop watering in June and allow the plants to go dormant over the summer. There is no need to do this because there is no additional benefit to the plant. Callas will continue to flower as long as they have sufficient water. The summer upsurge in weddings provides a good market for calla blooms.

Marketing

The flowers are cut when they are fully open. There is no loss in quality if the flowers remain on the plant for as much as a week after they open. Callas produce large amounts of ethylene gas. The grower must remember that callas cannot be stored or shipped with carnations, snapdragons, or other crops that are affected by the gas.

Problems

Diseases

Calla root rot is caused by *Phytophthora cryptogea* var. *richardiae*. This organism causes a dry rot of the rhizomes. The best control is to discard the diseased rhizomes.

Highly diseased old rhizomes can be saved by drying them, removing loose medium, scrubbing thoroughly, and then cutting out the rotted spots. The rhizomes are then soaked for 1 hour in a solution of 1 part formaldehyde diluted in 50 parts of water. The treated rhizomes can be grown using pasteurized medium, pots, and benches.

Erwinia aroidae or *E. caratovora* causes a soft rot that is controlled in the same manner as root rot.

Insects

Thrips, mealybugs, and red spider mites are common pests on callas. Use of the proper insecticides will control them. Sprays or smoke fumigants are most effective because systemic poisons do not work well on callas.

CROCUS ▒ *Crocus vernus, Iridaceae*

Crocus is a minor bulb crop. The bulbs are imported from the Netherlands in the fall. They are flowered for sales from the middle of December to the end of April.

Only large size bulbs are used, because they produce multiple flowers. The bulbs are precooled at 48°F (9°C) for 15 to 16 weeks. However, overcooling of the bulbs reduces the quality of the plants.

If the bulbs arrive before their scheduled cold storage date, they can be held at 63°F (17°C) in a well-ventilated area. They are removed from the packing bag and placed in shallow trays with wire mesh bottoms.

Crocus are often planted in novelty pots. Ten to 12 bulbs are planted in a 6-inch (15-cm) pot so that they just touch each other. Smaller pots are planted in the same way using fewer bulbs. The medium used was described earlier in this chapter.

When the roots are well developed and the tops are 1.5 to 2 inches long, the plants are ready to market. At this point, the stems are white and there are no signs of flower buds. There are several advantages to selling the plants at this stage of floral development. For one, the customer gets the most enjoyment from watching the plants come into bloom. Also, the grower's costs of production are reduced by moving the plants off the greenhouse benches. After the plants are brought into the greenhouse, they will flower after 2 weeks at a night temperature of 50°F to 55°F (10°C to 13°C). If the plants are sold at the full bloom stage, the blooms do not last long enough to give the customer full value for the money paid.

The following list gives the cultivars commonly used and the flower color when they are forced.

'Flower Record'—deep lavender
'Joan of Arc'—white
'King of the Striped'—lavender with small stripes
'Pickwick'—white with narrow lavender stripes
'Purpureus Grandiflorus'—deep lavender, fading to light at the petal edge
'Remembrance'—lavender
'Yellow'—golden yellow

Problems

Crocus have few problems. Temperatures above 60°F (15.5°C) should be avoided.

FREESIA ※ *Freesia refracta, Iridaceae*

A native of South Africa, the freesia is a cormous plant that has a very strong fragrance when it blooms (Figure 26-1).

The corms are planted directly in benches or flats or in pots or pans from August to November. They are placed so that the corms just touch each other. Ten to 12 corms can be placed in a 5-inch (13-cm) pot, and 15 corms can be placed in a 6-inch (15-cm) pot. If the corms are planted in flats, they are spaced on 2.5-inch (6.3-cm) centers. For ease of harvest in bed or bench plantings, they are spaced 1 inch (2.5 cm) apart in rows with 6 inches (15 cm) between the rows (Figure 26-2). The depth of planting is 1 inch (2.5 cm) below the surface of the medium.

The medium is usually a 2:1:1 mixture or 1:1:1 mixture by volume of soil, peat moss, and perlite or a commercially prepared peat-lite medium. The fertilizer requirement is 100 ppm each of N and K_2O applied with every other watering. Freesias are cool temperature crops. The best growth is made at a minimum night temperature of 50°F to 55°F (10°C to 13°C). Above 60°F (15.5°C), the plants produce weak stems and poor flowers.

FIGURE 26-1 A *Freesia* bloom.

FIGURE 26-2 A raised bed of *Freesia*.

Problems

There are few cultural problems with this species.

GLADIOLUS ✠ *Gladiolus* × *hortulanus, Iridacea*

Gladioli are also raised from corms. They produce a flower spike that is excellent for large cut flower arrangements. Miniature gladioli are also being grown. These smaller plants broaden the range of use of gladioli.

Gladioli are probably the most widely grown of all outdoor cut flowers in the United States. Florida has 4,200 acres (1,700 ha) in production. In addition, gladioli are grown in Alabama, Arkansas, Texas, and California. As spring moves northward, gladioli are grown as summer crops in more northern states. At one time gladioli were forced as cut flowers in northern greenhouses. This practice was stopped, however, when field production began in Florida.

In Norway, Sweden, and Finland, some gladioli are still forced in greenhouses. But again, this practice is declining because flowers are being sent to market from Israel and other Mediterranean countries.

More than 8,000 gladiolus cultivars have been developed so far. Commercial and amateur plant breeders introduce new cultivars every year. Of the total number of cultivars available, only 20 to 30 are grown for commercial use. The remaining cultivars are grown by amateurs.

Propagation and Growth

Gladioli are propagated from corms. The corm is the fleshy basal part of the plant. As the plant grows each year, new corms are formed and the old corm is gradually used up. **Cormels** are tiny corms that appear along the stolons, or underground stems that grow from the old corm (see Figure 18-21). Cormels are planted in beds, and after 2 or 3 years they are large enough to produce flowers.

Gladioli need a sandy soil that drains well. Clay or clay loam soils are not used because it is too difficult to harvest the bulbs from them. The soil in the production areas in Florida has a high sand content.

Gladiolus production in Florida takes place during the winter months. The corms are planted from September through February. The summer months are too warm for good flower production.

Fusarium, Curvularia, and *Stromatinia* fungi all attack the plants. The corms are soaked in hot water to reduce the number of disease organisms. If the corms are not treated, the fungi may become established in the fields where they will survive for many years.

The standard treatment is a 30-minute soak in water maintained at 136°F (57.8°C). The water should be stirred continuously to ensure that all of the corms receive the same treatment.

To prepare the corms to tolerate this high temperature, the corms and cormels harvested in warm weather are cured for 2 to 3 months at a temperature between 75°F and 85°F (24°C and 29.5°C). Smaller corms and cormels develop a deeper dormancy than do the large corms. As a result, large corms may be killed by the heat treatment.

As soon as the corms are removed from the hot water, they are cooled in cold running water or in ice water. The corms are then dried quickly in a current of warm air. The corms are then placed in cold storage at a temperature of 38°F to 42°F (3.5°C to 5.5°C). A storage period of 3 weeks is needed to break the dormant stage and make the corms sprout and flower evenly. Corms harvested in the summer are usually held in cold storage for at least 2 months.

Soils and Fertilization

The best blooms are obtained when gladioli are grown in soils having a pH value of about 6.0.

Growers once thought that gladiolus did not require fertilizers. However, current research shows that improved crops result from using fertilizers. The recommended preplant treatment is 500 to 750 pounds (227 kg to 340 kg) of 5-10-10 analysis fertilizer per row. Side dressings of nitrogen and potassium are made when the plants begin to develop flower spikes.

Planting

The corms are planted in rows and are placed 3 to 5 inches (7.5 to 13 cm) deep. A fungicidal dip is recommended before the corms are planted. The root side of the corm must face down as the corm is placed in the row. The corms cannot be thrown into the row because the probability is that at least half of them will land root side up. If this happens, the growth rate of the spike is delayed and the quality of the bloom is reduced. Growers in Florida use special tractors to plant the corms. Four people sit on each side of the tractor and place each corm into a separate cup on an endless belt. This belt deposits each corm in the planting furrow with the root side down at the rate of four to five corms for each foot of row.

Cormels or small, nonflowering corms are scattered in the furrow. Because cormels are several years away from flower production, this less costly and labor-saving practice can be used rather than the special tractor.

Before planting, the soil is fumigated to kill weeds, nematodes, and other pests. If the soil is not fumigated, herbicides (weed killers) are applied to control weeds. Shallow cultivation is one method of controlling the weeds until the plants grow too large to permit the tractor to move down the rows. Because the list of approved chemicals changes regularly, the local agricultural experimental station should be contacted to obtain recommendations for the weed-control chemicals that are safe to use.

Irrigation

Quality gladiolus cannot be grown without some form of irrigation. Irrigation supplies the needed moisture during dry weather and can also be used on hot days to cool the plants and prevent **tip burn**. If frosts are expected, the irrigation system can be used to protect the plants from freezing damage.

Irrigation water is applied by overhead sprinklers or by flooding the rows. If flooding is the method to be used, the land must be level so that the water can be distributed evenly throughout the field. Large machines called land planers are used to level the fields before the crop is planted.

Cutting for Market

Flowers to be used in local markets are cut when the first one or two **florets** open. Spikes that are to be shipped to distant markets are cut when the florets that are formed first are just beginning to show color. This condition is called the tight bud stage.

The gladiolus quickly displays a condition called **negative geotropism** when the spike is cut and placed on a flat surface; that is, the tips of the flower curve upward. To prevent this reaction, the spikes must be upright through all handling and shipping. The spikes are usually packed in large cardboard boxes with printed directions that caution that the boxes are to stand on end at all times.

Some growers place freshly cut gladiolus spikes in cool water in the packing shed. Other growers pack them first and then place them in cold storage at 40°F to 44°F (4.5°C to 7.5°C). Storage below 40°F (4.5°C) is not recommended, because the flowers of some cultivars do not open properly if they are cooled too much.

The cut flowers are graded and then are gathered into bunches of a dozen spikes each. The commercial grades by which the spikes are sorted are given in Table 26-1.

Handling Corms

When the flowers are harvested, the stem is cut to leave as much of the foliage on the plant as possible. Photosynthesis occurring in this foliage builds up a supply of stored food in the corms. The plants continue to grow for 6 to 8 weeks after the flowers are cut if the conditions are favorable. During this time, the new corms mature and young cormels appear.

The corms are removed from the field when the weather is dry. The old tops and weeds are cleared from the field using a rotary mower. The machine used to harvest the corms is a modified potato digger that lifts the corms from the soil, shakes off the excess soil, and places the corms into containers. The corms sunburn easily and must be covered immediately with burlap or other material. The digging machines can be adjusted so that the corms are bruised as little as possible.

For small plantings, a mold board plow can be used to bring the corms to the surface of the soil where they are gathered by hand.

The corms are then cleaned and dipped in a fungicide. For maximum disease protection, the corms should be treated within 3 days of digging. The corms are placed in trays for drying in a well-ventilated building. Following the drying period, they are stored in a frost-free area.

Any cormels recovered in the digging operation are treated in the same manner as the corms. In the next year, these corms, or planting stock, will be

Table 26-1

Commercial Grades of Cut Gladiolus		
Grade	Minimum Number of Florets on Each Spike	Minimum Length of Spike, in Inches (cm)
Extra Fancy	12	Over 42 (107)
Fancy	12	38-42 (96-107)
Special	12	34-38 (86-96)
A	10	30-34 (76-86)
B	7 or 8	24-30 (61-76)
C	6	24-30 (61-76)

Table 26-2

Grades of Gladiolus Corms	
Grade	Size Diameter in Inches (cm)
Jumbo	over 2 (5)
No. 1	1.5–2 (3.8–5)
No. 2	1.25–1.5 (3.2–3.8)
No. 3	1–1.25 (2.5–3.2)
No. 4	0.75–1 (1.9–2.5)
No. 5	0.5–0.75 (1.2–1.9)
No. 6	0.13–0.38 (0.9–1.2)

grown on to flowering size. The grades used by the industry to size corms are given in Table 26-2. Cormels usually are not graded.

Problems

Cultural Disorders

Bends in the stems may be due to a lack of boron. Another possible cause is wilting due to a lack of water followed by renewed growth. Short, stubby flower spikes are caused by a lack of water or root injury. Tip burn of the leaves may occur if the plants lack water when the temperature is high. Fluoride injury also results in tip burn.

Diseases

The high humidity and warm temperatures required to produce quality gladiolus blooms also promote a large number of diseases. If gladioli are planted close to other crops, such as beans, peas, and other members of the iris family, aphids carry **viruses** from one crop to another. Viruses that commonly infest gladiolus are cucumber mosaic, tomato ring spot, tobacco ring spot, and bean yellow mosaic.

Brown rot is caused by a fungus, *Fusarium oxysporum* f. *gladioli*. The disease can remain in the soil for several years even when gladiolus are not planted. Chemical treatment of the corms as soon as they are harvested and fumigation of the planting fields help to reduce losses from the disease.

Botrytis gladiolorum causes a leaf and flower spotting stem rot, and corm rot while the corms are in storage. The disease is not easy to control but measures that can be taken include roguing (removing) the plants from the field, removing all frost damaged tissue, and chemically treating the corms at various stages. A regular spray program in the field is recommended. During normal weather, the crop can be sprayed once a week. If the weather is rainy or damp, it may be necessary to spray every 2 or 3 days.

Curvularia trifolii is a warm-weather disease that can seriously threaten the crop during wet conditions. Young growing tissue is often attacked first, but all parts of the plant can be infected.

The control of *Curvularia trifolii* involves hot water dips, fumigation of field soils, and repeated fungicidal sprays.

Nematodes are parasites that live in the soil and attack the roots of plants. The most troublesome species is the root-knot nematode, *Meloidogyne incognita*,

Meloidogyne incognita var. *acrita*. These parasites cause small galls or "knots" to develop on the roots. Affected plants are stunted in size and wilt easily. Nematodes are most active when the soil temperature is high, as in late spring and summer.

Fields formerly planted with vegetables or other ornamentals may contain large populations of root-knot nematodes. Before such fields are planted with gladiolus corms, the soil must be fumigated.

Insects

The insects that infest gladioli—aphids, spider mites, thrips, and wireworms—can be controlled by the use of the proper insecticides. On small plantings, a mulch of aluminum foil is effective in repelling flying aphids that transmit viruses.

HYACINTH ⊠ *Hyacinthus orientalis, Liliaceae*

Hyacinths are flowered for sales from Christmas through late April. Easter is a major sales period for hyacinths.

Two types of hyacinths are sold by bulb suppliers—prepared bulbs and regular bulbs. Prepared bulbs are grown in fields where heat pipes increase the normal soil temperature. These bulbs are used for early forcing. Regular bulbs are given the standard treatment and are used for normal forcing.

Hyacinth bulbs are graded according to their circumference, or distance around the bulb. They range in size from 5.9 or 6.3 inches to 7.5 or 7.9 inches (15 or 16 cm to 19 or 20 cm). The larger sizes are used as prepared bulbs, and the smaller bulbs are used for standard forcing.

If the prepared bulbs cannot be planted immediately, they are held at 48°F to 55°F (9°C to 13°C). Thus, they receive partial precooling treatment while they are in storage. Because the bulbs may root prematurely if they are stored in cardboard boxes or paper bags, they are scattered loosely on trays.

Prepared hyacinths require 10 weeks of cold treatment. There are three stages of treatment. From the time of planting until December 1–5, the bulbs are held at 48°F (9°C). For the next 4 weeks, until January 1–5, the bulbs are held at 41°F (5°C). When the shoot reaches a length of 1.5 inches (3.8 cm), the temperature is reduced to 32°F to 35°F (0°C to 2°C) and maintained at this level until the bulbs are brought into the greenhouse for forcing.

Regular bulbs are stored at 63°F (17°C) until they are ready for precooling. The temperature program described for prepared bulbs is also used for regular bulbs, which need 13 weeks for precooling.

The number of hyacinth bulbs planted depends upon the size of the pot. Generally, one bulb is placed in a 4-inch (10-cm) pot, three bulbs are placed in a 6-inch (15-cm) pan, and seven or eight bulbs will fill a large bulb pan. A 1:1:1 mixture by volume of soil, peat moss, and perlite is the recommended potting medium. The bulbs must be well watered before they are stored.

The plants are removed from storage when the flower spike is 3 to 4 inches (7.5 to 10 cm) long. For early flowering, they are forced at temperatures of 73°F (23°C). For flowering from mid-January to late February, the temperature is held at 65°F (18°C). A temperature of 60°F (16°C) is maintained for the rest of the season.

The following list gives the cultivars commonly used for forcing and flower color.

'Amethyst'—light lavender
'Amsterdam'—very deep pink
'Anna Liza'—violet
'Anna Marie'—light pink
'Bismarck'—light blue
'Blue Blazer'—blue
'Blue Giant'—light blue
'Blue Jacket'—very deep blue
'Carnegie'—white
'Delft Blue'—blue
'Eros'—deep pink
'Jan Bos'—red
'Lady Derby'—light pink
'L'Innocence'—white
'Madame Krueger'—white
'Marconi'—deep pink
'Marie'—very dark blue
'Ostara'—deep blue
'Pink Pearl'—deep pink
'Pink Supreme'—pink
'Queen of the Pinks'—deep pink
'Violet Pearl'—violet
'White Pearl'—white

Growth Retardants and Height Control

The growth retardant that is active in controlling the height of potted hyacinths is ethephon (*eth-e-fon*), sold as Florel®. Ethephon has an influence on suppressing the formation of ethylene in the plant growing point thereby restricting stem stretch.

The material is applied as a spray to runoff. A spray concentration of 1,000 or 2,000 ppm is used. To prepare a solution of 1,000 ppm concentration, dilute 0.5 pint (237 ml) of Florel® to a final volume of 2.5 gallons (9.5 l). This amount of chemical is enough to treat 500 6-inch (15-cm) pots.

For a 2,000 ppm solution concentration, double the amount of Florel® used to 1 pint (474 ml) and dilute it to 2.5 gallons (9.5 l) to treat 500 6-inch (15-cm) pots.

To be effective, the best time to spray the solution is late afternoon. The leaves should not be wet for a period of 12 hours after the treatment to allow maximum absorption to take place. The plant (leaves/stalk) height should be 3 to 4 inches (7.5 to 10 cm) tall. If a second treatment is needed, apply it 2 or 3 days after the first spray. Do not wait longer than this, because these plants grow very quickly. A second treatment must not be applied if the flower bud is easily visible in the foliage.

Use freshly prepared solutions and apply immediately. Solutions older than 4 hours may break down chemically and not be effective. Follow all label directions regarding application, personal protection, and disposal of any solution left over.

Problems

One common problem is **spitting** of the flower spike. When the plants are ready to be forced, the stem may break above the soil line, or just below the flower head. The break is clean and looks like a knife cut. Several possible reasons for spitting can be listed: planting too early in the fall (before October 15); a high root pressure caused by fertilizing with ammonium sulfate (hyacinths do not require a fertilizer); poor drainage in the outdoor storage area; and exposure to cold when the plants are moved from the cold frame to the greenhouse. It is doubtful that the last reason listed is a factor, because the problem has occurred with hyacinths never exposed to a temperature less than the storage temperature.

IRIS � *Iris species, Iridaceae*

The iris cultivars used by growers for greenhouse forcing are the result of crosses of several species. *Iris xiphium preacox* × *Iris tingitana* × *Iris lusitanica* had their origins in Spain and North Africa (Tangiers and Morocco). The Pacific Northwest of the United States produces iris bulbs that are considered by growers to be the best available for forcing. The plants are field grown through winter.

When the bulbs are harvested in July and August, the flower bud has not formed in the new bulb. Within 5 days of digging, the bulbs are heat cured at 90°F (32°C) or given an ethylene treatment to hasten flower bud development. This is followed by 6 weeks of 48°F (9°C) storage, until they are shipped in late August or early September.

Bulbs not scheduled for early planting are held at 70°F to 80°F (21°C to 27.5°C) in open trays in a well-ventilated area. Bulbs to be precooled are held at 50°F (10°C) for 6 weeks. The precooling treatment can be given while the bulbs are in their shipping bags, or they can be planted and then placed in cool storage.

Most iris are forced for cut flowers. They are planted 4 to 5 inches (10 to 12.5 cm) deep in boxes or are placed directly in raised greenhouse benches. Iris planted in shallow flats dry out too quickly. As a result, the flower buds develop to a certain point and then dry up. This condition is known as **bud blasting**. Bulbs planted too close together are also subject to blasting.

The bulbs are spaced in a 3-inch by 3-inch (7.5-cm by 7.5-cm) pattern and are placed in a refrigerator where they are held at 50°F (10°C). They are brought into the greenhouse for forcing when the tops are 2 to 3 inches (5 to 7.5 cm) tall. At least 1 inch (2.5 cm) is allowed on all sides of the boxes for good air circulation. The plants may root into the medium beneath the boxes. If they do, the containers cannot be moved without breaking the roots. If this happens, the flowers will abort (go blind) from a lack of water. The plants require a full light intensity at 60°F (15.5°C) to produce high-quality plants.

Iris planted in greenhouse benches are grown at a temperature of 50°F (10°C) until they reach a forcing size. The temperature is then raised to 60°F (15.5°C). From the time of planting, nearly 3 months are required to bring the crop to flower. A succession of plantings is made so that the blooms are available for a longer period of time.

Iris are harvested for market when the buds are just beginning to show color. The buds open very quickly. If the blooms are to be sold locally, they are cut when the flowers are fully open. The bulbs are graded according to the circumference in centimeters. The size ranges are from 2.4 or 2.8 to 4.3 or 4.7 inches (6 or 7 cm to 11 or 12 cm). The larger bulbs are used for precooling and early forcing.

For forcing in midseason and later, smaller bulbs are used. Because of the better growing weather less time is needed to force the plants into flower. That is, the plants can be forced in 8 weeks rather than the 12 weeks needed to force flowers for Christmas and early January.

Problems

Cultural Disorders

Blasting of iris flowers can be caused by a number of factors: too little space between the bulbs, damage to the root system that prevents sufficient water uptake, very dry conditions, and forcing at too warm a temperature.

Blindness, or a failure of the flower bud to form, is the result of improper treatments during storage.

Diseases

Dry rot, virus disease, and leaf mottling caused by a virus affect the plants. *Fusarium* and *Penicillium* also affect the plants.

Insects

Pests include tulip aphids, bulb mites, bulb nematodes, and wireworms.

LILY ※ *Lilium longiflorum, Liliaceae*

Lilies are grown in the United States for cut flowers and as potted plants. In 2005, more than 8 million lilies were grown for sale at Easter as potted plants. The wholesale value was $35.7 million. The wholesale value for potted other lilies, Asiatics, and Orientals is no longer available. The wholesale value of lilies grown for cut flowers was $76.8 million in 2005.

Cultivars used for potted plants are 'Ace' and 'Nellie White.'

The ancestors of the lily used by growers were discovered on the Liu Chiu Islands south of Japan. Before World War II, Japan supplied the United States with almost all of the bulbs used for Easter forcing. When the war cut off these imports, the United States turned to lily production. Bulbs were also imported from Mexico, Bermuda, Cuba, Canada, and the Netherlands in an effort to supply the demand.

Lily production also occurs in Florida ('Creole' lilies) and along the Atlantic Coastal plain, especially in Georgia ('Georgia' lilies).

The commercial production of lily bulbs in Oregon and Washington State increased greatly after World War II. At one time several hundred growers produced bulbs in this area. At present, only a few growers produce bulbs to meet the commercial demand. Dahlstrom and Watt in Smith River, California, is one of the largest producers of lily bulbs. Many of these growers have been in business for over 40 years. As a result, they know the crop very well. The climate in Washington and Oregon is excellent for bulb production because of the mild winters, large amounts of rain, well-drained soils, and temperatures that rarely go below freezing.

Cultivars

Modern cultivars are the result of many years of development. Plant breeders are working continually to improve the stock.

The 'Ace' cultivar was developed in 1935 from the stock of a private breeder. The plant has excellent characteristics: a pyramidal shape, medium height, broad dark green leaves, and a large number of flowers. As many as 23 blooms have been counted on one plant. Less than 5 percent of the lily bulbs forced today are 'Ace.'

'Nellie White' is the most important bulb in terms of numbers forced for Easter. Approximately 95 percent of the lilies forced for Easter are 'Nellie White' cultivar. This cultivar is the result of a selection made by James Langlois of Oregon in 1955. The plant is short and column shaped, with heavyweight, dark green leaves. The leaves are medium in length and stand out horizontal to the axis of the stem. Medium size flowers are produced. There are fewer flowers per plant as compared to the 'Ace' cultivar.

Field Production

Propagation

Small, 0.5-inch (1.25-cm) diameter bulbs form in the axils of the stem. These bulbs are the main source of planting stock. Additional stock is obtained by scaling the bulbs. The lily bulb is somewhat like an onion. That is, it consists of a large number of close-fitting scales attached to a basal plate at the bottom. As soon as the plants have flowered in the spring, the bulbs are removed. The scales are broken off and scattered in rows. Tiny **bulblets** grow from the scales and sprout the following spring. After these "yearlings" have grown for 2 or 3 years, they are large enough to be forced.

Bulbs in the field are mature and ready for harvesting about 6 to 8 weeks after the plants have flowered. The success of the grower in forcing the bulbs into bloom depends upon the ripeness or maturity of the bulbs. Bulbs that are not fully mature cannot be forced into bloom according to schedule.

Harvesting

The bulbs are harvested in one operation by machines that lift the bulbs from the field and shake off the excess soil. The bulbs are then moved to a packing shed where they are further cleaned, graded, and packed into shipping boxes.

Grading

A lily bulb is graded according to its circumference. The commercial grades range in inches from 5–6, 7–8, 9–10, and 11–12. In the metric system of measurement, the grades are expressed in centimeters, such as 13–15, 18–20, 23–25, and 28–30.

Packing

The bulbs are packed in shipping boxes (Figures 26-3 and 26-4). Fine, humid peat moss is placed around the bulbs to protect them from damage during shipment. The peat moss also protects the bulbs from large temperature changes that may be harmful. The peat moss must be moist enough to prevent drying of the bulbs, but it must not be so wet that the bulbs sprout prematurely. For

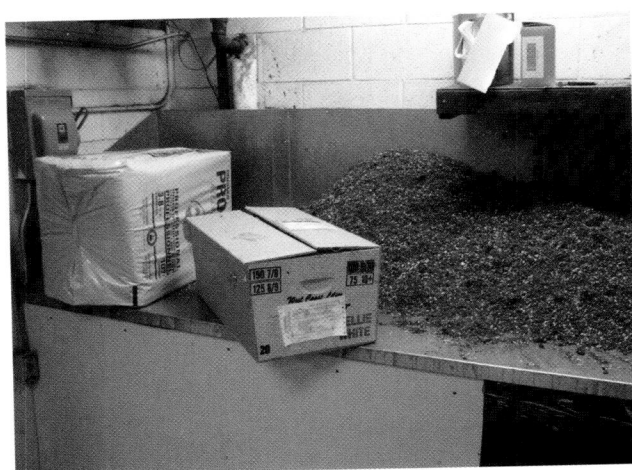

FIGURE 26-3 Shipping container of Easter lily bulbs.

FIGURE 26-4 Easter lily bulbs packed in sphagnum moss.

precooled bulbs, the cases are placed in storage at 31°F (−0.5°C) until they are shipped in late November.

Greenhouse Forcing

Precooling

Some growers prefer to precool the bulbs after they have been received from the producer. For these growers, the producer ships the cases of bulbs as soon as they are packed.

Precooling and cold storage treatment may be given to the lilies before they are unpacked from the shipping case. The lily bulb is then potted and placed in a rooting room for cold treatment. Other growers prefer to pot the bulbs and place them in protected cold frames. The natural changes in the temperature will give the plants the required treatment. Some growers claim that the natural or cold frame treatment (CFT) results in a higher-quality plant with more flowers. Both methods of treatment are discussed here.

Regardless of the method used, research has shown that the best temperature for cold treatment of bulbs grown on the West Coast is 35°F to 40°F (1.5°C to 4.5°C), and 45°F to 50°F (7°C to 10°C) for bulbs grown in the South.

Rooting Room Method

As soon as the bulbs are received, they are potted, watered in well, and placed in a rooting room at the required temperature. The plants are held in storage for at least 6 weeks (1,000 hours). After treatment, they are moved to the greenhouse for forcing. If the bulbs must be placed in storage later, because Easter is later in the spring, they can be held in refrigerated storage at 35°F (1.5°C) until they are ready for potting.

Cold Frame Treatment

The cold frame treatment method, known by many as natural cooling, is used more by western bulb forcers than by eastern growers. The use of CFT in the more severe winters in the eastern United States requires significantly more labor.

The bulbs are shipped as soon as they are removed from the soil. They are potted as soon as they are received. If the bulbs cannot be potted immediately, they can be held for a short time in the packing case at a temperature of 70°F to 80°F (21°C to 27.5°C). After potting, they are placed in cold frames or in a cold greenhouse in full light intensity. Any shading reduces both the number of flowers and the final plant height. The pots are placed so that the rims touch. In the cold frame they are usually plunged to the rim in sawdust or a similar material that acts as insulation. In this way, a more even temperature can be maintained around the developing root system. Where it gets very cold, the sawdust helps to prevent the roots from freezing.

When the night temperature goes below 25°F (−4°C) heat is needed. A minimum temperature of 33°F (0.5°C) is maintained. In a plastic or glass covered greenhouse, the pots are placed on raised benches. The minimum night temperature allowed is 30°F (−1.0°C) and the maximum day temperature is 50°F (10°C).

The pots must be watered as required but should not be kept too wet.

Controlled Temperature Forcing

The **controlled temperature forcing (CTF)** method is not to be confused with the cold frame treatment or with natural cooling. In the CTF method the grower pots nonprecooled bulbs as soon as they are received and waters them in well. This method requires that the medium be kept moist and that accurate temperature control be kept.

After potting, the bulbs are placed at a temperature of 63°F to 65°F (17°C to 18.5°C) for rooting. The warm temperature is thought to aid root growth. As a result, a very high flower bud count is obtained. After 3 weeks, the soil temperature is reduced to 35°F to 40°F (1.5°C to 4.5°C) for the 'Ace' cultivar and to 40°F to 45°F (4.5°C to 7°C) for 'Nellie White.' These temperatures are maintained until about 110 days before the scheduled blooming date. The bulbs are then forced at 63°F to 65°F (17°C to 18.5°C).

Timing and Flower Bud Formation

The time required to force lilies to flower from the time they are removed from storage is approximately 110 to 120 days at a night temperature of 63°F (17°C). Flower buds are formed when the stem is about 2 to 3 inches (5 to 7.5 cm) above the top of the bulb. When the bulbs are planted so that the top of the bulb is 2 inches (5 cm) below the surface of the medium, flower buds start to form just as the shoot breaks through the medium.

The number of flower buds formed is affected by several factors: temperatures that are too high or too low, too little light, a lack of water, too much water, and high soluble salt levels. The response of the plant to temperature means that the grower can increase the bud count by lowering the temperature.

The lily **inflorescence** (flowering head) consists of primary, secondary, and tertiary flowers (Figure 26-5). Primary flowers form regardless of the cooling technique used. Once the primary flowers form, the stem grows longer and the secondary buds develop. If the period of flower bud formation can be extended, a greater number of secondary buds will form. Tertiary buds develop in the axils of the bract leaves on the primary and secondary flower stems.

Temperature is the controlling factor in the timing of flower bud formation. When the bulbs are brought in for forcing, they are held at a night temperature of 63°F (17°C). At this temperature, bud formation always begins during

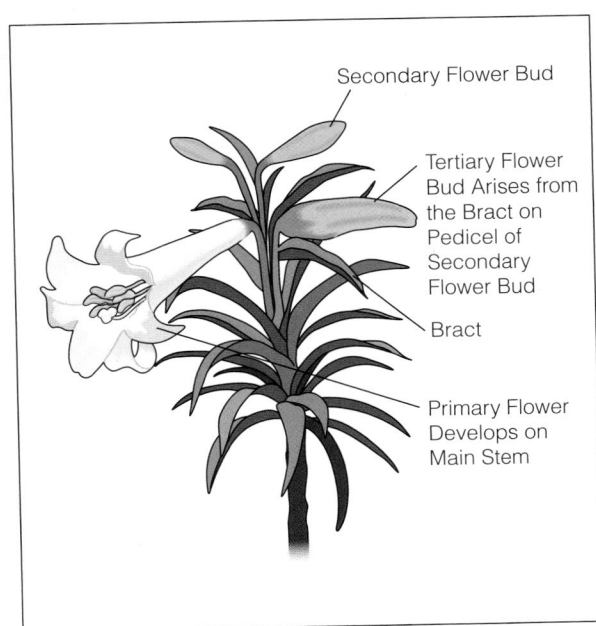

FIGURE 26-5 The primary, secondary, and tertiary flower buds of a lily.

Secondary Flower Bud

Tertiary Flower Bud Arises from the Bract on Pedicel of Secondary Flower Bud

Bract

Primary Flower Develops on Main Stem

the last week of January and the first week of February. Just as this period is ending, the night temperature in the greenhouse is lowered to 45°F to 50°F (7°C to 10°C) for 7 to 14 days. The cooler temperature prolongs the time during which flower buds form, resulting in more buds. The day temperature is kept below 60°F (15.5°C).

The grower determines when the temperature is to be lowered by counting the leaves on the plant. The leaves are counted by starting at the bottom of the plant and counting up the stem. For the cultivar 'Nellie White,' the temperature is lowered when the plants are in the bud-forming stage and 25 to 30 leaves have unfolded to a 45-degree angle with the stem.

After the period of lowered temperatures, the greenhouse is brought back to a temperature of 63°F (17°C).

There are several advantages to this system of forcing flower bud formation: A smaller bulb can be used to obtain the same number of flowers as can be obtained from large bulbs; a greater number of buds from a bulb of a given size results in a better-quality plant; and there is a small savings in the cost of fuel.

Potting

Bulb Size

The bulbs used for forcing are generally the 7- to 8-inch (17.5- to 20-cm) grade or the 8- to 9-inch (20- to 22.5-cm) grade. Larger bulbs are more expensive even though they produce more flowers. The very large bulbs are not used as often as before because of the reduced temperature method described in the previous paragraphs.

The recommended pot for lilies is a 5.5- or 6-inch (13.8- or 15-cm) three-quarter size (Figure 26-6). Clay pots are rarely used, but if used, they must be soaked overnight to prevent them from absorbing moisture from the medium after the bulbs are planted. A small piece of broken pot or an aero drain can be placed over the hole in the bottom of the pot.

Contrary to popular belief, drainage is not increased by adding a handful of gravel, sand, or perlite to the bottom of the pot. Actually, these materials shorten the length of the soil column and decrease the drainage.

The bulb is placed so that the nose or tip of the bulb is 2 inches (5 cm) below the surface of the medium (Figure 26-7). Lily plants develop strong stem roots that help support the plant. In the event that the bulb roots become diseased, the stem roots, which are more resistant to disease, can take up moisture and nutrients. Many growers will dip the bulbs in a fungicide prior to planting to prevent this (Figure 26-8).

Medium

The growing medium must have excellent drainage and aeration because lilies are sensitive to two root rot diseases. Even though the plants may be diseased, the crop can be saved if the medium is open and porous.

A 1:1:1 mixture of loam soil, peat moss, and perlite was once considered to be the best growing medium for lilies. Most growers now use a modified peat-lite mixture consisting of 2:1:1 parts by volume of sphagnum peat moss,

FIGURE 26-6 A standard 6-inch (15-cm) pot with 1 inch (2.54 cm) of medium at the bottom of the pot.

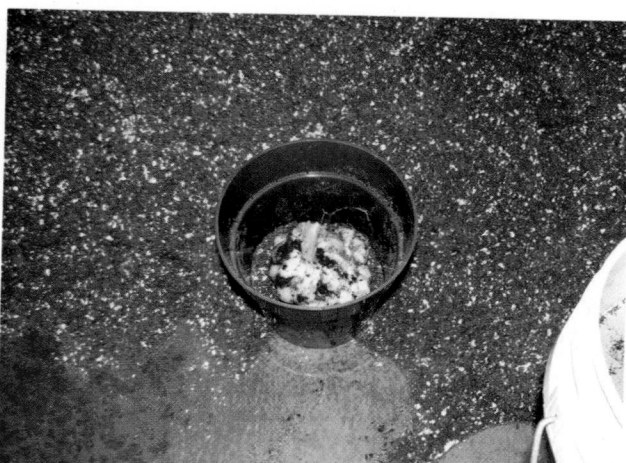

FIGURE 26-7 A standard 6-inch (15-cm) pot with an Easter lily bulb placed on 1-inch (2.54 cm) of medium at the bottom of the pot.

FIGURE 26-8 Easter lilies are often drenched in a fungicide prior to potting.

vermiculite, and perlite or a commercially prepared peat-lite medium. If toppling is an issue, some growers increase the weight of the mix using pasteurized sand added at the rate of 10 percent of the total volume.

Lilies are a limestone loving crop, and so 5 pounds (3.0 kg) of calcium limestone and 3 pounds (1.8 kg) of dolomitic limestone are added to the mix. A small amount of phosphorus is added to the medium because better-quality plants are obtained. However, too much phosphorus leads to leaf scorch. Regular superphosphate contains fluoride, which also causes leaf scorch. However, the fluoride should not be a problem because of the large amounts of limestone used. Fluoride is tied up by the calcium of the limestone and may cause leaf tip burn.

Regular superphosphate is added at a rate of 1 pound (0.6 kg) to a cubic yard (cubic meter) of mix. Treble superphosphate is used at the rate of 0.5 pound (0.4 kg) in a cubic yard (cubic meter) of mix.

Fertilization

The fertilization program begins when the new shoots first break through the surface of the medium. Lilies need large amounts of nitrogen and potassium. When fertilizing at every watering at a 1:100 dilution ratio, a stock solution is made consisting of 5 ounces (142 g) of potassium nitrate and 10 ounces (284 g) of calcium nitrate in 1 gallon (3.785 l) of water. This mixture supplies 168 ppm N and 148 ppm K at each watering. When the buds are 0.5 to 0.75 inch (1.25 to 1.91 cm) long, a stock solution for a 1:100 dilution is made using 10 ounces (284 g) of calcium nitrate in 1 gallon (3.785 l) of water. To obtain stock solutions for a 1:200 dilution ratio, double the amounts of the fertilizers.

If the plants are to be fertilized once a week, an alternating program of fertilizers is used. Three-quarters of a pound (0.34 kg) of potassium nitrate in 100 gallons (378.5 l) of water is applied 1 week. The next week, the plants are given 1.5 pounds (0.68 kg) of calcium nitrate in 100 gallons (378.5 l) of water. These fertilizers are alternated until the buds are 0.5 inch (1.25 cm) long. From this point until the plants are sold, a weekly application is made of calcium nitrate at the rate of 2 pounds (0.91 kg) in 100 gallons (378.5 l) of water.

Ammonium sulfate should not be used as a source of nitrogen. This material has an acid reaction that reduces the pH. A mixture that is neutral in reaction is one part of ammonium sulfate and four parts of calcium nitrate. This mixture is applied at the rate of 12 ounces (0.34 kg) in 100 gallons (378.5 l) of water at every irrigation; or 1.5 pounds (0.68 kg) in 100 gallons (378.5 l) every week; or 3 pounds (1.36 kg) in 100 gallons (378.5 l) every 2 weeks.

If the pH of the medium falls below 6.0, a limewater solution is applied. One-half to 1 pound (0.23 to 0.45 kg) of hydrated lime is dissolved in 100 gallons (378.5 l) of water. The mixture is stirred thoroughly and is allowed to settle overnight. When the mixture is to be used, it is not stirred. The clear liquid is applied in place of a regular irrigation. Normally, one treatment is enough to raise the pH of most media above 6.0.

Watering

Lilies grow best when the root ball is kept moist but not too wet. The plants are well watered when they are potted. Then they are allowed to dry out between waterings. Too much water can kill the roots quickly. In addition, a wet medium promotes root disease organisms. A Chapin watering system can be used to advantage on a lily crop.

Light, Temperature, and Height Control

One of the major problems a grower faces is controlling the height of lilies. The lilies are either too tall or too short. The use of growth retardants has become an important factor in controlling the plant height. A number of other factors also affect the growth of the plants.

Light Intensity

One of the most important environmental factors in controlling the height of the plants is the light intensity. If the plants are grown under dirty glass or in conditions that reduce the light intensity, they will grow tall. Crowding or shading also results in tall plants. In addition, a low light level reduces the amount of photosynthesis in the plant, resulting in poor leaf color. Lilies must be grown under full light intensity to obtain the highest quality plants.

The use of artificial light or black cloth can control the height of the plants. If the plants are given 18-hour days, they are twice as tall as those given 9-hour days. Tall lilies can be obtained by lighting them from 10 P.M. to 2 A.M. with 10 to 20 fc (108 to 216 lux) of incandescent light. For 'Ace' lilies, the lights are used for a period from 35 to 75 days after planting. This is the period when the 'Ace' cultivar is most responsive to a long-day treatment.

The black cloth treatment must be used from 5 P.M. to 8 A.M. daily to keep the plants short. For 'Ace' lilies, the treatment is given for a period from 35 to 95 days after planting.

Temperature

Lilies are forced to flower by keeping them at a night temperature of 63°F to 65°F (17°C to 18.5°C) and a day temperature 10°F (8°C) higher. At these temperatures, the crop will flower in about 120 days. At times, the crop may not grow according to the schedule. At other times, a period of bright, sunny weather in spring may bring the crop along too fast. A late Easter is always a problem. The grower may find it necessary to hold the lilies back to keep them from blooming too soon.

As stated earlier, Easter lilies require 120 days of forcing time in the greenhouse, but often there is less than 90 days because Easter falls on a different day each year. Liturgically, Easter Sunday falls on the first Sunday that follows the first full moon after the vernal equinox, which is any day between March 22 and April 25. Most growers target their ship date the week before Palm Sunday for retail sales the following week. Easter lilies have little value the Monday after Easter.

Range of Easter Dates and Key Forcing Dates		
Easter Classification	First Sunday of Lent	Palm Sunday
Early: March 26	February 13	March 19
Early: April 1	February 19	March 25
Medium: April 7	February 25	March 31
Medium: April 13	March 2	April 6
Late: April 18	March 7	April 11
Late: April 22	March 11	April 15

Lilies are very sensitive to temperature changes. Growth slows when the temperature decreases and speeds up as the temperature increases (Figure 26-9). There are limits to the temperature ranges at which lilies can be grown, and the grower must observe these limits.

If the crop is growing too fast, the temperature is lowered to a minimum night temperature of 55°F (13°C). During the day, the crop should be kept as cool as possible. In bright, sunny weather the daytime temperature may become very high. The plants can be kept cool by ventilating them. However, they must not be chilled. Shading is not used to reduce the temperature because the plants will stretch.

When the crop is behind schedule, the grower must increase the temperature to speed up the growth. As long as the light intensity is good and the humidity is high, the temperature during the day may be allowed to go to 85°F (29.5°C). When the plants are grown at a warm temperature, they must be syringed from overhead several times a day to keep the relative humidity high.

A sweating procedure may be used if the crop is far behind schedule. A plastic tent is placed over the plants. The air temperature inside the tent is allowed to go to 90°F (32°C) during the day. The plastic keeps the humidity around the plants at a high level. Sweating is a harsh treatment, which is used as a last resort only.

Regardless of the method used to hasten growth, the night temperature must not exceed 68°F to 70°F (20°C to 21°C). High night temperatures waste carbohydrates made during the day.

The extreme temperature treatments described for slowing or hastening growth are not to be used until the flower buds are at least 1 inch (2.5 cm) long. At this stage of development, they normally will not abort. Any loss of flower buds downgrades the quality of the plants.

The use of DIF is very effective in controlling lily plant height (see Chapter 12).

If the crop is being grown at a night temperature of 60°F to 63°F (15.5°C to 17.0°C), the flower buds are 0.75 inch (1.87 cm) long 6 weeks before Easter. At 2 weeks before Easter, they are 2.25 inches (5.7 cm) long and are starting to bend at an angle of 90 degrees to the stem. Five to 7 days before Easter they are 5 to 5.5 inches (12.5 to 13.75 cm) long. One day before they begin to open, they are puffy and white, known as the puffy white stage (Figure 26-10). It is at this point that the plants are placed in cold storage if they are 10 to 14 days early for sales.

FIGURE 26-9 These two Easter lily plants received the same precooling and the same time in the greenhouse. The shorter plant on the left came from a cool area of the greenhouse and the plant on the right came from a warmer area of the greenhouse.

FIGURE 26-10 The plant to the right has flower buds that are at what is known as the puffy white stage. It is at this point that the plants are placed in cold storage if they are 10 days to 2 weeks early for sales. The plant to the left needs to be in the greenhouse for a while longer.

Scheduling

Planting to Shoot Emergence

The first stage involves the bulb being panned and covered by the medium. Throughout the production period, the controlling factor is temperature. During the initial stages, as the bulb is developing a root system and shoot growth begins, the minimum substrate temperature is between 60°F and 63°F (15.6°C and 17.2°C) but should not exceed 70°F (21°C). Shoot emergence should take 2 weeks and should occur in early January.

Emergence to Flower Bud Initiation

During this period, Leaves are unfolding. It is at the terminal apex where flower buds are initiated. The night temperature at this point should be between 60°F and 63°F (15.6°C and 17.2°C) and the day temperatures should be kept cool. High day temperatures can delay flower initiation, which typically occurs around January 19.

Flower Bud Initiation to Visible Bud

During this period, one-half of total stem length will occur and leaves initiated below the flower buds will unfold exposing the flower buds. This stage should occur by the first Sunday of Lent.

Average Temperature Effect on Visible Bud to Flower		
Temp (°F)	VB to Flower (days)	Decrease in Days from VB-F due to 5°F Increase
55	42	–
60	38	4
65	34	4
70	31	3
75	27	3
80	25	2
85	24	1

Visible Bud to Open Flower

The time from visible bud to open flower is the time to fine-tune development and target sales for the week or so prior to Easter.

More advanced plants can be moved to cooler greenhouse sections and slower plants are moved to warmer greenhouse sections. The key is to market Easter lilies when older flower buds are in the puffy white stage. The time from visible bud to puffy white stage is 35 days at 63°F (17°C). The ideal temperatures should be between 60°F and 70°F (15.6°C and 21°C) as temperatures greater than 70°F (21°C) reduce plant quality. This stage should be reached by Palm Sunday.

Leaf Counting

Easter lilies are determinate plants and once the apical meristem initiates flower buds, no new leaves will be formed. The total number of leaves on the plant compared to the leaves that have not unfolded from the apex is used to determine the stage of development and scheduling. The leaf unfolding rate can be modified by temperature. Leaf counting begins at the time of flower initiation and continues up to visible bud.

The leaf counting procedure begins around January 15–20 after flower bud initiation occurs as well as the formation of stem roots. First, remove substrate to determine that stem roots have formed. Next, dissect and examine the shoot apex from several plants to determine that flower initiation has occurred. The vegetative apex is round and smooth and the floral apex will have a distinctive mound. The leaf number is fixed after flower initiation has occurred. Finally, count leaves on several plants to determine how many leaves are yet to unfold. The leaf unfolding rate is controlled by temperature (Figures 26-11 through 26-13).

Storing Plants

Lilies that are ready to flower can be held in a refrigerator at 31°F (−0.5°C) without lights for 10 to 14 days (Figure 26-14). The plants can be placed in storage when the flower bud closest to opening is puffy white and ready to

FIGURE 26-11 The Easter lily leaves to the right have unfolded and the leaves to the left are yet to unfold.

FIGURE 26-12 A closer view of the flower buds of an Easter lily stem.

FIGURE 26-13 Developing flower primordia of an Easter lily plant.

FIGURE 26-14 These Easter lily bulbs are at the right stage to go into storage at 31°F (0.5°C).

Effect of Average Daily Temperature on Leaf Unfolding Rate	
Leaves Uunfolded per Day	Average Daily Temperature (°F)
1.0	53
1.1	55
1.2	57
1.3	59
1.4	61
1.5	63
1.6	65
1.7	67
1.8	69

crack open. The plants are watered well and mist sprayed with fungicide to prevent *Botrytis* gray mold. The plants are removed from storage a day before they are to be sold. The plants cannot be moved directly from storage into a greenhouse where the light intensity is high. They should be warmed gradually to room temperature in a shaded location.

Growth Retardants and Height Control

One method of controlling the height of lilies is to vary the environmental conditions of temperature and light. Another method of height control involves the use of chemical growth retardants. Whichever method of height control is used, graphical tracking as described in Chapter 25 for poinsettias also should be used. One such growth retardant is A-REST®, a product of SePro Corporation. This substance can be used both as a soil drench and as a foliar spray.

The amount of retardant applied depends upon the cultivars. 'Ace' and 'Nellie White' need less treatment than 'Georgia' and 'Creole' lilies. Plants grown from small bulbs need less retardant than those from larger bulbs. Growers whose lilies are always too tall may use slightly more retardant than is recommended for a plant of average height.

If pine bark is a large part of the mix, more retardant must be used. The bark takes up some of the chemical before it reaches the roots. For plants in a pine bark medium, a spray treatment is preferred to a drench.

Drench Treatment

A-REST® is applied when the stems are 3 to 6 inches (7.5 to 15 cm) tall, as measured from the soil to the top of the plant. Treatment with A-REST® reduces the distance between the leaves (length of the internode).

The chemical is sold as a 0.026 percent solution. This solution is diluted at the rate of 1 pint (16 fluid ounces or 480 ml) in 24 gallons (90.8 l) of water. When the solution is applied at the rate of 6 fluid ounces (180 ml) per pot, 500 pots can be treated.

The soil must be slightly moist when the treatment is made. The moist soil ensures that the solution will wet the entire soil volume and will not run through to the bottom of the pot. If the soil is dry, some of the solution will run out of the pot and the rest will not be spread evenly among the roots.

Spray Treatment

A spray or foliar treatment is made using enough solution to cover 100 square feet (9.3 m²) of area. The plants should be spaced so that the leaves are touching. The plants are sprayed when they reach a height of 5 to 7 inches (12.5 to 17.5 cm). Each plant is sprayed with 10 ml of a solution diluted to a concentration of 50 ppm of active ingredient. That is, 1 pint (473 ml) of 0.026 percent solution is diluted in 4 pints of water to give a solution with a concentration of nearly 50 ppm. The foliage is sprayed until it is covered but there is no runoff. Too much retardant can have an undesirable effect on plant growth.

Problems

Cultural Disorders

There are a number of reasons for the failure to form flower buds: The bulbs are immature when they are harvested, the cold storage is inadequate, the light intensity is too low, the temperature is too high, and the plants are too dry at the critical bud formation stage.

Blasting of the buds may result from a lack of water under high-stress conditions, and a high temperature combined with a low relative humidity during forcing, or lack of calcium.

Bud splitting is caused by aphids feeding on the young growing buds.

Diseases

Root rot is caused by two types of fungi, *Rhizoctonia* and *Pythium*. *Pythium* usually develops when the medium is kept too wet. *Rhizoctonia* occurs more commonly as a dry rot. The best preventive measure is careful attention to sanitation and cleanliness. All media, pots, benches, and other items used in the propagation of the bulbs must be chemically or steam pasteurized. The bulbs are planted directly from the crate and are not piled on a dirty work table or the floor. All pots are filled to the same depth to allow the same amount of room for water and the fertilizer solution. As soon as the bulbs are potted, the containers are drenched with a recommended fungicide.

Botrytis gray mold caused by *B. cinerea* may cause petal blight or spotting. Good air movement and ventilation are effective in controlling this fungus. *Botrytis* is rarely seen in a well-ventilated greenhouse. To prevent *Botrytis* from starting while the plants are in cold storage, the flower buds can be given a light, mist type spray of the recommended chemical before they are placed in storage.

Several viruses are known to infest lilies. Unfortunately, little can be done to control them because the chemicals that kill the viruses also kill the cells of the plant. Tissue culture methods may result in virus-free plants. However, it is doubtful that the plants will remain disease free because aphids carry the viruses and spread them from plant to plant. The most important step that can be taken to control the viruses is to control the aphids.

Three viruses that attack lilies are named according to the symptoms they produce on the leaves: fleck, streak, and curly-stripe. The rosette virus causes the plant to remain short and develop a rosette appearance, similar to a dandelion.

Insects

The major insect pest is the aphid. Several systemic poisons are effective in controlling aphids.

A fungus or root gnat may appear in greenhouses used to grow lilies. This insect does not harm lilies directly because it feeds on dead organic matter in the medium and on the benches. The insect can be controlled by maintaining good sanitation and cleaning up any pools of water that could support algae growth. If the insects appear, the pots are drenched with a recommended insecticide.

Lower leaf yellowing is a condition caused by low fertility and cloudy weather, but also may occur during cold storage of plants. The PGR Fascination, which is a blend of cytokinin and gibberellins, effectively delays the aging of leaves. This leaf yellowing occurs from the bottom of the plant up and accelerates the aging of flowers. The use of Fascination® is now a standard treatment during the final stages of Easter lily production.

NARCISSUS (Daffodils) � *Narcissus species, Amaryllidaceae*

Flower growers force daffodils that are grown in Washington State and Oregon and in the Netherlands. The Netherlands Flower Bulb Institute and the Dutch Bulb Exporters Association have sponsored research on a number of bulbous crops. The results of this research enable growers to have a better understanding of how bulbs respond to environmental conditions. This means that growers can design more accurate programs for the handling of bulbs in both the storage and forcing stages.

Daffodils are forced as both potted plants and cut flowers.

Cut Daffodils

Plants are flowered from mid-January to early April for cut flowers. The best-quality flowers are obtained when the plants are forced in a rooting room. However, cold frames and buried storage can be used. The following guidelines for forcing are based on the use of a rooting room.

Table 26-3

Periods for Flowering Narcissus for Cut Flowers*		
Period	Approximate Time of Blooming	Main Holiday
1	January 12-24	—
2	January 25-February 9	—
3	February 10-March 3	Valentine's Day
4	March 4-19	—
5	March 20-April 8	Early Easter

* Adapted with modifications from the *Holland Bulb Forcers Guide*, 1996.

The information is presented for five flowering periods (Table 26-3). These periods can be varied, but they represent times that fit nicely with other schedules.

When narcissus bulbs are dug from the field, the flower parts are partially formed. The bulbs are stored at 90°F (32°C) for 4 days to enable the flower parts to develop fully. If the bulbs are stored too long at this temperature, both the flower size and the stem length are greatly reduced.

The bulbs are then held at 55°F (13°C) until the proper time for shipping. The bulbs are usually shipped to arrive in early September. As soon as they are received, the bulbs are ventilated, inspected, and stored at the proper temperature. Bulbs that are to be precooled are held at 48°F (9°C). The bulbs are stored at 55°F (13°C) if they are not to be planted or precooled for some time.

Generally, bulbs used for forcing are double-nosed bulbs. They are listed as Double Nose I and Double Nose II in suppliers' catalogs. The Double Nose I type produces more flowers than the Double Nose II bulb.

Precooled Bulbs

Bulbs to be precooled are planted in a 1:1:1 by volume mixture of soil, peat moss, and perlite. If the pH of the mixture is about 6.0, fertilizer is not required. Narcissus to be grown for cut flowers need room for good root development. Therefore, they are planted in bulb flats that are 4 to 5 inches (10 to 12.5 cm) deep. The bulbs are planted so that they just touch each other. The nose of the bulb is about 1 inch (2.5 cm) below the surface of the medium. The flats are then placed in the rooting room. Table 26-4 lists starting dates in the treatment schedule for precooled and nonprecooled bulbs.

Bulbs may be precooled dry in the shipping case. Brokers or dealers often use this method. The bulbs are held at 50°F (10°C) for a minimum of 6 weeks. They are shipped to reach the grower by mid-October. The bulbs are planted in flats, which are placed in a refrigerator at 50°F (10°C), or they are buried outside. Another 10 weeks of cold is required before the bulbs can be brought in for forcing. If the soil temperature is too warm when they are buried outside, the precooling effect may be partly offset. As a result, the flower quality is poor, and blind shoots may develop.

By placing the bulbs in a rooting room, the grower has more control of the storage temperature. From the time the bulbs are placed in the rooting room, the temperature is held at 48°F (9°C) until December 1–5. The temperature is held at 41°F (5°C) from December 1–5 to January 1–5. This is followed by a

Table 26-4

Scheduled Dates for Precooling, Planting, and Forcing of Narcissus for Cut Flowers[1]					
Flower Period	Start Precooling	Plant	Dates into Greenhouse	Approximate Flowering Date	Weeks of Cold
1 PC[2] at 48°F (9°C)	August 26–31	October 1–7	December 21	January 12	16.5
			December 28	January 16	17.5
			January 2	January 22	18
2 PC at 48°F (9°C)	September 1–5	October 1–7	January 5	January 25	17.5
			January 12	February 1	18.5
			January 19	February 7	19.5
3 NP[3]	None	September 20–25	January 20	February 10	17
			January 27	February 17	18
			February 4	February 24	19
4 NP	None	October 10–15	February 15	March 4	18
			February 22	March 10	19
			March 1	March 16	20
5	None	October 25–30	March 8	March 20	18.5
			March 15	March 27	19.5
			March 22	April 2	20.5
			March 29	April 8	21.5

1. Adapted with modifications from the *Holland Bulb Forcers Guide*, 1996.
2. PC = Precooled bulbs.
3. NP = Nonprecooled bulbs.

further period at 32°F to 35°F (0°C to 2°C) until the flats are brought in for forcing.

If cut flowers are to be harvested, the stem should be at least 14 inches tall when the flowers reach the **gooseneck stage**. The stem length is influenced by several factors, including the cultivar used, the cold treatment given, the light intensity to which the bulbs are first exposed when they are removed from storage, and the greenhouse forcing temperature.

As the plants grow, the stems stretch even in the darkness of the storage room. If the grower finds that the stems are slightly short when the plants are brought into the greenhouse for forcing, they can be covered with sheets of newspaper for a couple of days. The reduced light intensity promotes stem growth.

A greenhouse temperature of 55°F (13°C) results in the longest stem length. If the day temperature is above 60°F (15.5°C), the plants are forced into bloom quickly at the expense of stem length.

Harvesting for Market

Narcissus can be harvested for market when the flower bud is at the gooseneck stage of development. They should not be harvested earlier than this because they will not open properly. Harvesting them beyond the gooseneck stage reduces their useful life for the consumer.

The flowers are grouped into bunches of a dozen blooms each. The bunches are stored dry in an upright position so that the stem does not bend. The blooms can be held for 1 week when they are stored at 32°F to 35°F (0°C to 2°C). The relative humidity should be no more than 90 percent.

Some people are sensitive to the sap (juice) that drips from cut narcissus and must wear rubber gloves when working with them. The sap spots other flowers if it is allowed to drip on them.

Suitable Cultivars

The cultivars suitable for forcing as cut flowers are given in Table 26-5.

Pot Daffodils

Daffodils grown in pots have a longer flowering season than the cut flowers because the plants can be sold when the flower buds are still in the **pencil stage**. Less time is required to reach this stage as compared to the gooseneck stage. Suggested flowering periods are given in Table 26-6.

As soon as the bulbs are received, they are ventilated, inspected, and stored. A precooling temperature of 48°F (9°C) is used. If the bulbs are not to be planted immediately, they are held at 63°F (17°C).

Table 26-5

Narcissus Cultivars and Type of Bulb Needed for Forcing as Cut Flowers[1]			
	Flowering Period and Type of Bulb Needed		
Class 'Cultivar'–Color	1	2	3, 4, 5
Large Cupped			
'Carlton'–all yellow	PC[2]	PC	NP[3]
'Flower Record'–yellow and orange cup, white perianth	–	PC	NP
'Fortune'–yellow to orange cup, yellow perianth	PC	PC	NP
'Ice Follies'–light yellow cup, white perianth	PC	PC	NP
'Mercato'–deep orange cup, white perianth	–	PC	NP
'Yellow Sun'–all yellow	PC	PC	NP
Small Cupped			
'Barrett Browning'–deep orange cup, white perianth	PC	PC	NP
Trumpet			
'Brigton'–all yellow	–	–	NP
'Dutch Master'–all yellow	–	PC	NP
'Golden Harvest'–all yellow	PC	PC	NP
'Joseph MacLeod'–all yellow	PC	PC	NP
'Unsurpassable'–all yellow	PC	PC	NP

1. Adapted with modification from the *Holland Bulb Forcers Guide*, 1996.
2. PC = Precooled bulbs.
3. NC = Nonprecooled bulbs.

Table 26-6

Periods for Flowering Narcissus for Potted Plants*		
Period	Approximate Time of Blooming	Main Holiday
1	December 22–31	Christmas
2	January 1–16	–
3	January 17–February 2	–
4	February 3–24	Valentine's Day
5	February 25–March 21	–
6	March 22–April 9	Early Easter
7	April 10–20	Late Easter

* Adapted with modifications from the *Holland Bulb Forcers Guide,* 1996.

The recommended potting soil was once a 1:1:1 mixture by volume of soil, peat moss, and perlite. Most growers now use a peat-lite medium. The bulbs should be placed in bulb pans or in three-quarter size pots. White plastic pots provide a nice color contrast to the green foliage. The plants should be 10 to 14 inches (25 to 35 cm) tall when the flower buds reach the gooseneck stage.

Two rooting room temperature programs are used for potted daffodils. The cool temperatures given at an earlier date help bring the bulbs to the proper stage for early forcing. Table 26-7 lists the dates for the various stages in the forcing program.

After the plants are brought into the greenhouse, they are forced at a temperature of 60°F to 63°F (15.5°C to 17°C). This temperature range results in shorter plants than are obtained with cooler forcing temperatures. Temperatures above 65°F (18.5°C) are to be avoided at all times.

Cultivars

The cultivars that are preferred for potted plants are given in Table 26-8. Other cultivars tend to grow too tall and should not be used for potted plants.

Stage of Marketing

Potted daffodils are ready for market when the flower buds reach the pencil stage of growth. In other words, the flower bud is beginning to swell. There is a faint color in the bud at this stage. The stem is starting to bend from an upright (vertical) position to the right angle gooseneck position. Flowers sold at this stage of development give the customer a longer blooming period.

If the crop matures ahead of schedule, it must be placed in cold storage or its market value will be lost. The plants are placed in storage before the flowers show any color. To reduce the possibility of *Botrytis* starting during storage, the crop is mist sprayed with a fungicide. The foliage must be dried thoroughly before the crop is placed in the refrigerator. The storage temperature is 32°F to 35°F (0°C to 2°C). The plants can be held for 10 to 14 days.

Table 26-7

				Scheduled Dates for Treatment Program of Narcissus for Potted Plants[1]		
Flower Period	Start Precooling	Planting Date	Rooting Room Temperature Sequence[2]	Date into Greenhouse	Approximate Flowering Date	Weeks of Cold
1 PC[3] at 48°F (9°C)	August 26–31	October 1–7	A	December 1	December 22	13.5
				December 8	December 27	14.5
2 PC at 48°F (9°C)	August 26–31	October 1–7	A	December 10	January 1	15
				December 17	January 7	17
				December 24	January 13	17
3 PC at 48°F (9°C)	September 1–5	October 1–7	A	December 31	January 17	17.5
				January 7	January 23	18
				January 13	January 28	19
3 NP[4]	None	September	A	December 31	January 17	13.5
				January 7	January 23	14.5
				January 13	January 28	15.5
4 NP	None	October 1–5	A	January 16	February 3	15
				January 26	February 11	16.5
				February 4	February 18	17.5
5 NP	None	Oct. 28–Nov. 2	B	February 12	February 25	15
				February 20	March 5	16
				March 2	March 15	17
6 NP	None	November 8–12	B	March 10	March 22	17
				March 17	March 29	18
				March 25	April 5	19
7 NP	None	November 8–12	B	April 2	April 10	22
				April 7	April 15	22.5
				April 14	April 20	23.5

1. Adapted with modifications from the *Holland Bulb Forcers Guide*, 1996.
2. Rooting Room A Temperature Sequence:
    Planting to November 5–10, 48°F (9°C); to January 1–5, 41°F (5°C); then 32°F–35°F (0°C–2°C).
   Rooting Room B Temperature Sequence:
    Planting to December 1–5, 48°F (9°C); to January 1–5, 41°F (5°C); then 32°F–35°F (0°C–2°C).
Note: The change from 48°F (9°C) is made when the roots grow out of the bottom of the pot. The change from 41°F (5°C) to 32°F (0°C) is made when the shoots are about 1.5–2 inches (3.75–5 cm) long.
3. PC = Precooled bulbs.
4. NP = Nonprecooled bulbs.

The plants are removed from storage 2 days before they are to be sold. During this time, the medium warms up and the plants begin more active growth.

Growth Retardants and Height Control

The same chemical, ethephon (Florel®), and specific recommendations given for controlling the height of hyacinths are also used for controlling the height of potted daffodils.

Table 26-8

Narcissus Cultivars and Type of Bulb Needed for Forcing for Potted Plants[1]					
Class 'Cultivar'—Color	**Flowering Period and Type of Bulb Needed**				
	1	**2**	**3**	**4, 5**	**6, 7**
Trumpet 'Dutch Master'—yellow	–	PC[2]	PC	NP[3]	NP
'Explorer'—yellow	–	PC	PC	NP	NP
'Magnet'—yellow trumpet, white perianth	–	–	–	NP	NP
'Mount Hood'—white	–	–	PC	NP	NP
Large Cupped 'Carlton'—yellow	–	PC	PC	NP	–
'Flower Drift'—white with orange center	–	–	PC	NP	NP
'Ice Follies'—cream cup and white perianth	–	–	PC	NP	NP
Small Cupped 'Barrett Browning'—deep orange cup and white perianth	–	–	PC	NP	NP
Double 'Bridal Crown'—white with orange center	–	PC	NP	NP	NP
Small Trumpet 'February Gold'—yellow	PC	PC	NP	NP	NP
'Peeping Tom'—yellow	PC	PC	NP	NP	–
'TeteaTete'—yellow	PC	PC	NP	NP	NP
Tazetta 'Geranium'—white with orange cup	–	–	–	NP	NP
Split Corona 'Carrata'—all white	PC	PC	NP	NP	NP
'Chanterelle'—all yellow	–	–	–	NP	NP
Jonquilla 'Pippit'—lemon-yellow perianth with white cap	–	–	–	NP	NP

1. Adapted with modifications from the *Holland Bulb Forcers Guide*, 1996.
2. PC = Precooled bulbs.
3. NP = Nonprecooled bulbs.

Problems

Diseases

The major problem affecting daffodils is a *Fusarium* basal rot. This disease is carried from production fields. The basal plate of each bulb must be carefully inspected for soft tissue. Any bulbs suspected of being infected are discarded.

TULIPS ⊠ *Tulipa species and hybrids, Liliaceae*

It is estimated that more tulips are grown per square mile in the Netherlands than in any other country in the world. The Netherlands is the center of European commercial production. Most tulips grown in the United States come from Michigan, Oregon, and Washington.

Flower Formation

Unlike some other bulbous species, the flower parts of the tulip are present but not formed when the bulbs are dug from the field. The bulbs have to go through a series of growth stages before the floral parts are developed enough to permit the bulbs to be given the cold treatment. The bulbs are stored at a temperature of 63°F (17°C) to hasten the formation of the floral parts. If the tulips are subjected to high temperatures at this time, the floral parts may be killed. Heat damage may occur during shipment from the producer to the grower. When a shipment of tulips is received, the grower should cut a bulb in half lengthwise. If the center of the bulb is brown, the bulbs have been exposed to high temperatures. If the flower parts are dead, the bulbs cannot be forced. The grower should file a claim with the carrier.

Bulbs must be cut open frequently to determine when **Stage G** in the development of the bulb is reached. All of the flower parts are formed in Stage G, and the bulbs are ready for the cool temperature treatment.

To determine if Stage G has been reached, a sharp knife is used to cut the bulb through the center. A drop or two of ink can be placed on the tissue to mark the different parts. An experienced eye aided by a 10× hand lens is required to pick out the floral parts. For bulbs arriving in late September or early October, the shoots within the bulbs should be nearly 1 inch (2.5 cm) long. With the help of the magnifying lens, these shoots are relatively easy to pick out. Bulbs received late in October may even have visible shoots.

Certain cultivars require 10 days more of storage at 63°F (17°C) after Stage G is reached to prevent blasting when the flowers are forced. These cultivars are 'Apeldoorn,' 'Attila,' 'Bing Crosby,' 'Cantor,' 'Christmas Marvel,' 'Demetor,' 'Dix's Favourite,' 'Dreaming Maid,' 'Gander,' and 'Paul Richter.'

Precooling

Tulips are precooled in a rooting room in the same manner as other bulbous crops. Plants precooled for forcing as cut flowers or potted plants are held at 44.5°F (7°C). A temperature of 48°F (9°C) is used for bulbs that are to be forced as potted plants in period 2 and cut flowers in periods 2 and 3. The precooling period may be as long as 18 to 20 weeks.

For cut flowers, another precooling treatment program can be used. The bulbs are cooled for 12 weeks in the shipping cases at 41°F (5°C). They are then planted directly in greenhouse benches where they are forced into bloom. If the bulbs are not to be precooled immediately, they are stored at 55°F (13°C). Storage at this temperature helps to produce long stems.

Bulbs scheduled for forcing as potted plants are stored at 63°F (17°C) before cooling. This temperature helps to keep the stems short.

Cut Flowers

Tulips grown for cut flowers are programmed to flower from early January to late May. The periods of flowering are given in Table 26-9.

Tulip bulbs are graded according to their circumference (in centimeters). After the bulbs are removed from the field, they are cleaned. The bulbs are sorted by placing them on a series of screens with holes of different sizes. A bulb that passes through a 4.7-inch (12-cm) hole but is held on a 4.3 inch (11-cm) screen is graded as 4.3-4.7 inch (11–12 cm). Most tulip bulbs used for forcing are 4.7 inch (12 cm) and larger in size. In some cases, 4.3-4.7 inch (11–12 cm) bulbs may be used. Larger bulbs ensure the production of larger flowers.

Table 26-9

Periods for Flowering Tulips for Cut Flowers[1]		
Period	Approximate Time of Blooming	Main Holiday
1	January 3–22	–
2	January 23–February 2	–
3	February 3–23	Valentine's Day
4	February 24–March 15	–
5	March 16–April 4	Early Easter
6	April 5–25	Late Easter
7	April 26–May 3	Mother's Day

[1]Adapted with modifications from the *Holland Bulb Forcers Guide*, 1996.

A rooting room is used to provide the controlled temperatures needed. Tulips can be stored in cold frames outdoors, but a rooting room is preferred because the potted bulbs can be inspected frequently and the temperature can be controlled accurately. The rooting room schedule for the various flowering periods is given in Table 26-10.

Media and Fertilization

Tulips, like lilies, benefit from additions of fertilizer. The fertilization program begins with the planting medium. Formerly, a 1:1:1 mixture of equal parts by volume of soil, peat moss, and perlite was common, but most growers now use a peat-lite mix. Superphosphate is added at the rate of 2 pounds (1.2 kg) of regular superphosphate or 1 pound (0.6 kg) of treble superphosphate to each cubic yard (cubic meter) of mix. Ground limestone is added to bring the pH to 6.0. For the peat-lite mixes, at least 7.5 pounds (4.4 kg) of limestone are added to a cubic yard (cubic meter) of mix. Tulips are subject to a disease known as **tulip topple**. Calcium is an important factor in preventing this physiological stem disease that causes the flower to fall over.

If superphosphate alone is added to peat-lite media, the rate of tulip topple is twice as great as in media with added limestone or with limestone and superphosphate. The addition of both materials results in shorter stems as compared to the stem length when each material is used alone.

The bulbs are planted in flats 4 to 5 inches (10 to 12.5 cm) deep. The tip of the bulb is just covered with the medium to give the roots the maximum room for growth.

While the plants are being forced in the greenhouse, the type of fertilizer applied is alternated. One week the plants are fertilized with calcium nitrate at a rate of 2 pounds (0.9 kg) in 100 gallons (378.5 l) of water; the next week a complete analysis 20-10-20 fertilizer is applied at the rate of 14 ounces (397 g) in 100 gallons (378.5 l) of water.

Greenhouse Forcing

Plants to be forced for the first two flowering periods may require stretching by being exposed to a low light intensity. For several days after the plants are placed in the greenhouse, they are stored beneath the benches or are covered

Table 26-10

Flower Period	Start Precooling	Planting Date	Rooting Room Temperature Sequence[2]	Date into Greenhouse	Approximate Flowering Date	Weeks of Cold
1 PC[3] at 44.5°F (7°C)	August 23–28	October 1–7	A	December 10	January 3	15.5
				December 17	January 10	16.5
				December 24	January 17	17.5
2 PC at 48°F (9°C)	August 26–31	October 1–7	A	December 30	January 23	17.5
				January 7	January 30	18.5
3 PC at 48°F (9°C)	September 3–7	October 6–10	A	January 10	February 3	18
				January 17	February 10	19
				January 24	February 17	20
3 NP[4] at 55°F (13°C)	None	September 18–22	A	January 10	February 3	16
				January 17	February 10	17
				January 24	February 17	18
4 NP at 55°F (13°C)	None	September 25–30	A	February 2	February 24	18
				February 9	March 2	19
				February 16	March 10	20
5 NP at 55°F (13°C)	None	Oct. 16–20	B	February 23	March 16	18
				March 2	March 23	19
				March 9	March 30	20
6 NP at 55°F (13°C)	None	November 6–11	B	March 15	April 5	18
				March 23	April 12	19
				March 30	April 19	20
7 NP at 55°F (13°C)	None	November 6–11	B	April 6	April 26	21
				April 13	May 3	22

Title row: **Scheduled Dates for Treating of Tulips for Cut Flowers[1]**

1. Adapted with modifications from the *Holland Bulb Forcers Guide*, 1996.
2. Rooting Room A Temperature Sequence:
    Planting to November 5–10, 48°F (9°C); to January 1–5, 41°F (5°C); then 32°F–35°F (0°C–2°C).
  Rooting Room B Temperature Sequence:
    Planting to December 1–5, 48°F (9°C); to January 1–5, 41°F (5°C); then 32°F–35°F (0°C–2°C).
Note: The change from 48°F (9°C) is made when the roots grow out of the bottom of the pot. The change from 41°F (5°C) to 32°F (0°C) is made when the shoots are 2 inches (5 cm) tall.
3. PC = Precooled bulbs.
4. NP = Nonprecooled bulbs.

with newspaper. The reduced light intensity causes the stems to stretch to the required height. Plants forced for later flowering periods do not require this treatment. The stems reach the desired length in the rooting room.

The plants are forced at a night temperature of 63°F (17°C) and 68°F (20°C) during the day. Higher temperatures cause weak stems, and lower temperatures slow the rate of flower development.

Harvesting for Market

When buds are developed and the stems are at least 14 inches (35 cm) long, the tulips can be harvested. Tulips are cut in the bud stage because they are less easily damaged in handling. In addition, they have a longer lasting life for the consumer if they are cut at this stage.

To harvest cut flowers, most growers pull the entire plant. They slide a sharp blade along the side of the stem and cut through the bulb. In this way, an extra inch (2.54 cm) or more of stem length is gained. Tulip stems begin their growth just above the basal plate.

Cut tulips are graded and grouped in bunches of a dozen blooms. They are placed in water only if they are being stored for 24 to 48 hours and are going to market immediately. They can be stored dry for up to 1 week. The graded flowers are wrapped tightly in tissue paper and stored in a horizontal position.

If the market is slow, the bulb is not removed and the tulips are stored dry in an upright position to keep the stems from bending. They can be held for up to 2 weeks. The proper storage temperature for all three methods is 32°F to 35°F (0°C to 2°C). The relative humidity for tulips in storage should be greater than 90 percent.

Cultivars

Cultivars used for forcing as cut tulips are listed in Table 26-11.

Table 26-11

Selected Tulip Cultivars and the Type of Bulb Needed for Forcing as Cut Flowers[1]								
		Flowering Period and Type of Bulb Needed						
Color	Cultivar	1	2	3	4	5	6	7
Red	'Albury'	–	–	–	–	–	NP[2]	NP
	'Apeldoorn'	–	–	PC[3]	NP	NP	–	–
	'Bing Crosby'	–	–	PC	NP	NP	NP	–
	'Diplomate'	–	–	–	–	NP	NP	–
	'Dix's Favourite'	PC	PC	NP	NP	–	–	–
	'Frankfurt'	–	–	–	–	NP	NP	–
	'Henry Dunant'	–	–	PC	NP	NP	NP	–
	'London'	–	–	PC	NP	–	–	–
	'Oxford'	–	–	–	NP	NP	NP	–
	'Parade'	–	–	–	NP	NP	NP	–
	'Paul Richter'	–	PC	PC	NP	NP	–	–
	'Red Gander'	PC	PC	NP	NP	NP	NP	–
Pink or Rose	'Anne Claire'	–	PC	PC	NP	NP	NP	NP
	'Albury'	–	–	–	NP	NP	NP	NP
	'Apeldoorn'	–	–	–	–	NP	NP	NP
	'Bing Crosby'	PC	PC	NP	NP	–	–	–
	'Diplomate'	–	–	PC	NP	NP	NP	NP
	'Dix's Favourite'	–	–	–	NP	NP	NP	NP
	'Frankfurt'	PC	PC	NP	NP	NP	–	–
	'Henry Dunant'	–	–	PC	NP	NP	–	–
	'Peerless Pink'	–	–	–	–	NP	NP	NP

Color	Cultivar	Flowering Period and Type of Bulb Needed						
		1	2	3	4	5	6	7
	'Pink Supreme'	–	–	–	–	NP	NP	–
	'Rosario'	–	–	PC	NP	NP	NP	NP
	'Smiling Queen'	–	–	–	–	NP	NP	NP
	'Wim van Est'	–	–	PC	NP	NP	–	–
Yellow	'Bellona'	–	–	PC	NP	NP	–	–
	'Golden Apeldoorn'	–	–	PC	NP	NP	–	–
	'Golden Melody'	–	PC	PC	NP	NP	NP	–
	'Golden Oxford'	–	–	PC	NP	NP	NP	–
	'Jewel of Spring'	–	–	PC	NP	NP	–	–
	'West Point'	–	–	–	–	NP	NP	NP
White	'Hibernia'	–	–	PC	NP	NP	NP	–
	'Kansas'	–	–	–	NP	NP	NP	NP
	'White Triumphator'	–	–	PC	NP	NP	–	–
Lavender	'Attila'	–	–	PC	NP	NP	NP	NP
	'Blue Bell'	–	–	–	–	NP	NP	NP
	'Demeter'	PC	PC	NP	NP	–	–	–
	'Negrita'	–	–	–	–	–	NP	NP
	'Purple Star'	–	PC	PC	NP	NP	–	–
	'Queen of Night'	–	–	–	NP	NP	NP	–
Orange	'Fidelio'	–	–	–	NP	NP	NP	–
Apricot	'Apricot Beauty'	–	PC	PC	NP	NP	–	–
Bicolors Red and white or cream	'Gander's Rhapsody'	PC	PC	NP	NP	NP	NP	–
	'Lucky Strike'	–	PC	PC	NP	NP	–	–
	'Merry Widow'	–	–	PC	NP	NP	–	–
	'Princess Victoria'	–	–	–	–	NP	NP	NP
Red and yellow or cream	'Aladdin'	–	PC	PC	NP	NP	–	–
	'Beauty of Apeldoorn'	–	–	PC	NP	NP	–	–
	'Gudoshnik'	–	–	PC	NP	NP	–	–
	'Kees Nelis'	–	–	PC	NP	NP	NP	–
	'Queen of Sheba'	–	PC	PC	NP	NP	–	–
Lavender and white	'Dreaming Maid'	–	PC	PC	NP	NP	–	–
	'Maytime'	–	–	PC	NP	NP	–	–

1. Adapted with modifications from the *Holland Bulb Forcers Guide*, 1996.
2. NP = Nonprecooled bulbs.
3. PC = Precooled bulbs.

Tulips Precooled at 41°F (5°C)

Cut flowers for Valentine's Day can be obtained using a special method of precooling tulips dry. Very precise conditions are needed: Bulbs must have reached Stage G development (preferably beyond this stage), they must be precooled properly, and they must be planted and forced directly in the greenhouse.

Because of the precise schedule to be followed at each stage of handling, the grower should expect a crop loss of 10 percent when determining the number of flowers needed for market. An additional 5 to 15 percent loss may result if the grower fails to remove the tunic (dark brown skin) from the basal plate of the bulb. This tunic can be removed easily during the last week of storage. At this time the bulb is swollen slightly, and the tunic is loose. The basal plate must not be damaged. Any damage will affect flowering. Any bulbs suspected of infection with *Fusarium* disease are discarded.

Preplanting Treatments

Bulbs forced at 41°F (5°C) for cut tulips must be 4.7 to 5.5 inches (12 to 14 cm) in size. Smaller bulbs do not produce flowers of the necessary quality.

When the bulbs arrive from the producer in September, they are inspected for signs of overheating. If an automatic temperature recording device is included with the shipment and it shows that the bulbs were exposed to temperatures greater than 68°F (20°C), extra precooling is needed.

Between the time the bulbs arrive and the start of precooling, they are stored at 55°F (13°C). The 12-week precooling treatment is started at the proper time. Tables 26-12 and 26-13 list the cultivars that can be used with this production method. The relative humidity in the storage room is held at 80 to 90 percent.

Table 26-12

Scheduled Treatment of 41°F (5°C) Tulips for Valentine's Day[1]			
Cultivars	Storage Temperatue Before Precooling	Start of Precooling at 41°F (5°C)[2]	Planting Date in the Greenhouse
'Diplomate'	All cultivars are stored at 55°F (13°C)	September 20	December 13
'Kingdom'			
'Queen of Sheba'			
'Apeldoorn'		September 24	December 17
'Apeldoorn's Elite'			
'Beauty of Appeldoorn'			
'Don Quichotte'			
'Golden Apeldoorn'			
'Golden Melody'			
'Henry Dunant'			
'Hibernia'			
'London'			
'Oxford'			
'Parade'			
'Bing Crosby'		September 27	December 20
'Gander'			
'Paul Richter'			
'Apricot Beauty'		September 30	December 23
'Demeter'			

1. Adapted with modifications from the *Holland Bulb Forcers Guide*, 1996.
2. Bulbs must be at Stage G before they can be placed in precooling.

Table 26-13

Tulip Cultivars Precooled at 41°F (5°C) for Forcing as Cut Flowers for Valentine's Day[1]			
Color	Cultivar	Approximate Total Stem Length in Inches (cm)	Approximate Days to Flower
Red	'Apeldoorn'	18–20 (46–51)	50
	'Bing Crosby'	16–18 (40–46)	47
	'Diplomate'	15–17 (38–43)	55
	'Henry Dunant'	16–18 (40–46)	50
	'Kingdom'	16–18 (40–46)	55
	'London'	18–20 (46–51)	50
	'Orient Express'	16–18 (40–46)	55
	'Oxford'	18–20 (46–51)	50
	'Parade'	20–22 (51–56)	50
	'Paul Richter'	15–17 (38–43)	47
Rose	'Don Quichotte'	20–22 (51–56)	50
	'Gander'	16–18 (40–46)	47
Yellow	'Golden Apeldoorn'	18–20 (46–51)	50
	'Golden Melody'	15–17 (38–43)	50
White	'Hibernia'	16–18 (40–46)	50
Lavender	'Demeter'	16–18 (40–46)	50
Apricot	'Apricot Beauty'	15–17 (38–43)	43
Bicolor	'Apeldoorn's Elite'	18–20 (46–51)	43
	'Beauty of Apeldoorn'	18–20 (46–51)	50
	'Queen of Sheba'	22–24 (56–61)	50

1. Adapted with modifications from the *Holland Bulb Forcers Guide*, 1996.

Greenhouse Forcing

The planting medium in the ground beds is rototilled or spaded to a depth of at least 8 inches (20 cm) and preferably 10 inches (25 cm). Successful forcing of these special precooled bulbs requires a deep soil. They cannot be forced in bulb flats. Although fair results are obtained by forcing the bulbs in 6- to 8-inch (15- to 20-cm) deep benches, ground beds give the best results.

Limestone is added to bring the pH to at least 5.0. If the medium cannot be tested, 5 to 10 pounds (2.3 to 4.6 kg) of ground limestone are broadcast on 100 square feet (9.3 m²) of area before the ground is tilled. If the soil is a heavy clay, a 1-inch (2.5-cm) layer of perlite or sand and a 2-inch (5-cm) layer of sphagnum peat moss are added before the ground is tilled. Tulips do not grow well in heavy, clay soils. The medium must be steam or chemically pasteurized before planting.

A tulip bulb is planted with the tip, or nose, 1 inch (2.5 cm) below the surface of the medium. After all of the bulbs are planted, they are watered in well using cold water. This watering helps to reduce the soil temperature to the 55°F (13°C) level that must be held for 2 weeks. After this period the bulbs should be well rooted. At this time a second fungicide treatment is applied to prevent *Pythium*. The first treatment is applied when the bulbs are planted.

The air temperature is held at 58°F to 60°F (14.5°C to 15.5°C) day and night. Full ventilation is needed on warm days to keep the temperature down to this range. The greenhouse glass must be kept as clean as possible so that the plants will receive full light.

When the roots show good development, about 10 to 14 days after planting, calcium nitrate is added at the rate of 2 pounds (0.9 kg) in 100 gallons (378.5 l) of water. This fertilizer is alternated every other week with a full analysis 20-10-20 fertilizer applied at the rate of 14 ounces (397 g) in 100 gallons (378.5 l) of water. In place of the 20-10-20 liquid fertilizer, Osmocote® 14-14-14 may be added as a topdressing at the rate of 15 pounds (6.8 kg) to 100 square feet (9.3 m²) of area. Weekly applications of calcium nitrate are made in combination with Osmocote® 14-14-14. Cultivars suitable for forcing at 41°F (5°C) are listed in Tables 26-12 and 26-13.

Harvesting for Market

The procedures for harvesting and storing tulips are the same as those described previously for standard forced crops. Special handling is not required.

Pot Grown Tulips

Tulips forced as flowering potted plants are handled in much the same manner as those forced for cut flowers. The flowering season for potted bulbs is from early January to early May. Seven flowering periods are given in Table 26-14. Bulbs precooled for potted tulips need 15 weeks of cold temperatures. The bulbs used are generally 4.7 to 5.5 inches (12 to 14 cm) in size.

A rooting room is recommended, but good potted tulips can be grown using cold frames.

When the bulbs are received, those that are not to be precooled are held at 63°F (17°C). At this temperature, the final stem length is a little shorter than if the bulbs are held at 55°F (13°C). The schedule for precooling is given in Table 26-15.

Table 26-14

Periods for Flowering Tulips for Potted Plants[1]		
Period	Approximate Time of Blooming	Main Holiday
1	January 1–19	–
2	January 20–February 7	–
3	February 8–28	Valentine's Day
4	March 1–18	–
5	March 19–April 2	Early Easter
6	April 3–18	Late Easter
7	April 19–May 5	Mother's Day

1. Adapted with modifications from the *Holland Bulb Forcers Guide*, 1996.

Table 26-15

Flower Period	Start Precooling	Planting Date	Rooting Room Temperature Sequence[2]	Date into Greenhouse	Approximate Flowering Date	Weeks of Cold
1 PC[3] at 44.5°F (7°C)	August 26–31	October 1–7	A	December 10	January 1	15
				December 17	January 8	16
				December 24	January 15	17
2 PC at 48°F (9°C)	September 1–7	October 6–10	A	December 30	January 20	16.5
				January 5	January 28	17.5
				January 12	February 5	18.5
2 NP[4]	None	September 18–22	A	December 30	January 20	14.5
				January 5	January 28	15.5
				January 12	February 5	16.5
3 NP	None	October 1–7	A	January 18	February 8	15
				January 25	February 16	16
				February 1	February 23	17
4 NP	None	October 24–28	B	February 8	March 1	15
				February 15	March 8	16
				February 22	March 15	17
5 NP (early Easter)	None	November 6–11	B	February 28	March 20	16
				March 6	March 26	17
				March 13	April 1	18
6 NP (middle to late Easter)	None	November 10–15	B	March 16	April 4	18
				March 22	April 10	19
				March 28	April 17	20
7 NP (late Easter)	None	November 10–15	B	April 1	April 21	20.5
				April 8	April 28	21.5
				April 15	May 3	22.5
				April 22	May 10	23.5

Title: **Scheduled Dates for Growing Tulips for Potted Plants[1]**

1. Adapted with modifications from the *Holland Bulb Forcers Guide*, 1996.
2. Rooting Room A Temperature Sequence:
    Planting to November 5–10, 48°F (9°C); to January 1–5, 41°F (5°C); then 32°F–35°F (0°C–2°C).
   Rooting Room B Temperature Sequence:
    Planting to December 1–5, 48°F (9°C); to January 1–5, 41°F (5°C); then 32°F–35°F (0°C–2°C).
   Note: The change from 48°F (9°C) is made when the roots grow out of the bottom of the pot. The change from 41°F (5°C) to 32°F (0°C) is made when the shoots are 2 inches (5 cm) tall.
3. PC = Precooled bulbs.
4. NP = Nonprecooled bulbs.

Media and Planting

The medium may be a 1:1:1 mixture by volume of soil, peat moss, and sand or perlite. Limestone is added at the rate of 5 to 10 pounds (11 to 22 kg) per cubic yard (cubic meter). Twenty percent superphosphate is also added at the rate of 2 pounds (4.4 kg) per cubic yard (cubic meter). Peat-lite mix A can also be used with 10 percent sand added for weight.

The medium is pasteurized to prevent diseases. If the medium cannot be pasteurized, fungicide drenches are applied after the roots are well developed. Some fungicides restrict root growth. Therefore, the grower must be careful to apply the drench only after the roots are well developed.

Bulb pans or three-quarter size pots are used because they are more in scale with the height of the finished plants. Three or four bulbs are planted in a 4-inch (10-cm) pot, five or six bulbs are placed in a 5-inch (12.5-cm) pot, six or seven bulbs in a 6-inch (15-cm) pot, eight or nine bulbs in a 7-inch (17.5-cm) pan, and 11 or 12 bulbs in an 8-inch (20-cm) pan.

The first leaf develops from the flat side of the tulip bulb. When the potted plants are grown, they will have a pleasing appearance, if the flat side of the bulb is planted facing toward the rim of the pan. Bulbs planted within the outside row are randomly placed.

The pans should be half-filled with medium. The bulbs are set so that just the tips of the bulbs are even with the top of the pan. The medium covering the bulbs prevents them from pushing up out of the pot.

The bulbs should not be forced into the medium because the basal plate may be damaged. The bulbs are set gently in place, and the medium is filled in around them. A reservoir of at least 0.25 inch (0.6 cm) must be left at the top of the pot for water and fertilizer solution.

Labels must be used to identify the cultivars. Generally, wooden or plastic labels are used and are marked with a soft lead pencil. Ballpoint pens cannot be used because the ink is soon washed off the labels.

Many growers prepare a map of the rooting room to show where each cultivar is located. Such a map helps save time and labor when the pots are moved to the greenhouse for forcing.

Tulips may cause an allergic reaction in some people, resulting in a severe rash similar to poison ivy. Because this rash usually develops on the hands, the use of rubber gloves may reduce the severity of the reaction. However, the only effective preventive measure is not to work with tulips.

Cultivars that are suitable for forcing as potted plants are given in Table 26-16.

Greenhouse Forcing and Fertilization

The pots are brought into the greenhouse on the scheduled dates (Figures 26-15 and 26-16). If rooting is behind schedule, the pans should be held in the rooting room for several more days. If the plants root ahead of schedule, they are held in the rooting room until the proper time. It is better to hold the plants than to have them ready too early for markets.

The temperature in the greenhouse should be held at 63°F (17°C) at night and at 68°F (20°C) during the day. A higher temperature in the daytime may be used to hasten flowering. The night temperature should not be increased because the food reserves are reduced. Development will be slowed if the day and night temperatures are held a few degrees lower than recommended values.

The plants are fertilized using calcium nitrate at the rate of 2 pounds (0.9 kg) in 100 gallons (378.5 l) of water alternated weekly with 20-10-20 fertilizer at the rate of 14 ounces (397 g) in 100 gallons (378.5 l) of water.

Table 26-16

Tulip Cultivars for Standard Forcing as Potted Plants[1]

Color	Cultivar	Flowering Period and Type of Bulb Needed						
		1	2	3	4	5	6	7
Red	'Albury'	–	–	–	–	NP[2]	NP	NP
	'Arma'	–	–	–	NP	NP	NP	NP
	'Bing Crosby'	–	PC[3]	NP	NP	AR[2]	AR	–
	'Cassini'	–	PC	NP	NP	AR	–	–
	'Charles'	–	PC	NP	NP	AR	AR	–
	'Couleur Cardinal'	–	–	–	–	NP	NP	NP
	'Diplomate'	–	–	–	NP	NP	AR	AR
	'Frankfurt'	–	–	NP	NP	NP	AR	AR
	'Henry Dunant'	–	PC	NP	NP	AR	–	–
	'Ile de France'	–	–	NP	NP	NP	NP	–
	'Merry Christmas'	PC	PC	NP	NP	NP	NP	NP
	'Olaf'	–	PC	NP	NP	NP	–	–
	'Paul Richter'	PC	PC	NP	NP	NP	AR	–
	'Prominence'	–	–	–	–	NP	NP	NP
	'Robinea'	PC	NP	NP	–	–	–	–
	'Ruby Red'	–	PC	NP	NP	–	–	–
	'Stockholm'	–	PC	NP	NP	–	–	–
	'Topscore'	PC	PC/NP	NP	–	–	–	–
	'Trance'	–	–	NP	NP	NP	NP	–
Pink or Rose	'Angelique'	–	PC	NP	NP	AR	AR	–
	'Blenda'	PC	PC/NP	NP	NP	NP	–	–
	'Christmas Marvel'	–	–	NP	NP	NP	NP	AR
	'High Noon'	–	–	–	–	NP	NP	NP
	'Palestrina'	–	–	–	–	NP	NP	NP
	'Peerless Pink'	–	PC	NP	NP	AR	AR	–
Yellow	'Bellona'	–	PC	NP	NP	AR	AR	–
	'Christmas Gold'	–	PC	NP	NP	–	–	–
	'Kareol'	–	PC	NP	NP	AR	AR	–
	'Makassar'	–	–	–	–	NP	NP	AR
	'Monte Carlo'	PC	PC	NP	NP	AR	AR	–
	'Yellow Present'	–	–	–	–	NP	NP	NP
	'Yokohama'	–	PC	NP	NP	NP	NP	AR
White	'Hibernia'	–	PC	NP	NP	NP	AR	–
	'Inzell'	PC	PC	NP	NP	NP	NP	NP
	'Pax'	–	PC	NP	NP	NP	AR	–
	'Snowstar'	PC	PC	NP	NP	–	–	–

(continues)

Table 26-16 *(Continued)*

Color	Cultivar	Flowering Period and Type of Bulb Needed						
		1	2	3	4	5	6	7
Lavender	'Attila'	–	PC	NP	NP	NP	–	–
	'Negrita'	–	–	NP	NP	AR	AR	AR
	'Prince Charles'	–	PC	NP	NP	NP	AR	–
Orange	'Jimmy'	–	–	–	NP	NP	AR	AR
	'Orange Monarch'	–	PC	NP	NP	NP	NP	–
	'Orange Sun'	–	–	–	–	NP	NP	–
	'Princess Irene'	–	–	–	–	NP	NP	NP
Apricot	'Apricot Beauty'	PC	PC/NP	NP	NP	AR	–	–
Bicolors red and white	'Invasion'	–	–	–	NP	NP	NP	NP
	'Leen v.d. Mark'	PC	PC	NP	NP	NP	NP	NP
	'Merry Widow'	–	PC	NP	NP	AR	AR	–
	'Mirjoran'	–	PC	NP	NP	–	–	–
Red and yellow or cream	'Abra'	–	PC	NP	NP	NP	NP	AR
	'Coriolan'	–	–	NP	NP	NP	AR	–
	'Etude'	–	–	NP	NP	NP	NP	NP
	'Golden Eddy'	–	–	–	–	NP	NP	NP
	'Karel Doorman'	–	PC	NP	NP	AR	AR	–
	'Kees Nelis'	–	PC	NP	NP	NP	AR	–
	'Los Angeles'	–	–	–	NP	NP	AR	–
	'Margot Fonteyn'	–	–	NP	NP	NP	NP	NP
	'Mirjoran'	–	PC	NP	NP	NP	–	–
	'Thule'	–	PC	NP	NP	AR	AR	–
Greigii	'Authority'	–	–	NP	NP	NP	NP	–
	'Plaisir'	–	–	NP	NP	NP	NP	–
	'Red Riding Hood'	–	–	–	–	NP	NP	NP
Murillo Tulips (Double)	'Electra'	–	–	–	–	NP	NP	NP
	'Mr.VanderHoef'	–	–	–	–	NP	NP	NP
	'Orange Nassau'	–	–	–	–	NP	NP	NP
	'Peach Blossom'	–	–	–	–	NP	NP	NP
	'Schoonoord'	–	–	–	–	NP	NP	NP
	'Willemsoord'	–	–	–	–	NP	NP	NP

1. Adapted with modifications from the *Holland Bulb Forcers Guide*, 1996.
2. NP = Nonprecooled bulbs.
3. PC = Precooled bulbs.
4. AR = A-REST® required.

FIGURE 26-15 Tulips being forced into bloom in the greenhouse for Easter sales.

FIGURE 26-16 A striped leaf variety of tulip being forced for Easter sales.

Growth Retardants

One of the most commonly used growth retardants is A-REST® ancymidol, manufactured by SePro Corporation, of Carmel, Indiana. For tulips, A-REST® is applied in liquid form to the growing medium when the shoots are 3 to 4 inches (7.5 to 10 cm) long. The effect of the chemical is to reduce the distance between the nodes. That is, the growth of the basal part of the stem is reduced. As a result, the plant has a more pleasing appearance. The diluted solution is applied during the greenhouse phase of production. A-REST® cannot be used in the darkness of the rooting room.

Before applying A-REST®, the pots are watered 12 to 24 hours earlier so that the medium is moist. The amount of chemical supplied is controlled by the amount of solution added to each pot. Four fluid ounces (120 ml) of solution is put on 6-inch (15-cm) pots. Two fluid ounces (60 ml) is added to 4-inch (10-cm) pots. If the pots are larger than 6 inches (15 cm) in diameter, the grower must determine the strength of the solution and the proper volume to be used. Dilution rates are given in Table 26-17.

Not all cultivars show the same response when treated with A-REST®. To retard some cultivars, the chemical must be used at a stronger concentration. Other cultivars require no treatment at all. Generally, the plants treated with A-REST® are those scheduled to bloom in periods 5, 6, and 7. Table 26-18 lists

Table 26-17

	A-REST® Used as a Soil Drench in 1-Quart Dilutions[1]					
Pot Size	Rate Required in Milligrams Active Ingredient	Number of Pots to Be Treated with 3-Gallon (11.4 l) Dilution	Ounces of A-REST® Dissolved in 3 Gallons (11.4 l) of Water[2]	Number of Pots Treated per Quart (946 ml) of Retardant	One Quart of Retardant Is Dissolved in This Amount of Water in Gallons (l)	Liquid Ounces (ml) of Solution Applied per Pot
6 inch (15 cm)	0.125	96	1.5	2,000	64 (242)	4 (120)
	0.250	96	3.0	1,000	32 (121)	4 (120)
	0.500	96	6.0	500	16 (60.5)	4 (120)
4 inch (10 cm)	0.125	192	1.5	4,000	64 (242)	2 (60)
	0.250	192	3.0	2,000	32 (121)	2 (60)
	0.500	192	6.0	1,000	16 (60.5)	2 (60)

1. Adapted with modifications from the *Holland Bulb Forcers Guide*, 1996.
2. One liquid ounce of ancymidol A-REST® contains 7.8 mg of active ingredient. One liquid ounce is equal to 30 ml (3 gallons equal 11.4 l).

Table 26-18

Application of A-REST® in Selected Flowering Periods for Tulip Cultivars Used for Potted Plants[1]				
Flowering Period	Holiday	Rate Required (mg)	Time of Application in Greenhouse (days)	Cultivars
5 A-REST® used to control overall stem growth	Early Easter	0	—	'Abra,' 'Albury,' 'Angelique,' 'Arma,' 'Couleur Cardinal,' 'Edith Eddy,' 'Electra,' 'Etude,' Golden Mirjoram,' 'Jimmy,' 'Peach Blossom,' 'Princess Irene,' 'Red Riding Hood,' 'Robinea,' 'Schoonoord,' 'Stockholm,' and 'Willemsoord'
		0.125	1 to 2	'Bellona,' 'Christmas Marvel,' 'Diplomate,' 'Frankfurt,' 'Invasion,' 'Palestrina,' 'Pax,' 'Preludium,' 'Prominence,' 'Yellow Present,' and 'Yokohoma'
		0.250	1 to 2	'Apricot Beauty,' 'Bing Crosby,' 'Cassini,' 'Comet,' 'Coriolan,' 'Esther,' 'Hibernia,' 'Ile de France,' 'Inzell,' 'Karel Doorman,' 'Kees Nelis,' 'Merry Widovi,' 'Monte Carlo,' 'Olaf,' 'Orange Sun,' 'Oscar,' 'Page Polka,' 'Peerless Pink,' 'Prince Charles,' and Wirosa'
		0.500	1 to 2	'Attila,' 'Henry Dunant,' 'Los Angeles,' 'Madame Spoor,' 'Makassar,' and 'Paul Richter'

(continues)

Table 26-18 *(Continued)*

Flowering Period	Holiday	Rate Required (mg)	Time of Application in Greenhouse (days)	Cultivars
6 A-REST® used to control stem growth	Late Easter	0	–	'Couleur Cardinal,' 'Plaisir,' 'Princess Irene,' 'Red Riding Hood,' and 'Stockholm'
		0.125	1 to 2	'Abra,' 'Diplomate,' 'Etude,' 'Invasion,' and 'Robinea'
		0.250	1 to 2	'Albury,' 'Angelique,' 'Arma,' 'Bellona,' 'Bing Crosby,' 'Capri,' 'Christmas Marvel,' 'Esther,' 'Frankfurt,' 'High Noon,' 'Inzell,' 'Jimmy,' 'Kareol,' 'Merry Christmas,' 'Olaf,' 'Page Polka,' 'Palestrina,' 'Pax,' 'Preludium,' 'Prominence,' 'Shirley,' 'Wirosa,' 'Yellow Present,' and 'Yokohoma'
		0.500	1 to 2	'Attila,' 'Coriolan,' 'Hibernia,' 'Ile de France,' 'Karel Doorman,' 'Kees Nelis,' 'Los Angeles,' 'Makassar,' 'Merry Widow,' 'Orange Monarch,' 'Orange Sun,' 'Oscar,' 'Peerless Pink,' and 'Prince Charles'
7 A-REST® used to control stem growth	Mother's Day	0	–	'Electra,' 'Mr. Van de Hoef,' 'Orange Nassau,' 'Peach Blossom,' 'Red Riding Hood,' 'Schoonoord,' and 'Willemsoord'
		0.125	1 to 2	'Pax'
		0.250	1 to 2	'Abra,' 'Albury,' 'Capri,' 'Cardinal,' 'Couleur,' 'Diplomate,' 'Etude,' 'Invasion,' 'La Suisse,' 'Olaf,' 'Page Polka,' 'Palestrina,' 'Princess Irene,' 'Robinea,' 'Yellow Present,' and 'Yokohoma'
		0.500	1 to 2	'Arma,' 'Frankfurt,' 'High Noon,' 'Jimmy,' 'Leen v.d. Mark,' 'Makassar,' 'Merry Christmas,' 'Negrita,' Peerless Pink,' 'Shirley,' 'Varinas,' and 'Wirosa'

1. Adapted with modifications from the *Holland Bulb Forcers Guide*, 1996.

the concentrations of retardant to be applied to control the overall stem growth of tulips. Note in the table that the solution is added to the medium on the day the plants are removed from storage and placed in the greenhouse, or on the following day.

Problems

Cultural Disorders

If the bulbs are overheated during shipment or in storage, blind shoots may develop. When the stems fail to grow, the cold treatment may have been too short or the wrong temperature was used. Tulip topple is caused by a lack of calcium.

Disease

Botrytis blight is one of the most serious diseases affecting tulips. It causes small yellow spots on the leaves and flowers. The disease organism winters over on the bulbs. To control the disease, the bulbs are dipped in a fungicide before planting. Good ventilation and air circulation must be maintained in the greenhouse to prevent stagnant air conditions. Overhead fans are helpful in moving the air.

Fusarium disease is usually more a problem for the bulb producer than it is for the bulb forcer. The bulbs must be inspected upon arrival. Any bulbs that appear to be diseased are discarded. The remaining bulbs are dipped in the proper fungicide before planting.

Insects

The bulbs may be infested with bulb mites and aphids when they are received from the producer. These insects can be controlled with a dusting of insecticide.

Rodents

Rodents eat bulbs for food. To protect the crop, the rooting room must be made rodent proof. In addition, naphthalene moth flakes or crystals can be sprinkled on the pots and around the room.

CAUTION

Do not use flakes or crystals made from paradichlorobenzene. This substance seriously injures the bulbs and prevents them from growing.

ACHIEVEMENT REVIEW

Select the best answer or answers to complete each statement. List the appropriate letter(s).

1 Which of the following may be used to produce cut flowers?
   a. Corms
   b. Bulbs
   c. Tubers
   d. Rhizomes

2 The crown of the anemone plant is set above the surface of the medium to prevent:
   a. waterlogging.
   b. insect damage.
   c. crown rot.
   d. *Botrytis*.

3 Calla lilies are grown from:
   a. bulbs.
   b. corms.
   c. tubers.
   d. rhizomes.

4 Calla lilies should not be stored with carnations or snapdragons because calla lilies produce large quantities of:
   a. carbon dioxide.
   b. hydrogen.
   c. ethylene gas.
   d. ammonia fumes.

5 The amount of land in Florida devoted to gladiolus culture is:
   a. 500 acres (202.4 hectares).
   b. 1,000 acres (404.8 ha).
   c. 4,200 acres (1,700 ha).
   d. 10,000 acres (4,048.5 ha).

6 Tip burn of gladiolus leaves can be caused by:
   a. excess nitrogen.
   b. too much phosphorus.
   c. fluorine injury.
   d. lack of iron.

7 When gladiolus are shipped, they should be:
   a. placed flat in cardboard boxes.
   b. standing upright.
   c. placed in water.
   d. shipped in none of the above ways.

8 Which of the following tulip diseases causes ethylene gas to form?
   a. *Rhizoctonia*
   b. *Fusarium*
   c. *Botrytis*
   d. *Verticillium*

9 If new clay pots are used to force bulb crops, the pots should be:
   a. steam pasteurized before use.
   b. stored outside over winter.
   c. soaked in water for at least 4 hours before use.
   d. potted up immediately when they are received.

10 Most bulb crops that are forced for blooms require _____ amounts of fertilizer.
   a. zero
   b. medium
   c. low
   d. high

11 Bulb crops need a cold temperature treatment to:
   a. preserve carbohydrates.
   b. make them last longer.
   c. cause the tissue changes required if flowering is to take place.
   d. prevent diseases.

12 Hyacinth and tulip bulbs are graded according to:
   a. weight.
   b. number of flower buds.
   c. circumference in centimeters.
   d. diameter in inches.

13 The best Wedgewood iris bulbs used for forcing are grown in:
   a. the Netherlands.
   b. Spain.
   c. the Pacific Northwest of the United States.
   d. the Jersey Isles.

14 At a night temperature of 60°F (15.5°C), Easter lilies are forced in bloom in:
   a. 90 to 100 days.
   b. 110 to 120 days.
   c. 120 to 140 days.
   d. 130 to 140 days.

15 Lilies make their best growth when the potting medium contains a large amount of:
   a. manure.
   b. limestone.
   c. superphosphate.
   d. potassium.

16 An effective way to keep lilies from growing too tall is to:
   a. provide long days in the middle of the night.
   b. use the short-day treatment with black cloth.
   c. treat the soil with A-REST®.
   d. spray the plants with gibberellic acid.

17 Narcissus sold for cut flowers are harvested when they are:
   a. at the "gooseneck" stage.
   b. just beginning to show color in the flower.
   c. at the "pencil green" stage.
   d. half open.

18 Tulip bulbs must be stored at a temperature of _____ to hasten the formation of the flower parts.
   a. 45°F (7°C)
   b. 50°F (10°C)
   c. 55°F (13°C)
   d. 63°F (17°C)

19 Overheating of tulip bulbs in shipping:
   a. hastens bloom.
   b. delays bloom.
   c. kills the floral parts.
   d. has no effect on growth.

Student Project 1

Objectives

This exercise teaches the student:

■ how to observe the effects of different fertilizers on the growth and flowering of lilies.
■ how to observe the effects of long days on the growth of lilies.

Materials

Wooden box, inside dimensions of 12 inches by 12 inches by 12 inches (30 cm by 30 cm by 30 cm) (1 **cubic foot** [0.03 m^3] in volume).

30	6-inch (15-cm) diameter, three-quarter size plastic pots (new)
5 pounds (2.3 kg)	Ground limestone
5 pounds (2.3 kg)	20 percent superphosphate
5 pounds (2.3 kg)	Hydrated lime
1 pound (0.45 kg)	Aluminum sulfate
30	Lily bulbs 9- to 10-inch (22.5- to 25-cm) size, precooled, cultivar 'Ace'
3 cubic feet (0.08 m^3)	2:1:1 mixture of peat moss, perlite, and vermiculite

Note: Any fertilizer remaining may be used for other projects.

Procedure

1. Prepare the following mixes of fertilizer and potting medium.
   a. The control medium has no fertilizer added.
   b. A high pH effect is obtained by adding 4 ounces (113.4 g) of hydrated lime to 0.5 cubic foot (0.015 m^3) of medium. Mix the materials thoroughly.
   c. For a low pH effect, add 4 ounces (113.4 g) of aluminum sulfate to 0.5 cubic foot (0.5 m^3) of medium and mix the materials thoroughly.

d. A high phosphorus effect is obtained by adding 4 ounces (113.4 g) of 20 percent superphosphate to 0.5 cubic foot (0.015 m³) of medium. Mix the materials thoroughly.

e. To obtain the recommended levels of calcium and phosphorus, add 4 ounces (113.4 g) of ground limestone and 0.5 ounce (14.17 g) of superphosphate to 0.5 cubic foot (0.015 m³) of medium. Mix the materials thoroughly.

2. Add moisture to the media as needed until they have a good consistency for potting.

Note: An easy way of mixing 0.5 cubic foot (0.015 m³) of medium with added fertilizer and water is to place all materials in a 5-gallon (19-1) plastic bag. Blow into the bag to inflate it. The materials can then be mixed by tumbling. Practice will show how much mixing is required to obtain a good distribution of materials.

Planting Methods

Six pots can be planted using 0.5 cubic foot (0.015 m³) of medium.

1. Place a handful of medium in the bottom of each pot. When the bulb is placed on the medium, the tip should be 1.5 to 2 inches (3.75 to 5 cm) below the top of the soil.
2. Fill in around the bulb with medium. Press the medium firmly around the bulbs, but do not pack it.
3. Add medium to fill the pot to within 0.5 inch (1.25 cm) of the top of the pot.
4. Water the pots thoroughly and place them in a greenhouse at 60°F (15.5°C) in full light intensity.
5. Half of the pots from each treatment are to be given 18-hour days under incandescent lights providing 10 fc (108 lux) of light.
6. The remaining pots should be protected from light.
7. When the shoot of the bulb breaks the surface of the medium, ammonium nitrate fertilizer is used once a week at the rate of 1.5 pounds (1.5 kg) in 100 gallons (378.5 l) of water.

Observations

1. Record the date of emergence of the shoots.
2. Record, by weekly measurements, the rate of growth of the plants in each treatment.
3. Record the number of flowers that develop and the number of scorched leaves that appear on each plant in each treatment. Discuss the results with your instructor.

Student Project 2

Objectives

This exercise teaches the student:

■ how to observe the effects of the addition of fertilizer on the growth of tulips.
■ how to observe the effects of planting depth on tulips.

Materials

30   6-inch (15-cm) plastic bulb pans (new)
150 Precooled tulip bulbs, all the same cultivar

The same medium and fertilizers described in Student Project 1 are used in this project.

Procedure for Objective 1

1. Plant five bulbs in each of three pans using each medium. Use the recommended planting depth for tulips.
2. Water the plants thoroughly and place them in the proper growing environment.

Procedure for Objective 2

1. Plant three pans with five bulbs in each pan using each medium. Place the bulbs in the bottom of each pan before any medium is added. This type of deep planting is not recommended generally.
2. Water the plants thoroughly and place them in the proper growing environment. Water thoroughly.

Observations

1. Record the date of emergence of the shoots.
2. Measure the shoots weekly to determine the growth rate.
3. Record the date when the buds are first visible.
4. Record the date of first color.
5. Record the date when the flowers first open.
6. Discuss the results of this experiment with your instructor.

Tropical Foliage Plants

OBJECTIVES:

After studying this chapter, the student should have learned:

■ the most important families of tropical foliage plants.

■ how to identify a large number of foliage plants by their scientific names.

■ the cultural requirements for various types of foliage plants.

Tropical and semitropical foliage plants have always been popular. From the parlor palm and the cast-iron plant of the Victorian era to the present, the industry has shown incredible growth.

The total foliage plant sales reported for indoor and patio use in 2005 were $652 million. This value is 100 percent greater than the estimated net value of foliage sales in 1976. When the 2005 total includes foliage grown in hanging baskets, the value of sales was $721 million. Foliage plant sales in 36 states in 2005 surpassed those for cut flowers by more than $324 million. Sales of bedding and garden plants at roughly $2.6 billion were two and one-half times more than the sales of foliage plants.

Although a slowdown of sales has occurred the last few years, foliage plants make up over 47 percent of all the potted crops sold in the United States. Potted flowering plants had a wholesale value of $809 million in 2005.

Production Areas

The state of Florida produces more foliage plants than any other state (Table 27-1). California, the second ranking state in sales, has less than one-fourth that of Florida. Texas produces sales that are a little more than one-fourth that of California. Hawaii sales of foliage plants are more than 85 percent that of Texas. Many other states also produce foliage plants but not in so large volumes as the leading four states. This is because the plants must be grown in a heated greenhouse.

Table 27-1

| The Four Leading States in Sales of Potted Foliage and Foliage Hanging Baskets* ||
State	Sales (000)
Florida	$499,656
California	109,351
Texas	19,957
Hawaii	17,081

*2006 United States Department of Agriculture, Floriculture Crops, 2005 Summary

The climate in southern Florida, California, and Texas permits growers to produce foliage crops outdoors usually under Saran shade cloth. When a cold wave from the north threatens their crops, they must provide protective measures to keep the plants from freezing. Foliage growers in Hawaii do not have to worry about freezes.

Cold weather is not the only threat that southern and Hawaiian foliage plant producers face. Severe storms in the form of hurricanes have devastated foliage plant production areas in Florida, Texas, and Hawaii. Hurricanes Dennis, Katrina, Rita, and Wilma in 2005 caused over $1.12 billion in damage to the nursery and foliage industry in Florida alone. The production area in south Florida, around the Homestead area, was particularly hard hit. Many of the businesses destroyed still have not fully recovered from that hurricane.

Foliage propagation materials are imported into the United States from Puerto Rico, Central and South America, countries in the Caribbean area, South Africa, and Europe. Federal quarantine laws prohibit the import of any plant materials with soils attached to them.

Major production areas in Florida are found near Orlando and Apopka in north Florida and near southwest Miami in south Florida. Apopka is known as the "Foliage Capital of America."

Growers in the Orlando-Apopka area produce plants in heated glass and fiberglass-covered greenhouses. Freezing temperatures in the winter are common around Orlando.

Growers in southern Florida produce plants in unheated Saran plastic-covered houses. The primary function of the Saran covering is to reduce the high light intensity. In Miami, the temperature rarely falls below freezing. However, a general freeze in January 1977 caused millions of dollars worth of damage to tropical foliage plants, vegetables grown outdoors, and citrus crops.

World Distribution

Tropical foliage plants grow naturally around the world. Thousands of species of tropical and semitropical plants are known. A species is a subdivision of a genus. A group of genera (plural of genus) comprise a family. The genera of a family are thought to be related to each other. Of course, not all species in every genera are grown as ornamental plants. Many of the plants have the same kind of cultural requirements. However, plants that belong to the same species often have different cultural needs.

This text cannot describe the culture of all of the foliage plants used in the florist trade today. Only the most commonly grown plants are presented. The reader is urged to investigate other florists' crops not described here.

The following sections describe 26 families, including at least one genus and species in the family (but usually several are presented). The plants are listed alphabetically within a family rather than alphabetically by genus, species, or common name. The Liberty Hyde Bailey Hortorium at Cornell University in Ithaca, New York, publishes a formal botanical reference known as "*Hortus* Third." In this publication the plants are listed alphabetically by genera.

In several cases, plants with incorrect names are offered in the horticultural trade. In these cases the scientific name from "*Hortus* Third" is given first. The horticultural name is then given in parentheses, followed by the most commonly used name. Many plants have more than one common name.

Grouping the plants by families makes it possible to comment about conditions that may be of special importance. For example, members of the *Liliaceae* (lily family) often develop leaf tip burn. One probable cause for tip burn is fluoride toxicity.

General Culture

Tropical and subtropical foliage plants grow in a wide range of conditions in their natural surroundings. The light intensities required by various plants range from 12,000 fc (129.2K lux) at the brightest time of day to 10 fc (108 lux) in the deep shade of the jungle. Some plants grow as **epiphytes** (air-loving plants) high in the tops of tall trees under filtered sunlight, and others grow as **hydrophytes** (water-loving plants) among the flooded roots in jungle swamps.

The foliage plant grower tries to give the plants the cultural conditions needed for best growth. The grower must control factors such as the light intensity, moisture, temperature, and fertilization so that they are as similar to the natural growing conditions as possible.

The grower also tries to condition the plants to the environment they will be going into after they are sold. The plants need help in making the change from a tropical jungle plant to a tropical indoor plant. This process is called acclimatization (*ac-climb-a-ties-a-shun*) and is explained in more detail at the end of this chapter.

Media and Fertilizer

In their natural state, these foliage plants grow in a variety of media, ranging from the extremes of air alone for epiphytes to water alone for hydrophytes. Usually, the media contain a large amount of organic matter resulting from the decay of natural plant materials. In commercial greenhouse production, large amounts of sphagnum peat moss are used as the organic material in media. In addition, ground bark, weathered wood chips, sawdust, perlite, sand, vermiculite, and even soil are used.

A commonly used medium is a 2:1:1 by volume mixture of sphagnum peat moss, and perlite or sand. Some growers prefer to use a 1:1:1 by volume mixture of the same materials. Many growers continue to use some topsoil and make their own mixes.

Two variations of the standard Cornell peat-lite mixes are called Cornell Tropical Plant Mixes. For general use, the Foliage Plant Mix is preferred (Table 27-2). The Epiphytic Mix (Table 27-3) is used for plants requiring a great deal of aeration and dry conditions between waterings.

The Foliage Plant Mix is used for plants that need moist conditions for maximum growth. Some of these plants are as follows:

Amaryllis
Aphelandra squarrosa
Begonia
Beloperone guttata
Caladium
Cissus
Citrus
Coleus
Fern
Ficus
Hedera helix cultivars
Maranta

Table 27-2

Cornell Foliage Plant Mix[1]			
Material	Amount Used for 1 Cubic Yard[2]	Amount Used for 1 Bushel[3]	Amount Used for 1 Cubic Meter[4]
Sphagnum Peat Moss (Screened 0.5 inch)	13 bushels	0.5 bushel	16 bushels
No. 2 Horticultural Vermiculite	6.5 bushels	0.25 bushel	8 bushels
Perlite (Medium Grade)	6.5 bushels	0.25 bushel	8 bushels
Ground Dolomitic Limestone	8.25 lb	8 tbs[5]	4.9 kg
20 Percent Superphosphate (Powdered)	2 lb	2 tbs	1.2 kg
10-10-10 Fertilizer	2.75 lb	3 tbs	1.7 kg
Iron Sulfate	0.75 lb	1 tbs	0.4 kg
Potassium Nitrate (14-0-44)	1 lb	1 tbs	0.6 kg
Trace Element Materials (Use Only One)			
Fritted Trace Elements FTE 555	2 oz	–	74.1 g
or ESMIGRAN	4 lb	4 tbs	2.4 kg
or PERK®	4 lb	4 tbs	2.4 kg
Wetting Agent			
AquaGro® or Surfside® 30 Liquid	3 oz	–	111.2 g
AquaGro® or Surfside® Granular	1.5 qt	0.25 cup	1.4 l

1. Adapted with modifications from the *New York State Flower Industries Bulletin, 33*, April 1973.
2. One cubic yard equals 27 cubic feet or about 22 bushels. A 15 to 20 percent shrinkage occurs in mixing. Therefore, 4 bushels (5 cubic feet) are added to obtain a full cubic yard.
3. One bushel equals 1.25 cubic feet.
4. One cubic meter equals 35.31 cubic feet.
5. Level tablespoon equals 15 cc.

Table 27-3

Material	Amount Used for 1 Cubic Yard[2]	Amount Used for 1 Bushel[3]	Amount Used for 1 Cubic Meter[4]
Cornell Epiphytic Plant Mix[1]			
Sphagnum Peat Moss (Screened 0.5 inch)	8.66 bushels	0.33 bushel	11.8 bushels
Fir Bark (Douglas, Red or White) (0.13- to 0.25-inch size)[5]	8.66 bushels	0.33 bushel	11.8 bushels
Perlite (Medium Grade)	8.66 bushels	0.33 bushel	11.8 bushels
Ground Dolomitic Limestone	7 lb	8 tbs[6]	4.2 kg
20 Percent Super-phosphate (Powdered)	4.5 lb	6 tbs	2.7 kg
10-10-10 Fertilizer	2.5 lb	3 tbs	1.5 kg
Iron Sulfate	0.5 lb	1 tbs	0.3 kg
Trace Element Materials (Use Only One)			
Fritted Trace Elements FTE 555	2 oz	–	74.1 g
or ESMIGRAN	4 lb	4 tbs	2.4 kg
or PERK®	4 lb	4 tbs	2.4 kg
Wetting Agent			
AquaGro® or Surfside® 30 Liquid	3 oz	–	111.2 g
AquaGro® or Surfside® Granular	1.5 qt	0.25 cup	1.4 l

1. Adapted with modifications from the *New York State Flower Industries Bulletin, 33,* April 1973.
2. One cubic yard equals 27 cubic feet or about 22 bushels. A 15 to 20 percent shrinkage occurs in mixing. Therefore, 4 bushels (5 cubic feet) are added to obtain a full cubic yard.
3. One bushel equals 1.25 cubic feet.
4. One cubic meter equals 35.31 cubic feet.
5. Fresh Douglas fir bark, red or white, has a pH of 5.0. It becomes slightly more alkaline as it weathers.
6. Level tablespoon equals 15 cc.

Oxalis
Palm
Pelargonium
Pilea
Sansevieria
Soleirolia
Tolmiea menziesii

The Epiphytic Mix is used for plants that need excellent drainage, high aeration, and dry conditions between waterings. Some of these plants are as follows:

African violets
Aglaonema
Aloe
Bromeliad
Cacti

Crassula
Dieffenbachia
Episcia
Gloxinia
Hoya
Monstera
Nephthytis
Philodendron
Pothos
Peperomia
Schefflera (Brassaia)
Syngonium

The fertilization program for foliage plants is similar to that for other potted crops. When limestone and superphosphate are added as a preplant dressing, nitrogen and potassium are the only fertilizers added on a regular basis. Some growers prefer to add a small amount of phosphorus with the nitrogen and potassium. Most foliage plant growers use fertilizer injectors. Foliar analyses of experimental foliage plants have shown that the plants contain more potassium than nitrogen. As a result, growers use a 20-5-30 analysis fertilizer, which is applied to supply a concentration of 150 ppm nitrogen.

When slow-release fertilizers are used in tropical plant mixes, Osmocote® 14-14-14 or Magamp® 7-40-6 is also used at the rate of 5 pounds (2.98 kg) in a cubic yard (cubic meter) of mix. Foliage plants grown for indoor use cannot tolerate high soluble salt levels. The soil should be tested regularly to determine the salt levels. These tests will indicate when leaching is needed.

Foliage plants show the same types of nutrient deficiency symptoms as other plants. The reader should review Chapter 14 for a discussion of deficiency symptoms.

Light and Temperature

Light is probably the most important and least understood of the environmental factors affecting foliage plants. During the production phase, the grower may expose the plants to as much light as they can withstand without injury. This treatment results in rapid growth. A rapid growth rate means that the plants need to stay in the greenhouse for less time. Therefore, there is less cost to the grower to produce the crop. However, the grower must remember that when the plants are sold, they will be placed indoors in areas where the light intensity may range from 25 to 200 fc (0.27K to 2.16K lux). If the plants are grown in the highest light intensity they can tolerate, it will be much harder for them to adjust to the reduced light levels in the average home.

After the plants leave the greenhouse, the interior lighting provided should be enough to maintain the plants without promoting too much growth. If the plants become overgrown, they must be pruned or replaced.

In the greenhouse, most foliage plants grow well when the light level is 1,000 to 2,000 fc (6.75K to 21.5K lux). Some flowering plants, such as African violets and gloxinias, prefer a light level of from 850 fc (9.18K lux) to 1,200 to 1,500 fc (12.9K to 16.1K lux). Injury occurs at a greater light intensity. Most ferns require 1,000 to 2,000 fc (10.75K to 21.5K lux) of light. Some plants that prefer high light conditions, such as *Codiaeum variegatum*, produce their best growth and coloration in full light intensity.

In the following descriptions of foliage plants, any changes from the average light requirements are noted. Otherwise, the light intensity during the brightest part of the day should be no more than 2,500 to 3,000 fc (26.9K to 32.3K lux). A light meter is required to ensure that the proper light levels are maintained.

Tropical foliage plants require warm temperatures. Standard temperatures in the greenhouse are a minimum of 65°F (18.5°C) at night and 75°F to 85°F (24°C to 29.5°C) during the day. Northern growers who must watch energy costs should not attempt to produce tropical foliage plants in the winter if the temperature requirements cannot be met. Note in the following descriptions that temperature requirements of some crops differ from the standard conditions. Although some foliage plants can withstand low temperatures for short periods of time, the temperature normally should not be allowed to drop below 50°F (10°C). One exception is *Philodendron selloum*, which can be cooled to 28°F (−2°C) for several hours without injury.

Moisture and Humidity

Growing foliage plants need high moisture levels, good aeration, and high humidity. When water is applied, the medium must be wetted thoroughly.

The relative humidity in the greenhouse should be in the range of 70 to 80 percent. A lower humidity may cause leaf tip injury on sensitive plants. "Dry" houses should not be used for foliage plant production.

ACANTHACEAE

Fittonia, or silver nerve plant, grows wild in Colombia and Peru. This plant grows best in low light intensity and is often grown under the greenhouse benches. The temperature should not be allowed to go below 55°F (13°C). *Fittonia* is propagated by stem cuttings.

F. verschaffeltii has dark green leaves with rosy red veins. A variant, *F. verschaffeltii argyroneura*, has lacy, light green leaves with silver-white veins and midrib.

Hypoestes phyllostachya, polka dot plant, and *Pachystachys lutea*, golden shrimp plant, are also members of this family. *Hypoestes* is propagated from seed or terminal cuttings. *Pachystachys* is a recent introduction to the trade and is grown from terminal cuttings. The plants should receive full light intensity, or not more than 15 percent shade. The plants can be treated with a growth retardant, such as chlormequat, to reduce the stem length and increase the number of flowers produced.

AGAVACEAE

The genera of this family were removed from the *Liliaceae* and *Amaryllidaceae* families. Older floriculture textbooks may list several of the following crops as *Liliaceae*.

Several plants in this family are important producers of fiber, such as *Sansevieria*. In their natural habitats the plants grow in dry regions. Therefore, they need a well-drained growing medium with good aeration. The epiphytic mix is suitable. Plants in this family are propagated by seeds, suckers (offshoots), and cuttings, or by the division of rhizomes.

Cordyline

These plants are shrubby or treelike with long, narrow leaves. The leaves are often variegated or striped and are crowded at the ends of the stems. Propagation is by seeds, stem section cuttings, and root layering. The stems are cut into pieces 2 to 4 inches (5 to 10 cm) long. The pieces are placed on the propagation medium. Bottom heat promotes rooting.

C. terminalis is the Ti (*tee*) or good luck plant of Hawaii. The leaves of this plant are used to make hula skirts. The plant is also known as the *Dracaena* palm.

Growers propagate several brightly colored dwarf variants of *C. terminalis*. 'Baby Dolls' has brilliant reddish-purple foliage. It must be grown in a very high light intensity to keep the foliage color bright.

Dracaena

These shrubby or treelike plants produce leaves at the ends of the stems. The stems may vary in thickness from slender to very stout. Dracaena is often called the dragon tree (Figures 27-1A and 27-1B).

D. cincta has long, narrow, sword-shaped leaves with a reddish-brown marginal stripe. Plants sold as *D. marginata* are thought to be *D. cincta* or *D. concinna*.

D. deremensis 'Warnecki' is the most commonly grown cultivar. The leaves have two parallel white strips. 'Janet Craig' is a green form of *D. deremensis* (Figure 27-2).

D. fragrans has long, dark green leaves that curve gracefully. *D. fragrans* 'Massangeana' has green leaves with a central stripe of gold. It is known as the corn plant (Figure 27-3).

D. Sanderiana, or Sanders Dracaena, is also known as the Belgian evergreen. This plant has long, narrow leaves with broad white marginal stripes.

Sanseveria

This plant has several common names, including bowstring hemp and devil's tongue. It is a stiff, erect herbaceous plant found in the dry areas of Africa and

FIGURE 27-1A A nice, multiple-stemmed example of *Dracaena marginata*.

FIGURE 27-1B A specimen tree of Dracaena marginata growing on the grounds of the Coronada Restaurant in San Diego, California.

FIGURE 27-2 *Dracaena deremensis -* 'Janet Craig' is the green form of this species.

FIGURE 27-3 *Dracaena fragrans,* 'Massangeana' commonly called the corn plant because of its similar appearance. The yellow stripe down the center of the leaves is distinctive.

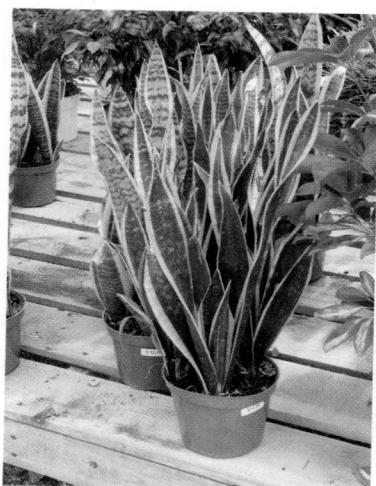

FIGURE 27-4 *Sanseverieria trifasciata* is a durable, low-light plant.

Southern Asia. It grows from thick, short rhizomes. It is one of the most durable house plants and can withstand low light levels and no water for weeks (Figure 27-4). It is grown commercially for its strong, white leaf fibers. The name, bowstring hemp, arises from this commercial use.

The plant is propagated by divisions of the rhizomes or by leaf section cuttings. Leaf section cuttings of *S. trifasciata* 'Laurentii' do not reproduce plants having the golden striped margin. Such plants are obtained only from rhizome divisions.

S. trifasciata, snake plant or mother-in-law's tongue, has swordlike leaves with whitish green, light green, and bluish-green crossbands. *S. trifasciata* 'Hahnii,' Hahn's bird's nest sansevieria, is a dwarf form of *S. trifasciata*. 'Golden Hahnii' is a variant of 'Hahnii' called golden bird's nest sansevieria.

The plants are injured by sudden chilling or if cold water is dropped on the foliage. An internal soft rot develops, and the appearance of the plant is ruined.

AMARANTHACEAE

Iresine herbstii (*I. reticulata* Hort.), or bloodleaf, is a small herbaceous plant in the greenhouse. The leaves are notched at the ends and are purple-red or green with prominent yellow veins. The plant is propagated by terminal cuttings. It is sometimes used as a bedding plant.

ARACEAE

The Arum or Aroid family contains over 2,000 species that grow in many tropical areas. Terrestrial (land), aquatic (water), and epiphytic (air-loving) plants are all found in the family. The plants may be stemless with leaves growing from corms or rhizomes. A number of plants have aerial stems that climb by means of aerial roots.

Aglaonema, or Chinese evergreen, is native to tropical Asia. The colorful leaves are often lance shaped (Figure 27-5). The plants may be less than 1 foot tall to 2 to 3 feet (0.6 to 0.9 m) in height. *A. commutatum* has dark, glossy green leaves with a gray-green area along the main vein. Other popular cultivars are *A. crispum*, *A. commutatum* cv. 'Pseudobracteatum,' and *A. crispum* (*A. Roebelinii* Hort.).

The plants grow well in a low light intensity. They must not be exposed to temperatures below 55°F (13°C).

Aglaonema is propagated by terminal cuttings or by single-node stem cuttings. The temperature in the propagation bench should be 70°F to 75°F (21°C to 24°C).

Anthurium, Anthurium andraeanum, and its close relative *Anthurium scherzeranum*, or flamingo flower, are grown widely in Hawaii. These plants are the second largest crop grown for cut flower exports. The plants produce an exotic flower that lasts 3 to 4 weeks (Figure 27-6). The color range of the blooms, from bright red through pink to cream to white, is the result of extensive breeding. The spadix is the central part of the flower and is supported on the colorful *spathe*.

A minimum temperature of 65°F (18.5°C) is needed. Generally, *Anthurium* are propagated from divisions of the old plant. Large numbers are propagated by tissue culture methods (Figure 27-7). Production from seed is difficult because the seed must be sown within 24 to 48 hours after it ripens. The percentage of germination is very low if the seed is sown after this period.

FIGURE 27-5 *Aglaonema* 'Silver Queen' in a commercial airport terminal building.

FIGURE 27-6 *Anthurium* flower.

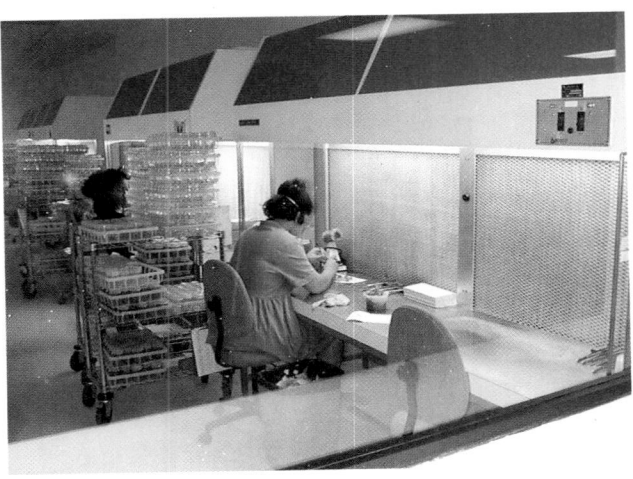

FIGURE 27-7 Many foliage species that are difficult to propagate from seed or vegetatively are propagated through tissue culture.

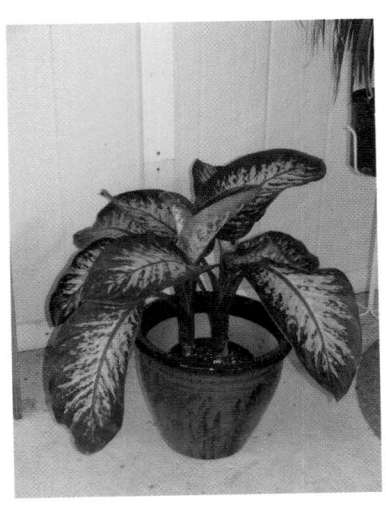

FIGURE 27-8 *Dieffenbachia* in a retail setting.

CAUTION

Never chew the leaves or stems of any ornamental plants.

The plants require a very porous medium. In Hawaii they are grown in beds of chopped osmunda fiber, wood chips, and sawdust. A light intensity of 1,500 fc (16.1K lux) is sufficient when combined with high humidity and frequent waterings.

Dieffenbachia, or 'Dumb Cane,' can cause paralysis of the tongue and throat, which may lead to suffocation, if the stems or leaves are chewed. The paralysis is due to the **oxalic acid** crystals contained in the plant tissue.

There are several species and cultivars of *Dieffenbachia* that are popular with consumers. New cultivars are being added as plant breeding programs continue (Figures 27-8 and 27-9).

D. maculata cv. 'Rudolph Roehrs,' *D. maculata* cv. 'Baranquiniana' (*D. picta* Hort.), and *D.* × *memoria corsii* (*D. picta* var. *memoria* Hort.) are commonly grown. The leaves of these plants are colored in various patterns of greenish yellow.

FIGURE 27-9 *Dieffenbachia* in a waiting lounge.

The plants are propagated by terminal cuttings, or by 2- to 3-inch (5- to 7.5-cm) long stem segments placed on the propagation medium. *Diffenbachia* can also be air layered. The cuttings must be protected from drafts and should not be exposed to temperatures below 60°F (15.5°C).

Monstera, the window leaf species, is a large epiphytic climber native to tropical America.

M. deliciosa (*Philodendron pertusum* Hort.), the breadfruit tree or Swiss cheese plant, is the most popular variety. It is also called the splitleaf philodendron, although it is not a true philodendron. The leaves are long ovals that may be up to 3 feet wide. The presence of the *geniculum*, a structure resembling a bent knee at the joint between the leaf blade and the leaf stem, identifies the plant as a *Monstera* and not a *Philodendron*. The philodendron has a smooth joint.

The culture of *Monstera* is the same as that of philodendrons.

Most of the material offered in the trade as *Nephthytis* is probably *Syngonium*. True *Nephthytis* has little ornamental value.

Philodendron is an herbaceous epiphyte of tropical America and other warm locations. These plants grow as vines with aerial roots. Some of them are tree-like in their mature form. Differences in the leaf form between the juvenile and adult stages of growth make identification difficult. There is a large number of hybrids in the trade (Figure 27-10).

Andreanum (Hort.), *Philodendron melanochrysum*, or black gold philodendron, has long leaves, with a velvet sheen. The upper surface of the leaves is black-green.

P. bipennifolium (*P. panduriforme* Hort.), or fiddle leaf philodendron, has violin-shaped, glossy, dark green leaves.

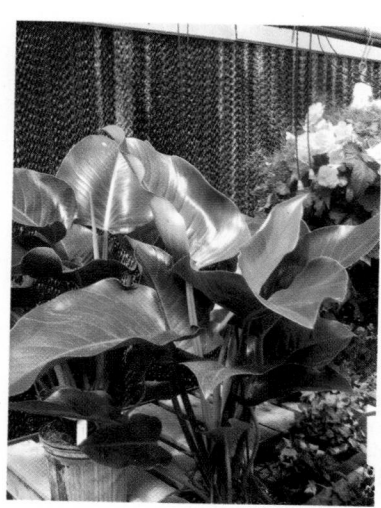

FIGURE 27-10 *Philodendron* 'Rojo Congo' is a hybrid of *Philodendron* 'Imperial Red' and *P. tatei* spp. *Melanochlorum*.

P. scandens subspecies *oxycardium* is often sold as *P. cordatum*. Also called parlor ivy, *P. cordatum* is probably the most widely sold of all of the philodendrons.

P. domesticum (*P. hastatum* Hort.), or spade leaf philodendron, has a long, triangular leaf. The petiole of the leaf is as long as the leaf.

P. pertusum is really *Monstera deliciosa*.

P. sagittifolium (*P. sagittatum* Hort.) has long, firm, triangular leaf blades. The color is a glossy bright green. It is difficult to distinguish *P. selloum* from *P. bipinnatifidum*, from which it has been separated. Cultivars of *P. selloum* include 'German Selloum,' 'Miniature Selloum,' 'Seaside,' and 'Uruguay.' 'Golden Selloum' is another cultivar that generally is too large for the average home. These plants grow well under poor conditions and can withstand a temperature of 28°F (−2°C) for several hours.

P. squamiferum, or anchor plant, has anchor-shaped leaves. The petioles of the leaves are red and have fleshy bristles or scales 0.19 inch (5 mm) long.

Philodendrons are propagated from seeds, terminal cuttings, divisions, and stem section cuttings. In addition, they can be air layered.

Pothos are climbing vines with nearly 70 species. Most of the plants grown are *Scindapsus*.

Scindapsus are climbing plants found in the Malay Archipelago. The leaf markings of these plants are **variegated** patterns ranging from stripes to small blotches. *Scindapsus pictus* cv. 'Argyraeus' (*Pothos argyraeus* Hort.) has spotted bluish-green leaves. *S. aureus*, or golden pothos, is *Epipremnum aureum*. The juvenile growth is glossy and bright green with irregularly splotched or marbled yellow or white leaves (Figures 27-11 and 27-12). 'Wilcoxii' is a cultivar of *E. aureum*.

The culture for *Scindapsus* is the same as the culture for *Philodendron*.

Spathiphyllum (*Spathiphyllum Clevelandii* or white Anthurium) is also known as the "Spathe flower" (Figures 27-13 and 27-14). Clevelandii is a listed name of no botanical standing. 'Mauna Loa' is a hybrid cultivar that is commonly available with several other cultivars. These plants prefer partial shade and will produce flowers even in very low light conditions.

Syngonium consists of 20 species of stout climbing vines with a milky juice. *S. podophyllum* (*Nephthytis liberiia* Hort.) is called the arrowhead vine because the leaves look like arrowheads (Figure 27-15). The leaves are glossy and green and may be variegated with silver, cream, or white markings. *S. wendlandii* (*Nephthytis wendlandii* Hort.) has three parted, dark green leaves

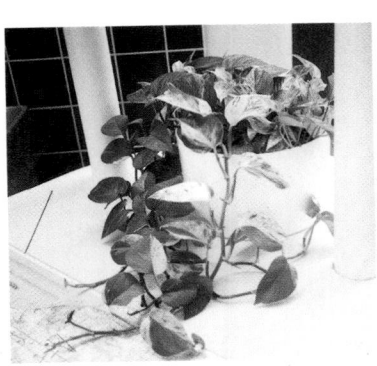

FIGURE 27-11 *Epipremnum aureum*, Golden Pothos thriving in a counter at a Traveler's Aid position in an airport lounge.

FIGURE 27-12 *Epipremnum aureum*, Golden Pothos on a retail bench.

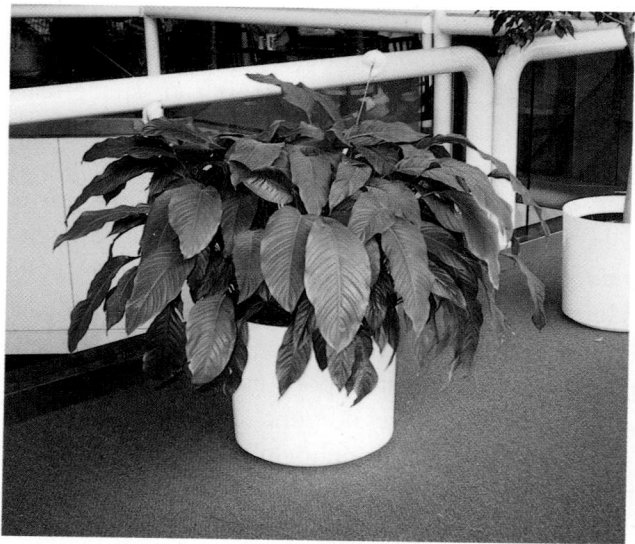

FIGURE 27-13 *Spathiphyllum* in an airport terminal building. This plant does well in low light areas.

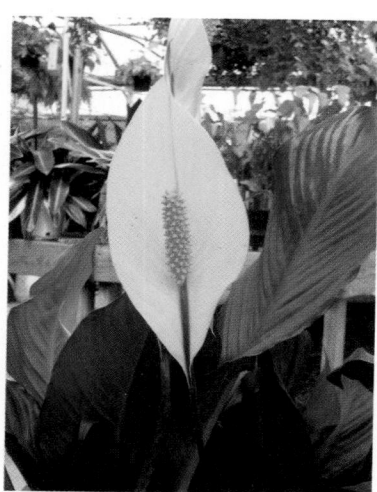

FIGURE 27-14 The flower of the *Spathiphyllum* is known as a spadix.

FIGURE 27-15 *Syngonium podophyllum* 'White Butterfly', Nephthytis or Arrowhead plant. (*Courtesy of PLANET, formerly the Associated Landscape Contractors of America.*)

with a velvety sheen. The culture of *Syngonium* is the same as that described for *Philodendron*.

ARALIACEAE

The Aralia or Ginseng family consists of herbaceous shrubs, trees, and vines that are widely distributed in both temperate and tropical climates. The juvenile growth habit and leaves may differ from the adult phase. Many of the species grown as pot plants never outgrow the juvenile phase. In their natural surroundings, some of the species grow to heights of 40 feet (12 m). *Brassaia actinophylla* (*Schefflera actinophylla* Hort.), or the umbrella tree or octopus tree, grow naturally from India to the Malay Peninsula, as well as in the Philippine Islands, northeastern Australia, and Hawaii. The leaves have long petioles and are **palmately compound** with five to seven or more leaflets (Figure 27-16). In the juvenile phase, each leaf rarely has more than three leaflets on a stem. In general, the flowering part, resembling an octopus with upside down tentacles, is seen only in the mature forms of the plant.

FIGURE 27-16 A well-grown *Brassaia actinophylla* in an airport connecting corridor where the plant receives the high light intensity it needs.

FIGURE 27-17 *Brassaia arboricola* is a smaller growing plant than *B. actinophylla*. This specimen is located in a lounge area.

Brassaia arboricola (Figure 27-17) is a low growing, shrublike plant. It is sometimes referred to as Hawaiian schefflera.

Brassaia will withstand dry conditions. A moderate light intensity should be provided. The plant is propagated from seeds and terminal cuttings and by air layering. Plants propagated by tissue culture methods are available from laboratories specializing in this means of propagation.

Dizygotheca, or false aralia, is native to New Caledonia and Polynesia. The plant has palmately compound leaves arranged in whorls at the end of the petiole. Other genera include *D. elegantissima* (*Aralia elegantissima* Hort.), *D. Kerchoveana*, and *D. Veitchii*.

The plants prefer a high light intensity, but not direct sunlight. They are grown in an epiphytic mix that is allowed to dry out between waterings.

Fatshedera, Ivy Tree, is also known as Botanical Wonder.

F. lizei is *Fatsia japonica* 'Moseri' crossed with *Hedera helix* var. *Hibernica*. This shrub was originally cultivated in France. The shrub has palmate leaves with three to five lobes. The stems are weak. The plant may grow to a height of 8 feet (2.4 m) or more. The best growth occurs in high light conditions.

Fatsia japonica (Thunb.), or Japanese fatsia, is native to Japan. It has thick stems with little branching. The palmate leaves are large and leathery, with 7 to 11 lobes. Propagation is by seeds, terminal cuttings, and root cuttings.

Polyscias is the true aralia. The shrubs and trees are native to Polynesia and tropical Asia. Frequently, the foliage is aromatic. These plants are often used as hedges in tropical areas. Propagation is by cuttings made from mature wood.

P. balfouriana (*Aralia balfouriana* Hort.) is known as the Balfour aralia. The leaves are 2 to 4 inches (5 to 10 cm) wide and have three leaflets. The stems are green, speckled with gray.

P. filicifolia (*Aralia filicifolia*), or fern leaf aralia, has purple branches with pinnate leaves. The leaflets vary in shape and size from 7-inch (17.5-cm) long oblongs to 1-foot (30-cm) long narrow, pinnatified, sharply toothed leaves. Both kinds of leaves are found on the same plant. The leaflets of the 'Marginata' and 'Variegata' cultivars have white margins. *P. filicifolia* is widely used for hedges in Key West, Florida.

P. fruticosa, or Ming aralia, has a feathery appearance. The leaflets vary in shape and size. *P. fruticosa* grows slowly and makes an outstanding specimen plant.

P. guilfoylei, or geranium leaf aralia, has pinnate leaves up to 16 inches (40 cm) long. The leaves have white margins or splashes.

Tupidanthus calyptratus is represented by one species native to tropical Asia. In cultivation it is a shrub with smooth leaves. In nature it is a climbing, woody vine with compound palmate leaves consisting of seven to nine leaflets. The plant looks very much like *Brassaia actinophylla*, but the leaves appear to be leathery. *Tupidanthus* can withstand an even lower temperature than *Brassaia*, as demonstrated in the freeze of 1977. It is very difficult to propagate from seed. The small quantity of seed produced has a low germination rate. The plant is generally propagated from cuttings. Several laboratories in California are supplying plants propagated by tissue culture methods.

ARAUCARIACEAE

Araucaria heterophylla, or Norfolk Island pine, differs from members of the Pinaceae family, the true pines (Figures 27-18 and 27-19). The Norfolk Island pine has cone scales without distinct bracts. Mature trees may reach a height of 200 feet (60 m). It makes an excellent specimen plant for indoor use.

Propagation is by tip cuttings of vigorous erect tips or seed. Plants showing spreading type growth are obtained from cuttings made from lateral branches.

ASTERACEAE (Compositae)

The Sunflower family is one of the largest families of plants. It contains roughly 950 genera with more than 20,000 species. This family is so large that it is divided into 12 or 13 tribes according to various characteristics. Two of its members are grown as foliage plants: *Gynura aurantiaca*, velvet plant; and *Gynura procumbens*, 'Purple Passion' vine. The major attraction of these plants is the velvety purple foliage. When the plants receive too little light, they lose their leaf color. The flowers are objectionable because they have a very bad odor. The plants are excellent for use in hanging baskets. Propagation is by terminal cuttings.

FIGURE 27-18 *Araucaria heterophylla*, Norfolk Island pine plants for sale in a greenhouse retail store.

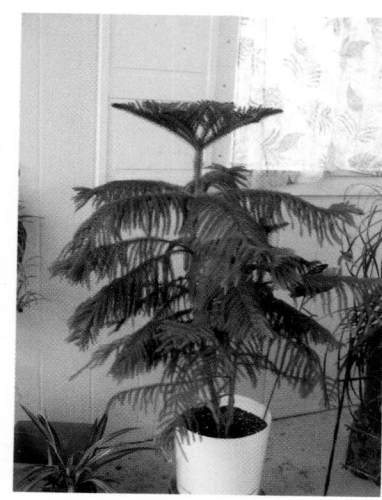

FIGURE 27-19 *Araucaria heterophylla*, Norfolk Island pine.

FIGURE 27-20 *Aechemea fasciata*, Bromeliad.

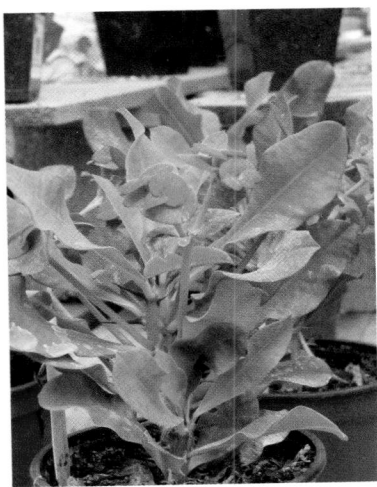

FIGURE 27-21 Crown of Thorns, *Euphorbia milii*, has flowers also known as cyathia, which are similar to poinsettia flowers.

FIGURE 27-22 Croton, *Codiaeum variegatum*.

BROMELIACEAE

Bromeliads are members of the pineapple family. A few plants are terrestrial in growth habit, but most are epiphytes. The leaves are strong and stiff. They grow in sheaths, one within another, to form a cup at the base of the plant. The flowers are borne on spikes, heads, racemes, or branching panicles (Figure 27-20).

Bromeliads are planted in an epiphytic mix because they need a well-drained soil with a pH ranging from acid to neutral. They are given frequent light waterings. In nature they grow in the filtered sunlight found high in the tops of trees. In the greenhouse, they need a high light level but must be protected from direct sunlight.

The most common genera grown are *Aechmea*, *Billbergia*, *Cryptanthus*, *Tillandsia*, and *Vriesea*.

Ananas, or pineapple, is grown commercially for its fruit. *A. comosus* is also grown as an ornamental. It requires a high light intensity at all times of the year.

Bromeliaceae are propagated by seeds and offsets. Offsets taken from pineapple plants are called **ratoons**.

COMMELINACEAE

The Spiderwort family consists of nearly 40 genera. Most of these are low growing herbaceous ground covers. Some of the plants are excellent for hanging basket culture. The most common basket type plants are *Tradescantia* and *Zebrina*, or wandering jew. Both types are very easy to propagate from seeds, terminal cuttings, or divisions.

Rhoeo spathacea, or Moses-in-the-cradle, is another member of this family. It is propagated easily from seed. The leaves are upright, with a dark green upper surface and a purple lower surface. Tiny white flowers appear in a boat-shaped envelope formed by two bracts growing at the bases of the leaves. This plant does well in full sunlight and in partial shade.

EUPHORBIACEAE

The Spurge family contains more than 283 genera and 7,300 species. The plants in this family are noted for their milky sap. A well-known member of the family is the poinsettia. An interesting member of the same genus as poinsettia is the crown of thorns *(Euphorbia milii)* (Figure 27-21).

Acalypha hispida, the chenille plant, and *Codiaeum variegatum*, or croton, are two members that are grown as foliage plants. Both need a high light intensity. The bright red and yellow colors of the croton fade to green if the light level is low. Many cultivars of *C. variegatum* are available to growers (Figure 27-22). One cultivar is the gold dust croton. New cultivars of *C. variegatum* need less light.

The plants are propagated from seeds and cuttings or by air layering.

GESNERIACEAE

Members of this family consist of herbaceous tropical plants that are terrestrial or epiphytic in growth habit. The roots of the plants may be fibrous rhizomes, scaly rhizomes, tubers, or stolons. The plants are commonly grown in greenhouses and in private homes under fluorescent lights. An open, well-drained soil is required. It must hold moisture without staying overly wet. A high humidity and warm temperatures are also required. The plants must be protected from direct sunlight. The maximum light intensity that can be tolerated without damage is 2,000 fc (21.5K lux).

The African violet (*Saintpaulia ionantha*) (see Figures 25-1 through 25-5), gloxinia (*Sinningia speciosa*), and hybrids of Streptocarpus belong to this family. The epiphytic *Aeschynanthus* and *Columneas* are well suited to hanging baskets. *Achimenes* and *Episcias*, or flame violets, are considered to be terrestrial plants.

LAMIACEAE (Labiatae)

The mint family is easily identified by the square stems of its members. The plants contain glands that secrete volatile oils with pungent aromas. Many members of the mint family are used as herbs in cooking.

Plectranthus australis, or Swedish ivy, is the most widely grown ornamental (Figure 27-23). Its trailing growth habit makes it an excellent plant for hanging baskets. The leaves are bright green with a waxy appearance. The leaves of *P. coleoides*, 'Marginatus,' have white edges.

LILIACEAE

The lily family contains 240 genera and more than 3,000 species. Many of its members are used in medicine; others have edible parts. Many members of this family are grown as ornamentals. Leaf tip burn is a problem with several of the foliage ornamentals in this family. Some relate the tip burn to excessive fluoride in the water or growing medium. When limestone is added to the medium to keep the pH above 6.5, tip burn does not occur. If fluoride is the problem, the high pH ties up the fluoride so that the plant cannot absorb it.

Tip burn is also caused by a low relative humidity, excessive drying of the plants, and high total soluble salt levels.

Asparagus setaceus (*Asparagus plumosus* Hort.), or asparagus fern, is used as a pot plant. The plumes are also cut for use in flower arrangements (Figure 27-24). Several cultivars are 'Cupresoides,' 'Nanus,' 'Pyramidalis,' and 'Robustus.'

Asparagus densiflorus (*Asparagus sprengeri* Hort.) is more woody than *A. setaceus* and has spines. The cultivars used are 'Myers,' 'Sprengeri,' 'Sprengeri Deflexus,'

FIGURE 27-23 Swedish ivy, *Plectranthus australis*.

FIGURE 27-24 *Asparagus densi-florus* 'Sprengeri', Asparagus fern. *(Courtesy of Ball Horticulture Co.)*

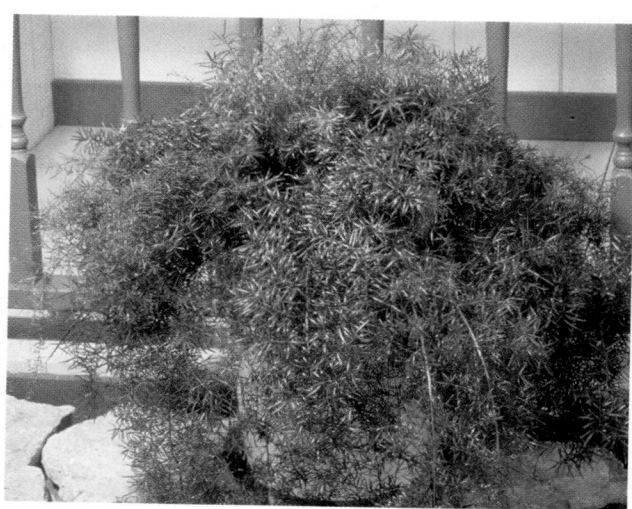

'S. Nanus,' and 'S. Robustus.' 'Robustus' has a faster rate of growth than 'Sprengeri.'

Both cultivars are propagated from seeds planted in the spring. The seed must be soaked in warm water for 24 hours before planting. It is sown 0.5 inch (1.25 cm) deep in the medium. When the medium is kept moist, the seed germinates in 3 weeks at 60°F to 65°F (15.5°C to 18.5°C). On the average, 50 percent germination is considered acceptable. However, newer cultivars have a better rate of germination. Eight to 9 months are needed to produce a 3-inch (7.5-cm) pot plant from seed. The plants grow best in a moderate light intensity. Full sunlight causes the foliage to turn yellow.

Aspidistra elatior, or cast-iron plant, has the ability to survive the poorest growing conditions. Aspidistra is native to Himalaya, China, Japan, and Taiwan. The cultivar 'Variegata' has variegated leaves. The plant is a slow grower. Propagation is by division in the early spring.

Chlorophytum comosum, or spider plant, is a rhizomatous herb. It is native to all continents except North America and Europe. It is easy to grow and can be propagated by divisions and seed. Small plantlets grow at the ends of the stems. These plantlets are commonly removed and started as new plants.

The production of the plants is controlled by the day length. When 8-hour days are given, the mother plant produces many plantlets. When the days are longer, few plantlets develop.

The leaves of *C. comosum* are all green. The leaves of the cultivar 'Mandaianum' have a bright yellow central strip. 'Variegatum' has leaves with a white margin, and 'Vittatum' produces leaves with a white central stripe. The variegated cultivars are often mixed in the trade.

To prevent tip burn in this member of the lily family, 1 pound (0.45 kg) of limestone is added to 1 cubic foot (28.3 l) of medium.

MARANTACEAE

This plant has the habit of folding its leaves at night, giving rise to its common name, the prayer plant. It is native to tropical America. Its colorful branching leaves form clumps of growth.

Maranta leuconeura Kerchoviana has variegated leaves spotted light and dark brown and green, and having a satiny luster (Figure 27-25).

Massangeana is a listed horticultural name for *M. leuconeura* var. *leuconeura*.

Calathea is similar in appearance to *Maranta*. It has brightly marked leaves that grow in clumps. Calatheas and Marantas require low to medium light conditions and must be shaded from direct sunlight. The temperature should be no less than 65°F (18.5°C) at night. The medium must be very moist and have good drainage. Calatheas are propagated by divisions of the crown, and tubers, and by terminal cuttings taken in the spring.

C. makoyana, the peacock plant, is native to Brazil. The upper leaf surface is cream or olive green with dark green oval spots along the veins and dark green borders.

MORACEAE

The mulberry family contains 1,400 to 1,850 species of trees, shrubs, and climbing plants in 53 to 75 genera. Most of the plants have a milky latex (sap), and

FIGURE 27-25 *Maranta leuconeura,* Prayer Plant. *(Courtesy of PLANET, formerly the Associated Landscape Contractors Association.)*

FIGURE 27-26 A well-grown specimen of variegated Java fig, *Ficus benjamina.* This plant is very sensitive to water stress and will readily drop its leaves if not watered properly.

FIGURE 27-27 A young plant of *Ficus elastica,* the rubber tree.

many have edible fruit. The most popular genus in the ornamental horticulture industry is *Ficus.* This genus contains nearly 800 species of trees, shrubs, and clinging vines with woody roots. Plants in this genus produce edible figs, fodder, natural rubber, and bark cloth.

Ficus benjamina, weeping fig or Java fig, is grown as a shrub or a tree (Figure 27-26). Early in its growth, it is epiphytic. Weeping figs have gracefully drooping branches and are popular as indoor specimens. Propagation is by seed and terminal cuttings.

A variegated form of *F. benjamina* has become popular. Ficus must be properly acclimated to interior environments or they will prematurely drop leaves in large numbers.

F. elastica, the rubber plant or India rubber tree, grows to be a large tree even in an indoor environment (Figure 27-27). The leaves are thick and smooth. In size, they are 5 inches (12.5 cm) wide by 12 inches (30 cm) long. The cultivar 'Abijan,' named for Abijan, South Africa, is a recent introduction. Its dark leaves are almost purple. 'Decora' has broad, dark, glossy green leaves with an ivory midrib and a red underside. 'Doescheri' has marbled grey-green leaves

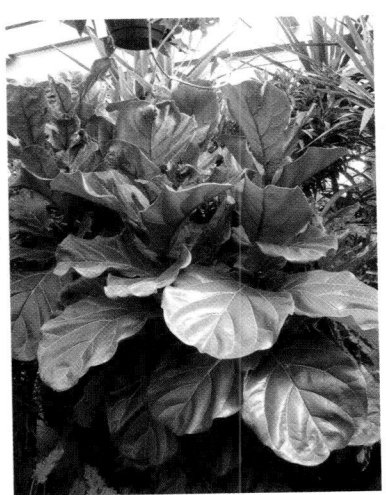

FIGURE 27-28 *Ficus lynata,* Fiddleleaf Fig.

with creamy yellow midribs and pink petioles. 'Variegata' has white or yellow leaves with light green margins.

F. lynata (*F. Pandunata* Hort.), or fiddle leaf fig, has large, violin-shaped leaves (Figure 27-28). In nature it grows to a height of 40 feet (12 m) and more. Two or three stems planted in a large tub make a more attractive specimen than a single plant.

F. pumila, or creeping fig, is a vine that clings to walls by its roots. 'Variegata' is a green and white cultivar.

F. retusa, or Indian laurel tree, is incorrectly called *F. nitida,* which is a variant of *F. benjamina.*

MYRSINACEAE

The most prominent ornamental member of the Myrsine family is *Ardisia crenata,* or coral berry. This plant is grown for its crisp foliage and long-lasting, bright red berries. *A. crispa* is often confused with *A. crenata.* The distinguishing feature is that the twigs of *A. crenata* are slightly pubescent (hairy) when young.

PALMAE (Arecaceae)

The members of the palm family are often called the aristocrats of the foliage world and are widely used as ornamental plants.

Palms are propagated from seed in October and November. The seed must be no more than 2 weeks old when it is planted or it will not germinate. Even when fresh seed is used, the germination percentage is very poor.

A partial list of dwarf or semidwarf palms suitable for pots or tubs consists of the following plants.

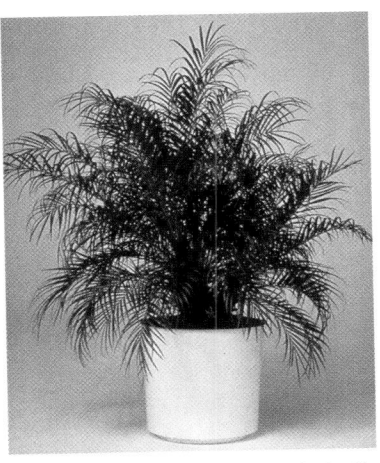

FIGURE 27-29 *Phoenix roebelenii,* Dwarf Date Palm. *(Courtesy of PLANET, formerly the Associated Landscape Contractors of America.)*

- *Chamaedorea cataractarum* and *C. costaricana,* or fish tail palms, tolerate cool temperatures.
- *C. elegans* (*Neantha Bella* Hort.) and *C. Ernesti-Augusti* are both known as the parlor palm.
- *C. erumpens* and *C. seifrizii* are known as the bamboo palm. Both types can withstand cooler-than-average temperatures.
- *Chamerops humilis* is the European fan palm.
- *Howea belmoreana* and *Howea forsterana* (Kentia palm Hort.) are the sentry palms.
- *Licuala grandis* and *L. spinosa* tolerate some cold. They need a preventive mite control. They are slow growers and require 4 to 5 years to reach a height of 18 to 24 inches (45 to 60 cm).
- *Livistona rotundifolia* is called the little footstool palm.
- *Phoenix roebelenii,* or dwarf date palm, is an excellent ornamental plant (Figure 27-29). There are 13 species. The palm has a very slow rate of growth and is somewhat cold tolerant.
- *Reinhardtia gracilis,* or window leaf palm, prefers deep shade and low fertility levels.
- *Rhapis excelsa* and *R. humilis* are known as lady palms. Both species tolerate cold temperatures and are slow growers. An unnamed *Rhapis* species known as "Thai dwarf" requires 5 to 6 years to reach a saleable size from seed.

The following list gives palm species that make good tub or pot plants when they are young. However, as they get older, they must be moved to larger containers, discarded, or planted in the ground.

Acoelorrhaphe Wrightii
Aiphanes caryotaefolia
Archontophoenix alexandrae
Archontophoenix Cunninghamiana
Arecastrum Romanzoffianum (Queen Palm)
Butia capitata
Caryota mitis
Caryota urens
Cocos nuncifera (Coconut palm)
Dictyosperma album
Heterospathe elata
Livistona chinensis (Chinese fan palm)
Phoenix canariensis
Phoenix reclinata
Phoenix rupicola
Thrinax radiata
Trachycarpus Takil (windmill palm)
Veitchii Merrillii (Christmas palm)
Washingtonia robusta

After they reach a saleable size, palms can be kept in the same container. When the palm is repotted each year, one-quarter of the root system is removed and the palm is replaced in the same pot. New medium is added to fill around the plant. New roots will grow into the potting medium. The plants should not be exposed to temperatures below 35°F (1.5°C).

PANDANACEAE

The members of the screw-pine family have stiff and leathery leaves. Two important species horticulturally are *Pandanus Veitchii*, which has variegated foliage, and *Pandanus utilis*, which has the spiral growth character that gives the family the name screw-pine. They are swordlike with pointed spines on the margins. Plants in the juvenile stage are used as ornamental specimens. The plants in this family need constant moisture and warmth. They are propagated by suckers that arise at the base of the old plants. These suckers are removed and planted to separate pots. The success of rooting depends upon bottom heat. The plants can be propagated from seed when it is available.

PIPERACEAE

The most commonly seen species of the pepper family is the *Peperomia*. Plants in this group require warm temperatures, a low light intensity, a high level of atmospheric humidity, and a low moisture level in the growing medium. Many species are epiphytic. Propagation is by stem cuttings, leaf or leaf-petiole cuttings, and division of the crowns.

Peperomia argyreia (*P. Sandersii* Hort.), or watermelon peperomia, has dark green leaves with silvery-gray radiating rays. The leaves look like small watermelons.

P. caperata, 'Emerald Ripple,' has glossy, dark green, pleated leaves. The leaves of the cultivar 'Variegata' have a broad white margin with a central zone of green.

P. griseoargentea 'Nigra' has dark leaves with a black-green coloring along the veins.

P. obtusifolia, or baby rubber plant, is available as the cultivar 'Variegata.' The leaves have broad, irregular, creamy white margins and central areas blotched with grey-green. The amount of nitrogen fertilization and the light intensity affect the amount and intensity of the variegation.

PITTOSPORACEAE

The Pittosporum family consists of trees, shrubs, and some woody climbers. *Pittosporum tobira,* mock orange or Australian laurel, has thick leathery leaves borne on coarse stems. The cultivar 'Variegata' has green leaves variegated with white. This plant is a slow grower indoors and holds its shape well.

POLYPODIACEAE

Ferns belong to the family Polypodiaceae of the order Filices. There are about 7,000 species of ferns in 180 genera. Most ferns used as ornamental plants are low growing, but some are tree ferns. Ferns normally are found on the floor of the forest or jungle where they receive filtered sunlight of low intensity. Therefore, in the home environment, they need low to medium light. They also require high moisture levels. Most ferns are grown as specimen plants because they have the potential to grow to a large size.

Ferns are propagated by divisions of the clumps, runners, and spores. The production of ferns from spores is an exacting procedure. The average grower should not attempt this type of propagation but should obtain plants from a specialty producer.

Tissue culture techniques are now used to propagate many ferns. Spores are sown on sterilized sphagnum peat moss in pots or flats. The containers are covered with glass or plastic and placed in 80 to 85 percent shade (2,000 fc) (21.5K lux) at 75°F to 86°F (24°C to 30°C). *Prothalli* appear in 2 weeks. These liver-shaped organs produce male and female structures that then produce the fern plant in 3 to 4 months.

The more popular species, varieties, and cultivars grown include the following: *Adiantum cuneatum,* maidenhair fern; *Asplenium nidus,* bird's nest fern; *Cyrtomium falcatum* 'Rochfordianum,' holly fern (so named because the leaves look like those of the holly plant).

A very popular plant is *Nephrolepis exaltata* 'Boston Express' (Figure 27-30). There are many cultivars of Boston including 'Compacta,' which is a dwarf Boston fern, 'Fluffy Ruffles,' 'Porters,' 'Roosevelt,' and 'Sword Fern.'

Platycerium, or staghorn fern, is an epiphytic fern (Figure 27-31). Seventeen species of *Platycerium* are grown commercially. In nature they grow in trees. *P. bifurcatum* and *P. vassei,* which have an erect growth habit, are the most popular species. They can be planted in sphagnum moss wired to a board.

FIGURE 27-30 *Nephrolepis exaltata bostoniensis*, Boston Fern. *(Courtesy H. Edward Reiley.)*

FIGURE 27-31 A large specimen of the staghorn fern, *Platycerium bifurcatum.*

An occasional fertilization with dried blood or bonemeal is sufficient. Tree fern slabs instead of a board are also used as a base.

The staghorn ferns prefer a high light intensity but not direct sunlight. Warm temperatures are needed.

Pteris cretica, brook fern, is a small fern that is often grown as a table specimen. The cultivars commonly grown include 'Albo-lineata,' 'Cristata,' 'Parkeri,' 'Roeweri,' 'Wilsonii,' and 'Wimsettii.' *P. ensiformis* 'Victoriae' is known as the silver leaf fern.

Polypodium aureum, the rabbitsfoot fern, has surface creeping rhizomes that look like the feet of a rabbit. *P. scolopendria* is the wart fern. It is an epiphytic fern with creeping rhizomes that are sea green in color with dark scales. The leaves have prominent son (singular, **sorus**), that look like warts.

Many other ferns are grown commercially. The student is encouraged to investigate in more detail this fascinating group of plants.

RUBIACEAE

The Madder family contains nearly 400 genera and 4,800 to 5,000 species. Coffee and gardenia are members of this family.

Coffea arabica, coffee, is grown both as a commercial food crop (coffee beans) and as an ornamental plant. The coffee plant is an attractive plant with lustrous, dark green foliage. The ruby red berries turn plum purple when they are ripe. The coffee plant is propagated by terminal cuttings and seed. The seed should be ripe and the surrounding husk removed to improve germination when it is sown. The best percentage of germination occurs when bottom heat is applied and a high moisture level is maintained. The plants do best in full light, but they will tolerate very light shade.

RUTACEAE

The Rue family contains 150 genera and nearly 1,600 species. Citrus fruits are members of this family. A familiar ornamental is *Citris × citrofortunella mitis*, the Calamondin orange. It is propagated from cuttings in full light intensity. Plants grown from seed seldom flower or fruit. The calamondin orange is a popular house plant because of its bright green leaves and attractive fruit. The juice is well flavored but very acidic.

C. Limon cv. 'Ponderosa' produces large lemons and is often grown at one end of the greenhouse. The fruit may be 4 to 5 inches (10 to 12.5 cm) in diameter. The fruit is quite pithy and can be used to make pies.

SAXIFRAGACEAE

Several members of the Saxifrage family are used as ornamentals. *Saxifragaceae stolonifera* (*S. sarmentosa* Hort.) produces long stolons or runners like strawberries. In fact, the plant is called the strawberry geranium. The leaves have long petioles and coarse teeth. The underside of the leaf is red, and the upper surface is veined with silver. The plants are used in hanging baskets. They are grown at a night temperature of 55°F (13°C). Propagation is by runners and division of the large plant.

URTRICACEAE

Soleirolia, baby's tears, is an attractive ornamental in the Nettle family. It is a perennial that grows to form a small, dense mat. In milder climates it is a useful ground cover as an ornamental, but it is usually grown in hanging baskets.

VITACEAE

FIGURE 27-32 *Cissus rhombifolia* 'Mandaiana', Grape Ivy. (*Courtesy of PLANET, formerly the Associated Landscape Contractors of America.*)

The Grape family contains woody vines that climb by means of **tendrils.** Two members of this family are grown for their ornamental value: *Cissus antarctica* (*Vitis antarctica* Hort.), the kangaroo vine, and *Cissus rhombifolia* (*Vitis rhombifolia* Hort.), the grape ivy (Figure 27-32). Both plants are compact growers that are excellent in hanging baskets. They are propagated by terminal cuttings and stem section cuttings.

A table (Table 27-4) summarizing the methods of propagation used for the species listed in the chapter is included in the section on acclimatization.

Foliage Plant Acclimatization

Many flower growers and retailers receive foliage plants from Florida and other southern locations for resale in the North. Some of these plants are in the form of rooted cuttings or small plants. At this early stage of development, they are potted and placed in the greenhouse until they grow to a saleable size. These plants should remain in the greenhouse at least 4 weeks before they are made available for sale. The grower is doing the customer a disservice if the plants are offered too early.

Other plants are obtained as larger specimens for use in interior decoration. These plants also require special handling. The expanded market for **interior foliage plants** has caused producers to force plants at an accelerated growth rate so that more crops can be delivered to market. This accelerated rate is achieved by using large amounts of fertilizer and water and exposing the plants to as much light as they can withstand. As a result, the plants shipped to northern growers are conditioned to an accelerated rate of growth.

If these plants are then placed in an indoor environment without a period of adjustment, they suffer from shock. The average indoor environment has a low light

Table 27-4

Methods of Propagation of a Selected Group of Foliage Plants			
KEY. The preferred method is given first. Alternate methods follow.			
A	Seeds	I	Ratoon crowns
B	Stem tip cuttings	J	Offshoots young plants
C	Stem segment cuttings	K	Tubercles formed near leaves
D	Leaf bud cuttings	L	Bulbs
E	Leaf segment cuttings	M	Air layering
F	Root cuttings	N	Spores
G	Rhizomes	O	Semimicropropagation-tissue culture
H	Divisions of plants		

Foliage Plant	Method of Propagation
Aglaonema	B, A, C, H
Anthurium andraeanum	A (must be fresh), H
Aphelandra squarrosa	A, D, C
Araucaria heterophylla	A, B (use only terminal cuttings)
Ardisia crispa	E, A
Asparagus setaceus	A (only fresh seeds), H
Aspidistra elatior	H
Aucubajaponica	B
Begonia argentea	E
Begonia masoniana	E
Begonia semperflorens	A, B
Brassaia actinophylla	A, B, M
Carissa grandiflora	B
Chlorophytum comosum	J
Cissus antarctica	B, D
Citro fortunella mitis	B
Clerodendrum thomsoniae	B, D, A
Codiaeum variegatum	B, M
Coffea arabica	A
Cordyline terminalis	B, C
Crassula arborescens	E, B
Cyperus alternifolia	A
Dieffenbachia amoena	B, C
Dizygotheca elegantissima	B, A, C
Dracaena australis	B, C
Dracaena deremensis	B, C
Dracaena marginata	B, C
Dracaena sanderiana	B, C
Epipremnum aureum (Pothos aureus) (Scindapsus aureus)	B, D (layering)
Fatshedera lizei	B, D
Fatsia japonica	A, B
Ficus benjamina	A, B
Ficus elastica 'Abijan'	A, M, B, D
Ficus elastica 'Decora'	A, M, B, D
Ficus lyrata (F. pandurata)	A, M, B, D

(continues)

Table 27-4 *(Continued)*

Foliage Plant	Method of Propagation
Ficus pumila	A, H (layering)
Ficus retusa (F. nitida)	A, B
Fittonia verschaffeltii	B, M
Gardenia jasminoides	B
Gynura aurantiaca	B, D
Hedera canariensis	B, C
Hedera helix	B, C
Hoya carnosa	D
Maranta leuconeura	H, B
Monstera deliciosa (P. pertusum)	A, B, D (layering)
Pandanus veitchii	B
Pellionia pulchra	B
Peperomia argyreia (P. sandersii)	B, D, E (layering)
Peperomia caperata 'Emerald Ripple'	B, D, E (layering)
Peperomia metallica	B, D, E (layering)
Peperomia obtusifolia	B, D, E (layering)
Peperomia verschaffeltii	B, D, E (layering)
Philodendron bipennifolium (P. panduraeforme)	B, D, M, A
Philodendron oxycardium (P. cordatum)	B
Philodendron scandens	B, D
Philodendron selloum, Saddle leaf	B, H, A
Philodendron wendlandii	H, A
Pilea cadierei	B, H
Pilea involucrata	B, H
Pilea microphylla	B, H
Pittosporum tobira	B, A
Plectranthus australis	B
Podocarpus gracilior	B
Podocarpus macrophyllus	B
Polyscias balfouriana	B
Polyscias fruticosa	B
Polyscias guilfoylei	B
Rhoeo spathacea (R. discolor)	A
Sansevieria trifasciata	F, D (does not come true)
Saxifraga stolonifera	B
Schlumbergera bridgesii	B, C
Schlumbergera truncatus	B, C
Senecio mikanioides	B, C
Soleirolia	H
Spathiphyllum 'Clevelandii'	H
Streptocarpus × *hybridus*	A, G
Syngonium wendlandii	H, A
Tolmiea menziesii	E
Tradescanti albiflora	B, D

(continues)

Table 27-4 *(Continued)*

PALMS

Caryota mitis (dwarf fishtail palm)	All palms are propagated by seeds.
Chamaedorea costaricana (showy bamboo palm)	
Chamaedorea elatior (Mexican rattan palm)	
Chamaedorea elegans (Neanthe bella palm)	
Chamaedorea erumpens (bamboo palm)	
Chamaedorea metallica (miniature fishtail)	
Chamaedorea stolonifera (climbing fishtail palm)	
Chrysalidocarpus lutescens (Areca palm)	
Howea belmoreana (sentry palm)	
Phoenix roebelenii (pigmy date palm)	
Rhapis excelsa (Flabelliformis) (lady palm)	

FERNS

Adiantum species	Ferns may be propagated by divisions of the clumps or spores. Propagation of ferns from spores is an exacting procedure requiring an expert. Plants from spores need 2 to 3 years to reach saleable size. Ferns may also be propagated by tissue culture methods.
Asplenium nidus (bird's nest fern)	
Cyrtomium falcatum 'Rochfordianum'	
Nephrolepis exaltata 'Boston Express' (Boston fern)	
Platycerium	
Polypodium vulgare virginianum (wall fern)	
Pteris species	

intensity in the range of 100 to 200 fc (1.08K to 2.16K lux). The relative humidity is low, and watering is done on a weekly basis. In addition, the temperature is variable. If the plant is not acclimatized to the new environment, it will drop from 30 to 75 percent of its leaves. In extreme cases, some plants have shed all of their leaves.

Acclimatization is a process by which a living organism adapts from one set of environmental conditions to a different set of conditions. Human beings adapted to the unfriendly environment of the moon by taking life support systems to the surface. Plants cannot do this. They must adapt to their new surroundings or die.

Acclimatization Factors

Light is the most important factor to be considered in the acclimatization of plants to new surroundings. To give plants the time to adjust to reduced light, they are held for several weeks at an intermediate light intensity. That is, the light level is lower than the level at which they are grown, but it is greater than the level of the indoor environment in which they will be placed. The plants are placed in a shaded greenhouse where the maximum light intensity during the brightest part of the day is 1,500 to 2,000 fc (16.1K to 21.5K lux).

The plants can also be placed in an area lighted by artificial light alone. The plants are exposed to 300 to 500 fc (3.23 to 5.4K lux) from cool-white fluorescent lamps for not less than 12 hours a day.

As soon as the plants are received, they are placed in the reduced light area. They are inspected for damage, insects, and diseases. If there are signs of insects, such as eggs or insects in other immature stages of development, the plants are sprayed. This treatment will prevent the spread of the insects to other plants.

During the acclimatization period, the first watering is heavy to reduce high soluble salt levels. The waterings are then reduced gradually until the plants are being watered twice a week. Then the watering is reduced to once a week.

The plants receive no fertilizer because they need none during this period of reduced growth. When the plants are kept in an indoor environment, they need about one-tenth the amount of fertilizer required to produce them. The plants can be kept healthy by applying fertilizer every 3 months.

Light Levels for the Indoor Environment

The question of how much light a plant needs to survive depends on many factors. The most important factor, of course, is the natural preference of the plant itself. Is it a low light plant, such as a fern, or is it a high light plant, such as *Ficus benjamina*? The consumer should be familiar with the light needs of all foliage plants purchased.

Another factor is whether the plants were grown in full sun or in partial shade. Plants grown in full sunlight need more light in the indoor setting than do plants grown in the shade. The following guidelines will assist the consumer in providing adequate light.

- Plants grown in the sun need a minimum of 150 fc (1.62K lux), 12 hours a day, 7 days a week.
- Plants grown in the shade need a minimum of 75 fc (0.81K lux), 12 hours a day, 7 days a week.

If the light intensity and the length of exposure to the light are lower than the recommended levels, the life of the plant is reduced.

Foliage plants in office buildings that are closed and unlighted on weekends will have a short life. However, if the plants receive natural daylight, or if the light is supplemented with artificial light, the plants should last for several years (if they are not abused in other ways).

In general, the number of weeks of acclimatization required is related to the type of growth of the plant and the light necessary to produce that growth. *Brassaia actinophylla* has a high light requirement for growth. Thus, it needs 6 to 8 weeks to become acclimated. *Spathiphyllum* 'Mauna Loa' grows in low light and needs only 1 to 2 weeks for conditioning.

Increased knowledge about the process of acclimatization has led many southern growers to begin the process while the plants are growing. As a result the plants are improved in quality, and less time is required by the northern grower to complete acclimatization.

Methods of Propagation

Table 27-4 summarizes the various methods used to propagate foliage plants. The details of these methods are covered in Chapters 17 and 18. The preferred method of propagation is identified in the table by the first letter given. Alternate methods that can be used follow in descending order of preference. In some cases, the choice of method depends upon the experience of the grower. It should be noted that more foliage plants are listed in the table than are covered in the detailed information given in this chapter.

Problems

The problems given are general to all foliage plants and can be summarized as follows.

Cultural Disorders

Leaves turn yellow and fall for one or more of the following reasons.

- Air pollution—gases
- In storage in the dark for too long a time in shipment
- Too little water
- Light intensity too low
- Failure to acclimatize plants properly
- Chilling injury
- Lack of oxygen from overwatering
- Root damage due to soil insects or diseases—any condition that damages roots and prevents the uptake of nutrients and water
- Light intensity too high for low light crops

Brown or burned leaf tips, especially on members of the lily family, are due to the following:

- Fluoride toxicity
- Low relative humidity and moisture stress
- Too high an air temperature
- High soluble salt levels
- Overfertilization

Weak growth may be the result of the following:

- Too low a light intensity
- Lack of fertilizer
- Large amount of soil nematodes
- Root rot organisms
- Poor root system due to a cold medium, disease infestation, or root-chewing insects

Diseases

Bacterial and fungal diseases are as common on foliage crops as on other crops. The warm, humid conditions that are often recommended for growth are also ideal for the development of disease.

The grower must inspect all plants received to ensure that as many as possible are disease free. Sanitary growing practices must be followed in each phase of culture. The foliage must be kept as dry as possible to limit the spread of disease, especially soft rot. Approved preventive sprays are applied at the rates recommended on the labels.

Insects

Foliage plants are attacked by a large number of insect pests. The use of the proper control measures will minimize infestations. Aphids, caterpillars, mealybugs, red spider mites and its many relatives, scales, thrips, and whiteflies all attack foliage plants.

ACHIEVEMENT REVIEW

Select the best answer or answers to complete each statement. List the appropriate letter(s).

1 Since 1976, the net value of the foliage industry has risen _____ times.
   a. two
   b. four
   c. five
   d. ten

2 In 1994, it was estimated that foliage plants accounted for approximately:
   a. one-fourth of all pot plants sold in the United States.
   b. one-third of all pot plants sold in the United States.
   c. one-half of all pot plants sold in the United States.
   d. over one-half of all pot plants sold in the United States.

3 Large numbers of foliage plants are grown in Florida and California because:
   a. they like foliage plants.
   b. plants can be grown outdoors without fear of frost.
   c. foliage plants do best in sandy soils.
   d. foliage plants require warm temperatures.

4 The highest light intensity under which foliage may grow outdoors is:
   a. 5,000 fc (53.8K lux).
   b. 7,500 fc (80.1K lux).
   c. 10,000 fc (107.7K lux).
   d. 12,000 fc (129.2K lux).

5 Epiphytic plants grow best in:
   a. highly aerated soil mixes.
   b. moderately aerated soil mixes.
   c. poorly aerated soil mixes.
   d. flooded soils.

6 Foliage plant soils usually contain a very large amount of:
   a. sandy material.
   b. small gravel.
   c. vermiculite.
   d. organic matter.

7 AquaGro® and Surfside® are added to soil mixes to:
   a. provide nitrogen to the plants.
   b. increase the potassium content of the media.
   c. improve the drainage of the media.
   d. increase the water-holding capacity of the media.

8 A 1-bushel measure of soil contains:
   a. 2 gallons (7.6 l).
   b. 1 cubic foot (28.6 l).
   c. 1.25 cubic feet (35.8 l).
   d. 3 cubic feet (85.8 l).

9 The preferred fertilizer for foliage plants has an analysis of:
   a. 10-64.
   b. 10-10-10.
   c. 20-10-20.
   d. 20-5-30.

10 The most important environmental factor affecting foliage and the one least understood is:
   a. soils.
   b. water.
   c. temperature.
   d. light.

11 The plant known as *Monstera deliciosa* is often called:
   a. *Philodendron selloum.*
   b. *Philodendron hastatum.*
   c. *Philodendron pertusum.*
   d. *Philodendron bipennifolium.*

12 The cast-iron plant gets its name because it:
   a. is rusty red in color.
   b. cannot be cut with a knife.
   c. is very heavy.
   d. can withstand very poor growing conditions.

13 Palms are propagated by:
   a. terminal cuttings.
   b. root divisions.
   c. seeds.
   d. layering.

14 Ferns are propagated by means of:
   a. spores.
   b. divisions.
   c. runners.
   d. all of the above.

15 The amount of fertilizer needed by a plant growing in an indoor environment is _____ that used to produce it outdoors.
   a. the same as
   b. one-half
   c. one-fifth
   d. one-tenth

16 Members of the *Lillaceae* family may suffer tip burn due to:
   a. overwatering.
   b. very low humidity.
   c. fluoride toxicity.
   d. potassium deficiency.

Bedding Plants

OBJECTIVES:

After studying this chapter, the student should have learned:

■ the most important bedding plants grown.

■ the proper conditions for seed germination.

■ correct methods for sowing seeds.

■ the practice of transplanting seeds.

■ cultural methods of growing high-quality bedding plants.

The branch of the horticulture industry known as the bedding plant business has shown a steady increase in sales for more than 40 years. Most homeowners include bedding plants in the landscaping around their homes.

The development of lightweight film plastics allowed growers to construct low-cost greenhouse structures to house the larger numbers of plants required. After World War II, the bedding plant business moved from cold frames into the greenhouse and has stayed there ever since. The bedding plant business is the least expensive way to get started on a lifetime career in horticulture.

The Production of Bedding Plants

Record Keeping

Success in any business requires organization and well-kept records. This statement is particularly true when growing bedding plants. The best source of information on schedules, culture requirements, and the response of plants to the specific conditions of the greenhouse is the record the grower keeps of previous crops.

The type of information recorded should include notes on the types of plant materials used (seed or transplants), the media, growing conditions, and marketing information.

Seed

If seed is used, the information recorded should include the company the seed was purchased from, the date ordered, date delivered, cost per pack, number of packs purchased, date sown, date germinated, number of seeds or packets used per seed flat, percentage germination, number of good seedlings transplanted, germination medium used, watering methods, bottom heat temperature, air temperature (day and night), and weather conditions during the germination period.

Transplants

The information to be recorded for transplants includes the number of transplants purchased, where they were purchased, the cost of the transplants, the date transplanted, growing medium used, type of growing container, source of growing container, price paid for containers, and the number purchased. Notes should be kept on how well the plants grow in the containers used.

Media

Information to be recorded concerning the media includes the cost if the media were purchased, the components and amounts used if the grower mixed the media (indicate units of soil, peat moss, perlite, or sand used and the amounts of fertilizer and limestone added in pounds or ounces), and type of sterilization (such as steam or soil fumigant). If a soil fumigant is used, record the amount applied.

Growing Conditions

The grower should record the type of fertilizers used, the method and schedule for applying the fertilizer, the greenhouse air temperatures (day and night), crop response to greenhouse conditions, and the weather. The rate of growth of the crop is affected by the amount of sunshine or cloudy weather and the amount of rainy or clear weather. The grower should also include notes on fungicide and insecticide treatments, giving the material used and the amount applied. These records are also necessary to be in compliance with new worker-protection-standard laws. If smoke fumigants are used, the information recorded should include the volume in cubic feet of each greenhouse. This information enables the grower to apply the proper amount of fumigant. An overdose can cause plant injury. If too little fumigant is applied, the insect population is unharmed. The date when each planting is ready for market is also recorded.

Marketing

Weekly records should be kept of the colors and varieties of plants that sell well. Items that are not selling must be recorded as well. Accurate records of sales help the grower plan the order for the following year.

Other Records

The grower must record all expenses, such as labor, heat, taxes, supplies, and vehicle operating and maintenance costs. Not only does the Internal Revenue Service require such records, but the grower also will find such information helpful in determining if a profit was made for the year.

Ordering Seeds

Good crops start with good seeds. Orders should be placed with a reputable seed supply house that will stand behind its products. Orders for seed are usually placed in the late summer or early fall. Bargains in seed may be more costly than good seed because of a poor percentage of germination and low-quality plants.

Some growers choose to only order seeds on an as-needed basis to reduce a large upfront, out of pocket expense; and others choose to order an annual supply based on experience and sales records. If an annual supply is ordered, proper storage facilities must be maintained to protect the investment. If seeds are ordered on an as-needed basis, shipping interruptions are common, which include bad weather and delivery service employee strikes.

Germination Methods

The correct methods for germinating seed are presented in Chapter 17. The peat-lite media are used to germinate bedding plant seeds. Germination usually occurs in 3 to 5 days. Begonias and petunias may need slightly more time to germinate. As soon as the seed germinates, the containers are removed from the mist spray or plastic cover used to maintain a high humidity. Sturdy plants will develop if they are given full sun, fresh air, and a night temperature of 50°F to 60°F (10°C to 15.5°C). Strong root systems develop in cool temperatures. The plants need good roots to withstand the shock of transplanting. The plants are hardened to help reduce damping off, a seedling disease.

Germination Medium

Some growers prefer to start seeds in a modification of the standard peat-lite mixes used to grow crops. This mix has a lower total soluble salt content than the regular peat-lite mixes.

Seeds of plants grown for spring sales contain all of the food needed to make them germinate. After the new leaves have formed, the seedlings must depend upon photosynthesis to make food. Nutrients taken up through the root system combine with the newly formed plant foods to give the plant the energy it needs to grow.

Seedling root systems are very sensitive to high soluble salt levels. While the seedling is in the germination flat, the greatest requirement is for nitrogen (Figure 28-1). A small amount of phosphorus is also taken up. These nutrients are provided by the germination medium described in Table 28-1.

Three to 4 gallons (14.8 to 20 l) of water are added to each cubic yard (cubic meter) of germination medium. The ingredients must be mixed thoroughly to ensure complete distribution.

The seed flats are filled with the medium and are wetted thoroughly. The flats are allowed to stand for several hours or overnight and are then rewetted. The seeds are sown after the second wetting (Figures 28-2 and 28-3). See Chapter 17 for plug seedling production.

Transplanting

Seedlings are ready for transplanting when they have two or three sets of true leaves. Some growers recommend that the seedlings be transplanted when the

FIGURE 28-1 Seedlings are germinated in a peat and vermiculite medium. Extra vermiculite has been added to the surface of the medium to help retain moisture.

Table 28-1

Modified Cornell Peat-Lite Medium for Germinating Seeds		
Materials	Quantity Used in 1 Cubic Yard[1]	Quantity Used in 1 Cubic Meter[2]
Sphagnum Peat Moss	13 bushels	0.56 m^3
Vermiculite, No. 3 or No. 4 Grade	13 bushels	0.56 m^3
Dolomitic Limestone	5 pounds	3.0 kg
Finely Ground 20% Superphosphate	2 pounds	1.2 kg
Ammonium Nitrate	0.75 pound	0.4 kg
Trace Elements *(use only one)*		
ESMIGRAN	2 pounds	1.2 kg
or PERK®	2 pounds	1.2 kg
or FTE 555	1 ounce	36.8 g
Wetting Agent		
Liquid:		
Aquagro®		
or Surfside®	3 ounces	110.5 g
Granular:		
Aquagro®		
or Surfside®	1.5 quarts	1.41 liters

1. A cubic yard equals 27 cubic feet or approximately 22 bushels. During mixing, 15-20 percent shrinkage takes place. Therefore 4 bushels (5 cubic feet) are added to obtain a full cubic yard.
2. To compensate for shrinkage (15-20 percent) during mixing, 0.12 m^3 extra are added.

first true leaves appear after the cotyledon leaves. However, most plants are still too small to be handled without damage at this stage.

If the seedlings are allowed to grow to a height of 2 to 3 inches (5 to 7.5 cm) in the seed flat (Figure 28-4), they become hard and are starved for nutrients. When these seedlings are transplanted, the plants are of inferior quality. Methods of holding seedlings in storage are covered in Chapter 17.

Only the number of seedlings that can be planted in a few minutes should be removed from a flat at one time. The seedlings should not be exposed to

FIGURE 28-2 Overhead mist or traveling booms are used to water seed flats.

FIGURE 28-3 A fine spray nozzle is used to water the seedlings after the flats are moved from the germination environment.

FIGURE 28-4 Seedlings sown in rows. The seedlings in the center flat are nearly at the proper size for transplanting.

direct sunlight because the roots are tender and easily killed. The transplanted seedlings are watered immediately. They are protected for the first 24 hours from conditions that cause excess moisture stress. In other words, the plants are placed under temporary shade or are misted from overhead several times a day to prevent wilting.

Medium

Bedding plants are normally grown in the Cornell peat-lite mixes and commercial products based on the Cornell mixes. High-quality bedding plants can be grown in many different kinds of media. There are several advantages in using a standardized growing medium: It is readily available, the nutrient content is always the same, it is easy to handle, and high-quality crops are produced.

The formulas for the peat-lite mixes are given in Chapter 12. Whenever a grower changes a peat-lite mix, there will be a trial-and-error period while the grower determines how the crops respond to the changes. Both the expert and the beginning grower will find that using the peat-lite mixes is one of the most important factors in growing high-quality bedding plants.

Figures 28-5 through 28-10 show various stages in the process of filling and handling bedding plant containers.

FIGURE 28-5 Potting medium is loaded into the hopper and blended.

FIGURE 28-6 The medium is fed onto trays after blending.

FIGURE 28-7 Medium can be mixed during the off season and stored until needed.

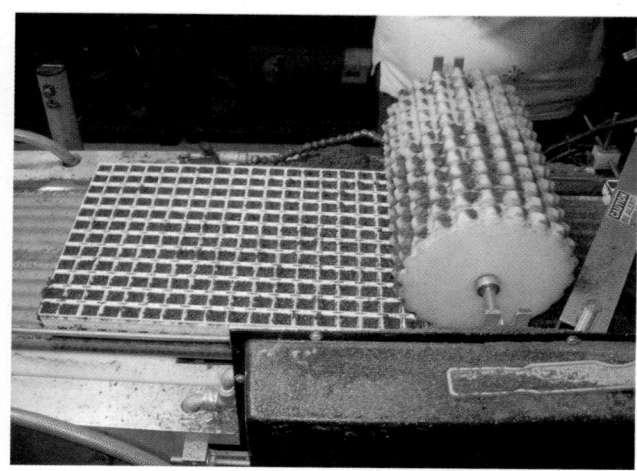

FIGURE 28-8 Rollers are common for seeding machines as well as flat filling machines for packing flats uniformly.

FIGURE 28-9 Another combination flat and pot filler.

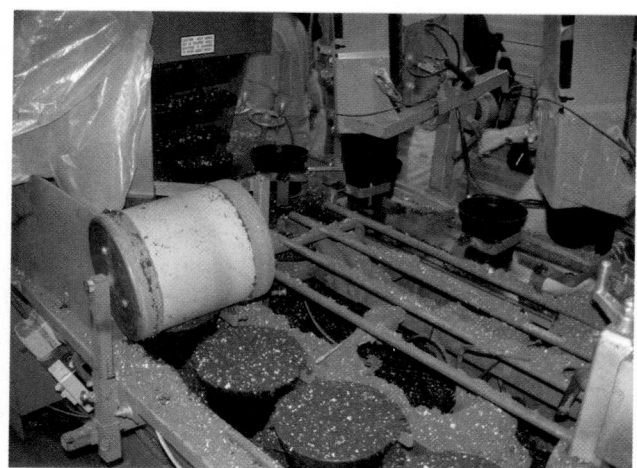

FIGURE 28-10 Pots as well as flats can be filled automatically.

Fertilization

Liquid fertilizers normally are applied by the injector method. Foliar analyses of bedding plants have shown that they contain about equal parts of nitrogen and potassium on a dry weight basis. The phosphorus content is about one-sixth to one-eighth that of nitrogen and potassium.

The amount of fertilizer added to the peat-lite mix will give the plants a good start. Seven to 10 days after transplanting, a regular program of fertilization is started. A 15-0-15 analysis dark weather fertilizer, or a 23-0-23 analysis fertilizer consisting of equal parts of ammonium nitrate and potassium nitrate, is preferred to a fertilizer with a high ammonium nitrogen and phosphorus content. The fertilizer is applied at the rate of 100 to 150 ppm of nitrogen at every watering. A greater concentration does not improve the quality of the plants. Some crops may be damaged by fertilizers applied in greater concentrations. Figure 28-11 shows the response of several annuals to fertilizer solutions of different strengths.

Magamp® 7-40-6 and Osmocote® 14-14-14 may be added when transplanting to carry the plants over if the grower forgets to fertilize the crop. Either material is added at the maximum rate of 4 pounds (2.4 kg) per cubic yard (cubic meter) of medium. The application of a greater amount means that the grower no longer controls the rate of growth. In addition, it may also be a waste of money.

Most bedding plants are grown at a cool temperature from 55°F to 60°F (13°C to 15.5°C). Neither Magamp® 7-40-6 nor Osmocote® 14-14-14 releases its nutrients efficiently at temperatures below 60°F (15.5°C).

Temperature

The night temperature maintained during production is 60°F (15.5°C). The day temperature should not exceed 70°F (21°C). At a higher temperature, the plant growth is too soft. High-quality crops require cool temperatures and fresh air. As the plants reach a saleable size, they are hardened off by exposure to a night temperature in the range of 40°F to 45°F (4.5°C to 7°C). Some growers merely protect the plants from freezing temperatures at the end of the season.

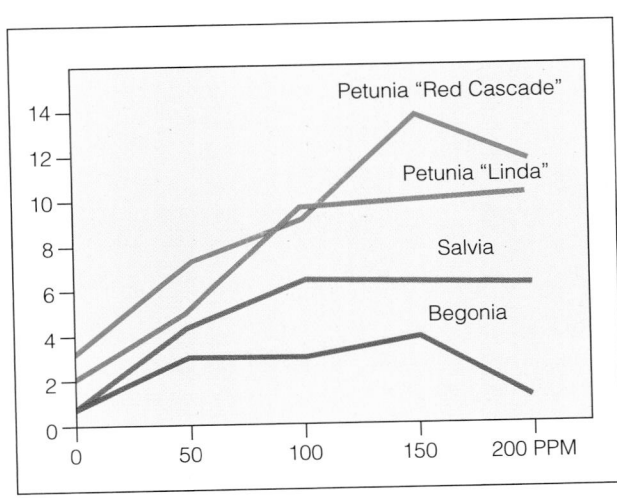

FIGURE 28-11 The growth of annuals fertilized with different rates of 20-20-20 fertilizer is measured as the dry weight, in grams.

Growth Retardants

The grower may follow all of the recommendations for growing the crop and still find that it is becoming too tall. In such cases, the grower can use chemical growth retardants, such as B-NINE SP®. A solution made from 5 level teaspoons (75 cc) of B-NINE SP® diluted with water to 1 gallon (3.785 l) provides 2,500 ppm of active ingredients, or a 0.25 percent concentration. This solution is sprayed on the crop from 3 to 4 weeks after transplanting. Generally, one spraying is sufficient. Some growers treat the plants once a week for 2 or 3 weeks.

Treatment with B-NINE SP® also delays the bloom and reduces the flower size of some plants. However, neither result is considered serious. One advantage to the use of B-NINE SP® is some added protection against air pollution injury. The effects of a B-NINE SP® treatment dissipate about 3 weeks after it is applied. In other words, there is no carryover of the treatment into the customer's planting. More information on growth retardants can be found in Chapter 21.

Containers

Many bedding plants are grown in multicell plastic trays. These "cell-paks" give the grower a master container that is easy to pull apart for sales. Quality bedding plants can be grown in several types of containers when there is enough media volume to ensure good root growth (Figure 28-12).

Marketing

High-quality plants are of no value if the grower does not sell them (Figure 28-13). Bedding plants can be marketed as a self-service operation with the use of individual or multicell packs. The grower displays the plants neatly on raised tables that are convenient for the customer. Self-service carts (Figure 28-14) allow the customer to move through wide aisles to select desired plants. Prices are prominently marked. Most growers place a marker in each pack indicating the name of the cultivar and listing essential culture information (Figures 28-15, 28-16, 28-17, 28-18, and 28-19).

FIGURE 28-12 Conventional bedding plant trays are called 10 × 20, but depending on the manufacturer, they range in size from 10 inches by 20 inches to 11 inches by 21 inches. The cell-pack insert pictured is called a 1204. That means there are 12 four-packs for 48 plants. The sales unit is the four-pack.

FIGURE 28-13 High-quality bedding plants in the greenhouse are of no value until they are sold.

FIGURE 28-14 A self-service hand cart for customer use. Use of such carts permits the customer to purchase several plants without the need for carrying them.

FIGURE 28-15 The tags placed in each pack identify crops and give some planting information.

FIGURE 28-16 Each neatly arranged container is marked with a tag.

FIGURE 28-17 Tags placed into each container specify the plant name, cultivar, and essential culture information. Many retail vendors also require a bar code.

FIGURE 28-18 Tags representing national programs will help the retailer market plants with logos and taglines reflecting national marketing campaigns. National marketing campaigns are often promoted in home and gardening magazines as well as home improvement television programming.

FIGURE 28-19 Local and regional plant promotion programs are designed to assist the retailer in marketing quality plants targeted for the local climate. The Plant Select® tags picture represents a program designed to seek out and distribute the very best plants for gardens from the high plains to the intermountain region of the United States.

Seed suppliers often provide growers with point of sales aids, such as large colored posters of the important new varieties and cultivars. In addition, planting charts and other types of cultural information are supplied to the grower to assist in answering questions from customers.

Bedding plant growers are assisted by a once state, but now national, organization called the Ohio Florists Association (OFA). With headquarters in Columbus, Ohio, OFA supplies growers with many sales and production aids that are helpful in stimulating the sale of bedding plants. Many seed companies and plant providers provide production and sales aids as well.

If bedding plants are sold at the retail level from the greenhouse, a large parking area is essential for customer use. The provision for adequate parking should be a primary consideration when planning the layout of a new greenhouse range.

Popular Crops

Many crops can be grown as bedding plants. The number of species and cultivars continues to grow at a rapid pace, although only a few crops seem popular with customers. These are the standard crops grown in the trade. The following sections describe the cultural details for these crops.

AGERATUM (Floss Flower) ✤
Ageratum houstonianum, Asteraceae

Ageratum is one of the most useful of the bedding plants. It is rarely called by its common name (floss flower). It grows equally well in full sunlight or in partial shade. At one time it was propagated by cuttings. Now growers propagate ageratum from seed. At 60°F (15.5°C), hybrids are grown from seed to the blooming stage in 2 to 12 weeks. The colors available are blue, pink, and white. A yellow hybrid is grown for cut flowers.

Among the better blue cultivars are 'Adriatic,' 'Blue Blazer,' 'Blue Danube,' 'Blue Lagoon,' 'Blue Triomphe,' and 'Pacific.' The Hawaii series are new offerings that include 'Blue Hawaii,' 'Royal Hawaii' (a deep lavender), and 'White Hawaii Improved.'

Ageratum is also grown as a specimen plant in 4-inch (10-cm) pots. Growing the plants in pots rather than packs requires an additional 2 weeks to produce a marketable flower. Ageratum is very sensitive to frost. This plant should not be offered for sale until there is no further danger of frost.

ALYSSUM (Sweet Alyssum) ✤
Lobularia maritima, Brassicaceae

Alyssum produces a compact carpet of flowers and is very popular for use in edging borders or beds. Individual plants are spaced on 12-inch (30-cm) centers. Their spreading growth habit soon forms a dense mass. Alyssum can withstand the hot dry conditions of summer very well.

The most popular cultivars offered are the 'Easter Bonnet' series (deep pink and violet), 'New Carpet of Snow' (white), 'Oriental Night' (violet purple),

'Royal Couple' (rich purple), and 'Minimum' (a pure white). Most cultivars grow to a height of 3 to 6 inches (7.5 to 15 cm). They make their best growth when the temperature is cool.

ASTERS ⬚ *Callistephus chinensis, Asteraceae*

Asters are grown for cut flowers and as bedding plant stock. The plants are subject to a disease called "Aster yellows," which makes them less desirable at times. The disease is carried by leafhoppers, and there is little that can be done to control it, except to screen the growing area to exclude their entry.

Seed is sown 5 to 6 weeks before the plants are needed for sale. Because asters are very sensitive to frost, they cannot be offered for sale until there is no further danger of frost. The plants are sold when they have no flowers. After several weeks in the garden, they will bloom.

The most popular cultivars for bedding plants are 'Dwarf Queen Mix' (12 inches [30 cm] tall), 'Milady Mix' (10 inches [25 cm]), 'Dwarf Sparkles' (10 inches [25 cm]), and 'Pinocchio Mix' (8 inches [20 cm]). Other cultivars are grown for cut flowers.

BEGONIAS ⬚ *Begoniacea*

The production of all types of begonias is covered in Chapter 23.

BROWALLIA ⬚ *Browallia speciosa, Solanaceae*

Browallia is an excellent plant for use in pots, borders, and hanging baskets. Seeds are sown in January to produce 4-inch (10-cm) pot plants by late April and May. The plants grow best in full sun at 65°F (18.5°C). If the temperature at which they are grown is too cold, they become chlorotic.

Speciosa major Bell series are the most popular. Included are 'Blue' (lavender blue), 'Heavenly' (Cambridge blue), 'Marine' (deep blue indigo), and 'Silver' (white). 'Jingle Bells Mix' is also available. Browallia flowers fall off the plants easily. The plants should be hardened off carefully before they are offered for sale.

CALADIUM ⬚ *Caladium hortulanum, Araceae*

The production of Caladium is covered in Chapter 23.

CELOSIA (Cockscomb) ⬚
Celosia argentea Var. cristata, Amaranthaceae

Celosia has been a favorite of gardeners for many years and is one of the showiest bedding plants. The seeds are sown 6 to 8 weeks before the plants are wanted for sale. They require a temperature of at least 65°F (18.5°C). If the plants become potbound, or if they are grown at a temperature cooler than the recommended minimum, they produce flowers too soon and remain stunted. The grower does not pinch celosia. It should not be allowed to flower before it is planted in the garden.

One type of celosia is the *plumosus*, which has feathery plume type flowers. The dwarf cultivars are used as pot plants. The second type of celosia is the *cristata*, or crested, type. The flowers of these celosias resemble cockscombs. Three popular low-growing (up to 8 inches [20 cm] in height) *cristata* type cultivars are 'Dwarf Empress Improved,' 'Kardinal,' and 'Jewel Box' mixed (includes pastels and bicolor, 'Jewel Box' red, and 'Coral Garden' mixed colors).

Taller globosa cultivars are used as cut flowers. The new 'Sparkler' series has carmine, cream, orange, red, yellow, and mixed colors.

Plumosa cultivars range in height from very dwarf to 27 inches (67.5 cm). Among this group are 'Apricot Brandy' (apricot orange; 16 inches [40 cm]), 'Castle' series (pink, yellow, scarlet, and mixed; semidwarf, 14 inches [35 cm]), 'Century' series (new in 1995; cream, fire red, red, rose, and yellow; 27 inches [67.5 cm]), 'New Look' (scarlet plume; 15 inches [37.5 cm]). 'New Look,' 'Castle' series pink, and 'Century' series mix were All American selection winners.

COLEUS ▨ *Coleus blumei Lamiaceae*

Coleus belong to the mint family, as indicated by the square stems. They are grown as bedding plants for shaded locations or as potted specimen plants. The colorful foliage makes the plants popular with customers. Extensive breeding programs have resulted in an expanded range of colors. Two new groups of coleus are the Rainbow class and the Carefree class.

Coleus are tropical plants that prefer a warm temperature. The minimum night temperature is 60°F (15.5°C). Most of the newer cultivars are grown from seed. High temperatures, 70°F to 80°F (21°C to 27°C), are required for proper germination of coleus seeds. Seed should be exposed to light and dark for best germination. Do not cover the seed. All cultivars are easy to propagate from terminal cuttings. Coleus normally root in 7 to 10 days.

Several series make up the offerings of seed suppliers of this crop. Among the more popular are the 'Fairway' series. These are small-leaved dwarf types that are very slow to flower. They make excellent 4- and 5-inch (10- and 15-cm) pots. 'Magic' has crimson red leaves that turn to tricolors of green, cream, and crimson. 'Mosaic' has an irregular mosaic pattern with cream and crimson leaves. 'Red Velvet' has deep maroon-red foliage. 'Rose' shows a bright rose with green edge leaf color. 'Salmon Rose' has a wide leaf margin of mottled green surrounding a salmon-rose center.

The 'Wizard' series produces plants 10 to 12 inches (25 to 30 cm) in height with heart-shaped leaves. They have excellent basal branching and are outstanding for baskets, packs, and pots. Cultivars in this series are 'Autumn' (autumn bronze shades), 'Golden Yellow,' 'Jade' (cream with a green leaf margin), 'Pineapple' (yellow-green with red blotches), 'Pink,' 'Red' (a deep red center with chocolate margin), 'Rose,' 'Scarlet,' and 'Sunset' (deep apricot). 'Velvet' and 'Mixed' round out this group.

'Red Poncho' and 'Scarlet Poncho' are in the 'Poncho' series, which have a cascading growth habit that makes them ideal for hanging baskets.

'Carefree Mix' and 'Saber Mix' are also widely grown.

GERANIUMS ▩ *Pelargonium xhortorum, Geraniaceae*

The geraniums used as bedding plants are grown from seed or from cuttings propagated asexually. The production of the geraniums from cuttings is covered in Chapter 25. This section is concerned with the production of bedding plant geraniums from seed.

Pot Plants from Seed

Cultivars

Certain geranium cultivars are bred to be flowered from seed. Improvements in plant breeding methods have shortened the time needed from sowing seed to plants in bloom ready for sales. Production time ranges from 12 weeks for plants in the 'Multibloom' series, which are among the earliest to bloom, to 16 weeks for those in the 'Picasso' series. Plants in the 'Elite' and 'Pinto' series and 'Hollywood Star' require 14 weeks to reach flowering from seed sowing. Colors available range from soft pink through deep red, rose, salmon, orchid, scarlet, violet, and white. The red cultivars are more widely grown than any other colors.

Media

The peat-lite medium described at the beginning of this chapter is used as the germination medium for geraniums.

Containers

To start seeds, only disease-free containers can be used. Reusable plastic flats are dipped in Green-Shield® or in a solution of 1 part formaldehyde in 50 parts of water. Diseases attack seedling geraniums as readily as plants grown from cuttings.

Seed

Geranium seed has a hard seed coat. The seeds must be specially treated before water will penetrate the seed coat. The seed must be scarified before it will take up moisture. To do this, the seed coat can be filed, it can be nicked with a knife, an end can be chipped off, or the seed can be soaked in acid. Commercial seed suppliers normally scarify the seed before it is packed for sale.

Geranium seed is large—about 0.19 inch (5 mm). A 0.3-ounce (1.1-g) seed packet of the 'Sprinter' cultivar contains roughly 175 seeds. The seeds are planted at least 0.38 inch (10 mm) apart in a row. Rows are spaced 1.5 to 2 inches (4 to 10 cm) apart. The seed is covered lightly with 0.13 inch (3 mm) of medium.

Moisture

Adequate moisture is a factor in obtaining a good percentage of germination. Before planting, the seed flats are given a thorough watering to soak the medium from top to bottom. To ensure a good wetting, the flats are watered the night before the seed is to be sown. After sowing, the medium is wetted again.

To prevent uneven drying of the medium, the flats are covered with opaque white plastic. Black plastic must not be used. When clear plastic is used, the flats cannot be placed in direct sunlight. A temperature buildup in the flat cooks the seed.

FIGURE 28-20 Inside view of a walk-in germination chamber that accommodates racks of seeded flats. These rooms are designed to be maintained at or near 100 percent relative humidity. This particular chamber also provides light, which may be required to germinate many species of bedding plants.

Low-pressure mist, if used, is applied on a very short schedule. Excessive mist cools the medium and promotes disease.

Many growers use controlled temperature and humidity germination chambers for starting seedlings (Figure 28-20).

Temperature

The minimum temperature recommended for the medium is 75°F (24°C). A probe type thermometer is used to monitor this temperature. The air temperature is maintained in the range of 65°F to 70°F (18.5°C to 21°C).

Transplanting

Most of the seed germinates in 6 to 10 days. The seed flats are held in the germinating environment to give any slow starting seeds a chance to germinate. However, after 2 weeks, there will be no further germination. At the end of 3 weeks, all of the seedlings are ready for transplanting (Figure 28-21). They are planted deep to prevent them from falling over as they grow larger. Deep planting means that the stem is inserted to the point where the leaves are attached.

The preferred growing medium is the peat-lite mix A. If more weight is needed, 10 percent clean white sand is added. Any other type of sand or soil should not be added until it is pasteurized.

Seedlings are transplanted directly to finish size pots, either 3.5 inches (9 cm) or 4 inches (10 cm). Approximately 100 3.5-inch (9-cm) pots or 75 4-inch (10-cm) pots can be set so that they touch in 1 square yard (0.75 m²) of bench area. When the leaves touch, the pots are moved to provide more room. The 3.5-inch (9-cm) pots are spaced 5 inches by 6 inches (12.5 cm by 15 cm) on centers. The 4-inch (10-cm) pots are spaced 6 inches by 6 inches (15 cm by 15 cm) on centers. At these spacings, 42 3-inch (7.5-cm) pots and 36 4-inch (10-cm) pots can be placed in 1 square yard (0.75 m²) of bench area.

Temperature

At night, a minimum air temperature of 62°F (17°C) is maintained at the pot level. A thermometer is placed among the plants to monitor the temperature. The plants will bloom earlier if they are grown at 65°F (18.5°C) for 4 to 6 weeks after transplanting. On sunny days the day temperature is held to 68°F to 70°F (20°C to 21°C). On cloudy days the night temperature is maintained through the day.

FIGURE 28-21 Seedling geraniums grown as plugs.

Two months after the seedlings are transplanted, the night temperature is lowered over a period of two or three nights to 55°F (13°C). Cooler temperatures will delay blooming. The day temperature is held at 65°F (23°C).

Supplemental Lighting

The plants can be lighted to hasten growth, either in the seedling flats or after they are transplanted to pots. Twenty-five watts of energy are required to light each square foot (0.1 m^2) of area using cool white fluorescent lights. Six square feet (0.6 m^2) of area can be lighted using four 40-watt tubes, 4 feet (1.2 m) long. The lamps are placed 12 to 15 inches (30 to 37.5 cm) above the plants. The lights are turned on for 16 hours daily and may be used to add to the natural daylight. This amount of lighting will hasten the blooms by 10 to 14 days.

Fertilization

Geraniums are vigorous growers, requiring more fertilizer than other bedding plants. High-quality plants are obtained by alternating 25-0-25 analysis fertilizer with potassium nitrate. Each substance is used at the rate of 2 pounds (90.9 kg) in 100 gallons (378.5 l) of water. In a constant fertilization program, 200 ppm of nitrogen and potassium are applied at each irrigation. The phosphorus added to the peat-lite mix before planting is enough to sustain the plants until they are sold.

Growth Retardants

One of the few growth retardants that is effective on geraniums is Cycocel®. Five and one-half weeks after the seed is sown, 1,500 ppm of Cycocel® is applied in a spray treatment. The timing of the application is critical to the success of the treatment. One and three-quarter ounces (50 g) of Cycocel® is diluted in 1 gallon (3.785 l) of water to make a solution with the desired 1,500 ppm concentration. Good coverage is obtained by adding a wetting agent. One-eighth teaspoon (0.6 ml) of wetting agent is added to the growth retardant solution. The plants are not watered for 24 hours after treatment. Two sprays are sufficient, and the second treatment can be given 12 to 14 days after the first.

The Cycocel® treatment may cause a slight yellowing around the edges of the leaves. The plants soon outgrow this condition.

A spray of A-REST® can also be applied to the crop. To obtain a 200 ppm concentration, 24 fluid ounces (720 ml) of A-REST® are mixed with 8 fluid ounces (240 ml) of water. The solution is applied as a very fine spray. One quart of solution treats 60 to 80 square feet (5.5 to 7.5 m^2) of small pots.

Wilted plants must not be treated because they do not take up the solution properly through the leaves. To ensure that the leaves are turgid before applying the growth retardant, water the plants several hours before the treatment. For more information on growth retardants, see Chapter 21.

Sowing Schedule

At 42° north latitude, seed is sown December 13–15 to obtain plants in flower by April 15. For a May 5 blooming date, seed is sown on January 6. When seed is sown January 26, 75 percent of the plants will be in bloom by May 25, or in time for the important Memorial Day holiday.

In southern growing regions, the sowing and selling dates occur 2 weeks earlier. The higher light intensities in the South cause more rapid growth and flower development.

Market Pack Production

The production of plants in **market packs** or trays is similar to the cultural methods for flowering plants. However, the plants to be sold in packs are started later, and they are sold as nonflowering, green plants. Because less growth is required for these plants, as compared to flowering plants, it is not necessary to start the seed so early.

When seed is planted January 20, the plants are ready for sale May 5. A sowing date of February 2 yields plants ready for sale by May 15. For June 1 sales, a sowing is made February 22.

Speedling® Geraniums

The novel method of producing geraniums from seed is to buy previously started seedlings and grow them to a saleable size. A franchise operation in Florida produces millions of vegetable and flower seedlings as transplants. The latest crop offering of this business is the **Speedling® geranium.**

The high light intensity and warm temperatures of Florida produce a top-quality seedling. The seedlings are shipped to northern growers who then bring them to a finish size. Speedling® geraniums are used to produce flowering plants.

For northern growers, the advantage in using Speedling® geraniums is that heat is not required in the greenhouse so early in the season. Heating costs in January and February are much greater than they are in March and April. It is expected that there will be a large increase in the number of geraniums produced from Speedling® as a result of increased energy costs.

Problems

Seed geraniums have the same cultural problems, insect pests, and diseases as geraniums produced from cuttings. These problems were described in detail in Chapter 25.

IMPATIENS ✺ *Impatiens wallerana, Balsaminaceae*

The development of new cultivars has made the impatiens the most important bedding crop grown for use in shaded locations. These cultivars provide a broad range of colors and a number of dwarf forms.

One seed supplier, in their 1994–1995 catalogue, offered 110 cultivars of impatiens in eight series, plus five separate ones.

Several years ago, most impatiens were propagated from cuttings. Now they are grown from seed. Impatiens seeds are not easy to germinate, but there will be few problems if the grower observes the light and moisture requirements.

To achieve maximum germination, a peat-lite mix is used because it has excellent aeration but still holds moisture.

Impatiens seed responds quickly to moisture stress. For this reason the seed should be covered lightly with the propagation medium. If too much medium is placed on the seed, it will not germinate because no light reaches it. Impatiens need light for germination.

The percentage of germination and seedling growth are improved if the seed flats are given artificial light. The same lighting arrangement described previously for geraniums can be used for impatiens with cool white fluorescent lamps providing light for 12 to 16 hours a day.

The optimum temperature of the medium is 70°F to 75°F (21°C to 24°C). After germination, the seedlings are grown at a night temperature of 65°F (18.5°C). Cold water and mist cannot be used on the seed flats. Cold water lowers the temperature of the medium and delays germination. Growth is also retarded if the air temperature exceeds 80°F (27.5°C) during the day.

Ammonium fertilizers retard germination. The same amount of calcium nitrate can be substituted for the ammonium nitrate in the special peat-lite germination medium. The use of calcium nitrate prevents the development of ammonia problems.

Impatiens are very sensitive to high soluble salt levels. The salt level of the germination medium should be less than 70 mhos using a 1:2 medium:water dilution. The seedlings are more uniform when they are grown at lower salt levels.

The seedlings must be watered early in the day. To reduce the amount of disease, the foliage must be kept dry. The damping off diseases that affect impatiens are *Rhizoctonia* and *Pythium*. The fungicides used to control damping off are harmful to seed germination. If one of these chemicals must be applied, it is used at one-half the recommended rate of application.

Fertilizer solutions are applied in concentrations not exceeding 100 ppm N. Frequent applications of low nutrient strength solutions are preferred.

Impatiens respond to foliar sprays of B-NINE SP®. A solution with a concentration of 0.05 percent (500 ppm) is applied 8 weeks after the seed is sown, if required.

Nine to 12 weeks after sowing the seed, the plants should be in bloom and ready for sale. If the plants grow too tall for sale, they can be pruned back. They will branch and make new growth that will bloom in 3 or 4 weeks.

Impatiens have shown a strong position in the marketplace. They have been the number one crop for many years in a row. In 1994, the wholesale value of bedding impatiens sold was nearly $82 million.

LOBELIA (Fairy Wings) ▒ *Lobelia erinus, Lobeliaceae*

Lobelia is an outstanding small plant that grows slowly. It is excellent for use in borders, pots, and hanging baskets. Brilliant blue is the most popular color, but white and carmine red are also sold.

The seed is extremely fine and should not be covered. Five to seven seeds are often placed in each cell of a pack. The small clumps are planted without being separated.

The seed is planted in January to produce flowering plants for Memorial Day sales. The plants are grown cool at temperatures of 45°F to 50°F (7°C to 10°C). Very light shade will help them to withstand the heat of summer.

Of the cultivars available, 'Fountain' series, 'Regatta' series, and 'Sapphire' are classed as trailing types, which are ideal for hanging baskets. Upright compact types, growing usually only 4 to 6 inches (10 to 15 cm) in height, are 'Blue Moon,' 'Cambridge Blue,' 'Cobalt Blue,' and 'Crystal Palace,' which are all blue. The 'Palace' series has blue, blue with a white eye, royal, and white.

'Rapid' series, which is excellent for growing in packs, offers blue, white, and violet-blue colors. 'Rosamond' is carmine-rose with a white eye.

MARIGOLDS ※ *Tagetes patula and Tagetes erecta, Asteraceae*

There are more sizes, flower types, and growth habits for marigolds than any other bedding plants. Marigolds range in size from the less than 6-inch (15-cm) tall *Tagetes patula* (French-type marigold) to the 3-foot (0.9-m) tall *Tagetes erecta* (African-type marigold) 'Crackerjack' cultivar. The dwarf forms are excellent as pot plants and for bedding plants in edgings and borders. The tall forms are used behind shorter plants in beds and as cut flowers. Both types of marigolds are native American species.

In the past, marigolds had a strong fragrance that many people found objectionable. The newer cultivars have a less pungent aroma but still smell like marigolds.

The plants are easily grown from seed. The best germination occurs at 70°F to 75°F (21°C to 24°C). As soon as the seed germinates, the seedlings are given a cooler temperature in the range of 60°F to 62°F (15.5°C to 16.5°C) to produce heavier growth. Very often, they are finished at a night temperature of 50°F to 55°F (10°C to 13°C).

The flowering of marigolds is affected by day length. Long days promote vegetative growth. Short days hasten blooming. Seeds sown in late January are exposed to 16 hours of cool white fluorescent light. The seed flats or plugs are placed 8 to 10 inches (20 to 25 cm) below the lamps. The most economical time to light the plants is in the seed flat stage when a large number of plants are close together.

After the seedlings are transplanted, they are placed in a greenhouse where they are given artificial short days by means of the black cloth treatment. The plants are shaded from 5:00 P.M. to 8:00 A.M. to produce a 9-hour day. Short-day conditions cause the plants to flower 6 to 10 weeks after the seed is sown. Not all cultivars require the black cloth, short-day treatment to hasten blooming. African cultivars appear to benefit most from providing a short-day treatment for 2 weeks when sown after mid-February. The treatment is started at germination.

The widespread use of automatic seed-sowing machines has forced seed suppliers of this crop to "de-tail" the seed—that is, remove the bristlelike projections on the basal end of the seed.

Marigolds should be grown in full light intensity to produce the best-quality plants. They have few cultural problems.

PANSY ※ *Viola × wittrockiana, Violaceae*

Pansies have long been a favorite as a bedding plant. Their happy moon-face flowers are available in many colors. Two methods of growing pansies are used: overwintering culture and greenhouse culture.

Overwintering Culture

Seed is planted in August. When the seedlings are large enough, they are transplanted to cold frames or they are planted in the field with 6 inches (15 cm) between them in each direction. The field beds are six to eight rows wide. There is a 12-inch (30-cm) wide path between the beds. The plants are fertilized once after they are established. A 20-10-20 analysis fertilizer is applied at the rate of 2 pounds (0.9 kg) in 100 gallons (378.5 l) of water.

The plants are not covered until the ground is frozen. Then straw is piled loosely over the field beds. For plants being held in cold frames, the glass or plastic cover is added. The temperature is kept as cool as possible to ensure minimum growth. When the temperature warms up in the spring, the plants are uncovered and allowed to bloom.

The plants are harvested from both field beds and cold frames and are marketed as groups of half a dozen or a dozen plants in baskets. The plants can be held in a refrigerator at 35°F (1.5°C) if the temperature becomes too warm during the selling period.

Producing pansies in southern climates for fall sales requires special handling of the seeded flats. Greenhouse temperatures in the South in late July and early August are much too warm to obtain good seed germination. To overcome this problem, some pansy growers use refrigerated germination rooms where temperatures are carefully controlled. Other southern growers rely on northern growers to produce their crops in plug culture. Plug-started pansy seedlings can be shipped south without too much loss. They can also be held by the receiving grower for 7 to 10 days if that grower is not ready to plant them.

Greenhouse Culture

Seeds are planted in December and January. They are germinated at 65°F to 70°F (18.5°C to 21°C). As soon as germination is complete, the seedlings are moved into a greenhouse where the night temperature is 50°F (10°C). After the seedlings are transplanted, they are given the full light intensity and are grown on at 50°F (10°C). Some growers shift the plants to a cold frame in early March, weather permitting (Figure 28-22).

Breeding of new pansy F_1 hybrids has been almost as prolific as for petunias. Many new series are offered in the latest seed catalogues. The 'Atlas' F_1 series is a large flowered type, as is the 'Bingo' series. Several of the colors offered in this series have blotches. The 'Delta' F_1 series is new and a heat-tolerant successor to the 'Roc' series. It also blooms 2 weeks earlier than plants in the 'Roc' series. Colors in this series range from blue through violet, yellow, pink, red, and white (Figure 28-23).

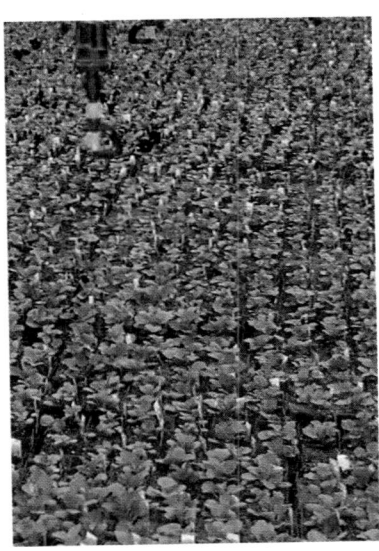

FIGURE 28-22 These pansies are being grown in a greenhouse and will be ready in 6 weeks for sale.

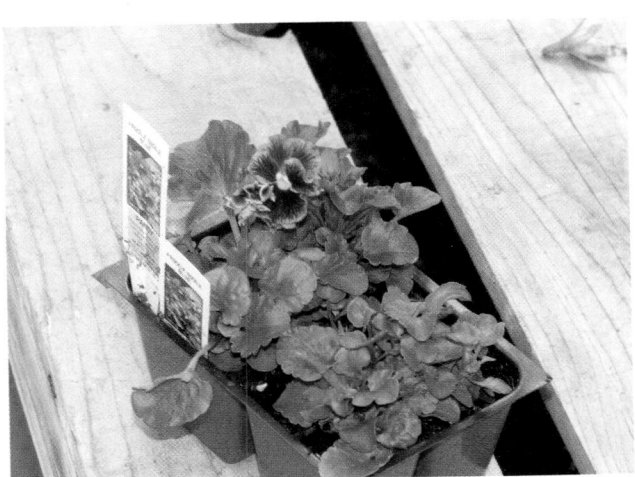

FIGURE 28-23 A four-pack of pansies.

PETUNIA ⌘ *Petunia hybrida, Solanaceae*

For more than 100 years, petunias have been one of the five leading annuals grown in the United States. According to one author, one of the reasons for the beginning of the modern bedding plant business was the demand for petunias.

Types of Petunias

Modern growers identify five types of petunias: grandiflora singles, grandiflora doubles, multiflora doubles, and multiflora singles (Figure 28-24).

Grandiflora Singles

More of these plants are grown than any other type of petunia. The plants produce a large number of flowers, and each flower may be up to 5 inches (12.5 cm) across. This type of petunia blooms early in the season. One seed company alone offers a total of 127 cultivars in 19 series or single offerings. Among the series listed are 'Daddy,' 'Dream,' 'Falcon,' 'Flash,' 'Picotee,' 'Supermagic,' and 'Ultra.'

The supergrandiflora or **Cascade petunia** is a popular mutation of the grandiflora type. It is an outstanding plant for use in hanging baskets, window boxes, or any location where the long, trailing branches can hang freely. The 'Supercascade' series is representative of this type.

Grandiflora Doubles

This type of petunia was the first to be offered as F_1 hybrid seeds. The plants are commonly grown in pots as specimens. Early January sowings produce good 3-inch (7.5-cm) pot plants that will flower by late May. These plants must be grown warm and must be fertilized well for best growth and good-quality flowers.

Multiflora Doubles

This type of petunia has smaller flowers than the grandiflora doubles. The plants are more compact in size and have a greater number of flowers on each plant. The flower has ruffled edges and looks like a miniature carnation.

FIGURE 28-24 Petunias ready for sale in a retail setting.

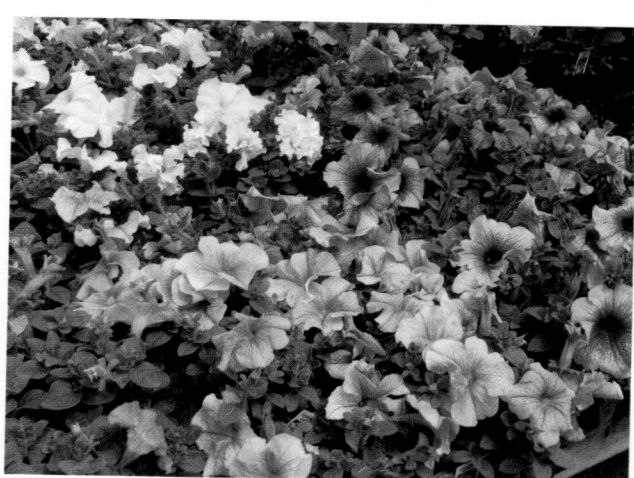

Popular cultivars include the "Tart" series: 'Cherry Tart,' 'Plum Tart,' 'Apple Tart,' 'Peach Tart,' 'Strawberry Tart,' and 'Snowberry Tart.'

Multiflora Singles

This type of petunia has many small flowers per plant. These plants mature faster than the other types of petunias, produce the most branches and flowers, and are available in the greatest range of colors of all the petunia classes.

Multiflora favorites are found in the following series: 'Carpet,' 'Celebrity,' 'Horizon,' 'Merlin,' 'Polo,' and 'Primetime.'

Spreading Types

This class of garden petunias was introduced in 1995. 'Purple Wave' is the first cultivar of this class, which has a ground hugging or procumbent growth habit. The growth is all horizontal. The stems grow along the ground and can spread 2 to 4 feet (60 to 120 cm) from where planted. It is recommended that 'Purple Wave' be produced in hanging baskets not market packs, because the vining growth habit weaves together like carpet.

Culture of Petunias

The culture of petunias is similar to that described for other bedding annuals. Several sowings of seed are made beginning in late December to provide a succession of plants for transplanting throughout the season. Some growers produce two or more crops in the spring growing season.

Sowing Seed

Petunia seed is very small. One ounce of seed contains about 285,000 seeds. The seed is placed in rows on the surface of a peat-lite medium. The seed is not covered because it is so small that it falls into the spaces between the particles of the medium. It is easy to oversow petunia seed, resulting in small, weak seedlings. Many growers use a 1/128-ounce packet to sow two 10-inch by 20-inch (25-cm by 50-cm) seed flats. Most petunias are started as plug tray seedlings today.

Germination Methods

Maximum germination requires a bottom heat of 75°F to 80°F (24°C to 27.5°C). The seed flats are placed on heating mats, overhead pipes, or on a concrete surface with heating cables embedded in it.

During germination, the seed must be kept evenly moist. A low-pressure mist may be used. However, caution should be exercised in using the mist because the cold water lowers the temperature of the medium by 10° or more. Many growers fog the flats by hand several times a day.

A lighted chamber can be used to germinate petunia seeds on a regular schedule. The seeds are sown in trays that are watered thoroughly and allowed to drain. The trays are then placed on shelves. The relative humidity in the chamber is kept at 98 to 100 percent.

Cool white, warm white, natural white, or daylight fluorescent lamps are used alone or in combination. The lamps are placed 6 inches (15 cm) above the flats. Artificial light is provided for 16 hours a day. The plants are not lighted 24 hours because this practice wastes energy. The light intensity should be

850 to 1,000 fc (9.2K to 10.8K lux). This intensity is obtained by using two four-tube fixtures mounted side by side, 6 inches (15 cm) above a 3-foot (1-m) wide bench. The recommended temperature is 72°F to 75°F (22°C to 24°C).

Petunia seed germinates in 4 to 5 days. Plants are ready for transplanting 2.5 to 3 weeks after germination. Immediately after transplanting, the containers are given reduced light until the seedlings adapt to the bright daylight.

Growing On

The Cornell peat-lite mix is commonly used as the transplanting medium. Commercial products may be used also, including Jiffy-Mix®, Jiffy-Mix Plus®, Redi-Earth®, and Pro-Mix®.

A quality crop is obtained if the grower is careful about following a schedule for fertilization, watering, and ventilation. As soon as the plants are growing well, a 20-10-20 analysis fertilizer solution is applied to provide 150 to 200 ppm of nutrients. The plants are fertilized at every watering until they are sold.

For several weeks after transplanting, the night temperature is held at 60°F (15.5°C). As the plants mature, the temperature is lowered gradually. During the last few weeks of growth the plants are grown at the natural night temperature, protected from freezing only.

It is recommended that the packs of plants be grown on benches that are at least 1 foot (30 cm) off the floor. The medium becomes too cold if the plants are placed on the floor, resulting in a reduced uptake of nutrients. Air movement under and around the plants helps to reduce problems from *Botrytis* and other diseases.

Sprays of B-NINE® at a concentration of 0.25 percent (2,500 ppm) are applied to prevent too much stem growth, especially if the spring is warm. When the plants reach the size of a half-dollar, they are sprayed. Following this, two or three additional sprays are applied at 1-week intervals.

The plants must be hardened before they are offered for sale. Hardened plants can withstand a light frost after they are planted in the garden.

The grower can select from dozens of colors for the various types of petunias. Before buying any seed, the grower should determine what types of flowers and what colors are most popular in the local market. Reds and pinks are always popular, followed by solid whites, blues, and pastels. Many mixed colors are also available and are very popular.

PHLOX (Annual Phlox) ▨ *Phlox drummondii, Polemoniaceae*

Annual phlox is native to the southwestern United States and Mexico. The seed is sown from early March to the middle of March to obtain bedding plants for sale in the spring. 'Chanal' was the first cultivar to be a truly double phlox. The opening flowers resemble miniature pink roses. The cultivar 'Globe Mix' produces a fine ball-like plant. 'Palona Mix' is an early flowering type that is more compact and more uniform than any other phlox. 'Petticoat Mix' and 'Twinkle Mix' both have star-shaped flowers. The former has a dwarf compact growth. The latter comes in brilliant colors.

The greenhouse culture of phlox is similar to that described for petunias. The plants are grown at 50°F (10°C). The growing period for potted specimens is from 11 to 13 weeks from seed sowing to bloom. Plants grown in packs require 9 to 11 weeks to bloom.

SALVIA (Scarlet Sage) ✼ *Salvia splendens, Lamiaceae*

Scarlet Sage is a very popular bedding plant. The upright, brilliant red, spike-shaped flowers provide an outstanding display in the garden (Figure 28-25). There are five groups of salvias that range in height from 8 to 10 inches (20 to 25 cm) to 26 to 30 inches (65 to 75 cm). Popular red cultivars are 'Red Vista,' 'Scarlet King,' and 'Scarlet Majesty.' The 'Sizzler' series grow to a height of 10 to 12 inches (25 to 30 cm). Red, lavender purple, salmon, and white are colors in this series.

The best percentage of seed germination occurs in a medium with a low salt content. The medium must not be fumigated with methylbromide because this substance is toxic to salvia. The seeds are sown on top of the medium. The flats are covered with white opaque plastic, or they are placed under mist to maintain a high humidity. As soon as the seedlings appear, the seed flats are moved to the growing area where the temperature is 55°F to 60°F (13.5°C to 15.5°C).

The plants require full light intensity. Salvia are very sensitive to high soluble salt levels. Therefore, the fertilizer solution can have a maximum concentration of only 150 ppm N.

From seed sowing, 7 to 8 weeks are required to produce flowering plants. Small plants in pots require an additional 2 to 3 weeks of growing time before they will bloom.

SNAPDRAGONS ✼ *Antirrhinum majus, Schrophulariaceae*

New hybrids developed by plant breeders have increased the popularity of snapdragons in recent years. The F_1 hybrid "Rocket" series was a major breakthrough in breeding snapdragons. These plants grow to be 30 to 36 inches (75 to 90 cm) tall. From this beginning, many other series were developed. Popular new cultivars developed especially for garden use are three dwarf series. The 'Chime' series produces finished plants that are only 6 to 8 inches (15 to 20 cm) tall. The plants are also suitable for 4-inch (10-cm) pot culture, cemetery urns, and color bowls. The 'Floral Showers' series grow to a finished

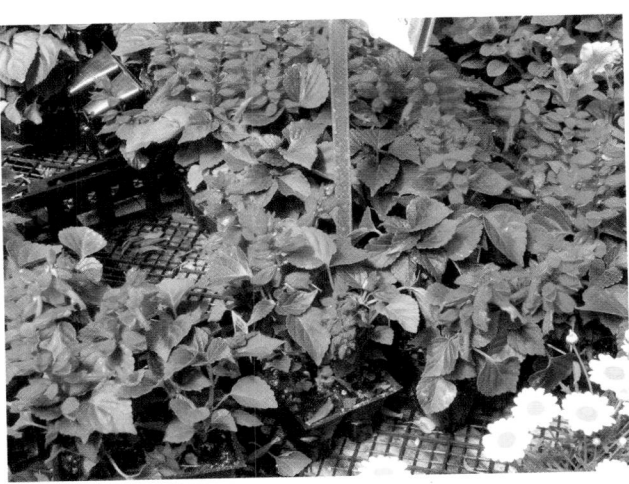

FIGURE 28-25 Salvias are popular retail bedding plants.

height of 6 to 8 inches (15 to 20 cm). It is extra early in bloom, needing a minimum of heat. This group is also excellent for 4-inch (10-cm) pots. The 'Bell' series produces strong basal branching plants that grow to 8 to 10 inches (20 to 25 cm). The semidwarf series 'Liberty' provides plants that reach a final height of 18 to 22 inches (45 to 55 cm). Colors in the dwarf and semidwarf series range from bronze through purple, yellow, pink, rose, red, and white. Each of these series offers mixed as well as pure colors.

Snapdragon seeds are very small. As a result, planting consists of pressing them lightly into the top layer of the peat-lite medium. Snapdragons are cool temperature crops. The temperature of the germinating medium should be 70°F (21°C). As soon as the seedlings are visible, the containers are placed in an area where the temperature is 50°F (10°C).

After transplanting to market packs, plants of the tall growing cultivars are allowed to grow until they have five or six sets of leaves. Then they are pinched or pruned to the point where three sets of leaves remain. This treatment causes each plant to branch until it has four or five stems.

Snapdragons do not grow well in a medium with a high soluble salt level. When a solution of 20-10-20 analysis fertilizer is applied, the concentration must be no more than 150 ppm N.

From seed sowing, 2 to 3 months are required to produce flowering plants. When tall cultivars are planted in the garden as early as possible, they often produce a second crop after the first crop is harvested for cut flowers. In milder climates, snapdragons may be planted outside as soon as the ground can be prepared.

VERBENA �ख *Verbena × hybrida, Verbenaceae*

Verbena is a popular bedding plant that is available in many colors. It is fairly easy to grow. Seed is planted in mid-February. Germination is slow and may require several weeks to achieve the average rate of germination, or 50 percent. The temperature should be in the range of 65°F to 70°F (18.5°C to 21°C) for germination. Better seedlings are produced under artificial lights than under normal greenhouse conditions. The seeds must not be overwatered.

After the seedlings are transplanted, they are grown at a night temperature of 50°F (10°C). When the main stem is 2 inches (5 cm) long, the plants are pinched by removing the tip only. Pinching causes the plants to branch, resulting in compact, bushy growth.

There are two types of verbenas: One type has a spreading growth habit, and the other is an upright form. Pot plants of the dwarf upright type require 10 to 11 weeks to grow to blooming size from the time seed is sown. When the plants are grown in packs, they are ready for sale 1 to 2 weeks earlier than the pot plants.

Semiupright types are represented by the 'Novalis' strain and the 'Amour' series. Both produce excellent compact plants that reach a height of 8 to 10 inches (20 to 25 cm) in the garden. 'Sandy Scarlet,' 'Sandy White,' 'Trinidad,' and 'Peaches and Cream' are also classified as semiupright types. The 'Romance' strain contains seven cultivars that range in color from deep rose through scarlet, violet, and white. This group has a spreading type habit that forms a dense mass of flowers when fully grown.

ZINNIA ⌧ *Zinnia elegans, Asteraceae*

Zinnias are native to Mexico and grow best under hot summer conditions. Extensive efforts by plant breeders have resulted in a relatively large number of different type zinnias that range from dwarf hybrids through tall inbreds. The 'Short Stuff' F_1 series grows to 8 to 10 inches (20 to 25 cm) tall. The 'Dreamland' series reaches a height of 8 to 12 inches (25 to 30 cm). They reach bloom in just 6 weeks from sowing seed.

The 'Small World' series are semidwarf, reaching a height of 12 to 14 inches (30 to 35 cm). In addition to being ideal for packs and bedding, they make great 4-inch (10-cm) pots. The 'Ruffles' series are used as cut flowers growing to 24 to 30 inches (60 to 75 cm) in addition to being grown as bedding plants. The 'Radiant' strain has large, double flowered Dahlia type blooms. It is outstanding as a cut flower growing to 30 inches (75 cm) or more. The tallest growing zinnias are the 'Giant Dahlia' type. They grow to 3.5 to 4 feet (150 to 120 cm). They are excellent for use as cut flowers.

Zinnias require warm temperatures. The seed is germinated at 70°F (21°C). After the seedlings emerge, they are held at a night temperature of 60°F (15.5°C). This is the lowest temperature to which the plants can be exposed. At a lower temperature, they may suffer shock, resulting in stunted growth. It is preferable to sow the seeds later in the season and grow them at a warm temperature rather than sowing them early and trying to grow the plants at a cooler temperature.

Vegetable Crops

Many bedding plant producers also grow vegetable transplants. The production of vegetable plants may account for 10 to 20 percent or more of the grower's crops.

Cultural Practices

In general, vegetable transplants are grown using the same methods as those described for bedding plants. The recommended germination temperatures for most vegetable crops range between 70°F and 80°F (21°C and 27.5°C). Crops requiring cool temperatures are usually grown at a temperature 10° lower during the day than that at which warm-temperature plants are grown. Broccoli, Brussels sprouts, cabbage, cauliflower, lettuce, and onions are grown at 60°F to 70°F (15.5°C to 21°C) during the day. The night temperature for these vegetables is reduced 10°.

Eggplant, peppers, and tomatoes are grown at 70°F to 80°F (21°C to 27°C) during the day and 10° cooler at night. However, eggplant prefers a minimum night temperature of 65°F (18.5°C).

Cucumber, muskmelon, summer squash, and watermelon grow best at a day temperature of 70°F to 90°F (21°C to 32°C) and a night temperature of 60°F to 70°F (15.5°C to 21°C).

Sowing Seed

Most vegetable seed is quite large and easy to handle. No more than 10 seeds are sown per inch of row in the seed flat. As a result the seedlings are not crowded, and they can grow at their maximum rate. Cucumber, muskmelon, squash, and watermelon have large seeds, which are planted directly into the growing packs. Three seeds are planted in each unit. The seed for these crops should not be planted until 2 weeks before the seedlings can be transplanted to the field. The growth of the plants in the field is greatly reduced if the seedlings become potbound in small containers before transplanting. It is difficult to transplant these vine type crops without disturbing their roots. Therefore, the seed should be started in thin-wall peat pots, Jiffy-7s®, or similar units that can be planted in the soil with the seedlings.

Tomatoes, peppers, eggplants, broccoli, cabbage, cauliflower, and lettuce are planted 6 to 8 weeks before they are needed. These crops are transplanted at least once before they are offered for sale. When they are transplanted, the roots are pruned. As a result, the plants develop stockier growth than if they were seeded directly into the final container.

Hardening Off

All vegetable transplants must be hardened before they can be planted in the ground. The growing temperature is reduced gradually over a period of 10 to 14 days. Two practices that are not recommended are reducing the temperature suddenly and restricting watering to keep plants dry. Such treatment may retard the growth so much that the plants cannot recover quickly.

Problems

Cultural Disorders

Bedding plants are subject to a large number of problems. Germination problems may be due to a variety of reasons including improper temperatures, over- or underwatering, preemergent or postemergent damping off due to disease, a high soluble salt level in the medium, methylbromide used to fumigate the medium, seeds planted too deep in the medium, and the use of old seed.

Problems that arise after transplanting are given in Table 28-2.

Diseases

Diseases in bedding plants can be minimized by pasteurizing any soils used in the medium or by using prepared peat-lite mixes. To prevent contamination of the plants and the medium, good sanitation must be practiced during all phases of culture. Loss from *Botrytis* can be reduced by spacing the containers properly and by providing good ventilation and air circulation.

Damping off and stem rots are caused by the *Pythium*, *Rhizoctonia*, and *Fusarium* disease organisms. To reduce these diseases, the seeds are treated with a fungicide before sowing. Then they are planted in a pasteurized medium or in a peat-lite type mix. Fungicidal drenches can be used if required, but they may reduce the percentage germination of many crops. More information on greenhouse diseases can be found in Chapter 20.

Insects

Insects that attack bedding plant crops include aphids, caterpillars, cutworms, flea beetles, fungus gnats, leafminers, mealybugs, scale insects, symphillids,

Table 28-2

Problems in Bedding Plant Production		
Problem	**Possible Cause**	**Suggested Remedy**
Light green, off-color plants	Lack of nutrients	Apply a complete liquid fertilizer
	High soluble salt level in medium	If the concentration of the soluble salts is over 150 on a 1:2 media:water test, leach the containers
	Medium temperature is too low	Raise packs off the cold floor; increase the air temperature
	Waterlogged medium	Improve drainage away from the packs; lighten the medium with perlite
Chlorosis, interveinal yellowing	Root injury due to several causes	Use chelated iron, try to correct the main problem
	Wrong media pH	Test the media before using so that limestone can be added appropriately
	Over- or underwatering	Correct watering practices
	High soluble salts	Leach media
	Herbicide residue	No known control; try an activated charcoal slurry[1]; avoid buying contaminated media
	Excessive trace elements	Apply the proper amounts of trace elements; try leaching the medium
Uneven growth	Poor mixing of media	Mix for longer periods
	Non-level flats resulting in non-uniform water-fertilizer application	Level the flats
	Uneven moisture levels in medium at transplanting	Use a surfactant to wet the medium in the containers 24 hours prior to transplanting
	High soluble salt levels	Leach the containers
	Poor grading of seedlings	Select for uniform size plants for each pack at transplanting. Check quality of plugs prior to transplanting
Purple leaves on tomatoes, marigolds, and petunias	Phosphorus deficiency	Fertilize with 10-52-17 analysis fertilizer or other liquid fertilizer with a high phosphorus analysis
Seedlings fall over	Grown too cold	Check the night air and media temperatures
	Damping off	Use pasteurized media; drench media with appropriate fungicide; improve air circulation around plants
Plants missing, holes in leaves	Cockroaches, slugs, snails, rodents	Use baits, traps, or appropriate pesticides
Lower leaves brown, rotting, and dying	*Botrytis* and/or other diseases	Improve air circulation around plants; increase the air temperature; use appropriate fungicide

(continues)

Table 28-2 *(Continued)*

Problem	Possible Cause	Suggested Remedy
Slow plant growth	Too little fertilizer	Test the soil and apply regular fertilizer
	Grown too cold	Raise the plants off the ground, increase air temperature
	Grown too dry	Increase frequency of irrigations
	Herbicide present	Know the history of the soil, medium, and/or compost; avoid using field soils; try an activated charcoal slurry[1]
Tall and spindly growth	Too much nitrogen fertilizer applied	Test the medium; check the proportioner ratio. Lower the temperature, especially night temperatures; improve air circulation using HAF fans or fan tubes
	High spring temperatures	Program use of chemical plant growth retardants at the proper stage
	Overwatered	Reduce the frequency of irrigations
	Low light intensity	Keep the greenhouse glazing clean; wash dirt from plastic coverings; paint wood work and benches white

1. An activated charcoal slurry is made by adding 1 pound (0.45 kg) of finely ground activated charcoal to 1 gallon (3.785 liters) of water. The mixture is stirred vigorously and the suspension is applied to the containers as evenly as possible.

springtails, tarnished plant bugs, thrips, and whiteflies. The insecticide recommended for a particular pest is used to control that insect. The initial treatments are made with small amounts of insecticide to determine if any plant injury occurs.

Slugs, snails, and cockroaches are treated by pesticides labeled specifically for their control. More information on greenhouse insects can be found in Chapter 19.

ACHIEVEMENT REVIEW

Select the best answer or answers to complete each statement. List the appropriate letter(s).

1 Bedding plant production expanded greatly after:
   a. World War II.
   b. the Korean War.
   c. the Vietnam War.
   d. the Desert Storm conflict.

2 The best source of information about sowing seeds and timing crops is:
   a. growers' catalogues.
   b. schedules from local growers.
   c. textbook programs.
   d. the growers' own records from previous years.

3 Orders for seeds for the following year are usually placed in:
   a. the spring.
   b. the late summer or early fall.
   c. early summer.
   d. late fall or early winter.

4 Seedlings are ready for planting when:
   a. the cotyledons appear above the germination medium.
   b. two or three true leaves have developed.
   c. they are 2 to 3 inches (5 to 7.5 cm) tall.
   d. they are 3 to 5 inches (7.5 to 12.5 cm) tall.

5 If a grower adds any material, such as sand or soil, to a prepared peat-lite mix, the material should be:
   a. screened to remove rocks and sticks.
   b. mixed thoroughly with superphosphate.
   c. steam pasteurized or fumigated.
   d. drenched heavily with captan.

6 Bedding plants are fertilized at every watering with a full analysis fertilizer at the rate of:
   a. 10 to 25 ppm.
   b. 50 to 75 ppm.
   c. 100 to 150 ppm.
   d. 200 to 225 ppm.

7 When Osmocote® 14-14-14 fertilizer is applied at a cold (50°F [10°C]) temperature:
   a. the uptake by the plants is increased.
   b. there is no effect on plant uptake.
   c. the uptake by the plants is decreased.
   d. uptake by the plants is prevented.

8 Growth retardants, such as B-NINE SP®, can be applied as a spray treatment:
   a. as soon as the seedlings are transplanted.
   b. 1 week after transplanting.
   c. 3 to 4 weeks after transplanting.
   d. when the flowers just begin to show color in the buds.

9 Market packs or containers holding six to a dozen plants are popular with growers for marketing plants because:
   a. they can be displayed neatly.
   b. they work well in self-service areas.
   c. it is easy to price the plants by inserting tags in the packs.
   d. of all of the above.

10 One of the most important items required if bedding plants are sold at retail is:
   a. a large supply of change in the cash register.
   b. not more than one cash register for checkout.
   c. a good supply of paper bags.
   d. a large parking area for customers' cars.

Cut Flower Crops

Major Cut Flowers: Antirrhinum, Dendranthema, Dianthus

OBJECTIVES:

After studying this chapter, the student should have learned:

- three major cut flower crops.
- the cultural needs of each crop.
- methods of harvesting and preparing each crop for market.
- diseases affecting each crop and methods of control.
- the insects that attack each crop.

Crops have been produced for cut flowers for many centuries. Drawings from ancient Egypt, Greece, and Rome show cut flowers arranged in vases around the home. Flower arrangement itself was viewed by the Romans as a noble art. Over the years many dramatic changes have occurred in the production of cut flowers. In its infancy the business was conducted outdoors. That is, all of the flowers were cultivated outside during the normal growing season. The development of protective glasshouses led to year-round production. Of course, the grower still grew seasonal crops.

An important advance in flower growing resulted when researchers determined how day length controls the flowering of many crops. This knowledge meant that the grower would no longer be restricted to the production of seasonal crops. During the late 1930s and early 1940s, the schedules were developed that permitted *Dendranthema* (chrysanthemums) to be flowered year-round. Knowledge of the techniques of day length control led to the development of the outdoor chrysanthemum industry in Florida. Starting from an area of less than 5 acres in 1949–1950, chrysanthemum production in Florida increased to around 400 acres in 1958. Chrysanthemum production as a cut flower in Florida today does not exist because most former chrysanthemum growers are now in the business of producing foliage plants. Offshore production of chrysanthemums in Central and South America has caused this change.

The success of the outdoor production of chrysanthemums led to a large increase in the production of gladioli in Florida. Beginning in southern states, gladioli were planted in successively more northern areas as spring moved north. At one time, they were widely grown by many producers during the summer months. Then the gladiolus business expanded in Florida due to more efficient methods of growing the crop. Now, as with chrysanthemums, gladiolus in Florida does not exist. The production of gladiolus remains in a few states in the northeast and elsewhere. Only a few commercial growers are producing gladiolus for local markets.

The cut flower market in the United States has been affected by a large increase in the production of carnations, chrysanthemums, and roses in Central and South America and Mexico. Carnations need a high light intensity and cool temperatures to form quality flowers with stiff stems. Denver, Colorado, at an elevation of 5,000 feet (1,524 meters), was the carnation capital of America. Environmental conditions in California also were favorable to carnation production.

Bogotá, Colombia, in South America is located at an elevation of nearly 8,600 feet (2,621 meters). The climate around Bogota is ideal for the production of carnations. There is plentiful, inexpensive labor in the area as well. Three carnation growers were in business in the mid-1960s. Now, there are well over 300 carnation producers in the vicinity of Bogotá. Millions of high-quality flowers are exported annually to the United States and Europe. These carnations are much lower in cost than crops produced in the northeastern United States. Because carnation production in the northeast became no longer profitable, growers have changed to other crops.

Some carnations are still being grown in Colorado and California. These crops are used locally or shipped to eastern markets. Stiff competition from the South American crops has all but eliminated this crop in the United States.

Cut chrysanthemums are also imported from Central and South America. These flowers are shipped to markets all over the United States. The number of chrysanthemums imported has greatly reduced the number of chrysanthemum growers in this country.

As production costs continue to rise, it is very likely that other areas of the world will become important producers of cut flowers because of ideal climatic conditions and cheap labor. However, once the crop is produced, it must be shipped to market. The cost of transportation may be more than the amount saved in producing the crops. In this situation, growers in the United States may still be competitive with imported crops.

Five major cut flower crops are grown in the United States: *Antirrhinum* (snapdragons), *Dendranthema* (chrysanthemums), *Dianthus* (carnations), orchids, and roses. The specific cultural details concerning light, temperature, water, gases, growing media, and fertilization were covered in earlier chapters. If special emphasis is required for any of these factors, it is indicated in the following sections for the specific crop.

ANTIRRHINUM (Snapdragons) ▧
Antirrhinum majus Scrophulariaceae

History

Snapdragons are native to the Mediterranean region. One of the earliest published accounts of snapdragon color variations appeared in 1578. The plants were first cultivated in Italy. Then they were introduced to other areas in Europe and Great Britain, where they were popular plants in outdoor gardens. Snapdragons are cool temperature crops and can be started early in the season. They provide good color through the summer and well into fall.

Early species were propagated by terminal cuttings. The use of seed did not become a general practice until the early 1900s. The appearance of a persistent rust disease stimulated breeding efforts to develop a resistant strain. Because the disease is not carried in the seed, this method of propagation gradually replaced cuttings. Rust is still a problem in California, but much less so now that fewer cuttings are made.

One of the first hybrid snapdragons to win an All-American award was introduced in 1957. The development of the 'Rocket' series in the early 1960s gave growers a new class of snapdragons for outdoor culture.

For many years, breeders have worked to improve snapdragons used as greenhouse crops. In the late 1930s, A. J. Snyder of Rockwood, Pennsylvania, developed one of the first summer flowering snapdragon cultivars, 'Lavender Rose.' There is now an entire series of Rockwood hybrids that flower at different seasons of the year.

Fred Windmiller of Columbus, Ohio, was one of the first breeders to produce F_1 hybrid snapdragons. His first hybrid, 'Christmas Cheer,' helped to lay the groundwork for the snapdragons now used for greenhouse forcing.

The peloric flower type snapdragons were introduced in 1973 by Glenn Goldsmith. The breeding program for this group of greenhouse snapdragons started in 1960. The plants in this group have flowers described as azalea type, butterfly type, and bell type. The new greenhouse group consists of eight butterfly type and four azalea type flower lines.

Classification

Snapdragons are classified into four flowering groups according to the response of the cultivars to temperature, light intensity, and day length. The groups are divided geographically into north and south groups using 40° north latitude as the dividing line. The flowering groups are given in Tables 29-1 and 29-2. Note in the spring/fall groups that two flowering periods are given. The conditions for flowering in these periods are nearly the same.

Suppliers' catalogues list the snapdragon cultivars included in each of these groups.

Seed Germination

Snapdragon seeds are very small. A trade packet of greenhouse snapdragon seed contains nearly 2,000 seeds. One packet is used to sow two standard 10-inch by 20-inch (25-cm by 50-cm) seed flats in rows spaced 1.5 to 2 inches (3.75 to 5 cm) apart. The germination medium is a special peat-lite mix using

Table 29-1

Classification of Snapdragon Flowering Response Groups North of 40° North Latitude[1]		
Flowering Group	**Dates to Sow Seed**	**Dates to Flower**
I Winter	August 15–28	Dec. 15–Feb. 15
II Late Winter	July 24–Nov. 14	Oct. 25–May 10
III Spring/Fall	Dec. 10–Mar.6	May 10–July 1
	June 18–July 16	Sept. 10–Oct. 25
IV Summer	Mar. 28–June 10	July 1–Sept. 10

1. Adapted with modifications from *Bedding Plant News*, September 1977.

Table 29-2

Classification of Snapdragon Flowering Response Groups South of 40° North Latitude[1]		
Flowering Group	**Dates to Sow Seed**	**Dates to Flower**
I Winter	Aug. 25–Sept. 18	Dec. 10–Feb. 17
II Late Winter	Aug. 22–Dec. 20	Dec. 1–May 4
III Spring/Fall	July 6–Aug. 16	Oct. 1–Dec. 7
	Jan. 7–Mar. 8	Apr. 30–June 14
IV Summer	Mar. 15–July 2	June. 14–Oct. 14

1. Adapted with modifications from *Bedding Plant News*, September 1977.

calcium nitrate rather than ammonium nitrate. The successful germination of snapdragons requires a medium with a low soluble salt level and excellent drainage and aeration. Rows 0.25 inch deep and 0.5 inch wide are pressed into the moistened surface of the medium. The seeds are sprinkled sparingly in the row and left uncovered. Some growers report that germination occurs faster if the seed is watered in with a solution of 1 teaspoon (0.5 g) of potassium nitrate in 1 gallon (3.785 l) of water.

The peat-lite mixes produce excellent seedlings. However, some growers prefer to add a small amount of soil to the mix. Any soil added must be steam pasteurized to destroy disease organisms. Snapdragon seedlings are very sensitive to damping off caused by the *Pythium* and *Rhizoctonia* species.

The medium may be drenched with a fungicide after the seed is sown. This treatment may reduce the percentage of germination. The fungicide captan cannot be used as a drench because it greatly reduces germination.

The seed flats are placed under low-pressure mist. They are misted for 3 to 5 minutes, every 12 to 15 minutes. In the winter, mist is used from 10 A.M. until 3:30 or 4:00 P.M. During the spring, summer, and fall, the length of time the flats are misted is increased as needed.

If mist is not used, the flats may be covered with opaque white plastic. Germination is hastened when bottom heat at 70°F (21°C) is applied. As soon as the seedlings appear, the plastic is removed. The seed flats are then placed in an area where the night temperature is 60°F (15.5°C). The seedlings are grown at this temperature until they are ready for transplanting. The transplanting stage is reached when the plants develop the first set of true leaves.

Most greenhouse growers producing cut snapdragons obtain their plants from commercial propagators who start the plants as plug seedlings.

Media and Fertilizer

The growing medium for snapdragons must have excellent drainage and aeration, but it must also hold moisture and nutrients. The preferred medium was once a 1:1:1 by volume mixture of loam soil, sphagnum peat moss, and perlite with a pH of 6.0 to 6.5, but many now use a peat-lite mix unless they are growing directly in the ground. The medium should be at least 6 inches (15 cm) deep for proper drainage. A low to medium soluble salt level is required. Good growth is obtained when the soluble salts are less than 150 mhos on a 1:2 media:water ratio test. Snapdragons are very sensitive to high soluble salt levels.

If snapdragons follow chrysanthemums in a planting, the soluble salt levels must be checked. The level of fertilization for chrysanthemums is usually much higher than that required by snapdragons. Leaf analyses to determine the nutrient content of snapdragons show that the levels of nitrogen, potassium, phosphorus, calcium, and magnesium are as high as the levels of these elements in chrysanthemums. However, the root systems of snapdragon plants are such that the plants cannot tolerate the addition of large amounts of fertilizer at one time. Thus, snapdragons are fertilized using lower concentrations of nutrient strength.

When a 1:100 dilution ratio injector is used, each 100 square foot (9.3 m²) area is treated with a preplant application of 5 pounds (2.3 kg) of 20 percent superphosphate and 1 pound 0.45 kg of 10-10-10 analysis fertilizer. Liquid fertilizer is applied at each watering at the rate of 60 to 90 ppm each of nitrogen, phosphorus, and potassium from a 20-10-20 analysis fertilizer. When the flower buds are well developed, only potassium nitrate is applied at the rate of 8 ounces (0.23 kg) in 100 gallons (378.5 l) of water. This solution supplies about 84 ppm of nitrogen and 240 ppm of potassium. The extra potassium improves the stem strength. Fertilizer is not applied after the flower buds show color. The crop is finished using tap water alone so that the soil will have a low soluble salt content for the next crop.

Snapdragons are very sensitive to a lack of boron. The symptoms can be confused with those caused by a cyclamen mite infestation. On seedlings, the new leaves are curled and twisted. On older plants, terminal growth stops and new side shoots form witches' brooms. Flower buds abort if the boron deficiency occurs as the flower buds are forming. A flower bud that aborts can be sliced in half to show the effect of too little boron. That is, the developing flower parts inside the bud turn brown. To prevent a boron deficiency, borax can be applied twice a year at the rate of 0.5 ounce (14.2 g) applied to 100 square feet (9.3 m²) of area. Sodium tetraborate ($Na_2B_4O_7 \cdot 4H_2O$) containing 11.8 percent actual boron can be purchased from a chemical supply house. Serious plant damage can occur if more than the recommended amount of boron is applied. Boraxo® can also be used. Polyborchlorate and other weed killers containing large amounts of boron should not be applied.

Watering

The best growth of snapdragons is obtained when the soil is kept moist. Overwatering in the winter must be avoided. During the spring, summer, and fall, a lack of moisture can severely check the growth, particularly when the flower buds are forming.

To minimize leaf disease problems, the plants should be irrigated without wetting the foliage. Snapdragons do not require any misting or wetting of the foliage.

Spacing

The plants are grown with single stems to produce high-quality flowers. Pinched plants require an additional 4 to 5 weeks to reach the flowering stage. **Single stem** crops are easier to time than pinched crops. When the crop is planted between September 15 and February 1, the plants are grown at a spacing of 4 inches by 5 inches (10 cm by 12.5 cm) or 20 square inches (125 cm^2) for each stem. Crops grown in benches for the rest of the year are given a 4-inch by 4-inch (10-cm by 10-cm) spacing, or 16 square inches (100 cm^2) per stem.

Support

Straight stems are produced by supporting the plants. A system of netting or a wire grid is used (Figure 29-1). The general practice is to install a single set of wires and string on a growing crop. As the height of the crop increases, the wires and strings are raised (Figure 29-2).

Light and Temperature

The full light intensity is required to produce high-quality crops during the fall, winter, and spring flowering periods. A light shade is applied to the greenhouse glass over summer flowering snapdragons to reduce any intense heat. If fan and pad cooling is used, then shade is not required.

Snapdragons respond to both day length and temperature. However, the type of response is used only to sort out the cultivars that belong in each response group. The response is not used in the culture of snapdragons as it is in the culture of chrysanthemums.

The seedlings grow faster in artificial light at an intensity of 10 to 15 **watts** for each square foot of area. Lighting is economical only during the seedling stage when the plants are close together. The light arrangement used on seedling geraniums is also used on snapdragons.

FIGURE 29-1 Newly transplanted snapdragon plugs with netting ready to be raised as the crop grows.

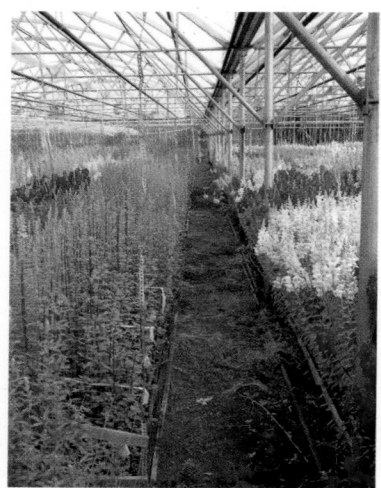

FIGURE 29-2 The wire and string support is moved up as the plants grow taller, as shown on this crop that is ready for harvest.

Snapdragons are produced at a night temperature of 50°F to 52°F (10°C to 11°C). On cloudy days the temperature is 5° higher. On clear days the temperature must not exceed 65°F (18.5°C). If necessary, ventilation is used to keep the temperatures in the proper range.

Carbon Dioxide

The quality of snapdragons is improved when carbon dioxide is added during the daylight hours. A concentration of 700 to 800 ppm is maintained from daylight until 4:00 P.M. Carbon dioxide is not used when the greenhouse is being ventilated. It is a waste of money to use carbon dioxide when the exhaust fans are on or when the ventilators are open.

Harvesting for Market

The plants are harvested when a minimum 10 florets are open on a stem or the lower third of the florets are open. The head of the snapdragon is called a **spike.** As soon as the spikes are cut, they are placed upright in containers with the stem ends in water. If cut snapdragons are placed in a horizontal position for a period of time, they show a negative geotropic response. In other words, the ends of the stems curve upward. The curving is permanent once it occurs. The cut flowers must be stored and shipped in an upright position. Newer snapdragon cultivars have a reduced geotropic response as compared to older cultivars. Yet, the issue of geotropism is one of the characteristics that makes this plant difficult to ship via air freight, thus making it unattractive to offshore producers. This keeps snapdragons as a mostly locally grown cut flower crop.

The flowers are sorted according to stem length and are grouped in bunches of a dozen blooms. At present there are no standard grading methods.

Problems

Cultural Disorders

Floret skips are caused by low temperature. Shelling, or dropping of florets, is due to pollination by bees and ethylene gas from air pollution, fruits, vegetables, and other flowers. Snapdragons cannot be stored or shipped in the same container as Calla lilies. In addition, they cannot be refrigerated with fruits or vegetables. Hollow stems are caused by low light levels, resulting in reduced carbohydrate formation. The addition of carbon dioxide to the greenhouse atmosphere can help overcome this problem.

Moisture stress causes wilting on sunny winter days following a long period of cloudy weather. In other words, the plants are transpiring faster than the roots can take up water from the cold soil. Temporary shade is applied over the plants to reduce the rate of transpiration. In a day or two, the plants usually adapt to the brighter light. Bottom heat can reduce this problem as well.

Diseases

Wilting may be caused by *Pythium* disease. If the roots are brown, the disease may be present. However, high soluble salt levels also cause brown roots. The presence of *Pythium* must be confirmed by laboratory analysis.

Rhizoctonia injures the stem at the soil line. This disease is controlled by drenching the flats with a recommended fungicide.

Botrytis gray mold is caused by *Botrytis cinerea*. Powdery mildew is caused by a fungus of the *Oidium* species. Both diseases attack the leaves and flowers and are promoted by high humidity and poor air circulation. Several steps can be taken to prevent these diseases, including not wetting the leaves, increasing the air circulation by means of overhead fans, picking off and cleaning up infested leaves, and using chemical smoke fumigants (rather than spraying).

Thielaviopsis basicola causes black root rot. To control this disease, the seed flats, media, and production benches are steam pasteurized.

Infection of the plants by the airborne fungus *Puccinia antirrhini* causes rust. Rust-resistant cultivars are used to reduce the incidence of this disease. As a further precaution, however, the preventive procedures used to combat *Botrytis* and mildew are also used to prevent rust.

Peronospora antirrhini and *Peronospora* sp. cause downy mildew. It is observed when the greenhouse conditions are cool and the relative humidity is high. The leaves will appear white to purplish-gray on the undersides as the downy mildew sporulates (Figure 29-3). Downy mildew can be controlled by keeping the humidity below 85 percent during the spring and fall months where the temperatures favor disease development. Preventative fungicide sprays are also effective.

Insects

The most troublesome insects are aphids, cabbage looper, garden symphylans, red spider mites (in warm weather), slugs, stalk borer, and whitefly. Thrips may be a problem in warm weather when they enter the greenhouse through the ventilators or the cooling pads. Bees cause floret drop in some cultivars. To overcome this problem, the vents are screened to keep the bees from entering the greenhouse, or shatterproof cultivars are grown.

CHRYSANTHEMUMS �ख *Dendranthema grandiflora, Asteraceae*

This crop was previously known as *Chrysanthemum morifolium*. It shall be referred to as chrysanthemum in this section.

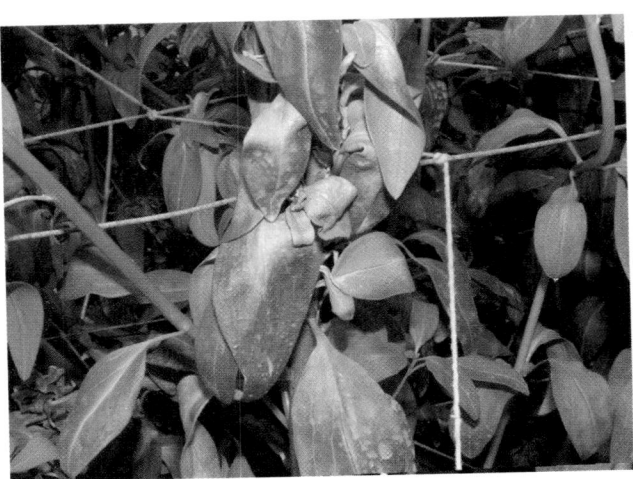

FIGURE 29-3 Downy mildew (*Peronospora antirrhini Peronospora* sp.) on snapdragons.

History

Chrysanthemums belong to the largest family of plants, the Asteraceae (formerly **Compositae**). The large number of plants in this family includes sunflowers, daisies, dandelions, and china asters. The many forms and types of modern chrysanthemums descended from two similar botanical species. It is thought that the first cultivated species of chrysanthemum originated in China over 2,000 years ago. In 1796, *Botanical Magazine* illustrated an important, large-flowered chrysanthemum modification. The beginning of modern record keeping and the development of new cultivars may date from this time.

At present, there are more than 700 named cultivars. Less than 50 of these cultivars are grown to produce the cut flowers harvested in the trade. The chrysanthemum is the third most important cut flower crop grown in the United States.

Flower Type

The chrysanthemum flower consists of many florets. The florets on the outer perimeter of the flower head are well-developed petals and are known as **ray florets**. These florets are female and normally do not have stamens. **Disk florets** are poorly developed, tubular in shape, and found in the center of the flower head. Disk florets have both male and female flower parts.

Classification

Several factors are used to classify chrysanthemums, including characteristics of the flower, commercial use and culture, and their response to day length or photoperiod.

Flower Characteristics

The National Chrysanthemum Society established a system of classification based on the following flower types: singles, anemones, pompons, decoratives, and large flowered (standard).

Singles are daisylike flowers having one or more outer rows of ray florets. The center of the flower consists of short, flat disk florets.

Anemones are similar to singles except that the disk florets are longer. These florets are often a different color, giving the flowers a pillow effect.

Pompons consist of short, broad incurving ray florets that form a small rounded head. Pompons are subdivided according to the diameter of the flower head: small or button (1.5 inches [3.75 cm] or less), intermediate (1.5 to 2 inches [3.75 to 5 cm]), and large (2.5 to 4 inches [7.75 to 10 cm]).

Decoratives are similar to pompons except the petals of the outer ray florets are longer than those of the center ray florets. This configuration gives the flower a flattened appearance.

Large-flowered standard plants have blooms that are larger than 4 inches (10 cm) in diameter. These plants are grown as "disbuds" normally. Disbudding involves removing all flower buds except the main bloom. In this type of flower, the disk florets are completely hidden by the large ray florets. The standard group has four subdivisions: incurved, reflexed, tubular ray, and miscellaneous.

Incurved flowers are glove shaped, and the ray florets curve inward toward the top.

Reflexed flowers have egg-shaped heads with overlapping ray florets curving downward. Tubular ray flowers have tube-shaped ray petals. There are four subgroups in this group:

1. The spider type has long, tubular ray florets that droop gracefully.
2. The fuji type is similar to the spider type except that the ray florets are shorter and stiffer.
3. The quill type has tubular ray florets. The florets are not as long as those of the spider or fuji types. The petals resemble feather quills.
4. The spoon type is similar to the quill type except that the open petal ends are flattened, resembling a spoon.

Flowers in the miscellaneous group are novelty types that have little or no commercial importance.

Commercial Use

The commercial classification contains two groups, cut flowers and potted plants. The cut flower group is further divided into two groups: single stem and **spray type** chrysanthemums.

1. Single stem chrysanthemums are produced by removing all of the flowers except the main flower. Standards, disbuds, and commercials are used to produce single stem flowers.
2. Spray chrysanthemums are produced by allowing all flowers to develop on a stem. Flower types that can be grown as sprays include pompons, singles, anemones, decoratives, and novelties.

The potted plant group is also divided into two groups: year-round pot mums and garden mums.

1. Year-round pot mums are produced from cultivars having the following flower types: standards, large and medium pompons, and decoratives. These plants are usually disbudded to a single flower on each stem.
2. Garden mums are grown using cultivars intended for garden planting. These cultivars usually flower late and may be hardy in mild winter climates.

Response Groups

The third method of classifying chrysanthemums sorts the plants according to the number of weeks between the start of short days and flowering. In this classification, the response groups are known as 7-, 8-, 9-, 10-, 11-, 12-, 13-, and 14-week groups. Cultivars in the longer response groups (12-, 13-, or 14-week groups) are used for winter flowering. Cultivars flowered in summer, especially garden mums, belong to the 7-, 8-, 9-, and 10-week response groups.

New Cultivars

New cultivars are introduced every year. In their 1993–1994 catalogue, Yoder Brothers Inc. of Barberton, Ohio, one of the largest specialty propagators of chrysanthemum cuttings in the world, listed 10 new spray type chrysanthemums. The total number of spray type cultivars given was 138. They also listed 40 standard type cultivars for sale. Several hundred million cuttings are sold as unrooted or rooted plants each year by Yoder Brothers Inc. to customers around the world.

One method of improving cultivars is **hybridization** or the combination of two parents to produce offspring with specific traits. Selection is a method by which the breeder picks only the best offspring or those with the most desirable traits for growing on.

Another way of obtaining new cultivars is sporting or spontaneous **mutation.** That is, natural forces cause a change in some characteristic of the plant. The change may be in the flower color or shape, or it may be in the leaf size and/or shape. Many modern cultivars are the result of natural mutations.

The plants can be treated with chemicals or exposed to radiation to produce artificial mutations.

New cultivars cannot be sold for forcing until they undergo a series of tests. These tests ensure that the new cultivar is stable in its genetic character. An unstable cultivar may revert to a plant that has undesirable characteristics and is unsaleable.

Stock Plants and Propagation

The large chrysanthemum growers maintain **stock plants** for propagation. It is recommended, however, that smaller growers purchase rooted cuttings from specialty propagators. These growers will find that maintaining stock plants consumes valuable greenhouse space that could be used to grow other saleable crops.

If a grower decides to propagate cuttings from stock plants, fresh plant material should be purchased from a specialty propagator at least every 2 years. The grower will find that if the same plant material is propagated year after year, the quality of the crops is gradually reduced. Fresh plant material for stock plants will reestablish a quality crop.

When the rooted cuttings are received from the propagator, they are planted in benches in medium with a minimum depth of 6 inches (15 cm). The ideal growing medium was once a 2:1:1 mixture by volume of loam soil, sphagnum peat moss, and perlite, but many modern growers now use a peat-lite medium unless they are growing directly in the ground. Regular superphosphate is added at a rate of 5 pounds (2.28 kg) to 100 square feet (9.3 m^2) of area. Treble superphosphate may be used instead of regular superphosphate at the rate of 2.5 pounds (1.14 kg) to 100 square feet (9.3 m^2). Limestone is added to bring the pH to a range of 6.0 to 6.5. The limestone and superphosphate are rototilled into the medium or are hand mixed to obtain a uniform distribution. The medium is steam pasteurized or chemically fumigated.

Rooted cuttings are planted in a 6-inch by 6-inch (15-cm by 15-cm) spacing. As soon as the new growth is 1 inch (2.5 cm) long, the plants are given a soft pinch. The plants must be kept in vegetative growth because flowers are not wanted at this point. The vegetative stage is maintained by using incandescent lights to break up the dark period. This period must not exceed 7 hours. The lighting schedule described for flowering plants is also used to keep chrysanthemums in the vegetative stage.

As soon as the new growth from the first pinch is 6 to 8 inches (15 to 20 cm) long, cuttings can be taken. If cuttings are not made at this time, the plants are given a second pinch, which causes them to produce more breaks (branches). Any cuttings made are rooted in a propagation bench, or they are placed in dry refrigerated storage at 31°F (−0.5°C). The cuttings can be held for 1 to 2 months in storage. Following this period, they are rooted. Some cultivars can be stored for 3 to 4 months. Rooted cuttings can also be stored dry at 31°F (−0.5°C) for 3 to 4 weeks. However, unrooted cuttings that were stored at 31°F (−0.5°C) cannot be rooted and returned to storage. The cuttings cannot survive this treatment because they lack sufficient food reserves.

The process of taking cuttings and allowing new growth to come back is repeated until five or six flushes (groups of cuttings) are obtained. The plants are then replaced with new planting stock.

Three- to 4-inch (7.5- to 10-cm) long terminal cuttings are made by snapping them from the plants with the fingers. If a knife is needed to cut the stems, they are too woody. Cuttings are taken from 14 to 21 days before they are to be planted in the production bench. Leaves that would be buried in the propagation medium are removed. No other trimming is done.

The basal end of each cutting is dusted with a rooting hormone (refer to Chapter 18). The use of the hormone improves the percentage and evenness of rooting. The temperature of both the rooting medium and the air is a minimum of 65°F (18.5°C). When the cuttings are first placed in the medium, low pressure mist is applied for 12 seconds every 3 minutes. After 3 or 4 days, the mist is applied every 8 to 10 minutes. The mist is used to keep the leaves moist; there should be little runoff. Too much mist chills the plants and leaches nutrients from the leaves (refer to Figure 18-7).

A **mist fertilizer** can be applied using 4 to 6 ounces (113.4 to 170 g) of 23-19-17 analysis fertilizer in 100 gallons (378.5 l) of water. The fertilizer is applied at the same rate as the tap water misting. The cuttings treated in this manner grow better and flower earlier than nonmisted cuttings. Cuttings that are fertilized by mist cannot be stored at 31°F (−0.5°C) after rooting because they will rot. Therefore, they must be planted in the production beds as soon as they are rooted.

As the cuttings root, the frequency at which they are misted, either with fertilizer or plain tap water, is reduced. This process hardens the cuttings and helps them to adjust to the new, drier environment of the production bench.

When the fertilizer is not applied as a mist, growers use a solution of 20-10-20 analysis fertilizer at the rate of 2 pounds (0.9 kg) in 100 gallons (378.5 l) of water. This solution is applied 7 days after the cuttings are placed in the propagation medium. Fertilizer is applied again 2 to 3 days before the cuttings are removed from the medium.

The cuttings are stuck at a spacing of 1 inch (2.5 cm) between plants and 2 inches (5 cm) between rows. The medium is a 1:1 mixture by volume of peat moss and perlite, or peat moss and vermiculite. A 1:1 mixture of perlite and vermiculite is used by many growers. Ground limestone is added to media containing peat moss at the rate of 5 pounds in each cubic yard.

The full light intensity is used at all times. In the middle of summer, light shade may be used on the glass if it is very hot. However, shade causes the cuttings to stretch and should be used sparingly.

As soon as the cuttings are rooted, they are planted in the production benches. If the plants remain too long in the propagation bench, they become too hard.

Media, Fertilizers, and Carbon Dioxide

Chrysanthemums grow well in a variety of media having the following characteristics: good drainage and aeration, good water- and nutrient-holding capability, and freedom from disease organisms and insects. The media described for growing stock plants is also used for cut flower production. Growers may add organic matter in the form of chopped straw between each crop. This material normally breaks down completely in 3 or 4 months. Extra nitrogen must be added to prevent a nitrogen deficiency.

Chrysanthemums are described as heavy feeders. Foliar analyses show that chrysanthemums normally contain 4 to 5 percent nitrogen and 5 to 6 percent potassium. Mums produce heavy growth in 3 or 4 months and require regular fertilization.

The preferred program is to apply fertilizer with every watering. A 1:100 fertilizer injector is used to apply 20-10-20 analysis fertilizer at a concentration of 200 ppm each of nitrogen, phosphorus, and potassium. Once the plants are established in the bench, and not later than 1 week after planting, the fertilization program is started. When color first shows in the flowers, the program is changed. From this point until the crop is harvested, only potassium nitrate is applied at the rate of 12 ounces (340 g) in 100 gallons (378.5 l) of water to supply about 125 ppm of nitrogen and 350 ppm of potassium.

During dark weather and winter, an alternative program is recommended. This program uses nitrogen and potassium from nitrate fertilizer. Six and three-quarter ounces (191 g) of potassium nitrate is combined with 11 ounces (312 g) of calcium nitrate in 100 gallons (378.5 l) of water. This solution supplies 200 ppm each of nitrogen and potassium. The fertilizer is added with every watering until the buds show color. From this point on, the crop is finished using potassium nitrate at the rate of 12 ounces (340 g) in 100 gallons (378.5 l) of water.

The grower may decide to use Osmocote® 14-14-14 or Magamp® 7-40-6 alone or in combination with a liquid fertilizer. Either material is applied before planting at the rate of 10 to 12 pounds (4.5 to 5.5 kg) to 100 square feet (9.3 m²) of area. Osmocote® must be added after the medium is steam pasteurized. Additional fertilizer is not required when Osmocote® is used, but 1 pound (0.45 kg) of potassium nitrate is added to Magamp®. This material provides nitrate nitrogen to the plants until the ammonium nitrogen in Magamp® is converted to nitrate nitrogen.

The grower has very little control over the amount of nitrogen received by the crop when all of the fertilizer is added before planting. If warm weather occurs at the wrong time, the fertilizer will release too much nitrogen and the plants will make soft growth. The rate at which both Osmocote® and Magamp® release their nutrients is directly proportional to the temperature and the moisture level.

Some growers apply 5 pounds (2.28 kg) of Osmocote® or Magamp® before planting and then apply liquid fertilizer to obtain more control over the fertility of the crop.

Both Osmocote® and Magamp® must be added again before each crop is planted. Some Osmocote® formulations last longer than 3 to 4 months. These materials release their nutrients too slowly to produce good-quality cut mums.

When the medium contains a large amount of soil, a lack of trace elements normally is not a problem. A boron deficiency may develop in soil to which only sphagnum peat moss and high-analysis liquid fertilizers were added and that was used for chrysanthemum production of 10 years or more. To correct this deficiency, the grower must use a regular liquid fertilization program containing trace elements or must apply fitted trace elements or similar products. Chapter 15 gives the details of these programs.

In the winter months, the greenhouse ventilators are closed for days or weeks at a time. Under these conditions, carbon dioxide may become a limiting factor for growth. The plants will benefit from carbon dioxide added during the daylight hours to maintain levels of 1,000 to 1,500 ppm. Normally, carbon dioxide is added from daylight until 2 hours before sunset. The advantages of adding carbon dioxide include less time required to harvest the crop, an increase in the dry weight of the plants, an increase in the number of plants sorted into the better grades, increased stem strength, reduced quilling, and better flower color.

When carbon dioxide is used in the production program, more water and fertilizer must be added. The increased growth of the plants demands higher moisture and nutrient levels.

Watering

Chrysanthemums need large amounts of water to produce quality flowers. Watering systems used successfully in the production of chrysanthemums include the Gates® perimeter system and Chapin® ooze tube. The foliage remains dry when these methods of irrigation are used. As a result, the incidence of leaf disease is reduced. When fertilizer is applied at each watering, an additional 10 percent by volume of solution is applied to cause leaching. This practice prevents a buildup of high soluble salt levels. When a soil test shows the total soluble salt level to be more than 200 mhos for a 1:2 soil:water sample, the growing medium requires leaching. The methods of reducing the soluble salt levels are described in Chapter 16.

Light and Temperature

These environmental factors are very important to growth and flowering of chrysanthemums. There is a direct relationship between the amount of light chrysanthemums receive and the quality of growth. In northern areas, the light intensity is not excessively high for chrysanthemums, except perhaps in the middle of summer. Some light shade is required on chrysanthemums growing in fields in Florida in the winter and in outdoor beds in the summer in the northern United States.

For plants growing in a greenhouse, light shade is used in the summer to reduce the temperature. Generally, chrysanthemums are shaded only when they are first planted and to reduce the temperature in the summer. Mums are shaded after planting to reduce moisture stress until the root system is fully active and taking up water.

The lighting requirements for the control of flowering are described later in this chapter in the section on "Year-Round Scheduling." The reader should review Chapter 8 for an explanation of how light affects plant growth.

Accurate temperature control is necessary in growing quality chrysanthemums. Growers may be tempted to lower the greenhouse temperature to reduce fuel costs. However, if the temperature is reduced below the recommended levels for chrysanthemums, a number of unwanted conditions result: Blooming is delayed; the time in the production bench is increased; the flower size and stem length are reduced; and foliar diseases increase, resulting in a higher cost for chemical sprays. The grower must compare any money saved in fuel costs to the cost of additional chemicals for disease control sprays, the increased labor costs in applying the sprays, the loss in crop value because of reduced quality, and the increased production costs due to the longer bench time. Any decision regarding the greenhouse temperature can be made only after the grower evaluates and compares the various cost factors.

Optimum flower bud formation and development of the buds occur when the night temperature is in the range of 60°F to 62°F (15.5°C to 16.5°C). During the day, the recommended temperatures are 65°F (18.5°C) on cloudy days and 70°F (21°C) on clear days. Temperatures in excess of 80°F (27.5°C) during the day cause a delay in the flower development and reduce flower quality.

Once the flowers show good color, the temperature can be reduced. The night temperature can be reduced to 55°F (13°C) to improve the color and lasting life of cut mums. The greatest improvement is shown in the reds, bronzes, and deep pinks.

Plant Spacing

The spacing of the plants in the bench depends upon the growing season and the cultural program. During the winter, when the days generally are darker, each plant needs as much light as possible. Therefore, they are given more space. Plants that are pinched to produce two or three stems per plant require more bench space than plants being grown for single stems.

Single-stem crops require less labor because there is no pinching and **pruning.** In addition, they are ready for harvest 3 to 4 weeks earlier than pinched crops. As a result, more single-stem crops can be grown in the same area in a year. A disadvantage to a single-stem crop is that more cuttings are needed when the crop is planted.

Plants used for single stems are spaced 4 inches by 6 inches (10 cm by 15 cm) apart for spray types and 5 inches by 6 inches (12.5 cm by 15 cm) apart for standards. Summer grown fast disbuds are spaced 4 inches by 6 inches (10 cm by 15 cm) apart. Disbuds grown in the winter are spaced 6 inches by 6 inches (15 cm by 15 cm) apart.

When the cuttings are to be grown as pinched plants, fewer cuttings are needed to plant a given area. However, these plants need more time in the bench, and the quality of the crop is less than that of single-stem plants. Large-flowered standards are planted 7 inches by 8 inches (17.5 cm by 20 cm) or 8 inches by 8 inches (20 cm by 20 cm) apart.

Spray type mums that will be pruned to two stems are planted at a spacing of 7 inches by 8 inches (17.5 cm by 20 cm). Plants pruned to three stems are planted 8 inches by 8 inches (20 cm by 20 cm) apart.

Pinching and Timing

The plants are pinched to increase the number of flowers per plant. Advantages of pinching include reduced costs for planting materials, control of the spray formation for spray type cultivars, and some control of the type of bud formed.

Pinching removes the terminal growing point of the plant, and causes growth of the lateral branches. To obtain the greatest number of lateral branches and to ensure that the growth is not checked, a soft pinch is used. A hard pinch, which removes more than 1 inch (2.5 cm) of the growing tip, may check the growth of the plant.

Extra stems develop on a pinched crop. To prevent the formation of a large number of short stemmed, low-quality blooms, the stems are pruned until two or three stems remain to produce flowers. Plants in the outside rows of the bench are allowed to keep one more stem per plant because these rows receive more light. Pruning must not be delayed because the energy of the plant will be wasted in growth that is to be discarded. When older, woody stems are pruned, the remaining scars may not heal. Disease may enter the plant through these scars.

The shape of the spray formation is controlled by removing the central flower bud after the side shoots develop. Some spray type cultivars tend to produce a **clubbed spray** head that can be opened up by the proper timing of the pinch.

Bud Types, Formation, and Disbudding

Chrysanthemums form crown type flower buds and terminal buds.

Crown buds form early in the season. They are surrounded by vegetative buds, which develop and flower. If the natural days are not short enough for terminal budding, the crown bud remains dormant. Crown buds may be formed if there is an interruption in the lighting schedule during off-season culture. Five short days followed by long days are sufficient to cause the formation of crown buds. These buds can be identified by the narrow, straplike leaves that are formed below the bud.

Terminal buds form naturally later in the season. They are surrounded by other flower buds (Figure 29-4). The flower that develops from a terminal bud commonly is smaller than the flower produced from a crown bud. Normal foliage is produced below terminal buds. Ten or more short days are needed to form terminal buds.

When plants are being grown for single-stem culture, one bud is selected to flower. All other flower buds below the one selected are removed, or disbudded. When the main bud is about 0.25 to 0.38 inch (6.4 to 9.5 mm) wide, the small disbuds are removed easily by a sideways motion of the thumb. All of the energy growth of the plant goes into the production of one large flower after the disbuds are removed.

Once the plants are disbudded, the buds are collected and removed from the greenhouse. If they are allowed to remain in the benches or in the aisles, they can serve as breeding areas for *Botrytis*.

The central bud of spray type flowers may be removed to open up clubbed sprays. In some cases, extra shoots with small undeveloped buds grow below the main spray head. These shoots should be removed as soon as possible.

Chrysanthemums are photoperiod responders. That is, flower buds are formed when the day length becomes less than a certain number of hours. In other words, chrysanthemums are short-day plants.

The critical day length for flower bud initiation or formation is nearly 14.5 hours. The day length must be reduced to 13.5 hours to ensure flower bud

FIGURE 29-4 A crown bud surrounded by stems with terminal buds.

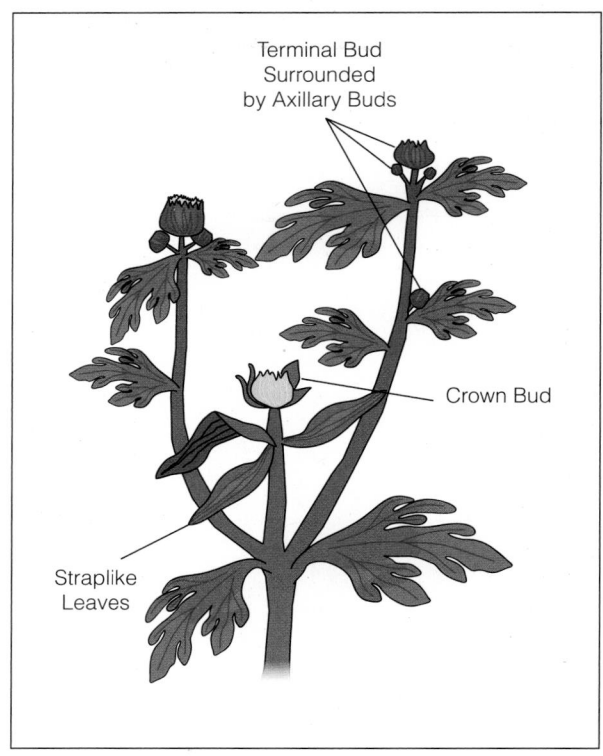

development. At 42° north latitude, the normal date of bud formation is between August 15 and September 5. Bud development continues as the days get shorter in the fall. By September 20, the day length is short enough, including morning and evening twilight, that bud development continues without delay.

At 42° south latitude the normal date of bud formation is between February 15 and March 5. Bud development continues as the days get shorter. By March 20, the day length is short enough that bud development continues without the need of black cloth treatment. As one travels closer to or farther from the equator, the dates of flower bud initiation and development are different from those given here.

To prevent flowering, chrysanthemums are given a light break in the middle of the night. A 9-hour dark period is sufficient to prevent a bloom. A dark period no longer than 7 hours is used. Growers provide the 2 extra hours of light to ensure that the plants do not bloom because of other factors. Knowing how to control the flowering of chrysanthemums permits the grower to force blooms year-round.

Year-Round Scheduling

Chrysanthemums bloom naturally in the Northern Hemisphere during the short days from August 15 to April 15. From April until August, the natural days are long enough to prevent flowering.

Lighting

As little as 2 fc (22 lux) of light will keep chrysanthemums in the vegetative stage of growth. To ensure that the plants do not flower, a minimum of 10 fc (108 lux) of light is supplied in the darkest part of the bench. Incandescent

Table 29-3

Lighting Schedule for Chrysanthemums at 42° N. Latitude		
Month	Hours of Light	Begin Lighting
June and July	2	11:00 P.M.
August, September, October, March, April, May	3	10:30 P.M.
November, December, January, February	4	10:00 P.M.

FIGURE 29-5 A lighting system used in southern California field production to keep chrysanthemums in a vegetative condition.

bulbs can provide this intensity by using 1.25 watts of light for each square foot of greenhouse area, including walkways (Figure 29-5).

For single benches up to 4 feet (1.2 m) wide, 60-watt bulbs in reflector fixtures are spaced 4 feet (1.2 m) apart, 2 to 3 feet (0.6 to 0.9 m) above the plants. If two benches are placed side by side and are no more than 4 feet (1.2 m) wide, a single line of 100-watt bulbs in reflector fixtures can be installed over the aisle. The fixtures are spaced 6 feet (1.8 m) apart and no more than 4 feet (1.2 m) above the plant height. A light meter is used to ensure that the correct light intensity is being provided.

During June and July, the days are naturally long enough to prevent flowering. However, light is used at night for several hours to prevent premature bud set. The schedule for lighting is given in Table 29-3.

Lighting is scheduled to divide the dark period into two equal parts. An automatic time clock is used to control the on–off cycle of the lights. The time clock can be set to the exact length of time required for lighting. The lights are on constantly for the entire lighting period.

Intermittent Lighting

Research has shown that it is not necessary to light the plants continuously during the light break to keep them vegetative. Intermittent lighting also stops the flowering response. This method of lighting calls for short intervals of light through the entire lighting period. As a result, the grower will see a significant decrease in the electrical energy used for lighting and does not require an upsizing in the electrical equipment to accommodate the electrical load required for continuous lighting.

To prevent flowering, the plants must be lighted for 20 percent of a 30-minute dark period. In other words, the lights are turned on for 6 minutes and turned off for 24 minutes in each 30-minute period. The hours of lighting are the same as those listed for continuous lighting.

Special programming units can be installed that will cycle the lights on and off. The same unit can be used to cycle the lights automatically so that as many as 50 benches are lighted in rotation.

Interrupted Lighting

Normally, lighting is started when the cuttings are planted. The lights are used for a certain number of weeks until the plants make the proper amount of stem growth. The lights are then turned off. Two standard chrysanthemum cultivars, 'Indianapolis' and 'Mefo,' have a low petal count in the period from December through March. This condition often results in small, flat flower heads. A schedule of **interrupted lighting** can overcome this problem. The method is effective only for the 'Indianapolis' and 'Mefo' cultivars. The plants require a minimum night temperature of 62°F (16.5°C). Once the flower buds are well formed and showing color, the temperature is reduced gradually over 3 or 4 nights to 55°F (13°C).

The grower uses the following method to determine when the lighting is to be interrupted.

1. The normal schedule for the cultivar is consulted to determine when the lighting is stopped.
2. The grower counts back 12 days.
3. On this date, the grower begins 9 consecutive short days.
4. At the end of the short-day period, the plants are given 12 long days.
5. This period is followed by short days until the plants flower.

The plants are stimulated to form buds by the first period of short days. The long-day period following the short days slows the development of buds, resulting in a 50 percent increase in the number of petals formed.

For example, 'Indianapolis White No. 4' is scheduled to flower on December 16. According to the program for this crop, the lights are turned off October 7 to give the plants the 10 weeks required for blooming. The grower counts back 12 days from October 7 to September 25. On this date, short days are given using the black cloth treatment. The treatment continues for 9 days up to and including October 3. On October 4, the lights are turned on to give long days. The lights are continued for 12 days. At the end of this time, on October 16, short days are started again and continued until the plants bloom.

Temperature control is an important factor in the success of this program. Interrupted lighting cannot be used in summer because it is not possible to control the temperature as required. Actually, the program is not needed in the summer because there is enough light to produce the rounded flower head desired.

Artificial Short Days

When the day length is longer than the critical number of hours, artificial short-day conditions must be provided. Artificial short days are also given whenever the flowering stage may be interrupted by stray light. For example, light may strike the crop from another bench where the plants are being given long days. Lights from shopping centers, high-intensity night lights, light from traffic, or light from any source will affect the flowering stage if it is more than 2 fc (32 lux). Light from a full moon has no photoperiodic effect on mums. Even on the clearest night, moon light provides only 0.05 fc (0.5 lux).

Short days are given by using opaque, black sateen cloth (Figures 29-6 and 29-7). The cloth is pulled over wire supports by hand or by means of a semiautomatic system (described in Chapter 8). Polypropylene plastic covers may be used also. Polyethylene should not be used because the relative humidity under it rises to a very high level, leading to an increased probability of disease problems.

The plants are shaded from 4:30 or 5:00 P.M. to 7:30 or 8:00 A.M. Twelve hours of continuous darkness are sufficient.

Semiautomatic systems using a time clock can be programmed to pull the cloth. Such a system is an advantage in summer during the daylight saving

FIGURE 29-6 Blackout curtains at about 50 percent closure for creating artificial short days.

FIGURE 29-7 For smaller growers, a manual shading system is adequate as shown with this bench top shade system.

period. If the black cloth is pulled at 4:30 P.M. DST (clock time), high temperatures of 120°F (49°C) or more can build up under the cover because it is really 3:30 P.M. (sun time). Such a high temperature can cause serious delays in bud formation and flowering. A semiautomatic system can pull the cover later in the evening after the workers have left and the heat of the day is reduced.

When the natural days are short enough to cause bud formation, the black cloth treatment is stopped as soon as the buds begin to show color. The plants then will go on to flower without delay. During the long-day conditions of spring and summer, the black cloth is pulled until the flowers are very well developed. If the black cloth treatment is stopped too soon, there may be some delay in bloom.

Harvesting for Market

The stage of bloom at which the crop is harvested depends upon several factors. Flowers cut for local markets are usually fairly open. Flowers that are to be shipped long distances to market are harvested when the flower bud is tighter. The flowers of some cultivars may be cut at a tighter stage than others because they open more quickly. Flowers harvested in the winter are allowed to open more than those cut during good light periods. An experimental method of opening flowers uses blooms cut in a fairly tight bud stage.

Flowers are cut to provide the longest stems possible. The Society of American Florists has defined grades for **standard mums** (Table 29-4).

Freshly cut flowers are placed in a container of room temperature water as soon as possible after cutting. Clean containers and clean water must be used because dirty materials greatly decrease the lasting life of cut flowers. The cut flowers are then placed in a refrigerator at 38°F to 40°F (3.5°C to 4.5°C) for at least 4 hours to allow them to take up a maximum amount of water. If a refrigerator is not available, then a cool basement or other cool area can be used. The flowers are then graded and packed in dozens.

When the flowers are to be shipped any distance, they are packed in large cardboard shipping boxes. Wooden cleats are used to hold the flowers in place to prevent bruising. Newspaper or other types of padding can be used also.

Table 29-4

Standard Chrysanthemum Grades Adapted by the Society of American Florists		
	Minimum Measurement in Inches (cm)	
Grade	Flower Diameter	Overall Stem Length[1]
Gold (trial only)	$6^1/_4$ (16)	30 (76)
Blue	$5^1/_2$ (14)	30 (76)
Red	$4^3/_4$ (12)	30 (76)
Green	4 (10)	24 (61)
Yellow	$3^1/_4$ (8)	18 (46)

1. Stem length includes the flower head.
"Flowers shall be clean and bright with firm petals and leaves. The flower shall be the normal shape for the cultivar, held upright. The head shall be fairly tight with center petals (disk flowers) unopened. The center of the flower shall not be exposed. Flowers shall be free of defects, dirt or other foreign material; discoloration, insects, diseases or spray residues. The stems shall be essentially straight and still, able to support the flower in an upright position. Foliage shall be stripped from not more than the lower one-third of the stem."

During the summer months, long-distance shipments are packed and pre-cooled in the box. The flowers must be packed dry to minimize diseases that may damage the foliage during shipment. In winter the plants are packed in insulated boxes for protection from freezing.

Bud Cut Standard Chrysanthemums

Growers are experimenting with the practice of harvesting standard mums in the bud stage. This method is practical and has several advantages: More crops can be grown in an area over a period of time because they are harvested early; more flowers can be packed in a shipping box, reducing the shipping costs; there is less damage to the flower heads; an outdoor crop can be harvested early if a frost threatens; and both the grower and the wholesaler need to provide less storage space.

Disadvantages to harvesting mums in the bud stage include the problem that more preservatives must be used at a higher cost; either the wholesaler or the retailer must provide space where the buds are opened; extra labor is required to open the buds; and there is a delay before the mums are ready for sale.

Procedures

Flowers to be harvested in the bud stage are taken from the Blue, Red, or Green grades established by the Society of American Florists. The buds must be 2 to 2.5 inches (5 to 6 cm) wide. The leaves are stripped from the lower third or the lower half of the stem. The bud mums are not put in water; they are shipped dry. The buds can be stored dry in a refrigerator at 33°F to 40°F (0.5°C to 4.5°C) until the time for shipment.

Opening the Buds

The retailer or wholesaler unpacks the buds and prepares them for opening. One-half to 1 inch (1.25 to 2.5 cm) of the basal end of each stem is cut off using a sharp knife. A smooth, clean cut is preferred to a ragged one. The stems are placed in glass, plastic, or ceramic containers filled to a depth of 5 inches (12.5 cm) with the opening solution. Metal containers cannot be used because there is a reaction with the chemicals in the solution.

The opening solution consists of sugar and 8-hydroxyquinoline citrate (**8-HQC**). A 200-ppm concentration of 8-HQC is made by dissolving 0.33 ounce (9 g) of the chemical in 12 gallons (45.4 l) of water. Two pounds (0.9 kg) of sugar are also added to the 12 gallons (45.4 l) of water to yield a 2 percent by weight solution of sucrose and 8-HQC.

The containers of flowers are placed in a room at a temperature of 70°F to 75°F (21°C to 24°C). Normally, the buds open to full market size in about 6 days. The containers can be kept in darkness or they can be lighted with 75 fc (810 lux) of fluorescent light for 12 hours a day. Lighting appears to have no effect on opening.

Once the flowers are open, they can be placed in storage at 33°F to 40°F (0.5°C to 4.5°C) until they are sold. If the flowers are transferred from the opening solution to containers of plain tap water, the solution can be reused. The solution can be used for 2 weeks before it must be replaced.

The opening solution discolors the lower part of the stems slightly. The discoloration does not affect their lasting life.

Several commercial opening solutions are available and are advertised in trade papers.

Problems

Cultural Disorders

Blindness, or the failure of buds to form, occurs because of incorrect temperatures at the critical bud initiation stage. In the winter the temperature may be too low, or in the summer it may be too high under black cloth.

Buds may not develop because **shading** is removed too soon, the plants are deficient in boron, or the plants are diseased.

Chlorosis of the foliage normally occurs because of root injury that reduces the uptake of nutrients (review Chapter 14 for nutrient deficiency symptoms). Overwatering in the early stages of growth also causes chlorosis. In addition, chlorosis may be due to an incorrect soil pH and to insects and diseases infesting the roots.

Bright sun can cause petal burn of standard flowers. Light shade may be required when the plants come into bloom. A boron deficiency can cause a halo of brown petals on 'Indianapolis' cultivars (see Chapter 14). *Botrytis* can be a problem when the humidity is very high and the air circulation is poor.

In the winter, certain cultivars tend to have flat flower heads and a low petal count. These problems are corrected by using the interrupted lighting program described earlier.

Diseases

Disease prevention is less expensive than controlling diseases after the plants are infested. Good preventive measures include using overhead fans or fan and tube units to maintain good air movement and keep the foliage dry, and watering and fertilizing the crops without splashing the foliage. Mums are misted from overhead only when they are propagated and for a few days after transplanting.

Ascochyta blight is caused by the fungus *Mycosphaerella ligulicola* (*Ascochyta chrysanthemii*). This blight usually appears on outdoor crops only. The buds and flowers must be misted with a fungicide two or three times a week. In rainy weather the spray must be applied as soon as possible following the rain. If the rain occurs as frequent showers, it may be necessary to apply the fungicide two or three times a day.

Botrytis petal spot is caused by *Botrytis cinerea*. Preventive measures include providing good ventilation and air movement and removing all disbuds and stripped leaves from the greenhouse.

Erwinia chrysanthemi is a bacterium that causes bacterial stem blight. The grower should order culture indexed or meristem cultured cuttings to ensure disease-free plants. If the disease starts in a planting, there is no control. All infested plants must be removed and burned. All materials used in the culture of the crop must be steam pasteurized.

Leaf nematodes produce a V-shaped area of dead tissue between the veins of the leaves. The nematodes travel from plant to plant in films of water. To prevent the spread of these organisms, the foliage must be kept as dry as possible. The recommended insecticides can be sprayed on the crop to control these pests.

Septoria leaf spot is caused by two fungi, *S. obesa* and *S. chrysanthemella*. Preventive measures include maintaining good ventilation and air movement through the plants and spraying with a fungicide.

Tomato spotted wilt virus has become a serious disease. It is spread by thrips.

Verticillium wilt is caused by a *Verticillium* soil fungus that is taken into a plant through the roots. The organisms enter the vascular system and plug the veins, resulting in wilting. The disease is controlled by sterilizing the growing medium and using wilt-resistant cultivars.

Insects

Chrysanthemums are popular hosts for a wide variety of insects. The appropriate chemical treatments are used to control the pests. The following insects attack chrysanthemums (in decreasing order of importance): thrips, aphids, spider and other two-spotted mites, leafminers, leafrollers, leaf tiers, cabbage loopers, cut worms, corn ear worms, earwigs, mealybugs, tarnished plant bugs, and sow bugs. Leafminers are primarily an issue when night photoperiod lighting is used, which attracts the adults.

Cultivars

No attempt is made to list the chrysanthemum cultivars that are available to the grower. Detailed information is available from the specialty propagators who supply cuttings to growers. Salespeople who service the commercial growers are another good source of information on available cultivars.

DIANTHUS (Carnation) ▩ *Dianthus caryophyllus, Caryophyllaceae*

History

Dianthus species are native to Europe, Asia, and Japan. Carnations were known to the Greeks and Romans who wove garlands of the flowers to crown their athletes. A French grower named Dalmais produced the first perpetual blooming carnation in fields along the Riviera coast of the Mediterranean. The light intensity and other environmental conditions in this area are ideal for the production of these plants. Large numbers of carnations are still grown in this region, as well as around San Remo in Italy.

In 1852, Charles Marc, a French grower who lived in Flatbush, Long Island, imported seedlings of the French carnations. The beginning of the American carnation industry dates from this point. Since this time, many people have been involved in improving carnation cultivars through selection and plant breeding.

William Sim of North Berwick, Maine, introduced a new cultivar around 1938. After his death, the cultivar was named 'William Sim' in his honor. Many experts consider the 'William Sim' carnation to be the finest carnation developed to date. The sports and variants bred from the original cultivars are grown all over the world today. More than 200 mutations originated from the 'William Sim' cultivar.

Many other cultivars have been introduced over the years. The color range of carnations is extensive. Only blue, green, and black are not produced naturally. The most popular colors are the reds, pinks, and white. There are many shades of red and pink flowers, as well as a large number of speckled, variegated, and multicolored flowers. Colors that are not produced naturally are made by dyeing white flowers. For example, the green carnations that appear on St. Patrick's Day are dyed. Tinting of carnations has become a standard practice over the past 10 to 15 years.

Crop Importance and Areas of Production

Of all the cut flower crops affected by offshore production, carnations have been impacted the greatest. In 1976, the total retail value of carnations sold in the United States was $160 million. At that time 72 percent of the carnations sold in this country were grown here. In 1994, the wholesale value of both standards and spray type minicarnations grown in the United States was a little over $44 million, and in 2005, the wholesale value was $1.8 million.

These sales were derived from the production of 8.95 million stems of both spray type minicarnations and standard carnations. In 2002, Colombia, South America, alone shipped $9.78 million of both mini spray types and standards into the United States. It is easy to see why cut carnations are no longer a major crop produced in the United States. As cited in Chapter 2, Kenya is becoming a major producer of carnations at minimal costs. Should Kenya begin exporting into the United States, it would seem obvious that it would no longer be profitable to produce carnations here.

As described in the Floriculture Crops 2005 Summary, only 24 major producers of these flowers are still surviving in the United States. The majority of these blooms are being produced in California. In 2004, Colorado producers grew about 6 percent of what are grown in California. Except for Hawaii where carnations are used mostly for making leis, the other states produce relatively few carnations.

Propagation

Many of the new carnation cultivars are intended for propagation by seed. Some growers use seed to experiment with growing selected cultivars as potted plants.

Potted carnations are not widely grown for commercial sales. However, this crop has a potential that can be developed.

Carnations are propagated by means of 4- or 5-inch (10- or 12.5-cm) long terminal cuttings. Stock plants are grown only to provide cuttings for propagation. The cuttings must be in a vegetative condition. As a result the stock plants are given 9-hour days. This means that they are shaded using black cloth from March through September.

Growers can purchase rooted terminal cuttings from several specialty propagators, if desired.

Leaf bud cuttings have been used to propagate mutations or sports of cultivars that often have a small amount of tissue that can be used for propagation. As compared to plants grown from terminal cuttings, those growing from leaf bud cuttings need an additional 5 weeks to come into flower.

Virus-free clonal carnations are produced by **meristem tissue culture** methods. The techniques of tissue culture are discussed in Chapter 18. Selected carnation plants are grown at temperatures of 100°F to 110°F (38°C to 43°C). Microscopic sections removed from the growing tip are placed on a sterile medium. The grower must perform a series of exacting cultural procedures to produce bacteria-, fungus-, and virus-free plants from these tissues. The plants are free of diseases initially, but they are not disease resistant. Even meristem cultured carnations can be infested with disease when they are grown in unsanitary conditions.

The carnation has a soft stem that breaks easily. The cuttings are snapped from the plant with a sideways motion of the fingers. A knife is not used. There is no need to trim any foliage from the cuttings. They are dusted lightly with a rooting hormone of the proper strength. Then they are stuck in the propagation medium, which is a 1:1 mixture by volume of perlite and sphagnum peat moss. Some growers use sand alone, but this medium dries out too quickly in warm weather.

Carnations can also be propagated in preformed media. Figure 29-8 shows carnation cuttings that have been rooted in Smithers Oasis Wedge® propagation media. Rooting occurs in 21 to 28 days after sticking. The units are then ready for planting in the production area.

The cuttings are usually propagated in March, April, and early May for direct planting to the production benches. They are stuck in the medium at a spacing of 1 inch (2.5 cm) between plants in the row and 2 to 3 inches (5 to 7.5 cm)

FIGURE 29-8 Carnation cuttings rooted in Smithers Oasis Wedge® propagation media.

between rows. A low-pressure mist is applied to maintain an even level of moisture. The frequency of the misting cycle is adjusted as needed.

Carnations cannot be fertilized using a mist type system. The waxy covering of the leaves prevents nutrient uptake. A solution of 2 pounds (0.9 kg) of 20-10-20 analysis fertilizer in 100 gallons (378.5 l) of water is applied to the media 2 weeks after the cuttings are placed in the propagation bench.

The air temperature is maintained at 50°F (10°C), and the medium temperature should be 5° to 10° warmer. In 2 to 3 weeks at these temperatures, the cuttings should be rooted and ready for transplanting. The cuttings are given the full light intensity.

Cutting Storage

Carnation cuttings may be held in cold storage if the propagation space is not ready. Dry unrooted cuttings are placed in polyethylene-lined boxes. The plastic is folded over the cuttings but is not sealed. The plants respire slowly and require an exchange of carbon dioxide and oxygen. If the plastic is sealed, the slow process of fermentation begins and soon renders the cuttings useless.

From 200 to 300 cuttings can be placed in one box. If too many cuttings are placed in the box, the process of fermentation produces heat, which ruins the cuttings before they are chilled.

The boxes are stacked using a minimum of 1-inch (2.5-cm) thick spacers between them to permit air circulation. At storage temperatures of 32°F to 33°F (0°C to 0.5°C), unrooted cuttings can be stored for a maximum period of 3 months. For short-term storage, from 10 to 14 days, a temperature of 40°F (4.5°C) is recommended.

Unrooted cuttings given cold storage do not require special treatment when they are rooted. The methods used to root fresh cuttings are also used to root cuttings taken out of cold storage.

Rooted cuttings can be stored as well. When they are well rooted, they are removed from the propagation medium and excess medium is removed. The cuttings are placed upright in plastic-lined boxes. The plastic is folded over but is not sealed. Rooted cuttings can be stored for as long as 3 months, but a shorter time is preferred. The storage temperature is the same as that used for unrooted cuttings, 32°F to 33°F (0°C to 0.5°C). Unrooted cuttings that are removed from storage and rooted cannot be returned to cold storage. These plants do not have the carbohydrate reserves needed to survive this kind of treatment.

Growing On

Planting Medium

Carnations grow well in many types of media. All of the media must have excellent aeration and drainage. Carnations do not grow well in an environment where the roots are poorly drained and oxygen starved. The medium must be heavy enough to support the young plants and hold moisture and nutrients.

A 2:1:1 or 1:1:1 mixture by volume of loam soil, sphagnum peat moss, and perlite or sand continues to be the commonly used medium. Other combinations of materials may be used if the grower can test the soil to determine the nutrient content.

The pH should be in the range between 6.0 and 6.5. Nutrients are provided by adding 20 percent superphosphate at the rate of 5 pounds (2.27 kg) to 100 square feet (9.3 m^2) of area, or treble superphosphate at the rate of 2.5 pounds (1.13 kg) to 100 square feet (9.3 m^2). Limestone, if required, and superphosphate are added before the cuttings are planted. The nutrient materials must be mixed thoroughly through the growing medium. For example, phosphorus must be evenly distributed because it does not move far from the point it is added to the medium.

The medium is steam pasteurized to eliminate disease organisms, nematodes, and other insects. If steam sterilization is not possible, certain chemicals may be used. However, any soil fumigant containing bromine, such as methyl bromide, cannot be used because the substance is toxic to carnations.

Raised benches are preferred to ground beds for growing carnations. It is easier to control disease and follow proper sanitation practices in the benches. The various production tasks, including planting, pinching, tying up, disbudding, and harvesting, are easier to do if the crop is in raised benches. The medium must be at least 6 inches (15 cm) deep. The benches must have good drainage so that thorough watering and leaching can be achieved.

Plant Spacing

The spacing of the plants affects the number of flowers obtained, the quality of the flowers, and the dollar return to the grower. If the plants are grown too close together, they compete for light, water, and nutrients. Although many flowers may be produced, often they are of reduced quality.

A plant spacing wider than that recommended may improve the quality of bloom, but fewer flowers will be cut. The spacing is too great if the number of flowers harvested per square foot is not enough to pay the costs of production plus a profit.

The cultivar used is a factor in spacing the plants. Some cultivars produce denser growth than others. The pinching and timing program used also affects the spacing. A grower may plan to produce a crop of red carnations for the Christmas market. As soon as the plants flower, they will be pulled from the bench. Therefore, the growers will place the plants close together to obtain a large number of blooms. If a grower plans to double pinch a crop and keep it in production for 18 months or longer, a wide spacing is used.

The normal spacing used to produce a pinched crop is 6 inches by 8 inches (15 cm by 20 cm), or three plants to a square foot (32/m^2). Other planting distances have been tried, but the spacing given is the one preferred and used by most growers.

Timing

When the plants flower depends upon several factors, including when the crop is planted, the pinching schedule used, and the cultural program followed.

Carnations are cool temperature crops. They also require a high light intensity to produce the best growth. Growth is adversely affected by temperatures above 65°F to 70°F (18.5°C to 21°C). The cuttings are planted in the benches in the early spring so that the plants are well established before the warm summer weather. Rooted cuttings are planted directly to the production benches in late April, May, and early June. Planting is delayed as long as possible to enable the grower to harvest flowers from the growing crop. Carnations are popular in the spring market.

If the carnations are planted too late, there is too little time between planting and pinching for the plants to develop the heavy growth desired before the warm summer weather begins. At one time, growers planted rooted cuttings outdoors early in spring. This practice allowed the growers to continue the production of a crop indoors for several weeks. After the old crop was pulled and the growing area prepared for a new crop, the plants started outdoors were brought in and benched. By this time they had been pinched and had developed several very strong breaks. However, this practice is no longer profitable because labor costs have increased and less expensive flowers can be imported from South America.

Single Pinch

Singe-pinch carnations usually show heavy growth in 2 weeks. Rooted cuttings are planted in late April or early May and are pinched approximately 30 days later. The pinch consists of removing the top of the plant. Although the amount of stem removed is almost enough to make a cutting, it is not used for this purpose but is discarded.

Removal of the main stem stimulates the rapid growth of side shoots that have already developed. These shoots are allowed to go on and flower. The first crop of flowers can be harvested in September. After this first crop, there is a long period without a large number of blooms. The second crop starts to come into bloom in December and continues through January. The return of the plants from this second flowering period is slow because of the poor light conditions in winter. The plants begin a third period of flowering in May. In many cases, growers would like to harvest this crop but must sacrifice it to make room for the new crop.

The general growth pattern of carnations in the northeastern states is characterized by high production peaks and nonflowering valleys. The peaks are caused by the high light conditions of summer, and the valleys are due to dark winter weather. In areas where the high light intensity is more uniform, such as in California, the number of flowers blooming each month is more equal.

Pinch and a Half

The **pinch and one-half** procedure permits the flowering of the crop to be spread out more evenly through the year. One month after planting, all of the plants are pinched by removing the top growth. Thirty days after the first pinch, one-half of the breaks that develop from the first pinch are pinched. As a result, there is a high rate of production for the first crop. This peak is followed by a long period of a slowly decreasing but steady yield of blooms. A second peak of flowering then occurs. Slightly more time is needed, using the pinch and one-half method, to reach a peak yield for the first crop as compared to a crop given a single pinch. Although the number of flowers harvested in the peak period is not as large as for the single-pinch crop, there is a large increase in the number of flowers cut between the production peaks.

Carnations should not be planted in greenhouses in the northeastern United States after June 15. After this date, the day length is decreasing too quickly.

The plants cannot build up the growth needed to go into the dark winter. A planting made June 15 requires a pinch not later than July 15. A general rule is that for every week the pinch is delayed after July 15, the crop will flower a month later in the spring. In California and other high light areas, carnations are planted 12 months of the year.

Fertilization and Carbon Dioxide

Over the years many recommendations have been made for fertilizing carnations. If carnations have been grown in the same production medium for many years, additional boron may be required. Excellent carnations are produced when phosphorus is added to the medium before planting and 200 ppm each of nitrogen and potassium are added at each watering. Half of the nitrogen should be in the nitrate form. During the winter months, at least 75 percent of the nitrogen should be in the nitrate form. Ammonium fertilizer applied when the light intensity is low results in carnations with weak stems. Fertilizers with a high sodium chloride or sulfate content should not be used because they contribute to a high soluble salt level in the medium. Carnations cannot tolerate high levels of soluble salts.

To prevent trace element deficiencies, fitted trace elements or other controlled-release materials are added in March and September. If a boron deficiency develops, sodium tetraborate ($Na_2B_4O_7 \cdot 4H_2O$) is added to the medium at a rate not to exceed 1 ounce (28.35 g) to 100 square feet (9.3 m^2) of area. Generally, just one treatment a year is sufficient. Overtreatment with boron can lead to severe plant damage.

Supplemental carbon dioxide applied to carnation crops results in stronger stems, larger flowers, and faster cropping times. Some growers use carbon dioxide, and others prefer not to use it in the greenhouse. Carnations appear to obtain the greatest benefit from a maximum level of 450 to 500 ppm of carbon dioxide. When carbon dioxide is injected into the greenhouse, the minimum temperature can be increased 5° above the normal level.

Watering

Carnations can be irrigated using the Gates perimeter system, gro-hose, ooze tubes, or a handheld hose. Regardless of the method used, the foliage should be kept as dry as possible to minimize disease.

When carnations are grown in raised beds, tensiometers are placed in the medium at a depth equal to two-thirds of the total depth. The plants are irrigated when the reading of the tensiometer dial is 10. When the reading is greater than 10, there is too little reserve moisture in the soil. If the plants are not irrigated immediately, they will be damaged by the lack of water.

In ground beds, where the medium is deeper, the tensiometer cup is placed 6 to 8 inches (15 to 20 cm) below the surface of the soil. When the dial indicates 30, the carnations must be irrigated. The timing of the irrigation of ground beds is not as critical as it is for raised benches. In ground beds, the roots of the plant extend well below the tensiometer depth and can take up water from the subsoil. If the ground beds have a solid base of concrete or other nonporous material, then the plants are irrigated as soon as they need water. This is the procedure used for raised benches also.

Light and Temperature

Carnations require full light intensity throughout their culture to make maximum growth. The one exception to this statement may be the case of a greenhouse crop growing in the summer in the North. It may be necessary to apply a light shade to the greenhouse to reduce the temperature. A high temperature is more harmful to the plants than the reduced light level caused by the shade. As soon as weather permits, the shading compound is removed from the glass and the plants are given full light.

The quality of carnation blooms is directly related to the light and temperature levels. If the temperature is too high when the light intensity is low, there are reductions in the flower size, stem strength, and lasting life. These changes do not appear quickly, and they cannot be corrected quickly by changing the environment. Carnations are slow to respond to improper temperatures. For example, the grower who tries to force the Christmas crop by keeping the temperature 5° too warm in December will find that the spring crop is of poor quality.

Table 29-5 lists suggested temperatures for growing carnations in different seasons.

Accurate temperature control is an important factor in reducing a problem that is particularly troublesome to carnation growers. The problem is **calyx splitting**, which can happen at any time but is more common in the fall and spring.

The calyx of the flower splits because the flower produces more petals than the calyx can hold. A number of reasons are given for this problem. The genetic makeup of some cultivars causes more splits than in other cultivars. The calyx of some cultivars is long and narrow. This shape cannot withstand the same amount of pressure from the growing flower as a short, wide calyx. Other reasons for splitting include high levels of phosphorus fertilization, a boron deficiency, and the presence of diseases.

Generally, calyx splitting is attributed to a wide variation between the day and night temperatures. Splitting is most severe when the day temperature is

Table 29-5

	Recommended Temperatures for Carnations Grown Between 35° and 45° North Latitude With and Without the Addition of Carbon Dioxide (CO_2)[1]									
	Night Temperature				Day Temperature					
							CO_2 Added			
	No CO_2 Added		CO_2 Added		No CO_2 Added		Cloudy Day		Sunny Day	
Period of Year	F°	C°	F°	C°	F°	C°	F°	C°	F°	C°
September 15–October 31	50–52	10–11	52–54	11–12	60	15.5	65	18.5	70	21.0
November 1–February 15	48–50	9–10	50–52	10–11	57	14.0	63	17.0	67	19.5
February 16–April 30	50–52	10–11	52–53	11–11.5	60	15.5	65	18.5	70	21.0
May 1–September 14	54	12	—[2]	—	65–70	18.5–21	—	—	—	—

1. Adapted with modifications from Holley 1963 and Langhans et al. 1961.
2. Carbon dioxide is not used in the summer.

allowed to exceed the minimum night temperature by 15°. To prevent the rise in temperature, the greenhouse is ventilated early in the day during the fall and spring when the bright sunny weather is likely to send the greenhouse temperature soaring. In the winter, fan and tube ventilation is used to minimize the chilling of the crop that can occur when cold outside air is suddenly introduced into the greenhouse.

Accurate temperature control in the greenhouse depends upon the thermometers used to measure the temperature and the thermostats that regulate the heat and ventilation. These devices must be checked for accuracy. If the thermometer is inaccurate by 5°, then the plants are exposed to incorrect temperatures. Any thermometers that are more than 2° off in temperature must be replaced.

Thermometers and thermostats should be located at the plant level not at eye level. Both types of devices are located in aspirated shelters to shield them from direct sunshine that causes false readings.

Harvesting for Market

The stage of development at which carnations are harvested is determined by the distance the blooms must be shipped to market, the cultivar grown, and the season of the year. The life of cut carnations is directly related to their total carbohydrate (sugar) content. As the flowers get older, their sugar content decreases. Carnations contain the highest percentage of sugar in the mid-afternoon. Therefore, the flowers are harvested at this time to ensure the maximum lasting life.

There is an estimated loss of nearly 20 percent of the total number of carnations cut because of improper care and handling at harvest and as the flowers are moved through the marketing chain. This loss has a value of nearly $30 million annually for flower cut and **bud cut** carnations.

The flowers are cut when the floral head is still a little tight or is barely opened. They are placed in clean water or in water to which a floral preservative is added. The purpose of the **floral preservative** is to improve the lasting life of cut flowers. There are many preservatives available to the commercial grower. Some preservatives are more effective than others. Therefore, the grower, florist, or other person working with the cut flowers is urged to experiment to find the preservative(s) best suited to specific conditions.

The containers are placed in a refrigerator at a temperature of 31°F (−0.5°C) and a relative humidity of 92 to 94 percent. The cut flowers are held at these conditions for a minimum of 4 hours while they take up the solution hydrate. After they are hardened, they are graded according to the flower size, stem length, flower form, and stem strength. Table 29-6 lists the **standard grades** for carnations recommended by the Society of American Florists.

The graded flowers are packed in bunches of 25 and shipped to market. In winter the shipping boxes are lined with insulation to protect the flowers from freezing temperatures.

Carnations can be held in **dry storage at 31°F (−0.5°C)** for several weeks without damage. The flowers are placed in polyethylene-lined boxes. The plastic is not folded over the blooms until they are at the storage temperature. This practice prevents the condensation of moisture inside the plastic. A high

Table 29-6

Standard Grades for Carnations Adopted by the Society of American Florists and the American Carnation Society			
Grade Designation	Minimum Flower Diameter in Inches (cm)	Overall Length in Inches (cm)	Stem Strength[1]
Blue	$2^3/_4$ (7)	22–26 (55–65)	10-2
Red	$2^1/_4$ (6)	17–22 (43–55)	10-2
Green	None (0)	10–17 (25–43)	Unrestricted

1. The flower head will not bend lower than a 10 o'clock or a 2 o'clock position when the base is held upright at a 6-o'clock position.

humidity in the box may lead to the start of *Botrytis* disease. The containers are stacked using 0.5-inch (1.3-cm) thick wooden spacers between the boxes to ensure free air movement.

Bud Cut Carnations

The bud cut method of harvesting carnations is being used by more and more growers. The advantages of cutting carnations in the bud stage are the same as those for harvesting chrysanthemums in the bud stage.

When one-half of the buds show color, the carnations are harvested. They are stored dry at 31°F (−0.5°C). The wholesaler or retailer processes the cut blooms by removing 0.5 inch from the base of the stem. The buds are then placed in a bud opening (preservative) solution. A typical solution consists of 200 ppm of 8-HQC and a 5 percent sugar solution. This solution is prepared by dissolving 0.3 ounce (9 g) of 8-HQC and 5 pounds (2.3 kg) of sugar in 12 gallons (45.4 l) of water. The containers are placed in a 70°F to 75°F (21°C to 24°C) room where the buds are lighted for 12 hours a day with 150 fc (1,620 lux) from cool white fluorescent lamps. The relative humidity of the room is in the range of 50 to 75 percent. The buds will open to full flower size in 4 or 5 days. After opening, they are given the same treatment as regular cut carnations.

Problems

Cultural Disorders

Weak stems result from inadequate light in the winter and higher-than-recommended night temperatures.

The causes of calyx splitting were described earlier in the "Light and Temperature" section.

Sleepiness, or failure to open, is caused by ethylene gas. Carnations cannot be stored with Calla lilies, fruits, or vegetables because of the large amounts of ethylene gas produced. If the grower expects the carnations to be exposed to ethylene gas, **ethylene scrubbers** can be packed with the flowers. These materials absorb the ethylene gas from the air. High temperatures cause small flowers in light colors. Boron and calcium deficiencies cause flower bud abortion. Petal burn is the result of moisture stress due to a lack of water and a large volume of air movement across the plants.

Disease

Carnations are affected by a large number of diseases. To minimize disease problems, the growing media are pasteurized; culture-indexed, tissue cultured cuttings are used; and rigorous sanitation procedures are maintained throughout the growth of the crop. If controls for a specific disease exist, they must be used as soon as the disease is identified.

Alternaria leaf spot and branch rot are caused by the fungus *Alternaria dianthi*. Dry foliage prevents the spread of the organism. Weekly sprays of the proper fungicide are effective in controlling the disease.

Bacterial wilt is caused by *Pseudomonas caryophylli*. The organisms grow inside the plant. The symptoms appear at a high temperature. The tissue just beneath the epidermis becomes very sticky. The disease spreads quickly in warm weather. All infected plants must be removed and burned. The production area must be steam pasteurized.

Botrytis cinerea is the common gray mold that causes leaf spotting. The use of heat and ventilation to keep the relative humidity low is effective in controlling the disease. Fallen leaves, flowers, and other debris must not be allowed to remain in the production area.

Three *Fusarium* organisms attack carnations. *F. oxysporum* and *F. dianthi* are soilborne fungi that cause wilt. They are also carried in cuttings from diseased mother plants. *F. roseum* is another soil fungus that causes a stem rot. Soil drenches are effective against this organism although they may check the growth of the plants temporarily. A bud rot is caused by *F. tricinctum poae*, which is carried to the buds by grass mites. The mites burrow into the buds and cause the inside petals to be affected first. Infested buds are picked and burned. Miticide sprays are also helpful.

Greasy blotch is caused by the fungus *Zygophiala jamaicensis*. The symptoms appear on the leaves in the form of greasy-looking spots that resemble a spider web. The disease is favored by the high humidity and free moisture found in leaky and poorly ventilated greenhouses.

Stem rot is caused by *Rhizoctonia solani*, one of the common soil fungi. Several soil drenches are effective in controlling this disease.

The fungus *Uromyces caryophyllinus* causes rust. The spores can be spread by air currents, but they are commonly spread by water splashing on the foliage. Prevention of rust requires good ventilation and dry foliage. If rust starts in a crop, there are effective fungicides that can be applied as a spray once a week.

Viruses

Meristem cultured planting material is free of viruses. However, aphids transmit viruses from plant to plant. Therefore, it is doubtful that plant material can remain virus free.

Insects

A very large number of insects attack carnations. Control of the insects requires regular inspections of the crop and the use of the appropriate chemicals. Sometimes, however, even these precautions cannot save a crop from predatory insects.

Insects that attack carnations include aphids, flower thrips, leafhoppers, leafminers, leafrollers, caterpillars, shoot mites, slugs, snails, sowbugs, spider mites, thrips, tortrix, and other moths.

Symphyllans and termites attack the root systems. Cutworms and cockroaches eat the flower buds and young tender shoots. These pests are active in the dark. Therefore, night inspections of the plants are recommended to detect insect activity that may not be apparent during the day.

Cultivars

Several hundred carnation cultivars are available to growers. Plant breeders are continuing to develop new cultivars. A reputable horticultural supplier is a reliable source of cultivar information.

ACHIEVEMENT REVIEW

Select the best answer or answers to complete each statement. List the appropriate letter(s).

1 The production of cut flowers for decorative use can be traced to:
   a. early 1900.
   b. pioneer days.
   c. early London.
   d. the ancient Egyptian and Roman civilizations.

2 The chrysanthemum industry in Florida was made possible by:
   a. the invention of plastic shade cloth.
   b. knowledge of how to control the photoperiod.
   c. the propagation of chrysanthemums as cuttings.
   d. hurricane forecasting.

3 U.S. carnation production has all but been eliminated due to competition from:
   a. Costa Rica.
   b. Puerto Rico.
   c. Brazil.
   d. Colombia.

4 Snapdragons grown as cut flower crops are native to:
   a. England.
   b. the United States.
   c. the Mediterranean region.
   d. North Africa.

5 The most recent type of snapdragon introduced by plant breeders is the:
   a. 'Rocket' series.
   b. 'Rockwood' series.
   c. 'Windmiller' types.
   d. peloric types.

6 The classification of snapdragons is based on their response to:
   a. temperature.
   b. light intensity.
   c. length of day or photoperiod.
   d. all of the above.

7 One trade packet of snapdragon seed is used to sow _____ 10-inch by 22-inch (25-cm by 50-cm) seed flats.
   a. one
   b. one and one-half
   c. two
   d. four

8 Snapdragon seedlings are very sensitive to damping off caused by:
   a. *Pythium* and *Rhizoctonia*.
   b. *Botrytis*.
   c. *Verticillium* wilt.
   d. bacterial disease.

9 Snapdragon seedlings should be transplanted when:
   a. cotyledon leaves are fully developed.
   b. the first set of true leaves is mature.
   c. the seedlings are 2 inches (5 cm) tall.
   d. the seedlings are 3 inches (7.5 cm) tall.

10 Before the crop is planted, 20 percent superphosphate is applied to 100 square feet (9.3 m^2) of area at the rate of:
   a. 1 pound (0.45 kg).
   b. 2.5 pounds (1.14 kg).
   c. 5 pounds (2.27 kg).
   d. 10 pounds (4.54 kg).

11 Witches' broom in snapdragons is caused by a lack of:
   a. nitrogen.
   b. calcium.
   c. phosphorus.
   d. boron.

12 Single-stem crops grown in the winter are planted at a spacing of:
   a. 3 inches by 5 inches (7.5 cm by 12.5 cm).
   b. 4 inches by 5 inches (10 cm by 12.5 cm).
   c. 5 inches by 5 inches (12.5 cm by 12.5 cm).
   d. 6 inches by 5 inches (15 cm by 12.5 cm).

13 The use of artificial light on snapdragons is economical:
   a. when the plants are in the seedling stage.
   b. when the plants are first transplanted.
   c. just before flower bud formation.
   d. any time during growth.

14 Chrysanthemums belong to the family:
   a. Asteraceae.
   b. Lamiaceae.
   c. Liliaceae.
   d. Ranunculacea.

15 Modern chrysanthemums are the descendants of flowers that originated more than 2,000 years ago in:
   a. China.
   b. Japan.
   c. Tibet.
   d. Vietnam.

16 Chrysanthemums classified in the 10-week response group:
   a. flower 10 weeks after planting.
   b. require 10 weeks of long days.
   c. flower 10 weeks after the start of short days.
   d. require 2 weeks of long days and 8 weeks of short days to bloom.

17 One of the largest chrysanthemum propagators in the United States today is:
  a. California-Florida.
  b. Paul Ecke, Jr.
  c. W. Atlee Burpee.
  d. Yoder Brothers.

18 Chrysanthemums are usually propagated by means of:
  a. seed.
  b. leaf bud cuttings.
  c. stem section cuttings.
  d. terminal cuttings.

19 Chrysanthemums propagated under a mist fertilizer system:
  a. can be stored at 31°F (−0.5°C) for 2 weeks.
  b. take longer to root than when tap water alone is used.
  c. must be planted in production benches as soon as they are rooted.
  d. root in less time than unmisted plants.

20 Chrysanthemums are fertilized:
  a. as soon as they are planted.
  b. 2 weeks after planting.
  c. 3 weeks after planting.
  d. when buds are visible.

21 Carbon dioxide is added to the greenhouse atmosphere to:
  a. improve the respiration of the plants.
  b. increase photosynthesis.
  c. reduce the amount of disease.
  d. decrease the number of insects.

22 Chrysanthemums form flower buds when the day length is 14.5 hours. The flower buds continue to develop when the day length is no longer than:
  a. 14.5 hours.
  b. 13.5 hours.
  c. 12.5 hours.
  d. 10.5 hours.

23 The minimum night temperature for flower bud formation is:
  a. 54°F to 55°F (12°C to 13°C).
  b. 58°F to 60°F (14.5°C to 15.5°C).
  c. 60°F to 62°F (15.5°C to 16.5°C).
  d. 64°F to 66°F (18.0°C to 19.0°C).

24 Chrysanthemums remain vegetative if the light intensity is greater than:
  a. 0.05 fc (0.55 lux).
  b. 2 fc (22 1ux).
  c. 10 fc (108 lux).
  d. 20 fc (216 lux).

25 Intermittent lighting of mums requires that the lights be on for _____ percent of 30-minute intervals.
  a. 10
  b. 20

c. 30

d. 50

26 Interrupted lighting of 'Mefo' and 'Indianapolis' cultivars:
   a. delays flowering.
   b. increases the petal number in the flower.
   c. decreases the petal number in the flower.
   d. develops flatter flowers.

27 Mums cut for shipment are placed in a refrigerator to allow them to:
   a. become crisp.
   b. take up a maximum amount of water.
   c. decrease the rate of transpiration.
   d. increase the rate of respiration.

28 A major advantage of bud cut chrysanthemums is that:
   a. harvesting requires more time.
   b. more flowers can be placed in a shipping container.
   c. fewer diseases attack the plants.
   d. insects are less of a problem.

29 The start of the American carnation industry began on Long Island in:
   a. 1767.
   b. 1852.
   c. 1900.
   d. 1938.

30 Due to flowers imported from South America and other countries, the percentage of total sales of carnations in the United States has _____ since 1976.
   a. decreased slightly
   b. decreased greatly
   c. increased
   d. remained the same

31 Carnations propagated from seed are mostly:
   a. 'William Sim' types.
   b. new cultivars from plant breeders.
   c. spring crops only.
   d. spray type carnations.

32 Meristem tissue-cultured plants are free of:
   a. bacterial disease.
   b. fungus disease.
   c. viruses.
   d. all of the above.

33 Unrooted carnation cuttings can be held in a refrigerator at 32°F (0°C) for a maximum of:
   a. 10 days.
   b. 30 days.
   c. 180 days.
   d. 1 year.

34 The proper pH range for a medium used to grow carnations is:
   a. 5.0 to 5.5.
   b. 5.5 to 6.0.
   c. 6.0 to 6.5.
   d. 7.0 to 7.5.

35 The chemical soil fumigant that cannot be used to treat media for carnations is:
   a. chloropicrin.
   b. Vapam.
   c. methyl bromide.
   d. formaldehyde.

36 The normal planting distance for cuttings of a pinched crop is:
   a. 4 inches by 4 inches (10 cm by 10 cm).
   b. 6 inches by 6 inches (15 cm by 15 cm).
   c. 6 inches by 8 inches (15 cm by 20 cm).
   d. 8 inches by 8 inches (20 cm by 20 cm).

37 To even out the peaks and valleys in the quantities of flowers harvested, the most effective procedure is to:
   a. grow the plants a single-stem crop.
   b. pinch them once.
   c. use the pinch and one-half method.
   d. use a half pinch only.

38 Ammonium fertilizers added to carnations in low light intensity conditions cause:
   a. large flowers.
   b. stiff stem growth.
   c. weak stems.
   d. soft growth.

39 Carnations often show a deficiency in the trace element:
   a. aluminum.
   b. boron.
   c. copper.
   d. molybdenum.

40 If carbon dioxide is added to the greenhouse environment for carnations, the minimum recommended temperatures:
   a. are lowered 5°.
   b. can be raised 5°.
   c. are lowered 5° with CO_2 injection.
   d. are maintained at the same level.

41 The lasting life of cut carnations is directly related to:
   a. the size of the bloom.
   b. the length of the stem.
   c. the protein content.
   d. the total carbohydrate sugar content.

42 Bud cut carnations are opened under _____ of artificial light from cool white fluorescent lamps.
   a. 10 fc (108 lux)
   b. 50 fc (540 lux)
   c. 75 fc (810 lux)
   d. 150 fc (1,620 lux)

43 Sleepiness of carnations is the result of:
   a. high temperatures.
   b. low light levels.
   c. ethylene gas.
   d. carbon dioxide gas.

Student Activities

I. Antirrhinums (Snapdragons)

To complete these exercises during the class year, it may be necessary for the instructor to sow seeds early to obtain seedlings in time for transplanting during the first practicum session.

Student Project 1—Sowing Seeds

Objective

This project will teach the student the proper methods of sowing snapdragon seeds. The need for sowing at the correct depth and spacing is demonstrated.

Materials

- Snapdragon seed
- Prepared peat-lite sowing medium such as Jiffy-Mix®, Redi-Earth®, Pro-Mix®
- Seed flats, or 5-inch by 7-inch (12.5-cm by 17.5-cm) market packs
- Plastic bags
- Wooden labels
- Soft lead pencil

Procedure

Depth of Sowing

1. Fill the seed containers with the peat-lite medium. Firm the medium at the edges. Water the containers well and let them stand overnight before planting.
2. Using a soft lead pencil, mark wooden labels with the date, cultivar used, and the treatment. Seeds are to be sown (a) on the medium surface, (b) covered by 0.25 inch (6 mm) of medium, and (c) covered with 1 inch (25 mm) of medium.

3. Use the edge of a 0.25-inch (6-mm) wide board to mark the rows and prepare them to the correct depth.
4. Sow seed in the rows at a rate of no more than 10 seeds to 1 inch (25 mm) of row.
5. Cover the seed according to the experimental treatment. Uncovered seed is sprinkled on the surface of the medium.
6. Water the seed flats with a fine FOGGIT® nozzle spray. The containers are allowed to drain for 1 hour. Then they are placed in plastic bags. Secure the ends of the bags with a tie or rubber band.
7. Place the seed flats in the proper germination environment. Follow the recommended germination procedures.

Observations

1. Examine the containers daily after 3 days.
2. Record the date when the seedlings first emerge in each container.
3. Record observations on the appearance of seedlings; that is, indicate if the growth is strong or weak, and good, fair, or poor.
4. The seedlings are allowed to grow until the first true leaves appear. The seedlings can be used for growing on.

Procedure

Depth of Sowing

1. Prepare the seed flats as described under "Depth of Sowing." Mark the rows on the surface of the medium only.
2. Line each row with a narrow strip of fine white mountain sand. The seeds are easier to see against the sand as compared to the medium. Ocean sand cannot be used because it has high soluble salt levels.
3. The seed is sown in three ways. Thin sowing is difficult because the seeds are very tiny.
   a. Thin sowing: 6 to 8 seeds in 1 inch (2.5 cm) of row.
   b. Medium sowing: 10 to 15 seeds in 1 inch (2.5 cm) of row.
   c. Thick sowing: 20 to 30 seeds in 1 inch (2.5 cm) of row.
4. Water with a fine mist nozzle and place the containers in plastic bags in the germination environment.

Observations

1. After 3 days, begin daily observations.
2. Record the date when the seeds first germinate.
3. The seeds are allowed to grow until the first true leaf appears.
4. Record observations of the stem thickness and leaf size of the seedlings for each method of sowing.
5. Remove some seedlings as if they were to be transplanted. Observe the difficulty of separating the seedlings without injuring them.
6. Discuss with the instructor the reasons for the differences noted in the various treatments.

Student Project 2—Planting Distances

This exercise requires greenhouse space and media prepared for snapdragons grown for cut flowers. Small plots consisting of four to six rows of plants are used. Seedlings started for this project or those from the sowing project can be used.

Objective

This project will teach the student how to determine the effect on growth of the density of the plant per square foot of area.

Procedure

1. Seedlings are planted at spacings of:

2 inches by 2 inches (5 cm by 5 cm)   3 inches by 4 inches (7.5 cm by 10 cm)
2 inches by 4 inches (5 cm by 10 cm)   4 inches by 4 inches (10 cm by 10 cm)
3 inches by 3 inches (7.5 cm by 7.5 cm) 4 inches by 5 inches (10 cm by 12.5 cm)

2. Four plots are planted using even spacing so that averages can be obtained.
3. Crops grown in the fall and winter require 4 to 5 months to flower. In this period, the plants are measured weekly. The measurements are plotted on a chart or a line graph.
4. At the end of the study, record the following data for each plot.
   a. Overall length of the stems
   b. Spike length
   c. Number of flowers on each spike
   d. Fresh weight of each stem and flower
5. Add the figures for each factor measured and determine the average value for each plot.
6. Discuss with the instructor the possible reasons for any differences obtained.
7. A variation of this study can be made by using different fertilizers. Smaller plots are prepared using fewer plant spacings. The plots are treated with 20-10-20 analysis fertilizer at concentrations of 10 ppm each of nitrogen, phosphorus, and potassium; 50 ppm each; and 100 ppm each.
8. Record the same observations required in steps 3 and 4.

II. Dendranthema (Chrysanthemums)

Student Project 1

Grow standard mums with and without disbudding to observe the effect on final flower size.

Objectives

The objectives of both chrysanthemum projects are to acquaint the student with the rate of growth of chrysanthemums and with some of the commercial procedures used in growing the crop.

Materials

10 feet (3 m) of greenhouse bench space is prepared for cut chrysanthemum production.
Rooted cuttings of standard mums (large single flowers) are obtained from a commercial propagator or from material previously propagated from stock plants.

Procedure

1. The cuttings are planted according to recommended procedures. Information published by George J. Ball, Inc. or Fred C. Gloeckner Co. is used to schedule the crop.
2. Weekly measurements of growth in height are made during the entire production period. These figures are plotted on a graph to show the rate of growth during long days and short days.
3. After starting short days and buds develop, one-half of the crop is not disbudded. All of the flower buds are allowed to develop.
4. The other half of the crop is disbudded following recommended procedures.
5. When the crop flowers, record the following information.
   a. The height of the plant from soil line to the top of the bloom
   b. The number of flowers on each stem of the nondisbudded plants
   c. The fresh weight of each flower (stem plus flower)
6. Find the average value for each treatment. Record this information on a chart or graph to show the differences in the growth rates for long days and short days.
7. Discuss with the instructor any differences in each group and their possible causes.

Student Project 2

Grow a group of 'Mefo' and 'Indianapolis' cultivars for blooming in February. Use the method of interrupted lighting on one-half of the crop. The other half of the crop is to be grown on a regular schedule.

Record final growth measurements for the stem length, flower size, and fresh weight of each flower. Determine the average value for each treatment. Discuss the differences obtained and the possible reasons for them. Does interrupted lighting improve the quality of the flowers?

III. Dianthus (Carnations)

The production of carnations is slow. To produce a flowering crop in the northeastern United States, the plants must be benched in May for growth through the summer. This schedule can present problems because of summer vacation periods, unless special arrangements are made for someone to care for the crops.

The students will benefit from an exercise in opening bud cut carnations. Such an exercise will show the methods used and the response of the plants.

The instructor is encouraged to obtain bud cut carnations from a nearby grower or wholesale house. The procedures recommended for opening buds are given in this chapter in the section on carnations.

Orchids and Roses

OBJECTIVES:

After studying this chapter, the student should have learned:

■ why orchids are not easily grown from seeds.

■ the different types of orchids grown commercially.

■ nutrient requirements of orchids.

■ methods of weed control in orchids.

■ methods of propagating roses.

■ how rose growing media are prepared.

■ methods of pinching and cutting roses to time the crop for different holidays.

ORCHIDS ▨ *Orchidaceae*

At present, 500 genera with more than 15,000 species of orchids are known. Before World War I there was so much interest in orchids that $8,000 was paid for a single plant of one species. Orchids have long been regarded as the royalty of the flower world. However, the popularity of orchids has decreased greatly over the past 20 years. The reasons for the decline are difficult to pinpoint. Cost is not a factor because the flowers are less expensive now than several years ago.

Orchid species grow naturally in every country of the world. Those species considered to be florists' orchids normally are found in the deep jungles and the tropical rain forests. Brazil is noted for the beauty of its orchids. More than 1,765 species native to Brazil are listed in *Flora Brasiliensis*.

There are two types of orchids. **Epiphytes** grow on trees and other forms of support. They have large **aerial roots** and obtain their nutrients from dust in the air and the rain water that washes down the tree trunks and over the roots. Epiphytes grown commercially include *Brassavola, Cattleya, Dendrobium, Laelia, Miltonia, Oncidium, Odontoglossum,* and *Phalaenopsis*. The most important commercial crops are the *Cattleya* and *Laelia* orchids and the many hybrids derived from them.

Terrestrial orchids grow in a soil type medium. Commercially grown terrestrial orchids include *Calanthe, Cymbidium, Cypripedium* (also called

Paphiopedilum), and *Vanda*. The last three groups are the most important commercial crops.

Orchids have two habits of growth. In **sympodial growth**, the stem grows along the surface of the medium (Figures 30-1 and 30-2). After one or two seasons, the shoot stops growing and forms a pseudobulb and leaf or a **flower sheath.** The next shoot grows in the axil of a leaf scale at the base of the shoot. The flower develops from an upright sheath. Orchids having a sympodial growth habit are *Cattleya*, *Calanthe*, and *Laelia*.

Monopodial growth is the more familiar form of growth. As in most flower crops, this growth habit is upright. The terminal bud develops in a vertical position, and new roots develop from the stem. *Phalaenopsis* and *Vanda* orchids have a monopodial habit of growth (Figure 30-3).

Other Characteristics of Orchids

The rate of growth of orchids is slow compared to chrysanthemums or roses. Orchids generally produce a few, thick leaves in each growth cycle. There may be more than one growth cycle in a year if the conditions are favorable. The leaves remain on the plant for several years before they abscise (drop off).

FIGURE 30-1 Sympodial growth habit of *Cattleya* orchids showing aerial roots covered with white valamen.

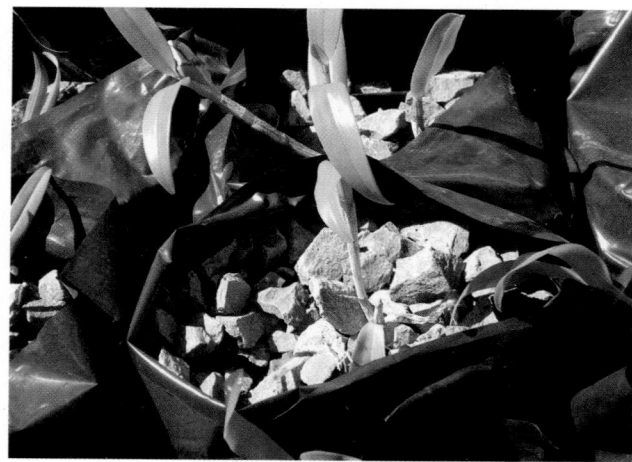

FIGURE 30-2 Close-up of developing lead on newly divided plants. (Note that these plants are growing in stones rather than bark or osmunda fiber.)

FIGURE 30-3 Monopodial growth habit of *Phalaenopsis* orchid with emerging aerial roots.

Many orchid species produce a pseudobulb (Figure 30-4). *Pseudo* means false and the bulblike growth is not a true bulb. The pseudobulb is like a true bulb in that it stores food and moisture in its fleshy tissue.

The leaves may drop from the pseudobulb, but the bulb itself remains alive for several years. While it is alive, the pseudobulb serves a life support function for the rest of the plant.

The aerial roots of orchids are fleshy and thick. Unlike other plants, epiphytes rarely have root hairs. In place of these structures, the roots are covered with a thick layer of cells. The epidermis of this layer is called the **velamen** (Figure 30-5). Directly below the velamen is the cortex, which contains chloroplasts.

It has been determined that the velamen allows an exchange of oxygen and carbon dioxide to take place. Water and nutrients are not taken up unless the aerial roots enter the potting medium or fasten themselves to a solid surface. When this happens, the structure of the root is changed. Water and nutrients

FIGURE 30-4 Pseudobulbs of the *Oncindium* orchid are the swollen stem sections near the rim of the pot.

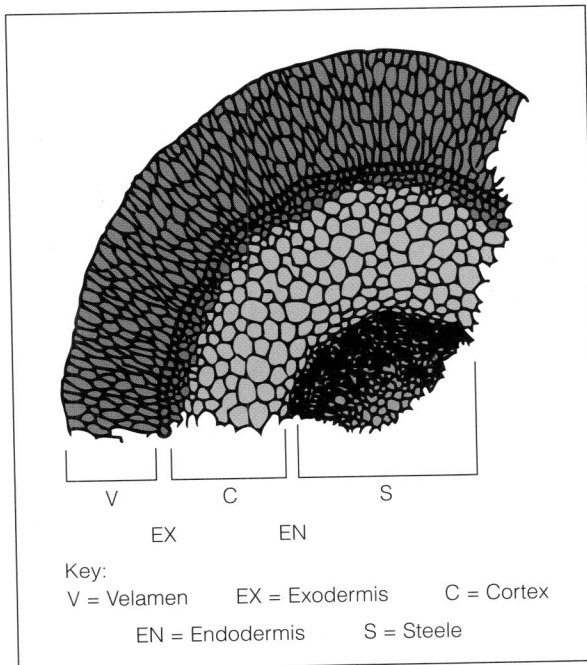

FIGURE 30-5 Cross section of aerial root.

Key:
V = Velamen   EX = Exodermis   C = Cortex
EN = Endodermis   S = Steele

can be taken up and translocated to the rest of the plant. Orchid growers should direct the aerial roots into the potting medium.

Propagation

Orchids are propagated by seeds, divisions of the mother plant, and tissue culture methods.

Seeds

The propagation of an adult orchid plant from seed is a long process. From 5 to 7 years are required to grow a flowering plant from seed. Six months to a year or longer may pass from the time of pollination until the seed is ripe. When the seed pod splits open, the seed is ripe. The pod is usually enclosed in a plastic bag so that the seed is not lost if the pod opens during the night.

Orchid seed is extremely small. Seed of the *Cattleya* species and hybrids is 0.006 inch (0.1 mm) long and 0.004 inch (0.09 mm) wide. A *Cattleya* seed pod measuring 3 inches (7.5 cm) in length and slightly more than 1 inch (2.5 cm) in width contains 500,000 or more seeds.

Orchid seed has an embryo, but it does not contain any stored food in the form of an endosperm or cotyledons. To ensure that the seed develops properly, all of the required nutrients must be supplied. A **nutrient agar** solution was developed to supply the necessary materials.

Dr. Lewis Knudson, a botanist at Cornell University, developed the agar culture method in 1922. The original nutrient solution was modified by Dr. Knudson in 1943 and 1946. The solution developed in 1946 is still used by orchid propagators. The constituents of this solution are as follows:

Material	Grams
Calcium nitrate, $Ca(NO_3)_2 \cdot 4H_2O$	1.00
Monobasic potassium phosphate, KH_2PO_4	0.25
Magnesium sulfate, $MgSO_4 \cdot 7H_2O$	0.25
Ammonium sulfate, $(NH_4)_2SO_7$	0.50
Ferrous sulfate, $FeSO_4 \cdot 7H_2O$	0.025
Manganous sulfate, $MnSO_4 \cdot 4H_2O$	0.0075
Sucrose	20
Agar	2–15
Distilled water	1 liter

After sterilization, 1/10 normal hydrochloric acid is added to reduce the pH to a range of 4.8 to 5.2. The preferred pH is 5.0. At a pH of 4.8, some *Cattleya* hybrid embryos do not survive. If the pH goes above 5.5, the embryos may not develop chlorophyll.

Orchid growers normally use one or more of the numerous commercial tissue culture and propagation media available.

Sowing Seed

Aseptic conditions are required to germinate orchid seeds successfully. The propagator works in an enclosed and disinfected room. If an entire room cannot be set aside, the propagator may work in a small unit where sterilized air blows over and away from the operator. This unit is known as a laminar flow hood and keeps the area spore free (Figure 30-6).

FIGURE 30-6 A laminar flow hood used for sowing orchid seeds and general tissue culture work.

FIGURE 30-7 Banks of seedling orchids grown in aseptic glass bottles.

Glass containers used to hold the media are sterilized. This procedure calls for the nutrient agar solution to be poured into each container. The open end of the bottle or flask is plugged with cotton, or a porous plastic stopper is inserted. The containers and media are autoclaved (sterilized) for 20 minutes under 15 pounds of pressure. The sterilized bottles and media are allowed to cool slowly after they are removed from the autoclave. The stoppers are removed from the containers in the sterile transfer chamber or in the area where the seed is added.

The seed is sterilized by soaking it for 15 minutes, with shaking every 5 minutes, in a 1:15 solution of sodium hypochlorite (bleach) and water. The seeds are transferred from this solution to the surface of the sterilized agar medium using a sterilized medicine dropper. When the stopper is removed from the test tube or flask containing the medium, the mouth of the flask is passed through an alcohol flame to kill any spores that may be present. After the seed is placed on the medium, the mouth of the flask is passed through the flame again and the flask is restoppered.

The seeded flasks are placed in the germination environment, such as an artificially lighted room or heavily shaded greenhouse (Figure 30-7). The flasks are provided with a light intensity of 850 fc (9.18K lux) for 16 hours a day from cool white fluorescent lights. The seed may be lighted for 24 hours a day to hasten growth after germination occurs. The air temperature should be in the range of 75°F to 80°F (24°C to 27.5°C).

In the greenhouse the light intensity is reduced by heavy shading. At the brightest part of the day, 1,000 fc (10.75K lux) is the maximum light level permitted on the containers. Artificial lighting can be used in the greenhouse to add to the natural daylight.

After the seeds germinate and begin to develop, they are exposed gradually to the greenhouse environment by loosening the stoppers on the containers a small amount each day. When the plants are large enough to be handled with forceps or tweezers, they are transferred to an agar medium in larger flasks or other containers. When the plants become crowded in these containers, they are moved to a community pot or to a flat. The flat is filled with **osmunda** fiber that is clipped with shears to form an even surface. The seedlings are spaced about 1 inch (2.5 cm) apart. A 5- to 6-inch (12.5- to 15-cm) diameter pot can hold 50 to 75 seedlings; a 3-inch (7.5-cm) pot can hold 20 to 30 seedlings.

The seedlings are removed from the agar by slowly dissolving it in warm water. Hot water is not used because it may injure the tiny seedlings. The combined agar and seedling liquid is poured into the osmunda fiber. The seedlings are handled carefully using large tweezers or forceps.

Care of the Young Plants

The seedling containers may be kept in the germination environment. The temperature is lowered to 70°F to 72°F (21°C to 23°C). Long days are provided to hasten growth. Carbon dioxide may be applied in a concentration of 500 to 800 ppm to hasten the growth of the small plants. As they develop, the small plants are fertilized with 50 to 75 ppm of nitrogen from a 30-10-10 analysis fertilizer. Orchids require large amounts of nitrogen.

The light intensity can be increased gradually as the plants grow larger. After 1 year the young plants should be able to withstand 1,200 to 1,500 fc (12.7K to 16.1K lux) of light. The plants are shifted into larger pots as they become crowded. Eventually they are large enough to be placed in separate 2-inch (5-cm) pots.

Division

The plants are propagated by division when the large plants overgrow the pots. *Cattleyas* are separated so that at least four strong pseudobulbs are attached to a rhizome. If fewer pseudobulbs are included, the plant usually does not survive.

Divisions are made before new growth begins in the form of a **lead**. The leads are tender and easily broken. A year's growth is lost if the lead is broken from the plant.

The basal end of the rhizome is placed at the far side of the pot to allow room for growth for 2 or 3 years before the plant must be repotted. Orchids that grow more rapidly may need repotting every year. The potting medium is firmed around the roots to hold the plant in place.

Meristem Tissue Culture

The basics of tissue culture, or micropropagation, are covered in Chapter 18. Georges M. Morel of Versailles, France, began in 1956 to experiment with meristem tissue culture of orchids. Previously, Morel had used the technique to free potatoes, dahlias, and carnations from viruses. Similarly, a virus threatened the commercial orchid industry and led growers to seek the assistance of Morel and others in combating the disease. The virus is no longer a threat to the orchid industry.

Meristem tissue culture led to the clonal propagation of plants. A group of plants that originate from the same cell are said to be clones. Each plant has an identical genetic makeup. In other words, the plants will produce flowers of the same type, shape, and color. All of the plants flower at the same time. This feature is particularly important to the grower because it allows the crop to be programmed accurately for any market. The cloning procedure is being used more often, but a large number of orchids are still propagated by means of regular propagation methods.

General Culture

Media and Fertilizers for Epiphytes

Epiphytic orchids require a very open and porous medium. For many years, a popular medium consisted of the roots of the Osmunda fern. To pot an orchid properly in osmunda fiber requires a strong wrist and 30 to 45 minutes for each plant.

The nutrient composition of osmunda fiber is very much like that of the orchids that grow in it. As the fiber breaks down slowly, it releases these nutrients to the plants.

The high cost of labor involved in repotting orchids in osmunda fiber led growers to investigate other media. In the early 1950s, it was determined that tree bark was an effective medium. Redwood, red cedar, tanbark, and red and white **fir bark** are now used as orchid media. Each material is used alone, or it may be combined with sphagnum peat moss and perlite. The amount of each material used to prepare the medium depends upon the grower's preferences. The bark is available in fine to medium particle sizes.

It may be necessary to repot orchids from an osmunda medium to a bark medium. Before potting, the bark is soaked overnight in a large container to allow it to take up as much moisture as possible. The bark is then mixed with moistened sphagnum peat moss. Some growers prepare a mix consisting of one-third bark and two-thirds sphagnum peat moss. Other growers prefer a 1:1:1 mixture by volume of bark, sphagnum peat moss, and perlite.

Plants to be repotted are prepared by trimming off dead roots, old osmunda fiber, and any dead pseudobulbs or other stem tissue. If the osmunda fiber is firmly attached to the roots, it is allowed to remain.

A full depth standard pot is used. To keep the media from sifting through the drain hole of the pot, an aero-drain or a piece of broken pot is placed over the drain hole. A handful of the potting medium is placed in the bottom of the pot. The plant is placed in the pot and is held with the top of the rhizome about 1 inch (2.5 cm) below the rim of the pot. The back pseudobulb is pressed lightly against the edge of the pot. The plant will be able to make two or three seasons' growth before it must be repotted. More potting mixture is poured around the roots. The pot is tapped on the work bench to remove air pockets and firm the medium. The grower may invest in vibrating machines, which are very useful in repotting orchids in bark mixes. The vibrating action quickly settles the media around the roots.

Plants placed in a bark medium need some support. Special wire supports are fastened to the rim of the pot to hold the plant upright. These supports can also be used to hang the plant from an overhead pipe structure.

As soon as the plants are potted, they are watered thoroughly. The plants should not be allowed to dry out during active growth. Plants growing in bark mixes must be watered more often than those growing in osmunda fiber. Bark mixes store less water than osmunda fiber.

Plants growing in bark mixes need more nitrogen than those growing in osmunda fiber. A small amount of nitrogen is released by osmunda fiber as it decomposes. However, bark mixes require extra nitrogen to make them decompose. If the grower does not supply the necessary nitrogen fertilizer, the orchid plant becomes nitrogen deficient. The organisms that break down the bark take up the nitrogen before the plants do. A 30-10-10 analysis fertilizer is used at the rate of 1 pound (0.45 kg) in 100 gallons (378.5 l) of water. Fertilizer is applied once a month in winter and twice a month during the summer or active growth periods. Plants growing in osmunda fiber are fertilized with a 10-10-10 analysis fertilizer applied at the rate of 1 pound (0.45 kg) in 100 gallons (378.5 l) of water. The fertilizer is used once a month in the winter and every 2 weeks in the summer or active growth periods.

Media and Fertilizers for Terrestrials

Commercial growers on the mainland United States normally produce two terrestrial type orchids: *Cymbidiums* and *Cypripediums*. *Vanda* orchids are grown in Hawaii.

Bark mixes are also used as the medium for terrestrial orchids. However, other materials are usually added such as sphagnum peat moss, leaf mold, ground bark, sawdust, or perlite. High-quality terrestrial orchids can be grown only when good moisture conditions are maintained around the roots. When *Cymbidium* orchids are grown in large pots or tubs, an overhead watering system is needed.

One recommended medium for terrestrial type orchids is a 1:1:1 mixture of sphagnum peat moss, bark, and perlite. The pH range must be 5.5 to 6.5. An alternative medium consists of three parts of coarse sphagnum peat moss (with 1 pound of ground limestone added to each bushel of peat, three parts of fir bark or another bark, or unrotted oak leaves), three parts of mulching or insulation grade redwood bark fiber, and one and one-quarter parts of sandy loam or sand. To each cubic yard (cubic meter) of this mixture, the following ingredients are added.

   2 pounds, 3 ounces (2.8 kg) of urea formaldehyde nitrogen (fine particle size)
   2 pounds, 3 ounces (2.8 kg) of finely ground 20 percent superphosphate
   8.8 ounces (0.32 kg) of potassium sulfate (sulfate of potash)
   1.0 ounce (28.35 g) of fritted trace elements (FTE 555)

The fertilizer materials are blended thoroughly with the medium. Water is added to minimize the dust and to aid in the mixing. Roughly 4 gallons (18 l) of water is added to a cubic yard (cubic meter).

Cymbidiums and *Cypripediums* are repotted when the plants grow too large for the containers. The pseudobulbs are separated into units of not less than four bulbs. The pseudobulbs are placed above the medium surface so that only the bases of the bulbs and the roots are covered (refer to Figure 30-4).

The orchids are fertilized every 3 to 4 weeks using a 30-10-10 analysis fertilizer at the rate of 12 to 16 ounces (340.2 to 454 g) in 100 gallons (378.5 l) of water. During active growth cycles, the plants may require fertilization more often. If a weaker solution is used, such as 6 ounces (170 g) of 30-10-10 analysis fertilizer in 100 gallons (378.5 l) of water, it is applied every 7 to 10 days.

Controlled-release fertilizers such as Magamp® or Osmocote® are not satisfactory for orchids. When they are used alone, they do not release their nutrients quickly enough, or the nutrients released are not held long enough by the very porous medium to serve the needs of the plants. Further research on the use of these materials must be completed before any general recommendations can be made.

Light and Temperature

Research and experience have shown that there is a relationship between the level of fertilization used and the maximum light intensity to which the plants can be subjected. This relationship is applicable to foliage plants also. In a high light intensity, the plants grow better with high levels of fertilization. If the light levels are low, the amount of fertilizer used must be reduced or the plants may be damaged.

This does not mean that higher and higher light levels can be used on *Cattleya* and *Cypripedium* orchids when the plants are given more fertilizer. There is a limit to the amount of light these orchids can tolerate without damage. For plants less than 2 years old, this level is around 1,200 to 1,500 fc (12.9K to 16.1K lux). The maximum light level increases for older, more mature plants to 3,000 to 3,500 fc (32.3K to 37.6K lux). Of course, certain cultivars can withstand slightly higher levels, and others need slightly lower intensity to prevent damage. *Cattleya* and *Cypripedium* orchids that are grown north of 42° north latitude must be shaded from full sunlight from April 1 to the middle of October. For the remainder of the year, they can be grown in the full light intensity.

Cymbidium and *Vanda* orchids are grown in the full light intensity year-round. In the summer, these plants should be removed from the greenhouse and placed outdoors. One important requirement for successful outdoor culture is to satisfy the water requirement of the plants. The growth of *Cymbidiums* is checked if they are allowed to dry out.

The flowering of orchids is controlled by the temperature and the photoperiod. Unfortunately, one set of optimum environmental conditions cannot be established for all orchids because of the complex genetic makeup of the many hybrids on the market.

Cattleya gaskelliana normally makes two successive growths in one year. When growth starts in November, December, and early January, the minimum night temperature is held at 65°F (18.5°C). This temperature prevents the flowering that normally occurs in the summer but encourages the development of the second growth. To hasten the formation of flower sheaths in the second growth, a short-day treatment using black cloth is started February 1. This treatment is continued until September 30. At this time, the night temperature is decreased to 55°F (13°C). The low temperature and the short days cause the two sets of growth to form flower buds. The plants bloom 3 to 4 months after the buds are initiated. Flower bud development occurs at a faster rate when the minimum night temperature is increased to 65°F (18.5°C) in late November and early December.

C. warscewiczii (*C. gigas*) also produces two sets of growth in a year in the commercial greenhouse. To control flowering for Easter, the plants are grown under natural conditions when the first set of growth matures in May, June, and July. The temperature in the greenhouse is kept below 55°F (13°C) until October when it is raised to 65°F (18.5°C) to force the development of the second growth. As soon as the new shoots begin to develop, the temperature is returned to 55°F (13°C). The second growth forms flower buds at these low-temperature, short-day conditions. As soon as the buds are well developed, the temperature can be raised to 65°F (18.5°C) to hasten blooming for Easter sales.

A group derived from *C. labiata* normally flowers in September. To delay the flowering until the period from Christmas through Easter, the plants are given long days beginning June 1. Lights are used in the middle of the night for 2 hours each night in June and July, 2.5 hours each night in August, 3 hours each night in September, 3.5 hours each night in October, 4 hours each night in November and February, and 4.5 hours each night in December and January. Incandescent lights are used (consult the schedule of lighting for chrysanthemums). The long-day treatments are stopped 3 months before flowers are wanted. When the plants are being lighted, the minimum temperature permitted is 65°F (18.5°C).

C. mossiae hybrids normally flower for Easter. This characteristic makes them the most important orchid species for Easter sales. The plants are given long-day conditions before November 1 to delay the crop. The night temperature is in the range of 55°F to 65°F (13°C to 18.5°C). To obtain early flowering, black cloth is used starting November 1 at a night temperature of 55°F (13°C). When the buds are clearly visible in the flower sheath, the temperature can be raised to 65°F (18.5°C) to hasten bloom. Flower buds form in 2 to 3 weeks of short days at the cool temperatures. Development of the buds to the flowering stage requires another 10 to 12 weeks at 65°F (18.5°C).

Various hybrids of members of these groups have been developed to flower between the blooming periods of the parents. If desired, a grower can have *Cattleya* type orchids in bloom year-round by selecting the hybrids carefully.

Cymbidiums normally flower in the spring. The flowering of *Cymbidiums* does not show a photoperiodic response. However, it is interesting to note that *Cymbidiums* flower in Australia when the environmental conditions are the same as the conditions required for flowering in the United States, except that the seasons are 6 months apart.

Temperature may be more of a factor in flowering *Cymbidiums*. The plants are grown at a night temperature of 50°F (10°C) after April 1. The plants are placed outdoors in summer in the full light intensity. In the winter, the night temperature is held at 50°F to 55°F (10°C to 13°C).

Vanda orchids are grown in open fields in Hawaii. They are used to make leis and are exported to mainland United States.

Harvesting for Market

Orchids are harvested when the last flower on the stem is fully open. As soon as the flowers are cut, they are placed into tubes of water. Orchids are not graded. The minimum storage temperature for orchids is 50°F (10°C). Temperatures below 50°F (10°C) damage the flowers.

Cymbidium flowers are harvested as a spike. *Oncidium* and *Phalaenopsis* flowers are harvested as a spike or individually (Figures 30-8A and B and 30-9). The flowers are removed and packed as separate blooms on a bed of shredded wax paper to prevent damage. The stems of individual flowers are placed in tubes of water to keep them fresh.

Problems

Cultural Disorders

Yellow foliage develops when the light levels are too high for the plants. Failure to flower may be due to the wrong temperature, improper light levels, or other factors.

FIGURE 30-8A Spikes of *Oncidium* flowers ready for harvest and will be used to manufacture Hawaiian leis.

FIGURE 30-8B A close-up view of *Oncidium* orchids.

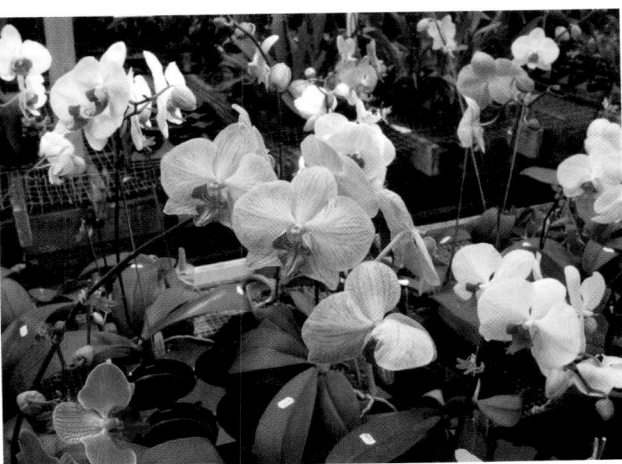

FIGURE 30-9 *Phalaenopsis* orchid blooms ready for harvest.

The sepals of the flowers become dry because of air pollution. There is no control for this problem except to install activated charcoal filters on all of the greenhouse ventilators. This is a great expense that the grower may want to compare against the cost of relocating to a pollution-free area.

Disease

Cattleya mosaic or black streak is caused by a virus. Virus-free, tissue culture propagated plants can be used, but the viruses are spread by aphids, and a virus-free crop cannot be guaranteed for long.

Botrytis causes a spotting of Cymbidium flowers and stem rot. Improved air movement through the plants keeps the foliage dry and reduces the likelihood of this disease.

Insects

Orchid weevils produce small holes in the leaves. Slugs, snails, and cockroaches feed on the flowers. Various types of baits can be set out to eliminate these pests. Scales may develop on the plants. These insects can be controlled by using the recommended insecticides.

Sow bugs and millipedes in the medium are controlled by insecticide drenches.

Weed Control

Few chemicals may be used to control weeds in orchid plantings.

ROSES ▧ *Rosa sp., Rosaceae*

History

The roses modern growers force for cut flowers are all hybrids. The hybrid tea roses and the floribunda type roses used by growers are the result of a series of crosses between various species followed by double and triple crosses.

Cultivars of hybrid tea roses produce large flowers with long stems. The most popular color for tea roses is red, followed by pink, orange, yellow, white, and variegated colors. The popularity of pink is increasing, largely because of a rose called 'Sonia.'

Rose breeders continue to produce new cultivars by the time-consuming process of hybridization. As an illustration of the amount of time required to introduce a new cultivar, a brief history of the cultivar 'Forever Yours' is traced.

'Forever Yours' was one of the two most popular red cultivars grown for cut flowers in the 1980s and early 1990s. The first crosses that ultimately resulted in this cultivar were made in the spring of 1959. Fifty to 60 cultivars were used to make 150 to 200 individual crosses. The plants from these crosses produced 35,000 to 40,000 seeds. The rose produces an average of eight to ten seeds in each pod that develops. The germination rate for the seeds planted was 50 percent, that is, half of the seeds germinated. The seedlings were planted to individual pots. When these plants bloomed, nearly one-quarter had to be discarded because they produced only single flowers. The greenhouse rose is a multipetalled bloom. A well-developed flower has 50 to 60 petals.

In the first year of the selection process, the number of plants selected was reduced again to about 1,500. Sixty or 70 of the best plants were grafted to the normal **understock** used. This practice speeds up growth and gives the plant breeder a chance to see how the new material performs as a graft.

By the second year, only six or seven of the grafted plants remained after the rigorous selection process used. Finally, one plant was selected to become 'Forever Yours.' Then it was necessary to build up the quantity of stock plants. Three years after the first cross was made, the breeder had about 150 plants. By the next year, over 6,000 stock plants were available. From this point the patent rights to the cultivar were sold to commercial propagators. For the next 2 years, the propagators increased the stock to 3 million plants that were sold to commercial growers. Beginning in the spring of 1959, it was not until 1967 that commercial rose growers received the first 'Forever Yours' plants to be forced for cut flowers.

Roses are the most important cut flower crop grown in the United States. Commercial rose growers usually do not mix other crops with roses but produce roses exclusively. The high light intensity needed for quality roses means that they must be grown under glass or high light transmitting plastics. An exception to this statement can be made for California and Colorado where the light intensity normally is so high that roses can be grown in fiberglass greenhouses. In Hawaii, the light intensity is so great that roses must be grown under Saran shade cloth to prevent damage to the flowers. In many offshore producing countries, roses are grown under polyethylene plastic covers. On a side note, only a few greenhouses in Colorado now grow roses in greenhouses.

Propagation

Roses are propagated from cuttings, grafting, or budding. Seeds are used only in the development of new hybrids.

Cuttings

Cuttings rooted from one-, two-, or three-eye stem sections are rarely used by commercial producers. Plants obtained from budded material are usually superior to stock rooted from stem sections.

Flowering wood is used to take cuttings from October to March. Three-eye cuttings are preferred because more shoots develop. Generally, there is more success in rooting three-eye cuttings than one- or two-eye cuttings. A single leaf at the top of the cutting is left on the stem. Most of the energy for making new roots comes from the food stored in the stem. The single leaf at the top contributes to the rooting process by keeping the xylem and phloem tissue active. If this leaf is damaged and falls off, the rooting process is slowed considerably.

Sand and vermiculite are commonly used rooting media. Roses also root well in Smithers Oasis Wedge® media. Peat moss is not a recommended addition to the medium because the pH is too acid. The cuttings are inserted 1 inch (2.5 cm) apart in rows spaced 3 inches (7.5 cm) apart. A rooting hormone is used to help root development. Rooting is promoted by providing bottom heat of 70°F to 72°F (21°C to 22°C). The recommended air temperature is 50°F to 55°F (10°C to 13°C). A low-pressure mist is used to maintain high relative humidity around the plants.

The cuttings should root in 4 to 7 weeks. Then they are planted in separate rose pots that are deeper than standard pots. The extra depth permits the maximum growth of the root system. New shoots develop from the axillary buds, or eyes, on the stem. When the shoots are 6 to 8 inches (15 to 20 cm) long, they are given a soft pinch to promote branching and the growth of more shoots. This process is continued until the grower decides the plant is large enough to go into the production bench.

Grafted Plants

At one time, grafting was the most widely used method for producing new cultivars. However, the procedure is rarely used now for several reasons: It is an expensive process, it is difficult to make the grafts, and a low percentage of the grafts are successful.

Budding

Commercial growers require millions of rose plants to produce cut flowers. These plants are propagated by specialty growers using the process of budding. Most of the budded roses for greenhouse use are grown in California, Oregon, and Texas. These areas have a naturally long growing season that allows the plants to complete growth in one year.

There are two types of budded plants: **started eye** and **dormant eye**. A started-eye plant is budded early in the season. The stock plant is broken just above the bud, shortly after the bud forms. The top stock is not removed completely. Because it is partially attached, there is a high exchange of nutrients and water past the location of the bud. As a result there is a more rapid union of the tissues, and the bud is forced into growth sooner. As soon as the bud starts good growth, the top stock is cut away.

Dormant-eye buds are propagated late in the season. The bud is inserted into the stem. The top stock is not removed, and the bud remains dormant. When the plants are harvested at the end of the growing season, the top stock is cut away.

The percentage of success in starting dormant-eye plants is less than 100 percent. Thus, growers prefer to use started-eye plants with growth that began in the field. If dormant-eye plants are benched and do not grow, the grower loses money because of the cost of the plants, the labor used to bench them, and the wasted bench space.

Understock

Roses budded for greenhouse forcing use a different understock than those grown for outdoor gardens. A good rose understock has several characteristics: it is easy to bud, is capable of making strong unions, does not send out suckers or extraneous shoots, is free of root diseases and insects, is resistant to root diseases, and promotes perpetual growth; that is, the understock does not go dormant or cause dormancy in the wood budded to it.

The understocks commonly used in the United States are *Rosa chinensis*, variety Manetti for field grown plants. Grafted miniroses are grown on *Rosa indica*, variety Odorata stock. *Rosa canina*, L. is used as the understock in Europe. *Rosa Natal briar* is used as the understock in Europe for miniroses.

Commercial propagators of roses maintain large blocks of *R. manetti* stock for the cuttings that serve as the understock. These stocks are maintained for 8 to 10 years. After this time, they are renewed because the vigor of the stock runs out due to cropping and an increase in virus diseases.

After a period of reduced watering to harden them, the understock canes are cut into pieces 8.5 inches (22 cm) long. A sharp knife is used to remove all but three eyes at the top of each cutting. A rooting hormone is applied to the basal end of each cutting to stimulate rooting. A mulch of asphalt paper or black plastic is placed on the ground, and the cuttings are planted in rows through holes in the paper or plastic. The mulch helps to prevent weed growth and warms the soil.

Fields where rose stock is to be grown are leveled so that irrigation ditches can be used. As soon as the cuttings are planted, the fields are irrigated. By early February the cuttings root, and top growth begins. As soon as the new shoots are 6 to 8 inches (15 to 20 cm) long, the budding process starts.

Two people are required to do the actual budding. A T-shaped cut is made on the stem. One person inserts the bud, which is shield shaped and is known as a shield, in the cut. The "T and shield" method of budding is quick and has a high percentage of success. The second person wraps the bud with a rubber strip to hold it in place securely. The buds are inserted on the north side of the stock. In this location, they are not desiccated (dried out) by the sun before the bud and stock tissue unite. The rubber used to tie the bud to the stem disintegrates in about the same amount of time required for the union to knit together. If the rubber does not disintegrate in this time, it is cut away so that it does not strangle the plant and prevent the translocation of nutrients and water. The bud and scion should unite in 2 weeks to form a continuous conducting tissue.

When the buds are inserted, the top stock is bent over but is not removed. After 30 days, the top stock is cut off if the **bud union** has formed properly.

The budded plants are grown on for the remainder of the season to produce strong branches. In early December the stems are cut back until they are about 14 inches (35 cm) high. A special machine is used to trim the stems and make chips of the wood removed. The chips are returned to the field to add organic matter.

A U-shaped unit mounted on a tractor digs the plants from the field. Excess soil is removed from the plants, which are then taken to the packing house for grading. Roses are grouped into XXX, XX, and X grades. The XXX grade must have at least four strong stems. Most growers use the XXX and XX grades for forcing.

After the plants are cleaned and graded, the tops of the canes are dipped in a paraffin or plastic sealing compound. This treatment holds moisture in the plants. They are stored at 32°F to 33°F (0°C to 0.5°C) until they are removed to fill orders. Plants removed from the field early require at least 3 weeks of storage to mature properly.

When orders are received, the roses are placed into polyethylene-lined cardboard boxes. One box holds from 250 to 350 plants. The number of plants packed depends on the size of the plants because cultivars differ in size. The plastic is pulled around the plants and taped to form a moisture-proof wrapping.

A small amount of air exchange is required to allow the escape of ethylene gas. However, at temperatures below 40°F (4.5°C) very little ethylene is produced by rose plants. Some growers report that a buildup of ethylene gas in the shipping probably is responsible for the failure of roses to force shoots. Such a buildup of gas may occur if the plants are overheated in shipment.

Recent demands for rose plants for hydroponic culture have increased the demand for miniature plants. Here, rose understocks are taken as an unrooted stem and a dormant bud is chip budded onto the understock. The budded understock is then rooted while the chip bud union is formed (Figure 30-10).

FIGURE 30-10 Greenhouse grown chip budded rose plants ready for transplant.

Preplant Handling

The preplant handling of roses is discussed in the section on pot roses in Chapter 25. The reader is referred to this section for review.

Greenhouse Planting

Media

Rose plants grow well in a wide range of media if the media provide good aeration and drainage, hold nutrients and water, and are free of disease organisms and insect pests.

Commercial rose growers that choose to grow roses in ground beds must meticulously prepare the medium for their roses. The base for the medium is a good field of sod that has been growing for 3 years or longer. The sod should be a mix of grasses such as timothy, redtop, and orchard grass. Perennial ryegrass roots deeper than other grasses and provides more organic matter. Clover is not required in the grass mix. The field is fertilized at least twice a year with nitrogen to promote good grass growth. The field is mowed at least once a year. The clippings are left to add to the organic content of the sod.

One year before the topsoil is to be stripped from the field, the soil is tested. Limestone is added to bring the pH up to 6.0 to 6.5. The field is plowed and seeded with rye in the fall. The rye is allowed to grow as late as possible in spring. Before it forms heads, it is plowed into the soil. This green manure crop adds further organic matter to the soil. The ground is tilled two or three times during the summer to reduce the weeds.

In the fall, 20 percent superphosphate is added at the rate of 1 ton to an acre (2.24 mt/ha). If treble superphosphate is used, it is added at the rate of 0.5 ton to an acre (1.12 mt/ha). If manure is available in quantity, it is added at the rate of 10 to 15 tons to each acre (22.4 to 33.6 mt/ha). Cattle manure is often fortified with superphosphate in the barn to prevent the loss of ammonia. In this case, less phosphorus is used in a separate application when manure is added. The manure is plowed into the soil, and then the soil is stripped to a depth of 6 inches (15 cm). The soil is piled to a depth of 6 to 8 feet (1.8 to 2.4 m) and finished with a flat top. This shape allows the fall rains and winter snow to soak through the pile. In the spring the soil is ready for use.

Rose growers who cannot prepare their own soil must purchase it from some other source. Before making any purchases, the grower must obtain a sample of the soil and have it tested for its nutrient content. The soil can be analyzed at any state agricultural college or commercial soil testing laboratory. The results of the test guide the grower in adding limestone and superphosphate when the medium is prepared.

The grower can conduct a test in the greenhouse to determine if the soil being considered for purchase contains residues of chemical weed killers. Weed killers are used so often in farming today that any soil obtained from recently cropped farmland must be suspected of contamination. Some weed killers leave a residue that is active for 2 or 3 years after the material is used. A simple test can be made to determine if weed killers are present in the soil. Three 5- or 6-inch (12.5- or 15-cm) pots filled with the soil in question are planted with oats, beans, and cucumbers. Each crop is planted in a separate pot. Three more pots are filled with soil that is known to be free from the chemicals. These pots are planted with the same crops for comparison.

All of the pots are placed in the greenhouse until the seed germinates. A minimum night temperature of 60°F (15.5°C) is maintained. If the weed killers atrizine or simazine are present, the oat seedling leaves are straw colored. The beans and cucumbers will grow to a height of 4 to 5 inches (10 to 12.5 cm) and then stop. The seedlings in the test soil are compared with those in the known soil. If the seedling growth is not the same for both types of soil, the grower should not purchase the test soil. Another, uncontaminated source of soil must be found.

Preparing the Production Media

The basic soil obtained from the field composting of sod is mixed with other materials to make the production medium. Roses are grown from 4 to 6 years or longer in the same medium. During this time, there is no opportunity to modify it. Therefore, the rose medium must be formulated to ensure that the good drainage and aeration characteristics persist for a long time.

At least half of the medium should be a good loam soil. One good medium is a 2:1:1 mixture by volume of loam soil, sphagnum peat moss, and perlite or sand. When field composted soil is used, a 3:1:1 mixture of the materials listed previously is prepared. Some growers prefer this medium because of its greater volume of soil.

Other materials can be added, including wood chips, sawdust, calcined clays, peanut hulls, and corncobs. The advantages and disadvantages of these materials were described in Section 4.

The pH of the medium is adjusted by adding limestone until it is in the range of 6.0 to 6.5. The soil is tested to determine if superphosphate is needed. If there is any doubt about the need for phosphorus, it is recommended that superphosphate be added at the rate of 5 pounds (2.28 kg) to 100 square feet (9.3 m²) of area. Some growers also add 5 pounds (2.28 kg) of gypsum calcium sulfate to each 100 square feet (9.3 m²) of area. The soil pH is not altered, but calcium is added and the soil structure is improved.

Roses grow best in a medium that is at least 6 inches (15 cm) deep after it settles. To provide adequate room for root growth, an 8- to 10-inch (20- to 25-cm) deep medium is preferred. Ground beds with solid, V-shaped bottoms are used because the rose plants may reach a height of 6 feet (1.8 m) or more. It is more difficult and time consuming to produce a rose crop on raised benches.

If the ground beds do not have solid bottoms, the roots may grow below the level of soil preparation and sterilization. As a result, the plants may become infested with diseases and nematodes from the soil below the medium. In addition, the grower loses some control over the water and nutrient uptake of the crop if the roots can grow into the subsoil.

The medium is steam pasteurized. Few rose growers lack the facilities to steam pasteurize their soils. Aerated steam is preferred because it is faster, thus reducing energy costs and the beneficial soil microbes are not killed by the low-temperature heat. The soil is steamed for 30 minutes at 140°F (60°C).

Planting

The budded plants are shipped to reach the grower in mid-December. They may be planted from this time until June or July. Most growers plant during the spring months to give the plants a good start before the heat of summer. A planting schedule could be set up easily if roses were replanted every 4 years. In this case, 25 percent of the greenhouse range would be planted every year. However, growers now keep plants in production only as long as they are profitable. This means that a cultivar that does not sell well may be removed after 2 years. The grower should review the performance of each cultivar to determine if it is to be replaced.

The budded plants are inspected, and any broken roots and stems are pruned. Some growers cut back all of the stems to an even 6-inch (15-cm) length. On some cultivars, this pruning may sacrifice one or more strong top buds.

The depth of planting depends upon the use of a mulch. If a mulch is used, the plant is placed so that the bud union is about 2 inches (5 cm) above the soil surface. The grower can then add a 3-inch (7.5-cm) mulch, which settles to a final 2-inch (5-cm) depth.

If a mulch is not used, the plant is placed so that the bud union is just above the soil surface. At this level, excess moisture does not settle around the union and cause it to rot. In addition, the bud union is not set so low that roots start to grow from the scion wood. If roots do form, the plants are said to be on their own roots. In most cases, this condition is not desirable. However, there have been instances where plants have gone on their own roots and production was greatly improved for a year or two. The plants are placed at a spacing of 12 inches by 12 inches (30 cm by 30 cm). Closer spacing reduces the quality and growth of the plants.

The new plantings are misted from above several times a day to maintain high humidity around the buds and force faster growth. For late spring in warm weather, a low-pressure mist helps to start the breaks.

If it is suspected that the plants will be slow to develop new shoots, they are covered with white plastic to produce high-temperature and high-humidity conditions that will hasten growth. The grower must monitor the temperature under the plastic, particularly on a bright sunny day. The tender shoots are damaged by very high temperatures. As soon as the shoots are 1 to 2 inches (2.5 to 5 cm) long, the plastic is removed.

The plants require extra care and attention for the first 5 to 6 weeks of growth. The recommended night temperature is 55°F (13°C); however, it may not be possible to maintain this temperature if production plants are in the same house. A cool night temperature slows vegetative growth, allowing the root system to develop at a rate of growth that is more nearly equal to the top growth. The top growth for the first 5 to 6 weeks is nourished by the food stored in the stems and large roots. The new roots absorb water but few nutrients during this period of time. Therefore, a high level of nutrients in the medium can damage tender roots. After 6 weeks, the root system is large enough to permit a regular fertilization program.

Rose Production in Rock Wool or Coir Coconut Fiber

Most rose growers in the United States and Canada produce roses in hydroponic systems using rock wool or coir coconut fiber (Figures 30-11 and 30-12). The production of roses as cut flowers in rock wool media is practiced more so in Europe, but is gaining popularity in the United States. Rock wool culture of roses is a highly sophisticated technique of growing plants hydroponically.

FIGURE 30-11 Budded rose plants newly transplanted into hydroponic troughs filled with coir coconut fiber.

FIGURE 30-12 A close-up view of newly transplanted budded rose plants.

The rock wool medium is basically inert as far as supplying plant nutrients is concerned. It is simply an anchor system for the plant. It also serves as a short-time reservoir for nutrients and water, which must be supplied on a regular basis.

Because rock wool media provide no buffering or cation-exchange capacity as does soil, the nutrient solutions supplied must be carefully prepared, and their application controlled and monitored frequently. The total soluble salt content and the pH of the solutions must be checked several times daily to ensure that they are in the proper range for the crop. Failure to do so may result in extensive damage to the root system and the subsequent flower production.

Roses are generally grown on slabs of rock wool that are 3 by 8 by 39 inches (7.5 by 20 by 100 cm) in size. The slabs are usually covered with white polyethylene, which helps to retain moisture and also reduces the growth of algae.

In recent months, trough production of roses using rock wool or a product called coir has been tried. Coir is fiber from the mesocarp tissue of coconuts. The product is light in weight and similar to sphagnum peat moss in color. It has a greater water-holding capacity than that of peat moss with a similar cation-exchange capacity. One unique characteristic of coir is that it is more easily rewetted than sphagnum peat moss if allowed to become excessively dry.

Growers who wish to switch their rose production from soil to these substitutes are cautioned to do so on a small scale first. After a couple of years' experience with the new media and system, they can then change over 100 percent.

Fertilization and Carbon Dioxide

Greenhouse roses have a medium-high nutrient requirement compared to other cut flower crops. Chrysanthemums require more fertilization than roses, which require more than carnations. When roses are planted, the nutrient levels are kept low. When the plants are cut back each year, the nutrient levels are again kept low until the new growth is large enough to use added nutrients.

Roses need a balanced fertilization program of N-P-K plus trace elements. Foliar analyses of rose leaves show that their nutrient content is 3 percent nitrogen (N), 0.20 percent phosphorus (P), 1.8 percent potassium (K), 1 percent calcium (Ca), and 0.25 percent magnesium (Mg). To maintain these levels, the nutrient solution must supply approximately 170 ppm N, 34 ppm P, 150 ppm K, 120 ppm Ca, and 12 ppm Mg at each watering. Table 30-1 lists the ingredients of a stock solution that supplies the required amounts of nutrients.

The fertilizer stock solution is fortified with a trace element solution (Table 30-2). In some cases, the amounts of trace elements added to the solution are very small. An accurate metric scale is used to weigh the ingredients.

The advantages of applying fertilizers at every irrigation include a savings in time and labor because the irrigation and fertilization are accomplished at the same time; a regular schedule of fertilization is followed; the soluble salt concentrations are neither too low nor too high; accurate soil and foliar tests can be made at any time to determine the true nutrient content of the soil and plant; the grower has greater control of crop growth; and there is less chance of an accidental overdose of nutrients.

Table 30-1

Nutrient Carriers and Quantities Added to 100 Gallons (378.5 l) of Water to Prepare a Concentrated Stock Solution[1]						
Nutrient Carrier	Pounds (kg) in 100 Gallons(378.5 l) of Water	PPM in Final Solution Applied at Each Irrigation				
		N	P	K	Ca	Mg
Calcium Nitrate	60 (27.3)	114	–	–	120	–
Potassium Nitrate	16.66 (7.6)	27	–	73	–	–
Potassium Chloride (60%)	12.5 (5.7)	–	–	79	–	–
Ammonium Nitrate	7 (3.2)	28	–	–	–	–
Magnesium Sulfate	10 (4.5)	–	–	–	–	12
Phosphoric Acid (85% H_3PO_4)[2]	3.33 (1.5)	–	34	–	–	–
Totals		169	34	152	120	12

1. This solution is applied at a 1:100 dilution ratio to supply the stated nutrient concentrations at each irrigation.
2. Phosphoric acid is a liquid. The amount needed is determined by weight. Caution: Handle the acid with care, wear eye protection, and use rubber gloves. Slowly add the acid to the water after all of the other nutrient carriers are added.

Table 30-2

Trace Element Materials Added to 100 Gallons (378.5 l) of Stock Solution Prepared in Table 30-1[1]			
Trace Element[2]	Percent Element Contained	Grams in 100 Gallons (378.5 l) of Water	PPM in Final Solution Applied at Each Irrigation
Iron	13.0	1,453.00	5.00
Manganese	12.0	204.75	0.65
Copper	9.0	8.40	0.02
Zinc	14.5	13.03	0.05
Boron	17.5	74.83	0.35

1. When this trace element solution is applied at a 1:100 dilution ratio, it supplies the stated concentration of nutrients at each irrigation.
2. Iron, manganese, copper, and zinc are chelates. Boron is boric acid H_3BO_3.

High-quality crops can also be obtained using other fertilizer programs if they are carried out on a regular schedule. In these programs, several nutrient carriers are alternated to supply the needed elements. Regular soil testing helps the grower determine how well the nutrient levels are being maintained.

Foliar or leaf analysis is a very useful procedure for checking the nutritional status of rose crops. Leaf samples are taken from several plants. The sample is usually the first and second five-leaflet leaves from the upper stem of a flower bud just beginning to show color. A commercial laboratory or a state agricultural college laboratory has the facilities for foliar analysis. The results of the analyses are interpreted for the grower by an expert. Laboratories normally charge for both soil tests and foliar analyses.

The symptoms of nutrient deficiencies in roses are the same as those described in Chapter 14. In many cases, a foliar analysis is required.

Carbon Dioxide

Rose growers add carbon dioxide to the greenhouse atmosphere from late October to April. Carbon dioxide is added as long as it is not necessary to open the ventilators to reduce the temperature in the greenhouse. A concentration of 1,000 to 1,200 ppm of carbon dioxide is maintained.

Foliar analyses of roses grown with and without high levels of carbon dioxide showed that the plants had the same nutrient content over a 2-year period. However, plants grown in a high carbon dioxide atmosphere require additional fertilizer and water because the plants make more growth compared to plants grown without carbon dioxide.

Carbon dioxide is colorless, odorless, and tasteless. In concentrations less than 10,000 ppm it is not considered to be harmful to humans.

Light and Temperature

Roses require full light intensity to produce quality flowers having a long lasting life. The greenhouse may require some shade on the glass in the summer to reduce the heat load on the plants. Studies of roses show that their growth curve follows the light intensity curve. During the high light periods of the year, growth is increased. When the light levels are low in the winter, growth occurs at a slower rate.

A low light level is the limiting factor for growth during the winter in the Pacific Northwest and in the North. Experiments with high-intensity discharge HID lamps show a 100 percent increase in the production of certain cultivars of roses during the winter months (Figure 30-13). Roses given 9 hours of 1,200 to 1,500 fc (12.9K to 16.1K lux) from HID lamps during the night show improved growth and flowering compared to roses grown with natural days. Even greater production is achieved using the same light intensity for 18 hours each day.

When the lights are on, less heat is needed to maintain the proper temperature because the lights and ballast generate a large amount of heat.

HID lighting units are priced at several hundred dollars each. When 1,000 or more units are placed in a greenhouse range, the cost of the installation is high. The grower must compare the cost of buying and installing the lights with the dollar return from an improved crop. The use of HID lights helps the grower make more efficient use of the greenhouse space.

FIGURE 30-13 A large gutter connected rose range using HID (high-intensity discharge) lights to improve rose growth in the winter months. (Note: The pink roofed sections were installed as an experiment to control powdery mildew spore germination by altering the wavelength of the light entering the greenhouse.)

FIGURE 30-14 A thermal (heat) blanket is installed in this greenhouse. In this picture, the heat blanket is gathered at each truss and is pulled over the crop each evening.

The normal production temperatures for greenhouse roses are 60°F to 62°F (15.5°C to 16.5°C) at night. The day temperature is 10° higher on cloudy days and 15° higher on clear days. When carbon dioxide is added to the atmosphere, the day temperature may be increased to 80°F to 85°F (27.5°C to 29.5°C) with no loss in the quality of the flowers. If HID lighting is used, a minimum temperature of 65°F (18.5°C) is maintained to get the most benefit from the treatment.

Energy conscious growers may be tempted to reduce the night temperatures by a few degrees to conserve fuel. However, research has shown that a reduction of the night temperature to 55°F (13°C) reduces the number of flowers and delays maturity from 1 to 3 weeks. Any money saved in fuel costs may be lost in a smaller crop.

Another suggested energy conservation method involves heating the soil and lowering the air temperature. Studies of this practice show that heating the soil above 67°F (19.5°C) reduces the growth of the roots and the tops.

Energy can be conserved in the greenhouse by the use of thermal blankets (Figure 30-14). These blankets are pulled over the plants at night to prevent heat loss through the glass. They are removed in the morning. If HID lights are used, move the ballasts below the thermal blanket to capitalize on the heat energy that is not converted to light.

Timing and Pinching

Roses are timed to produce flowers for the major holidays, such as Christmas, Valentine's Day, Easter, and Mother's Day. The pinching methods used on the plants control the timing.

When the roses are first planted and the new growth is 6 to 8 inches (15 to 20 cm) long, the plants are pinched. The terminal growing point and 1 to 2 inches (2.5 to 5 cm) of the stem are removed. Axillary shoots develop resulting in more leaves. Pinching is continued to cause the development of a large amount of leaf area that carries on photosynthesis in the low light conditions of winter.

After the plant reaches a good bushy stage, pinching for further development is stopped. The flowers are allowed to reach market size and are cut. Pinching after this point is used to time the crop.

Both hard and soft pinches are used on roses. A soft pinch is made by removing the growing tip of the shoot when the flower bud is less than 0.25 inch (6 mm) wide. A hard pinch is made when the flower bud is about 0.38 to 0.5 inch (9.5 to 12 mm) wide. The shoot tip and stem tissue are removed to the second five-leaflet leaf.

When a soft pinch is made, the tissue removed is usually very succulent. This means that the axillary bud that develops just below the area pinched grows into a straight stem. Normally, a soft pinch results in a single stem. Mature flowers are ready for harvest on growth produced from a soft pinch 3 to 7 days later than on growth from a hard pinch.

A hard pinch usually produces two stems because of the higher carbohydrate reserve of the stem. In addition, the axillary buds present at the time of the pinch are at a more advanced stage of growth. If a single shoot develops following a hard pinch the stem usually contains a crook.

General Timing

Following a harvest of flowers, the time required to produce a new crop varies from 6 to 8 weeks, depending upon the season of the year. As the days get longer in the spring and summer and the light intensity increases, a crop comes back in 6 weeks. Seven to 8 weeks are required in the late fall and through the winter to produce a new crop. This general timing information is useful to the grower in scheduling flowering for holidays.

Pinching for Holidays

A modest number of stems are given a hard pinch on October 1–4. As a result there is an increase of 15 to 30 percent in the number of flowers harvested at Thanksgiving (late November). The plants harvested at this time are ready for harvest again in mid-January. Regarding the so-called **Christmas pinch**, at 42° north latitude, a soft pinch is made about October 17–21 for the Christmas crop. In areas where the light intensity is higher, the soft pinch may be made a few days later.

The plants are given a hard pinch October 20–25. Some of the plants are given a hard pinch on December 21–22 to help increase the Valentine's Day (February 14) crop. A number of the plants harvested at Christmas will produce another crop for Valentine's Day.

Easter varies from early to late in the season. If Easter is late, flowers will be ready if the plants are given a hard pinch 7 weeks before the holiday. If Easter is early, the crop is pinched 4 to 5 days earlier because of the poorer light conditions. When Easter occurs in the last week in March, the plants harvested at this time are often ready for harvest again at Mother's Day (early May), if the weather in April is warm and bright. The timing of the crop can be adjusted somewhat by raising or lowering the night temperature by 5°.

Cutting Flowers Back

Roses must be cut back once a year to reduce the overall height. The time selected for cutback usually occurs when flower sales are poor and the growing conditions will aid the new growth of the plants. Two methods of cutback are used: gradual pruning or direct pruning.

Gradual Pruning

The method of gradual pruning ensures that the crop is never completely out of flower. This method is also called knife pruning because the cuts are made with a knife. Knife pruning starts with the Easter harvest. After the flower is removed, the stem is cut back into the hardwood to reduce the overall height. In this method, the plants are pruned to a minimum height of 24 inches (60 cm). As the next flower comes into bloom, the steps are repeated. The flower is cut with a maximum stem length. The remaining stem is then trimmed back to the desired length. By Mother's Day, all of the roses in the bench are cut back. Some of the shoots cut at the beginning of the pruning are ready to flower again. Any stems that are still too long are cut back at this time.

Direct Pruning or Brush Cut

Direct pruning (**brush-cut pruning**) is started after harvest for Mother's Day and before the weather warms up too much. All of the roses in the greenhouse are cut back to an even height. For first-year plants, the stems are cut to 18 inches (45 cm). The cut for each succeeding year is raised 6 inches (15 cm). If the cuts are made into wood that is too hard, the new shoots that develop are greatly reduced in number. A disadvantage of direct pruning is that flowers are not produced for 5 to 6 weeks or until the crop comes back. Therefore, this type of pruning is done in the summer when the demand for cut roses is low.

Regrowth of Cutback Plants

After the plants are cut back, they must be built up again to prepare for new harvests. A light shade may be applied to the greenhouse glass to reduce the heat. High humidity is maintained by syringing the plants from above several times a day. The humid environment helps to force the new growth. Fertilizer is not applied until the new growth is 6 to 8 inches (15 to 20 cm) long. The shoots are given a hard pinch to force bottom canes to develop and to increase the number of stems on the plants. At this stage, the plants are treated as if they had just been benched.

Bending of Arching Roses

A novel technique for pruning roses, called bending or arching, continues to gain in popularity since its introduction in the mid-1990s (Figure 30-15). Bending or arching is a method of pruning, where rather than removing shoots with lower quality flower buds or blind shoots, the shoots are bent down and left attached (Figure 30-16). The stem is broken slightly, which disrupts the flow of photosynthates from the foliage to the roots, providing photosynthates from leaves to supply flowering branches (Figure 30-17). From the horizontal branch, new flowering stems arise from existing nodes, which produce new flowering stems (Figure 30-18). Because this is done low on the plant, it reduces plant height and increases worker safety as well. Additional benefits of bending or arching include an increase in air movement in the plant canopy which reduces foliar diseases, an increase in light penetration into the plant canopy, and an increase in quality of flowering shoots which allows for harvest of larger stems.

FIGURE 30-15 Rose plants grown with the bending or arching technique.

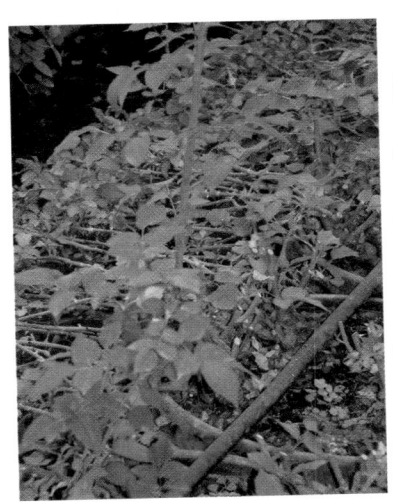

FIGURE 30-16 A closer view of the bending or arching technique. (Note: Many stems have already been harvested. The upright stem will be harvested soon.)

FIGURE 30-17 A newly bent rose cane with strong lateral growth.

FIGURE 30-18 Rose stems near ready for harvest from bent or arched canes.

Supporting the Plants

Roses must be supported to produce straight stems. At one time, a large wire stake was placed within a plant. Each stem was tied to the stake with raffia or string. However, the present cost of labor makes this method of support too expensive. Growers now stretch plastic mesh on frames above the plants (Figure 30-19). As the stems increase in length, they are tucked into the openings in the mesh. When the plants are removed for replanting, the plastic mesh is raised out of the way.

Harvesting for Market

Normally, roses first produce three or four three-leaflet leaves. These are followed by several five-leaflet leaves. As the plant reaches the flowering stage, more three-leaflet leaves are produced.

During low light periods, the plant must carry as much foliage as possible. For this reason, when flowers are harvested from August through December, they are cut to leave two five-leaflet leaves per stem. When flowers are cut in

FIGURE 30-19 Tall rose plants require support for straight stems.

the remainder of the year, only one five-leaflet leaf is left. To obtain longer stems for market, the plants may be given a "knuckle cut" in the spring. This means that cuts are made within 0.5 inch (1.2 cm) of a previous cut. Two or three new shoots normally develop from a knuckle cut. Ideal growing conditions are required to ensure that all of the shoots bloom. Normally, one or two shoots may go blind while the rest bloom.

When a shoot grows for a time and then stops it is said to be blind wood. It is thought that a lack of carbohydrates causes blind wood. Usually, more blind wood occurs in the low light periods of winter than in the spring and summer. Growers who use carbon dioxide and HID lights report less blind wood in their crops. Under these conditions, more carbohydrates are produced.

Grading and Packing

When the flowers are cut in the greenhouse, they are wrapped in plastic sheets or mesh (Figure 30-20). They are placed immediately in water or in a hydrating solution and then are moved directly into refrigerated storage.

The flowers are allowed to take up some water and then they are graded (Figure 30-21). The flowers are inspected for deformed blooms or other imperfections. Then they are placed into the trays of a mechanical sorter, which speeds up the process of grading. The machine sorts the flowers according to stem length. After sorting, the flowers are bunched.

The grower may use grading boards, which are marked in 3-inch (7.5-cm) intervals. After the plants are graded, they are wrapped in a spiral pack or are bunched and tied (Figures 30-22 and 30-23). The packs are returned to the refrigerator where they are placed in a holding or a preservative solution for storage until they are packed for shipment (Figures 30-24, 30-25, and 30-26). Some large rose growers also wholesale other flowers (Figure 30-27). When an order is received, the entire group of flowers is packed together for delivery.

The refrigerator temperature is maintained at 33°F to 35°F (0.5°C to 1.5°C). This temperature brings the respiration and transpiration rates nearly to a standstill, thus conserving carbohydrates in the flowers. The flowers are held in refrigerated storage for no more than a few days before they are sent to market.

FIGURE 30-20 Freshly cut roses are wrapped in plastic and immediately placed in hydrating solution. They are then moved immediately to refrigerated storage.

FIGURE 30-21 Each bin of this mechanical sorter used to grade roses holds a different stem length or flower color.

FIGURE 30-22 Roses after grading and sorting are hand separated.

FIGURE 30-23 Roses ready for final processing and storage.

FIGURE 30-24 Graded and sleeved roses ready for shipment.

FIGURE 30-25 End view of red roses in a spiral wrap.

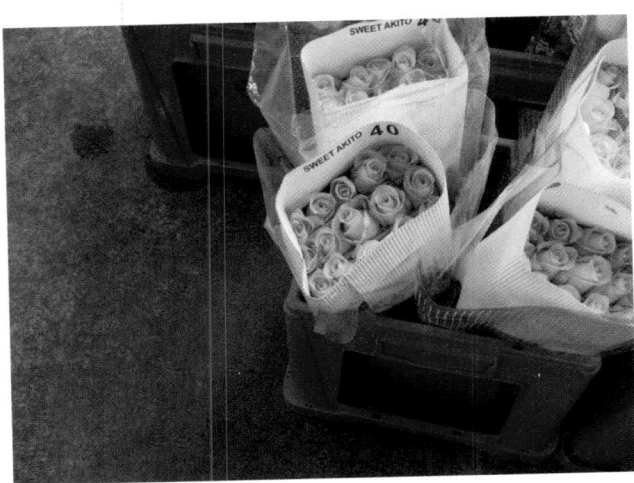

FIGURE 30-26 End view of pink roses in a bunch.

FIGURE 30-27 A small order being assembled by a wholesaler to be delivered to a retailer.

Roses can be held for several days in dry storage at 31°F (−0.5°C). This treatment is described in Chapter 32.

Lasting Life

The lasting life of a cut rose is one of the shortest for any cut flower. The methods of prolonging the cut life of a rose until it reaches the consumer are described in Chapter 32.

Problems

Cultural Disorders

Nutrient deficiency symptoms and the effects of excessive fertilization are described in Chapters 14 and 15.

Nematodes may infest the roots of the plants and cause stunting and chlorosis.

Malformed flowers are caused by an infestation of thrips and boron deficiency.

Leaf drop is due to overwatering of poorly drained soils, the application of too much fertilizer resulting in high soluble salt levels, and air pollution. In addition, if certain insecticides are applied when sulfur burners are being used to control mildew, a gas is formed that causes leaf drop.

An infestation of black spot increases the amount of ethylene in the greenhouse air, resulting in leaf drop.

Toxic fumes from certain rust-inhibiting paints may cause leaf injury and leaf drop.

The improper use of herbicides in a rose range may cause leaf drop.

Some cultivars of roses are affected with **bent neck**. The cause of this problem is not understood completely. One suggestion for overcoming the problem in the cultivar affected is to cut the flowers after they open somewhat more than usual. The additional time allows the tissue in the neck of the stem to mature and conduct water more readily to the flower.

Diseases

Powdery mildew is the largest problem, but the proper control of ventilation helps to minimize it (Figure 30-28). Rust diseases are common on roses as well

FIGURE 30-28 Powdery mildew, *Sphaerotheca pannosa var. rosae*, is a major disease of greenhouse roses. It is characterized by white, powdery lesions.

(Figure 30-29). Fungicide sprays are used as needed. Many growers will volatilize sulfur to control many foliar diseases (Figure 30-30).

Various canker organisms affect the cut stubs of stems. Control of these organisms is difficult. Viruses cause a mottling of the leaves but generally do no other damage.

Insects

The most serious insect problem faced by rose growers is the red spider mite (Figure 30-31). The insects soon develop resistance to the chemicals when insecticides are used regularly.

Aphids are an occasional problem, but they are easy to control. Infestations of leaf tiers and leafrollers are more serious. Control is difficult because plants are damaged by the larva stage, which is inside a rolled-up leaf and thus is protected from sprays.

FIGURE 30-29 Rust disease, *Phragimidium mucronatum*, is best controlled by increasing air circulation and reducing humidity. Infected leaves can be removed without damaging the plant, but is labor intensive.

FIGURE 30-30 This device is a heater that is used to volatilize sulfur for control of foliar diseases on greenhouse roses.

FIGURE 30-31 Spider mite infestation is typically identified on greenhouse roses with chlorotic stippling.

ACHIEVEMENT REVIEW

Select the best answer or answers to complete each statement. List the appropriate letter(s).

1 The two types of orchids grown are:
   a. epiphytes.
   b. hydrophytes.
   c. terrestrials.
   d. xerophytes.

2 Sympodial growth can be described as:
   a. vertical.
   b. at a 60-degree angle.
   c. at an 80-degree angle.
   d. horizontal.

3 Which of the following have a sympodial growth habit?
   a. *Brassavola*
   b. *Cattleya*
   c. *Laelia*
   d. *Vanda*

4 Compared to chrysanthemums, the growth rate of orchids is:
   a. very fast.
   b. moderately fast.
   c. fast.
   d. slow.

5 The function of the pseudobulb is to:
   a. separate the leaf from the stem.
   b. produce a flower.
   c. store food and moisture.
   d. start new leads.

6 Aerial roots of orchids that do not reenter the growing medium can:
   a. absorb water only.
   b. absorb nutrients.
   c. exchange oxygen and carbon dioxide only.
   d. do all of the above.

7 Germinating seed of orchids requires a very special procedure because the seed:
   a. is very small.
   b. has no seed coat.
   c. has no embryo.
   d. has no stored food.

8 Seed flasks may be placed under cool white fluorescent lamps that provide an intensity of:
   a. 500 fc (5.4K lux).
   b. 850 fc (9.2K lux).
   c. 1,500 fc (16.1K lux).
   d. 2,000 fc (21.5K lux).

9 Meristem tissue culture of orchids was started to eliminate the problem of:
   a. bacteria.
   b. fungi.
   c. red spider mites.
   d. viruses.

10 Plants growing in a bark medium require _____ water than do plants in osmunda fiber.
   a. less
   b. the same amount of
   c. more
   d. twice as much

11 Orchids growing in bark mixes have need for more:
   a. nitrogen.
   b. phosphorus.
   c. potassium.
   d. iron.

12 The proper temperature for refrigerating orchids is:
   a. 39°F (4°C).
   b. 45°F (7°C).
   c. 50°F (10°C).
   d. 55°F (13°C).

13 The rose cultivar 'Forever Yours' took _____ years to reach commercial production after the first crosses were made.
   a. 3
   b. 8
   c. 10
   d. 12

14 Roses are the _____ most important cut flower crop in the United States.
   a. first
   b. second
   c. third
   d. fourth

15 The most important types of rose plants used by commercial growers today are:
   a. own root.
   b. grafted.
   c. started eye.
   d. dormant eye.

16 The understock commonly used on roses is:
  a. *R. canina.*
  b. *R. chinensis* 'Manetti.'
  c. *R. multiflora.*
  d. *R. indica* 'Odorata.'

17 The preferred grade of rose stock used for greenhouse forcing is:
  a. No. 1.
  b. X.
  c. XX.
  d. XXX.

18 Large rose growers begin to prepare rose growing soil by:
  a. producing a good field sod.
  b. purchasing top soil.
  c. stripping old corn fields.
  d. using peat-lite mixes.

19 The proper pH for a rose growing medium is:
  a. 5.5 to 6.0.
  b. 6.0 to 6.5.
  c. 6.5 to 7.0.
  d. 7.0 to 7.5.

20 A simple test to determine if there is a weed killer residue in a soil to be purchased is to plant seeds of:
  a. beans.
  b. cucumbers.
  c. oats.
  d. all three of these.

21 The planting distance for roses is:
  a. 6 inches by 8 inches (15 cm by 20 cm).
  b. 8 inches by 8 inches (20 cm by 20 cm).
  c. 8 inches by 12 inches (20 cm by 30 cm).
  d. 12 inches by 12 inches (30 cm by 30 cm).

22 The bud union should be planted so that it is:
  a. 1 inch below the medium surface.
  b. just at the medium surface.
  c. 1 inch above the medium surface.
  d. 2 inches above the medium surface if a mulch is to be used.

23 Carbon dioxide is added to the greenhouse during the daylight hours to maintain a level of:
  a. 600 to 800 ppm.
  b. 800 to 1,000 ppm.
  c. 1,000 to 1,200 ppm.
  d. 2,000 to 2,500 ppm.

24 The most production is obtained when HID lights are used:
  a. during the daytime only.
  b. for 9 hours at night at 800 fc (8.64K lux).
  c. for 9 hours at night at 1,200 to 1,500 fc (12.9K to 16.1K lux).
  d. for 18 hours at 1,200 to 1,500 fc (12.9K to 16.1K lux).

25 Roses are given a hard pinch to:
  a. slow the growth.
  b. increase the number of stems.
  c. time the crop.
  d. remove excess foliage.

26 The hard pinch for the Christmas crop is given:
  a. September 20–25.
  b. October 17–21.
  c. October 20–25.
  d. November 1–5.

27 The direct pruning brush cut type of cutback used on roses is usually started after the:
  a. Christmas cut.
  b. Valentine's Day cut.
  c. Easter cut.
  d. Mother's Day cut.

Minor Cut Flower Crops

⊠ **Alstroemeria**

⊠ **Anthurium**

⊠ **Calendula**

⊠ **Callistephus**

⊠ **Centaurea**

⊠ **Dahlias**

⊠ **Freesias**

⊠ **Gerbera**

⊠ **Gypsophila**

⊠ **Iris**

⊠ **Matthiola**

⊠ **Narcissus**

⊠ **Peonies**

⊠ **Ranunculus**

⊠ **Schizanthus**

⊠ **Stephanotis**

⊠ **Tulips**

⊠ **Zantedeschia**

OBJECTIVES:

After studying this chapter, the student should have learned the cultural needs of 18 minor cut flower crops in regard to:

■ propagation.

■ media and fertilization.

■ spacing and support.

■ light and temperature.

■ harvesting for market.

■ disease.

■ insects.

ALSTROEMERIA ⊠ *Alstroemeria species, Amaryllidaceae*

Alstroemeria is native to Chile and the Andes Mountains of South America, and it is found in Brazil, Colombia, Ecuador, and Peru. It is popular as a cut flower in Europe and grown to a lesser extent as a cut flower in the United States, although it has the potential to become popular. Alstroemeria is field grown in the western part of the United States and shipped east for sales. It is greenhouse grown in California, Colorado, Minnesota, and New York.

The plants are herbaceous perennials. The flowers are lilylike in appearance with five or six small blooms on each stem. The most common species grown is *A. pelegrina* with lilac-colored flowers spotted with red-purple. The plants grow to a height of 2 to 5 feet (0.6 to 1.5 m). *A. aurantica* has reddish flowers and is used as an outdoor cut flower (Figures 31-1 and 31-2).

Propagation

The plants are propagated by dividing the mother plants. They are cut into clumps of three to five fleshy rhizomes with crowns attached. If the divisions are planted in early August and September, they will bloom from March to May when grown at a night temperature of 50°F (10°C). Bloom can be hastened by lighting the plants for 13 hours each day with incandescent bulbs.

FIGURE 31-1 Alstroemeria growing in a ground bed and supported by wire and string nearing harvest.

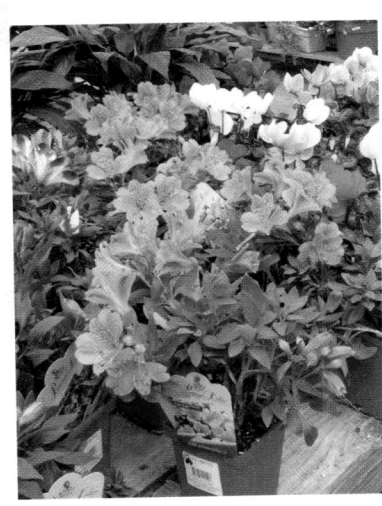

FIGURE 31-2 Alstroemeria blooms. New smaller selections are now available for pot production.

Note that this schedule is the same one used for cut mums. Under these conditions, October plantings will flower in late January and early February.

Media and Fertilization

The once preferred medium was a 1:1:1 mixture by volume of soil, peat moss, and perlite with a pH of 6.0 to 6.5. Most modern growers use a peat-lite medium in raised beds. The medium must have a good moisture-holding capacity. The plants are fertilized at every watering with a solution supplying approximately 175 ppm nitrogen, 18 ppm phosphorus, and 200 ppm potassium.

Support

The plants must be supported with wires and string or plastic mesh. If they are allowed to grow unsupported, they become tangled and are difficult to harvest. Stems are harvested by pulling directly up and out of the bed. They are not harvested by cutting.

Light and Temperature

Alstroemeria grow best in cool temperatures. Quality flowers are produced if the night temperature is 50°F (10°C). They may be grown through the summer if light shade is used on the greenhouse glass. Heat reduces the quality of the flowers.

Carrying Plants Over

After flowering, the foliage dies. The plants can be cut back to a height of 12 inches (30 cm). They are kept rather dry until fall. At that time, they start a second year of growth. If they are not cut back severely, they tend to grow too tall and produce vegetative growth, resulting in fewer flowers.

There is an alternate method of carrying the plants until the fall growing season begins. After the foliage dies, the roots are removed from the bench and stored at 60°F (15.5°C) until the fall planting.

FIGURE 31-3 Anthurium blooms packed and ready for shipment from Hawaii.

Problems

Cultural Disorders

Overcrowding causes the plants to remain vegetative. A low light intensity may have the same effect.

Diseases and Insects

The plants seem to be free of most diseases. Aphids and whiteflies are an occasional problem.

ANTHURIUM ⧆ *Araceae*

The production of *Anthurium* for cut flowers uses the same techniques described for pot plants (Figure 31-3).

CALENDULA (Pot Marigold) ⧆ *Calendula officinalis, Asteraceae*

Calendula is a cool temperature crop that was widely grown at one time. The increased costs of heating greenhouses may lead to a comeback for this crop. Calendula is native to southern Europe. Although it is called the pot marigold, calendulas are grown for cut flowers. They require temperatures of 45°F to 50°F (7°C to 10°C).

Propagation

The seeds are planted in August and are benched in September. Flowering begins in November. A September sowing produces blooms in late January and February. A January sowing will flower in the spring. When the spring temperatures begin to warm up, the plants start to produce single flowers, which have little sales value.

Media and Fertilizer

If the grower plans to rotate crops, calendula can follow a crop of chrysanthemums or snapdragons. Fertilizer is not required until the buds become visible. The plants are then fertilized with 100 ppm N, using a 20-10-20 analysis fertilizer.

A conventional peat-lite medium is used. Superphosphate is added, as is limestone to increase the pH to 6.5.

Ground beds are preferred to raised benches for calendula. In ground beds, the medium stays cooler and there is more available moisture, resulting in larger flowers with longer stems.

Spacing and Support

The plants are grown at a spacing of 12 inches by 12 inches (30 cm by 30 cm). They must be supported to ensure that the stems grow straight. The plants must be disbudded to obtain a single large flower.

Harvesting for Market

The plants are cut when the buds are about two-thirds open. The lower foliage on the stem is removed. The blooms are grouped in dozens. The flowers should be hardened in a refrigerator before they are graded and packed for market.

Problems

Cultural Disorders

Some cultivars produce more single type flowers than others. In addition, single flowers are caused by high temperatures.

A low light intensity leads to a low level of carbohydrate synthesis, resulting in weak stems and poor lasting life. The dollar return to a grower for calendulas is not large enough to permit an investment in HID lights to overcome this problem.

Diseases

Yellows is a disease that is spread by leafhoppers. For many years, this problem was considered to be caused by a virus. In 1967, Japanese researchers working on a mulberry dwarf disease similar to yellows discovered that there were no viruses present. Instead, they found mycoplasma organisms that cause certain diseases in birds, animals, and human beings. Plants infected with this organism produce yellowish-green stems and flowers. There is no control other than to screen the plants to keep leafhoppers away from them.

Powdery mildew may be a problem. Control consists of improved ventilation, low humidity, and the use of a fungicide applied at the rates recommended on the label.

Insects

Insect pests that attack this crop are aphids, cabbage loopers, garden flea hoppers, leafhoppers, leaf tier, mealybugs, sow bugs, tarnished plant bugs, and the variegated cutworm.

CALLISTEPHUS (China Aster) ▧
Callistephus chinensis, Asteraceae

Asters grown for cut flowers are native to China.

They rank second in importance to chrysanthemums as crops grown in cloth-covered houses. Their lasting life is almost equal to that of chrysanthemums.

Propagation

Asters are started from seed. They germinate in 8 to 10 days at a temperature of 70°F (21°C). If the medium contains soil, it should be sterilized. The peat-lite mixes are excellent for asters.

Two to 3 weeks after sowing, the seedlings are ready for transplanting to bands or small pots. The seedlings should not be allowed to become crowded. In addition, they cannot be held in the small containers for too long a period.

General Culture

Asters are grown and flowered as a year-round crop. In the winter and spring they are grown in greenhouses. In the summer they can be grown in a cloth-covered house.

The medium must be pasteurized to ensure that it is free of disease organisms. A peat-lite medium is used. Limestone and superphosphate are added at the standard rates to this mixture. Asters require good drainage. In a cloth-covered house, peat moss and perlite are worked into the soil to improve the drainage and aeration.

Asters require a medium strength fertilization program. A preplanting application is made of 10-10-10 analysis fertilizer at the rate of 2 pounds (0.9 kg) to 100 square feet (9.3 m²) of area. This is mixed thoroughly into the soil. Three weeks after planting, 20-10-20 analysis fertilizer is applied at the rate of 2 pounds (0.9 kg) in 100 gallons (378.5 l) of water. Normally, additional fertilization is not required.

Light and Temperature

Asters respond to both light and temperature in forming flower buds. The plants form flower buds only when the main stem elongates or grows vertically. Stem elongation takes place when the daylight period is longer than 15 hours or the temperature is greater than 70°F (21°C). It is this knowledge of how the plants respond to environmental conditions that permits year-round production. Growers use less expensive electricity compared to the cost of heating fuel to light the plants with incandescent lighting similar to the program used on chrysanthemums.

Asters are much more sensitive to light than chrysanthemums. A level of 0.3 fc (3.3 lux) of light stimulates growth of the axis. Normally, 2 fc (22 lux) are given for 4 to 6 hours. This amount of light is obtained using 25- to 40-watt incandescent bulbs spaced 6 feet (1.8 m) apart, 2.5 feet (0.8 m) above the tops of the plants. Some growers use only 5 fc (54 lux) of light in the darkest part of the bench. Other growers use chrysanthemum lighting programs.

Lighting is applied to divide the dark period into two equal parts. Lights are used before and after midnight for equal lengths of time. From August 15 until November 30, and from February 1 to May 10, lights are turned on for 4 hours each night from 10 P.M. until 2 A.M. Between December 1 and January 31, 5 hours of light are used. From May 10 until August 15, lights are not required because the natural days are long enough to cause stem elongation. Lights are used from seed sowing until the seedlings are planted in the cloth-covered house. For greenhouse crops, the lights are used until the plants are 2 feet (0.6 m) tall.

Cloth House Culture

Seeds for asters to be grown in cloth houses are sown around April 15. They are lighted and held in pots or bands after transplanting until May 20. They may be planted earlier if the last frost date is before May 20. The soil in the cloth house should be prepared so that it is ready to receive the plants. They are planted at a spacing of 12 inches by 12 inches (30 cm by 30 cm) in beds 4 feet (1.2 m) wide.

To minimize insect populations and aster yellows, the plants are sprayed every week with an all-purpose insecticide. A fungicide may be used also if the weather is rainy and damp. *Botrytis* petal blight may be a problem. Irrigation water should be available if needed during the growing season.

The plants must be supported with plastic mesh or wires and string. Normally, one layer of support is sufficient.

The plants are not pinched to produce branching because they branch naturally. The side shoots are disbudded to promote the maximum growth of the flower. If a very large number of side shoots develop, some of the lower branches may be removed. This practice prevents the grower from overcropping the plant, resulting in short stems and small flowers.

Greenhouse Year-Round Culture

Researchers at Ohio State University have developed a year-round schedule for flowering asters. This schedule is given in Table 31-1. The night temperature is maintained at 55°F (13°C). The plants are lighted from the time of seed sowing until they are 24 inches (60 cm) tall.

The plants are benched at a spacing of 8 inches by 8 inches (20 cm by 20 cm). The fertilizer program described for the cloth house crop is also used on the greenhouse crop.

Insect problems in the greenhouse are also not as severe as those in the cloth house. However, the plants still must be sprayed. Spraying every 10 days

Table 31-1

Schedule for Production of Year-Round Asters at a Minimum Night Temperature of 55°F (13°C) at 42° North Latitude			
Approximate Flowering	Sow Seed	Pot Seedlings	Bench Plants
January	July 15	August 10	September 15
February	August 15	September 10	November 1
March	September 10	October 1	December 1
April	October 20	November 15	January 1
May	December 20	January 10	March 1
June	February 1	February 20	April 1
July	March 15	April 1	May 15
August	April 20	May 10	June 15
September	May 20	June 10	July 15
October	June 10	July 1	August 1
November	June 20	July 15	August 20
December	July 10	August 1	September 1

should be satisfactory. Aphids and red spider mites may be troublesome in the late spring and early summer.

Harvesting for Market

The flowers are cut when they are three-quarters open. There are no standard grades for sorting the flowers. Growers normally sort them by flower size and stem length. They are bunched in dozens and hardened in the refrigerator before shipment. The lasting life of asters is equal to that of chrysanthemums.

Problems

Cultural Disorders

The disease aster yellows causes the plants to turn yellow-green and the flowers to remain green. Because this disease is transmitted by mycoplasma organisms carried by leafhoppers, the only control is to spray regularly to reduce the leafhopper population. The plants are screened from the insects by means of a cloth house.

Disease

In addition to aster yellows, *Fusarium* wilt attacks the plants. The major control methods include soil sterilization and strict sanitation procedures throughout the greenhouse. Wilt-resistant cultivars should be planted.

The plants may be infected with a root rot caused by *Phytophthora cryptogea*. Control is obtained by sterilizing the medium.

Septoria callistephi, Ascochyta asteris, and *Botrytis* all cause leaf spots or tip dieback. Fungicide sprays may control these organisms.

Insects

Asters may be infested by aphids, cyclamen mite, leafhoppers, leafrollers, mealybugs, red spider mites, and thrips.

CENTAUREA (Bachelor's Button) ▨
Centaurea cyanus, Asteraceae

Greenhouse growers often produce several species of *Centaurea. C. cineraria* and *C. gyninocarpa*, are used as bedding plants and in combination plantings where a silvery leaf is wanted for accent. Both species are called dusty miller.

C. cyanus is the species grown for cut flowers. Common names for this plant include blue bottle, bluet, bachelor's button, and cornflower. *C. cyanus* may be grown in the field during the spring and summer or in the greenhouse for flowering in the late winter and early spring. The cut flowers are a popular addition to bouquets of spring flowers.

Propagation

The plants are propagated from seed using standard propagation methods. They are grown as annuals. Seeds planted in September for greenhouse flowering from March to May need a long-day treatment starting in January. Seed sown in February is field planted in April and May for blooms from July through September.

Media and Fertilization

Quality crops are produced in a general peat-lite mixture. The fertilization program commonly used calls for 20-10-20 analysis fertilizer to be applied at a concentration of 150 ppm N at each watering. The plants should be kept moist but not overly wet.

Support

The plants require the same type of support as chrysanthemums. Pinching is not required. Plants grown in the field do not require support unless the wind conditions are extreme.

Light and Temperature

Long days hasten flowering, but they should not be started too early. After germination, the seedlings are planted and grown under natural day conditions. As a result numerous branches form at the base of the plant. Once these side shoots are 5 to 6 inches (12.5 to 15 cm) long, the long-day treatment is started. The plants will then go on to flower. The lighting schedule used for chrysanthemums can be used with *Centaurea*. Lights are started in October and used through March. Earlier flowering is achieved if full light intensity is required.

The minimum night temperature is 55°F (13°C). The day temperature may be 65°F (18.5°C) or slightly higher. Stockier growth is obtained at cooler temperatures.

Harvesting Market

The stems are cut when four or five flowers are open fully. They are sorted for uniformity and are tied in bunches of 25 stems.

Problems

Cultural Problems

Centaurea have few cultural problems.

Diseases

Aster yellows is a serious threat to *Centaurea*. This crop should not be planted in areas where asters were grown. The soil must be sterilized.

Basal rot is due to common soil-borne disease organisms such as *Rhizoctonia* and *Pythium*.

Leaf spots are caused by *Botrytis cinerea*.

Fusarium wilt is caused by *Fusarium oxysporum callistephi*, which infects the plant through the roots and enters the vascular tissue.

Verticillium wilt is another soil-borne disease. *Centaurea* should not be planted in areas previously planted to chrysanthemums, tomatoes, potatoes, and eggplants—ALL of these crops are infected by the disease.

Insects

Troublesome insects include aphids, leafrollers, red spider mites, and thrips.

DAHLIAS ✤ *Dahlia pinnata, Asteraceae*

Dahlias are grown as a cut flower crop for the summer and fall (Figure 31-4). Dwarf cultivars are sold in the spring for bedding use and as potted plants.

The plants are native to the high altitudes of Mexico. Therefore, the areas of greatest production have similar climates, such as coastal plains and higher altitudes where the conditions include bright sunshine, warm days, and cool nights.

Propagation

Dahlias for cut flowers are grown from terminal cuttings or divisions of the parent clumps. Cuttings are obtained from tuberous roots that start growth in January and February. The clumps of roots are planted in a moist peat-lite medium. The night temperature is held at 60°F (15.5°C). When the new shoots are 8 to 10 inches (20 to 25 cm) long, 3- to 4-inch (7.5- to 10-cm) long terminal cuttings are made. These cuttings are rooted at 65°F (18.5°C) using bottom heat. Cuttings made in March produce flowering plants in May.

The clumps are divided so that part of a stem section is included with each tuberous root. Unlike potatoes, which have true tubers, dahlias do not have adventitious buds called "eyes" from which new growth starts. The "eyes" of the dahlia are located at the base of the stem. The tuberous root serves as a source of stored food. If a stem section and eye are not attached to the tuber, a plant cannot develop.

Transplanting

Dahlias grow best in deep, sandy soils containing some organic matter. They do not grow well in poorly drained, heavy clay soils. Before planting, a 10-10-10 analysis fertilizer is tilled into the soil at the rate of 2 pounds (0.9 kg) per 100 square feet (9.3 m²) of area, or about 800 pounds to an acre (898 kg/ha). The plants may be given a side dressing of 5-10-10 fertilizer when flowering starts. No other fertilization is needed.

The tuberous roots are planted 4 to 5 inches (10 to 12.5 cm) below the surface of the soil. This depth promotes the growth of stem roots that help to support the plant. If cuttings are transferred from pots to the field, they are planted in the field at the same depth of planting as they were in pots.

FIGURE 31-4 Dahlias grown in a ground bed (lilies are planted in the row to the left)

Plants grown for cut flowers must be staked. The stake should be placed at the time of planting. If it is inserted next to the plant after root growth starts, the roots may be damaged. The plants are tied to the stake as they grow in height.

The plants are spaced 2 feet (0.6 m) apart in a row with rows 3 feet (0.9 m) apart. This spacing allows the grower to cultivate between the rows to reduce weeds. Black plastic mulch also prevents weed growth and helps to retain moisture in the soil. The black plastic can be placed on the ground and the seedlings planted through it.

Pinching and Disbudding

To promote branching, the stems are given a soft pinch when at least six nodes have formed. Pinching hastens the growth of the side shoots. As soon as buds appear, the side buds are removed. Therefore, only the main bud remains on the plant. Removing the side buds allows all of the energy of the plant to go into the development of one high-quality bloom.

Harvesting for Market

Dahlias do not open very much after they are cut. Therefore, they are harvested when the center florets of the blooms are almost fully open (Figure 31-5). Pompon types are cut with 12- to 18-inch (30- to 45-cm) long stems. The large cactus flowered types are cut with stems from 1 to 3 feet (0.3 to 0.9 m) long. The smaller flowers are sold in bunches of 12 blooms. The large flowers are sold as separate blooms.

To prevent the flowers from wilting, the ends of the stems are placed in boiling water for 1 minute. This treatment seals in the latex that otherwise plugs the water-conducting xylem vessels. The flowers last for only 4 to 5 days after cutting.

Storage of Tuberous Roots

As soon as the tops are killed by frost, they are cut off close to the crown. The tubers are removed from the soil, and all excess soil is removed. The tubers are air dried for 2 or 3 days at room temperature. Then they are stored at 35°F to 50°F (1.5°C to 10°C) in a relative humidity of 80 percent. If the humidity is too low, the tubers are stored in dry sand, peat moss, or vermiculite. This material,

FIGURE 31-5 Dahlia blooms ready for harvest.

which acts as insulation, and a temperature below 50°F (10°C) prevent the tubers from shriveling and drying.

Problems

Cultural Disorders

Stunted plants with many side branches may have been "stung" in the growing tip by tarnished plant bugs, leafhoppers, or thrips. The feeding of these insects injures the meristematic cells. Stunt disease is another cause of this problem. The disease is caused by a virus that cannot be controlled. Infected plants must be destroyed.

Diseases

Other viruses that infect the plants include mosaic, oak leaf, ring spot, and yellow ring spot.

Botrytis can cause rotting in storage. *Fusarium* may develop in storage because it is carried in on the tuberous roots.

Insects

Leafhoppers, spider mites, tarnished plant bugs, and thrips attack the plants.

FREESIAS ⊠ *Hybrida, Iridaceae*

The culture of Freesias is discussed in Chapter 25.

GERBERA (Transvaal Daisy) ⊠
Gerbera jamesonii, Asteraceae

The common name of this plant is rarely used. It is normally called gerbera. The plants provide excellent cut flowers in bright and pastel colors (Figure 31-6). The plants are grown in northern greenhouses for winter and spring cut flowers.

Propagation

Gerberas are propagated by seeds, divisions, and tissue culture techniques. California researchers first developed clonal plants. A number of cultivars that

FIGURE 31-6 Gerbera flowers ready for harvest.

produce single-color flowers are available from a commercial source in California.

Seeds are planted in January to produce flowering crops in 11 to 12 months. A crop is obtained in less time by planting divisions in the late spring. These plants begin to flower in the fall. The plants should be divided once a year to prevent the clumps from becoming too large, causing a decrease in flower quality.

Media and Fertilizers

A light sandy soil with a pH of 6.0 to 6.5 is preferred. The plants do not grow well in heavy soils that stay too moist. They also grow well in hydroponic systems called nutrient film technique (Figures 31-7A and 31-7B).

An adequate fertility level is maintained by adding a 20-10-20 analysis fertilizer at a concentration of 100 ppm.

Light and Temperature

Quality blooms require a high light intensity and cool night temperatures. The recommended night temperature is 50°F to 55°F (10°C to 13°C), and the day temperature is 10°F (8°C) higher.

Harvesting for Market

The flowers are cut when they are open fully. The stems show a negative geotropic effect similar to snapdragons if they are placed in a horizontal position. Therefore, they should be shipped upright but rarely are because of the problems of handling the flowers this way.

The flowers are sold in bunches of a dozen blooms. They have a relatively long lasting life.

Problems

Diseases

A number of diseases cause the leaves to wilt suddenly and turn a deep violet-red color. *Phytophthora cryptogea* causes root rot. There is no control except to sterilize the growing medium and propagate healthy plants only.

FIGURE 31-7A Gerbera is easily grown in hydroponic troughs.

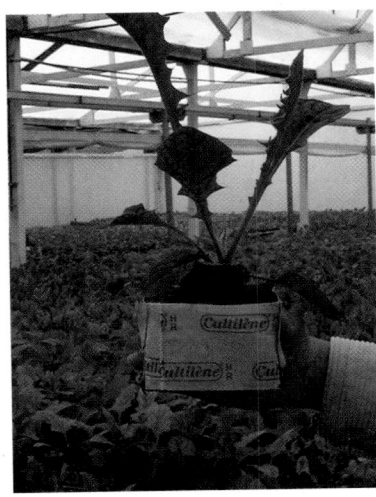

FIGURE 31-7B Prior to placing Gerbera plants into the hydroponic trough, they are established in rock wool cubes.

Mildew caused by *Erysiphe polygoni* stunts the plants. The disease can be controlled with the appropriate chemical.

Insects

Cyclamen mites, leafminers, leafrollers, mealybugs, thrips, and whitefly all attack the plant.

GYPSOPHILA (Baby's Breath) ※
Gypsophila species, Caryophyllaceae

Two species of *Gypsophila* are grown by florists. *G. elegans* is known as the annual baby's breath, and *G. paniculata* is the perennial baby's breath. Retail growers often produce *Gypsophila* as a summer crop for use as a filler in flower arrangements. In milder climates, *G. elegans* may be grown outdoors for 12 months a year. Most of the *G. paniculata* grown is used for dried flowers. *G. elegans* is grown from seed planted in sandy soil in rows 6 inches (15 cm) apart. The plants are thinned to stand 1 inch (2.5 cm) apart in the row.

G. elegans flowers under long-day conditions. Continuous production can be obtained by using a chrysanthemum lighting program from September 1 to April 1. A new crop is seeded every 7 to 10 days. The night temperature is held in the range of 45°F to 50°F (7°C to 10°C). When the flowers are fully developed, the plants are pulled from the soil. The roots are cut off, and the plants are tied in bunches of six to eight stems.

G. paniculata is propagated by root grafting desirable cultivars to seedlings during the early summer. The majority of the perennial baby's breath grown comes from the high-lime soils in the western United States. The heavy clay soils in the east cause the plants to die out after 2 or 3 years.

The plants are spaced 3 to 4 feet (0.9 to 1.2 m) apart to give them maximum light and ventilation. Nitrogen levels are kept low to reduce vegetative growth. A 5-10-10 analysis fertilizer is applied in May at the rate of 2 to 3 pounds (0.9 to 1.35 kg) to 100 square feet (9.3 m²). Additional fertilizer is not required.

The stem is cut when most of the flowers are open. The branches are hung up to dry in a cool drying shed. Forced hot air drying desiccates the plants too much. After drying, the plants may be dyed different colors.

Problems

The perennial baby's breath is subject to the aster yellows disease. The annual type is subject to damping off of the seedlings. Both diseases can be reduced by sterilizing the growing medium.

IRIS ※ *Iridaceae*

The production of iris as a cut flower was described in Chapter 23.

MATTHIOLA (Florists' Stock) ※
Matthiola incana, Brassicaceae

Florists' stock was grown as a cut flower crop in greenhouses in the eastern United States until it was learned that outdoor production in California produces a better flower at less cost. A few eastern retail growers produce crops for local sales.

Stocks are true biennials in that they must have a cold period called vernalization to initiate flowers. Therefore, they are grown for blooming in the winter and spring.

Propagation

Stocks are grown from seed that is planted directly in the production bench. Three seeds are placed in each hole. A marking board can be made by nailing 0.5-inch (12-mm) long by 0.5-inch (12-mm) thick wooden dowels 3 inches (7.5 cm) apart on a board that is the width of the bench. This board is pressed into the soil to mark an entire row at one time. The rows are spaced 6 inches (15 cm) apart.

After the seed is planted in the hole, the hole is filled with sand or coarse size vermiculite. These materials provide maximum aeration of the seed for quick germination.

Planting directly in the production bench may lead to the development of more single flowers. The flowers sold in the trade are the double type. By planting three seeds to a hole, the grower can increase the chance for doubles slightly by selective thinning.

Thinning is delayed until the plants have six leaves. At this stage, plants that will produce double flowers develop a notched or **lobed** leaf. Single-flowered plants are weaker in growth. The grower pulls the weaker seedling and those without notched leaves, leaving one plant per hole.

Media and Fertilization

Stocks require a high moisture level and good aeration. The once preferred medium was 2:1:1 mixture by volume of soil, peat moss, and perlite with a pH of 6.0 to 6.5. Peat-lite media with similar water-holding capacities are also used. Superphosphate is added at the standard rate. A preplanting application of potassium chloride is made at the rate of 0.25 pound (112 g) to 100 square feet (9.3 m^2) of area. Stocks are especially sensitive to potassium deficiency. It appears that a large amount of potassium is translocated into the developing flower. In media containing a marginal supply of potassium, the loss of the lower leaves is very severe when the plants flower.

During active growth, the plants are fertilized at every watering with a 20-5-30 analysis fertilizer applied at the rate of 13.3 ounces (377 g) in 100 gallons (378.5 l) of water. This amount supplies 200 ppm of nitrogen and 256 ppm of potassium at each irrigation.

Light and Temperature

The plants are grown under full light intensity at all times. As soon as the crop is planted, all shading compound must be removed from the greenhouse glass.

Flower buds are formed after the plants develop 10 or more mature leaves at temperatures below 60°F (15.5°C). Buds are initiated after 21 days of cool temperatures in the range of 50°F to 60°F (10°C to 13°C). Before and after this period the temperature is held at 60°F (15.5°C) to hasten growth. If seedlings are started below 60°F (15.5°C), they may remain dwarfed.

Problems

Cultural Disorders

Blind plants result if the temperature exceeds 60°F (15.5°C) during the critical bud formation process.

Short plants are caused by temperatures that are too low when the plants are seedlings.

Diseases

Rhizoctonia causes stem rot at the soil line. Gray foliage and wilting are other symptoms of *Rhizoctonia*.

Bacterial blight caused by *Phytomonas incanae* causes a sudden wilting and collapse of one or both cotyledons. To destroy this seed-borne bacterium, the seeds are heat treated in 129°F (65°C) water for 10 minutes followed by a quick cooling.

Botrytis cinerea causes the common gray mold. To control this disease, the humidity is decreased by providing better control of the heat and ventilation.

Insects

Aphids, leafrollers, and thrips are troublesome insects on stocks.

NARCISSUS ▨ *Amaryllidaceae*

The production of narcissus as a cut flower crop is described in Chapter 25.

PEONIES ▨ *Paeonia lactiflora, Ranunculaceae (Paeoniaceae)*

Peonies are one of the oldest flowers in cultivation. They were grown in China before AD 900. The peony is native to Siberia, China, and Japan.

The plants are hardy perennials that are grown outdoors in areas north of 35° north latitude. Flower buds develop after the plants experience temperatures below freezing for 2 or 3 months. The plants survive severe winters with very little protection.

Two types of peonies are grown: tree peonies and herbaceous peonies. Tree peonies have tall woody stems. Herbaceous, or garden, peonies are bushy plants that grow 2 to 4 feet (0.6 to 1.2 m) tall. Garden peonies are grown for cut flowers. There are five types of garden peonies, grouped according to their petal shape.

1. Single or Chinese peonies have one row of broad petals surrounding a cluster of yellow, pollen-bearing stamens. On the other flower types, the central stamens are replaced by petals.
2. Semidouble peonies have broad central petals.
3. Double peonies have central petals that are as wide as the outer petals.
4. Japanese peonies have long, thin central petals.
5. Anemone peonies have broad central petals.

Other minor types of petal shapes are recognized, but they are not of great importance in the cut flower market.

Propagation

Peonies are grown from tubers and are propagated by dividing the clumps of tubers. New growth develops from the "eyes" or buds of the tuber. To produce flowering plants the second year after planting, a single tuber should have not less than three eyes.

Peonies develop a straight, central taproot plus many short, thin roots. The taproot often extends 15 inches (37.5 cm) into the soil.

Media and Fertilization

The long taproot requires a thorough preparation of the growing medium. Peonies grow well in many different types of soils if they are deep and well drained. Many acres are devoted to peony cultivation in Illinois because of the excellent soil conditions.

The tubers are planted in September and October. The soil is thoroughly prepared to a depth of 18 inches (45 cm). Then it is mounded into raised beds to improve drainage away from the crowns. Well-rotted manure, peat moss, leaf mold, or other organic matter is mixed into the soil. One ounce (28.35 g) of 10-10-10 analysis fertilizer may be mixed thoroughly into the soil used to backfill the hole.

The tubers are planted with the topmost eye 2 inches (5 cm) below the soil surface. Deeper planting is not recommended. The hole is backfilled, and the soil is tamped firmly around the roots to eliminate air pockets. A thorough watering helps to firm the soil and remove any remaining air.

The plants are usually fertilized once in the spring. A handful of 10-10-10 analysis fertilizer is sprinkled in a circle around the plants. If the fertilizer is sprinkled too close to the crown, root damage may result.

Mulching the Plants

As soon as the ground freezes to a depth of 1 to 2 inches (2.5 to 5 cm), a mulch of 1 to 2 inches (2.5 to 5 cm) of straw or peat moss is applied. The winter mulch is used to prevent alternate freezing and thawing cycles that force the roots out of the ground.

In spring after the last frost, the winter mulch is removed. It is burned to destroy any insects or disease organisms that may have wintered over in it. A summer mulch of straw or peat moss is applied to a depth of 1 to 2 inches (2.5 to 5 cm). The summer mulch keeps the ground cool and conserves moisture. It is removed and burned when the winter mulch is applied.

Light

Peonies require full light intensity. They do not grow well in shaded locations. They must be protected from strong winds, which break the stems. If necessary, a windbreak may be used to protect the plants.

Harvesting for Market

The plants are harvested in the bud stage because the flowers open naturally after cutting. Early morning cutting is preferred. However, during warm weather, it may be necessary to harvest the flowers throughout the day and night to keep them from opening too far. The flowers are moved quickly into cold storage to keep the buds tight.

The stems are cut so that three leaves remain on the plant. The removal of too many leaves reduces the plant area where photosynthesis can take place. As a result there is a gradual decline in the vigor of the plant.

There are no formal standards for grading the flowers. Growers who sell cut peonies to wholesale markets ship only their best flowers and usually ask one price.

The wholesale production of outdoor peonies is popular with growers who cater to farmer's markets. However, the number of growers is dwindling.

Problems

Cultural Disorders

Plants that fail to flower may not have been exposed to sufficient cold during the winter. If the plants were divided the previous year, they may not develop enough in a year to reach a flowering condition. They should bloom after another year of growth.

Diseases

Bud blasting is due to *Botrytis paeoniae*. The plants are sprayed regularly with a fungicide to control this disease.

Phytophthora blight affects the plants in the same manner as *Botrytis*. The control methods are the same.

Stem rot is due to *Scierotinia sclerotiorum*, which causes sudden wilting and death of the stem. When the stem is cut lengthwise, small, black sclerotia are revealed. These are the fruiting bodies of the disease. Infested stems are removed carefully and burned to destroy the sclerotia.

Verticillium alboatrum causes the plants to wilt. The disease is contained in the roots and tubers of the plants. Infected plants should be destroyed. If the soil cannot be sterilized, a new planting area must be used because the disease organisms live in the soil.

Several viruses affect peonies. Mosaic is the most common virus. Crown elongation and leaf curl are also caused by viruses. The only known control is to remove and burn the infested plants.

Insects

The plants may be attacked by flower thrips, rose chafers, and stalk borers. However, these pests are not serious. Ants are found in large numbers on peonies when they are coming into bloom. The ants are gathering the sweet sap from the buds and do not damage the plants.

RANUNCULUS ▨ *Ranunculus asiaticus, Ranunculaceae*

The cultural methods for *Ranunculus* are the same as those described for Anemones.

SCHIZANTHUS (Butterfly Flower) ▨
Schizanthus species, Solanaceae

Schizanthus is called the poor man's orchid because the small flowers resemble orchids. The species commonly grown are native to Chile and are all annuals. Popular hybrids include *S. pinnatus*, *S. wisetonensis*, and *S. Grahamii*.

Propagation

The seeds are sown in September. The plants are grown in shallow benches with a maximum depth of 4 inches (10 cm), or they are grown in flats. The shallow depth forces the roots to grow into a solid mass. If the plants are grown in pots, they must be allowed to become potbound. The plants do not flower if a root mass is not formed.

They are spaced 6 to 8 inches (15 to 20 cm) apart in all directions. They are given one pinch to make them branch.

The plants start to bloom in February if they are given 4 hours of light beginning December 1. They are grown at a night temperature of 50°F (10°C).

The plants are cut when they are in full bloom. Generally, the flowers are not shipped because they shatter easily.

Problems

Insects

The plants may be infested with aphids, whiteflies, and red spider mites. Treatment with certain insecticides may cause flower drop.

STEPHANOTIS ☒ *Stephanotis floribunda, Asclepiadaceae*

Stephanotis is a sturdy, woody vine that produces an abundance of white star-shaped flowers having a strong fragrance. It is highly prized as a flower for wedding bouquets (Figure 31-8).

Propagation

The plants can be grown from seed, but the most common means of propagation is half-matured woody cuttings taken from December to March. The cuttings are from 4 to 6 inches (4 to 6 cm) long. They are treated with a rooting hormone and then are stuck into the propagation medium. The best rooting conditions include bottom heat, an air temperature of 70°F (21°C), and high humidity. If low-pressure mist is not available, a grafting case or a plastic tent is used.

FIGURE 31-8 Stephanotis flower clusters, which are cut as a unit.

Medium Light and Temperature

The plants are transplanted to large tubs or boxes after they are well rooted. They are grown for several years in these containers in the same soil. The preferred medium is a 3:1:1 or 2:1:1 mixture by volume of a good loam soil, peat moss, and perlite. This mix is similar to the medium used for growing roses. Stephanotis are often grown in rose houses because they need the same warm temperature, 60°F (15.5°C), and high humidity provided for roses.

The containers are placed where they are easy to water. The plants are trained over a trellis or an arbor type support. They normally flower from April through September. To bring the plants into bloom in late February and March, they are lighted with floodlights for 4 hours a night in December and January.

Fertilization

The plants are fertilized during periods of active growth using one-half the amount applied on roses.

Harvesting for Market

The flowers are cut as clusters. Fifty flowers are packed in a plastic bag. The bags are placed in cold storage at 33°F to 36°F (0.5°C to 2°C). The flower clusters are separated by the retailer as they are used.

Problems

Insects

Scale insects and mealybugs are the most troublesome insects on stephanotis. Few other insects attack the plants.

TULIPS ✖ *Liliaceae*

The production of tulips as a cut flower crop is discussed in Chapter 26 (Figure 31-9).

ZANTEDESCHIA (Calla Lily) ✖ *Araceae*

The production of the calla lily as a cut flower crop is discussed in Chapter 26.

FIGURE 31-9 Cut tulips ready for sale.

ACHIEVEMENT REVIEW

Select the best answer or answers to complete each statement. List the appropriate letter(s).

1 *Alstroemeria* is a popular cut flower in:
   a. California.
   b. China.
   c. Europe.
   d. the United States.

2 Propagation of *Alstroemeria* is by:
   a. seeds.
   b. terminal cuttings.
   c. division of crowns.
   d. root graft.

3 Anemones grow best if a 20-10-20 analysis fertilizer is used at a concentration of:
   a. 100 ppm.
   b. 150 ppm.
   c. 175 ppm.
   d. 200 ppm.

4 If Calendulas follow a crop of snapdragons or chrysanthemums, they must be fertilized:
   a. when they are first planted.
   b. when the buds first appear.
   c. when the flower color is visible.
   d. never.

5 Weak stems and the poor lasting life of Calendulas are due to:
   a. a lack of moisture.
   b. a lack of carbohydrates.
   c. a lack of proteins.
   d. an excessive light intensity.

6 Aster yellows is a disease caused by:
   a. bacteria.
   b. fungi.
   c. mycoplasma.
   d. viruses.

7 Asters respond to a light intensity as little as:
   a. 0.01 fc (0.11 lux).
   b. 0.05 fc (0.55 lux).
   c. 0.30 fc (3.3 lux).
   d. 2.00 fc (22 lux).

8 Asters will flower if they are given additional lighting until:
   a. the first true leaves are formed.
   b. flower buds are visible.
   c. they are 24 inches tall.
   d. they are cut off.

9 Dahlias grown for cut flowers are propagated from:
   a. seeds.
   b. terminal cuttings.
   c. divisions.
   d. root grafts.

10 Dahlias grown for cut flowers are harvested when the flower is:
   a. at the loose bud stage.
   b. half-open.
   c. two-thirds open.
   d. almost fully open.

11 The ends of Dahlia stems must be seared in boiling water to:
   a. kill disease.
   b. kill insects.
   c. seal their latex vessels.
   d. seal the xylem vessels.

12 Florists' stocks are widely grown as a field crop in:
   a. California.
   b. Florida.
   c. New York.
   d. Pennsylvania.

13 Seedlings of *Matthiola incana* that produce double flowers:
   a. are stronger growers than those producing single flowers.
   b. have a smooth leaf margin.
   c. have a notched leaf margin.
   d. do not differ from those that produce singles.

14 Stocks require large amounts of:
   a. nitrogen.
   b. phosphorus.
   c. potassium.
   d. boron.

15 Peonies come into flower following:
   a. 10 weeks of long days.
   b. 2 months of temperatures below freezing.
   c. 10 weeks of short days.
   d. all of the above.

16 Stephanotis flowering can be increased by:
   a. giving a cold treatment for 10 days.
   b. lighting the crop for 4 hours a day in the winter.
   c. giving a short-day treatment.
   d. growing the crop dry for 2 months.

Postharvest Handling and Marketing of Potted Plants and Cut Flowers

CHAPTER

32

Harvesting and Storage Methods

OBJECTIVES:

After studying this chapter, the student should have learned:

■ the causes of cut flower death.

■ the need for proper growing methods in relation to the lasting life of cut flowers.

■ three methods of cold storage used to hold cut flowers.

■ methods used to prolong the life of cut flowers.

Growers produce flower crops to sell in the market at a price that will give the grower a profit on the investment made. Flowers are a very perishable product. They are fragile and easily bruised by mishandling. Flowers have a relatively short life compared to vegetable crops such as potatoes. They cannot be canned or frozen and then reprocessed for use later. Some flowers are dried and used for their aesthetic value.

The majority of cut flowers and potted plants are offered for sale as fresh products. As such they have a very short useful life. It is estimated that the loss in fresh crops as a result of improper handling is equal to 20 percent of the total farm value of the crop. In 2005, the value of the shrinkage loss was estimated to be over $79 million. Any steps that can be taken to reduce this waste will mean a more satisfied consumer and a lower dollar loss to the industry.

Factors That Influence Cut Flower Life

When the flower is removed from the plant, it loses its life support system. The roots of the plant can no longer supply water and nutrients to the leaves and flowers. The life processes in the harvested blooms continue, however. Transpiration takes place, and water is lost through the leaves.

734

Food reserves are used up by respiration. The lasting life of cut flowers is greatly influenced by these two important plant growth processes.

The question is often asked, "Why do flowers die?" Those that remain on the plant follow a natural course of senescence. Senescence is the change from maturity to death. After the flower blooms, it is usually pollinated by insects that transfer the pollen from the stamens to the stigma. The brightly colored petals attract the insects and then fade and wilt. At the same time, the seed parts of the flower develop to ensure another generation of plants.

The basic reasons why cut flowers die are described in the following sections.

1. *Food reserves are used up.* A cut flower is still a living organism. Transpiration and respiration continue, although photosynthesis does not occur to replace the food used in these processes. Both processes require energy that is supplied by the food reserves in the plant or by nutrients added to the water in which the cut stems are placed. This is the major reason for the use of cut floral foods (nutrients). The flower dies when all of the food reserves are depleted or the flower can no longer use the nutrients supplied from outside.

2. *When there is too little water uptake, the flower dries out.* Movement of water through the stem to the leaves and the flowers is very important in prolonging the life of the flowers. If turgidity is not maintained, the plant wilts and dies regardless of the amount of food reserves present. A great deal of research has concentrated on the regulation of water uptake and the movement of water through plants.

3. *The flower continues to develop toward maturity.* Although cut flowers are not pollinated, they continue to develop toward maturity. When the flower is fully open and starts to decline, its useful life is over. The decline of carnations, chrysanthemums, and similar flowers is not as dramatic as that of roses and snapdragons.

4. *Diseases kill flowers.* Diseases may start when the flowers are in transit or in storage. Botrytis is a common and troublesome disease.

5. *Ethylene gas, other gases, or fume injury affect flower life.* Air pollution shortens the life of flowers. Regardless of where the pollution occurs—in the greenhouse, in the shipping environment, in storage, or in the home—flowers are affected. Ethylene is produced by fruits, vegetables, and flowers themselves. The grower must ensure that flowers are not shipped in the same container with fruits and vegetables. Calla lilies produce large amounts of ethylene. When snapdragon and calceolaria flowers are exposed to small amounts of ethylene gas, they quickly drop from the stem. Yet, snapdragon stems generate ethylene gas if their foliage is submerged in water.

6. *Flowers lose color as they age.* High temperatures hasten the loss of flower color. Fading also results from the reduction of stored food reserves in the plant. Some flowers, such as certain red roses, acquire an objectionable blue color as they age, due to a change in pH within the petals.

Production and Harvesting Methods

A chrysanthemum blooms 3 to 4 months or more after it is planted. The cut flower normally lasts 3 to 4 weeks if it is properly handled. It is estimated that one-third of the postharvest life of a cut flower is influenced by how it is grown.

The remaining two-thirds of its cut life are influenced by how it is handled after it is cut.

Preharvest Handling

There is a direct relationship between the lasting life of a cut flower and the amount of stored carbohydrates and sugars in the plant tissues. A high sugar content means a long lasting life. To obtain maximum cut flower life, the grower must ensure that certain cultural practices are carried out. These practices include the following:

- Fertilizing for quality yield and growth.
- Supplying adequate water so that the plants are not under stress because of lack of water. Automatic watering systems are an excellent investment in meeting the moisture needs of crops.
- Maintaining accurate temperature control at all stages of growth.
- Keeping the greenhouse glass clean at all times so that the plants can get the maximum light available.
- Controlling the environmental and cultural factors to produce flowers and plants with high carbohydrate content.
- Harvesting and shipping the flowers at the proper stage of development so that the customer receives the maximum value from them.

Harvesting

For each of the cut flowers described in the text, the proper stage for harvesting the flowers is discussed. However, some general comments about harvesting can be made.

Before the crop is ready for harvest, the grower should plan how the crop will be handled. The harvest should be organized so that the flowers are moved from the growing area to the packing area in the least time.

For chrysanthemums or other crops grown in the field, carts should be ready at the field with buckets of fresh, clean hydrating solution so that the freshly cut blooms can be placed immediately in the solution (Figure 32-1). If the weather is bright and sunny, the cut blooms must be shaded to prevent sun scald.

FIGURE 32-1 Farm wagons are used by many growers in Colombia to transport freshly cut flowers from the field or greenhouse to the packing shed. Buckets filled with water and flower food are kept under shade.

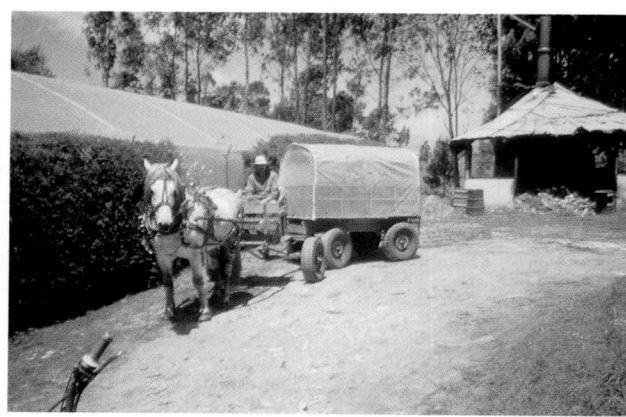

In the rose range, each house must contain buckets of fresh, clean, 100°F (38°C) hydrating solution. The freshly cut flowers are placed in the solution immediately. If the roses are going into dry storage at 31°F (−0.5°C), they are not placed in water but are moved quickly into the refrigerator.

When flowers are cut in the bud stage, the customer obtains the greatest benefit from a longer lasting life. Roses are always cut in the bud stage. The advantages of bud cut mums and carnations are convincing more growers to use this method of harvest. Tulips, iris, narcissus, and daffodils last longer if they are cut in the bud stage. Many of the bulbous cut flower crops are allowed to open too far before they are offered for sale.

Some crops do not open properly if they are harvested in the bud stage. Two crops that cannot be harvested in the bud stage are dahlias and *Anthuriums*. More research is required to determine if orchids can be harvested as buds and opened.

Some crops require special treatment that must be given as soon as they are harvested. For example, dahlias and poinsettias that are sold as cut flowers must be treated. That is, the ends of the stems are seared in an open flame or dipped in boiling water for 30 to 60 seconds. This treatment seals the latex-producing vessels so the latex does not plug the water-conducting xylem tubes.

Following harvest the crops must be moved quickly to the refrigerator for cold storage.

Cold Storage Treatments

Three methods of cold storage are used to prolong the keeping life of cut flowers: **wet-cold storage**, usually at 33°F to 35°F (0.5°C to 1.5°C); dry storage at 31°F (−0.5°C); and **freeze drying**.

Cut flowers are put into cold storage to delay the maturation process and prolong their lasting life. The respiration and transpiration rates are greatly reduced at cold temperatures. At a temperature of 32°F (0°C), respiration is said to be three times slower than it is at 50°F (10°C). This reduction in the rate of respiration conserves carbohydrates within the plant.

Wet-Cold Storage at 33°F to 35°F (0.5°C to 1.5°C)

This method of cold storage is used for the day to-day handling of cut flower crops (Figure 32-2). There are a few exceptions to the recommended temperatures.

FIGURE 32-2 Cold storage of cut roses.

Generally, gladiolus and most cultivars of orchids are not cooled below 40°F (4.5°C). Because of the danger of chilling injury, *Vanda* orchids are not cooled below 55°F (13°C).

The following guidelines are presented to help the grower provide the wet-cold storage conditions that will benefit the lasting life of cut flowers.

Equipment

1. The storage area must be insulated so that the correct temperatures can be held without a waste of energy.
2. Accurate thermostats and thermometers must be used. The accuracy of the devices can be checked every 6 months.
3. Direct blasts of cold air must not be allowed to strike the flowers. Baffles can be added as needed to protect the flowers.
4. A warm cabinet is to be provided inside for orchids.
5. An ethylene scrubber can be installed to remove any ethylene gas that may be present in storage. Although flowers generally do not produce ethylene at temperatures below 40°F (4.5°C), an ethylene scrubber is recommended in the event ethylene does develop from living or dead plant tissue.

Containers

1. The containers must be absolutely clean. They should be made of plastic rather than metal because some cut flower food solutions react with metal containers. White is preferred as it shows dirt more clearly.
2. The containers should be scrubbed and sanitized in an approved disinfectant solution or in a 1:9 chlorine bleach and water solution by volume. The containers are cleaned outside the refrigerator so that the fumes do not damage the cut flowers inside. The containers should be rinsed at least three times with clear water.
3. The containers should be washed and disinfected at least twice a week.

Storage of Plant Material

1. All cut stems are inspected for signs of disease on the leaves and flowers. Do not store any plant materials showing symptoms of disease.
2. Only top-quality, fresh plant material should be stored.
3. The flowers are prepared by cutting 0.5 inch (1 cm) from the base of stems of material that is not freshly cut, such as material received from another grower.
4. Strip any leaves from stems that will be under the surface of the solution. Leaves placed in the **cut flower food** solution will rot. The bacteria that cause the leaves to rot plug the water-conducting stem tissues and shorten the lasting life. Therefore, all leaves are removed from the part of the stem that is to be submerged in the preservative.
5. Plant materials must not be stored for long periods of time. The crops must move through storage rapidly. Old flowers are discarded.

Sanitation

1. Refuse and litter must not be allowed to accumulate in the greenhouse or in the storage area. All fallen leaves must be cleaned up as well.

2. A floor drain with a bucket trap that can be removed is useful for washing the floors on a regular basis. The bucket trap collects leaves that would otherwise enter the drain and rot. A drain may be a source of ethylene or other sewer gases.

3. The refrigerator must be cleaned thoroughly and washed down with a disinfectant at least once a month. The preferred frequency for such a cleaning is every 2 weeks.

Flowers placed into wet-cold storage should be in cut flower food solutions. When the flowers are first placed in the solution, it should be at a temperature of 100°F (38°C). There are a few exceptions to this statement, which are discussed in the section on cut flower foods. Wet-cold storage is used for short periods of time.

The USDA Agricultural Handbook No. 66, *The Commercial Storage of Fruits, Vegetables, and Florist and Nursery Stocks*, contains specific storage information on many floriculture crops. The most up–to-date version can be acquired from the USDA website, http://www.ba.ars.usda.gov/hb66/.

Dry-Cold Storage at 31°F (−0.5°C)

Cut flowers can be held for longer periods of time by placing them in dry-cold storage at 31°F (−0.5°C) (Figure 32-3). There are certain advantages to long-term storage. Flowers usually sell for higher prices at major holidays. In anticipation of holiday sales, the grower can build up a larger-than-normal supply of flowers by placing them in dry-cold storage. However, there are limits to the amount of time a grower can store cut flowers and still sell a product that offers full value to the consumer.

At times, too many flowers may appear on the market, driving the price down. The grower can place some flowers in dry-cold storage until the prices improve. At other times, excellent weather conditions may result in a very large number of flowers that are ready for harvest and market. Three or 4 days of above-normal temperatures in the spring may double the flower harvest. It may not be possible to grade and pack the flowers quickly enough. The extra flowers are placed in dry-cold storage until they can be processed. Generally, this type of storage is short term only.

FIGURE 32-3 Cut flowers boxed and stored dry. Here the boxes are stacked against a plenum that pulls chilled air through the box to remove heat and humidity.

Temperature

In dry-cold storage, the temperature must be held at 31°F (−0.5°C) exactly. At this temperature the life processes of the flower nearly come to a halt. As a result, carbohydrates and proteins are preserved in the plant tissues for very long periods of time.

Methods of Storage

The flowers are not put into water when they are first cut. No water is used during the storage process. The plants may lose a very small amount of water as a vapor. Therefore, the storage containers must be vapor proof or they will cause the flowers to lose moisture, resulting in dehydration. The flowers stop losing moisture as soon as the relative humidity inside the container reaches 100 percent. By filling the entire container with flowers, the moisture loss from each flower is kept to a minimum and no injury occurs.

The containers should prevent moisture loss, but they should not stop gas exchange. Carbon dioxide building up inside the container may reach a toxic level and damage the flowers. The most commonly used containers are boxes lined with plastic that can be sealed with tape. "Leverpak" drums are more rigid containers that give greater protection. The containers are stacked in the refrigerator with space between them to permit air circulation. The movement of air helps maintain a uniform temperature in the boxes or drums.

Another type of storage container can be made from a standard plastic stock bucket and a plastic bag. The bucket is filled with flowers standing upright. The flowers must not be crushed as they are placed in the bucket. A plastic garbage bag is placed over the flowers, and it is sealed to the bucket with tape. Polyethylene allows a limited exchange of gases but holds in moisture. The buckets must be supported by some means or weighted because they are top heavy when filled with flowers.

Materials Stored

The flowers placed in storage must be high quality, undamaged, disease free, and insect free. Dry-cold storage cannot improve the quality of poor flowers.

After the crop is cut, it should be moved into storage as quickly as possible. If the plants are warm because of greenhouse heat, they are placed dry in a refrigerator for cooling before they are packed into the dry-cold storage containers. Roses placed in water for hardening before they are put into dry storage turn blue during storage.

The grower may sort the flowers by stem length before they are put into storage. They can be graded according to appropriate standards before or after the storage period. Grading after storage means that the flowers can be inspected. If the flowers are stored ungraded for a holiday, preparing them for market may put a strain on an already overloaded grading room. Each grower must evaluate the advantages and disadvantages of grading blooms before or after dry-cold storage.

After Storage Treatments

After the flowers are removed from dry-cold storage, 1 inch (2.5 cm) is cut from the base of the stem. They are placed into a hydrating solution warmed to 100°F to 110°F (38°C to 43°C). The solution is about 4 to 5 inches (10 to 12.5 cm) deep in the container. This amount of solution should be enough to permit the flowers to absorb all they need without draining the container. A container must not be filled with too many flowers. During storage, the flowers become slightly wilted. During the freshening procedure, they need room to become fully turgid as they absorb the hydrating solution.

The containers are placed at room temperature out of drafts and out of direct sunlight for at least 1 hour, or as long as it takes to fully hydrate them. After the flowers are fully turgid, it is recommended that they then be placed into a cut flower food solution prepared at label recommended strength. This solution is prepared using clean water heated to 100°F to 110°F (38°C to 43°C). The hydration treatment is recommended before the cut flower food treatment to maximize solution uptake. If the flowers are removed from dry storage and are placed into a cut flower food solution, the sugar in the cut flower food solution prevents maximum uptake of water. This is due to the higher viscosity of the sugar solution than the hydrating solution. See Appendix IV for instructions on preparing a hydrating solution.

The containers are returned to a refrigerator at 33°F to 35°F (0.5°C to 1.5°C). The flowers are allowed to harden for 6 to 24 hours. After they are hardened, they can be handled like freshly cut flowers. Table 32-1 gives the optimum storage period for several crops with comments on special problems if any exist. All flowers cannot be placed safely in dry-cold storage at 31°F (−0.5°C).

Freeze Dry Flowers

The process of freeze drying has been used for many years in the fields of biology, museum specimen preservation, and salvage. It has only been in the last few years, since 1988, that research and engineering efforts have resulted in a reliable freeze drying method for flowers. The fragile structure of flowers has been a major hurdle to overcome. To achieve a dried flower that has retained its colors in true detail as well as its shape has been a major accomplishment.

Unlike the normal process used for drying flowers, where many of them are cut in the bud stage and open as they dry, freeze dried flowers are harvested at their peak beauty. The blooms are placed on trays and loaded in the drying

Table 32-1

Flower Crops and the Maximum Number of Days They Can Be Held in Dry-Cold Storage at 318F (20.58C)[1]		
Crops	Maximum Storage Time in Days	Comments
Carnations	28	Red and pink cultivars may develop scorched tips if they are stored longer than 12 to 15 days.
Chrysanthemums	28-30	Field grown mums must be completely free of *Botrytis*.
Roses	15	The cultivar tested was 'Better Times.' Other cultivars may store better or not as well.
Snapdragons	12	The first floret must be open no more than 10 days when the spikes are harvested. Because ethylene builds up in storage, shatter-resistant cultivars must be used.
Tulips	42-56	Cut in green bud stage. Store upright.

The flowers listed in this table are the only ones recommended for dry storage at 31°F (−0.5°C). Cymbidium orchids tolerate the temperature, but they can be "held" better on the plants in the greenhouse.

1. Modified from *Living Flowers That Last*, University of Missouri, 1963.

machine. They are immediately flash frozen to maintain their shape and form. After this a vacuum is applied to the chamber, which slowly withdraws the remaining moisture from the tissue. This is done over a period of several days. The amount of time it takes to dry the blooms varies with the type of flowers used and the moisture content. After the flowers are dried, they are treated with a special preservative that further protects them. This treatment also results in a longer life of the blooms when they are made up into arrangements.

The freeze dry process can also be applied to decorative herbs, fruits, and vegetables as well as seasonal cut flowers. All of these can be used as accent or focal points in flower arrangements. Flowers given for functional or sentimental occasions such as birthdays, bar mitzvahs, weddings, and dances can also be freeze dried, thus preserving their natural beauty for many years of memories.

Cut Flower Foods (Nutrients)

The first cut flower foods (nutrients) appeared shortly after World War II. These materials were designed to extend the lasting life of cut flowers. Some products are intended for use on a specific crop. Other products are broad spectrum formulations that can be used on several types of flowers.

Cut flower foods act in several ways to prolong the life of cut flowers. Most cut flower foods contain a large amount of sucrose (common table sugar), which serves as a food source for the cut flower. When the stems of the flowers are placed in the solution, bacteria and fungi are introduced. The sugar in the solution acts as a food source for these organisms as well. If no control is provided, the organisms multiply so rapidly that they soon plug the water-conducting vessels in the stems, resulting in premature wilting. A biocide is added to the solution to control microorganism growth.

Another organism control is the addition of a weak acid to reduce the pH to 3.5 to 4.0. Research studies show that the flowers last longer because at a low pH the xylem vessels do not plug so quickly.

Blockages of the xylem vessels of the stem also occur because of the buildup of natural substances that begins when the stems are cut. Other materials are added to the cut flower food solutions to slow down the development of such blockages. Effective chemical additions include 8-hydroxyquinoline sulfate (**8-HQS**) and 8-hydroxyquinoline citrate (8-HQC).

Cut flower foods are most effective if they are added to the water in which the flowers are placed when they are first cut. The materials should be used again at the wholesale level and by the retail florist when the flowers are sold to the consumer. A definite improvement in the lasting life of cut flowers is obtained with the use of cut flower food solutions. The cost of treatment at the grower, wholesaler, and retailer levels amounts to roughly 6 cents for 25 flowers. This is a very small price to pay for the benefits obtained.

The **Cornell cut flower food solution** was developed for use on roses. The original formulation has been modified and is now the basic solution used to open bud cut carnations and chrysanthemums. The Cornell formulation in various strengths is given in Table 32-2.

One of the greatest advancements in postharvest flower care in the past 20 years has been the introduction of a novel product called EthylBloc®, which is marketed by Floralife®, Inc. of Walterboro, South Carolina. The active ingredient of EthylBloc® is 1-methylcyclopropene, which is a nontoxic gas that

Table 32-2

Formulation of the Cornell Solution for Various Crops[1]				
	Quantities Added to 12 Gallons (45.4 l) of Water			
	Concentration Desired		8-HQC Stock Solution[2] in Fluid Ounces (ml)	Sucrose in Pounds (kg)
Crop	8-HQC (ppm)	Sucrose (%)		
Roses[3]	200	3-4	6 (180)	3-4 (1.4-1.8)
Snapdragons	300	1.5	8 (240)	1.5 (0.68)
Carnations	400	4	12 (360)	4 (1.8)
Mums	200	2	6 (180)	2 (0.9)
Gladiolus	600	4	18 (540)	4 (1.8)

1. This solution is to be used on a trial basis first for the listed crops.
2. 8-HQC stock solution is prepared by dissolving 1.75 ounce (49.5 g) of 8-HQC in 1 quart of water to provide a 10x concentration. The solution is stored in a plastic container.
3. Red roses only. For other colors, reduced concentrations of sucrose may be required to prevent injury. Trials are recommended.
NOTE 1. The solution is stored and used in nonmetallic containers.
NOTE 2. 8-hydroxyquinoline sulfate (8-HQS) can be substituted for 8-HQC.
NOTE 3. Deionized water is preferred to tap water, well water, pond water, or water from other sources. Fluoride levels greater than 1 ppm reduce the lasting life of gladiolus, lilies, and roses.

increases flower life, inhibits premature wilting, helps eliminate leaf yellowing, reduces flower browning, and stops flower and leaf drop. It is effective at both refrigerated and room temperatures.

EthylBloc® is applied as a gas to plant material, which means that the flowers or plants must be inside a chamber for treatment. Major flower shippers use this product by adding EthylBloc® to an enclosed truck and others add this product to a closed cooler. It is used by mixing a powder with water, and then the active ingredient, 1-methylcyclopropene, volatilizes into the confined atmosphere.

EthylBloc® works by blocking the site of action within plant cells where ethylene is active. A drawback is that EthylBloc® is applied as a gas and quickly dissipates. Also, as new cells develop and metabolize, EthylBloc® is not recycled in the plant system, thus it only serves as a temporary protection from ethylene damage, yet is very effective.

ACHIEVEMENT REVIEW

Select the best answer or answers to complete each statement. List the appropriate letter(s).

1 Cut flowers die because:
   a. the processes of transpiration and respiration continue.
   b. they no longer have a life support system.
   c. their food reserves are used up.
   d. of all of the above.

2 Ethylene gas may come from:
   a. air pollution.
   b. dry ice.
   c. calla lilies.
   d. orchids.

3 Freshly cut flowers should be:
   a. held at room temperature for 1 hour.
   b. placed immediately in 31°F (−0.5°C) wet storage.
   c. placed immediately in a 100°F (38°C) hydrating solution.
   d. placed immediately in ice water.

4 Which of the following crops cannot be harvested in the flower bud stage?
   a. Carnations
   b. Chrysanthemums
   c. Dahlias
   d. Roses

5 The proper storage temperature for wet-cold storage is:
   a. 31°F (−0.5°C).
   b. 33°F to 35°F (0.5°C to 1.5°C).
   c. 35°F to 38°F (1.5°C to 3.5°C).
   d. 38°F to 40°F (3.5°C to 4.5°C).

6 Gladiolus and certain orchids should not be stored at a temperature below:
   a. 31°F (−0.5°C).
   b. 35°F (1.5°C).
   c. 38°F (3.5°C).
   d. 40°F (4.5°C).

7 At a temperature of 50°F (10°C), respiration is _____ times faster than at 32°F (0°C).
   a. two
   b. three
   c. four
   d. six

8 The base of the stem of a cut flower is cut again to:
a. remove any rot.
b. expose fresh vascular tissue.
c. expose fresh pith.
d. shorten the distance water must travel.

9 Freeze drying is:
a. used in commercial floriculture.
b. experimental at present.
c. less beneficial than 31°F dry-cold storage.
d. the storage of flowers under pressure.

10 One of the most important chemicals used in cut flower foods to prevent stem blockages is:
a. sugar.
b. aspirin.
c. silver ions.
d. 8-HQC.

The Role of the Wholesaler in the Marketing and Distribution of Floriculture Crops

OBJECTIVES:

After studying this chapter, the student should have learned:

■ the role of the wholesale distributor in the sale of flowers.

■ how the Dutch clock auction works.

■ how consignment selling works.

■ the importance of careful handling of flowers.

Flower growers are located in nearly every country of the world. Retail growers normally produce flower crops that are sold in a retail store for a local market. A wholesale grower sells the crops at the wholesale level only; there are no retail sales at all.

The wholesale grower may have a very large business located some distance from urban centers. Many years ago, large growers were located on the edges of cities because the transportation was much slower than it is today and they had to be near their markets. Even the modern transportation network of highways does not make it practical for the wholesale grower to sell flowers directly to the retailer. In most cases, the retailers are widely scattered and the costs of delivery are considered too high for some, especially cut flower growers. Once, only growers sold directly to retailers because they operated a trucking service from city to city along a given route. Most of these routes are not more than 200 miles in length. Yet, now with more emphasis on service, almost all wholesale growers deliver, at least in part, flowers and plants directly to retailers.

The wholesale grower who does not sell crops directly to the retailer normally sells to a **wholesale distributor**.

Wholesale Distributor

The role of the wholesale distributor developed because the wholesale grower needed a buyer who would purchase crops in large quantities. The wholesale distributor then breaks up the crop into smaller units that are sold to several retailers. The wholesale distributor is also known as the break bulk distributor.

The wholesale distributor purchases some flowers from local growers. The majority of the crops sold today are obtained from more distant growers, many of whom are offshore. One wholesaler in Ohio reported that 95 percent of the flowers sold were imported. Only 5 percent were from local sources. In 2004, 59 percent of the cut flowers imported into the United States were from Colombia, 19 percent from Ecuador, 9 percent from the Netherlands, 3 percent from Canada and Costa Rica each, and 2 percent from Mexico. Cut flowers imported into the United States in 2004 were valued at $705.6 million in sales at wholesale compared to $396.7 million locally grown.

Modern air transportation enables offshore producers to move their bulky and highly perishable cut flower crops rapidly to market. Air time from Bogota to Miami is only 3 hours and 40 minutes compared to 3 hours and 45 minutes from Denver, Colorado. Flowers cut in the morning on a Colombian farm can be in the wholesale market in Miami, Florida, or New York City in 24 hours.

Wholesale distributors who do business in large cities such as New York City, Philadelphia, and Boston rely on retailers to come in early in the morning to personally inspect the flowers that they purchase and take them back to the shop. The retailers may visit the market as early as 6:00 A.M. to get the best quality.

Wholesale distributors who do business in semiurban or more rural areas rely on truck routes to deliver their flowers. Such purchases are made by telephone through in-house salespersons working in the wholesale house. These wholesalers usually deliver within a 100-mile radius of their location.

Several wholesalers may be grouped together in an industrial park located near a major airport. This makes it easy to obtain shipments from distant points as rapidly as possible. It also enables retailers to shop around among several suppliers to get the products they need. The longer cut flowers are in transit from the place of production to the location of sale, the shorter is their vase life in the hands of the consumer.

Clock Auction

European growers and distributors use a more modern, streamlined method of selling cut flowers and potted plants. The Dutch flower trade was one of the first to use a **clock auction** system to sell flowers. In and around Aalsmeer, Holland, there are about 11,000 flower growers. Their potted plants and cut flower crops are sold by auction in three large markets.

The merchandise is displayed on carts in a large hall before the auction begins. The lots are identified by number. The buyers are allowed to visit the floor and examine the flowers offered for sale. A bell rings to announce the beginning of the sales. The bell is the signal for all the buyers to return to the auction room.

Each buyer sits in a numbered chair that is equipped with an electrical button. A large clock faces the buyers. The bidding starts when the flowers being sold are wheeled in on the cart. The auctioneer starts the clock at a fair price for the flowers offered. The hand moves rapidly from the starting price to lower figures. When the price is right, the buyer presses the button on the chair to stop the clock. At the same time, the code number of the buyer, the flower lot number, and the sale price are recorded in a computer. A sale is completed in about 8 to 10 seconds.

As soon as a sale is completed, a new cart of flowers is brought in and the auction continues. At the close of business for the day the buyer is billed automatically for the purchases made. The buyer must pay before the flowers can be removed from the hall. The grower is paid for the crop on the day it is sold.

The clock auction is fast, and a large quantity of flowers can be moved through the market in a day. Because the auction is run on a cash basis, growers do not have to wait weeks for the money from the sale of their crops. A small percentage of each sale is used by the auction management to pay the workers who handle the flowers.

There is an active clock auction in Toronto, Montreal, and Vancouver, Canada. The only clock auction in the United States is in San Diego, California (Figure 33-1).

Wholesale Structure

In a survey of 109 companies in 2006 in North America, conducted by the Wholesale Florist and Florist Supplier Association (**WF&FSA**), the average annual sales volume was nearly $4 million. Thirty-two companies had annual sales that averaged $14.3 million. Forty-three companies showed average sales of $4 million. Thirty-four companies showed average sales of $2.2 million. As can be seen from these figures, the wholesale florist business turns a lot of dollars each year. Eighty-two percent of the income was derived from the sale of fresh floral products. The remaining 18 percent was from the sale of hard goods in the form of ribbons, accessory items, baskets, and other items (Figures 33-2, 33-3, 33-4, and 33-5). The hard goods category also includes a large selection of permanent flowers made from silk, other fabrics, and plastic (Figures 33-6, 33-7, and 33-8). The majority of these permanent flowers are imported from suppliers in China, Taiwan, Hong Kong, and other far eastern countries.

The four main cut flower crops sold by one wholesaler were standard carnations, roses, pompons, and minicarnations in that order.

FIGURE 33-1 Electronic clock used at the San Diego, California, clock auction for selling flowers.

FIGURE 33-2 Ribbons display at wholesale florists supply house for purchase by retail florists.

FIGURE 33-3 A fine selection of brass and ceramic containers for retail florists.

FIGURE 33-4 Plastic and woven straw baskets for retail florists to use in making.

FIGURE 33-5 Ceramic containers and a few woven baskets. All are used by retail florists in making floral arrangements.

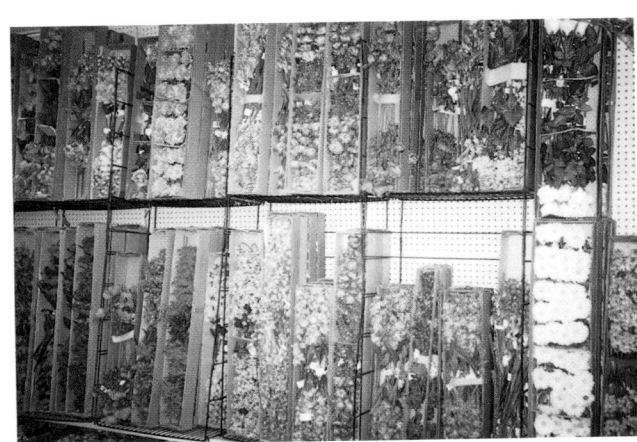

FIGURE 33-6 A wide array of single-stem artificial flowers. These are made of silk fabric.

FIGURE 33-7 A display of artificial flowers made from plastic and fabric.

FIGURE 33-8 Dried flower seed pods and other plant parts used for flower arrangements.

A modern-day wholesale distributor, in a semiurban area, may have as many as 400 or more accounts. These are generally with retailers and growers. They may also have a few accounts with other wholesale distributors in their areas. They buy and sell flowers among themselves as the need arises.

Methods of Selling

There are two types of business arrangements between the wholesale distributor and the suppliers. The oldest of the two types is the **consignment** for sale or commission sale. The grower sends a shipment to the wholesaler on consignment. The wholesale distributor then sells the flowers. The wholesale distributor charges the grower-consignor a fee for making the sale. This commission charge is approximately 25 percent of the gross wholesale selling price. For selected special crops, the commission may be closer to 30 percent.

For the commission fee, the wholesale distributor assumes the responsibility of selling the flowers after receiving them from the grower. This responsibility includes breaking the cartons and repacking the crop into smaller units, the costs of selling and delivering the flowers to retailers, and billing the retailers who buy the product.

The grower normally receives money once every 1 to 2 weeks for the sale of the crop, less the commission charge. Any flowers that are not sold are discarded or sold to vendors at distress merchandise prices. The pushcart peddlers usually buy their flowers at these prices.

Outright purchase is the second method of sales. About 30 percent of the flowers sold are purchased by the wholesale distributor who buys them from the grower, adds a fair markup, and sells them to the retailer. When flowers are in large supply, the wholesaler tries to sell the flowers purchased outright before those received on consignment.

Methods of Shipping Flowers

Most of the flowers sent to wholesale distributors are shipped by air and by truck. Cut flowers from Central and South America and Mexico are shipped by air to Miami, Florida, Houston, Texas, or other ports of entry where they undergo customs inspections. Importers who specialize in bringing in flowers from offshore have built large, state-of-the-art, refrigerated facilities to process these flowers. After customs inspection and processing, the shipments may go directly to wholesale distributors in other large or small cities who sell them to retail florists or other sales outlets. Some importers bypass the wholesaler and ship directly to the central processing facilities of large chains such as Kroger, Edwards, Home Depot, and others.

Cut flowers from California, Texas, and Florida are usually shipped by **airfreight.** The reduced costs of shipping by air that have taken place in the period since deregulation of the airlines and the advantages of rapid transport to the market make this the preferred method of transport from distant markets. Large numbers of cut flowers and cut greens are also shipped by overnight truck transport from Florida. The trucks are refrigerated to keep the flowers cool in summer and heated to keep them from freezing in the winter. Many of these trucks have dual chambers allowing two shipping temperatures so as to not damage tropical plant material.

Cut flowers are shipped in reinforced cardboard boxes. Years ago flowers were shipped in wooden boxes that were returned to the grower. These boxes were bulky and heavy and difficult to handle. The cardboard boxes used now are strong and dependable for lightweight flowers.

Flowers to be sent long distances are typically shipped dry. In the winter the cartons must be insulated with newspapers or commercial insulating liners. A box lined with 10 sheets of newspaper can protect the flowers from freezing for 2 hours when the outside temperature is 30°F (-1.0°C). A 0.25-inch-thick wrap of Styrofoam insulation protects the plants just as well. Regardless of how well insulated the shipping boxes are, they must never be left in below-freezing temperatures for any extended periods. Some boxes have integrated reservoirs for water and maintain the flowers upright to prevent curling due to geotropism.

In the summer the long-distance shipment of flowers may require the addition of crushed ice to the boxes to keep the flowers cool. Crushed ice melts in less than 24 hours, so its effect is short term. Crushed ice is recommended for shipments that take less than 24 hours to complete. Some growers also use dry ice, which is compressed carbon dioxide. It is so cold that if the cake of ice touches any flower stems, it can kill them. The carbon dioxide gas given off by the sublimation of the dry ice does not seem to harm the plants and slows respiration rates, further preserving the vase life of the flower. (Sublimation is a process by which a material is changed from a solid to a gas with no intermediate step.)

Local shipments to wholesalers are often made with the flowers standing in cut flower food solutions. This method of transporting the flowers ensures that their lasting life is as long as possible.

Handling upon Arrival

Flower shipments generally reach the wholesale distributor at 3:00 or 4:00 A.M. The schedule ensures that fresh material is available at the start of business. The wholesaler usually opens for business at 6:00 A.M. By 2:30 or 3:00 P.M., the flower handlers and salespeople have left, and only a few bookkeeping employees remain.

When the flowers arrive, they are checked against the invoices so that the wholesaler receives the proper materials. If any damaged or frozen flowers are found, the wholesaler files a claim against the carrier.

The wholesale distributor would prefer not to unpack the flowers from their shipping boxes. On occasion the wholesaler may sell the flowers in case lots to large retailers for special jobs requiring hundreds of blooms.

Most of the time, however, the flowers must be unpacked from the cases and placed on tables for display. Orders are filled as they are received. The wholesaler does not rely entirely on sales to walk-in retailers. Normally, three or four salespeople sell to retailers by telephone. When flowers are in large supply, each salesperson may spend the better part of the day making telephone calls.

Flowers that are unsold at the end of the day are put into holding solutions and placed in refrigerators at the recommended storage temperature. Flowers still unsold after 3 or 4 days are discarded or sold to street peddlers or convenience stores.

The proper handling of the flowers at the wholesale distributor level is just as important to the lasting life of cut flowers as the treatment given by the grower and the retailer. Far too often, cut flowers are handled roughly. They are dropped, tossed on hard tables, thrown through the air, and generally abused. Cut flowers are fragile and easily bruised by improper handling. They should always be handled with care.

Cut flowers that arrive from distant locations have probably been out of water for 48 hours or more (Figure 33-9). Because of this they require special handling to maximize their lasting life in the hands of the consumer. An air embolism (bubble of air) develops in the first 1 inch (2.5 cm) of the basal end of the stem. It was once recommended to remove this by cutting 1.5 inches (3.7 cm) from the base of the stem underwater. Cutting underwater prevents any more air from getting into the stem (Figure 33-10). Although this continues to be the most effective means for recutting a flower stem that has been in transit, research from the University of Florida and the University of California at Davis has shown that the practice of continuous use of the cutting water becomes very dirty. If the water is not changed frequently, the water becomes a very cloudy, bacterial soup. The antibacterial products added to the rehydrating solution are not adequate to maintain sanitation; therefore, the recommendation now is to recut the stems dry.

Immediately after cutting, the flowers are placed into a hydrating solution. A prepared hydrating solution is made with citric acid and using 100°F (38°C) warm water (see Appendix IV). The solution lowers the pH of the water to 3.5 to 4.0. It is at this pH that the maximum uptake of solution occurs. Many years of research have also shown that this pH keeps the xylem water-conducting vessels of the stem open for the longest period of time.

While in the hydrating solution, the flowers are held at room temperature of 60°F to 70°F (15.5°C to 21°C) for 1 hour or until they are fully turgid (Figure 33-11). They may then be placed into the refrigerator for further hardening (Figure 33-12). During the period of hydration the flowers should not be exposed to hot or cold drafts or be placed in direct sunlight. Water uptake is a slow process. A cut flower placed in a rapidly moving airstream could become desiccated before it has absorbed all the water it needs.

FIGURE 33-9 Tropical cut flowers as received from an off-shore producer.

FIGURE 33-10 A FELLY® underwater flower cutter. This unit can cut a dozen rose stems at one time.

FIGURE 33-11 Flowers in hydrating solution for 1 hour before being placed in flower food and then placed in the refrigerator.

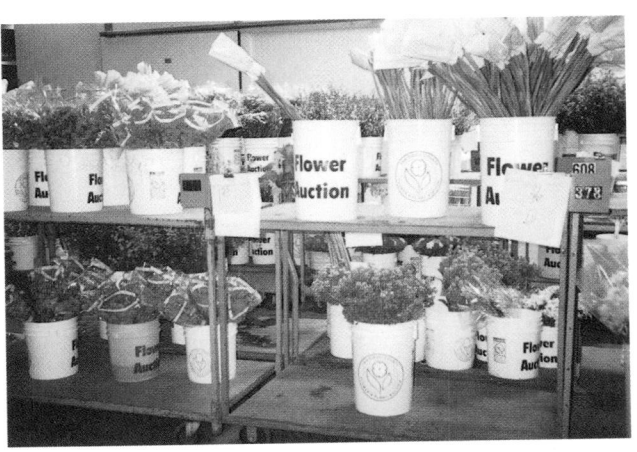

FIGURE 33-12 Flowers in flower food continuing the conditioning (hardening) treatment in the refrigerator.

Badly wilted flowers are wrapped with plastic until they become turgid. The plastic wrap maintains a high humidity around the blooms.

Dry pack flowers that have had the stem ends cut underwater should never be placed directly into a cut flower food solution. The major ingredient in most flower foods is a carbohydrate source, which is usually some kind of sugar. Sugar solutions have a higher osmotic concentration than water or hydrating solutions. As a result, flower stems that have been out of water for several hours to several days, even though they may be recut underwater, will take up less solution than if placed in a hydrating solution first.

After the flowers have been fully hydrated, they can then be placed into a cut flower food solution. This is usually done by the retail florist when the arrangement or bouquet is made up. Cut flowers that arrive at the wholesalers in holding or hydrating solutions should remain in those solutions until sold. If not sold in 3 days, the flowers are placed into newly prepared cut flower food solutions (Figure 33-13).

A room where bud cut chrysanthemums and carnations are opened is an ideal location for the conditioning treatment. Direct sunlight is not allowed to shine on the flowers. Any overheating of the flowers leads to severe moisture stress.

Strict sanitation measures are required in the refrigerators. Fallen leaves, broken flowers, and stems must not be allowed to accumulate. The containers must be washed thoroughly and disinfected following the same schedule used by the grower (Figure 33-13).

Flowers placed on tables should be handled carefully. Foam rubber padding covering the tabletop protects the heads of roses. Orchids are never removed from water. They are shipped normally with the stems inserted in small, water-filled tubes. Orchids should be displayed in racks where they can be held upright for maximum protection.

The last responsibility of the wholesale distributor in the marketing chain is to ensure that the flowers are delivered in excellent condition to the retailers (Figure 33-14). All flowers must be packed carefully (Figure 33-15). Orchids are placed on a bed of shredded wax paper. In addition, they are taped or wired to the bottom of the box to keep them from moving.

FIGURE 33-13 Assorted cut flowers in a large walk-in cooler.

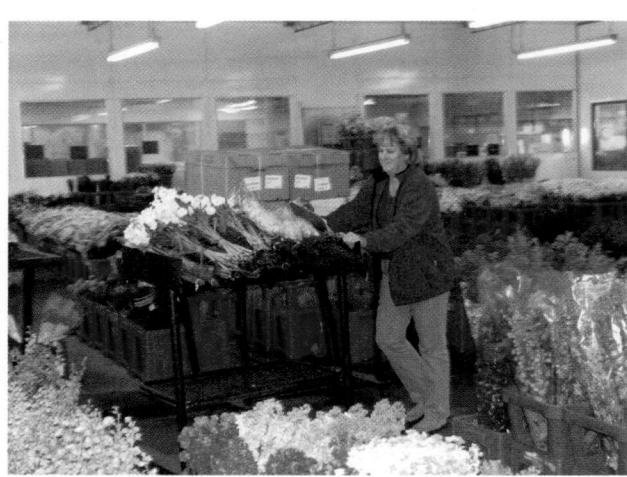

FIGURE 33-14 Wholesale sales associate gathers orders from the cooler.

FIGURE 33-15 Cut flower orders are assembled and then boxed for delivery.

When heavy greens are included in an order, they are placed on the bottom of the box to serve as a cushion for the lightweight flowers. In warm weather, ice is added to roses to keep them in better condition. Pads of rolled-up newspaper are placed under the heads of mums. Each mum flower may be wrapped in wax paper to keep the petals from sticking together. An order of flowers in a box should be tied down to keep them from shifting in the box when the cartons are handled roughly. The cleats used to anchor the flowers must not be installed so tightly that the stems are broken.

Approximately 85 percent of the flowers that pass through the wholesale distributor are sold to retailers. Ten to 11 percent of the flowers go to other wholesalers, and 4 to 5 percent are sold to other types of retailers.

Chain stores, supermarkets, and other types of mass market outlets purchase flowers directly from the growers. Wholesale distributors are rarely involved in such sales.

ACHIEVEMENT REVIEW

Select the best answer or answers to complete each statement. List the appropriate letter(s).

1 The wholesale distributors of flowers operate between the:
   a. consumer and the retailer.
   b. consumer and the grower.
   c. grower and the retailer.
   d. grower and the mass market outlet.

2 The wholesale distributor handles flowers from:
   a. local sources only.
   b. distant sources only.
   c. both distant and local sources.
   d. none of the above.

3 Most wholesalers are located in:
   a. small towns.
   b. medium size towns.
   c. large cities.
   d. farm districts.

4 The Dutch clock auction system is also used in the wholesale market in:
   a. New York City.
   b. San Diego, California.
   c. Toronto, Canada.
   d. Vancouver, Canada.

5 Sixty-six percent of the dollar income of the wholesalers is from the sale of:
   a. bows and ribbons.
   b. cut flowers.
   c. potted plants.
   d. permanent flowers.

6 In consignment selling, the:
   a. grower sells flowers outright to the wholesaler.
   b. grower sends flowers to the wholesaler and hopes they can be sold.
   c. wholesaler charges a commission for selling the grower's flowers.
   d. retailer buys flowers on credit.

7 Flowers shipped in the winter months are protected from freezing by:
   a. packaging them in insulated boxes.
   b. shipping them in heated trucks.
   c. putting heaters in the boxes.
   d. wrapping the boxes with newspapers.

8 The best way to make local flower shipments to wholesale distributors is to:
   a. pack the flowers with crushed ice.
   b. pack the flowers with dry ice.
   c. ship the flowers in containers with their stems in ice water.
   d. ship the flowers in containers with their stems in fresh cut flower food solutions.

9 The wholesaler sells flowers by means of:
a. walk-in buyers.
b. colorful picture catalogues.
c. door-to-door sales.
d. telephone contacts.

10 Bud cut chrysanthemums and carnations are:
a. not practical for shipment.
b. becoming more popular with florists.
c. opened at 31°F.
d. placed in direct sunlight to speed their opening.

The Retail Florist, Mass Market, and Other Outlets

OBJECTIVES:

After studying this chapter, the student should have learned:

■ how cut flowers are handled at the retail shop.

■ the importance of the mass market outlets to retail flower sales.

■ light needs for tropical foliage plants in the retail shop.

At one time the traditional retail florist was the last link in the marketing chain of the florist industry. This picture has changed considerably over the last 20 years. Supermarkets, discount chain stores, department stores, home improvement hardware chains, and other types of stores have all taken flower sales away from the retail florist.

There are close to 22,750 traditional retail florists in the United States. A traditional retail florist is defined as a business that buys flowers from a wholesaler and sells them to consumers. The main sources of income are from sales of cut flowers and potted plants. Other items that may also be sold are accessories, containers for the flowers and plants, ribbons and bows, and greeting cards. The specialty offered by retail florists is the artistic skill in design and arrangement of the flowers that they sell.

The majority of the retailer's business is the sale of flowers for social occasions, such as weddings, births, birthdays, and funerals, as well as for get-well wishes, special occasions, and holidays. About 10 percent of a retailer's business is from weddings and 22 percent from funerals. Very few impulse buyers visit retail flower shops. In fact, men buy from traditional florists more so than any other retail outlet, whereas women typically buy flowers from a supermarket flower outlet.

Many of the flowers sold for special occasions are done so by means of a wire service. When a customer wishes to send flowers to someone in a distant city, the order is placed with a local florist. This florist, who is a member of a wire service, then telephones, faxes, or e-mails the order to a retail florist in the city or town where the flowers are to be sent. The two primary wire services active in the United States include Florist Transworld Delivery, Inc. (FTD) and Teleflora. Within recent years a new service, 1-800-FLOWERS, offers customers a 24-hour service to send flowers. This service has taken business away from retail florists because of its ready accessibility. This company's busiest time for receiving orders is from 6:00 to 9:00 P.M., a time period when most florists are closed and do not answer their telephones.

As reported in information circulated by the Society of American Florists, the total flower sales through all sales outlets was $13 billion in 2002 and the projected online sales volume for 2006 is $2.4 billion in flowers. The traditional florists' share of these sales decreased from 62 percent to 42 percent. To remain competitive in the future, the traditional retail florist is going to have to become very aggressive in seeking new sales.

Buying flowers over the Internet has changed the floristry industry. Between $990 million and $1.5 billion of flowers were purchased using the Internet in 2004, which was about 1.5 percent of all online sales.

Types of Retail Outlets

The **retail outlet** located in the business district of a city caters to all types of customers and depends on the cash-and-carry trade. These outlets also conduct a large telephone order business. These stores maintain charge accounts and provide delivery services.

In high-rent districts, retail stores may be exclusive shops that cater to people interested in a selection of expensive, exotic flowers as well as the standard crops.

Neighborhood retail stores rely heavily on the walk-in trade, but much of their sales are handled by telephone. The neighborhood retailer knows many customers by name and can provide more personal service.

Shops in small towns are similar in some respects to neighborhood stores. Their sales consist of the walk-in trade and telephone orders. A delivery service is provided as many sales are made to people living outside the town in rural areas. The small-town retail florist often enjoys a high social position in the town (Figures 34-1 through 34-7).

FIGURE 34-1 Richards Flowers and Gifts is a small-city retail florist shop located on a main street in Kent, Ohio.

FIGURE 34-2 Refrigerator display at retail florist. Notice made-up arrangements for quick sales.

FIGURE 34-3 Made-up Christmas arrangements and other accessories on display.

FIGURE 34-4 Santa Claus welcomes Christmas shoppers to the retail shop.

FIGURE 34-5 Various selections of poinsettias. A small decorated Christmas tree is also displayed.

FIGURE 34-6 Permanent flowers offered for sale by the stem. Notice the unique way of displaying them in wooden barrels.

FIGURE 34-7 Gift corner of the retail shop. Jams, jellies, cheeses, and some candy on display.

Other Types of Retail Outlets

In the past 10 years, other types of retail outlets have appeared and increased in popularity. For example, the European type of flower market is becoming more common in the United States. Small kiosks or mini refrigerated sales carts can be found in airport lounges, train stations, large department stores, and shopping malls. Generally these units are satellite operations of a large retailer (Figure 34-8). One person can handle the sales activity at a unit. Some units are self-service. All sales are cash-and-carry. Makeup work or flower arrangements are not done at the cart. The units are serviced at least once a day by the retailer. On busy holidays, it may be necessary to service the units two or more times a day if the sales are high.

The recent interest in tropical foliage plants led to the introduction of a number of boutiques in many cities and towns. In many cases, the boutiques were operated by people who were more interested in making a profit than providing quality plants to customers. Again, sales were limited to cash-and-carry. Many of these shops closed because the operator had little knowledge about the care of the plants being sold.

Around the major holidays, a large number of nonretail shops begin to sell flowers, usually pot plants, but some cut flowers are offered. These items usually are purchased at a discount or in a large volume from a wholesaler or grower. The quality of these plants may not be high.

Mass Market Outlets

In most regions of the United States, a supermarket chain store may be listed in the yellow pages of the telephone directory under "Florists." This may seem strange until it is realized that **mass market outlets** have captured more than 40 percent of all flower and plant sales in the United States.

Supermarkets and other mass market sales outlets for potted plants and cut flowers have become major players in the floral business. Home Depot, Edwards, Wal-Mart, and K-Mart stores, just to mention a few, are the latest to see an opportunity to make money selling floral products (Figures 34-9 through 34-14).

Many of these retail outlets have expanded their offerings from holiday potted plants and cut flowers to a full line of services, including delivery, equal to retail flower shops. In addition to these, spring sales of annual bedding plants, woody ornamentals, and small trees, plus a full line of garden supplies are carried by many stores. The old observation that only cheap flowers are sold in

FIGURE 34-8 A kiosk at a major airport terminal building. This unit is self-service, making flowers available 24 hours a day.

FIGURE 34-9 The interior of a florist shop in a large supermarket chain store.

FIGURE 34-10 The flower coolers are maintained with prepared arrangements for ready impulse sales.

FIGURE 34-11 Colorful display of blooming plants are positioned in high traffic areas. Notice the vegetables in the background, which are a source of ethylene gas. Many supermarket florists separate their floral displays from produce.

FIGURE 34-12 Product price promotions drive impulse sales.

FIGURE 34-13 Readymade bouquets are typically prepared and packed at the farm for ready sales. They are maintained in easily accessible racks. All of the flowers are kept in flower food solutions to preserve freshness.

FIGURE 34-14 Container displays easily accessible to nearby flower displays improve sales.

mass market outlets is not true. These stores are serviced by some of the largest and highest-quality plant growers in the country.

Many of these growers now are paid for their products using the concept of pay by scan. Even though many parts of the retail industry have used this concept for years, such as soda, bread, and the like, this is new to the greenhouse industry. The basic premise is that the grower is not paid by the retailer for the product until it is paid for by the consumer. In other words, the grower owns the plant until the bar code on the pot passes through the scanner and is paid for. Much nervousness has abounded in the industry with this novel approach, but growers have coped by placing their employees at the retail site to maintain plant quality. In the end, the grower sells a quality plant, the retailer sells a quality plant, and the consumer takes home a quality plant.

Many supermarkets started out on a small scale with flowers and plants. This segment of the industry grew approximately 4 percent from 1992 to 2002. Initially these products were part of the produce section and were the responsibility of the produce manager. The losses were very large in the beginning because no one knew how to handle flowers and plants. Through experience, management learned to hire trained people to run the flower departments. As the chain stores expand and add to their floral and plant sections, improvements continue.

Employees now attend classes to learn the latest information on handling cut flowers and potted plants. The Fresh Produce & Floral Council (FPFC) is a division of the **Produce Marketing Association** with headquarters in La Mirada, California. The **FPFC** also sponsors sessions where supermarket floral buyers can experience the latest offerings of various suppliers (Figures 34-15, 34-16, 34-17, and 34-18).

Supermarket sales are increasing significantly each year. These new retail outlets will provide many fine job opportunities for anyone interested in floriculture.

Handling Flowers at the Retail Outlet Level

The retailer must follow the recommendations listed in Chapter 33 regarding the careful handling of flowers. Using cut flower food solutions at the label

FIGURE 34-15 New ideas pavilion at the 1995 Floral Marketing Association Convention. The display was sponsored by Smithers Oasis Company, Kent, Ohio —the world's largest and leading manufacturer of phenolic foam (Oasis®) used for making floral arrangements.

FIGURE 34-16 Curvilinear display refrigerator as seen at the 1995 FMA Convention.

FIGURE 34-17 *Stephanotis* grown as a potted plant for supermarket sales, 1995 FMA Convention.

FIGURE 34-18 Bromeliads of various types. They are rapidly becoming a popular flower for supermarket sales, 1995 FMA Convention.

recommended rates is critical to success. If used at less than these amounts, there is not enough of the biocide present to restrict the growth of those organisms whose presence plugs the water-conducting vessels of the flowers. The stock containers must be kept absolutely clean. At least once and preferably twice weekly these containers should be scrubbed with hot, soapy water and disinfected. Use chlorine bleach at 1 tablespoonful per gallon of water as a final rinse and then let the containers drip dry by inverting them.

The refrigerator also must be cleaned and disinfected once weekly. Special attention must be paid to the drain because leaves, stems, and other plant debris can collect here. When they rot, they can produce ammonia and other gases that reduce the life of cut flowers.

Do not use ammonia-based glass cleaners when cleaning the refrigerator glass—inside or outside surfaces. The fumes will injure sensitive flowers.

When cut flowers are received at the retail shop, unless they arrive in buckets with cut flower food solution, cut 1 inch (2.5 cm) off the base of the stem, preferably underwater. This cut removes any air blockages that may have developed and exposes fresh vascular tissue for water absorption. Although this continues to be the most effective means for recutting a flower stem that has been in transit, research from the University of Florida and the University of California at Davis has shown that the practice of continuous use of the cutting water becomes very dirty. If the water is not changed frequently, the water becomes a very cloudy, bacterial soup. The antibacterial products added to the rehydrating solution are not adequate to maintain sanitation; therefore, the recommendation now is to recut the stems dry.

For maximum flower life, dry pack flowers should first be hydrated in a hydration solution as described in Chapter 33. After the flowers have become fully turgid, they are placed in a cut flower food solution made up with warm 100°F (38°C) water. The containers are then placed in the refrigerator at the proper storage temperature for the crop.

An excellent publication on the proper care and handling of cut flower crops and potted flowering plants, *SAF Flower and Plant Care Manual,* is available from the Society of American Florists (SAF), 1601 Duke St., Alexandria, VA 22314. There is a charge for this publication.

ACHIEVEMENT REVIEW

Select the best answer or answers to complete each statement. List the appropriate letter(s).

1 In the United States, there are about _____ retail florists.
a. 10,000
b. 23,000
c. 40,000
d. 100,000

2 Part of the services offered by retailers is:
a. the sale of cut flowers and potted plants.
b. the artistic arrangement of cut flowers.
c. the delivery of plants and flowers.
d. all of the above.

3 Mass market outlets now sell more than _____ percent of all the flowers sold at the retail level in the United States.
a. 10
b. 15
c. 28
d. 40

4 The retailer has some control over the production of flower crops by:
a. ordering early.
b. telling the wholesaler what is needed.
c. refusing to buy poorly grown flowers.
d. growing his or her own plants.

5 Foam materials used by retailers to hold cut flowers should be:
a. presoaked thoroughly before use.
b. soaked after the arrangements are made.
c. refrigerated.
d. watered by the customer in the home.

6 Tropical foliage plants should receive not less than _____ of light for _____ hours a day.
a. 10 fc (108 lux); 10
b. 25 fc (270 lux); 10
c. 25 fc (270 lux); 12
d. 50 fc (540 lux); 12

APPENDICES

Appendix I: A Partial List of Boom Irrigation Manufacturers and Suppliers

Appendix II: Association of Natural Biocontrol Products List of US ANBP Supplier Members

Appendix III: University Plant Disease Laboratories for North America

Appendix IV: Formula for Preparing a Citric Acid Hydrating Solution

APPENDIX

I

A Partial List of Boom Irrigation Manufacturers and Suppliers

Agroponic Industries Ltd.
908 Ranchview Cres. NW.
Calgary, Alberta, Canada, T3G 1P9
Tel: (403) 241-8234
Fax: (403) 241-8238
http://agroponic.com

Cherry Creek Systems
2675 Akers Drive
Colorado Springs, CO 80922
Tel: (719) 380-8373 or (877) 558-3246
Fax: (719) 380-8695
http://www.cherrycreeksystems.com

Growing Systems, Inc.
2950 N. Weil Street
Milwaukee, WI 53212
Tel: (414) 263-3131
Fax: (414) 236-2454
http://www.growingsystemsinc.com

Solaris, a division of McConkey & Co.
1615 Puyallup St.
P.O Box 1690
Sumner, Washington 98390-0369
Tel: (253) 863-8111 or (800) 426-8124
Fax: (253) 863-5833
http://www.mcconkeyco.com

U.S.Global Resources
10242 59th Ave. South
Seattle WA 98178 USA
Tel: (206) 722-3999
Fax: (206) 721-1140
http://www.usgr.com

Association of Natural Biocontrol Products List of US ANBP Supplier Members

Ag Attack, Inc.
PO Box 81
Visalia CA 93279
Tel: (559) 625-4598
Fax: (559) 625-4598
http://www.agattack.com

ARBICO-Organics
PO Box 8910
Tuscon AZ 85738-0910
Tel: (800) 827-2847
Fax: (520) 825-2038
http://www.arbico-organics.com

Associates Insectary
PO Box 969
Santa Paula CA 93061-0969
Tel: (805) 933-1301
Fax: (805) 933-1304
http://www.associatesinsectary.com

Becker Underwood
801 Dayton Avenue
Ames IA 50010
Tel: (515) 232-5907
Fax: (515) 817-0722
http://www.beckerunderwood.com

Beneficial Insectary
9664 Tanqueray Ct.
Redding CA 96003
Tel: (800) 477-3715
Fax: (888) 472-0708
http://www.insectary.com

Bio Ag Services, Corp.
4218 W. Muscat, Fresno, CA 93706
Tel: (559) 268-2835
Fax: (559) 268-7945
http://www.bioagservices.com

Biobest Canada Ltd.
2020 Fox Run Road, RR 4
Leamington Ontario N8H 3V7
Tel: (519) 322-21-78
Fax: (519) 322-12-71
http://www.biobest.ca/

Biofac, Inc.
P. O. Box 87
Mathis, Texas
Tel: (361) 547-3259 or (800) 233-4914
Fax: (361) 547-9660
http://www.biofac.com

Biological Control of Weeds, Inc.
1418 Maple Dr.
Bozeman MT 59715
Tel: (406) 586-5111
Fax: (406) 586-1679
http://www.bio-control.com

Biotactics
25139 Briggs Road
Romoland Ca 92585
Tel: (951) 928-0215
Fax: (951) 928-2041
http://www.benemite.com

EcoSolutions, Inc.
2948 Landmark Way
Palm Harbor FL 34684
Tel: (727) 787-3669
Fax: (727) 787-3669

Garden's Alive
5100 Schenley Place
Lawrenceburg, IN 47025
Tel: (513) 354-1482
http://www.gardensalive.com

Harmony Farm Supply & Nursery
3244 Hwy.116 North
Sebastopol, CA 95472
Tel: (707) 823-9125
Fax: (707) 823-1734
http://www.harmonyfarm.com

Hydro Gardens-HGI Worldwide
PO Box 25845
Colorado Springs CO 80936
Tel: (800) 634-6362
Fax: (800) 634-6362
http://www.hydro-gardens.com

IPM Laboratories
980 Main Street
Locke NY 13092
Tel: (315) 497-2063
Fax: (315) 497-3129
http://www.ipmlabs.com

Koppert Biological Systems, Inc.
28465 Beverly Road
Romulus MI 48174
tel: (734) 641 3763
fax: (734) 641 3793
http://www.koppert.com

Kunafin "The Insectary"
P.O. Box 190
13955 U.S. Hwy 277
Quemado TX 78877
Tel: (830) 757-1181
Fax: (830) 757-1468
http://www.kunafin.com

M & R Durango, Inc.
P O Box 886
Bayfield CO 81122
Tel: (970) 259-3521
Fax: (970) 259-3857
http://www.goodbug.com

ORCON
5132 Venice Blvd
Los Angeles CA 90019
Tel: (323) 937-7444
Fax: (323) 937-0123
http://www.organiccontrol.com

Peaceful Valley Farm & Garden
Supply
P.O. Box 2209
125 Clydesdale Court
Grass Valley, CA 95945
Tel: (530) 272-4769
Fax: (530) 272-4794
http://www.groworganic.com

Pacific Biocontrol Corp.
PO Box 529
Placerville CA 95667
Tel: (530) 622-4866
Fax: (530) 621-3213
http://www.pacificbiocontrol.com

Plant Products Co. Ltd.
6299 Meadowsweet Ave NW
Canton OH 44718
Tel: (330) 418 7966
http://www.plantprod.com

Plant Protection Sciences, Inc.
PO Box 1690
Freedom CA 95019
Tel: (831) 728-3322
Fax: (831) 728-4967
http://www-plantprotection
   sciences.com

Rincon-Vitova Insectaries, Inc.
P.O. Box 1555
Ventura CA 93001
Tel: (805) 643-5407
Fax: (805) 643-6267
http://www.rinconvitova.com

Sespe Creek Insectary
PO Box 176
Lindsay CA 93247
Tel: (559) 562-6464
Fax: (559) 562-5565

Spalding Laboratories, Inc.
760 Printz Rd.
Arroyo Grande CA 93420
Tel: (805) 489-5946
Fax: (805) 489-0336
http://www.spalding-labs.com

Sterling Insectary
PO Box 1987
Delano CA 93216
Tel: (661) 792-6810
Fax: (661) 792-6880
http://www.sterlinginsectary.com

Sweetbriar Bio-Systems
151 Silliman Road
Watsonville CA 95076
Tel: (831) 722-5577
Fax: (831) 722-0191

Syngenta Bioline, Inc.
PO Box 2430
Oxnard CA 93034-2430
Tel: (805) 986-8265
Fax: (805) 986-8267
http://www.SyngentaBioline.com

West Valley Farm Enterprises
2925 Hempstead Way
Stockton CA 95207
Tel: (559) 693-1930

APPENDIX III

University Plant Disease Laboratories for North America

Plant pathogens pose a constant battle for the greenhouse grower. It is important to remember that symptoms can easily mislead us, resulting in improper treatment or subsequent crop losses. To make an accurate assessment of a suspected disease condition, a diagnostic facility is the most dependable course of action. With proper testing and appraisal, the grower can initiate the most effective treatment and better ensure a healthy saleable crop.

Below is a state by state list of university related plant disease laboratories in the United States and Canada. If you suspect a disease condition, it is recommended that you use the services of a laboratory in your area. Check with the laboratory prior to sending a sample to determine the type of sample and information to be included, proper packaging method and cost, if any.

US University Laboratories

Alabama
Plant Diagnostic Lab
Auburn University
ALFA Agricultural Services Bldg.
961 South Donahue Drive
Auburn, AL 36849-5624
Phone: (334) 844-5508, 5507
Fax: (334) 844-4072

Plant Diagnostic Lab at Birmingham
C. Beaty Hanna Hort. and Environ.
 Center
2612 Lane Park Road
Birmingham, AL 35223-1802
Phone: (205) 879-6964 ext. 19
Fax: (205) 414-3906

Alaska
Department of Plant, Animal, &
 Soil Sciences
University of Alaska
Ag. Forestry Exp. Station
Fairbanks, AK 99775-7200
Phone: (907) 474-7431
Fax: (907) 474-7439

Arizona
Contact your local County Farm
 Advisor or Extension Specialist at
 your nearest University.

Arkansas
Plant Disease Clinic
Lonoke Agricultural Center
P.O. Box 357; Hwy 70 East
Lonoke, AR 72086
Phone: (501) 676-3124
Fax: (501) 676-7847

California
Contact your local County Farm
 Advisor or Extension Specialist at
 your nearest University.

Colorado
Plant Diagnostic Clinic
Colorado State University Extension -
 Jefferson County
15200 W. 6th Avenue, Suite C
Golden, CO 80401
Phone: (303) 271-6620
Fax: (303) 271-6644

Plant Diagnostic Clinic
E 215 Plant Sciences Building
Department of Bioagricultural
   Sciences & Pest Management
Colorado State University
Fort Collins, CO 80523-1177
Phone: (970) 491-6950

Connecticut
Contact your local County Farm
   Advisor or Extension Specialist at
   your nearest University.

Delaware
Plant Diagnostic Clinic
Plant and Soil Sciences Department
151 Townsend Hall
University of Delaware
531 S College Ave
Newark, DE 19716-2170
Phone: (302) 831-1390
Fax: (302) 831-0605

Florida
Plant Disease Clinic
U.F., Bldg. 78 Mowry Road
PO Box 110830
University of Florida
Gainesville, FL 32611-0830
Phone: (352) 392-1795
Fax: (352) 392-3438

Three Regional State Labs:
Florida Extension Plant Diagnostic
   Clinic
IFAS/NFREC
155 Research Road
Quincy, FL 32351
Phone: (850) 875-7140
Fax: (850) 875-7188

Florida Extension Plant Disease Clinic
SW Florida Research & Education Ctr.
2686 State Road 29 N
Immokalee, FL 34142
Phone: (239) 658-3400
Fax: (239) 658-3469

Florida Extension Plant Disease Clinic
Tropical Research and Education
   Center
18905 SW 280th Street
Homestead, FL 33031
Phone: (305) 246-7000
Fax: (305) 246-7003

Georgia
Extension Plant Disease Clinic
Department of Plant Pathology
Miller Plant Sciences Building
University of Georgia
Athens, GA 30602
Phone: (706) 542-9140
Fax: (706) 542-4102

Hawaii
Plant Disease Clinic
Agricultural Diagnostic Service Center
1910 East-West Road
Sherman Lab 134
Honolulu, HI 96822
Phone: (808) 956-6706
Fax: (808) 956-2592

Idaho
Plant Pathology Laboratory
University of Idaho
29603 U of I Lane
Parma, ID 83660-6699
Phone: (208) 722-6701 Ext. 218 or 217
Fax: (208) 722-6708

Nematode Testing Service
University of Idaho
29603 U of I Lane
Parma, ID 83660-6699
Phone: (208) 722-6701 Ext. 237
Fax: (208) 722-6708

Illinois
(May-Sept)
Plant Clinic
1401 W. St. Mary's Road
University of Illinois
Urbana, IL 61802
Phone: (217) 333-0519

(Oct-Apr)
N-533 Turner Hall
1102 S. Goodwin Avenue
University of IL
Urbana, Illinois 61801
Phone: (217) 333-2478
Fax: (217) 333-1289

Indiana
Plant and Pest Diagnostic Laboratory
Purdue University
LSPS 101
915 W. State Street
West Lafayette, IN 47907-2054
Phone: (765) 494-7071
Fax: (765) 494-3958

Iowa
Department of Plant Pathology
323 Bessey Hall
Iowa State University
Ames, IA 50011
Phone: (515) 294-0581
Fax: (515) 294-9420

Kansas
Plant Disease Diagnostic Lab
Extension Plant Pathology
4032 Throckmorton Hall
Kansas State University
Manhattan, KS 66506-5504
Phone: (785) 532-5810
Fax: (785) 532-5692

Kentucky
(serving Central and Eastern
   Kentucky):
Plant Disease Diagnostic
   Laboratory
Agricultural Science Building-North
University of Kentucky
Lexington, KY 40546-0091
Phone: (859) 257-8949
Fax: (859) 323-1961

(serving Western Kentucky):
Plant Disease Diagnostic Lab
Department of Plant Pathology
UK Research and Education
   Center
P.O. Box 469
1205 Hopkinsville Street
Princeton, KY 42445
Phone: (270) 365-7541 Ext. 228
Fax: (270) 365-2667

Louisiana
Plant Disease Diagnostic Clinic
Department of Plant Pathology and
   Crop Physiology
302 Life Science Building
Baton Rouge, LA 70803
Phone: (225) 578-4562
Fax: (225) 578-1415

Maine
Pest Management Office
Cooperative Extension
University of Maine
491 College Avenue
Orono, ME 04473-1295
Phone: (207) 581-3880
Fax: (207) 581-3881

Maryland
Plant Diagnostic Laboratory
Department of Entomology
University of Maryland
3171 Plant Sciences Bldg
College Park, MD 20742-4454
Phone: (301) 405-1611
Fax: (301) 314-9290

Home and Garden Information
   Center
Maryland Cooperative Extension/
University of Maryland
12005 Homewood Road
Ellicott City, MD 21042-1542
Phone (410) 531-5556
Fax (410) 531-5567

Massachusetts
Commercial samples are handled by
   individual Extension specialists at
   the University.

Michigan
Diagnostic Services
101 CIPS
Michigan State University
East Lansing, MI 48824-1311
Phone: (517) 355-3504
Fax: (517) 432-0899

Minnesota
Department of Plant Pathology
495 Borlaug Hall
1991 Upper Buford Circle
University of Minnesota
St. Paul, MN 55108
Phone: (612) 625-1275
Fax: (612) 625-9728

Mississippi
Plant Pathology Lab
Room 9, Bost Extension Ctr.
Box 9655
Mississippi State, MS 39762-9655
Phone: (662) 325-2146
Fax: (662) 325-8336

Missouri
Extension Plant Diagnostic Clinic
23 Mumford Hall
University of Missouri
Columbia, MO 65211
Phone: (573) 882-3019
Fax: (573) 884-4288

Montana
Schutter Diagnostic Lab
119 Ag Bioscience Facility
Department of Plant Sciences &
  Plant Pathology
Montana State University
Bozeman, MT 59717-3150
Phone: (406) 994-5150
Fax: (406) 994-7600

Nebraska
Department of Plant Pathology
448 Plant Sciences
University of Nebraska
Lincoln, NE 68583-0722
Phone: (402) 472-2559
Fax: (402) 472-2853

Nevada
No university disease diagnostic
  services.

New Hampshire
Plant Diagnostic Lab
241 Spaulding Hall
Plant Biology Department
University of New Hampshire
Durham, NH 03824
Phone: (603) 862-3841
Fax: (603) 862-2717

New Jersey
Plant Diagnostic Lab
Rutgers University
P.O. Box 550
Milltown, NJ 08850
Phone: (732) 932-9140
Fax: (732) 932-1270

New Mexico
Plant Diagnostic Clinic
P.O. Box 30003, MSC. 3AE Plant
  Sciences
Cooperative Extension Service
New Mexico State University
Las Cruces, NM 88003
Phone: (505) 646-1621
Fax: (505) 646-8085

New York
Plant Disease Diagnostic Clinic
Department of Plant Pathology
334 Plant Science Bldg
Cornell University
Ithaca, NY 14853-4203
Phone: (607) 255-7850
Fax: (607) 255-4471

Long Island Horticultural Research
  and Extension Center
Cornell University
3059 Sound Ave.
Riverhead, N.Y. 11901
Phone: (631) 727-3595
Fax: (631) 727-3611

North Carolina
Plant Disease and Insect Clinic
Campus Box 7211
Room 1104 Williams Hall
North Carolina State University
Raleigh, NC 27695-7211
Phone: (919) 515-3619
Fax: (919) 515-7716

North Dakota
Plant Diagnostic Lab
306 Walster Hall
NDSU Plant Diagnostic Lab
P.O. Box 5012
North Dakota State University
Fargo, ND 58105

Ohio
C. Wayne Ellett Plant and Pest
  Diagnostic Clinic
110 Kottman Hall
2021 Coffey Road
The Ohio State University
Columbus, OH 43210-1087
Phone: (614) 292-5006
Fax: (614) 292-4455

Oklahoma
Plant Disease and Insect Diagnostic
  Lab
Department of Entomology & Plant
  Pathology
127 Noble Research Center
Oklahoma State University
Stillwater, OK 74078
Phone: (405) 744-9961
Fax: (405) 744-7373

Oregon
Plant Disease Clinic
Extension Plant Pathology
Cordley Hall, Room 1089
Oregon State University
Corvallis, OR 97331-2903
Phone: (541) 737-3472
Fax: (541) 737-2412

Plant Pathology Lab
H.A.R.E.C.
Oregon State University
P. O. Box 105
Hermiston, OR 97838
Phone: (541) 567-8321
Fax: (541) 567-2240

Pennsylvania
Plant Disease Clinic
220 Buckhout Laboratory
Pennsylvania State University
University Park, PA 16802
Phone: (814) 865-2204
Fax: (814) 863-7217

Rhode Island
URI Plant Protection Clinic
University of Rhode Island
CE Education Center
3 East Alumni Avenue
Kingston, RI 02881-0804
Phone: (401) 874-2967
Fax: (401) 874-5826

South Carolina
Agricultural Service Laboratory Plant
  Problem Clinic
Clemson University
171 Old Cherry Road
Clemson, SC 29634-0114
Phone: (864) 656-3125
Fax: (864) 656-2069

South Dakota
Plant Disease Clinic
Department of Plant Science
South Dakota State University
P.O. Box 2108 PSB 119
Brookings, SD 57007-1090
Phone: (605) 688-5157
Fax: (605) 688-4024

Tennessee
Plant and Pest Diagnostic Center
University of Tennessee
5201 Marchant Dr.
Nashville, TN 37211-5112
Phone: (615) 832-6802
Fax: (615) 781-2568

Texas
Texas Plant Disease Diagnostic Lab
Texas A&M University
1500 Research Parkway, Suite A 130
College Station, TX 77845
Phone: (979) 845-8032
Fax: (979) 845-6499

Utah
Utah Plant Pest Diagnostic Lab
Department of Biology
Utah State University
Logan, UT 84322-5305
Phone: (435) 797-2515
Fax: (435) 797-1575

Vermont
Plant Diagnostic Clinic
Department of Plant and Soil Science
University of Vermont
235 Hills Building
Burlington, VT 05405-0086
Phone: (802) 656-0493
Fax: (802) 656-4656

Virginia
Plant Disease Clinic
106 Price Hall
Department Of Pant Pathology,
   Physiology & Weed Science
VPI & SU
Blacksburg, VA 24061-0331
Phone: (540) 231-6758
Fax: (540) 231-7477 (3221)

Nematode Assay Lab
115 Price Hall
Department Of Pant Pathology,
   Physiology & Weed Science
VPI & SU
Blacksburg, VA 24061
Phone: (540) 231-4650

Washington
Plant Diagnostic Clinic
WSU-Puyallup Research and Extension
   Center
7612 Pioneer Way East
Puyallup, WA 98371-4998
Phone: (253) 445-4582
Fax: (253) 445-4569

West Virginia
Plant Disease Diagnostic Clinic
414 Brooks Hall
Downtown Campus
West Virginia University
Morgantown, WV 26506
Phone: (304) 293-3911
Fax: (304) 293-2872

Wisconsin
Plant Disease Diagnostic Clinic
Department of Plant Pathology
1630 Linden Drive
University of Wisconsin-Madison
Madison, WI 53706-1598
Phone: (608) 262-2863
Fax: (608) 263-2626

Wyoming
Extension Plant Pathology Laboratory
Department of Plant Sciences
University of Wyoming
P.O. Box 3354
Laramie, WY 82071-3354
Phone: (307) 766-5083
Fax: (307) 766-5549

Canadian Provinces

Alberta
Brooks Diagnostics Ltd.
PO Box 1701
Brooks, Alberta, T1R 1C5
Phone: (403) 362-5555

Alberta Agriculture Regional Crop
   Laboratory
Provincial Building
PO Box 159
Fairview, Alberta T0H 1L0
Phone: (403) 835-2291

British Columbia
British Columbia Ministry of
   Agriculture Plant Diagnostic
   Laboratory
Fisheries and Food
Abbotsford Agriculture Centre
1767 Angus Campbell Road
Abbotsford, British Columbia
   V3G 2M3
Phone: (604) 556-3126

Manitoba
545 University Crescent
Room 201
Winnipeg, Manitoba R3T 5S6
Phone: (204)945-7706

New Brunswick
New Brunswick Department of
   Agriculture & Rural Development
Plant Industry Branch
PO Box 6000
Fredericton, New Brunswick E3B 5H1
Phone: (506) 453-2108

Newfoundland

Nova Scotia
Nova Scotia Department of
   Agriculture and Marketing
Plant Pathology Laboratory
Plant Industry Branch
Kentville Research Station
Kentville, Nova Scotia B4N 1J5
Phone: (902) 679-6040

Nova Scotia Department of
   Agriculture and Marketing
Plant Industry Branch
PO Box 550
Touro, Nova Scotia B2N 5E3
Phone: (902) 893-6562

Ontario
Laboratory Services Division
University of Guelph

95 Stone Road W.,
Guelph, Ontario, N1H 8J7
Phone: (519) 767-6256
Fax: (519) 767-6240

Pest Diagnostic Clinic
Agriculture & Food Lab Services
   Center
PO Box 3650 - 95 Stone Road West -
   Zone 2
Guelph, Ontario N1H 8J7
Phone: (519) 767-6256

Quebec
Laboratoire de Diagnostic
Le Service de Phytotechnie du Quebec
2700 Rue Einstein - Room D.1.200H
Sainte Foy, Quebec G1P 3W8
Phone: (418) 643-4925

Saskatchewan
Crop Protection Laboratory
Ministry of Agriculture and Food
346 McDonald Street
Regina, Saskatchewan S4N 6P6
Phone: (306) 787-8130
Fax: (306) 787-8803

Formula for Preparing a Citric Acid Hydrating Solution

STOCK SOLUTION

Use 5 $\frac{1}{3}$ ounces (151 g) technical grade citric acid in 1 gallon (3.785 l) of warm tap water. Store in plastic or glass container in refrigerator at 35°F (1.5°C). Will last three to four weeks. Renew as soon as it turns cloudy.

WORKING SOLUTION

Dilute 2 ounces (60 ml) of the stock solution in each gallon (3.875 l) of warm tap water. The desired pH is 3.5 to 4.0. Discard solution at the end of each day.

PROCEDURE

Recut stems under warm water. Remove 1–2 inches (2.5–5.0 cm) of stem base. Immediately place into hydrating solution of 4-inch (10-cm) depth. Keep blooms at room temperature, out of draft and direct sunlight, for 1 hour or until fully turgid.

After flowers are turgid, place them in flower preservative made up to a 2 percent concentration and place them in the refrigerator until ready to use.

CAUTION: *Do not use metal containers. Use only plastic or other non-metallic buckets.*

A-REST®. A growth retardant used to keep plants short.

Abscission. Term used to describe the natural process of leaves falling from the plant.

Absorption. The uptake of nutrients from the soil solution.

Acclimatization (or acclimation). The process by which plants adapt to another climate or environment.

Acid organic matter. An organic material that leaves an acid residue in the growing medium; an example is sphagnum peat moss.

Acid reaction fertilizer. A fertilizer that causes the soil to become more acid; this type of fertilizer is also called acid residue fertilizer.

Acrylic. A plastic greenhouse covering consisting of two layers of material separated by air space to provide insulation.

Adult stage of growth. Advanced period of plant growth, leads to flowering and production of seed.

Aerated steam. Steam into which a stream of air is injected to hold the temperature at a constant value, usually 140°F (60°C).

Aeration. A characteristic condition of soils. Good aeration occurs in loose, open soils; poor aeration occurs in heavy, tight soils.

Aerial roots. Roots that remain in the air and exchange oxygen and carbon dioxide; they do not absorb water unless they enter a medium or fasten themselves to a solid surface.

Air freight. Transporting products by commercial air carriers; cut flowers are often shipped by airfreight.

Air pollutant. Material in the air that is polluting; a few pollutants are sulfur dioxide, ozone, and PAN.

Alkaline reaction fertilizer. A fertilizer that causes the soil to become more alkaline; this type of fertilizer is alkaline residue fertilizer.

Alloy. A combination of several metals. Aluminum alloys are used in the greenhouse framework.

Anaerobic. Lives without oxygen.

Anaphase. The period when the chromosomes divide into two equal groups and each half moves to opposite ends of the cell.

Anion. A negatively charged ion; nitrate is an anion NO_3^-.

Anion exchange capacity. The chemical activity of soils associated with negatively charged ions.

Annual plant. A plant that completes its life cycle in one growing season.

Anther. Pollen-bearing structure that is found at the end of the filament.

AquaGro®. A surfactant (wetting agent) used to reduce the surface tension of water.

Asexual (vegetative) propagation. This method of propagation uses sections of plants that are rooted.

Aspen fibers. Shredded wood fibers used to make up an evaporative pad.

Aspirated thermostat. Thermostat over which air is blown by a small fan; this type of thermostat is usually more accurate than a nonaspirated thermostat.

Autoclave. A method of sterilizing objects under pressure.

Automation. Mechanical or electronic means of doing work that formerly was done by hand—for example, thermostatic control of a heating system versus manual control.

Auxins. Control many plant growth functions including cell enlargement, apical dominance, and stimulation of roots.

Axillary bud. A bud that develops in the angle (axil) the leaf makes with the main stem.

Azalea pot. A three-quarter size pot; the height is three-quarters of its width.

Bent neck. A condition of roses that occurs a day or two after cutting; the stem fails to transport water to the flower, and the bud bends over without opening.

Biennial plant. A plant that requires two growing seasons to complete its life cycle.

B-I-N paint. A paint that is sometimes used successfully to cover pentachlorophenol-treated woods.

Black cloth. A means of providing artificial short days so that short-day plants will bloom.

Blindness. The failure to form flower buds because of a number of factors.

B-NINE SP®. A growth retardant applied as a foliar spray.

Bogotá. The capital of Colombia, South America, is the center of carnation production in that area.

Boiler horsepower. Method of rating the size of a boiler. One boiler horsepower is equal to the heat output of 33,475 Btu per hour.

Botrytis. A disease of the foliage caused by *Botrytis cinerea*.

Bottom heat. Heat applied to the rooting or propagation medium to hasten growth.

Bouyoucos bridge. An instrument that measures the moisture content of a growing medium; it is used with a Bouyoucos block.

Bract. A modified leaf that may be highly colored; poinsettias have very colorful red bracts.

Broadcast sowing. A method of sowing seed by sprinkling it randomly over the surface of the medium, as opposed to sowing in rows.

Brush-cut pruning. Roses are cut back to a uniform height in one cutting; the pruning is usually done after the Mother's Day cut.

Btu (British thermal unit). Amount of heat needed to raise the temperature of 1 pound of water 1 degree Fahrenheit, at, or near, its point of maximum density (39.1°F of heat). (Calorie. The amount of heat needed to raise the temperature of 1 gram of water 1°C, at 15°C.)

Bud blasting. The death of the developing flower bud, usually because of water stress.

Bud-cut. Flowers cut in the bud stage and then opened in a special solution.

Budding. A method of grafting used to propagate roses.

Bud union. The location where the bud was inserted into the understock.

Bulb cellar. Room used to give bulb crops a cold temperature treatment; also called a rooting room.

Bulblet. A small daughter bulb of a bulb-producing crop.

Bushel. A volume of media containing 1.25 cubic feet (0.03 m^3).

Calcined clay. Clay mineral heated to 800°C and then ground into granules.

Calyx. Made up of the sepals.

Calyx splitting. The calyx of the carnation breaks open, causing the petals to hang out; "splits" bring low prices in the market.

Cambial meristem. Tissues within the stem that divide and cause it to increase in circumference.

Capillary mat. A fiber mat that is used to distribute water to potted plants; the plants take up the water through capillary action.

Capillary tubes. Fine water-conducting tubes found naturally in soils.

Carbon dioxide. A colorless, odorless, tasteless gas used by plants to make food sugars (carbohydrates).

Carbon monoxide. A colorless, odorless poisonous gas that is formed by burning acrylic plastics in insufficient air.

Carotenoids. Yellow, orange, or red pigments found in the plants.

Cascade petunias. Petunias that have a trailing habit of growth; these plants are especially suitable for window boxes and hanging baskets.

Cation. A positively (+) charged ion; calcium is a cation (Ca).

Cation-exchange capacity. A chemical property of soils; cations have a positive (+) charge.

CEC (cation exchange capacity). Chemical activity in which soils pick up and release nutrient elements having a positive charge.

Cell. The smallest basic unit of life.

Cellulose. The wood fraction of bark products.

Cfm (cubic feet per minute). A measure of the amount of air moved by a fan; for example, an 18,000-cfm fan moves 18,000 cubic feet of air per minute.

Chamber steaming. This process is the same as vault steaming; items to be steamed are put in a closed chamber.

Chelated iron. A chemical form of iron that does not become tied up in the soil but remains available for plant uptake.

Chemical plant growth regulator (PGR). A natural or synthetic substance that in very small quantities regulates or controls some aspect of plant growth.

Chimney effect. Movement of air in a greenhouse created when both the top ridge and side ventilators are open.

Chlorophyll. The green pigment within the cell that is needed to carry on photosynthesis.

Chloropicrin (tear gas). A chemical used to fumigate soils.

Chloroplast. A plastid containing chlorophyll.

Chlorosis. Loss of normal leaf color due to a deficiency of an element or because of root injury.

Christmas pinch. Rose plants are pinched to bring the crop into flower for Christmas sales; the soft pinch is made October 17–21; the hard pinch is made October 20–25.

Chromosomes. The structures in the nucleus that contain the genes.

Clay pot. A pot made of fired clay; it is burnt orange in color and fragile (it is easily broken if dropped); it can be steam sterilized, has porous sidewalls, and is heavy in weight.

Clock auction. A wholesale floral auction where the sales are controlled by means of a large clock; clock auctions are located in Aalsmeer and other cities in the Netherlands. In North America, clock auctions are located in Toronto,

Montreal, and Vancouver, Canada, and in San Diego, California.

Clubbed spray. A spray type chrysanthemum having flowers too close together.

Coated seed. Seed that is coated by the supplier with clay or other substances; seeds treated in this manner are also known as pelleted seed.

Cold frame. Protective structure used to start plants early in the spring; usually covered with glass or plastic.

Cold storage. The treatment given to plants and bulbs to cause certain internal chemical changes that enable them to respond to forcing treatments.

Community pot/flat. A pot/flat where a large number of small seedings are transplanted for a short period before they are moved to other containers.

Complete fertilizer. A fertilizer that contains nitrogen, phosphorus, and potassium, the three major nutrient elements.

Compositae (Asteraceae). One of the largest plant families, including sunflowers, daisies, dandelions, chrysanthemums, and many other plants.

Connected greenhouse. Greenhouses that are connected at the side wall.

Consignment. One method used to sell flowers; the grower "consigns" a crop to the wholesaler who sells it and deducts a percentage of the selling price as a commission.

Constant fertilization. The application of a small amount of fertilizer at each watering.

Container. The object in which plants are grown, such as a pot, basket, bucket, or market pack.

Controlled-release fertilizer. A fertilizer that releases its nutrients slowly over a predetermined amount of time.

Controlled temperature forcing (CTF). A method used to precool lilies.

Convection. Movement of heated air due to natural forces.

Cormels. Small daughter structures produced by a corm; gladiolus produce cormels.

Cornell cut flower food solution. A standard bud-opening solution consisting of 8-HQC and sugar.

Cornell peat-lite mix. A combination of vermiculite or perlite and peat moss; developed at Cornell University and used as a growing medium.

Corolla. All of the petals together.

Cortex. Plant tissue just below the epidermis of the root and stem.

Cotyledon leaves. The first seed leaves that develop.

Creosote. A wood preservative that is phytotoxic and should never be used in the greenhouse.

Critical day length. The number of hours at which a plant responds to the day length by undergoing a change in its morphology (changing from vegetative growth to flowering growth).

Crown. A group of closely grouped stems or plantlets; African violets form multiple crowns.

Cubic foot. A volume of medium with dimensions of 1 foot by 1 foot by 1 foot.

Cubic yard. A volume of medium that is 1 yard long, 1 yard high, and 1 yard wide; 1 cubic yard contains 27 cubic feet and is equal to 21.6 bushels.

Cultivar. A propagated variety of a plant.

Curtain wall. A low sidewall, 2 to 3 feet high, made from wood, poured concrete, cinder block, glass, or plastic.

Curvilinear. Greenhouse with curved roof bars, usually at the lower part of the roof.

Cut flower food. Combination of chemicals made up as solution to increase the lasting life of cut flowers.

Cuticle. The protective coating on the leaf.

Cutin. Waterproof, waxy material that makes up the cuticle.

Cycocel®. A growth retardant that is applied as a soil drench or a foliar spray.

Cytokinins. Naturally occurring hormones produced in plant roots.

Cytoplasm. The total cell except the nucleus.

Damping off. Death of seedlings due to a soil-borne microorganism.

Day-neutral plant. A plant that flowers regardless of the length of day.

Day temperature. The temperature used in the greenhouse from 7 or 8 A.M. until 5 P.M.

Deficiency. A level of nutrition that is below that needed for good growth.

Deionized water. Water from which all ions are removed by means of the ion-exchange process.

Desiccation. Excessive drying of flowers; they do not recover from this extreme drying.

Detached greenhouse. A structure that stands apart from another greenhouse.

DIF. An abbreviation to designate the difference between day and night temperatures. DIF is calculated by subtracting the night temperature from the day temperature. A positive DIF is when the night temperature is less than the day temperature. A negative DIF is when the night temperature is greater than the day temperature.

Differentiation. See *Maturation*.

Diffusion. The movement of water into the roots; diffusion may be active or passive.

Disbudding. Removing the unwanted axillary bud to allow one bud to flower.

Disk florets. Florets in the center of the chrysanthemum head.

Disease organism. An organism that causes disease; the organisms may be found in the soil, water, or air.

Disinfectant. A chemical that kills microorganisms such as algae, bacteria, fungi, and viruses.

Diurnal cycle. An up-and-down fluctuation of temperature, day and night.

Division. A method of propagation similar to separation.

Dormancy. The resting stage of seed.

Dormant-eye rose. A budded rose where the "eye" is still dormant.

Drip gutter. Located below the gutter. This is a small channel that carries away the condensation that runs down the roof bar.

Dry storage at 31°F (–5°C). A method of storing certain flowers without water for extended periods.

Eave. The point at the top of the greenhouse side wall to which the roof bars are fastened.

Eave height. The height of the greenhouse side wall from the floor to the eave.

8-HQC. 8-hydroxyquinoline citrate is an additive in cut flower foods that helps to reduce stem blockages.

8-HQS. 8-hydroxyquinoline sulfate is an alternative additive in cut flower foods that performs the same function as 8-HQC.

Elbow arm. Mechanical leverage connector for opening and closing sash type ventilators.

Embryo. A miniature plant contained in the seed.

Endosperm and cotyledons. Food storage tissues in seed.

Epidermis. The bark or skin of the plant.

Epinasty. A growth condition shown by certain plants (tomatoes, marigolds) to excessive amounts of gas in the atmosphere.

Epiphyte. An air-loving plant that grows best in a very porous, highly aerated medium.

Equinox. When the natural day length is equal to the night; March 21 and September 21 are the spring and fall equinoxes, respectively.

ESMIGRAN. A trace element fertilizer produced by Mallinckrodt Chemical Corporation.

Essential element. A mineral element that is necessary for the growth of an organism; 16 elements are considered necessary for the optimum growth of plants.

Ethephon. A compound carried in liquid form as 2-chloroethyl-phosphonic acid.

Ethrel®. A growth-regulating chemical that is used to stimulate flowering in certain crops.

Ethylene. A gas formed from decaying plants, apples, and also produced by all flowers; it usually reduces the lasting life of cut flowers.

Ethylene scrubber. A material that absorbs ethylene gas from the atmosphere; this material is usually packed in shipping boxes when ethylene is a problem.

Etiolated seedlings. Stretched, leggy seedlings resulting from exposure to a very low light intensity.

Exclusion. The prevention of pest entry into a greenhouse through either screening or by not allowing infested plant material to be placed into the greenhouse.

Exhaust fan. Thirty-six to 48-inch (1 to 1.2 m)diameter fan used to exhaust air from the greenhouse.

Family. A subdivision of the plant kingdom; a family is subdivided into genera.

Fan and pad cooling. A method of reducing temperatures in the greenhouse. Outside air is pulled into a greenhouse through a wet pad by means of a fan.

Fertilizer. A material that contains nutrient elements.

Fiberglass. A plastic that contains fibers of glass that give it strength. Fiberglass may be flat or corrugated. It is used as a greenhouse glazing material. It burns very quickly. (FRP)

Fiber pot/pack. Pots/packs made of compressed fiber.

Filament. Stalk-like unit holding the anthers.

Finned pipe. Heat pipe to which closely spaced metal fins are fixed to increase the radiating surface of the pipe. One foot (0.3 m) of finned 2-inch (5-cm) pipe has the same radiating capacity as 5 feet (1.5 m) of nonfinned 2-inch (5-cm) pipe.

Fir bark. The medium used to grow orchids; it is usually mixed with peat moss.

Fire tube boiler. A boiler in which the heat is conducted through tubes surrounded by the water to be heated.

First true leaves. The first true leaves that develop after the cotyledon leaves.

Five leaflets. A rose leaf that has five leaflets.

Flaccid. Only partially filled with fluid, as in a wilted seedling.

Floral bud development. The next stage in the growth of the flower after bud initiation.

Floral bud initiation. The process of changing from a vegetative growing condition to a flowering stage of growth.

Floral preservative. A special solution used to increase the lasting life of cut flowers and to open bud-cut chrysanthemums and carnations. Also called cut flower food.

Floret. An individual flower on a spike, such as a gladiolus floret.

Flower sheath. The part of the orchid from which the flower stalk develops.

Fluorine. An element that exists as a gas or in combination with other elements; in excess amounts it causes plant injury.

Foam material. Retail florists use this material as a base for cut flower arrangements.

FOGGIT® nozzle. A special nozzle that produces a very fine, fog like spray; it is used for hand watering cuttings or seedlings.

Foliar analysis tissue testing. A method of determining the nutrient content of plant tissue by means of a chemical analysis of the leaves or other parts of the plant.

Footcandle. A measure of light intensity. Illumination at a point on a surface which is 1 foot from and perpendicular to a uniform point source of one candle.

Forced into bloom. The procedure whereby plants are made to bloom according to a schedule; for example, azaleas are forced for Easter sales.

Formalin. Commercial formaldehyde used to treat seed flats.

Foundation. The underground structure that supports sidewalls; also includes the footing that supports the load-bearing posts. The footing must be installed below the frost line.

Freeze drying. A method of preserving cut flowers, fruits, and vegetables in a permanent condition.

Frit. Controlled-release fertilizers, usually made of soft glass infused with nutrient elements and then ground to a fine powder.

FRP. Glass reinforced plastic (fiberglass)

FTE 555. Fritted trace element fertilizer.

Fumigation. Injection of chemicals into the soil or other medium to kill diseases and insects.

Fungicidal drench. A chemical drench used to combat certain soil-borne diseases.

Fungicide. A chemical used to combat fungus infections.

Genes. The messengers in the chromosomes that carry hereditary information.

Germination. Development of a seed from a resting stage to a stage of growth.

Germination media. The media (soil or soillessmixtures) in which seed is germinated.

Gibberellic acid. A growth-stimulating chemical.

Gibberellin. A naturally occurring plant hormone, which is responsible for the promotion of growth and cell wall elongation.

Glazing sill. Sash sill mounted at the top of the curtain wall, the unit on which the side glass rests. If a side ventilator is used, it is called a sash sill.

Gooseneck stage. The stage of growth at which the daffodil flower assumes a position horizontal to the ground.

Gothic arch. Greenhouse structure shaped like a church arch.

Gradual pruning. A method used to cut backroses that does not take them out of production completely.

Grafting. A method of propagation that is used for difficult-to-root plants; it is also used to take advantage of special characteristics of the rootstock.

Gravity flow return system. Gravity causes the return flow of cold water to the boiler.

Greenhouse atmosphere. The air inside the greenhouse.

Greenhouse effect. Process by which the rays of the sun entering the greenhouse as light are changed to heat energy and trapped inside.

Growing medium. A mixture of several materials in which plants are grown.

Guard cell. Specialized cells that control the opening and closing of the stomata; found mainly on the undersides of the leaves.

Guttation. The minute loss of water during the night at the tips and edges of the leaves.

Gutter. Section between two attached (ridge and furrow) greenhouses. It carries off rain water and melted snow.

Hanging basket. A growing container with attached supports that can be hung from a hook or rack.

Hardening off. A toning and strengthening process in which the temperature is lowered and/or water is withheld from the plants.

Hard pinch. A pinch that removes enough terminal growth that it can be used as a cutting.

Hardwood barks. Barks obtained from deciduous trees such as maple and oak.

Hardwood mulch. A hardwood bark product that contains less than 85 percent lignin.

Heat exchanger. Part of a heating system that is similar in action to an automobile radiator. Steam or hot water is circulated through a fin coil, and a fan blows air over the coil to force heat into the surrounding area. Unit heaters contain heat exchangers.

Heat transfer coefficient. A numerical value given to the relative amount of heat transmitted or transferred through a material. The heat transfer coefficient of glass is 1.15.

Herbaceous perennial. A plant that lives from year to year; the top growth dies back to the roots after a frost, but the roots live from year to year; apeony is an herbaceous perennial.

Herbicide. Weed killers; only certain types of chemicals are safe to use inside the greenhouse; 2, 4-D types should never be stored or used in a greenhouse.

HID (high-intensity discharge lamps). Used to supplement the daylight.

Homogeneous. Plants propagated vegetatively that are all the same.

Hotbed. A heated cold frame.

HOZON®. A simple fertilizer proportioner.

Humidistat. A control for the operation of a fogging or misting unit that operates on the relative humidity content of the air.

Humus. Organic matter in an advanced state of decay.

Hybridization. Plant breeders combine two hybrids to produce an improved cultivar.

Hydathode. The water of guttation exudes from this organ at the leaf tip.

Hydrophyte. A water-loving plant.

Hydroponic solution. Nutrient solution in which plants are grown.

Hypnum peat. A peat moss derived from *Hypnum* moss.

Hypocotyl. The first true stem of the seed.

Incandescent lamps. Standard bulb type lights that are high in red and infrared energy.

Indoleacetic acid (IAA). A naturally occurring auxin, but is not used commercially as it breaks down rapidly in light.

Indolebutyric acid (IBA). A synthetic auxin with commercial use.

Infiltration. Movement of outside air into the greenhouse through laps in the glass, holes, and other small openings.

Inflorescence. The flowering head of a plant.

Infrared radiation. Long-wavelength radiation from the sun, felt as heat.

Inhibitor. A substance in the seed coat that prevents or slows the germination process.

Inorganic material. Material that does not contain carbon; sand, perlite, and vermiculite are inorganic materials.

Insecticide. A chemical used to control insects.

Intercalary zone. In monocotyledonous plants, an area of meristematic tissue located just above the node.

Interior foliage plants. Tropical foliage plants that are grown in a special environment to condition them to life indoors.

Intermittent mist. A low-pressure system that applies a fine, fog type mist over cuttings; the frequency of the cycle of misting is controlled by mechanical or electronic means.

Interrupted lighting. A schedule of long and short days used to develop better flower heads on 'Indianapolis' and 'Mefo' cultivars in winter.

Interveinal chlorosis. Yellowing of the tissues between the veins.

Integrated pest management (IPM). The use of all possible means to control greenhouse pests. This includes diseases, insects, and weeds. Excluding insects by screening openings, ventilating to reduce relative humidity to control certain diseases, mechanically removing weeds by hand pulling or hoeing, using biological methods to control insects and diseases are all part of the IPM approach. Chemical pesticides are used only as a last resort.

Irrigation. The act of putting water on a crop.

Jiffy-Mix®. Commercial formulation of the Cornell peat-lite mixes made and sold by Jiffy Products of America.

Jiffy-7®. A commercially prepared, compressed peat moss disk used to propagate plants.

Juvenile stage of growth. Early stage of plant growth.

Kool-cel®. Trade name for a commercial, paper-based evaporative pad.

Lath house. A structure covered with wood or metal slats. It is used to protect plants from a high light intensity.

Layering. A method of propagation that produces a new plant while the stem being rooted is still attached to the mother plant; the mother plant serves a life support function. There are several methods of layering with air layering and mound layering popular choices.

Leaching action. The dilution and flushing away of nutrients or high soluble salt levels by means of heavy irrigations.

Lead. Newly developing shoot of an orchid (*Cattleya*).

Leaf bud cutting. A small section of the stem with a leaf attached; the new shoot arises from the axillary bud.

Leaf cutting. A single leaf from which new plants arise.

Leaf mold. The rotted remains of leaves.

Leaf scorch. A burning or necrosis of leaves; this condition is common to 'Croft' cultivar lilies.

Leaf tip burn. A necrosis or "scorching" of tips and edges of leaves; in members of the lily family, leaf tip burn is thought to be caused by fluorine toxicity; other causes include dry air, high soluble salt levels, and pesticide injury.

Lean-to. Greenhouse with a one-sided roof, usually built against another building or a greenhouse.

Light compensation point. The lowest light intensity at which the amount of food produced by photosynthesis just equals the energy used up in respiration.

Light meter. A meter used to measure light intensity; the measurement is usually in footcandles.

Light saturation point. The intensity of light beyond which the plant cannot use a brighter light for photosynthesis.

Lignin. The main component of bark, which decomposes slowly.

Lily pot. A three-quarter size pot.

Limiting growth factor. A factor that limits growth in the greenhouse; for example, a low light intensity in the winter is the most important limiting factor.

Liquid fertilizer. A fertilizer applied as a liquid.

Lobed. Deep indentation in the margins of a leaf, for example, a five-lobed leaf.

Long-day plant. A plant that flowers only when the day length is longer than a critical number of hours.

Louver. A mechanically or electrically operated unit that opens and closes to allow air to enter or exit a greenhouse.

Lumen. A measure of the light flux falling on a surface 1 square foot in area, every part of which is 1 foot from a point source of light of 1 footcandle in all directions.

Luxury consumption. The uptake of nutrient elements in amounts greater than is required for growth.

Macro element. Minerals needed in large quantities include carbon, hydrogen, oxygen, nitrogen, phosphorus, potassium, calcium, magnesium, and sulfur.

Magamp®. Controlled-release magnesium ammonium phosphate fertilizer.

Makeup air. Air needed to cause the burner of a heater to operate at 100 percent of its efficiency, usually drawn from outside the greenhouse.

Manure. Animal waste that is mixed with bedding material such as straw, sawdust, peat moss, wood chips, or similar products.

Marginal necrosis. Death of the margin (edge) of the leaf tissue; see *Leaf tip burn*.

Market pack. A fiber or plastic container used to grow bedding plants; the container holds from 6 to 12 plants.

Mass market outlet. A supermarket chain or discount stores that sell flowers.

Maturation. Area of the root where the cells reach maturity and form tissues.

Maximum. The highest level or the greatest value, such as the maximum temperature for growth.

Mechanical conditioning. A practice that is used to control plant height without the use of any chemicals.

Mechanical support. Support given by the growing medium to the plant.

Medium. The term used to describe the "soil" in which plants are grown; the plural of medium is *media*.

Meristematic. Actively dividing tissues.

Meristem tissue culture. A method of micropropagating plants using single cells; special techniques are required to grow plants from the cells; the method is also known as cloning.

Mesophyll. The middle tissue layers of the leaf, below the palisade layers.

Metabolic processes. Growth processes, or processes that carry on metabolism, developing tissues.

Metaphase. The second phase of cell division in which the divided chromosomes line up at the center, or equator, of the cell.

Mho. A measure of electrical conductance; the term is used to report the level of soluble salt concentration.

Micro element. Minerals needed in extremely small or trace amounts are iron, manganese, copper, boron, zinc, molybdenum, and chlorine.

Microorganism. A microscopic living organism in the soil or medium.

Micropropagation (tissue culture). A new propagation technique that uses a single cell of the meristematic tissue of a plant to produce a new plant; antiseptic techniques and specialized growth media are required; the procedure is also known as cloning.

Milliliter. A unit of measurement for liquids; one milliliter (1 ml) = one cubic centimeter (1 cm^3) = one-thirtieth ounce (0.03 oz).

Millimho. A measure of conductance, one-thousandth of a mho; a term used in reporting total dissolved solids or total soluble salts in solutions.

Millimicron. A measure of length; one millimicron equals one twenty-five millionth of an inch.

Mineral element. A substance used by an organism as a source of energy.

Minimum. Lowest or least, such as the minimum temperature for growth.

Mist fertilizer. A low-nutrient solution applied to newly propagated vegetative cuttings.

Mitochondria. Units within the cell that carry on respiration.

Mitosis. Cell division that occurs in several phases.

Mobility. The ability of mineral elements to move from one part of the plant to another; usually the elements move from the older tissue to newly developing tissues.

Moisture stress. A physiological condition caused by a shortage of water.

Monocotyledon. A single cotyledon plant; corn is a monocotyledon.

Monopodial growth. An upright type of growth shown by some orchids; *Vanda* orchids have this habit of growth.

Motorized accelerated system. Heating system that uses a pump or pumps to return cold water to the boiler.

Mottle leaf. A speckling and crinkling of the leaf tissue due to a lack of zinc.

Multicell pak. A plastic tray with many cells used to grow bedding plants as individual plants.

Mutation. A change in the character of an organism; mutations may be spontaneous, as in nature, or they may be induced artificially by chemicals or radiation.

Nanometer. A measure of wavelength that is equal to a millimicron.

Naphthalene acetic acid (NAA). A synthetic auxin with commercial use.

Naphthenate, copper and zinc. Wood preservatives (Cuprinol is one tradename) that are usually safe to use to treat greenhouse wood.

Negative geotropism. A reaction in which a plant bends upward away from gravity.

Nematode. An eel worm that lives in the soil and attacks the roots of plants.

Night temperature. The temperature used in the greenhouse from 5:00 P.M. until 7:00 or 8:00 A.M.

Nitrobacter **bacteria.** Bacteria that change the nitrite form of nitrogen to the nitrate form.

Nitrosomonas **bacteria.** Bacteria that change the ammonium form of nitrogen to the nitrite form.

Nucleus. Large central part of the cell that is surrounded by a membrane; it contains the chromosomes.

Nutrient agar. A medium used in tissue culture and seed sowing; agar is derived from seaweed (kelp) by a special process and nutrients are added to it.

Nutrient mist. A mist solution containing a small amount of fertilizer.

One-half strength. One-half the recommended amount.

Optimum. Best level or rate, such as the optimum rate of fertilization.

Organic fertilizer. A fertilizer material that contains carbon.

Organic material. Material containing carbon; peat moss, sawdust, and wood chips are organic materials.

Osmocote®. Controlled-release plastic coated fertilizer.

Outright purchase. A wholesaler buys flowers outright from the grower and sells them to retailers.

Ovary. The flower part that contains the ovule, or the seed.

Overfertilization. The application of too much fertilizer.

Overpotting. Putting too small a plant into too large a container.

Overwatering. The application of too much water.

Oxalic acid. A compound found in plants that can cause serious internal injury.

Palisade layers. Tissues just below the epidermis of the leaf; most photosynthesis takes place in the palisade layers.

Palmately compound. A leaf form consisting of several leaflets that are attached at the same point on the petiole.

Panning. Potting poinsettia plants or bulbs.

Paradichlorobenzene. A moth repellant that should not be used on tulips or other bulb crops to repel mice.

Parasite (or parasitoid). An organism that attacks and kills a single host. Typically the parasite completes all or part of its lifecycle in or on this individual pest.

Parenchyma cells. Tissue that is made up of loosely packed cells.

Pasteurization. Process that kills only harmful disease organisms; aerated steam at 140°F to 160°F (60°C to 72°C) is used to pasteurize soils.

Pathogen. Typically a microorganism that brings about a disease on a host animal or pest.

Peat bog. The area from which peat is dug; a natural deposit of peat moss.

Peat moss. Organic material used to grow plants.

Peat pots. Pots made of compressed peat moss and paper.

Pencil stage. At this stage of daffodil flower stem development, the bud points upright.

Pentachlorophenol. Phytotoxic wood preservative that should never be used in the greenhouse.

Perched water table. An area that contains 100 percent water at the bottom of a pot or bench filled with growing medium.

Perennial plant. A plant that lives from year to year; see *Herbaceous perennial*.

Perforated polyethylene tubing. Plastic tubing 18 to 24 inches (45 to 60 cm) wide, with regularly spaced holes on each side. Used to distribute air in a greenhouse.

PERK®. A trace element fertilizer manufactured by Kerr McGee Fertilizer Company.

Perlite. A volcanic mineral expanded by heating to 1,800°F (982°C); used as a soil additive, it is white in color.

Permanent wilting point. The point beyond which a wilting plant will not recover even when it is given water.

Petals. The showy, colored parts of the flower.

Petiole. The lower part of the leaf where it is attached to the plant; the petiole is often a stalk.

pH. The term used to denote the acidity or alkalinity of media or solutions.

Phloem. The vascular tissue that carries organic foods from the leaves to the roots and from the roots to the leaves.

Photoperiod. The length of the daily light period.

Photoperiod responder. A plant whose flowering depends upon the day length.

Photosynthates. Foods produced by the plant as a result of photosynthesis.

Photosynthesis. The conversion of light energy into chemical energy by means of chlorophyll; process by which plants change the energy of sunlight into usable food (carbohydrates and proteins).

Phototropism. Response shown by plants to light coming from one direction.

Phytochrome. An enzyme that controls the response of plants to day length.

Phytotoxic. Causing plant damage; certain gases are phytotoxic.

Pinch and one-half. A carnation crop is pinched 1 month after planting; 30 to 50 days after the first pinch, one-half of the shoots that develop from the first pinch are pinched. The process spreads out flowering over a longer period.

Pinching. The procedure by which a small part of the growing tip of the plant is removed to cause the development of axillary buds.

Pinnately compound. A leaf form consisting of several leaflets attached to a long petiole; the arrangement of leaflets may be opposite or alternate on the petiole.

Pistil. Female reproductive part of the plant consisting of the stigma, style, and ovary.

Pitch. The angle the roof makes with the horizontal.

Plant nutrient. Material used by plants to make growth.

Plastic pot. A pot made of any one of several types of plastics; it is light in weight, has nonporous sidewalls, and is available in many sizes and colors; plastic pots usually cannot be steam sterilized.

Plastids. Small bodies in the cell that contain the chloroplasts.

Plumule. The first terminal bud of the embryo; the first shoot that emerges from the seed.

Polyethylene. A petroleum-based flexible plastic used for many purposes; greenhouses can be covered with polyethylene.

Postemergent damping off. Soil organisms kill seedlings shortly after they emerge from the medium.

Postharvest life. The lasting life of a flower after it is harvested from the plant.

ppm (parts per million). A measure of the concentration of a fertilizer solution, for example, 200 ppm nitrogen. One ppm equals 1 milligram per liter.

Precooling. A cold temperature treatment given to tulips and other bulbs to make them flower.

Predator. Typically an insect or insect-related animal that attacks or feeds on a host or individual pest.

Preemergent damping off. Soil organisms kill seedlings shortly before they emerge from the medium.

Preharvest handling. Conditions under which the crop is grown before harvest.

Premature sprouting. A condition that affects lilies, causing them to sprout in the shipping case.

Prepared hyacinths. Hyacinth bulbs from plants that were grown in heated soils.

Preservative, wood. A fluid used to treat wood to prevent rotting. Creosote and pentachlorophenol are wood preservatives, but they must not be used in a greenhouse because their fumes are poisonous to plants.

Primary wall. The first wall laid down in the cell.

Produce Marketing Association. The Produce Marketing Association is the trade association for floral retailers and suppliers conducting business in the mass-market side of the floral industry.

Pro-Mix®. Commercial formulation of the Cornell peat-lite mix made and sold by Premier Brands.

Propagate. To reproduce by seed or other means.

Propagation. The increase of a plant species through generation after generation.

Propagation bench. The bench containing the rooting medium into which the cuttings are stuck to develop roots (rooting).

Propagation environment. The area where the propagation bench is located.

Propane gas. A fuel used to heat greenhouses; it is also burned to add carbon dioxide to the greenhouse atmosphere.

Prophase. The first phase of cell division.

Proportioner. A mechanical unit that dilutes a small amount of concentrated fertilizer stock solution into a large volume of water.

Protective environment. Surroundings for growing plants that are protected from the weather. The environment may be provided in a glasshouse, plastic covered greenhouse, cold frame, or hot bed.

Protoplasm. The organic contents of the cell.

Pruning. Removal of extra side shoots.

Pseudobulbs. The specialized storage tissues of orchids that are above the surface of the medium.

Pseudomonas species. Bacterial organisms that cause bacterial diseases in several plants; there are different species of *Pseudomonas*.

Psi (pounds per square inch). A method of expressing steam or water pressure.

Puddled soil. A clay soil that is worked when it is too wet; its structure is ruined.

Purlin. A structural unit that is used to support the sash bars. It is built lengthwise along the greenhouse. Purlins are supported by purlin posts and purlin braces.

PVC (polyvinyl chloride). A sheet plastic film.

PVF (polyvinyl fluoride). A fluorinated plastic that resists weathering. "Tedlar" is a commercial product used to coat fiberglass to make it last longer outdoors. PFV is resistant to ultraviolet radiation.

Pythium. A root rot disease of ornamentals; a preemergent damping off disease organism.

Quonset. A type of curved greenhouse that takes its name from the Quonset hut of World War II.

Rack and pinion. Gear type device for opening and closing sash type ventilators.

Radiant energy. Energy that comes from a radiant source such as the sun.

Radiational cooling. The loss of heat from plant leaves by means of radiation through the greenhouse glass; the loss is most intense on clear, cloudless nights.

Radiator. A type of heat exchanger.

Radicle. The first true root that emerges from the seed.

RA-PID-GRO®. A soluble fertilizer with an analysis of 23-19-17; in vegetative propagation it is used at the rate of 4 ounces (113 g) in 100 gallons (378.5 l) of water and is applied as a nutrient mist.

Ratoon. The offshoot pineapple that is used to propagate new plants.

Ray floret. Flowers around the perimeter of the chrysanthemum head.

Receptacle. The swollen part of the stem or stalk; the flower parts are attached to the receptacle.

Redi-Earth®. Commercial formulation of the Cornell peat-like mix made and sold by The Scotts Company, Marysville, OH.

Reed-sedge peat. Peat derived from the various reeds and sedges that usually grow together.

Relative humidity. The amount of moisture actually contained in the air compared to the total amount that can be held at a specific temperature with no condensation.

Repotting or shifting up. Replanting plants into another, usually larger, container.

Respiration. The process of exchanging oxygen and carbon dioxide through the stomata.

Response group. When applied to chrysanthemums, refers to the number of weeks after the start of short days that are required before the plant will bloom; for example, a chrysanthemum may belong to a 10-week response group.

Retail outlet. A place where flowers are sold to the public; it may be a traditional florist or a mass market outlet.

Rhizoctonia. A soil-borne disease that causes damping off of seedlings; the injury occurs at the soil line and is usually postemergent; a preemergent damping off disease organism.

Rhizome. An underground specialized stem.

Ridge. The top of the greenhouse; a ridge cap can be used as a walkway on the outside of the roof. Sash bars and the ridge ventilator fasten to the ridge.

Ring spot. A discoloration of African violet leaves resulting from cold water spotting the leaves; the cold water kills the chlorophyll.

Roof bar or sash bar. A specially grooved bar of wood or aluminum, which is used to hold the greenhouse covering.

Roof bar cap. A special unit that is fastened to the top of the roof bar to prevent glass slippage and leaks and to reduce weathering of the glazing compound.

Root cap. A group of cells at the tip of the root that protect the root.

Rooted cutting. The vegetatively propagated part of the plant that develops roots.

Root hairs. Specialized root cells in the epidermis of the root that absorb water and nutrients.

Rooting hormone. A root-stimulating substance that is applied to the base of a cutting; rooting hormones are available in several strengths.

Rosa manetti. The understock commonly used for grafting roses.

Salt index. A measure of the amount of salts a fertilizer adds to the soil.

Saran. A polypropylene mesh material used to cover structures to reduce the light intensity. Mostly used in Florida and other high light areas.

Sash bar. See *Roof bar.*

Scarification. The process of scratching a hard seed coat so that it can take up water.

Scion. The upper part of a grafted plant; the scion is fastened to the root stock.

Secondary roots. The branched roots that develop from the initial roots formed.

Seed geraniums. Geraniums grown from seed.

Seedling. Newly developing plant, started from a seed.

Semi-iron. A greenhouse framework that combines metal and wood as structural units.

Senescence. The change of cut flowers from maturity to death.

Sepals. Outermost parts of the flower.

Separation. A method of propagation in which clumps of roots or bulbs are separated.

Sexual propagation. This procedure uses seeds of plants for producing new plants.

Shading. The treatment given to mums to cause them to flower when the natural days are longer than the critical day length.

Shearing. A method of cutting back, or heading, plants and shaping them; azaleas are sheared.

Sheet film plastic. Flexible plastic that comes in very thin sheets; also known as flexible plastic. Polyethylene is a sheet film plastic.

Short-day plant. A plant that flowers only when the day length is shorter than a critical number of hours.

Sideposts. The weight-bearing units of the greenhouse.

Single-pinch carnations. Carnations that are given one pinch.

Single stem. A crop grown for single stems produces only one flower to a stem.

Sleepiness. The premature closing of carnation flowers, usually due to ethylene gas exposure.

Social necessity. A social event that is considered necessary to life, such as a wedding, birth, death, or illness.

Softened water. Water in which the calcium and magnesium ions and carbonates are exchanged chemically for sodium ions.

Soft pinch. The method of removing a small part (0.5 inch [1.25 cm]) of the top of the plant.

Softwood barks. Barks obtained from conifers such as fir, pine, and redwood.

Softwood mulch. A softwood bark product that contains less than 85 percent lignin.

Soil solution. The liquid part of the growing medium that contains the dissolved nutrients.

Soil structure. Refers to the arrangement of the soil particles.

Soil testing. A method of determining the nutrient content of a growing medium.

Soil texture. Refers to the size of the soil particles; gravel, sand, silt, and clay are classes of texture.

Solar energy. Radiant energy from the sun.

Soluble salts. The residue of fertilizers in the soil; soluble salts may be left by other materials, such as manure.

Solubridge®. An instrument that measures the total soluble salt content of growing media or solutions.

Sorus. A fruiting body on the fronds of ferns that emits spores; *son* is the plural of *sorus*.

Speedling® geraniums. Seed geraniums grown by the Speedling Corp. of Florida.

Sphagnum moss peat. A particular type of peat moss derived from the *Sphagnum moss* species.

Spike. A type of floral inflorescence.

Spitting. The breaking out of the flower stalk of the hyacinth.

Spray-type (mum). Chrysanthemums grown with many stems attached to the single large stem; also known as pompons.

Stage G. The stage in the development of tulip bulbs when all of the floral parts are visible.

Stamen. The male reproductive part of the plant, consisting of the filament and stamen.

Standard grades. Designations used to sort cut flowers to standard stem lengths or other measures of quality.

Standard mum. A large-flowered chrysanthemum, the "football" mum.

Standard pot. A full-size pot for which the height is equal to the width, measured at the top of the pot.

Started-eye rose. A budded rose plant with an "eye" growing as a branch.

Steam pasteurization. The use of steam to pasteurize the soil, pots, tools, and other materials used to produce a crop. A temperature of 160°F (57°C) is held for 30 minutes.

Stem section cutting. A section of the stem, usually with a leaf attached.

Sterilization. The process used to kill all organisms in a medium; may be accomplished by steam or chemicals.

Stigma. The part of the pistil that receives the pollen grain.

Stipules. Leaflike growths at the base of petioles; geraniums and roses have stipules.

Stock plants. A group of plants maintained to produce vegetative propagation material (cuttings).

Stoma. An opening in the leaf or epidermis that is controlled by guard cells; the stomata (plural of stoma) are the sites of gaseous exchange between the inside of the leaf and the atmosphere.

Style. The stalk through which the pollen tube grows to fertilize the ovary.

Subirrigation. Flats and/or pots are watered from the bottom, and the water rises by capillary action.

Sublimed. The change of substance from a solid state to a gaseous state without passing through an intermediate state; carbon dioxide sublimes from dry ice to the gas.

Surfactant. A chemical used to reduce the surface tension of water.

Surfside®. Surfactant (wetting agent) used to reduce the surface tension of water.

Sympodial growth. The horizontal growth of a rhizome; *Cattleya* orchids have this habit of growth

Telophase. The formation of new daughter cells.

Tendrils. Long, twining appendages used by plants for support.

Tensiometer. A mechanical device that measures the moisture content of a growing medium.

Terminal cutting. The most commonly used method of propagation, in which the stem tip is rooted.

Terrestrial. A plant that grows in soil as compared to an epiphytic plant, which grows in air.

Testa. The outer part of the seed coat.

Thermal blanket. A movable cover that is pulled at night to reduce the heat loss from the greenhouse.

Thermostat. A device that measures the temperature and controls the operation of a heating unit.

Thielaviopsis basicola. An organism that causes root rot in pansies, poinsettias, and geraniums.

Tip burn. See *Leaf scorch*.

Topsoil. Soil high in organic matter, usually the top 6-inch (15-cm) layer of soil.

Totipotency. The ability of each cell in a plant to reproduce the entire plant as an exact copy of the parent; in tissue-culture techniques, this ability is known as "cloning."

Traditional retail florist. A florist whose major business consists of the sale of cut flowers, arrangements, pot plants, and some accessories.

Translocation. The movement of water and organic foods through the vascular tissues.

Translucent. A term used to describe the light-transmitting property of a material. Translucent white plastic allows light to shine through, but the light is diffused and images cannot be seen clearly through the plastic.

Transmittance (light). The passage of light through a greenhouse covering. Black plastic has a light transmittance of 0 percent. Glass may transmit 92 percent of the light that strikes it, fiberglass about 90 percent.

Transpiration. The loss of water in the form of vapor from the leaves; transpiration takes place through the stomata.

Transplanting. Moving plants from one container to another; usually the container size increases; also, shifting seedlings from seed flats to packs.

Trimming for steam. Procedure whereby the water level in a hot water boiler is lowered slightly so that steam can be produced for steam sterilization.

Trombone heat system. A continuous line of heat pipes that look like a trombone.

Tropical foliage plants. Plants originating in tropical regions that are grown for their foliage.

Truss. A member used to support the greenhouse.

Tuber. A specialized stem structure that serves primarily as a food storage organ; white potatoes are tubers.

Tuberous root. Fleshy root that acts as a storage tissue.

Tulip topple. A physiological disease of tulips in which a lack of calcium causes the stem to collapse.

Turgid. Filled or swollen with fluid.

Ultraviolet inhibitor. A chemical added to polyethylene film that causes the plastic to resist breakdown due to the ultraviolet rays of the sun.

Ultraviolet radiation. Short-wavelength radiation of the sun; ultraviolet radiation causes sunburn in humans.

Understock (rootstock). The lower part of the grafted plant.

Uneven span greenhouse. The roof bars on one side are longer than those on the other; as a result, the roof has an uneven span.

Unicellular. Made of one cell.

Unit steam heater. A small heating unit usually mounted overhead in a greenhouse.

Vacuoles. A large empty chamber in a cell that is surrounded by a membrane.

Vapam. A soil fumigant.

Variegated. Shades of dark and light colors.

Vascular tissue. The specialized tissues that make up the pipelines that conduct water and nutrients through the plant.

Vegetative. Nonflowering stage of growth.

Vegetative meristem. Nonflowering growing point of the plant.

Velamen. The covering of aerial orchid roots.

Ventilator. A movable unit fastened to the ridge and sidewall, used to ventilate the greenhouse on hot days or whenever temperatures go above recommended levels.

Ventilator header. The ventilator rests on the ventilator header.

Vermiculite. A mica mineral expanded by heating to 1,400°F (760°C); used as part of the peat-lite mixes.

Verticillium. A fungus-producing organism that grows in the vascular system of the plant and causes wilting.

Verticillium **wilt.** Soil-borne disease.

Virus. Extremely small organisms that cause a variety of plant diseases.

Volatiles. Fumes given off from paints or other materials; some are damaging to plants.

Waterlogged soil. Soil that is overwatered.

Water tube boiler. A boiler in which the water to be heated is conducted through tubes that are surrounded by the heat rising from the fire.

Watt. The unit of measurement of electrical energy consumed.

Wet-cold storage. Storage of cut flowers with the stems inserted in a holding or cut flower food solution; the storage temperature is 33°F to 35°F (0.5°C to 1.5°C).

Wetting agent. Material added to water to reduce the surface tension and make the water wetter.

WF&FSA. Wholesale Florists and Florist Suppliers of America.

Wholesale distributor. Buys flowers from the grower and sells them to the retailer.

Witches' broom. The unchecked growth of many shoots caused by a lack of boron in chrysanthemums, carnations, and roses; the resulting growth is broomlike in appearance.

Xylem. The vascular tissue that carries water from the roots to the leaves.

Year-round flowering. The practice of flowering certain plants year-round; for example, azaleas and chrysanthemums are flowered year-round.

Zoned heating. A large greenhouse is equipped with several thermostats to control the temperatures by "zones." A more uniform temperature is possible with zoned heating than if one thermostat is used.

2, 4-Diphenyloxzoline derivatives, 360*t*

8-HQC, 657

8-HQS, 742

10-10-10 fertilizer, 205, 206*t. See also* Peat-lite mixes

A

Abamectin, 362*t*

Above ground plant structures, 136–139

Abscission, 138–139

Absorption
  mineral elements, 223–228
  water, 273–274

Abutilon, 420

Acalypha hispida, 589

Acanthaceae, 580

Accelerated heating systems, 92, 92*f*

Acclimatization, 68, 597–602

Acephate, 362*t*

Acephate + fenpropathrin, 365*t*

Acequinocyl, 362*t*

Acetamiprid, 362*t*

Achimenes, 420–421, 589

Acidic organic matter, 262

Acidity, 262–263, 291–292

Acrylic, 56*t*, 62–63, 64*f*

Action threshold, 349

Acylalanine + phenylpyrrole, 388*t*

Acylamine, 385*t*

Add reaction, in nitrogen, 244–245

Adiantum cuneatum, 595

Administrative skills, 29

Adult stage of growth, 301, 301*f*

Aerated steam, 212–213

Aeration, 167–168, 184, 310

Aerial roots, 679

Aerosol application, 369–370

Aeschynanthus, 589

A-frame framing method, 53, 54, 54*f*

African violet, 5*t*, 287*f*, 326*f*, 336*f*, 470–474

Agavaceae, 580–582

Ageratum, 306*t*, 614

Aglaonema, 583

Agricultural Bioterrorism Act, 26

Air conditioning, mechanical, 124

Air currents, 160–161

Air layering, 337

Air movement, 112, 112*f*, 113, 277

Air pollutants, 174–175

Air temperature
  metabolic processes and, 161
  photosynthesis and, 159–160
  reduction, 111–112
  respiration and, 160
  transpiration and, 160–161
  See also Temperature

Air-freight, 750

Alkaline reaction, in nitrogen, 245

Alkalinity adjustments, 261–262

Alloys, 50

Alstroemeria, 712–714

Alternative crops, 358–359

Aluminum, 238

Aluminum framework, 50

Aluminum sulfate, 263

Aluminum trioxide, 65

Alyssum, 306*t*, 614–615

Amaranthaceae, 582

Ammonia, 166, 172, 253*t*

Ammonium nitrate, 244–245, 253*t*, 260, 308, 608*t*, 698*t*

Ammonium phosphate, 245

Ammonium polyphosphate, 253*t*

Ammonium polysulfide, 253*t*

Ammonium sulfate, 244, 253*t*, 260*t*

Ammonium thiosulfate, 253*t*

Anaerobic organisms, 167

Ananas, 589

Anaphase, 324, 324*f*, 325

Ancymidol, 401–402, 402*t*

Andean Free Trade Act of 2002, 23

Andreanum, 584

Anemones, 521–522

Animal and Plant Health Inspection Service (APHIS), 26

Anion, 224, 225, 226, 232

Anion exchange capacity, 225–226

Annette Hegg Poinsettia, 491

Annual phlox. *See* Phlox

Annual plants, 301

Anther, 139, 139*f*

Anthurium, 583, 714

Antirrhinum. See Snapdragons

Aphids, 353*f*, 370–371, 371*f*, 394*t*

Aphrodite series, 434, 436

Apical meristem, 134*f*, 135*f*, 136*f*

Apple leaf, 137*f*

Application methods, for fertilizers, 252–260*t*

AquaGro, 202*t*, 294*f*, 608*t. See also* Peat-lite mixes

Arabian violet. *See* Exacum

Araceae, 582–586

Araliaceae, 586–588

Araucariaceae, 588

Arching, 702, 703*f*

Ardisia, 592

A-REST, 565, 566–567*t*, 568. *See also* Growth retardants

Aromatic hydrocarbon, 385*t*

Arrangements
  benches, 81–83
  greenhouses, 80–81
  piping, 102–103

Artificial light
  African violets, 473–474
  Begonias, 435*t*
  Chrysanthemums, 654–655

Artificial light intensity, 143–144, 145*f*

Artificial soils
  industry regulation, 208
  need for, 201
  standard growing media, 201–207

Asexual propagation, 300–301, 323
  basis for, 324–325
  divisions, 335–337
  environment, 328–335
  grafting, 338–339
  layering, 337–338
  methods of, 325–328
  micropropagation, 339–340
  separations, 335
  terminal cuttings (student project), 343–345
  See also Propagation

Asparagus, 590*f*, 598*t*

Aspidistra elatior, 591, 598*t*

Aspirated thermostat, 125–126

Asplenium nidus, 595

Asteraceae (Compositae), 588

Asters, 174*f*, 306*t*, 615

Atmospheric carbon dioxide, 168–169

Atmospheric oxygen, 167

Autoclave, 683

Automatic black cloth systems, 150

Auxins, 404

Axillary buds, 137

Azadirachtin, 361*t*

Azalea, 3*t*, 5*t*, 9*t*, 421–422
  forcing, 427, 428*t*, 431*t*
  origin, 422–423
  problems, 431–432
  production, 424–431
  propagation, 423–424

Azalea pots, 413

Azoxystrobin, 387*t*

B

Baby's breath. *See* Gypsophila

Baccto peat-lite mixes, 207

Bachelor's Button. *See* Centaurea

Bacillus subtillis GB03, 385*t*

Bacillus subtillis QST713, 385*t*

Bacillus thuringiensis Israelensis, 360*t*

Bacillus thuringiensis Kurstaki, 360*t*

Bacteria, 392

Bar caps, 47*f*, 48, 48*f*

Bark, 190, 262

Barrel vault greenhouses, 44, 44*f*

Bays, 50

Bean yellow mosaic, 394*t*

Beauveria bassiana, 360–361*t*

Bedding and garden plants, 3*t*, 4, 5*t*, 6*f*, 82*f*, 605
  Ageratum, 614
  Alyssum, 614–615
  Asters, 615
  Browallia, 615
  Celosia, 615–616
  Coleus, 616
  Geraniums, 617–620
  Impatiens, 620–621
  Lobelia, 621–622
  Marigolds, 622
  Pansy, 622–623
  Petunia, 624–626
  Phlox, 626–627
  popular crops, 614
  production of, 605–614
  Salvia, 627
  Snapdragons, 627–628
  temperature and, 162, 162*f*
  vegetable crops, 629–630, 631–632*t*
  Verbena, 628
  Zinnia, 629

Bees, 353*f*

Beet curly top, 394*t*

Begonia, 306*t*, 432–437

Begonia semperflorens, 437

Begonia socotrana, 433–434

Benches, 81, 81*f*, 82*f*, 83, 83*f*, 84, 84*f*, 85, 85*f*, 287*f*, 329

Bending, 702, 703*f*

Bent neck, 706

Benzamides, 385*t*

Benzoyl urea insect growth regulators, 360*t*

Bicarbonate, 385*t*

Biennial plants, 302

Bifenazate, 361*t*

Bifenthrin, 363t
Biological control, 357–358
Biopesticide, 360–361t, 385t
Biopesticide insect growth regulators, 361t
Black cloth systems, 150, 150f, 151, 151f, 655f
Blindness, 488
Bloom forcing. See Forcing
Blue Hydrangea, 485–486
Boilers, 103, 104, 104f, 105, 105f
Bone meal, steamed, 247
Boom irrigation systems, 285–286, 609f
Borden's, 38, 248
Boric acid, 253t, 361t, 385t
Boron, 223, 227f, 228f, 229, 235, 236, 237, 239t, 251, 698t
Botanical extract, 385t
Botanicals, 361t
Botrytis, 29, 384t
Botrytis blight, 384
Bottom heat, 159, 310f, 311f, 330f
Bougainvillea, 438–439
Boutiques, 760
Bouyoucos blocks, 279
Brassaia, 586
Brazil, 23
Brightpoints Poinsettia, 492
Broadcast sowing, 313
Bromeliaceae, 588
Bromeliads, 437–438, 588
Browallia, 306t, 615
Brush-cut pruning, 702
Btu (British thermal unit), 90, 90t
Bud blasting, 532
Bud cut carnations, 666, 667
Bud cut standard chrysanthemums, 656–657
Budding, 338–339, 691–693
Bud formation. See Flower bud formation
Bud union, 693
Bulb cellars, 67
Bulblets, 534
Bulbs, 516
    Anemones, 521–522
    Calla lilies, 522–524
    Crocus, 524–525
    forcing, 516–521
    Freesia, 525–526
    Gladiolus, 526–530
    Hyacinth, 530–532
    Iris, 532–533
    Lily, 533–546
    Narcissus (daffodils), 546–552, 552t
    soil temperature and, 163
    student projects, 571–573
    Tulips, 552–568
Bulb size, of potted lilies, 537–538
Bulb suppliers, 31
Buprofezin, 360t
Buried pipe steaming, 214, 214f
Business plan, for greenhouse enterprises, 74–75
Butterfly flower. See Schizanthus

C

Caladium, 439–440
Calathea, 591
Calceolaria, 289f, 306t, 440–441
Calcined clays, 202
Calcitic limestone, 263
Calcium, 223, 225, 227f, 228f, 233, 233f, 234f, 236, 239
Calcium carbonate, 253. See also Limestone
Calcium hydroxide, 261
Calcium nitrate, 202t, 206t, 245, 253t, 260, 698t. See also Peat-lite mixes
Calcium oxide, 65, 261
Calcium sulfate, 253t, 261–262
Calendula, 306t, 714–715

California
    cut flowers, 637
    plant production in, 5, 5t, 6t, 7t, 8, 8t, 9, 9t, 10, 10t, 11, 11t, 12f
    tropical foliage plants, 575, 575t
    white rust and, 26
Calla lilies, 522–524
Callistephus, 715–718
Callistephus chinensis. See Asters
Calyx, 139
Calyx splitting, 665–666
Cambial meristem (cambium), 134, 134f
Canada, 18t, 26, 185, 186f
Capillary mat, 289–291
Capillary tubes, 180, 181
Captan, 387t
Carbamate, 361t, 385t
Carbamate insect growth regulators, 361t
Carbaryl, 361t
Carbazates, 361t
Carbon, 223–24
Carbon dioxide, 43, 126, 166, 168–171
    Carnations, 664, 665f
    Chrysanthemums, 648–649
    photosynthesis and, 141–142
    Roses, 697–699
    Snapdragons, 642
    ventilation and, 112
    water loss and, 274–275
CARE plant modeling software system, 127
Carnations (Dianthus)
    Colombian production, 22f
    cut flowers, 637, 658–669
    dry-cold storage, 740
    ethylene and, 171f
    imports of, 18t, 19t
    Kenya production, 24f
    propagation environment, 332f
    seed germination, 306t
    student project, 678
    U.S. cut flower production of, 20t
Carotenoids, 133
Cascade geraniums, 476
Cascade petunia, 624
Cation, 194, 224, 225, 225f, 232, 239
Cation exchange capacity (CEC), 224–225
Cauliflower mosaic, 394t
Ceiling. See Interior ceiling
Cellar-rooting room, for bulbs, 520
Cell division zone, 135f
Cell enlargement zone, 135f
Cells, of plants, 132–133, 133f
Cellulose, 190
Celosia, 306t, 615–616
Centaurea, 718–719
cfm (cubic feet per minute), 116
Chain stores, 760, 760f, 761f, 762, 762f, 763f
Chamber steaming, 215–216
Chelated iron, 236
Chelates, 251
Chemical control, 358
Chemical fumigation, 309
Chemical plant growth regulator, 399, 404–405
Chemical plant growth retardants. See Growth retardants
Chemical reaction, of photosynthesis, 141–142
Chemicals, acidifying, 262–263
Chemical soil treatment, 216–217
Chimney effect, 49
China aster. See Callistephus
Chlorfenapyr, 364t
Chlorides, 225
Chlorine, 223, 235, 238, 292
Chlormequat chloride, 400–401
Chloronitriles, 385t

Chlorophyll, 133, 133f, 141, 152
Chlorophytum comosum, 591–598t
Chloropicrin, 216, 217
Chloroplasts, 133, 133f, 138f
Chlorosis, 162, 230, 233–234, 234f, 235, 235f, 237, 238, 239t, 432, 631t
Chlorothalonil, 385t
Chlorpyrofos, 363t
Chlorpyrofos + cyfluthrin, 365t
Christmas cactus, 444–445
Christmas cherry, 446
Christmas pinch, 701
Chromosomes, 133
Chrysanthemums, 3t, 5t, 8t, 14f, 454–455
    Colombian production, 23f
    cut flowers, 636–637, 643–658
    dry-cold storage, 740
    general culture, 455–462
    growth retardants and, 399f
    imports of, 18t, 19t
    irrigation, 287f
    problems, 462–463
    student project, 677–678
    U.S. cut flower production of, 20t
Cineraria, 306t, 446–447
Cinnamic acid amides, 385t
Cissus Antarctica, 597, 598t
Cissus rhombifolia, 597, 597f
Citrus, 596
Clarified hydrophobic extract of neem oil, 362t, 387t
Classification
    Chrysanthemums, 644–645
    Snapdragons, 638, 639t
Clay, 183f, 183t
Clay pots, 410, 411
Clean Air Act, 90
Clerodendrum, 448–449
Cleveland cherry. See Christmas cherry
Climate, 8t, 10, 11, 21, 24, 76–77. See also Weather
Clock auction, 747–748
Clofentezine, 365t
Cloth house culture, for Callistephus, 717
Clothianidin, 362t
Clubbed spray, 651
Coal, 90, 90t
Coated seed, 313
Cockscomb. See Celosia
Codiaeum variegatum, 580, 588t, 598f
Coffee, 596
Coir, 191, 192f
Coir coconut fiber, rose production in, 696–706
Cold frames, 66–67, 71–72, 518–520, 535–536
Cold period, for bulbs, 520–521
Cold soils, 277
Cold storage, for cut flowers, 737–743
Coleus, 306t, 616
Colloid, 225, 225f, 232, 238
Colocasia, 234f
Colombia, 13, 17, 18t, 21–23, 637, 736f
Color. See Flower color
Colorado, 7, 11t, 13, 637
Colorado series, 452
Columneas, 590
Commelinaceae, 589
Complete fertilizers, 248–250
Compositae, 644
Composition, of glass, 65
Compound layering, 338
Computerized climate control systems, 126, 127
Computer skills, 29–30
Concerto series, 452
Condensation, 161
Connected greenhouses, 43f, 43–46
Consignment, 750

Constant fertilization, of Chrysanthemums, 457
Construction
  cold frames, 71–72
  curtain walls, 52
  foundations, 51–52
  greenhouse, 46–49
  greenhouse benches, 84–85
  hotbed, 67
Containers and container grown crops, 410–411
  bedding plants, 612
  bulbs, 518
  clay pots, 411
  Geraniums (potted), 617
  growing media, 413–414
  marketing preparation for, 416–417
  peat pots and multicell paks, 412
  plastic pots, 411–412
  pot sizes and shapes, 413
  transplanting, potting, and repotting, 414–416
  wet-cold storage, 738
Control, of greenhouse environment, 124–127
Controlled-release fertilizers, 248–250, 411, 414, 457
Controlled temperature forcing (CTF), of Lilies, 536
Convection, 94
Cooling
  fan and pad, 119–122
  flowing water shade, 124
  greenhouse environment control, 124–127
  high-pressure fog, 122–123
  mechanical air conditioning, 124
  positive-pressure, 353, 354f
  radiational, 161
  shading, 123–124
  ventilation and, 111–112
Copper, 223, 227f, 228f, 235, 236, 237, 239t, 698t
Copper (complex), 386t
Copper (fixed), 386t
Copper hydroxide, 386t
Copper sulfate, 386t
Co-RAY-VAC System, 106, 107
Cordyline, 581
Cormels, 527
Corms, 526, 528–529, 529t
Corr embryo of, 303f
Cornell cut flower food solution, 742, 743t
Cornell Epiphytic Mix, 597, 577f
Cornell Folliage Plant Mix, 577, 577t
Cornell peat-lite mixes, 202–207, 208, 608t
Corolla, 139
Cortex, 135, 136f
Costa Rica, 18–19, 23
Cotyledons, 303
Creosote, 173
Crocus, 524–525
Crop fertilizers. See Fertilizers
Crops
  alternative, 358–359
  greenhouse enterprises and, 78
  See also specific flowers and plants
Crowns, 336
Cucumber mosaic, 394t
Cultivars, 148
  Carnations, 669
  Chrysanthemums, 645–646, 658
  Geraniums, 475–476, 617
  Hyacinths, 531
  Kalanchoe, 489
  Lilies, 534
  Narcissus (daffodils), 549t, 550, 552t
  Poinsettia, 491–492
  resistant, 358–359
  Roses, 512
  Tulips, 556–557t, 563–564t, 566–567t
Cultural control, 357

Cultural disorders
  Achimenes, 421
  African violets, 474
  Alstroemeria, 714
  Anemones, 522
  Azalea, 431
  Caladium, 439
  Calceolaria, 441
  Calendula, 715
  Callistephus, 718
  Carnations, 667
  Centaurea, 719
  Christmas cherry, 446
  Chrysanthemums, 461, 657
  Cineraria, 447
  Clerodendrum, 448
  Cyclamen, 454
  Dahlias, 722
  Gardenia, 465
  Genista, 466
  Geraniums, 481–482
  Gladiolus, 529
  Gloxinia, 484
  Hydrangea, 488
  interior foliage plants, 601–602
  Iris, 533
  Kalanchoe, 490
  Lilies, 545
  Matthiola, 726
  Orchids, 688–689
  Peonies, 728
  Roses, 706
  Snapdragons, 642
  Tulips, 568
  vegetable crops, 630, 631–632t
Curtain walls, 39, 47f, 52, 52f
Curved eave greenhouse, 38, 42
Curvilinear greenhouse, 38, 42
Cutback roses, 701–703
Cut flowers, 636–637
  Alstroemeria, 712–714
  Anthurium, 714
  benches, 81, 81f, 84f, (see also Benches)
  Calendula, 714–715
  Callistephus, 715–718
  Carnation, 658–669, 678
  Centaurea, 718–719
  Chrysanthemums, 643–658, 677–678
  cold storage treatments, 737–743
  Colombia, 23
  Dahlias, 720–722
  Gerbera, 722–724
  Gladiolus, 527–528, 528t
  Gypsophila, 724
  handling, 751–754
  harvesting, 736–737
  Kenya, 25
  life support factors, 734–735
  Matthiola, 724–726
  Narcissus (daffodils), 546–549, 549t
  offshoring and, 17–20
  Orchids, 679–690
  Peonies, 726–728
  preharvest handling, 736
  production and harvesting methods, 735–736
  Ranunculus, 728
  retail outlets, 763
  Roses, 690–707
  Schizanthus, 728–729
  shipping, 750–751
  Snapdragons, 638–643, 675–677
  Stephanotis, 729–730
  student projects, 675–678
  top crops grown, 3t
  Tulips, 553–556, 556t, 559t, 730f
  wholesale value (2004), 4

See also off shore flower Production; specific flowers
Cuticle, 138, 138f, 274
Cutin, 138
Cuttings
  Azalea, 423–424
  Carnation 660, 660f
  Chrysanthemums, 455–456, 456f, 457f
  Geraniums, 474–481
  leaf, 326
  leaf bud, 326
  Poinsettias, 493–494, 494f
  Roses, 691
  stem section, 326
  terminal, 325, 343–345
  See also Propagation
Cyathin, 587f
Cyclamen, 306t, 449–454
Cyclodiene organochlorines, 361t
Cycocel, 503, 503t
Cyfluthrin, 363t
Cyfluthrin + imadacloprid, 365t
Cyromazine, 365t
Cyrtomium falcatum, 595
Cytisus canariensis. See Genista
Cytokinins, 404–405
Cytoplasm, 133, 133f

D

Daffodils. See Narcissus
Dahlias, 360t, 336f, 720–722
Daisies, 22f See also Gerber daisies
Daminozide, 400, 401t
Damping off, 310
Day length
  Chrysanthemums, 460–461
  conditions, detection of, 152
  controlling, 149
  decreasing, 150–151
  effects of, 147–148, 151–152
  extending, 149–150
  Poinsettias, 500–501
  See also Light; Photoperiod
Day-neutral plants, 148
Day-night cycle, 159
Day-night temperature relationships, 157–159
Day temperature (DT), 157, 429
Deficiency, 222
  boron, 236, 239t
  calcium, 233, 233f, 234f, 239t
  chlorine, 238
  copper, 237, 239t
  iron, 235, 235f, 236, 237, 239t
  magnesium, 234, 234f, 239t
  manganese, 237, 239t
  molybdenum, 237
  nitrogen, 230–231, 239t
  phosphorus, 231–232, 239t
  potassium, 232, 233f, 239t
  sulfur, 239t
  zinc, 237, 239t
Deficiency symptoms, 239t
Delphiniums, 3t
Dendranthema. See Chrysanthemum
Dendrobium orchids, imports of, 19t
Desert coolers, 124, 124f
Desiccation, 157
Detached greenhouse, 38–40, 42–43
Diammonium phosphate, 245, 247, 253t, 260t
Dianthus. See Carnation
Dicalcium phosphate, 253t
Dichlorvos, 363t
Dicoarboximides, 386t
Dieffenbachia, 584
DIF day-night temperature relationships, 157–159, 399

Differentiation, 135
Diffusion, 170, 273, 275
Diflubenzuron, 360t
Dilution ratios, 255–257, 259t
Dimethoate, 363t
Dimethomorph, 385t
Dinotefuran, 362t
2,4-Diphenyloxzoline derivatives, 360t
Direct pruning, 702
Direct rooting, 334, 495–496
Disbudding
    Chrysanthemums, 459–460, 651–652
    Dahlias, 721
Disease organisms, 309
Diseases
    Achimenes, 421
    African violets, 474
    Alstroemeria, 714
    Anemones, 522
    Azalea, 432
    Begonias, 437
    Caladium, 440
    Calceolaria, 441
    Calendula, 715
    Calla lily, 524
    Callistephus, 718
    Carnations, 668
    Centaurea, 719
    Christmas cactus, 445
    Chrysanthemums, 461, 657–658
    Cineraria, 447
    Clerodendrum, 448–449
    cut flowers, 735
    Cyclamen, 454
    Dahlias, 722
    Gardenia, 465
    Geraniums, 482
    Gerbera, 723–724
    Gladiolus, 529–530
    Gloxinia, 484
    greenhouse diseases, 382–395
    growing skills and, 29
    Gypsophila, 724
    Hydrangea, 488
    imports and, 26
    interior foliage plants, 602
    Iris, 533
    Kalanchoe, 490–491
    Lilies, 545–546
    Matthiola, 726
    Narcissus (daffodils), 552
    Orchids, 689
    Peonies, 728
    Poinsettia, 504–506
    Primrose, 508
    Rieger elatior hybrids, 436
    Roses, 512, 706–707
    Snapdragons, 642–643
    Tulips, 568
    vegetable crops, 630
Disinfectant, 351, 412f
Disk florets, 644
Distribution, of tropical foliage plants, 575–576.
    See also Wholesale distributors
Dithiocarbamates, 386–387t
Diurnal cycle, 159
Division, 335–337, 684
Dixie series, 452
Dizygothea, 586
Dolomite limestone, 202t, 253t, 261, 263, 608t
Dome shaped greenhouse, 38
Dominican Republic, 23
Dormancy. See Seed dormancy
Dormant eye, 692
Dosotron, 256f
Double layer plastic, 108–109, 109f

Double phosphate, 246
Downy mildew, 384, 389t, 390, 643f
Dracaena, 581, 581f, 582f, 598t
Drenches, 309–310, 545
Dried cut flowers, imports of, 18t, 19t
Drip gutters, 47f, 48
Dry application fertilizers, 248, 252
Dry-cold storage, 666–667, 739–741
Dry ice, 169
Dutch Venlo greenhouses, 44f, 44–45
Duties, 23
Dwarf plams, 593
Dyed cut flowers, imports of, 18t, 19t

E
Easter lilies, 3t, 5t, 10t
Eave greenhouse, 38
Eave height, 39
Eave plate, 47, 47f
Ebb and flood benches, 84, 84f, 287f,
    330f, 413f, 490f
Ebb and flood irrigation, 286–289
Eckespoint Celebrate Poinsettia, 491
Eckespoint 'Lilo' Poinsettia, 491
Ecuador, 18t, 19t, 23
Elbow arm, 113, 113f
Electricity, 90
Elegance Geraniums, 475
ELISA, 393–394
Embryo, 303, 303f
Emergency heating equipment, 105
Endosperm, 303, 303f
Endosulfan, 361t
Energy issues, 8
Environmental control, 357
Epidermis, 134, 138f
Epinasty, 171
Epiphytes, 576, 578, 679, 685–686
Episcias, 589
Epsom salts. See Magnesium sulfate
Equinox, 147, 147f
Erwinia, 494f
ESMIGRAN, 202t, 206t, 250, 251, 608t. See also
    Peat-lite mixes
Essential macro elements, 223, 228–234
Essential mineral elements, 222–223
Essential trace elements, 228, 235–239
Established greenhouse businesses, acquisition
    of, 78–79
Ethephon, 404
EthylBloc, 742–743
Ethylene, 166, 171–172, 172f, 404
Ethylene scrubbers, 667
Ethylene vinyl acetate (EVA) copolymers, 58
Ethyl phosphonates, 387t
Etiolated, 312
Etoxazole, 360t
Etridiazole, 387t
Euphorbia pulcherrima. See Poinsettia
Euphorbiaceae, 589
Even span greenhouse, 38–40
Exacum, 463
Exchange capacity, 194
Exclusion, 350–353, 358
Exhaust fans, 110
Experimental heat conservation methods, 111
Expert knowledge requirements, for plant
    growers, 28–30
Exporters, 20–26
External gusset framing method, 53, 54, 54f

F
F₁ hybrids, 451–452
Facultative parasites, 384
Facultative saprophytes, 384
Fafard peat-lite mixes, 207, 207f

Fairy wings. See Lobelia
Fall equinox, 147, 147f
Family, 419
Fan and pad cooling, 109, 110f, 119–122
Fancy-leaved caladium. See Caladium
Fans, 96, 96f, 97, 97f, 110
Fan ventilation, 115–119, 115–119f
Fatshedera, 587, 598t
Federal Worker Protection Standards, 366
Fenamidone, 387t
Fenbutatin-oxide, 363t
Fenhexamid, 387t
Fenoxycarb, 361t
Fenpropathrin, 363t
Fenpyroximate, 363t
Ferns, 600t
Ferric iron, 235
Ferrous iron (iron sulfate), 235, 262–263
Fertilizer arithmetic, 257–258
Fertilizers/fertilization, 224, 242
    African violet, 471–472
    Alstroemeria, 713
    Anemones, 521
    application methods, 252–260
    asexual propagation environment and,
        333–334
    Azalea, 425
    bedding plants, 611, 611f
    Calceolaria, 441
    Calendula, 714–715
    Calla lily, 523
    Carnations, 664
    Centaurea, 719
    Chrysanthemums, 457, 647, 648–649
    complete, 248–250
    controlled-release, 248–250, 411, 414, 457
    Cyclamen, 450, 453
    Gardenia, 464–465
    Geraniums (potted), 619
    Gerbera, 723
    Gladiolus, 527
    inorganic, 243–248
    Lilies, 539
    Matthiola, 725
    Orchids, 685–686
    organic, 242–243
    Peonies, 727
    Petunias, 626
    plant nutrient need determination, 264–265
    Poinsettia, 498–500
    Roses, 697–699
    sexual propagation and, 317
    Snapdragons, 640
    soil acidity and alkalinity adjustments, 261–263
    soluble salts, 265–268, 269t
    Stephanotis, 730
    trace elements, 250–251
    tropical foliage plants, 576–579
    Tulips, 554, 562, 565
Fiberglass, 56t. See also Glass fiber–reinforced
    polyester
Fiber pots, 415
Ficus, 592, 592f, 593f, 598t, 599t
Field production, Lilies, 534–535
Field soil, 413–414
Filament, 139, 139f
Final shift, of Cyclamen, 450
Financing, for greenhouse enterprises, 74–76
Finish pans, Poinsettias in, 495–496
Finish size pots, direct rooting in, 334
Finned pipe, 101, 101f, 102
Fir bark, 685
Fire molds, 212, 213f
Fire tube boilers, 103, 104, 104f, 105
Fisher Pelfi Star Poinsettia, 491
Flaccid, 136

Flexible plastics, 60–65
Flexible polyvinyl chloride. *See* Polyvinyl chloride
Flonicamid, 364*t*
Flood irrigation. *See* Ebb and flood irrigation
Floral bud development, 163
Floral bud initiation, 163, 542–543
Floral initiation, effect of temperature on, 163–164
Floral preservative, 666
Florets, 527–528, 644
Floriculture business
    exporters, 20–26
    grower knowledge requirements, 28–30
    grower types, 30–31
    importing hazards, 26
    job opportunities, 31–34
    offshoring, 17–20
    overview, 2–6
    plant production areas, 6–14
    wholesale value (2004), 4
    See also Retail outlets; Wholesale distributors
Floriculture sales, wholesale values in, 4*f*
Florida
    cut flowers, 636–637
    importing and, 26
    plant production in, 5*t*, 6, 6*t*, 7*t*, 8, 8*t*, 9*t*, 10*t*, 11*t*, 14
    tropical foliage plants, 575, 575*t*
Florist greens, 3*t*, 4
Florists. *See* Retail outlets
Florists' stock. *See* Matthiola
Flower bud formation
    Azalea, 425–426
    Chrysanthemums, 651–652
    Gardenia, 465
    Hydrangea, 486–487, 488
    Lilies, 536–537, 542–543
    Narcissus (daffodils), 550
Flower color, 164, 485–486, 735
Flower crops, exporters of, 20–26. *See also* Bulbs; Cut flowers; *specific flowers*
Flower formation
    Azalea problems, 432
    Chrysanthemums, 644–645
    Tulips, 552–553, 560
Flowering
    Chrysanthemums, 460
    Azalea, 429–430, 431, 432
    See also Forcing
Flowering growth, temperature and, 163–164
Flowering maple. *See* Abutilon
Flowering plants, 419
    Abutilon, 420
    Achimenes, 420–421
    African violet, 470–474
    Azalea, 421–432
    Begonia, 432–437
    Bougainvillea, 438–439
    Bromeliads, 437–438
    Caladium, 439–440
    Calceolaria, 440–441
    Christmas cactus, 444–445
    Christmas cherry, 446
    Chrysanthemum, 454–463
    Cineraria, 446–447
    Clerodendrum, 448–449
    cultural information format, 419–420
    Cyclamen, 449–454
    Exacum, 463
    Fuchsia, 463–464
    Gardenia, 464–465
    Genista, 466
    Geraniums, 474–482
    Gloxinia, 483–484
    Hydrangea, 484–488
    Kalanchoe, 488–491

Poinsettia, 491–507
Primrose, 507–508
Roses (potted), 508–512
    See also Bulbs; Cut flowers; *specific flowers*
Flower markets (European style), 758, 758*f*, 759, 759*f*
Flower production, offshore versus U.S., 17–20. *See also* Production
Flower sheath, 680
Flowing water shade, 124
Fludioxonil, 387*t*
Fluorescent lights, 150, 473*f*
Fluoride, 292, 293*f*
Fluoride toxicity, 175
Fluorinated plastic. *See* Polyvinyl fluoride
Fluorine, 166, 239, 292, 293
Flurprimdol, 402
Flutolanil, 385
Fluvalinate, 363–364*t*
Foam, 111
Fog, 122–123, 330, 330*f*
Foliage plants, 3*t*, 4, 7. *See also* Tropical foliage plants
Foliar analysis, 264–265
Food storage tissue, in seeds, 303
Footcandles, 68
Forced into bloom, 487
Forcing
    Azalea, 427, 428*t*, 429–430
    bulbs, 516–521
    Hydrangea, 487, 488
    Lilies, 535–536
    Narcissus (daffodils), 548*t*, 549*t*, 552*t*
    Roses, 509–511
    Tulips, 554–555, 556–557*t*, 559–560, 562, 563, 564*t*, 565*f*
    See also Precooling
Foreign markets, shipping to, 26
Formalin, 217
Fosetyl-Al, 387*t*
Foss flower. *See* Ageratum
Foundations, 46, 51–52
Framework, of greenhouses, 46, 47
Framework components, 46–49
Framing
    plastic-covered, 53–55
    wooden, 52–53
Framing materials, for greenhouses, 49–50, 50*f*
Framing methods, for greenhouses, 49*f*, 51, 54*f*
Freesia, 525–526
Freeze drying, 737, 741–742
Fresh Produce & Floral Council (FPFC), 762, 762*f*, 763*f*
Frits, 250–251
FTE 555, 202*t*, 206*t*, 250, 680*t*. *See also* Peat-lite mixes
Fuchsia, 463–464
Fuel, types of, 90, 90*t*, 91
Fuel oil, 90, 90*t*, 91
Fuel storage tank, EPA approved, 91, 91*f*
Fumigants. *See* Soil fumigants
Fumigation, 211, 309
Fungi, 383–391
Fungicidal drenches, 309–310
Fungicides, 328, 333, 359, 538*f*. *See also* Pesticides
Fungus gnats, 371–372, 372*f*
Furnaces, hot air, 106
Fusarium, 191

G

Gable ends, in heat requirement calculations, 99
Gardenia, 464–465
Garden plants *See* Bedding and garden plants
Garden roses, 511–512
Gas (fuel), 90, 90*t*, 91

Gases, 166
    air pollutants, 174–175
    carbon dioxide, 168–171
    harmful, 171–173
    herbicides, 176
    oxygen, 167–168
    soil fumigants, 175–176
    wood preservatives, 173
Gas phase, of growing media, 182
Gates perimeter watering system, 284–285
Genes, 133
Genista, 466
Geographical location. *See* Plant production areas
Georgia Horticultural Growing Media Act, 208
Geraniums, 82*f*, 137*f*, 231*f*, 287*f*, 288*f*, 306*t*, 327*f*, 330*f*, 413*f*, 474–482, 617–620
Gerbera, 722–724
Gerber daisies, 3*t*, 24*f*
Germination, 154. *See also* Seed germination
Germination chambers, 315–316
Germination media, 307–308
Gesneriaceae, 589
Gibberellic acid, 430–431
Gibberellins, 305, 399, 404
Gladioli, 3*t*, 335*f*, 526–530
Glass, 56*t*, 65, 65*t*, 66, 99–100
Glass fiber–reinforced polyester (FRP), 60–62
Glass greenhouses, 44*f*, 91–94
Glazing materials, 55
    heat requirement calculations, 99–100
    light transmittance, 55, 55*f*, 56, 56*t*
    Saran, 66
    type I, 56–60
    type II, 60–65
    type III, 65–66
Glazing sill, 47
Glorisoa Daisy, 306*t*
Gloxinia, 306*t*, 483–484
Glycol-modified polyethylene terephthalate. *See* PETG
Glycosides, 362*t*
Gnats. *See* Fungus gnats
Gooseneck stage, 548
Gothic arch greenhouse, 38, 41–42, 42*f*
Grades, of glass, 65–66
Grading
    Carnations, 666, 667*t*
    Chrysanthemums, 655, 656*t*
    Lilies, 534
    Roses, 704, 705*f*, 706
Gradual pruning, 702
Grafting, 338–339, 339*f*, 423, 691
Grandiflora doubles, 624
Grandiflora singles, 624
Granular matrix sensors (GMS), 279, 279*f*
Granulation, 168
Gravel, 193
Gravity flow return system, 91–92, 92*f*
Gray mold. *See* Botrytis blight
Greenhouse atmosphere, 167
Greenhouse benches, 81–85
Greenhouse diseases, 382–383
    bacteria, 392
    fungi, 383–391
    phytoplasmas, 395
    plant disease, 383
    viruses, 392–395
Greenhouse effect, 89
Greenhouse enterprises, 73
    benches, 81–85
    financing, 74–76
    orientation and arrangement, 80–81
    service buildings, 85–86
    site selection, 76–79
    structures, 79–80

Greenhouse environment, 89
    control of, 124–127
    cooling, 111–112, 119–127
    heating, 89–111
    ventilation, 111–119
Greenhouse production, 32–33
Greenhouses, 38
Greenhouse range, aerial view of, 80f
    bulb cellars, 67
    Callistephus, 717–718
    carbon dioxide in, 169–170
    cold frames, 66–67, 71–72
    Colombia, 21–22
    construction of, 46–49
    curtain walls, 52
    foundations, 51–52
    framing materials, 49–50
    framing methods, 51
    glazing materials, 55–66
    Gloxinia, 483–484
    growing rooms, 68
    hotbeds, 66–67
    Kenya, 24
    Lilies, 535–536
    Mexico, 24
    organizational chart for, 32f
    orientation and arrangement of, 80, 80f,
        81, 81f
    Pansy, 623
    pest management, 353–356
    plastic-covered framing, 53–55
    Roses, 694–696
    Surface area calculation, 98f
    Tulips, 554–555, 559–560, 562, 565
    types of, 38–46
    volume calculations, 116, 117f
    wooden framing, 52–53
    greenhouse screen configurations, 352f
Greenhouse structures, 79–80
Gro-hose, 284, 248f
Ground limestone. See Limestone
Growers, 28–31, 33
Growing media, 180
    Alstroemeria, 713
    Anemones, 521
    artificial soils, 201–207
    bedding plants, 606, 609–610f
    bulbs, 518
    Calendula, 714–715
    Calla lily, 523
    Carnations, 661–662
    Centaurea, 719
    chemical soil treatment, 216–217
    Chrysanthemums, 456, 648–649
    composition of, 182
    containers, 413–414
    functions of, 181–182
    Geraniums, 476–477, 617
    Gerbera, 723
    Gladiolus, 526–527
    Lilies, 538–539
    Matthiola, 725
    Orchids, 685–686
    Peonies, 727
    perched water table, 181, 181f
    Petunias, 626
    Poinsettia, 497–498
    Roses, 694–695
    Snapdragons, 640
    steam pasteurization, 211–216
    tropical foliage plants, 576–579
    Tulips, 554, 561–562
    See also Artificial soils; Macro elements;
        Mineral elements; Soil; Trace elements
Growing methods
    Carnations, 661–666

Geraniums, 477, 478
Gloxinia, 483–484
Growing rooms, 68
Growing skills, 28–29
Growth retardants, 399–403
    bedding plants, 612
    Chrysanthemums, 459
    Geraniums (potted), 619
    Hyacinths, 531
    Lilies, 544–545
    Narcissus (daffodils), 551
    Poinsettia, 501–504
    Tulips, 565–568
Guard cells, 136f, 138
Guatemala, 18t, 19t, 23–24
Gutbier 'V-l0' Poinsettia, 491
Gutbier 'V-l4' Poinsettia, 492
Gutbier 'V-l7 Angelika' Poinsettia, 492
Guttation, 275–276
Gutters, 47
Gynura, 588
Gypsophila, 306t, 724
Gypsum. See Calcium sulfate

H
Handling
    cut flowers, 736
    Gladiolus, 528–529
    retail outlets, 762–763
    rooting and, 334
    Roses, 509–510, 694
    wholesale distributors, 751–754
Hanging baskets, 410
Hardening off, 329, 630
Hard pinch, 458
Hard water, 122–123, 291, 292f
Hardwood barks, 190
Hardwood mulch, 190–191
Harmful gases, 171–173
Harvesting
    Anemones, 522
    Calendula, 715
    Callistephus, 718
    Carnations, 666–667
    Centaurea, 719
    Chrysanthemums, 655–656
    cut flowers, 735–737
    Dahlias, 721
    Gerbera, 723
    Lilies, 534
    Narcissus (daffodils), 548–549
    Orchids, 688
    Peonies, 727–728
    Roses, 703–704
    Snapdragons, 642
    sphagnum peat moss, 186–187
    Stephanotis, 730
    Tulips, 555–556, 560
Hawaii, 7t, 8t, 10t, 12–13, 575, 575t, 714f
Head growers, 33
H.E. Anderson injection system, 254–255, 255f, 256f
Heat conservation, 79
Heat conservation methods, 108–111
Heat distribution, 94–97
Heat exchanger, 94
Heating, 89–90
    boilers, 103–105
    emergency heating equipment, 105
    fuel types, 90, 90t, 91
    glass greenhouses, 91–94
    heat conservation methods, 108–111
    heat distribution, 94–97
    heat requirements, calculation of, 98–100
    hot water and steam, 101–103
    plastic-covered greenhouses, 106–108
    solar, 111

Heating plant boilers, 103, 104, 104f, 105, 105f
Heat requirements, calculation of, 98, 98f, 99, 100
Heat transfer coefficient, 98
Height control
    Hyacinths, 531
    Lilies, 540–541, 542, 544–545
    Narcissus (daffodils), 551
Herbaceous perennials, 5t, 6f, 302
Herbicides, 176, 359. See also Pesticides
Hexythixox, 365t
High intensity discharge (HID), 41f, 83f, 143–144,
    144f, 149f
High-pressure fog, 122–123
High-volume spray application, 368–369
Holding, of Azalea, 431
Holiday forcing
    Azalea, 427–428t
    Hydrangea, 487, 488
    Roses, 509–511
Holiday pinching, for roses, 701
Hollyhock, 306t
Homogeneous plants, 323
Hoop house, 45f
Horizontal air flow (HAF) fans, 96–97, 97f
Horizontal unit heaters, 97
Hose watering, 281, 281f, 282f
Hot air furnaces, 106
Hot air heaters, 97
Hot water heating system, 91, 92, 92f, 93, 93f,
    101, 101f, 102, 103f
Hot water piping, 101–103
Hotbeds, 66–67, 67f
HOZON, 254, 255, 255f, 259t
Humidistat, 122
Humidity, for tropical foliage plants, 580.
    See also Relative humidity
Humus, 184
Hyacinth, 530–532
Hybridization, Chrysanthemums, 646
Hydathodes, 275
Hydrangea, 484–488
Hydrogen, 223, 225, 226
Hydrogen dioxide, 387t
Hydrophytes, 576
Hydroponic solution, 181–182
Hydroxyanilide, 387t
Hypnum peat moss, 187
Hypocotyls, 303

I
Illinois, plant production in, 7t, 8t, 9t, 10t,
    11t, 14
Imidacloprid, 362t
Imidazole, 387t
Impatiens, 306t, 620–621
Impatiens necrotic spot virus, 394t
Importing, 18t, 26
Incandescent lamps, 144, 149f
Indiana, plant production in, 11t
Indoleacetic acid (IAA), 404
Indolebutyric acid (IBA), 404
Industry regulation, for artificial soils, 208
Infiltration, 93
Inflatable interior ceiling, 111
Inflorescence, 536
Infrared heating system, 106, 107, 107f
Infrared radiation, 142
Inhibitor, 449
Inorganic fertilizers, 243–248
Inorganic materials, 182, 192–196
Insect control, for fan and pad cooling, 121
Insecticides, 223, 359. See also Pesticides
Insects, 29, 370–377
    Achimenes, 421
    African violets, 474
    Alstroemeria, 714

Anemones, 522
Azalea, 432
Begonias, 437
Caladium, 440
Calceolaria, 441
Calendula, 715
Calla lily, 524
Callistephus, 718
Carnations, 668–669
Centaurea, 719
Christmas cactus, 445
Christmas cherry, 446
Chrysanthemums, 462, 658
Cineraria, 447
Cyclamen, 454
Dahlias, 722
Fuchsia, 464
Gardenia, 465
Genista, 466
Geraniums, 482
Gerbera, 724
Gladiolus, 530
Gloxinia, 484
Hydrangea, 488
interior foliage plants, 602
Iris, 533
Kalanchoe, 491
Lilies, 546
Matthiola, 726
Orchids, 689
Peonies, 728
Poinsettia, 506–507
Primrose, 508
Rieger elatior hybrids, 436
Roses, 512, 707
Schizanthus, 729
Screen sizing, 353f
Snapdragons, 643
Stephanotis, 730
Tulips, 568
vegetable crops, 630, 632
Insulation, 111
Integrated pest management (IPM), 348–350,
    356–357, 382–383
bacteria, 392
biological control, 357–358
chemical control, 358
cultural control, 357
environmental control, 357
fungi, 383–391
greenhouse environment, 353–354
mechanical control, 357
monitoring, 354–356
pesticide selection, 359–370
pest types, 370–377
phytoplasmas, 395
plant disease, 383
prevention, 350
resistant cultivars and alternative crops,
    358–359
sanitation and exclusion, 350–353, 358
viruses, 392–395
Intercalary zone, 134, 134f
Interior ceiling, 109, 110, 110f, 111
Interior foliage plants, 597–602
Intermezzo series, 452
Interrupted lighting, for Chrysanthemums,
    653–654
Intermittent low-pressure misting, 311
Interveinal chlorosis, 233–234
Invertebrate pests, 370–377
Iowa, plant production in, 8t, 9t
Iprodione, 386t
Ireland, 186
Iris, 3t, 532–533
Iron, 223, 227f, 228f, 232, 235, 236, 237, 239t, 698t

Iron sulfate. *See* Ferrous iron
Iron trioxide, 65
Irrigation, 254, 269t
    boom, 285–286, 609f
    ebb and flood, 286–289
    Gladiolus, 527
    surface, 284
    timing of, 277–280
    See also Water
Irrigation control devices, 282, 283f
Israel, 25
Ivy Geraniums, 475

J
Japanse betele, 353f
Jerusalem cherry. *See* Christmas cherry
Jiffy-7, 450
Job opportunities, in floriculture industry,
    31–34
Juvenile stage of growth, 301, 301f

K
Kalanchoe, 306t, 488–491
Kenya, 24–25, 26
s-Kinporene, 361t
Kool-cel, 121
Kresoxim-methyl, 387t

L
Labor issues
    greenhouse site selection and, 77
    Kenya, 24–25
Lambda-cyhalothrin, 364t
Lamiaceae (Labiatae), 590
Laminated beam framing method, 53
Larkspur, 306t
Laser series, 452
Lasting life, 164, 706
Lath houses, 45, 45f
Laws, knowledge of, 30
Layering, 337–338
Leaf blade structure, 138–139
Leaf bud cuttings, 326
Leaf chlorosis, 329t
Leaf counting, for Lilies, 543
Leaf cuttings, 326, 326f
Leaf drop, in Azalea, 432. *See also* Abscission
Leafhoppers, 394t
Leaf miners, 353f, 372, 373f
Leaf mold, 189, 262
Leaf scorch, 412
Leaf shapes, 137, 137f
Leaf structure, 274f
Leaf tip burn, 292, 293
Lean-to greenhouse, 38, 40, 41, 41f
Leaves, 136f, 137, 233, 233f, 234, 234f,
    235, 235f, 274
Leave trimming, of Anemones, 522
Life cycle, of plants, 301, 301f
Life support, for cut flowers, 734–735
Light, 141
    African violet, 472–474
    Alstroemeria, 713
    Anemones, 521–522
    asexual propagation environment, 330–331
    Azalea, 427
    Begonias, 435, 435t
    Calla lily, 523
    Callistephus, 716
Light compensation point, 144–145
Light intensity, 41f, 126, 142, 144, 145f, 146f,
    277, 501, 540
Light levels, low, 146, 146f, 147
Light meters, 143, 143f
Light quality, 152–154

Light saturation point, 145, 145f, 146
Light transmittance, of glazing materials, 55,
    55f, 56, 56f
    Carnations, 665
    Centaurea, 719
    characteristics of, 142–147
    Chrysanthemums, 460–461, 649–650,
        652–655
    Cyclamen, 450–451, 453
    Gardenia, 465
    Geraniums (potted), 619
    Gerbera, 723
    germination environment and, 312
    Gloxinia, 483
    interior foliage plants, 600–601
    Matthiola, 725
    Orchids, 687–688
    Peonies, 727
    photoperiod, 147–152
    photosynthesis, 141–142
    Poinsettia, 500–501
    Rieger elatior hybrids, 435
    Roses, 699–700
    seed germination and, 305, 306t
    Snapdragons, 641–642
    Stephanotis, 730
    tropical foliage plants, 579–580
Lignin, 190
Liliaceae, 580
Lily, 3t, 533
    calcium deficiency, 233f
    cultivars, 534
    DFF and, 158f
    field production, 534–535
    flower bud formation and timing,
        536–537
    greenhouse fording, 535–536
    growth retardants, 544–545
    height control, 540–541, 542, 544–545
    leaf shape, 137f
    light intensity, 540
    potting, 537–540
    problems, 545–546
    scheduling, 542–543
    storage, 543–544
    temperature, 540–541, 542
Lima bean, embryo of, 303f
Limestone, 202t, 206t
    as fertilizer, 253t, 261–262, 263
    seed germination and, 308, 608t
    See also Peat-lite mixes
Limit controls, 125
Liquid application fertilizers, 248, 252–254
Liquid fertilizers, 457
Liquid phase, of growing media, 182
Loam, 182, 183f, 183t, 414
Lobelia, 306t, 621–622
Long-day plants (LDP), 148
Louisiana, plant production in, 7t, 11t
Low-volume spray application, 369
Low-volume water application methods, 282
Lumens, 143
Luxury consumption, 229

M
Macleaya extract, 385t
Macro elements, essential, 223, 228–234
Magamp, 205, 206t, 249, 252, 414
Magic flower. *See* Achimenes
Magnesium, 223, 227f, 228f, 233, 234,
    234f, 235, 239t
Magnesium ammonium phosphate, 249
Magnesium oxide, 65, 253t
Magnesium sulfate, 253t, 698t
Makeup air, 108
Malathion, 363t

Mancozeb, 386t
Manganese, 223, 227f, 228f, 235, 236, 237, 239t, 698t
Manganese sulfate, 253t
Manganese zinc, 386–387t
Manual ventilating sash system, 113, 113f, 114f
Manures, 189
Marantaceae, 591
Marginal necrosis, 232
Marigolds, 306t, 622
Marketing (market preparation)
    bedding plants, 606, 612, 613f, 614
    Calendula, 715
    Calla lily, 524, 527–528
    Callistephus, 718
    Carnations, 666–667
    Centaurea, 719
    Chrysanthemums, 655–656
    Dahlias, 721
    Gerbera, 723
    Peonies, 727–728
    Roses, 703–704
    Snapdragons, 642
    Stephanotis, 730
    Tulips, 555–556, 560
Marketing manager, 34
Marketing positions, 33–34
Marketing skills, 29
Market pack production, of Geraniums, 620
Market preparation
    container grown crops, 416–417
    Orchids, 688
    Poinsettia, 504
Market preparation/marketing, Narcissus (daffodils), 548–549, 550
Markets, for greenhouse enterprises, 76
Marvel series, 452
Massachusetts, plant production in, 9t
Mass market outlets, 760, 760f, 761f, 762, 762f, 763f
Materials Safety Data Sheet (MSDS), 366
Matthiola, 724–726
Maturation, 135
Maxi Flat Filler, 205f
Maximum temperature, 156
Mealybugs, 372, 373, 373f
Mechanical air conditioning, 124
Mechanical conditioning, 405
Mechanical control, 357
Mechanical heat distribution, 95–97
Mechanical skills, 29
Mechanical support, 181
Media
    asexual propagation, 331–333
    seed germination, 307–308
Medium, 180. See also Growing media
Mefanoxam, 385t
Mefanoxam + fludioxonil, 388t
Mercury, 172–173
Meristematic tissues, 133–134, 134f
Meristem tissue culture, 339, 660, 684
Mesophyll, 138, 138f
Metabolic processes, air temperature and, 161
Metaphase, 324, 324f, 325
Methiocarb, 361t
Methyl bromide, 217, 217f, 309
Metro Mix peat-lite mixes, 207
Mexico, 18t, 19t, 23, 24, 637
Michigan, plant production in, 5, 5t, 6t, 7t, 9t, 10t
Micro elements. See Trace elements
Micromax, 251
Microorganisms, soil, 161–162, 163f
Micropropagation, 339–340, 340f
Milbemectin, 362t
Mildews, 29

Mill finish, 50
Mineral elements, 222
    absorption, 223–228
    distribution in plants, 223
    essential, 222–223
Mineral soils, Solubridge readings for, 269t
Mini Flat Filler, 204
Miniature carnations, imports of, 18t, 19t
Miniature Cyclamen, 452
Miniature roses, 3t, 5t, 10t, 511f
Minimate series, 452
Minimum temperature, 156
Minnesota, plant production in, 9t, 10t, 11t
Miracle series, 452
Missouri, plant production in, 8t
Mist fertilizer, 647
Misting, 328f, 329–330, 609f
Mites, 373, 374, 374f, 707f
Mitochondria, 133, 133f
Mitosis, 324–325
Mix Maker 10, 204f
Mix Maker 30, 204f
Mix Maker 50, 203, 204f
Mobility, 230–231
Model Legislative Bill for Horticultural Growing Media Labeling, 208
MOils, 362t
Moisture
    asexual propagation environment and, 329–330
    calceolaria, 441
    Geraniums (potted), 617–618
    germination environment and, 311–312
    tropical foliage plants, 580
    See also Water
Moisture probes, 279–280, 280f
Moisture stress, 160, 276–277
Molybdenum, 223m, 227f, 235, 237–238, 251
Monitoring, for pests, 354–356
Monoammonium phosphate, 245, 247, 253t, 260t
Monocalcium phosphate, 247, 253t
Monopodial growth, 680
Monopotassium phosphate, 253t, 260t
Monstera, 584
Montreal Protocol on Substances that Deplete the Ozone Layer, 217
Moraceae, 591–593
Morning Glory, 306t
Moth flies, 371–372
Motorized accelerated heating systems, 92, 92f
Mottle leaf, 237
Mound layering, 337, 338f
Muck soil, 183
Mulch and Soil Council (MSC), 208
Mulching, for Peonies, 727
Multicell paks, 412
Multichamber greenhouses, 44–45
Multiflora doubles, 624–625
Multiflora singles, 625
Multilayer flexible plastics, 63–65
Mums. See Chrysanthemums
Mutation, 646
Mycolbutanil, 388t
Myron Meter, 260f, 293f
Myrsinaceae, 592

N
Naphthalene acetic acid (NAA), 404
Naphthenate, 52
Napthoquinone derivatives, 362t
Narcissus (daffodils), 546–552, 552t
Nasturtium, 306t
National Bark and Soil Production Association, 208

National Greenhouse Manufacturers Association (NGMA), 51
Natural air movement, 112, 112f, 113
Natural gas, 90, 90t, 91, 171
Natural heat distribution, 94–95
Natural light, for African violets, 472
Necerosis, 232, 233f, 235, 238, 239t
Negative geotropism, 528
Nematodes, 191, 357–358
Neonicotinoids, 362t
Nephrolepis exaltata, 595
Netherlands, 9, 18t, 19t, 25, 25f, 26
Neutral reaction, in nitrogen, 246
New Jersey, plant production in, 5t, 6t, 7t, 8t, 9t, 10t
New York, plant production in, 5t, 6t, 7t, 8t, 9t, 10t, 11t
Nicotine, 361t
Night temperature (NT), 157, 417, 717t
Nitrates, 225
Nitrification process, 229, 229f
Nitrobacter bacteria, 229
Nitroform, 248
Nitrogen, 227f, 228f, 228–231, 239, 243–246, 253t, 259t, 260t
Nitrogen cycle, 228f
Nitrogen deficiency, 230, 231, 239t
Nitrogen dioxide, 174
Nitrogen growth curve, 230f
Nitrosomonas bacteria, 229
North Carolina, plant production in, 5t, 6t, 7t, 8t, 9t, 10t
Novaluron, 360t
Novella series, 452
Nucleus, 133, 133f
Nuclear membrane, 133f
Nutrient agar, 682
Nutrient mist, 330
Nutrients. See Plant growth; Plant nutrients; specific nutrients

O
Oak leaf, shape of, 137f
Oat straw, 192
Obligate parasites, 383
Obligate saprophytes, 383
Offsets, 336
Offshore flower production, 17–20
Oglevee Poinsettia, 492
Ohio, plant production in, 5t, 6t, 7t, 9t, 10t
Ohio State skinner system, 285
Oil, 387t
One-half strength, 496
Ooze tubes, 282, 283f
Optimum temperature, 156
Orchids, 3t, 5t, 8t, 679–680
    characteristics of, 680–682
    general culture, 685–688
    harvesting for market, 688
    imports of, 18t, 19t
    problems, 688–690
    propagation, 682–684
Ordering
    bedding plant seeds, 607
    bulbs, 517
Oregon, plant production in, 9t, 9–11
Organic fertilizers, 242–243
Organic matter, 182
    acidic, 262
    pH and, 227f
    in soil, 184–185
    types of, 185–192
Organizational chart, for greenhouse, 32f
Organophosphate, 362, 363t
Organophosphate + pyrethroid, 365t
Organotins, 363t

Orientation, of greenhouses, 80–81
Origin, Azalea, 422–423
Osmocote, 205, 206t, 249, 250f, 252, 414
Outright purchase, 750
Ovary, 139, 139f
Oversteaming, 212
Oversupply, of water, 273–274
overwatering, 212
Overwintering culture, of Pansies, 622–623
Oxygen, 166, 167–168, 223–224, 238, 305, 310f
Ozone, 175

P
Pacific Northwest, plant production in, 9–11
Packing
    Lilies, 534–535
    Roses, 704, 705f, 706
Paclobutrazol, 401t, 402, 403, 403t
Pad cooling. See Fan and pad cooling
Paints, 173
Palisade layer, 138, 138f
Palmae (Arecaceae), 593–594
Pandanaceae, 594
Panning, for Poinsettias, 496, 497
Pansy, 306t, 622–623
Paraffinic oil, 362t
Parasites, 357–358
Pasteurization
    growing media and, 211–216
    sexual propagation and, 308–310
    See also Steam pasteurization
Pathogens, 358
PCNB, 385t
Peat bogs, 185
Peat lite mixes, 202, 207, 208, 269t, 414, 480f
Peat moss, 180, 185–187, 188f, 189f, 262, 307,
    308, 414. See also Sphagnum peat moss
Peat pots, 412
Pelargonium hortorum. See Geraniums
Pelargonium shortorum. See Geraniums
Pelleted seed, 313
Pencil stage, 549
Pennsylvania, 5t, 6t, 7t, 8t, 9t, 10t, 11t, 26
Pentachlorophenol, 173
Peonies, 726–728
Peperomia, 594–595, 599t
Perched water table, 181, 181f
Perennial plants, 5t, 6f, 302
Perforated polyethylene tubing, 95–96
Periwinkles, 306t
PERK, 202t, 206t, 251, 608t. See also Peat-lite
    mixes
Perlite, 180, 193, 203f, 307, 307f, 308, 414.
    See also Peat-lite mixes
Permanent interior ceiling, 109, 110, 110f
Permanent tissues, 134
Permanent wilting point, 136
Permethrin, 364t
Peroxyacetyl nitrate (PAN), 174, 174f
Personal preferences, for greenhouse site
    selection, 78
Pesticide resistance, 368
Pesticides
    application of, 368–370
    phytotoxicity, 366, 368
    safe use of, 366t
    selection of, 349, 359–370
    signal words and toxicity categories, 367t
Pest management. See Integrated pest
    management
Pest prevention, 350
Pests, types of, 370–377. See also Diseases;
    Insects
Petals, 139, 139f
PETG (glycol-modified polyethylene
    terephthalate), 63

Petiole, 137
Petunia, 234f, 306t, 624–626
pH
    fertilizers and, 260, 262, 263
    nutrient absorption and, 226–228
    water application and, 291–292
Phenoxypyrazoles, 363t
Phenylpyrrole, 387t
Philodendron, 584–586, 584f, 599t
Phloem, 135, 136f, 138f
Phlox, 306t, 626–627
Phosphate rock, 247
Phosphite, 387t
Phosphoric acid, 698t
Phosphorus, 232, 227t, 228f, 231, 232, 233, 235,
    238, 239f, 246, 247, 253t, 260t
Phosphorus acid, 387t
Photoperiod, 147–152, 500–501. See also Day
    length; Light
Photosynthates, 132
Photosynthesis, 108, 141–142, 159–160
Photosynthetic light, 152
Phototropism, 147
Phthalimides, 387t
Phytochrome, 154
Phytophthora, 191
Phytoplasmas, 395
Phytotoxicity, of pesticides, 366, 368
Pinch and one-half carnations, 663–664
Pinching
    Begonias, 436
    Chrysanthemums, 458–459, 650–651
    Dahlias, 721
    Poinsettia, 496–497
    Roses, 700–701
Pine needles, 262
Pink Hydrangea, 485–486
Pipe steaming. See Buried pipe steaming
Piperaceae, 594t
Piperadine, 387t
Piperalin, 387t
Piping, hot water, 101, 101f, 102, 103, 103f
Pistil, 139, 139f
Pitch, 39
Pittosporaceae, 594
Plain pipe, 101–102
Plant disease, 383. See also Diseases
Plant growth, 222
    macro elements, 228–234
    mineral elements, 222–228
    temperature and, 157
    trace (micro) elements, 235–239
    water and, 272–273
    See also Fertilizers; Growing media; Soil; Water
Plant growth regulation, 398–399
    chemical plant growth regulator, 399, 404–405
    chemical plant growth retardants, 399–403
    mechanical conditioning, 405
Plant growth regulators (PGRs). See Rooting
    hormones
Plant growth stage, nitrogen fertilizer effects
    on, 244
Planting
    Carnations, 661–662
    Gladiolus, 527
    Lilies, 542
    Narcissus (daffodils), 548t
    Roses, 694–696
    Tulips, 561–562
Plant life cycle, 301, 301f
Plant nutrients, 185
    cut flowers, 742–743
    fertilizers and, 264–265
    nitrogen, 243–244
    See also Fertilizers; Growing media;
        Specific nutrients

Plant production areas. See Production areas
Plant propagation. See Sexual propagation
Plant Quarantine Act 37 (Q-37), 3t
Plants, 132
    cells, 132–133
    harmful gases and, 171
    light requirements of, 144–146
    tissues, 133–139
    See also specific plants and plant types
Plastic-covered framing, 41f, 53–55
Plastic-covered greenhouses, 100, 106–108
Plastic pots, 410, 411–412
Plastics
    double layer, 108–109
    flexible, 60–65
    rigid, 54
    Saran, 66
    thin films, 54–55, 56–60
Plastids, 133, 133f
Platycerium, 595–596
Plectranthus, 590, 590f, 599t
Plug seedling culture, 316–317
Plug seedling production, 313–316
Plumule, 303
Pocketbook plant. See Calceolaria
Poinsettia, 3t, 5t, 7t, 491
    calcium deficiency, 234f
    cultivars, 491–492
    leaf shape, 137f
    production (regular program), 496–507
    propagation, 493–496
    Pythium infection, 391f
    stock plants, 492–493
Political issues, in Colombia, 23
Pollutants. See Air pollutants
Polycarbonate, 56t, 63
Polyethylene, 56, 56t, 57, 58
Polymethyl methacrylate (PMMA), 62–63
Polypodiaceae, 595–596
Polypodium, 596, 600t
Polyscias, 587
Polyvinyl chloride (PVC), 59
Polyvinyl fluoride (PVF), 59–60
Pompon chrysanthemums, imports of, 18t, 19t
Portulaca, 306t
Positive displacement pump, 254, 255f
Positive-pressure cooling, 353, 354f
Postemergent damping off, 310
Potassium, 223, 224, 227f, 228f, 231, 232, 233f,
    234, 235, 236, 239t, 247, 248, 253t, 260t
Potassium bicarbonate, 385t
Potassium chloride, 248, 253t, 260t, 698t
Potassium nitrate, 202t, 206t, 246, 247, 253t,
    260, 698t. See also Peat-lite mixes
Potassium salts of fatty acids, 364t
Potassium sulfate, 247, 253t, 260t
Potassium thiosulfate, 253t
Pothos, 585
Pot marigold. See Calendula
Pots
    clay, 410, 411
    fiber, 415
    peat, 412
    plastic, 410, 411–412
    sizes and shapes, 413
Potted plants and flowers, 3t, 4, 4f, 5t,
    8t, 9t, 10t
    Chrysanthemums, 456–457
    Gradenias, 464f
    Geraniums, 477f, 480f, 617–620
    Lilies, 537–540
    Narcissus (daffodils), 549–552, 552t
    retail outlets, 763
    temperature and, 162, 162f
    Tulips, 560–568
    See also specific plants and flowers

Potting, 414–416
Potting soil, for roses, 510
Powdery mildews, 390, 706f
Power ventilating sash system, 114–115, 115f
ppm, 223
Precooling
    Azalea, 426–427
    bulbs, 518
    Easter lilies, 541f
    Lilies, 535
    Narcissus (daffodils), 547–548, 548t
    Tulips, 553, 557–560
    See also Forcing
Predators, 357
Preemergent damping off, 310
Preharvest handling, cut flowers, 736
Prepared hyacinths, 530
Preplant handling
    Roses, 509–510, 694
    Tulips, 558–559
Preservatives, 52
Pretreatment, for seed sowing, 312
Primary wall, 133
Primrose, 507–508
Primula malacoides, 306t. See also Primrose
Production
    Azalea, 424–431
    bedding plants, 605–614
    Carnations, 659
    cut flowers, 735–736
    Geraniums, 474–481, 620
    Lilies, 534–535
    Poinsettia, 496–507
    Roses, 696–706
Production areas, 6–9
    California, 11–12
    Colorado, 13
    Florida, 14
    Hawaii, 12–13, 13f
    other areas, 14
    Pacific Northwest, 9–11
    Texas, 13–14
    tropical foliage plants, 574–575
Production managers, 33
Pro-Mix peat-lite mixes, 207
Propagation
    African violet, 470–471
    Alstroemeria, 712–713
    Azalea, 423–424
    Calendula, 714
    Callistephus, 716
    Carnations, 659–661
    Centaurea, 718
    Chrysanthemums, 646–647
    Dahlias, 720
    Geraniums, 478–481
    Gerbera, 722–723
    Gladiolus, 526–527
    Hydrangea, 484–485
    interior foliage plants, 597–602
    Lilies, 534
    Matthiola, 725
    Orchids, 682–684
    Peonies, 727
    Poinsettia, 493–496, 505
    Roses, 691
    Schizanthus, 729
    Stephanotis, 729
    tropical foliage plants, 597–600t
    See also Asexual propagation; Sexual
        propagation
Propagation benches, 329
Propagation environment, 328–335
Propagator, 32–33
Propamocarb, 385t
Propane, 91, 170, 170f

Prophase, 324, 324f, 325
Proportioning equipment, 245, 254–257,
    258–259, 260f, 260t
Protective environment, 38. See also
    Greenhouses
Pruning
    Azalea, 425
    Chrysanthemums, 650
    Roses, 701–702
Pseudobulbs, 337
Pseudomonas gardeniae, 465
psi (per square inch), 94
Pteris cretica, 596
Puddle soil, 184
Puerto Rico, 575
Purlins, 47f, 48
Pymetrozine, 364t
Pyraclostrobin, 387t
Pyrethrins, 361t
Pyrethroid, 363–364t
Pyrethroid + neonicotinoid, 365t
Pyridaben, 364t
Pyridazinones, 364t
Pyridine azomethines, 364t
Pyridine carboxamides, 364t
Pyridine insect growth regulators, 364t
Pyripoxyfen, 364t
Pyrroles, 364t
Pythium, 191, 212f, 391f

Q
Q-37. See Plant Quarantine Act 37
Quantitative long-day (QLD) plants, 148
Quantitative short-day (QSD) plants, 148
Quonset, 38, 41, 41f, 54, 54f, 57f, 60f

R
Rack and pinion, 113, 114f
Radiant energy, 89
Radiant energy spectrum, 142, 143f
Radiational cooling, 161
Radiators, 101
Radicle, 303
Ralstonia solanacearum race 3 biovar 2, 26
Range superintendent, 33
Ranunculus, 728. See also Anemones
RAPID-GRO, 330
Ray florets, 644
Raynoutria sachalinesis, 385t
Receptacle, 139, 139f
Record keeping, for bedding plants, 605–606
Records, for pest management, 349–350
Redi-earth peat-lite mixes, 207
Reed peat moss, 187, 189f
Regal Geraniums, 475
Regulation, of artificial soils, 208
Regulations, 30
Relative humidity, 126
    Azalea, 426–427
    reduction of, 112
    water loss and, 275
Reporting, 483–484
Repotting, 414–416
Representative Designer Geraniums, 476
Representative Ivy Geraniums, 476
Representative Showcase Geraniums, 476
Representative Zonal Geraniums, 476
Resetting, 237
Resistant cultivars, 358–359
Respiration, 138, 160
Response groups, 459, 645
Restricted entry intervals (REI), 366
Retail outlets, 757–760
Retardants. See Growth retardants
Retractable roofs, 58, 58f
Rhizoctonia, 191, 391f

Rhizomes, 336
Rhododendron species. See Azalea
Rhoeo spathacea, 589, 599t
Ridge, 47f, 48–49
Rieger elatior hybrids, 434–436
Rigid plastics, 54
Rigid polyvinyl chloride (PVC), 62. See also
    Polyvinyl chloride
Ring spot, 474
Rock wool, 194, 195f, 196, 333f, 480f, 696–706,
    723f
Rodent problems, for Tulips, 568
Roll-up walls, 113, 114f
Roof area, in heat requirement calculations,
    99–100
Roof bars, 41
Root cap, 134, 135f
Rooted cuttings, of Chrysanthemums, 455
Root hairs, water absorption trough, 136f
Rooting
    direct, 334
    handling after, 334
    storage after, 335
Rooting hormones, 326–328
Rooting room forcing, of Lilies, 535
Roots, 134–135, 136f, 273, 276
Root substrate application, 370
Root system, mineral element absorption
    through, 224
Rosemary plant, 231f
Roses, 3t
    budding 338–39, 339f
    cut flowers, 690–707
    cold storage, 737f
    dry-cold storage, 740
    leaf shape, 137f
    imports of, 18t, 19t
    mound layering, 338f
    potted, 508–512
    rock wool and, 194, 195f
    U.S. cut flower production of, 20t
Row sowing, 313
Royalty Geraniums, 475–476
Rubiaceae, 596
Rubeetia, 306t
Rust disease, 707f
Rutaceae, 596–597

S
Safety, for pesticides, 359–366, 366t, 367t
Safe, 306t
Saintpaulia ionantha. See African violet
Sales, of Azalea, 431
Sales positions, 33–34
Salt accumulation, 265–267
Salt fertilizers, 251, 252, 253t, 254
Salt index, 252, 253t, 254
Salt injury, action of, 266
Salts. See Soluble salts
Salvia, 306t, 627
Sand, 180, 183f, 183t, 185, 192–193
    container grown crops, 414
    limestone and, 262
    peat lite mixes and, 203f
    seed germination and, 307f, 308
    See also Peat-lite mixes
Sanitation
    asexual propagation environment, 333
    pest management, 350–353, 354f, 358
    wet-cold storage, 738–739
Sansevieria, 581–582
Saran, 45, 66
Sash bars, 48, 48f
Sash sill, 47, 47f
Saturated-media extract (SME), 268
Sawdust, 180, 189–190, 262

Sawtooth greenhouses, 43–44, 44f, 53
Saxifragaceae, 596
Saxifragaceae stolonifera, 596
Scale insects, 374
Scarification, 304
Scheduling
    Chrysanthemums, 652–655
    Lilies, 542–543
Schizanthus, 306t, 728–729
Schlumbergera brudgesii. See Christmas cactus
Schwabenland series, 434, 435–436
Scindapsus, 585, 598t
Scion, 444
Scissors truss framing method, 53, 54, 54f
Scouts, 355, 356f
Screening, 351–353
Secondary roots, 334
Sedge peat moss, 187, 189f
Seed coat, 303, 303f
Seed dormancy, 303–304
Seed flats, 317
Seed geraniums, 620
Seed germination, 301f, 304–305, 306t
    bedding plants, 607
    environment, 310–312
    fertilizers and, 317
    media, 307–308
    Petunias, 625–626
    Snapdragons, 638–640
    temperature and, 159
Seedlings, 135
Seed parts, 302–303
Seed propagation. See Propagation; Sexual
    propagation
Seeds
    bedding plants, 606, 607
    Geraniums, 482, 617
    Gloxinia, 483
    Orchids, 682
Seed Size 302f
Seed sowing, 312–317, 320–322
    Cyclamen, 449–450, 453
    Geraniums, 619–620
    Orchids, 682–684
    Petunias, 625
    Snapdragons, 675–676
    vegetable crops, 629–630
Seed suppliers, 31
Selling methods, of wholesale distributors, 750
Semiautomatic black cloth systems, 150–151
Semidwarf palms, 593
Semi-iron pipe framework, 50
Senecio cruentus. See Cineraria
Senescence, 735
Sepals, 139
Separations, 335, 139f
Serpentine layering, 338
Service availability, as greenhouse site selection
    factor, 78
Service buildings, 85–86
Sexual propagation, 300–301
    annual plants, 301
    biennial plants, 302
    fertilizer requirements, 317
    perennial plants, 302
    plant life cycle, 301, 301f
    seed dormancy, 303–304
    seed germination, 304–305, 306t
    seed germination environment, 310–312
    seed germination media, 307–308
    seed parts, 302–303
    seed sowing, 312–317, 320–322
    soil pasteurization, 308–310
    student project, 320–322
    transplanting, 317–318
    See also Propagation

Shade cloth. See Saran
Shading, 123, 123f, 124, 331f, 655f
Shamerock, 306t
Shearing
    Azalea, 425
    Genista, 466
Sheet film plastic, 41
Shifting up, 483–484
Shipping
    foreign markets, 26
    wholesale distributors, 750–751
    See also Transportation costs
Shipping room manager, 34
Shoot emergence, from Lilies, 542
Shore flies, 371–372, 372f
Short-day plants (SDP), 148, 149, 153f
Sideposts, 46–47
Sierra series, 452
Silica, 192
Silica dioxide, 65
Silicon, 238–239
Silt, 183f, 183t
Simple diffusion, 273
Simple layering, 337–338
Single detached greenhouse, 43
Single-element materials, 243
Single-pinch carnations, 663
Sinningia speciosa. See Gloxinia
Siphon hose proportioner, 254, 255f
Site selection, for greenhouses, 76–79
Size
    fans, 116
    greenhouse structures, 79
    pads, 120t, 121, 121t
Sizer, pads, 120–121
Skinner system, 285
s-Kinporene, 361t
Sleepiness, 667
Slugs, 375
Small seeds, sowing, 313
Smith Measure-Mix, 254, 256f
Smoke application, 370
Snails, 375
Snapdragons, 3t, 306t, 627–628
    cut flowers, 637, 638–643
    downy mildew, 389f
    dry-cold storage, 740
    soil oxygen and aeration, 168f
    student projects, 675–677
Snow ball plant. See Hydrangea
Soaps, 364t
Sodium, 238
Sodium nitrate, 245, 253t, 260t
Sodium oxide, 65
Sodium tetraborohydrate decahydrate,
    361t, 385t
Soft pinch, 458
Soft water, 122–123, 291, 292f
Softwood barks, 190
Softwood mulch, 190–191
Soil, 180
    African violet, 471–472
    carbon dioxide in, 170–171
    chemical treatment of, 216–217
    container grown crops, 413–414
    Gardenia, 464–465
    Geraniums, 476–477
    Gladiolus, 527
    as growing media, 180–182
    inorganic materials added to, 192–196
    moisture stress and, 277
    nitrgen cycle and nitrification process, 229f
    nitrogen fertilizers and, 244
    organic matter in, 184–192
    Roses, 510, 512
    Solubridge readings for, 269t

    student project, 198–200
    See also Artificial soils; Growing media
Soil acidity, 262–263
Soil aeration, 184
Soil air, 182
Soil alkalinity, 261–262
Soil-borne diseases, 390–391
Soil fumigants, 175–176
Soilless media, for Geraniums, 476–477
Soil microorganisms, soil temperature and,
    161–162, 163f
Soil oxygen and aeration, 167–168
Soil pasteurization
    growing media and, 211–216
    sexual propagation and, 308–310
Soil solution, 265
Soil structure, 183–184,
Soil temperature, 161–162, 163, 163f
Soil testing, 264
Soil texture, 182–183, 183f
Soil triangle, 183f
Solanum pseudocapsicum. See Christmas cherry
Solar energy, 90
Solar heating, 111
Soleirolia, 597, 599t
Solid phase, of growing media, 182
Solubility, of fertilizers, 252, 253t, 260t
Soluble salts, 244, 260, 265–268, 269t,
    276–277, 293
Solubridge readings, 269t
Sorus, 596
South Carolina, plant production in, 5t
Sowing. See Seed sowing
Spacing
    Calendula, 715
    Carnations, 662
    Chrysanthemums, 458, 650
    Poinsettia, 501
    Snapdragons, 641, 677
Spadix, 585f
Spathiphyllum, 586f, 599t, 601
Specialty producers, 30–31
Seedling geraniums, 620
Sphagnum peat moss, 185, 185f, 186, 187, 188f,
    203f, 262, 307, 308, 414, 535f, 608t.
    See also Peat-lite mixes
Spinosad, 364–365t
Spinosyns, 364–365t
Spiromesifen, 365t
Spitting, 532
Spray stakes, 285, 285f
Spray treatment, for Lilies, 545
Spray type chrysanthemums, 645
Spreading petunias, 625
Spring bulbs, 3t, 5t, 9t. See also Bulbs;
    Specific bulbs
Spring equinox, 147, 147f
Stage control, 125, 125f, 126
Stage G, 553
Stamens, 139, 139f
Standard chrysanthemums. See Chrysanthemums
Standard grades, 666, 667t. See also Grading
Standard mums, 655
Standard pots, 413
Stardom-floribundas, 475
Started eye, 692
Starting seeds, for Gloxinia, 483
Steamed bone meal, 247
Steam heating systems, 93–94, 94f, 95f, 101,
    101f, 102, 103, 103f
Steam pasteurization, 211–216, 308, 309, 333f
S.T.E.M., 250, 250f
Stems, 136, 274
Stem section cuttings, 326
Stem tip cuttings. See Terminal cuttings
Stephanotis, 729–730

Sterilization
   chemical soil treatment, 216–217
   salt accumulation, 266
   steam pasteurization, 211–216
Stigma, 139, 139f
Stipules, 137
Stock plants, 306t
   Chrysanthemums, 646–647
   Geraniums, 474–475
   Poinsettia, 492–493
Stoma, 136f, 138, 138f, 274
Storage
   Azalea, 424, 427
   bulbs, 67, 518–519
   Carbon dioxide, liquified, 170f
   Carnations, 661, 666–667
   cut flowers, 737–743
   Dahlias, 721–722
   Lilies, 543–544
   rooting and, 335
   transplanting and, 318
Straw, 192
Streptomyces lydicus WYEC, 385t
Strobilurins, 387t
Student projects
   bulbs, 571–573
   cut flowers, 675–678
   growing media, 198–200
   seed sowing, 320–322
   terminal cuttings, 343–345
Style, 139, 139f
Styrofoam, 111
Suberin, 274
Subirrigation, 311–312
Sublimed, 169
Submergence, effects of, 273–274
Sulfate, 225
Sulfate of potash-magnesia, 253t
Sulfur, 223, 227f, 234, 239t, 253t, 263, 707f
Sulfur dioxide, 175
Summer equinox, 147, 147f
Sun scorch, 472f
Sunshine peat-lite mixes, 207
Supermarkets, 760, 760f, 761f, 762,
   762f, 763f
Superphosphate, 202t, 206t, 234, 246, 247, 253t,
   308, 608t. See also Peat-lite mixes
Suppliers, of seeds and bulbs, 31
Support
   Alstroemeria, 713
   Calceolaria, 441
   Calendula, 715
   Centaurea, 719
   Roses, 703
   Snapdragons, 641, 642f
Surface irrigation methods, 284
Surface steaming, 214–215
Surfactants, 294
Surfside, 202t, 608t
Sweden, 186
Sweet Alyssum. See Alyssum
Sweetheart roses, imports of, 18t, 19t
Sweet pea, 306t
Sympodial growth, 680
Syngonium, 584–586, 586f, 599t

T

Tagetes erecta. See Marigolds
Tagetes patula. See Marigolds
Takii cultivars, 452
Taxes, 30
Telophase, 324, 324f, 325
Temperature, 156–157
   African violet, 472
   Alstroemeria, 713
   Anemones, 521–522

   asexual propagation environment, 331
   Azalea, 429
   Asters, 717t
   bedding plants, 611
   bulbs, 520–521
   Calla lily, 523
   Callistephus, 716
   Carnations, 665–666
   Centaurea, 719
   Chrysanthemums, 460, 649–650
   Cyclamen, 450–451, 453
   DIF day-night relationships, 157–159
   dry-cold storage, 740
   Easter lilies, 542, 544
   flowering growth, 163–164
   Gardenia, 465
   Geraniums (potted), 618
   Gerbera, 723
   germination environment and, 310–311
   glass house heating, 92, 93f
   hot water heating systems, 92
   Lilies, 540–541, 542
   Matthiola, 725
   moisture stress and, 277
   Orchids, 687–688
   Petunias, 626
   plant growth and, 157
   Poinsettia, 495, 495f, 500
   Roses, 699–700
   seed germination, 159, 305, 306t
   Snapdragons, 641–642
   Stephanotis, 730
   tropical foliage plants, 579–580
   Tulips, 558t, 559t
   vegetative growth, 159–163
   See also Cold sterage; Cooling; Heating;
      Procooling
Temporary interior ceiling, 110–111
Tendrils, 596
Tennessee, plant production in, 10t
Tensiometer, 278, 278f
Terminal cuttings, 325, 343–345
Terrestrial orchids, 679–680, 686
Tetrazines, 365t
Tetronic acids, 365t
Texas, plant production in, 5t, 6t, 7t, 8t, 9t,
   10t, 13–14, 575, 575t
Thailand, 18t, 19t, 25
Thanksgiving cactus, 444, 445f
Thermal fog application, 370
Thermistor, 126
Thermomeler, 215, 216f, 309f
Thermostats, 92, 118–119, 119f, 125–126
Thiadizole, 387t
Thiamethoxam, 362t
Thiazolidinones, 365t
Thielaviopsis, 191
Thielaviopsis basicola, 505
Thin plastic films, 54–55, 56–60
Thiophanate, 388t
Thiophanate + chloronitrile, 388–389t
Thiophanate + dicorbxamide, 389t
Thiophanate + dithiocarbamate, 389t
Thiophanate + iprodione, 389t
Thiophanate + thiadiazole, 389t
Thiophanate-methyl, 388t
Thiophanate-methyl + chlorothalonil,
   388–389t
Thiophanate-methyl + etridiazole, 389t
Thiophanate-methyl + mancozeb, 389t
Thrips, 353f, 375–376, 394t
Timing
   Carnations, 662–663
   Chrysanthemums, 650–651
   Hydrangea, 487, 488
   Lilies, 536–537

   Roses, 700–701
   See also Forcing
Tip burn, 293f, 527
Tip cutting, 326t
Tip layering, 338
Tissue, of plants, 133–139
Tissue culture, 339–340
Tissue testing, 264–265
Tobacco mosaic, 394t
Tomatoes, 195f
Tomato spotted wilt virus, 394t
Topography, for greenhouse site selection, 77
Topsoil, 201
Torenia, 306t
Totipotency, 324
Trace elements, 202t, 206t, 228, 235–239,
   250–251, 608t, 698t. See also Peat-lite
   mixes; specific trace elements
Tradescantia, 588
Traidimefon, 388t
Translocation, 135
Transpiration, 136, 160–161, 274–275
Transplanting, 317–318
   bedding plants, 606, 607–609
   Christmas cherry, 446
   container grown crops, 414–416
   Cyclamen, 450
   Dahlias, 720–721
   Geraniums (potted), 618–619
   Gloxinia, 483
Transportation costs, 23, 91
Transvaal daisy. See Gerbera
Treble superphosphate, 202t, 206t, 246.
   See also Peat-lite mixes
Trench layering, 338
Triazine insect growth regulators, 365t
Triazole, 388t
Trichodema harzianum T22, 385t
Trichoderma virens GL21, 385t
Trifloxystrobi, 387t
Triflumizole, 387t
Trimming, of Anemones, 522. See also Pruning
Trimming for steam, 93
Triple superphosphate, 253t
Trombone, 103
Tropical foliage plants, 574
   Acanthaceae, 580
   acclimatization, 597–602
   Agavaceae, 580–582
   Amaranthaceae, 582
   Araceae, 582–586
   Araliaceae, 586–588
   Araucariaceae, 588
   Asteraceae (Compositae), 588
   Bromeliaceae, 589
   Commelinaceae, 589
   Euphorbiaceae, 589
   general culture, 576–579
   Gesneriaceae, 589–590
   Lamiaceae (Labiatae), 590
   light and temperature, 579–580
   Liliaceae, 590–591
   Marantaceae, 591
   moisture and humidity, 580
   Moraceae, 591–593
   Myrsinaceae, 593
   Palmae (Arecaceae), 593–594
   Pandanaceae, 594
   Piperaceae, 595
   Pittosporaceae, 595
   Polypodiaceae, 595–596
   production areas, 574–575
   Propasation methods, 597, 600t
   Rubiaceae, 596
   Rutaceae, 597
   Saxifragaceae, 597

Urtricaceae, 597
Vitaceae, 597
world distribution, 575–576
Throughs, 84, 85, 723t
Truss, 48
Truss rafter framing method, 53, 54, 54f
Tuberous-rooted Begonias, 436–437
Tuberous-rooted Dahlias, 721–722
Tuberous roots and stems, 336
Tubers, 336, 484
Tube watering, 282, 283f
Tulips, 3t, 552
    cultivars, 556–557t
    cut flowers, 553–556
    dry-cold storage, 740
    flower formation, 552–553
    pot-grown, 560–568
    precooling, 553, 557–560
    problems, 568
Tulip topple, 554
Tupidanthus, 588
Turgid, 136
Turnip mosaic, 394t
Twister batch soil mixers, 203, 203f
Type I glazing, 56–60
Type II glazing, 60–65
Type III glazing, 65–66

U

Ultraviolet light, 152
Ultraviolet radiation, 152–153
Underground plant structures, 134–135
Understock, 690, 692–693
Uneven span greenhouses, 38, 40, 79–80
Unicellular, 133
Uniconazole, 402–403, 403t
United States flower production, offshore
        impact on, 17–20. See also Exporters;
        Importing
Unit heaters. See Horizontal unit heaters
Unit steam heater, 94
Unrooted Azalea cuttings, storage of, 424
Uramite, 248
Urea, 253t
Urea formaldehydes (UFN), 248–249
Urea nitrate, 246, 253t
Urtricaceae, 597

V

Vacuoles, 133, 133f
Vapam, 216
Vapor pressure deficit (VPD), 126
Vascalar bundle, 138f
Vascular system, 135, 136
Vascular tissue, 135
Vault steaming, 215–216
Vavra Geraniums, 476
Vegetable crops, 629–630, 631–632t
Vegetative axillary buds, 137

Vegetative growth, temperature and, 159–163
Vegetative propagation. See Asexual
        propagation
Velamen, 681
Venlo green house, 44, 44f, 45
Ventilating sash operation, 113, 113f, 114, 114f,
        115, 115f
Ventilation, 111
    air temperature reduction, 111–112
    carbon dioxide supply, 112
    fans, 115–119, 115–119f
    natural air movement, 112–112f, 113
    relative humidity reduction, 112
    thermostat controls, 118–119, 119f
    ventilating sash operation, 113, 113f, 114,
        114f, 115, 115f
Ventilator header, 49
Ventilators, 49
Venting heaters, 107–108, 188f
Venturi principle, 254, 255f
Verbena, 306t, 628
Vermiculite, 180, 193–194, 202t, 203f, 307f,
        308, 310f, 608f, 608t. See also
        Peat-lite mixes
Verticillium wilt, 216
Vinea, 306t
Viola x Wittrockiana. See Pansy
Virginia, plant production in, 8t
Viruses, 392–395, 668
Visible light, 152
Vitaceae, 597
Volatile products, 173

W

Wall area, in heat requirement calculations, 99
Washington, plant production in, 6t, 9–11
Water, 272
    application of, 277–294
    Calla lily, 523
    Carnations, 664
    Chrysanthemums, 458, 649
    Lilies, 539–540
    Petunias, 626
    photosynthesis and, 141, 142
    Poinsettia, 501
    role of, 272–277
    Snapdragons, 640–641
    Solubridge readings for, 269t
    See also Moisture
Water absorption (uptake), 273–274, 304–305
Water acidity, 291–292
Water flow shade, 124
Water injection bench, 287f
Water loops, 282, 283f, 284
Water losses, 274–276
Water movement, 136–137
Water quality, 291–293
Water softening, 122–123
Water stress. See Moisture stress

Water supply, to pad, 121–122
Water tube boilers, 103, 104, 105, 105f
Water wetters, 294
Watts, 641
Weather, 7t, 280, 575. See also Climate
Weather monitoring system, 126, 126f
Weeds, 350, 350f, 351, 354, 358, 359, 374, 690
Weights, of glass, 65–66
Wet-cold storage, 737–739
Wetting agents, 202t, 206t, 294, 608t. See also
        Peat-lite mixes
Wheat straw, 192
Whiteflies, 353f, 376–377, 506–507, 507f
White light, 152
White rust, 26
Wholesale distributors, 746–747
    clock auction, 747–748
    handling, 751–754
    selling methods, 750
    shipping methods, 750–751
    wholesale structure, 748, 748f, 749f, 750
Wholesale Florist and Florist Supplier
        Association (WF&FSA), 748
Wholesale growers, 30
Wholesale structure, 748, 748f, 749f, 750
Wilting, 275, 431–432
Winter equinox, 147, 147f
Wisconsin, plant production in, 9t
Witches' broom, 236
Wolmanizing, 53
Wood, as energy source, 90
Wood byproducts, 189–190. See also Sawdust
Wood chips, 180
Wooden framing, 50f, 52–53
Wood preservatives, 173
World distribution, of tropical foliage plants,
        575–576
Worldwide distribution. See also Wholesale
        distributors
Woven polyethylene, 58

X

Xylem, 135, 136f, 138f

Y

Year-round flowering
    Azalea, 429, 430, 431
    Callistephus, 717–718
    Chrysanthemums, 460, 652–655

Z

Zantedeschia. See Calla lily
Zebrine, 588
Zinc, 223, 227f, 228f, 235, 236, 237, 239t, 698t
Zinnia, 233f, 306t, 629
Zodiac series, 452
Zonal Geraniums, 475
Zoned heating, 92–93, 93f, 127
Zoning laws, 77